AF693003

Vehicle and Traffic Safety

Vehicle and Traffic Safety

Editors

Guzek Marek
Rafał Jurecki
Wojciech Wach

MDPI • Basel • Beijing • Wuhan • Barcelona • Belgrade • Manchester • Tokyo • Cluj • Tianjin

Editors

Guzek Marek
Warsaw University of Technology
Poland

Rafał Jurecki
Kielce University of Technology
Poland

Wojciech Wach
Institute of Forensic Research
Poland

Editorial Office
MDPI
St. Alban-Anlage 66
4052 Basel, Switzerland

This is a reprint of articles from the Special Issue published online in the open access journal *Energies* (ISSN 1996-1073) (available at: https://www.mdpi.com/journal/energies/special_issues/Vehicle_Traffic_Safety).

For citation purposes, cite each article independently as indicated on the article page online and as indicated below:

LastName, A.A.; LastName, B.B.; LastName, C.C. Article Title. *Journal Name* **Year**, *Volume Number*, Page Range.

ISBN 978-3-0365-5603-1 (Hbk)
ISBN 978-3-0365-5604-8 (PDF)

Contents

About the Editors

Guzek Marek

Marek Guzek (Ph.D., D. Sc.) is an associate professor at the Warsaw University of Technology. He is the Head of the Motor Vehicle Safety and Operation Team at the Faculty of Transport. His professional experience is closely related to the research of vehicles and drivers on issues of the safety of the driver–vehicle–environment system: vehicle dynamics, vehicles stability and handling, working and diagnostic processes motor vehicles, analysis of road traffic accidents taking into account the role of drivers, uncertainty in accident analysis, the use of EDR recorders for vehicle motion reconstruction, psychophysical properties of drivers in accident situations. He participates in the construction of car driving simulators used in the research of drivers and for their training. He is the author of over 120 publications in the abovementioned areas, including the monograph *Uncertainty in the Analysis of Road Accidents*. He is a member of SAE International (Society of Automotive Engineering) and the Association of Experts of Automobile Technology and Road Traffic (SRTSIRD), which is a member of FIEA (International Federation of Automotive Experts).

Rafał Jurecki

Rafał Stanisław Jurecki (Ph.D. D. Sc.) is an associate professor at Kielce University of Technology in Kielce, Poland. He was born in Kielce, Poland, in 1969. He obtained his M.S. degree in automotive engineering in 1994 and Ph.D. degree in mechanical engineering in 2006, both from Kielce University of Technology in Kielce, Poland. In 2017, he obtained the degree of habilitated doctor in transport at the Warsaw University of Technology in Warsaw, Poland. Between 1996 and 2006, he held the position of assistant lecturer at the Department of Automotive Engineering and Transport of the Kielce University of Technology. He has worked as an assistant professor in the same department since 2006 and as an associate professor since 2017. He is the author or co-author of about 150 articles and books. His research interests include driver behavior in emergency situations, vehicle dynamics, collision analysis, and motor vehicle testing and diagnostics. He holds 13 patents. Since 2019, he has served as Vice Dean for Teaching and Student Matters in the Faculty of Mechatronics and Mechanical Engineering of Kielce University of Technology, Poland.

Wojciech Wach

Wojciech Wach (Ph.D., D.Sc.) is an associate professor at the Institute of Forensic Research in Kraków, Poland, where he is the Head of the Department of Road Accident Analysis. He received his Ph.D. and habilitation degree in vehicle dynamics and vehicle accident reconstruction. His research interests include all issues related to vehicle dynamics and traffic safety in perspective of the accident reconstruction (vehicle dynamics, multi-body dynamics, collision analysis, road infrastructure equipment, visibility at night, uncertainty, modeling and simulation, biomechanics, photogrammetry, computer applications, etc.). He is the author or co-author of several books (inter alia *Simulation of Vehicle Accidents Using PC-Crash, Structural Reliability of Road Accidents Reconstruction, and Fundamentals in the Reconstruction of Road Accidents*) and over 250 articles. Wojciech Wach is a member the ENFSI (European Network of Forensic Sciences) Road Accident Analysis Working Group and of the EVU (European Association for Accident Research and Analysis) Scientific Board.

energies

Editorial

Vehicle and Traffic Safety

Marek Guzek [1,*], Rafał S. Jurecki [2] and Wojciech Wach [3]

1 Faculty of Transport, Warsaw University of Technology, 00-662 Warsaw, Poland
2 Faculty of Mechatronics and Mechanical Engineering, Kielce University of Technology, 25-314 Kielce, Poland; rjurecki@tu.kielce.pl
3 Institute of Forensic Research in Cracow, 31-033 Cracow, Poland; wwach@ies.gov.pl
* Correspondence: marek.guzek@pw.edu.pl

Citation: Guzek, M.; Jurecki, R.S.; Wach, W. Vehicle and Traffic Safety. *Energies* **2022**, *15*, 4573. https://doi.org/10.3390/en15134573

Received: 14 June 2022
Accepted: 20 June 2022
Published: 23 June 2022

1. Introduction

The role of the road transportation of people and goods is increasing. Despite significant progress in the development of car design and road safety improvement programs, the number of accidents and the casualties caused by them, as well as the related economic and social costs, remains unsatisfactory. In this context, it is extremely important to develop and exchange knowledge regarding contemporary car safety problems, interactions in the vehicle–human–road system, the development of car construction, and methods of analyzing road accidents.

This book contains submissions [1–29] that were invited to be published a Special Issue of *Energies* entitled "Vehicle and Traffic Safety". We invited researchers, specialists, and industry representatives to publish their achievements regarding various topics in the fields of ground vehicle transport and automotive engineering, active and passive vehicle safety, vehicle dynamics and stability, accident analysis and reconstruction, vehicle (and its assemblies) testing, the safety of alternative vehicle drives, including electric and hybrid vehicles, and the impact of aspects such as traffic control systems, road infrastructure (including street and road lighting), autonomous and connected vehicles, and human factors on road safety.

Detailed topics of interest in the call for papers included, but were not limited to:

- Ground vehicle safety;
- Vehicle active safety;
- Vehicle passive safety;
- Accident analysis;
- Accident reconstruction;
- Vehicle dynamics;
- Vehicle stability and handling;
- Autonomous vehicles;
- Autonomous and connected vehicles;
- Vehicle testing;
- The safety of electric/hybrid cars;
- Driving simulators;
- The visibility (recognizability) of pedestrians and obstacles;
- The lighting of vehicles and roads;
- Crashworthiness;
- Road safety infrastructure;
- Traffic control systems;
- Traffic organization.

2. A Brief Overview of the Content of This Special Issue

The response to our invitation may be considered broad. The related basic statistics are as follows:

- Submissions: 45;
- Published papers: 29 (all research articles);
- Rejections: 16.

In total, 115 authors contributed to the preparation of these 29 publications. The authors were mainly affiliated to universities (there were 31 such units among the affiliations), research institutes (8), and enterprises (2).

The authors came from 15 countries, representing a wide territorial cross-section:

Austria	2 Authors/1 Article
China	6/1
Croatia	3/1
Czechia	10/1
France	3/1
Greece	4/1
Germany	3/1
India	4/1
Italy	6/2
Korea	2/1
Poland	56/18
Saudi Arabia	3/1
Spain	4/1
UK	3/1
Ukraine	6/1

The topics of the articles varied. The authors presented the results of their research which covered many aspects that directly or indirectly affected issues relating to vehicle and road safety and referred to the previously mentioned topics. Various research methodologies were used (experimental research on real objects, simulation research, questionnaire research, etc.).

Some articles directly related to vehicle safety in the field of car passive safety [1,14], active vehicle safety [7,11,12,20,26], or comfort [25]. Several papers were devoted to issues related to the operation and safety of electric vehicles [10,19,29] and autonomous vehicles [2,8,9,13,16,18]. Several studies referred to selected problems in the reconstruction of road accidents [5,22,28]. A separate group of papers analyzed selected aspects of road infrastructure and traffic control [3,6,23,24]. All of the data show that the weakest link in the road safety system is man, and more specifically, vehicle drivers. Tests involving drivers were the topic of [4,15]. Car driving simulators are an essential tool in testing drivers, as well as design solutions. The use of such tools in research can be found in [21,27]. The study [17] was slightly different, as the authors analyzed the state of road traffic safety in a specific country and the impact of non-technical methods on shaping this state.

The overview presented above is, of course, simplified. The abovementioned links between topics and articles are not unambiguous. The works were often multi-threaded, indicating various influences in the human–vehicle–environment system.

The review presented above shows that a relatively large number of articles were in line with contemporary trends in the development of the automotive industry (e.g., electric vehicles and traffic automation and driver assistance systems). Relatively few papers focused on passive safety, and no articles considered crashworthiness.

3. Conclusions

Most of the topics indicated in the invitation were reflected in the submitted and accepted articles, and as Guest Editors, we are very satisfied with this. To conclude, we would like to thank all of the authors for their efforts in preparing the articles. We would also like to thank the reviewers and the publishing house staff for their efficient processing.

We are convinced that the presented collection of high-quality articles on the above-mentioned topics in this Special Issue of the journal will bring a lot of valuable experience and expand our knowledge in this field.

Author Contributions: M.G., R.S.J. and W.W. contributed equally to this article. All authors have read and agreed to the published version of the manuscript.

Funding: This research received no external funding.

Conflicts of Interest: The authors declare no conflict of interest.

References

1. Jaśkiewicz, M.; Frej, D.; Matej, J.; Chaba, R. Analysis of the Head of a Simulation Crash Test Dummy with Speed Motion. *Energies* **2021**, *14*, 1476. [CrossRef]
2. Prochowski, L.; Ziubiński, M.; Szwajkowski, P.; Gidlewski, M.; Pusty, T.; Stańczyk, T. Impact of Control System Model Parameters on the Obstacle Avoidance by an Autonomous Car-Trailer Unit: Research Results. *Energies* **2021**, *14*, 2958. [CrossRef]
3. Budzynski, M.; Gobis, A.; Guminska, L.; Jelinski, L.; Kiec, M.; Tomczuk, P. Assessment of the Influence of Road Infrastructure Parameters on the Behaviour of Drivers and Pedestrians in Pedestrian Crossing Areas. *Energies* **2021**, *14*, 3559. [CrossRef]
4. Jurecki, R.; Stańczyk, T. A Methodology for Evaluating Driving Styles in Various Road Conditions. *Energies* **2021**, *14*, 3570. [CrossRef]
5. Wach, W.; Zębala, J. Striated Tire Yaw Marks—Modeling and Validation. *Energies* **2021**, *14*, 4309. [CrossRef]
6. Ostrowski, K.; Budzynski, M. Measures of Functional Reliability of Two-Lane Highways. *Energies* **2021**, *14*, 4577. [CrossRef]
7. Babić, D.; Babić, D.; Fiolić, M.; Eichberger, A.; Magosi, Z. A Comparison of Lane Marking Detection Quality and View Range between Daytime and Night-Time Conditions by Machine Vision. *Energies* **2021**, *14*, 4666. [CrossRef]
8. Masood, K.; Morales, D.; Fremont, V.; Zoppi, M.; Molfino, R. Parking Pose Generation for Autonomous Freight Collection by Pallet Handling Car-like Robot. *Energies* **2021**, *14*, 4677. [CrossRef]
9. Stoma, M.; Dudziak, A.; Caban, J.; Droździel, P. The Future of Autonomous Vehicles in the Opinion of Automotive Market Users. *Energies* **2021**, *14*, 4777. [CrossRef]
10. Szumska, E.; Jurecki, R. Parameters Influencing on Electric Vehicle Range. *Energies* **2021**, *14*, 4821. [CrossRef]
11. Lee, S.; Kim, B.; Lee, S. Study on Image Correction and Optimization of Mounting Positions of Dual Cameras for Vehicle Test. *Energies* **2021**, *14*, 4857. [CrossRef]
12. Dębowski, A. Analysis of the Effect of Mass Parameters on Motorcycle Vibration and Stability. *Energies* **2021**, *14*, 5090. [CrossRef]
13. Anastasiadou, K.; Gavanas, N.; Pitsiava-Latinopoulou, M.; Bekiaris, E. Infrastructure Planning for Autonomous Electric Vehicles, Integrating Safety and Sustainability Aspects: A Multi-Criteria Analysis Approach. *Energies* **2021**, *14*, 5269. [CrossRef]
14. Shirur, N.; Birkner, C.; Henze, R.; Deserno, T. Tactile Occupant Detection Sensor for Automotive Airbag. *Energies* **2021**, *14*, 5288. [CrossRef]
15. Gazdzinski, S.; Binder, M.; Bortkiewicz, A.; Baran, P.; Dziuda, Ł. Effects of All-Night Driving on Selective Attention in Professional Truck Drivers: A Preliminary Functional Magnetic Resonance Study. *Energies* **2021**, *14*, 5409. [CrossRef]
16. Dudziak, A.; Stoma, M.; Kuranc, A.; Caban, J. Assessment of Social Acceptance for Autonomous Vehicles in Southeastern Poland. *Energies* **2021**, *14*, 5778. [CrossRef]
17. Alonso, F.; Faus, M.; Fernández, C.; Useche, S. "Where Have I Heard It?" Assessing the Recall of Traffic Safety Campaigns in the Dominican Republic. *Energies* **2021**, *14*, 5792. [CrossRef]
18. Montani, M.; Ronchi, L.; Capitani, R.; Annicchiarico, C. A Hierarchical Autonomous Driver for a Racing Car: Real-Time Planning and Tracking of the Trajectory. *Energies* **2021**, *14*, 6008. [CrossRef]
19. Alosaimi, W.; Ansari, M.; Alharbi, A.; Alyami, H.; Ali, S.; Agrawal, A.; Khan, R. Toward a Unified Model Approach for Evaluating Different Electric Vehicles. *Energies* **2021**, *14*, 6120. [CrossRef]
20. Jackowski, J.; Żmuda, M.; Wieczorek, M.; Zuska, A. Quasi-Static Research of ATV/UTV Non-Pneumatic Tires. *Energies* **2021**, *14*, 6557. [CrossRef]
21. Choromański, W.; Grabarek, I.; Kozłowski, M. Integrated Design of a Custom Steering System in Cars and Verification of Its Correct Functioning. *Energies* **2021**, *14*, 6740. [CrossRef]
22. Guzek, M.; Lozia, Z. Are EDR Devices Undoubtedly Helpful in the Reconstruction of a Road Traffic Accident? *Energies* **2021**, *14*, 6940. [CrossRef]
23. Krukowicz, T.; Firląg, K.; Suda, J.; Czerliński, M. Analysis of the Impact of Countdown Signal Timers on Driving Behavior and Road Safety. *Energies* **2021**, *14*, 7081. [CrossRef]
24. Llewellyn, R.; Cowie, J.; Fountas, G. Solar-Powered Active Road Studs and Highway Infrastructure: Effect on Vehicle Speeds. *Energies* **2021**, *14*, 7209. [CrossRef]
25. Kernytskyy, I.; Yakovenko, Y.; Horbay, O.; Ryviuk, M.; Humenyuk, R.; Sholudko, Y.; Voichyshyn, Y.; Mazur, Ł.; Osiński, P.; Rusakov, K.; et al. Development of Comfort and Safety Performance of Passenger Seats in Large City Buses. *Energies* **2021**, *14*, 7471. [CrossRef]

26. Parczewski, K.; Wnęk, H. Influence of Clearance on the Rocker Arm Pin on the Steerability and Stability of the Vehicle Motion. *Energies* **2021**, *14*, 7827. [CrossRef]
27. Chen, L.; Xie, J.; Wu, S.; Guo, F.; Chen, Z.; Tan, W. Validation of Vehicle Driving Simulator from Perspective of Velocity and Trajectory Based Driving Behavior under Curve Conditions. *Energies* **2021**, *14*, 8429. [CrossRef]
28. Mikulec, R.; Semela, M.; Bradáč, A.; Tokař, S.; Bilík, M.; Křižák, M.; Belák, M.; Kledus, R.; Haring, A.; Rábek, V. Obsolete or Viable? Revision of Lane-Change Manoeuvre Duration Empirical Calculation. *Energies* **2021**, *14*, 8439. [CrossRef]
29. Dukalski, P.; Będkowski, B.; Parczewski, K.; Wnęk, H.; Urbaś, A.; Augustynek, K. Analysis of the Influence of Motors Installed in Passenger Car Wheels on the Torsion Beam of the Rear Axle Suspension. *Energies* **2022**, *15*, 222. [CrossRef]

energies

MDPI

Article

Analysis of the Head of a Simulation Crash Test Dummy with Speed Motion

Marek Jaśkiewicz [1], Damian Frej [1,*], Jan Matej [2] and Rafał Chaba [3]

[1] Department of Automotive Engineering and Transport, Kielce University of Technology, Avenue Tysiąclecia Państwa Polskiego 7, 25-314 Kielce, Poland; m.jaskiewicz@tu.kielce.pl
[2] Institute of Vehicles, Warsaw University of Technology, 02-524 Warsaw, Poland; jan.matej@pw.edu.pl
[3] Globtrak Polska Sp. z.o.o., ul. Wincentego Witosa 65/2, 25-561 Kielce, Poland; r.chaba@globtrak.pl
* Correspondence: frejdamian@gmail.com; Tel.: +48-41-342-4718

Abstract: The article presents a model of an anthropometric dummy designed for low velocity crash tests, designed in ADAMS. The model consists of rigid bodies connected with special joints with appropriately selected stiffness and damping. The simulation dummy has the appropriate dimensions, shape, and mass of individual elements to suit a 50 percentile male. The purpose of this article is to draw attention to low speed crash tests. Current dummies such as THOR and Hybrid III are used for crash tests at speeds above 40 km/h. In contrast, the low-speed test dummy currently used is the BioRID-II dummy, which is mainly adapted to the whiplash test at speeds of up to 16km/h. Thus, it can be seen that there is a gap in the use of crash test dummies. There are no low-speed dummies for side and front crash tests, and there are no dummies for rear crash tests between 16 km/h and 25 km/h. Which corresponds to a collision of a passenger vehicle with a hard obstacle at a speed of 30 km/h. Therefore, in collisions with low speeds of 20 km/h, the splash airbag will probably not be activated. The article contains the results of a computer simulation at a speed of 20 km/h vehicle out in the ADAMS program. These results were compared with the experimental results of the laboratory crash test using volunteers and the Hybrid III dummy. The simulation results are the basis for building the physical model dummy. The simulation aims to reflect the greatest possible compliance of the movements of individual parts of the human body during a collision at low speed.

Keywords: computer simulation; dummy; crash test

Citation: Jaśkiewicz, M.; Frej, D.; Matej, J.; Chaba, R. Analysis of the Head of a Simulation Crash Test Dummy with Speed Motion. *Energies* **2021**, *14*, 1476. https://doi.org/10.3390/en14051476

Academic Editor: Chunhua Liu

Received: 15 January 2021
Accepted: 2 March 2021
Published: 8 March 2021

1. Introduction

Anthropometric dummies are specialized research devices that simulate both dimensions, weight proportions, and joints in the human body [1,2]. Dummies are used by vehicle manufacturers and aircraft to predict potential human injuries in an accident. Current crash test dummies have built-in sensors to record data such as impact speed, crush force, bending, torque, and crash braking pressure. Anthropometric dummies are cataloged by purpose [3–5]. Other dummies are used for the rear, side, and front tests. Moreover, they are tailored to a specific test in such a way as to collect as much information as possible. Therefore, the side test dummies have the upper limbs removed, which could interfere with the measurements during a crash. First of all, it should be noted that dummies are used to increase security [6,7].

Research issues concerning the protection of the driver and passengers of passenger vehicles against the consequences of any road collisions belong to the issues of impact biomechanics. An important element is the choice of the method used as well as the criteria for assessing the risk of sustaining an injury [8,9]. It should be noted that the result of the complicated nature of the loads acting on the driver or passenger during a collision is the main reason for the model testing of the applied vehicle passive safety systems. In this case, there are two types of model analyzes [10–12]:

- Computer simulation tests that are performed in simulation programs with the use of virtual models representing experimental dummies,
- experimental crash-test studies that are performed using material human models reflecting human behavior during a collision.

Experimental research is characterized by high testing costs, which significantly reduces the frequency of testing. The present times associated with sudden and rapid technological growth contribute to the increased popularity of simulation research. This type of research is characterized by [12–14]:

- Low costs of building simulation models,
- ease of conducting research on virtual models.

The process of correlating the results of computer simulation tests with the use of virtual manikins is called validation. The validation is to ensure that the results obtained with the use of the simulation model reflect to some extent the physical model [15,16]. On the other hand, the process of creating a simulation model requires discretization of geometry, appropriate selection of elements, defining connection elements, selection of material specifications and its physical properties, as well as knowledge of loading conditions. In addition, the selection and determination of all the above variables and parameters significantly affects the accuracy and correctness of the simulation result. Therefore, mechanical engineering in the field of modeling reliable models of anthropometric dummies can only be verified by comparing the results with actual crash tests [17,18]. Figure 1 shows the THOR anthropometric dummy, both in the real form (Figure 1a) and in the simulation form (Figure 1b), intended for frontal crash tests.

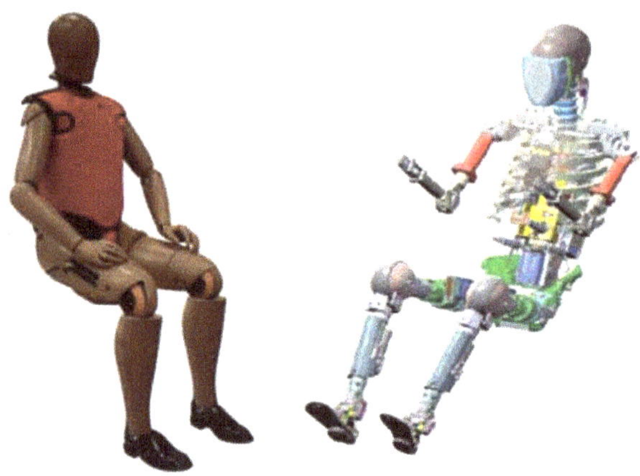

(a) (b)

Figure 1. Anthropometric dummy for the THOR crash test [18]: (**a**) THOR-50th Dummy; (**b**) THOR-FT dummy CAD (Computer aided design)-model.

Crash test dummies consist of the spine element assemblies, the head assembly, the spine assembly, the thoracic assembly, the lower limb assembly, and the upper limb assembly. The ideal structural design of an anthropometric crash test dummy assumes a simple structure that does not require frequent calibration, while maintaining high durability and repeatability of results [19–21]. Currently, in the case of frontal collisions, the most advanced dummy is the THOR [19,20]. Its structure is similar to the human body. It has a developed spine with flexible joints in both the thoracic and lumbar sections. Its predecessor and competitor is the Hybrid III dummy, which is used all over the world in crash tests. It was on its basis that the side and rear collision dummies were created [21]. Hybrid III dummies have a high rate of compatibility with the human body. Research on the behavior of dummies during crash tests led to the development of an assessment of head, face, chest, abdomen, lower limb, and upper limb injuries, which are used by all countries around the world [21–23]

Crash tests of various types of security systems aimed at increasing passenger safety require significant financial and time expenditure [21,24,25]. Therefore, simulations are performed first. The simulation programs of anthropometric manikins include such programs as: Madymo, Dytran, Dassault Systèmes SIMULIA, ANSYS Engineering Simulation, SIMPACT Caterham Super 7, MSC Software ADAMS [26,27]. These programs allow you to perform simulated crash tests with minimal financial burden. Crash tests are carried out only after collecting the appropriate amount of data confirming the effectiveness of the simulation tests. The demand for simulation programs and computer simulations before experimenting influences the development of specialized software networks [28–30].

When reviewing the literature, it should be noted that simulation dummies are also available for lateral, frontal, and rear tests. The simulation dummies are a reflection of the existing anthropometric dummies [18,31]. Child simulation dummies are also available. They are designed to assess the safety of vehicles and child seats or booster cushions [32,33]. Figure 2 shows the children's simulation dummies. The simulation dummies are intended for the same crash tests as their physical counterparts, because before performing a given experiment, simulation tests should first be carried out, which are less expensive for experimental tests [18,34].

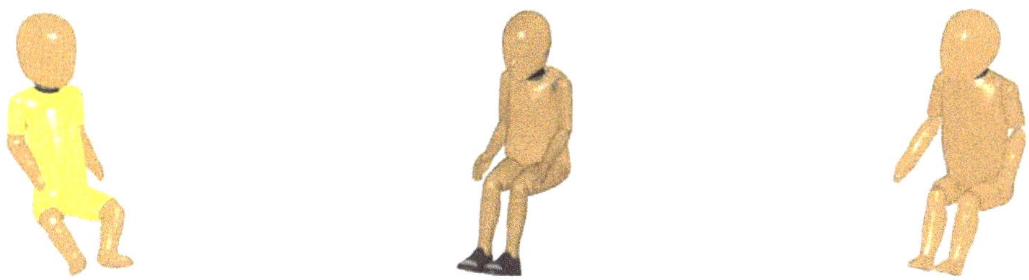

(a) (b) (c)

Figure 2. Children's dummies [18]: (**a**) Seria Q FE, (**b**) Hybrid III Children Series FE, (**c**) CRABI-12MO FE.

The development of technology and the growing popularity of simulation programs resulted in an increasing interest in virtual crash tests. The article [35] presents a simulation model of a person sitting in a wheelchair. This model can be used for wheelchair crash response testing or for assessing the effect of various factors on the safety of passengers in a collision. However, the articles [36–38] present simulations of crash tests aimed at increasing the passive safety of a passenger vehicle. The articles [39,40] present simulations of crash tests with the use of airbags. These simulations are designed to reduce the failure rate of the airbag system, because if the airbag is deployed in inappropriate conditions, it may cause bone injuries and fractures. The articles [41–43] present the use of computer simulation to analyze the injuries of rear seat passengers in a frontal collision. These studies showed that a suitable belt load limiter could reduce dummy injuries. On the other hand, in [44], simulation tests were used to develop a system for restraining bus passengers during emergency braking. These simulation studies have shown that three-point seat belts can better protect the passenger, and that increasing the angle of the airbag can reduce the amount of injury to passengers.

Low speed crash tests are used to test the strength of vehicle bumpers, and they are performed at speeds of 10 km/h and 5 km/h [45]. These tests do not include the study of the movement of the individual parts of the dummy's body. Moreover, it should be noted that in the literature, there are scientific studies aimed at determining the minimum speed at which permanent damage to the human body occurs. Only volunteers are used in tests at speeds of up to 20 km/h. Dummies for such tests are not constructed [46–48].

Therefore, it should be noted that the current anthropometric dummies are constantly being improved and modified. An ideal dummy should be highly resistant to damage during crash tests and highly reliable compared to the human body [49,50]. Since the development of the first crash test dummy, it was known that the dummy could not be used for all types of crash tests. Therefore, a division into rear, side, and front crash test dummies was created. The first side test dummies, such as EUROSID 1, BIOSID, were developed in 1989, while the BioRID side test dummy was developed in 2000 by engineers from General Motors and NHTSA [17,51]. The dummy was quickly upgraded to the BioRID II version. The new version of the rear crash test dummy was mainly used for the Whiplash tests at speeds up to 16 km/h. A characteristic feature of this dummy is an extensive spine consisting of the same number of vertebrae as a human. In contrast, the side test dummies were devoid of upper limbs, which could interfere with data collection during the crash test [51–53].

Most of the constructed anthropometric dummies have their counterparts in a digital version. Simulation dummies have very well correlated components and subassemblies with the human body [54,55]. Thanks to the development of simulation dummies, it was possible to compare physical dummies with simulation dummies and with volunteers. The first simulation dummies such as BIOSID or Hybrid III have been validated, and their authors assure that the models are able to predict injuries that may occur in a vehicle accident [56,57]. However, it should be noted that the current anthropometric dummies are used for specific crash tests. These tests are characterized by a specific speed and angle of the vehicle hitting an obstacle. Therefore, there is no anthropometric dummy that could be used for any type of crash test using low speeds [58,59].

2. Research Object

The model of a physical-point crash test dummy is a set of interrelated lumps with adequate damping and stiffness. Each element in the construction of a dummy has the appropriate shape and mass. All elements of the dummy are connected with each other by means of elaborate joints that reflect the range of human body movement. The model of the simulation dummy was made in the MSC Adams program [44,60]. This program examines the dynamics of the movement of the various parts of the dummy's body. At the same time, it allows the transformation of a rigid body into a flexible model using the finite element method. The program environment makes it possible to modify and change the parameters of individual dummy elements. Moreover, it allows for obtaining information on the exact position of all parts of the dummy's body at a given moment in time and then presenting them in a graphical manner. The following assumptions were made for the construction of the mathematical model of the anthropometric dummy:

- System of rigid bodies.
- Known dimensions, masses, and moments of inertia.
- Model in which the movement takes place in three-dimensional space.
- Connections of solids using hinges.
- The only force acting on the system is the initial speed Vx (chair speed).
- Belts and seat modeled on the basis of experimental research.

The dummy is designed in the manner of the Hybrid III manikin representing the 50th percentile male. It consists of 17 elements connected by joints. Table 1 shows the mass of the individual body parts of the designed dummy. The dummy was placed on the vehicle seat. The seat is made of one block with appropriately selected stiffness and damping characteristics. Figure 3 shows a simulation dummy designed in the MSC Adams program.

Table 1. The masses of individual body parts of the simulation dummy.

The Name of the Block	Mass (kg)
forearm	3.60
arm	0.50
hand	4.60
food	9.4
shank	0.60
thigh	13.80
neck	0.95
head	3.70
hips	20.40
chest	9.80
stomach	10.00
(Σ)	**78.7**

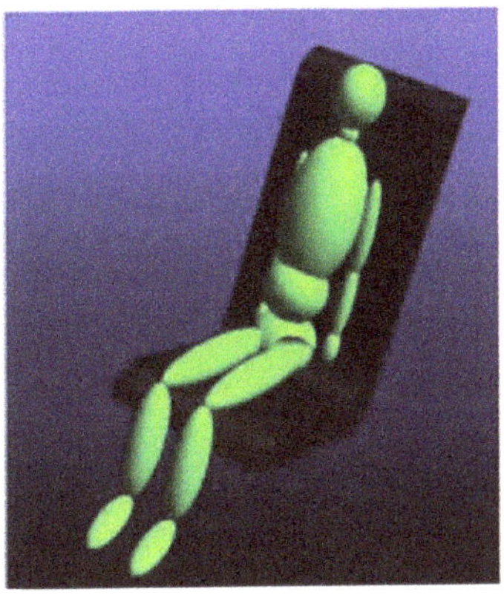

Figure 3. Anthropometric dummy for crash tests made in ADAMS program.

The MSC ADAMS program allows you to create and connect individual dummy solids using complex joints imitating joints. In addition, ADAMS allows you to introduce restrictions related to the movement of individual elements, select the appropriate articulation, and the number of degrees of freedom.

The simulation results of the crash test with the dummy carried out for the speed of 20 km/h were compared with the results of the experiment with the Hybrid III dummy at the same speed. Moreover, in order to check the reliability of the results, they were compared with the results of an experimental crash test in laboratory conditions with the participation of volunteers. An experimental crash test using the Hybrid III dummy was carried out at the Automotive Industry Institute in Warsaw [48]. The experimental test stand consists of a trolley to which a vehicle seat with a dummy is attached. The trolley is accelerated to a speed of 20 km/h using flexible ropes with a total length of 25 m. The expected speed is obtained by pulling the rope cart to a certain length. Releasing the ropes causes the trolley to move in a chaotic manner, which is braked with the help of 60 cm polyurethane sleeves. The experiment with a Hybrid III dummy representing a 50th centile male was recorded with a high-speed Phantom V310 camera at 600 frames/second [43,48]. Figure 4 shows the test stand located at the Automotive Industry Institute in Warsaw.

Experimental test conditions were replicated in MSC Adams. The computer simulation uses a built-in software camera operating at 2500 frames per second. The duration of the simulation was 5 s. The seat with the dummy was accelerated from zero to the assumed speed. In the next stage, the chair was moved by the force of inertia for 2 s, and then it was decelerated to zero. Braking the vehicle seat simulates a collision with a stationary obstacle. Figure 5 shows a simulation of the dummy's movement at appropriate times.

The simulation dummy was equipped with seat belts modeled as linear Kelvin-Voigt susceptibility elements. Due to the accuracy of displacements of individual parts of the dummy's body, the seat belts were not visible during the simulation.

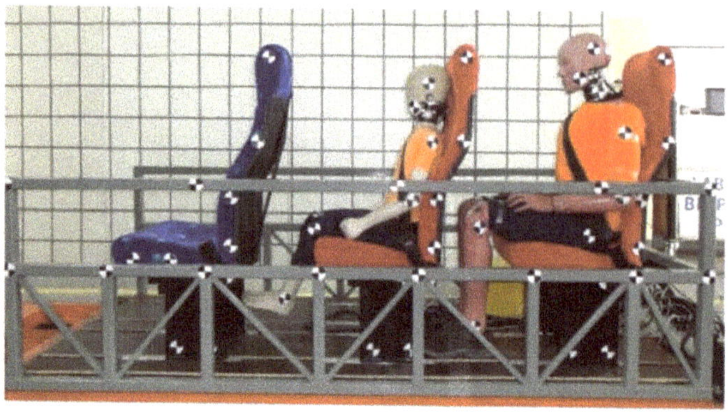

Figure 4. Test stand located at the Automotive Industry Institute in Warsaw.

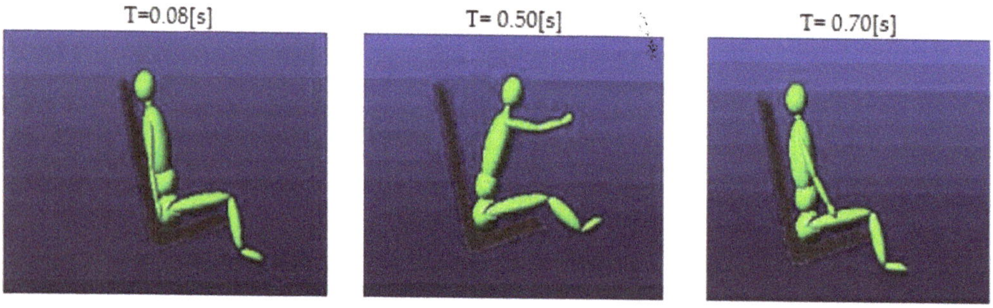

Figure 5. Displacement of the dummy at selected moments of time.

Experimental tests with the use of volunteers were carried out in compliance with all safety standards on a test stand intended for crash tests at low speed. The stand for simulation testing of dynamic stress, especially in the driver's safety elements, was submitted for protection at the Patent Office of the Republic of Poland on 27 May 2019 under the number P.430055 [60]. An experimental crash test was carried out in the laboratory of motor vehicles and tractors at the Kielce University of Technology. The stand consists of a 10m long track along which a vehicle seat moves. At the end of the track, there are shock absorbers that simulate a collision with a stationary obstacle. The speed of the collision depends on the road length and the track gradient. The vehicle seat is on rollers that allow it to move freely along the tracks of the station. The measurement takes place when the electro valve is released, which keeps the chair with the dummy or the volunteer at a given height. Measurements on the experimental stand were recorded with a high-speed Phantom camera at 2500 frames per second. Figure 6 shows the view of the crash test stand at low speeds.

Experimental studies with the participation of volunteers were performed on a sample of 20 people. The results of the experimental studies were averaged. The averaged dimensions of the people participating in the experiment are presented in Table 2. The averaged results showed a picture of a person representing the 50th percentile of the population.

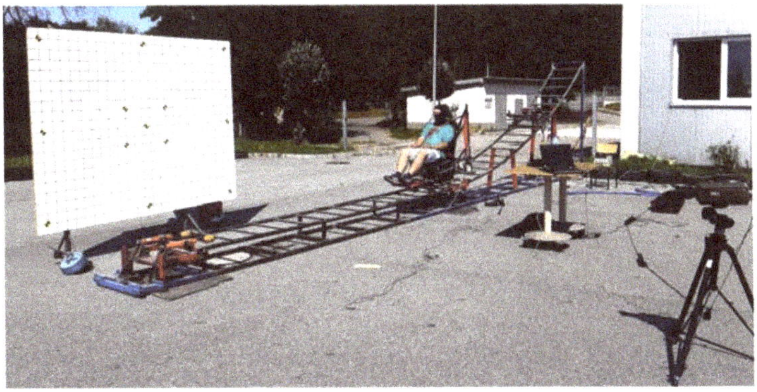

Figure 6. Test stand for crash tests at low speed.

Table 2. Volunteer anthropometric dimensions.

Parameters	Values	Allocation to the Population Percentile
Mass (kg)	90	C95
Height (cm)	181	C95
Head circumference (cm)	59	C95
Torso length (cm)	60	C95
Chest circumference (cm)	108	C95
Arm circumference (cm)	37	C50
Arm length (cm)	33	C50
Forearm circumference (cm)	31	C50
Wrist circumference (cm)	20	C50
Wrist width (cm)	10	C50
Hand width (cm)	25	C95
Thigh circumference (cm)	57	C50
Circumference of the lower leg (cm)	40	C50
Ankle circumference (cm)	26	C50
Foot length (cm)	24	C50

Comparative tests of the low speed crash test dummy were conducted with the Hybrid III dummy and volunteer. It should be noted that in the case of the test with the Hybrid III dummy and volunteer, other vehicle seats were used, therefore the stiffness of the backrest and the headrest are different. Moreover, the designed vehicle seat in the ADAMS program consists of a single body having the same stiffness, which makes it difficult to compare the head displacement of the simulation dummy with the Hybrid III and volunteer.

3. Results of Simulation Tests

The analysis of the film and the elaboration of the results were done with the aid of the TEMA program. This program allows us to analyze the movement of objects recorded with the camera. The program included a head movement trajectory for both the Hybrid III dummy and the volunteer. Figure 7 shows a comparison of the simulated dummy head displacement with the research experiments. It should be noted that in the case of comparing the results of computer simulation studies with the results of Hybrid III and volunteers, the compliance is at the level of 88%. The result in the first phase of the head movement caused by the sudden stopping of the seat is characterized by a perfect mapping of the simulation to real conditions, while in the case of the head returning to the vehicle seat, a difference of 12% is visible, which is caused by the difference in the stiffness of the seats in the computer simulation and the experiment.

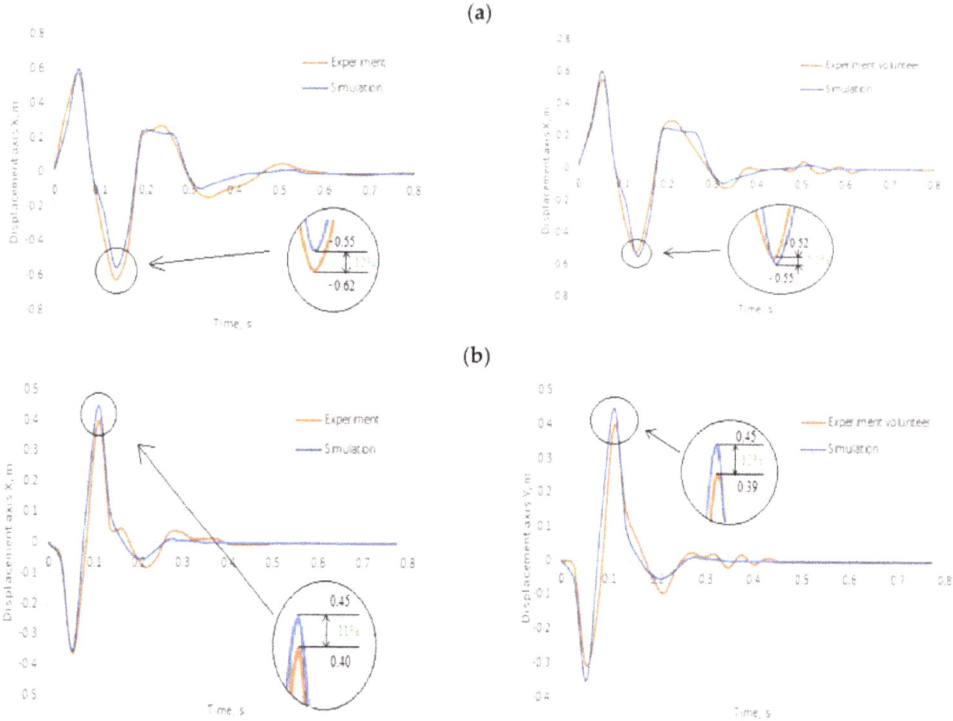

Figure 7. Displacement of the dummy head (**a**) relative to the X axis, (**b**) relative to the Y axis.

Table 3 shows the percent accuracy of the dummy's head displacement, in the first phase of the impact, when the dummy's head tilts the most forward and in the second phase, when the dummy's head returns to the vehicle seat and tilts its most rearward. It can be seen that for both the X-axis and the Y-axis in the first phase of the head movement, the difference between the simulation dummy and the hybrid III dummy is up to 5%. The simulation dummy is characterized by greater accuracy with the hybrid III dummy. It should be noted that the volunteer does not have anthropometric dimensions typical for the 50th percentile of the population, which may result in a lower accuracy of data replication in the comparison of the simulation of the dummy with the volunteer.

Table 3. Comparison of the accuracy of the dummy head displacement.

Comparison	The First Phase of the Head Movement		The Second Phase of Head Movement	
	Relative to the X Axis	Relative to the Y Axis	Relative to the X Axis	Relative to the Y Axis
Simulation dummy—Hybrid III dummy	95%	98%	88%	89%
Simulation dummy—Volunteer	94%	90%	94.5%	82%

Figure 8 shows a comparison of the first phase of the movement of the computer simulation and the experiment with the Hybrid III dummy. The difference in the results is 5%.

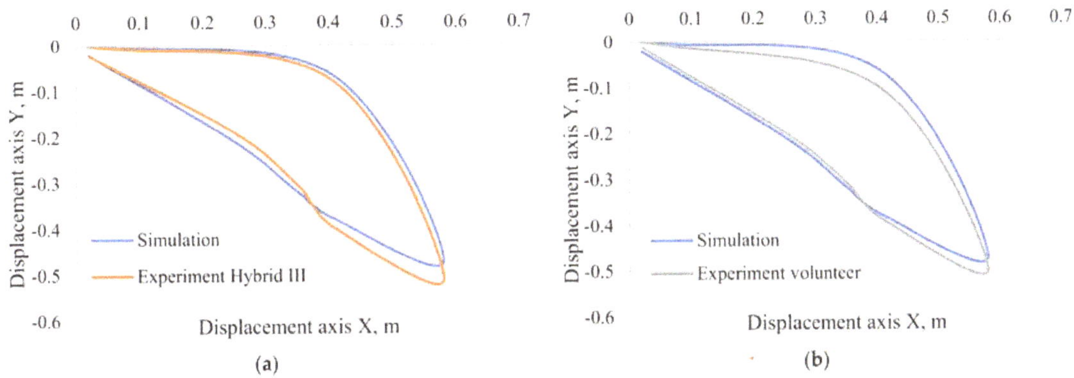

Figure 8. Head movement trajectory; (**a**) head displacement comparison for the simulation dummy and the Hybrid III dummy, (**b**) head displacement comparison for simulation dummy and volunteer.

Using TEMA software, the accelerations acting on the head of the Hybrid III dummy and the volunteers were determined. However, for the simulation dummy, the accelerations acting on the head were determined in the ADAMS program. The results of the dynamic analysis for the three types of studies showed a slight difference in the course of the accidental head acceleration. The resultant acceleration of the simulation dummy's head was compared with the result of the Hybrid III dummy and the volunteer in Figure 9.

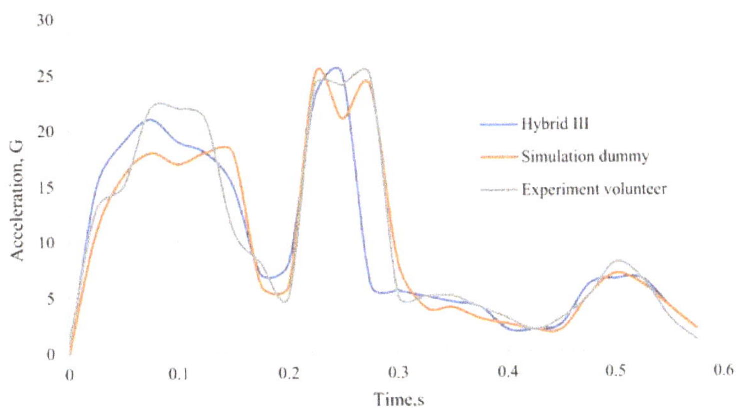

Figure 9. The resultant head acceleration.

4. Conclusions

The anthropometric dummies and simulation dummies used are used for crash tests from a speed of 40 km/h. It should be noted that at this speed, the human body behaves differently than in a collision with a speed of 20 km/h. Low speed collisions occur especially in the case of rear collisions as a result of the vehicle speed and distance not being adjusted. In addition, rear collisions are becoming more frequent due to the development of the infrastructure of expressways and highways, where traffic flows in one direction. It should be noted that there is a need to build a test dummy at a low speed of impact, as the human body will behave differently in a collision at 20 km/h than in a collision at 50 km/h. Serious damage to the spine can occur even in a collision with a speed of 20 km/h. In addition, at this speed, the airbag will not be activated in a passenger vehicle, so a collision could seriously damage occupants' health [61,62].

The current crash test dummies have been designed to best reflect the human body. In addition, the dummies were divided into rear, side, and front crash tests. Because they are constructed in such a way as to collect as much information as possible to help in the detection of injuries and the improvement of passenger vehicle safety. In the scientific literature, there are many scientific publications on the comparisons of anthropometric mannequins among themselves, thanks to which we know perfectly well which mannequin better replicates a given part of a human. For example, the BIORID II manikin reproduces a human spine better than the Hybrid III manikin. Therefore, it is recommended for rear crash tests. It should be emphasized that the current dummies are a specialized research unit equipped with thousands of sensors. The authors would like to point out that each of the dummies has its own purpose and there is no one universal dummy that could be used in any type of crash test at any speed which would perfectly replicate the human body.

The ADAMS program enables modification of the dummy's construction and adopted model parameters. Comparing the computer simulation with the experiment at 20 km/h showed that the computer simulation data differed from the experimental data by 12%. Moreover, by modifying the stiffness and damping parameters as well as the mass of individual parts of the dummy's body, it is possible to obtain crash test data taking into account individual percentiles of the human population.

The model made in the ADAMS program is a conceptual model on the basis of which the physical model will be made. The simulation model has knee and shoulder joints, which are patent applications [63–65]. These joints were first tested using the ADAMS program, and then, after successful attempts to reproduce the ranges of human body motion, they were made and incorporated into a physical model of dummy. The comparison of the simulation model with the actual experiment shows that such selected parameters of stiffness and damping in individual joints are able to reproduce individual movements of the human body.

Work on the physical structure of the simulation dummy involves the use of elements corresponding to the shape, mass, and dimensions of individual parts of the human body, as well as the use of special joints reflecting the range of motion of individual human joints [44].

The tests carried out to compare the constructed dummy with the experiments confirm the validity of the construction of the crash test dummy at low speed. In addition, it marks the beginning of work on an anthropometric dummy used in low speed crash tests for both front and rear collisions and with different impact angles.

Author Contributions: Conceptualization, D.F.; formal analysis, D.F. and M.J.; funding acquisition, D.F.; methodology, D.F. and M.J.; project administration, D.F. and M.J.; resources, D.F.; software, D.F. and M.J.; writing—original draft, D.F. and R.C.; Writing—review & editing, M.J. and J.M. All authors have read and agreed to the published version of the manuscript.

Funding: The research was financed by the National Centre for Research and Development, grant number POIR.04.01.04 00 0004/19 00.

Acknowledgments: The research was carried out as part of the Innovative system research project supporting the motor vehicle insurance risk assessment dedicated to UBI (Usage Based Insurance) No. POIR.04.01.04 00 0004/19 00 financed by the National Centre for Research and Development.

Conflicts of Interest: The authors declare no conflict of interest.

References

1. Nie, B.; Sathyanarayan, D.; Ye, X.; Crandall, J.R.; Panzer, M.B. Active muscle response contributes to increased injury risk of lower extremity in occupant–knee airbag interaction. *Traffic Inj. Prev.* **2017**, *19*, 76–82. [CrossRef]
2. Corazza, M.; Trincone, S.; Virgili, A. Effects of Airbag Deployment. *Am. J. Clin. Dermatol.* **2004**, *5*, 295–300. [CrossRef]

3. THOR-50M Qualification Procedures Manual. Available online: https://www.nhtsa.gov/sites/nhtsa.dot.gov/files/thor-50m_qualification_august2016.pdf (accessed on 1 February 2021).
4. Description and Performance of the Hybrid III Three-Year-Old, Six-Year-Old and Small Female Test Dummies in Restraint System and Out-Of-Position Air Bag Environments. Available online: https://www.nhtsa.gov/sites/nhtsa.dot.gov/files/1998esvpaper.pdf (accessed on 1 February 2021).
5. Dębecki, R. Kiedy Działa Airbag? [When Does Airbag Work?]. Available online: https://www.auto-swiat.pl/porady/eksploatacja/kiedy-dziala-airbag/cfbvlwt (accessed on 1 February 2021).
6. Kostek, R.; Aleksandrowicz, P. Simulation of car collision with an impact block. In Proceedings of the 11th International Congress of Automotive and Transport Engineering 2017, Pitesti, Romania, 8–10 November 2017; pp. 1–6. [CrossRef]
7. Cunningham, K.; Brown, T.D.; Gradwell, E.; Nee, P.A. Airbag associated fatal head injury: Case report and review of the literature on airbag injuries. *Emerg. Med. J.* **2000**, *17*, 139–142. [CrossRef]
8. Liu, Z.; Zhang, X.; Chen, H.; Shi, Y. Optimization of calibration parameters for hybrid III 5th female dummy neck calibration based on taguchi method. In Proceedings of the 2010 Third International Conference on Information and Computing, Wuxi, China, 6 June 2010; Volume 4, pp. 97–100. [CrossRef]
9. Goo, A.; Laubscher, C.A.; Farris, R.J.; Sawicki, J.T. Design and Evaluation of a Pediatric Lower-Limb Exoskeleton Joint Actuator. *Actuators* **2020**, *9*, 138. [CrossRef]
10. Lemmen, P.; Hynd, D.; Carroll, J.; Davidsson, J.; Bernard, B.; Song, E.; Steeger, B. Thoracic Injury Assessment for Improved Vehicle Safety. *Procedia Soc. Behav. Sci.* **2012**, *48*, 1649–1661. [CrossRef]
11. Frej, D.; Zuska, A.; Cadge, K. Analysis of vertical vibrations affecting a child transported in a child seat during a vehicle passing over the release speed bump. *Arch. Automot. Eng.* **2019**, *86*, 111–125. [CrossRef]
12. Amiri, S.; Naserkhaki, S.; Parnianpour, M. Modeling and validation of a detailed FE viscoelastic lumbar spine model for vehicle occupant dummies. *Comput. Biol. Med.* **2018**, *99*, 191–200. [CrossRef]
13. Nycz, D. Influence of selected design parameters of the composite-foam cover rail on the course of the TB11 crash test of a road safety barrier forming a horizontal concave arc. *Arch. Automot. Eng. Arch. Motoryz.* **2017**, *76*. [CrossRef]
14. Szada-Borzyszkowski, W.; Szada-Borzyszkowska, M. Wpływ prędkości pojazdów na wielkość uszkodzeń podczas zderzenia (Influence of vehicle speed on the magnitude of damage during a collision). *Autobusy Tech. Eksploat. Syst. Transp.* **2015**, *16*, 224–227.
15. Brolin, K.; Stockman, I.; Andersson, M.; Bohman, K.; Gras, L.L.; Jakobsson, L. Safety of children in cars: A review of biomechanical aspects and human body models. *Iatss Res.* **2015**, *38*, 92–102. [CrossRef]
16. Żuchowski, A. Seatbelt impact on a child during a frontal collision. *Arch. Automot. Eng. Arch. Motoryz.* **2017**, *75*. [CrossRef]
17. Żuchowski, A. Analysis of the influence of the vehicle impact velocity on the loads of the dummies in the front and rear seats. *Arch. Automot. Eng. Arch. Motoryz.* **2018**, *81*, 159–176. [CrossRef]
18. Available online: https://www.humaneticsatd.com (accessed on 1 January 2020).
19. Final Publishable Report. Available online: https://ec.europa.eu/transport/road_safety/sites/roadsafety/files/pdf/projects_sources/fid_final_report.pdf (accessed on 1 January 2020).
20. Stańczyk, T.L.; Zuska, A. Review of anthropodynamic dummies used to evaluate the effect of vibrations on sitting human (vehicle driver). *Arch. Automot. Eng. Arch. Motoryz.* **2014**, *65*, 65–74.
21. Jaśkiewicz, M.; Jurecki, R.; Witaszek, K.; Więckowski, D. Overview and analysis of dummies used for crash tests. *Sci. J. Marit. Univ. Szczec.* **2013**, *35*, 22–31.
22. Kent, R.; Bass, C.; Woods, W.; Salzar, R.; Lee, S.; Melvin, J. The role of muscle tensing on the force-deflection response of the thorax and a reassessment of frontal impact thoracic biofidelity corridors. *J. Automob. Eng. Proc. Inst. Mech. Eng.* **2006**, *220*, 853–868. [CrossRef]
23. Jaśkiewicz, M.; Więckowski, D.; Witaszek, K. Przegląd i analiza manekinów stosowanych do testów zderzeniowych (Review and analysis of crash test dummies). *Zesz. Nauk. Mech. Politech. Opol.* **2013**, *101*, 47–48.
24. Lemmen, P.; Been, B.; Hynd, D.; Carroll, J. Development of an advanced thorax/shoulder complex for the THOR dummy. In Proceedings of the Symposium on International Automotive Technology (SIAT), Pune, India, 9–12 January 2013; Volume 17, pp. 25–26. [CrossRef]
25. Mohan, P.; Marzougui, D.; Kan, C. *Development and Validation of Hybrid III Crash Test Dummy*; SAE Technical Paper 2009-01-0473; SAE International; The George Washington University: Washington, DC, USA, 2009. [CrossRef]
26. Zhang, W.; Kapoor, T.; Tot, M.; Altenhof, W. *A Comparison of the Kinematics of a Child Finite Element Model and the HYBRID III 3-Year-Old Dummies in Frontal Crashes*; SAE Technical Paper 2007-01-0977; SAE International: Warrendale PA, USA, 2007. [CrossRef]
27. Wang, W.; Zhang, D.; Ji, J.; Tian, H.; Zhao, H. Seat belt protection of train driver during secondary impact. In Proceedings of the 2016 IEEE International Conference on Intelligent Transportation Engineering (ICITE), Singapore, 20–22 August 2016; pp. 85–88. [CrossRef]
28. Jaśkiewicz, M.; Więckowski, D.; Poliak, M.; Frej, D. Shoulder joint design of the anthropometric dummy used for crash tests. In Proceedings of the 2020 XII International Science-Technical Conference Automotive Safety, Kielce, Poland, 21–23 October 2020; pp. 1–5. [CrossRef]

29. Nordhoff, L.; Freeman, M.; Siegmund, G. Human Volunteer and Anthropomorphic Dummy Tests of General Motors Driver Air Cushion System, SAE Paper No. 740578, Society of Automotive Engineers, Warrendale, PA, 1974. In *Human Subject Crash Testing: Innovations and Advances;* SAE International: Warrendale, PA, USA, 2007; pp. 307–321.
30. Biofidelity Impact Response Requirements for an Advanced Mid-Sized Male Crash Test Dummy. Available online: https://hal.archives-ouvertes.fr/hal-01490941/document (accessed on 1 January 2020).
31. Forman, J.; Poplin, G.S.; Shaw, C.G.; McMurry, T.L.; Schmidt, K.; Ash, J.; Sunnevang, C. Automobile injury trends in the contemporary fleet: Belted occupants in frontal collisions. *Traffic Inj. Prev.* **2019**, *20*, 607–612. [CrossRef]
32. Han, Y.; Tang, H.; Tian, F.; Huang, H.; Mizuno, K. Analysis of the equation of motion on the chest of hybrid III 3YO dummy in dynamic loads. In Proceedings of the 2018 International Conference on Robots and Intelligent System (ICRIS), Changsha, China, 26–27 May 2018; pp. 529–533. [CrossRef]
33. Ludwinek, K.; Jurecki, R.; Jaskiewicz, M.; Szumska, E.; Sulowicz, M. A test stand for the experimental analysis of physical quantities during crash test at low speeds. In Proceedings of the 2018 XI International Science-Technical Conference Automotive Safety, Casta, Slovakia, 18–20 April 2018; pp. 1–7. [CrossRef]
34. Manary, M.A.; Klinich, K.D.; Orton, N.R.; Reed, M.R.; Rupp, J.D. *Comparing the CRABI-12 and CRABI-18 for Infant Child Restraint System Evaluation;* (Report No. DOT HS 812 156); National Highway Traffic Safety Administration: Washington, DC, USA, 2015. Available online: https://www.nhtsa.gov/sites/nhtsa.dot.gov/files/documents/11316-crabi_12and18_060215_v4_tag.pdf (accessed on 1 February 2021).
35. Bertocci, G.E.; Szobota, S.; Hobson, D.A.; Digges, K. Computer simulation and sled test validation of a powerbase wheelchair and occupant subjected to frontal crash conditions. *IEEE Trans. Rehabil. Eng.* **1999**, *7*, 234–244. [CrossRef]
36. Thuong, L.N.P. Vehicle frontal impact to pole barrier simulation using computer finite element model. In Proceedings of the 2018 4th International Conference on Green Technology and Sustainable Development (GTSD), Ho Chi Minh City, Vietnam, 23–24 November 2018; pp. 273–277. [CrossRef]
37. Sun, Y.; Sun, X.; Rong, J.; Tai, Y. Development of energy absorbing refuge island using crash simulation. In Proceedings of the 2011 International Conference on Transportation, Mechanical, and Electrical Engineering (TMEE), Changchun, China, 16–18 December 2011; pp. 1012–1015. [CrossRef]
38. Zhang, Y.; Song, J.; Qian, Y.; Gao, Q. Simulation and optimization of vehicle side airbag. In Proceedings of the 2018 International Computers, Signals and Systems Conference (ICOMSSC), Dalian, China, 19–21 December 2018; pp. 320–323. [CrossRef]
39. Shameem, S.; Prasad, G.R.K.; Ch, V.T.; Kumar, K.B.; Yaswanth, M.; Ch, M. Design simulation and analysis of crash sensor for air bag system. In Proceedings of the 2018 3rd International Conference on Inventive Computation Technologies (ICICT), Coimbatore, India, 15–16 November 2018; pp. 718–723. [CrossRef]
40. Sharma, A.; Sharma, S.; Chhabra, M. Recent Developments in Airbags for Passenger Safety in Automobile Engineering. *Int. J. Res. Advent Technol.* **2017**, *5*, 2321–9637.
41. He, P.; Jiang, X.; Yang, J. A Study on rear seat occupant injuries in side impact. In Proceedings of the 2012 Third International Conference on Digital Manufacturing and Automation, GuiLin, China, 31 July–2 August 2012; pp. 144–147. [CrossRef]
42. Zhang, Y.; Ju, C.; Yue, G.; Chen, X.; Sun, H. Simulation analysis of the rear seat female in front impact. In Proceedings of the 2012 Third International Conference on Digital Manufacturing and Automation, GuiLin, China, 31 July–2 August 2012; pp. 767–770. [CrossRef]
43. Ruhai, G.; Manjiang, H.; Dong, X. Simulation study of bus occupant restraint system in emergency brake. In Proceedings of the 2010 International Conference on Optoelectronics and Image Processing, Haikou, China, 11–12 November 2010; pp. 392–395. [CrossRef]
44. Jaśkiewicz, M.; Frej, D.; Tarnapowicz, D.; Poliak, M. Upper limb design of an anthropometric crash test dummy for low impact rates. *Polymers* **2020**, *12*, 2641.
45. Sonawane, C.R.; Shelar, A.L. Strength Enhancement of Car Front Bumper for Slow Speed Impact by FEA Method as per IIHS Regulation. *J. Inst. Eng. India Ser. C* **2018**, *99*, 599–606. Available online: https://doi.org/10.1007/s40032-017-0365-y (accessed on 2 March 2021). [CrossRef]
46. Jakobsson, L.; Norin, H.; Bunketorp, O. In-depth study of whiplash associated disorders in frontal Impacts: Influencing factors and consequences. In Proceedings of the International IRCOBI Conference on the Biomechanics of Impact, Munich, Germany, 18–20 September 2002; pp. 211–222.
47. Croft, A.C.; Eldridge, T.R. Human subject rear passenger symptom response to frontal car-to-car low-speed crash tests. *J. Chiropr. Med.* **2011**, *10*, 141–146. [CrossRef]
48. Krafft, M.; Kullgren, A.; Ydenius, A.; Boström, O.; Håland, Y.; Tingvall, C. Rear impact neck protection by reducing occupant forward acceleration—A study of cars on Swedish roads equipped with crash recorders and a new anti-whiplash device. In Proceedings of the International IRCOBI Conference, Graz, Austria, 22–24 September 2004; pp. 221–231.
49. Jaśkiewicz, M.; Frej, D.; Šarkan, B. Construction of the knee joint of the dummy designed for crash tests. *Trans. Res. Procedia* **2020**, *44*, 121–128. [CrossRef]
50. Nordhoff, L.; Freeman, M.; Siegmund, G. Kinetic and kinematic responses of the RID2a, hybrid III, and human volunteers in low-speed rear-end collisions. *Stapp Car Crash J.* **2001**, *45*, 239–256, (in Human Subject Crash Testing: Innovations and Advances, SAE 2007, 865–882).

51. Nordhoff, L.; Freeman, M.; Siegmund, G. TRL Rear Impact Volunteer Testing: Methods and Measurements. In *Human Subject Crash Testing: Innovations and Advances, Proceedings of the 1st APSN Workshop on Biomechanical Experiments, Advanced Passive Safety Network, Graz, Austria, 21 September 2004*; SAE International: Warrendale, PA, USA, 2007; pp. 611–627.
52. Nordhoff, L.; Freeman, M.; Siegmund, G. Human Volunteer Head-T1 Response for Oblique Impact Conditions. In *Human Subject Crash Testing: Innovations and Advances, Proceedings of the International IRCOBI Conference on the Biomechanics of Impact, Graz, Austria, 22–24 September 2004; pp. 53–68*; SAE International: Warrendale, PA, USA, 2007; pp. 269–284.
53. Nordhoff, L.; Freeman, M.; Siegmund, G. Relationship between localized spine deformation and cervical vertebral motions for low speed rear impacts using human volunteers. In *Human Subject Crash Testing: Innovations and Advances, Proceedings of the International IRCOBI Conference on the Biomechanics of Impact, Stiges, Spain, 23–24 September 1999; pp. 149–164*; SAE International: Warrendale, PA, USA, 2007; pp. 773–788.
54. Yamazaki, K.; Ono, K.; Ishii, M. Biofidelity of rear impact dummies in low speed rear-end impact: Comparison of rigid seat and mass production car seat in human volunteers. International Phantom data sheet. In Proceedings of the 2008 International Ircobi Conference on the Biomechanics of Injury, Bern, Switzerland, 17–19 September 2008. Available online: http://www.komiweb.co.kr/data/v310.pdf (accessed on 22 February 2020).
55. Castro, W.H.; Meyer, S.J.; Becke, M.E.; Nentwig, C.G.; Hein, M.F.; Ercan, B.I.; Thomann, S.; Wessels, U.; Du Chesne, A.E. No stress—No whiplash? Prevalence of "whiplash" symptoms following exposure to a placebo rear-end collision. *Int. J. Leg. Med.* **2001**, *114*, 316–322. [CrossRef] [PubMed]
56. Nordhoff, L.S., Jr.; Freeman, M.D.; Siegmund, G.P. *Human Subject Crash Testing: Innovations and Advances (PT-134)*; SAE International: Warrendale, PA, USA, 2007; ISBN 9780768019315.
57. Hynd, D.; Willis, C.; Roberts, A. TRL rear impact volunteer testing: Method and measurements. In Proceedings of the D16-WG 5.2 Workshop on Biomechanical Experiments, Berks, UK, 21 September 2004; pp. 18–28.
58. Lopez-Valdes, F.J.; Kent, R.; Arbogast, K.; Higuchi, K. Analysis of spinal motion during frontal impacts. Comparison between PMHS and ATD. *Ann. Adv. Automot. Med.* **2010**, *54*, 61–78. [PubMed]
59. Noureddine, A.; Eskandarian, A.; Kennerly, D. Computer modeling and validation of a Hybrid III dummy for crashworthiness simulation. *Math. Comput. Model.* **2002**, *35*, 885–893. [CrossRef]
60. Bailey, A.; Chirstopher, J.; Henderson, K.; Brozoski, F.; Salzar, R.S. Comparison of hybrid-III and PMHS response to simulated underbody blast loading conditions. In Proceedings of the 13th International Research Council on the Biomechanics of Injury Conference (IRCOBI), Gothenburg, Sweden, 11–13 September 2013; pp. 58–171.
61. Stand for Simulation Testing of Dynamic Stresses, Especially in the Elements Securing the Driver. Available online: https://patents.google.com/patent/PL430055A1/pl?oq=430055+ (accessed on 1 January 2020).
62. Zellmer, H.; Muser, M.; Stamm, M.; Walz, F.; Hell, W.; Langweider, K.; Philippens, M. Performance comparison of rear impact dummies: Hybrid III (TRID), BioRID and RID 2. In Proceedings of the International IRCOBI Conference on the Biomechanics of Impact, Munich, Germany, 18–20 September 2002.
63. Cappon, H.J.; Philippens, M.; van Ratingen, M.R.; Wismans, J.S.H.M. Evaluation of dummy behaviour during low severity rear impact. In Proceedings of the IRCOBI Conference on the Biomechanics of Impact, Montpeller, France, 20–22 September 2000; pp. 53–66.
64. Frej, D.; Ja'skiewicz, M. Przegub Kolanowy Manekina Antropometrycznego do Testów Zderzeniowych [Knee-Joint of the Anthropometric Crash Test Dummy]. PL. Patent P.431522, 21 October 2019.
65. Frej, D.; Ja´skiewicz, M. Przegub Barkowy Manekina Antropometrycznego do Testów Zderzeniowych [Shoulder Joint of the Anthropometric Crash Test Dummy Application]. PL. Patent P.433041, 27 February 2020.

energies

MDPI

Article

Impact of Control System Model Parameters on the Obstacle Avoidance by an Autonomous Car-Trailer Unit: Research Results

Leon Prochowski [1,2], Mateusz Ziubiński [1,*], Patryk Szwajkowski [3], Mirosław Gidlewski [1,2], Tomasz Pusty [4] and Tomasz Lech Stańczyk [5]

1 Institute of Vehicles and Transportation, Military University of Technology (WAT), gen. Sylwestra Kaliskiego 2 Street, 00-908 Warsaw, Poland; leon.prochowski@wat.edu.pl (L.P.); miroslaw.gidlewski@wat.edu.pl (M.G.)
2 Łukasiewicz Research Network—Automotive Industry Institute (Łukasiewicz-PIMOT), Jagiellońska 55 Street, 03-301 Warsaw, Poland
3 Electromobility Department, Łukasiewicz Research Network—Automotive Industry Institute (Łukasiewicz-PIMOT), Jagiellońska 55 Street, 03-301 Warsaw, Poland; patryk.szwajkowski@pimot.lukasiewicz.gov.pl
4 Vehicle Tests Laboratory, Łukasiewicz Research Network—Automotive Industry Institute (Łukasiewicz-PIMOT), Jagiellońska 55 Street, 03-301 Warsaw, Poland; tomasz.pusty@pimot.lukasiewicz.gov.pl
5 Department of Automotive Engineering and Transport, Kielce University of Technology, Ave. 1000—lecia Państwa Polskiego 7, 25-314 Kielce, Poland; stanczyk@tu.kielce.pl
* Correspondence: mateusz.ziubinski@wat.edu.pl

Citation: Prochowski, L.; Ziubiński, M.; Szwajkowski, P.; Gidlewski, M.; Pusty, T.; Stańczyk, T.L. Impact of Control System Model Parameters on the Obstacle Avoidance by an Autonomous Car-Trailer Unit: Research Results. *Energies* **2021**, *14*, 2958. https://doi.org/10.3390/en14102958

Academic Editor: Islam Safak Bayram

Received: 2 April 2021
Accepted: 17 May 2021
Published: 20 May 2021

Abstract: The introduction of autonomous cars will help to improve road traffic safety, and the use of a cargo trailer improves the energy efficiency of transport. One of the critical (collision) road situations has been considered, where immediate counteraction is required in a space that has been only partly defined. This research work was aimed at determining the impact of the trajectory planning method and the values of some parameters of the control system on the feasibility of safe avoidance of an obstacle that has suddenly appeared. The obstacle is assumed to be a motor vehicle moving on a road intersection along a collision path in relation to the autonomous car-trailer unit (CT unit) travelling at high speed. Analysis of cooperation between several non-linear models (representing the car, trailer, tyre–road interaction, and driving controller) has been carried out. Mathematical models of the control system and the CT unit have been built. The process of selection of temporary and variable parameters, applied to the control system for the time of the critical situation under consideration, has been shown. The research work carried out has made it possible to recommend appropriate parameter values for the control system.

Keywords: safety of autonomous motor vehicle; autonomous motor vehicle with a trailer; critical road situations; vehicle trajectory planning; mathematical modelling motor vehicle with a trailer; control system of an autonomous motor vehicle

1. Introduction

Autonomous vehicles are considered as a solution expected to improve the efficiency of transport processes. An important good point of the introduction of autonomous vehicles will be an improvement in road traffic safety. Such an effect, however, will not be produced automatically. It may be achieved through research on vehicle behaviour, including the selection of a vehicle controlling method that would be adequate for difficult road situations. The problem of adapting the vehicle control process to special road situations has been raised, e.g., in [1–7]. A key factor is here the programming of the vehicle control system, in which the algorithms responsible for planning the obstacle-avoiding trajectories are of significant importance. At present, the research works on the control systems are predominantly focused on the planning of vehicle trajectories for lane-change maneuvers (obstacle avoiding, overtaking) [6,8–11].

A lane-change maneuver of this kind is usually planned in advance and performed in predictable conditions, where the restrictions arising from the necessity of avoiding collisions with other vehicles participating in the traffic or from the kinematic and dynamic properties of the vehicle in question can be easily met. In such a situation, a planning stage can be introduced to the control system, at which time the appropriate trajectory would be chosen from a library of solutions pre-programmed in the system controller [12].

The addition of a trailer to the autonomous vehicle will result in better economic and energy efficiency of the transport processes. However, this will also bring about a change in the dynamic properties of the CT unit compared with those of the motor vehicle alone. The presence of a trailer radically affects the dynamics of the towing vehicle and reduces the stability of the vehicle combination as a whole. An extensive review of such research works on CT units has been presented in [13,14]. The attaching of a trailer to a motor vehicle may also cause oscillations of the CT unit in the final phase of the obstacle avoidance process (in the initial phase, the trajectories of both vehicles are almost identical). This can be observed, e.g., in the profile of the trajectory of the CT unit's center of gravity (CG) for rising vehicles' yaw angles from the carriageway centerline [15]. The research on the stability of CT unit's motion, reported in [13,14], has shown that the instabilities occurring in the trailer and vehicle's motion strongly depend on the mass and moment of inertia of the vehicles and on the drawbar length. Experimental research on the stability of a car-trailer unit within a sped range of 48–90 km/h, with a rapid turn of the steering wheel, has confirmed decisive impact of the parameters mentioned above on the behaviour of the vehicle combination [16]. These findings have been taken into account in the modelling described hereafter. The controlling of a motorcar-trailer unit is a far more complex issue in comparison with the controlling of a motor vehicle alone [15]. The values and ranges of the input parameters applied to the control system model must be different as well.

The participation of autonomous motor vehicles in the road traffic may be described with using a few elementary vehicle trajectory models: following of the preceding vehicle, following-up of predesigned reference models, and driving to follow up models that would represent the planned (theoretical) vehicle trajectory [5,6,17]. The design of such models is based on analyses of specific traffic situations and the most frequent drivers' behaviors [18,19]. This is of critical importance for the safe operation of autonomous vehicles in the road traffic where cars driven by human drivers will remain predominate for many upcoming years.

Regardless of the driving model chosen, autonomous vehicles move to follow up a pre-planned trajectory. This is also the case when the obstacle avoidance takes place. A critical review of the trajectory planning methods has been presented, e.g., in [1,11]. The trajectory planning is based on determining the curve that describes the lateral displacement of the center of vehicle mass (e.g., when the vehicle changes the lane to the adjacent one) in the form of algebraic equations, which may represent sequences of circular arcs, polynomial splines, clothoid splines, Bézier curves, etc. [1,5,9,20]. When the desired vehicle trajectory is determined, it is important that the basic limitations dictated by the properties of real cars and road surface should be taken into account [21]. As an example, a method of generating the vehicle trajectory has been presented in [8], where the maximum acceptable values of the lateral (centripetal) acceleration of the car were taken into account. In the work reported in [9], the vehicle trajectories and their curvatures were planned taking into account the comfort of vehicle occupants, e.g., a requirement was adopted that the lateral acceleration should not exceed 1.6 m/s^2 (the limit for good comfort) or 3.6 m/s^2 (the limit for medium occupant's comfort). In the vehicle control system, calculations are carried out to plan the desired vehicle trajectory and to track the actual one. As another option, a controller provided with a library of pre-programmed solutions suitable for plannable situations may be used [12]. An example of planning a trajectory for a safe lane-change and obstacle avoidance maneuver, taking into account the current traffic situation and the dynamic properties of the CT unit, is shown in [10]. In the vehicle control process, the goal is to minimize the distance between the current position of the center of vehicle mass and the planned vehicle trajectory, and to minimize the difference between the angular positions of the longitudinal vehicle axis and the tangent to the said trajectory [22–24].

In this role, PID (proportional–integral–derivative) controllers and controllers based on fuzzy logic predominate. As an example: in [20,24], the control process is based on the follow-up of the preset vehicle trajectory by a PID controller and the effective use of fuzzy logic in the controllers of mobile robots and vehicles is shown in [18,23,25]. A good result of controlling the drive of mobile platforms in [21] was achieved by using two different control techniques. The controlling of a car with a trailer in typical road situations has been analyzed in [10], where the current obstacle position and the static space limitations posed by the road infrastructure have been taken into account. An interesting method of planning the vehicle trajectory, taking into account the field of "obstacle repulsion" potential during the lane-change maneuver, has been proposed in [6]. In [26], on the other hand, the trajectory was planned based on the preset direction of vehicle movement and the positions of the centers of front and rear axles of an articulated wheel loader relative to the optimum trajectory. The motion of such a machine in a predetermined environment has been described by a trajectory composed of circular arcs and line segments.

At present, the vehicle trajectory is predominantly planned on the grounds of the limitations dictated by the structural vehicle's properties [7,27], and the basic vehicle control methods include the fuzzy logic algorithms [11,27].

Most of the reported methods of generating the desired vehicle trajectory apply to typical maneuvers often performed in road traffic. In contrast, there is a lack of research works and models applicable to the critical situations where autonomous vehicles towing trailers with a high speed would be involved. Particularly dangerous situations take place when the vehicle and the obstacle move along collision paths and in an environment that has been only partly defined.

The study presented includes an analysis of the problem of avoiding an obstacle in a critical road situation that may arise from, e.g., another vehicle driver's failure to yield the right of way on a road intersection. This usually develops into front-to-side collisions of moving vehicles; the percentage of such collisions in the total number of road accidents in Poland shows an upward trend. At present, the said percentage amounts to 32% [28].

It is peculiar to the critical situations that they require difficult defensive (accident-avoiding) maneuvers to be performed, which are often based on very aggressive vehicle control. If this is the case, the vehicle trajectory is planned without being impeded by any limitations related to the possible occurrence of excessive lateral acceleration, tire sideslip, or development of forces exceeding the lateral tire-road adhesion. The analysis applies to a road situation where immediate counteraction is required in a space that has been only partly defined. The autonomous vehicle's control system is expected to plan a safe vehicle trajectory based on information received from an environment perception system. It has been assumed that the algorithm of controlling the vehicle will not change in spite of the occurrence of a critical situation. Nevertheless, the following factors may change according to the information received from the environment perception system:

- vehicle trajectories planning (calculation) method;
- values of the parameters that are treated as variables in the vehicle control algorithm during the obstacle avoidance maneuvers.

Therefore, a temporary trajectory is planned in the critical situation under analysis in order to avoid a collision with the obstacle.

The objective of this study is to determine the impact of the trajectory planning method and of the values of some control system parameters on the feasibility of the safe avoidance of an obstacle having suddenly appeared. In this study, the obstacle is a motor vehicle whose driver has violated traffic regulations. The obstacle is moving on a road intersection with poor visibility along a collision path in relation to an autonomous CT unit travelling with a high speed (Figure 1). After hard braking, the said motor vehicle has stopped with blocking one lane for the CT unit. A trajectory planning method and desirable values of the temporary parameters of the control system, which is based on an anticipating model and fuzzy logic, will be shown. The areas of advantageous choice of the temporary parameter values for the critical situation under analysis will be indicated. The problem is explored using computer simulation based on a model of CT unit's dynamics in curvilinear motion. The model was subjected to a validation process, in which results of experimental tests of dynamic lane changing by the CT unit were used.

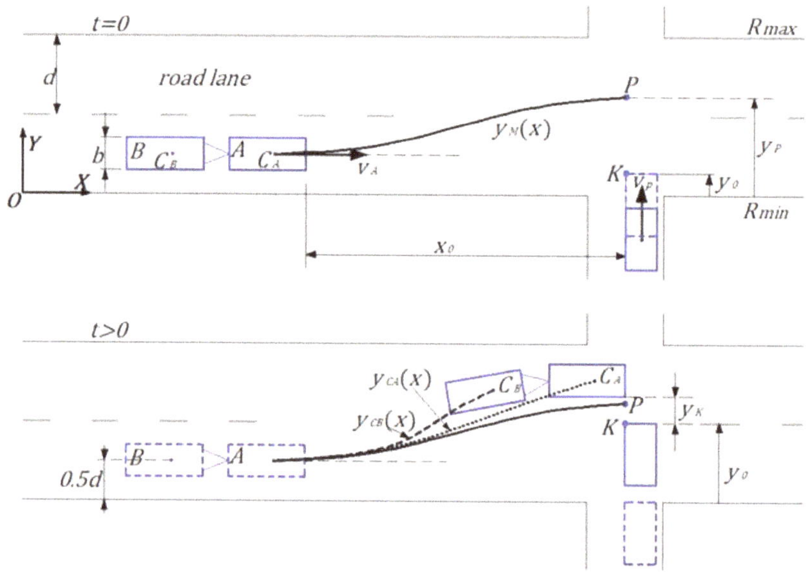

Figure 1. Road situation under analysis.

In this study, a situation is analyzed in which the vehicle's environment perception system has detected an obstacle that suddenly appeared at a distance that may be shorter than the stopping distance of the autonomous vehicle with a trailer. For such a situation, an assumption has been made that the information received from the environment perception system will cause the settings of the control system of the CT unit to be adjusted as appropriate. The new settings will be introduced temporarily (only for the time of avoiding the obstacle) and their values will differ from those required at the stable vehicle drive before and after the obstacle avoidance maneuver. The necessity of local trajectory planning in critical situations has been pointed out, e.g., in [29], where the limitations additionally arising from excessive tire slip and from development of forces exceeding the lateral tire-road adhesion have been highlighted.

In this problem, the control system must cope with a very difficult task. This is not only due to the very short time available for the perception of a specific road situation and for the trajectory planning, but also because of the dilemmas that would arise from the possible lack of any non-collision solution. Such a sudden situation, where only very few of the practicable defensive maneuvers may result in the successful avoidance of a road accident, may be defined as a critical one.

The behavior of a CT unit on the 0–60 m road section under analysis, i.e., before and beyond the obstacle, has been analyzed in [2]. In particular, the trajectory of the center of mass of the vehicle combination has been examined. The analysis presented herein is more detailed and the motion of outermost points on the external edges of the vehicle and trailer has been observed. Simultaneously, the analysis has been reduced to a 0–30 m road section, i.e., to the situation before the point where a collision between the vehicles involved may occur.

2. Scenario of the Road Situation under Analysis

In this study, the motion of the combination of an autonomous motorcar with a cargo trailer (CT unit) on two lines road with right of way is analyzed. During the motion, the vehicle's perception system has just detected the sudden appearance of another vehicle moving along a collision path (Figure 1). The said other vehicle may be expected to block within a short time the whole width of the lane used by the CT unit and thus to become an obstacle for the latter.

The following notation will be used in this study:

A, B—autonomous motorcar and trailer, respectively of center of mass C_A, C_B;
R_{min}, R_{max}—symbols indicating the outermost edges of the lanes involved;
L_a—anticipation radius, used when trajectory $y_T(x)$ is generated;
$y_M(x)$, $y_T(x)$—planned and preset vehicle trajectory;
y_0—instantaneous obstacle position in relation to lane edge R_{min};
$y_{CA}(x)$, $y_{CB}(x)$—trajectories of the centers of mass of vehicles A and B;
K, P—characteristic points: trace of the obstacle edge and target point for the planning of a safe trajectory $y_M(x)$;
y_W—clearance margin, necessary for safe obstacle avoidance;
y_K—clearance between the vehicle and the obstacle at the instant when the latter is being passed by;
b, d—widths of the vehicle combination (CT unit) and the lane;
δ_H, δ, α, β—steering wheel angle, steering angle, tire sideslip angle, and angle of position of the tangent to the planned or preset trajectory of the vehicles;
R_0, κ—radius and curvature of the vehicle trajectory;
ω_u; $v_{Tu} = \omega_u r_{Du}$—angular velocity of the uth vehicle wheel and circumferential velocity of the tyre;
r_{Du}—dynamic tire radius of the uth vehicle wheel;
v, v_A—velocity of the center of mass of vehicle A;
$\Delta\psi = \psi_A - \psi_B$—trailer drawbar turning angle;
ψ_A, ψ_B—yaw angles of the motor vehicle and the trailer;
F_Q, a_y—centrifugal inertia force and lateral acceleration of center of mass of the vehicle body.

The OXY coordinate system is attached to the R_{min} lane edge.

3. Control System and Model of Dynamics of the Vehicles

3.1. Structure of the Autonomous Vehicle's Control System

The CT unit's motion results from the cooperation of the following three major system components:

- environment perception system;
- vehicle control system;
- model adopted to represent the dynamics of the vehicle combination (CT unit).

The cooperation of the above system components has been illustrated in Figure 2.

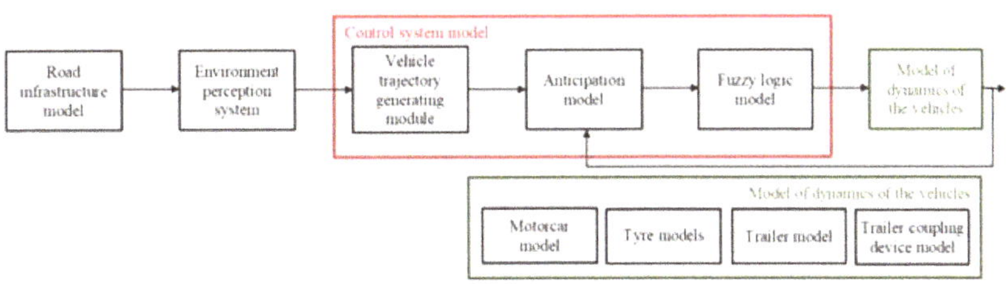

Figure 2. Structure of the control system.

3.2. Procedure of Determining the Control Signal Using a Fuzzy Logic Model

The sudden intrusion of an obstacle onto the lane used by the autonomous vehicle, as analyzed herein, means that an obstacle ($y_0 > 0$) appears at a distance of x_0 ahead of the vehicle (Figure 1). A case is addressed where this distance may be shorter than the stopping distance S_Z for the CT unit. Then, a safe solution may be to avoid the obstacle with using the adjacent road lane. In such a case, the method detecting the obstacle edge K by the environment perception system is very effective [30]. Based on an analysis of the current position of edge K relative to the R_{min} lane edge, the predicted position of the target point P (Figure 1), i.e., the y_P value, is calculated from Equation (1):

$$y_P = y_0 + 0.5b + y_W \tag{1}$$

This makes it possible to calculate trajectory $y_M(x)$ in the control system (Figure 3); the trajectory is treated as the desired (planned) path of the center of mass C_A within the x_0 road section [3]. The selection of an algorithm to generate this trajectory has been described in a subsequent part of this paper. The said trajectory is applied as an input to the anticipating model, where trajectory $y_T(x)$ is calculated based on equations:

$$\begin{gathered} y_T(x) = 0.5d \text{ for } x \in \langle 0; L_a) \\ y_T(x) = y_M\left(\frac{x-L_a}{x_0-L_a}x\right) \text{ for } x \in \langle L_a; x_0) \\ y_T(x) = y_0 \text{ for } x = x_0 \end{gathered} \tag{2}$$

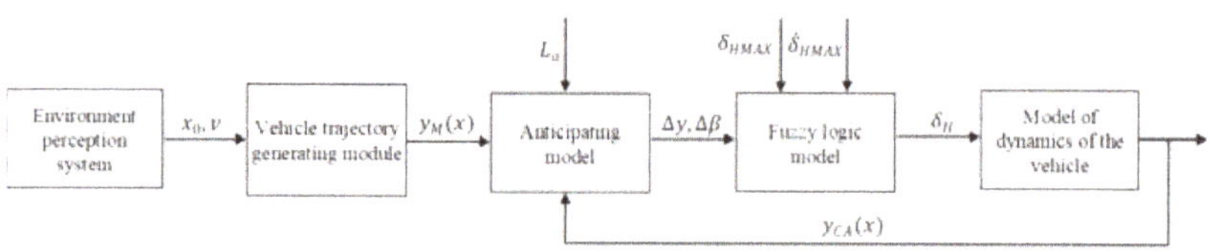

Figure 3. Schematic diagram of the processing of trajectory $y_M(x)$ in order to determine the steering wheel angle δ_H settings.

In Equation (2), the properties of the anticipating model and the value of the anticipation radius L_a are made use of. The determination of the anticipation radius value, together with the method of planning the vehicle trajectory for the critical situation under analysis, has been presented in Section 6. Trajectory $y_T(x)$ and path $y_{CA}(x)$ of the center of mass of the motor vehicle make a basis for determining the Δy and $\Delta\beta$ values.

The Δy and $\Delta\beta$ values describe the divergence between the trajectory planned and the actual vehicle path. They have been shown in Figure 4:

$$\begin{aligned} \Delta y &= y_T(x) - y_{CA}(x) \\ \Delta\beta &= \beta_T(x) - \beta_{CA}(x) \end{aligned} \tag{3}$$

The fuzzy logic model used in the control system minimizes the Δy and $\Delta\beta$ values by immediate and ongoing adjustment of the steering wheel angle δ_H in the model of vehicle dynamics. The fuzzy logic model is shown in Figure 5. It represents the connection between the input signals (Δy and $\Delta\beta$) and the steering wheel angle values δ_H necessary for the obstacle to be avoided. The model includes inference rules based on associating the input signal values with possible logic states of these parameters. The inference rules are based on functions with trapezoidal profiles. In the model, the limit values of the input signals have been adopted as limitations δ_{HMAX} and $\dot{\delta}_{HMAX}$, determined by motor vehicle construction. To pass from the logic value of the input signal to the resultant value of the steering wheel angle δ_H, the analytical centrode model was used.

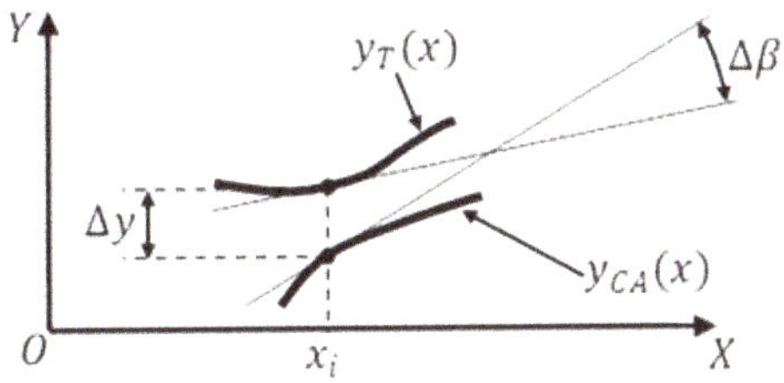

Figure 4. Illustration for determining Δy and $\Delta\beta$ based on the course of the $y_T(x)$ and $y_{CA}(x)$ curves at point x_i.

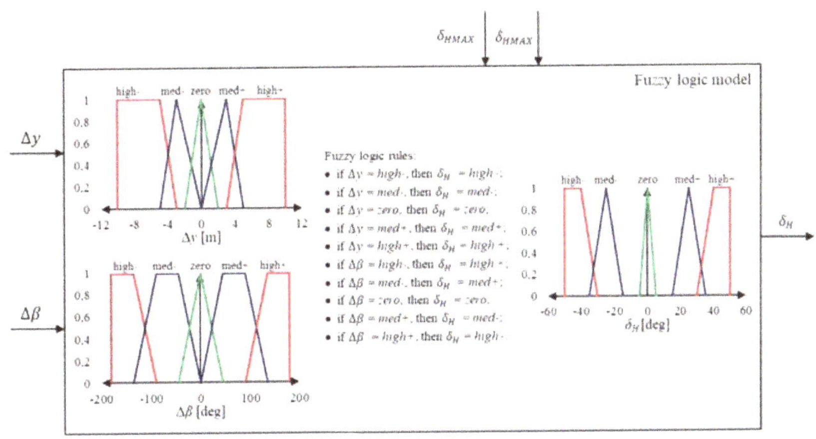

Figure 5. Structure of the fuzzy logic model used.

3.3. Model of Dynamics of the CT Unit

3.3.1. Generalized Coordinates; Equations of Motion of the CT Unit

The model of a CT unit is shown in Figure 6. It is generated in the PC-CRASH computer program, which is used to model the movement of vehicles in collision situations [31,32]. To describe and analyze the CT unit's motion, global and local coordinate systems have been used, pursuant to ISO 8855:

- $OXYZ$—Cartesian coordinate system $\{O\}$, attached to the road infrastructure;
- $C_sX_sY_sZ_s$—local coordinate systems $\{C_s\}$, with their origins at the centres of mass of vehicle bodies; $\{s\} = [A, B]$;
- $O_{Tu}x_{Tu}y_{Tu}z_{Tu}$—local coordinate systems $\{O_{Tu}\}$, each having its origin at the centre of the tyre-road contact area of the uth wheel ($u = 1, 2, \ldots, 6$; 4 vehicle wheels and 2 trailer wheels, the latter being assumed as twin wheels).

The vehicle bodies are solid with 6 degrees of freedom. Each of the wheels has a degree of freedom related to its rotational motion, which means that it has a moment of inertia relative to the axis of wheel rotation. Road wheels are linked with vehicle bodies by spring and damping elements with non-linear characteristics. They can move parallel to the C_sZ_s axis relative to the vehicle body.

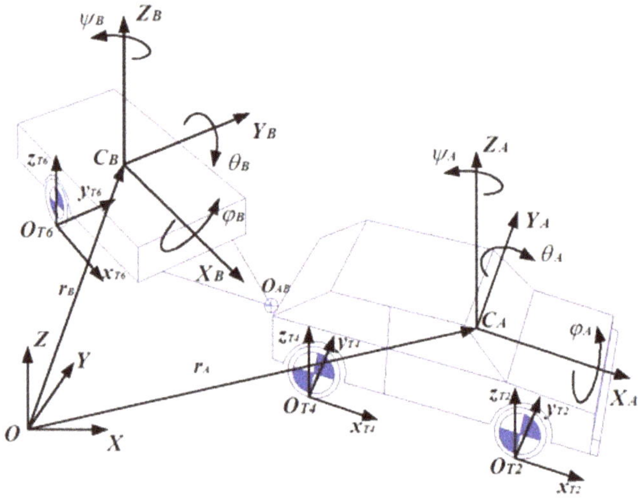

Figure 6. Model of a CT unit and the coordinate systems.

The interdependences between the coordinate systems may be described as follows:

$$\begin{bmatrix} X \\ Y \\ Z \end{bmatrix} = N_{OCs} \begin{bmatrix} X_s \\ Y_s \\ Z_s \end{bmatrix} \tag{4}$$

(global system $\{O\}$ vs. local system $\{C_s\}$ as defined above)

$$N_{OCs} = \begin{bmatrix} \cos\psi_s \cos\theta_s & \cos\psi_s \sin\theta_s \sin\varphi_s - \sin\psi_s \cos\varphi_s & \cos\psi_s \sin\theta_s \cos\varphi_s + \sin\psi_s \sin\varphi_s \\ \sin\psi_s \cos\theta_s & \sin\psi_s \sin\theta_s \sin\varphi_s + \cos\psi_s \cos\varphi_s & \sin\psi_s \sin\theta_s \cos\varphi_s - \cos\psi_s \sin\varphi_s \\ -\sin\theta_s & \cos\theta_s \sin\varphi_s & \cos\theta_s \cos\varphi_s \end{bmatrix} \tag{5}$$

where:
ψ_s, θ_s, φ_s— quasi-Euler angles, defining the orientation of the local system $\{C_s\}$ relative to the global system $\{O\}$, i.e.:
ψ_s— yaw angle;
θ_s— pitch angle;
φ_s— roll angle;

$$\begin{bmatrix} X_s \\ Y_s \\ Z_s \end{bmatrix} = N_{CsOTu} \begin{bmatrix} x_{Tu} \\ y_{Tu} \\ z_{Tu} \end{bmatrix} \tag{6}$$

(local system $\{C_s\}$ vs. local system $\{O_{Tu}\}$, as defined above);

$$N_{CsOTu} = \begin{bmatrix} \cos\delta_u & -\sin\delta_u & 0 \\ \sin\delta_u & \cos\delta_u & 0 \\ 0 & 0 & 1 \end{bmatrix} \tag{7}$$

where:

δ_u—steering angle of the front left and front right wheel of the motor vehicle (u = [1,2]).

The steering angles of individual wheels are in conformity with the Ackermann model and are determined by the steering wheel angle δ_H. The values of this angle are calculated in the control system (Figure 3).

For the non-steerable wheels, the coordinate systems $\{O_{T3}\}$ and $\{O_{T4}\}$ are parallel to the $\{C_A\}$ system and the $\{O_{T5}\}$ and $\{O_{T6}\}$ systems are parallel to the $\{C_B\}$ system. The transforms describing the interdependence between these systems are unit matrices (e.g., Equation (7) for $\delta_u = 0$).

The physical model of the motorcar consists of a vehicle body and 4 road wheels; for the trailer, the physical model consists of a trailer body and 2 wheels. The models of dynamics of the car and the trailer, if treated separately, have 10 and 8 degrees of freedom, respectively. When the motorcar and the trailer are coupled together by means of a ball joint O_{AB} (Figure 6), constraints are imposed on the trailer's motion and the number of the degrees of freedom of the CT unit is thus reduced to 15.

The set of the generalized coordinates, which completely define the car and trailer's positions, may be written as follows:

$$q_A = [x_A y_A z_A \psi_A \theta_A \varphi_A \omega_1 \omega_2 \omega_3 \omega_4]^T \tag{8}$$

$$q_B = [\psi_B \theta_B \varphi_B \omega_5 \omega_6]^T \tag{9}$$

The CT unit's motion can be described by vectorial equations [30,31]:

$$m_s(\ddot{\boldsymbol{r}}_s + \boldsymbol{\Omega}_s \times \dot{\boldsymbol{r}}_s) = \sum_i^n \boldsymbol{F}_{si} \tag{10}$$

$$\boldsymbol{T}_s \dot{\boldsymbol{\Omega}}_s + \boldsymbol{\Omega}_s \times \boldsymbol{T}_s \boldsymbol{\Omega}_s = \sum_j^k \boldsymbol{M}_{sj} \tag{11}$$

where:
m_s—vehicle mass;
$\boldsymbol{r}_s$—vector from the origin of the global coordinate system to the center of mass C_s in the global coordinate system $\{O\}$; $\boldsymbol{r}_s = \begin{bmatrix} x_s & y_s & z_s \end{bmatrix}^T$ (cf. Figure 6); $\dot{\boldsymbol{r}}_s = \boldsymbol{v}_s$; $\ddot{\boldsymbol{r}}_s = \boldsymbol{a}_s$;
$\boldsymbol{F}_{si}$, $\boldsymbol{M}_{sj}$— generalized external forces and moments acting on vehicle s;
$\boldsymbol{T}_s$—tensor of inertia of vehicle s relative to the vehicle center of mass in the local coordinate system $\{C_s\}$;
$\boldsymbol{\Omega}_s$—vector of the yaw velocity of the body of vehicle s in the local coordinate system $\{C_s\}$; $\boldsymbol{\Omega}_s = \begin{bmatrix} \dot{\varphi}_s & \dot{\theta}_s & \dot{\psi}_s \end{bmatrix}^T$.

3.3.2. Forces in the Vehicle-Trailer Coupling Device

The generalized external forces F_{si} also include the coupling device force R_s. The coupling device equation, being at the same time an equation of the constraints imposed upon the trailer, has been written as follows:

$$\begin{aligned} R_{AX} + R_{BX} &= \Delta R_{1zad} \\ R_{AY} + R_{BY} &= \Delta R_{2zad} \\ R_{AZ} + R_{BZ} &= \Delta R_{3zad} \end{aligned} \tag{12}$$

where:
R_{AX}, R_{AY}, R_{AZ}, R_{BX}, R_{BY}, R_{BZ}—components of the coupling device force for motorcar A and trailer B, as appropriate, calculated in every step of the integration of model equations and expressed in the global coordinate system $\{O\}$;
ΔR_{1zad}, ΔR_{2zad}, ΔR_{3zad}—acceptable values of the differences between components of the coupling device force.

3.3.3. Model of the Tire–Road Interaction

The tire–road interaction has been described with the use of the non-linear TMeasy model [33–36]. This model makes it possible to determine the external forces F_{xT} and F_{yT}, generated in the tire-road contact area and acting from the road via the suspension system onto the vehicle. These forces are functions of the normal tyre-road contact force $F_{zT}(t)$ and of the longitudinal tyre slip ratio s_{xT} and lateral tyre slip ratio s_{yT}; for individual wheels, they are calculated from equations:

$$F_{xT}(s_{xT}, t) = \mu\, \mu_x(s_{xT})\, F_{zT}(t) \tag{13}$$

$$F_{yT}(s_{yT}, t) = \mu\, \mu_y(s_{yT})\, F_{zT}(t) \tag{14}$$

where:
μ—local tire-road adhesion coefficient;
$\mu_x(s_{xT})$, $\mu_y(s_{yT})$—unit longitudinal and lateral forces as characteristics describing the properties of a specific tire model as functions of tire slip ratio;
$F_{zT}(t)$—current value of the normal tire-road contact force for each wheel.

Figure 7 shows, inter alia, the tire velocity vectors, which are necessary for determining the tire slip ratio. The position of the center of the tire-road contact area (point O_T) has been defined using the local coordinate systems attached to vehicle bodies $\{C_s\}$ and to vehicle wheels $\{O_T x_T y_T z_T\}$, shown in Figure 6.

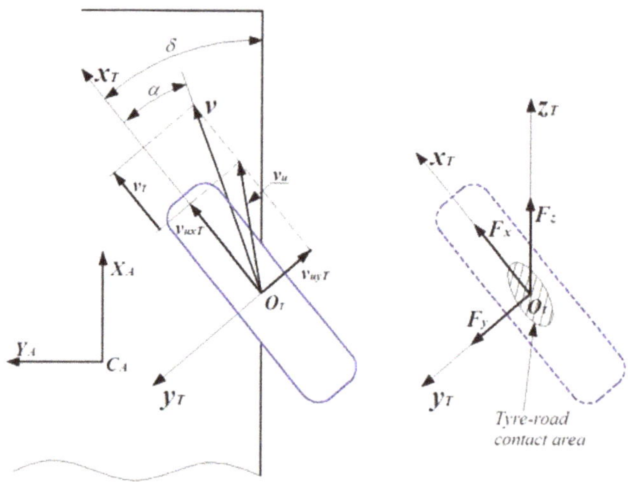

Figure 7. Kinematics and dynamics of the steerable wheel.

Based on Figure 7, the longitudinal tire slip ratio s_{xT} and lateral tyre slip ratio s_{yT} have been determined for the uth wheel:

$$s_{xT} = \frac{v_{uxT}}{(v_T + v_{uxT})} \tag{15}$$

$$s_{yT} = \frac{v_{uyT}}{(v_T + v_{uxT})} \tag{16}$$

where:
v_u—uth wheel slip velocity, resulting from longitudinal and lateral slip velocities, i.e., v_{uxT} and v_{uyT}, respectively;
$v_T + v_{uxT}$—longitudinal component of the wheel center velocity vector;
v_T—circumferential velocity of the tire, resulting from the rotational wheel motion.

The tire sideslip angle has been determined from the equation:

$$\alpha = -arctan\frac{v_{uyT}}{(v_T + v_{uxT})} \tag{17}$$

The values of the tire slip ratios according to (15) and (16) are necessary to calculate the forces F_{xT} and F_{yT} tangential to the road surface. The elastodynamic tire characteristics have a significant impact on the CT unit's behavior in the road situation under consideration (high traveling speed with high values of longitudinal slip ratio and sideslip angle of vehicle tires).

The tire model parameter values taken for this study have been based on the results of the experimental testing of tires 185R14C and 235/60R16 [37–39].

Figure 8 shows an example comparison of tangential force curves $F_x(s_x)$ and $F_y(s_y)$ for tests carried out on road surfaces with adhesion coefficients of $\mu_1 = 0.8$ (dry asphalt concrete) and $\mu_2 = 0.8$ (wet road). The $F_y(\alpha)$ curve has been plotted for only one road surface type. The curves presented show that the maximum values of the tangential reactions at the tire-road contact area occur at a slip ratio of about 0.15 and a tire sideslip angle of about 8 deg. Further growth in the slip ratio causes a reduction in the tangential reactions and, in consequence, increasing deviation of the vehicle's motion from the trajectory planned.

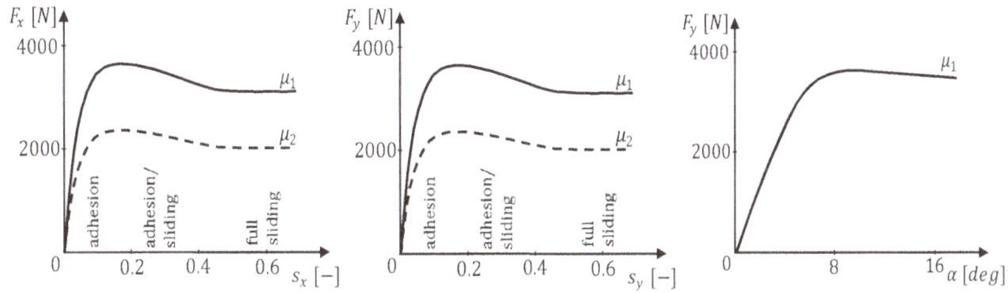

Figure 8. Non-linear characteristics of the dependence of tangential reactions on tire slip ratio and sideslip angle.

4. Validation of the Model of a CT Unit

The model validation was preceded by a parametrization process based on the results of measurements of mass distribution, as well as dimensions and characteristics of the suspension system and tires, carried out on the vehicle combination under test. The vehicle combination under test can be seen in Figure 9, when it was performing a dynamic lane-change maneuver according to the ISO 3888-1 standard [40]. Results of the measurements carried out were compared with results of simulation of a double lane-change maneuver.

Figure 9. Experimental tests.

During the validation tests, a road infrastructure model according to the ISO 3888-1 standard and a CT unit control system model according to Figures 2, 3 and 5 were also used. The trajectory of the center of mass of the motorcar model $y_M(x)$ was specially selected for the steering wheel angle $\delta_H(x)$ obtained to be in conformity with the curve recorded during the experimental tests. The result of such a model validation procedure has been presented in Figure 10.

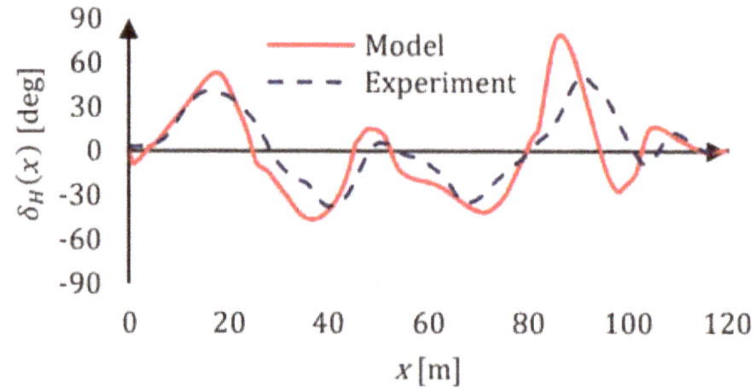

Figure 10. Comparison of the steering wheel angle curves $\delta_H(x)$ obtained in the model tests and experimental tests carried out to validate the CT model.

Example comparisons of results of the experimental and simulation tests have been presented in Figure 11.

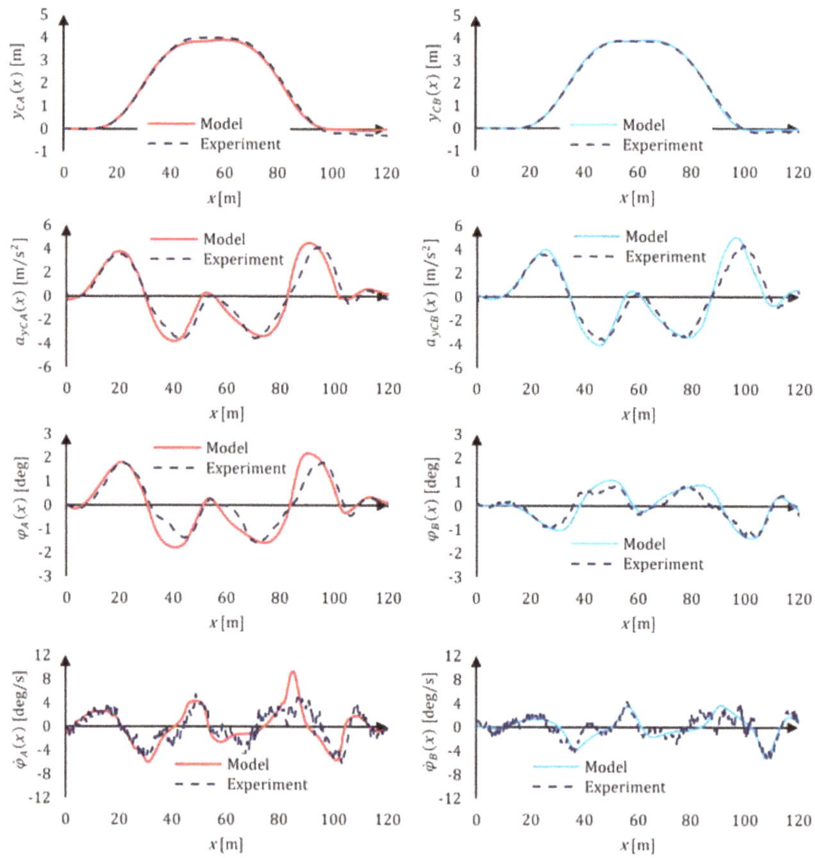

Figure 11. *Cont.*

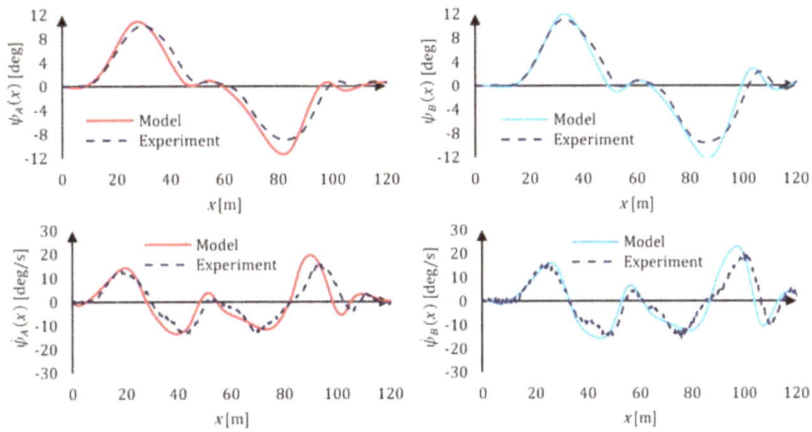

Figure 11. Comparison of results of the experimental and simulation tests of the CT unit performing a double lane-change maneuver with a speed of v = 60 km/h (left: motorcar; right: trailer).

Figure 11 shows the validation results. For model validation purposes, the following results of the model and experimental tests have been compared with each other, separately for the motorcar ($s = A$) and for the trailer ($s = B$): trajectory $y_{Cs}(x)$ and lateral acceleration $a_{yCs}(x)$ of the centre of vehicle mass, vehicle roll angle $\varphi_s(x)$ and velocity $\dot{\varphi}_s(x)$, and vehicle yaw angle $\psi_s(x)$ and velocity $\dot{\psi}_s(x)$. The conformity between the extreme and mean characteristic values at various stages of the maneuver performed was analyzed. The validation process, based on the results of parametrization of the CT unit and control system models, resulted in the obtaining of good consistency between the kinematics of the model and the actual motion of the real object (i.e., the motorcar and the trailer). The results of this highly multi-aspect assessment may be considered as confirming good agreement between the profiles of individual physical quantities, which means good agreement between the characteristics of the object and its model.

5. Calculation of Trajectory $y_M(x)$

5.1. Data Necessary to Plan the Obstacle-Avoiding Trajectory

At first, the control system calculates trajectory $y_M(x)$ (Figures 2 and 3). The calculations are based on the information received from the environment perception system that a critical situation has arisen. The process of analyzing the critical situation begins at the instant when the perception system identifies the appearance of an obstacle ($y_0 > 0$) on the lane used by the vehicle at a distance of x_0 ahead (Figure 1). If $x_0 < S_Z$ then the obstacle avoidance procedure is started. S_Z is the length of the stopping distance of the CT unit in the current local road conditions. The trajectory $y_M(x)$ necessary for safe obstacle avoidance is calculated with taking into account the following:

- The trajectory $y_M(x)$ planned is to be a continuation of the preceding path of the vehicle and to include smooth transition from straight-line motion to curvilinear motion (Figure 1), with the obstacle position being taken into account.
- Trajectory $y_M(x)$ is designed in a two-dimensional space, based on the information obtained from the system of perception of the situation in front of the vehicle.
- The positions of point K and target point P (see (1)) are predicted based on the information provided by the environment perception system and on the method pre-programmed in the control system for critical situations.
- An assumption is made that the curvature of the trajectory section where $0.5x_0 < x < x_0$ should be opposite to that of the section where $0 < x < 0.5x_0$ and its absolute value should not exceed that of the latter.

When trajectory $y_M(x)$ is calculated, the following limitations posed by the design features of the CT unit must be taken into account:

$$\delta \leq \delta_{MAX} = 30\ldots35 \text{ deg} - \text{confinement of the steering angle;} \tag{18}$$

$$R_{A\ \mathrm{MIN}} \geq \frac{L}{tg\delta_{MAX}} \quad R_{BMIN} \geq \sqrt{R_{AMIN}^2 + l_h^2 - l_{hp}^2} \tag{19}$$

where:
$R_{A\ MIN}$, $R_{B\ MIN}$—minimum radii of curvature of car and trailer's trajectories, respectively;
L—motorcar's wheelbase;
l_h and l_{hp}—distances from the coupling device centerline to the rear axle of the towing vehicle and to the trailer axle, respectively.

5.2. Methods Considered and Their Calculational Models

In result of an analysis of the methods of planning a safe vehicle trajectory [1,9,20], the functions based on a cosine curve, circular arcs, and parabola segments were selected for further consideration. Thus, the following functions, going through point $P(x_0; y_P)$, have been used for designing the trajectory in the global coordinate system based on lane edge R_{min} (Figure 1):

1. Cosine curve (dark blue in Figure 12):

$$y_M(x) = 0.5(y_P - 0.5d)\left(1 - \cos\left(\frac{x}{x_0}180^\circ\right)\right) + 0.5d \ for\ x \in \langle 0;\, x_0\rangle \tag{20}$$

2. Circular arcs tangent to each other at point $M_1(0.5x_0; 0.5(y_P - 0.5d) + 0.5d)$, with identical radius R_0 (red in Figure 12):

$$y_M(x) = R_0 - \sqrt{{R_0}^2 - x^2} + 0.5d \ for\ x \in \langle 0;\, 0.5x_0\rangle$$

$$y_M(x) = \sqrt{{R_0}^2 - (x - x_0)^2} + y_P - R_0 \ for\ x \in \langle 0.5x_0;\, x_0\rangle \tag{21}$$

$$R_0 = 0.25\frac{{x_0}^2 + (y_P - 0.5d)^2}{y_P - 0.5d}$$

3. Two parabolas tangent to each other at point $M_2(0.1x_0; 0.1(y_P - 0.5d) + 0.5d)$, (green in Figure 12, defined by equations):

$$y_M(x) = a_1 x^2 + 0.5d \ for\ x \in \langle 0;\, 0.1x_0\rangle$$

$$a_1 = \frac{0.1(y_P - 0.5d)}{(0.1x_0)^2}$$

$$y_M(x) = a_2(x - x_0)^2 + y_P \ for\ x \in \langle 0.1x_0;\, x_0\rangle \tag{22}$$

$$a_2 = \frac{-0.9(y_P - 0.5d)}{(0.1x_0 - x_0)^2}$$

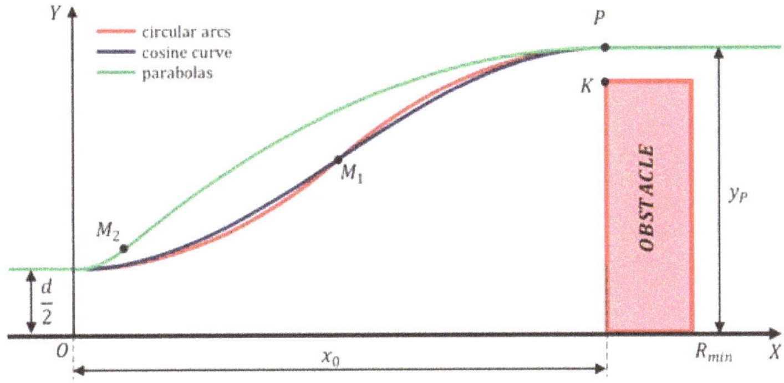

Figure 12. Shapes of trajectory $y_M(x)$, obtained from the three trajectory calculation methods.

6. Test Results Concerning the Selection of a Method to Calculate the CT Unit Trajectory in a Critical Situation

6.1. Assumptions Adopted for The Simulation Tests at the Method Selection Stage

The properties of the control system depend on many factors. These factors were divided into three groups: constant, temporary, and variable ones. The group of temporary factors included the trajectory planning algorithm and the L_a value. As a variable factor, the clearance margin (y_W) value, applied as a parameter to the control system, is considered. Tests were prepared to select the said factors for the critical situation under consideration. The following main assumptions were adopted for the simulation tests:

- The CT unit consisted of a motorcar with a mass of 1800 kg and a centre-axle trailer with a mass of 1800 kg as well.
- Before the critical situation arose, the CT unit moved rectilinearly with a constant speed (v = const.) in the middle of the right lane of a carriageway.
- The level and flat road had two lanes with a width of $d = 4$ m each and a shoulder 1 m wide.
- The width of the CT unit was $b = 2$ m.

When the model parameter values were selected, the maximum possible trailer weight was assumed. Such a choice has a favorable impact on transport efficiency but adversely affects the stability of motion of a CT unit along a curvilinear path [14].

At this stage, the tests were carried out for the following solution alternatives examined:

- three trajectory planning methods (computationally described by Equations (20)–(22));
- four values of the CT unit's speed ($v = 40$ km/h, 60 km/h, 70 km/h, and 80 km/h);
- five values of the anticipation radius ($L_a = 4$ m, 6 m, 8 m, 10 m, and 12 m);
- seven clearance margin values ($y_W = 0$ m, 0.25 m, 0.5 m, 0.75 m, 1.0 m, 1.5 m, and 2.0 m);
- two states of the asphalt concrete road surface (dry and wet).

While the simulation tests covered so many solution alternatives, only one critical situation was addressed, where another motor vehicle suddenly appeared on a road intersection with poor visibility and blocked the whole width of the lane used by the CT unit (Figure 1).

6.2. Example of Calculation Results

Simulation tests were carried out for 840 trajectory planning alternatives, as described above. Fragments of the calculation results have been presented in Figures 13–15 and in Table 1; a complete set of the results will be used in the procedure of selecting a method to calculate the CT unit trajectory and the L_a value. Figure 13 shows the courses of the trajectories $y_M(x)$ and $y_T(x)$ determined for different L_a values (according to (2)) and for three vehicle trajectory calculation methods. Trajectory $y_M(x)$ has been plotted with a dotted line. Figure 13A,C,E (on the left) show examples of trajectories of the CT unit moving with a speed of $v = 60$ km/h; on the right (Figure 13B,D,F), there are model responses obtained for each of the trajectory planning methods and for three vehicle speed values. The obstacle avoidance process obtained for three trajectory calculation alternatives and for vehicle speeds $v = 60$ km/h, 70 km/h, and 80 km/h has been presented in Figure 14. Figure 15 shows a comparison of animations of the CT unit's motion on dry and wet road surfaces.

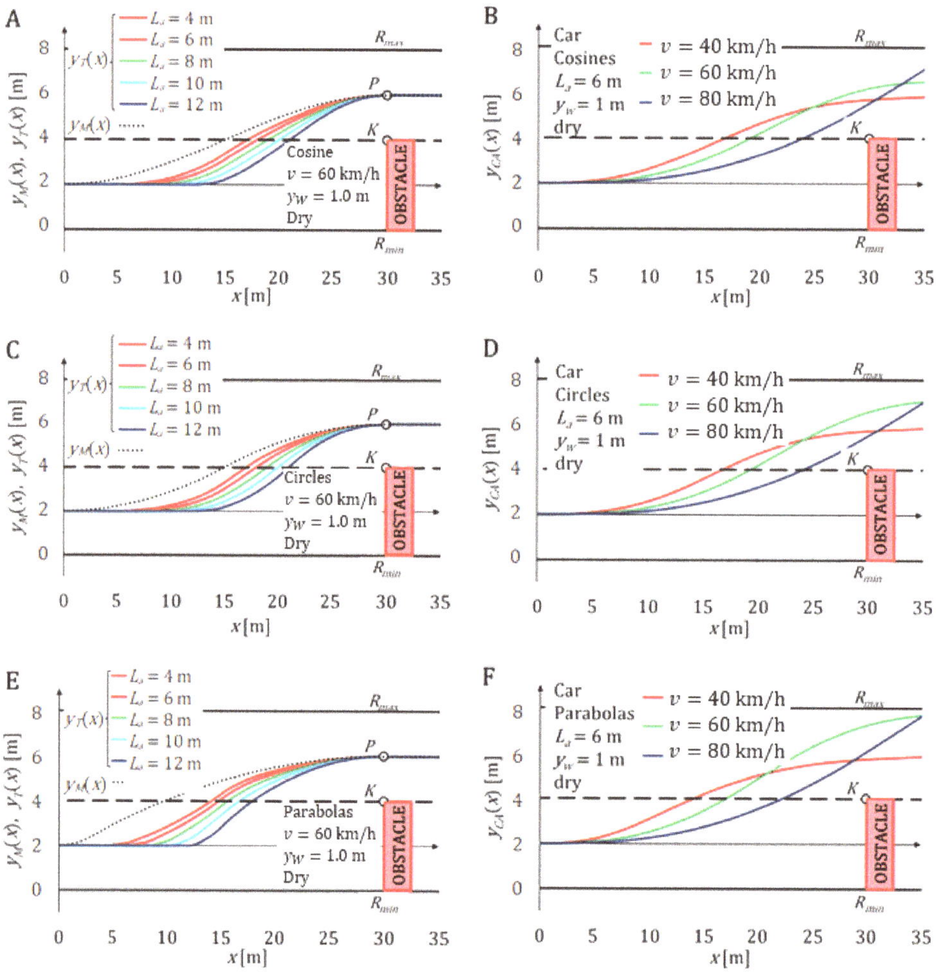

Figure 13. Comparison of trajectories $y_M(x)$ and $y_T(x)$ for different trajectory calculation methods (left) and of car paths $y_{CA}(x)$ obtained (right) for dry road surface; (**A**,**B**)—method with a cosine curve; (**C**,**D**)—method with circular arcs; (**E**,**F**)—method with parabolas.

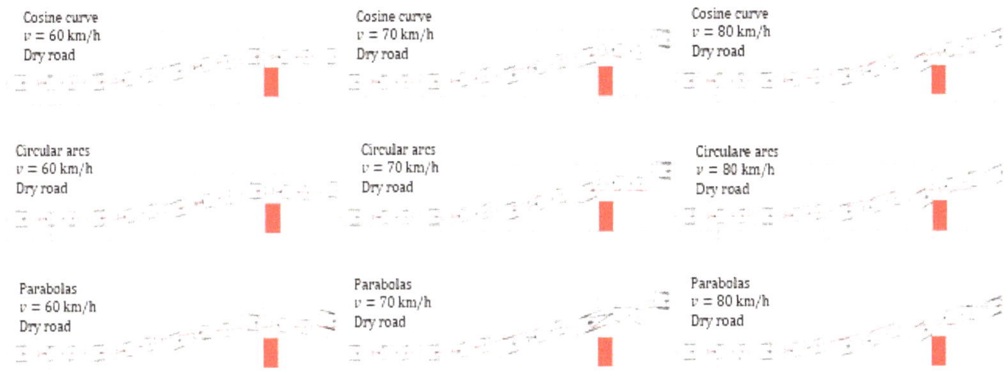

Figure 14. Comparison of animations of the CT unit's motion for $v = 60$ km/h, 70 km/h and 80 km/h and for three $y_M(x)$ determination methods; dry road surface.

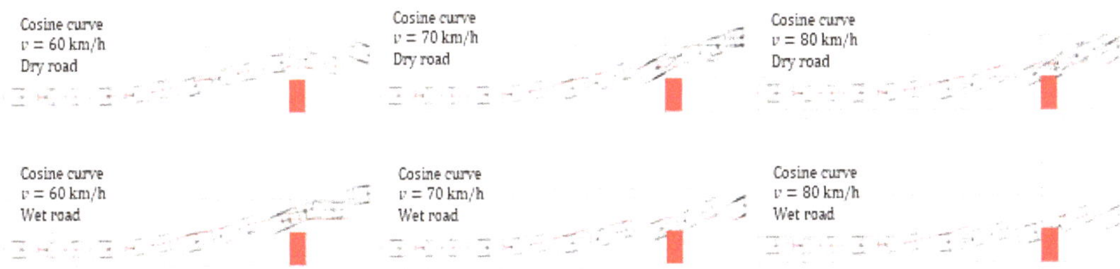

Figure 15. Comparison of animations of the CT unit's motion on dry and wet road surface for $v = 60$ km/h, 70 km/h and 80 km/h; trajectory calculation method with a cosine curve.

This brief summary of research results shows the impact of the $y_M(x)$ calculation method on the obstacle avoidance process. The example presented in Figure 13 indicates the favorable course of the process when the methods with a cosine curve or circular arcs are used to calculate $y_M(x)$. In Figure 14, we can see that the cosine method produced an advantageous effect for $v = 70$ km/h, but for $v = 80$ km/h, the trajectory calculated by the method with parabolas is better. For the wet road (Figure 15), good results were obtained for $v = 60$ km/h and the cosine method. The selection of L_a, in turn, has an impact on the trajectory curvature in each of the methods under consideration. For rising vehicle speed values, increasing the impact of the trajectory planning methods on the course of the obstacle avoidance process can be observed in the analysis.

This means that to select the optimum method of planning the trajectory $y_M(x)$ and value L_a, a lot of obstacle avoidance alternatives and curves representing changes in various physical quantities, obtained as simulation results, must be taken into consideration, because each of them helps to describe a different aspect of vehicles' behavior in a critical situation.

Table 1. Fragment of a set of results of simulations of CT unit's motion during an obstacle avoidance maneuver.

Vehicle s	v [km/h]	y_W [m]	y_P [m]	$x(y_{C_s}=y_P)$ [m]	Safe Avoidance of the Obstacle [-]	Stay in the Lane [-]	α_{Ekstr} $x\in(0;0.5x_0>$ [deg]	α_{Ekstr} $x\in(0.5x_0;x_0>$ [deg]	α_{Ekstr} $x\in(x_0;1.33x_0>$ [deg]	a_{yEkstr} $x\in(0;0.5x_0>$ [m/s^2]	a_{yEkstr} $x\in(0.5x_0;x_0>$ [m/s^2]	a_{yEkstr} $x\in(x_0;1.33x_0>$ [m/s^2]	$\Delta\psi_{Ekstr}$ $x\in(0;x_0>$ [deg]	y_K [m] N_{FR}	y_K [m] N_{RR}
1	2	3	4	5	6	7	8	9	10	11	12	13	14	15	16
A	40	0.0	5.0	>40	NO	YES	1.26	1.17	0.38	1.99	2.00	0.68	0	−0.21	−0.23
A	60	0.0	5.0	28.95	YES	YES	4.24	7.88	9.06	4.29	4.64	4.66	0	0.01	0.49
A	60	0.25	5.25	29.79	YES	NO	4.46	8.42	10.23	4.39	4.79	4.79	0	0.12	0.66
A	70	1.75	6.75	25.36	YES	NO	6.52	12.57	11.96	5.71	5.77	4.26	0	2.40	1.33
A	70	2.0	7.0	25.36	YES	NO	6.57	12.66	12.16	5.71	5.76	4.26	0	2.43	1.34
B	40	0.0	5.0	>40	NO	YES	1.32	1.20	0.36	1.91	1.71	0.71	4.01	−0.24	−0.24
B	60	0.0	5.0	26.61	YES	YES	3.96	6.46	12.94	4.82	6.02	6.55	13.28	0.46	0.92
B	60	0.25	5.25	26.60	YES	YES	4.05	6.52	15.12	4.92	6.02	6.69	14.42	0.62	1.09
B	70	1.75	6.75	27.74	YES	NO	3.77	14.57	19.07	5.01	6.85	6.85	12.39	1.30	0.01
B	70	2.0	7.0	27.74	YES	NO	3.78	14.48	19.41	5.02	6.76	6.85	12.44	1.32	0.02

In the table, the vehicle corners have been given symbols according to the N_{is} system, where $i = \{FR, RR\}$ (**F**ront **R**ight, **R**ear **R**ight) and $s = \{A, B\}$ (car and trailer, respectively).

6.3. Procedure of Analysing the Calculation Results

When the best trajectory planning method and the desirable L_a value were selected, individual simulations were treated as successive alternatives of solving the obstacle avoidance problem. The selection was based on an analysis of a set of simulation results, which included numerical values, curves recorded, and logical values.

A procedure was established, which was taken in apart from the AHP (analytic hierarchy process) method [41,42]. In that method, logical values may be used apart from numerical ones, which was of considerable importance in the assessment process. The following steps were adopted in the procedure:

1. Selection of a decision goal (referred to as "goal") as the primary criterion of assessment of the solutions available.
2. Hierarchical arranging of the criteria of analysis, which would be considered as partial goals or sub-criteria.
3. Determining of weights for individual sub-criteria.
4. Calculating of partial assessment results for individual physical quantities in the set of results of successive simulation alternatives (solutions) adopted in examining the obstacle avoidance process. The partial assessment results would be based on the product of the weights referred to in item 3 of this list and the degree of satisfying a specific criterion by the solution alternative under analysis.
5. The final assessment result and selection of the values of the target solution would be determined by aggregation of the partial assessment results.

The results of experimental and simulation tests [43] of similar processes have made it possible to formulate the following criteria of recognizing the performance of the obstacle avoidance maneuver (i.e., the criteria and sub-criteria) as correct:

- Successful avoidance of a collision with the obstacle and keeping of the CT unit within the road lanes as planned.
- Smooth growth in the lateral displacements $y_{CA}(x)$ and $y_{CB}(x)$ until the obstacle is passed by.
- Short length of the distance travelled where the "adjustment" of the vehicle position (measured by the yaw angle $\psi_s(x)$) took place and limited value of the extreme vehicle trajectory overshoot, i.e., preventing of the vehicle from leaving the lane planned.
- The lowest possible extreme values of the steering wheel angle δ_{HEkstr}, tyre sideslip angles of vehicle's rear axle and trailer's axle (α_{AEkstr} and α_{BEkstr}, respectively), lateral acceleration ($a_{yCAEkstr}$ and $a_{yCBEkstr}$), and trailer drawbar turning angle $\Delta\psi$, informing that the CT unit's motion was kept stable.

The above was taken into account when the results obtained for individual solutions (i.e., the simulation results for individual alternatives) were assessed. In consequence, the following has been decided:

A Goal: safe avoidance of the obstacle.

B Criteria as partial goals:

(a) safe vehicle path, i.e., CT unit's motion between lines R_{min} and R_{max} (Figure 1) or $y_{CA}(x)$, $y_{CB}(x) \in \{0.5b; 2d - 0.5b\}$;

(b) vehicle's motion stability maintained, i.e., $a_y < a_{yMAX}$ and $\Delta\psi < \Delta\psi_{MAX}$.

C Sub-criteria:

(a) direct:

- clearance between the car side and the obstacle when the latter is being passed by (y_{KA});
- clearance between the trailer side and the obstacle (y_{KB});

(b) indirect:

- car yaw angle relative to the preset vehicle trajectory ($\Delta\beta$);
- departure of the actual car trajectory from the planned one ($|y_M - y_{CA}|$);

- departure of the actual trailer trajectory from the planned one ($|y_M - y_{CB}|$);
- stable growth in the lateral displacement ($y_{CA}(x)$).

D Alternatives:

(a) length of the imaginary anticipation radius (L_a) from 4 m to 12 m;
(b) methods of calculation of the trajectory planned: with a cosine curve, circular arcs, and parabolas;
(c) vehicle speed (v) from 40 km/h to 80 km/h on dry and wet road surface.

E Criterion (limit) values:

(a) Sub-criteria (Ca)

- 2.0 m > y_{KA} > 0 m;
- 2.0 m > y_{KB}> 0 m;

(b) Sub-criteria (Cb)

- $|\alpha| < 16$ deg; $|\Delta\beta| < 90$ deg;
- $|y_M - y_{CA}| < 2$ m;
- $|y_M - y_{CB}| < 2$ m;
- $x(y_{CA} = 3\text{ m}) < 25$ m.

The most desirable values of the parameters taken as the sub-criteria may be described as follows:

- for (Ca), the highest weights are to be assigned to the values exceeding 0.5 m;
- for (Cb), the lower the values, the higher weights should be assigned.

6.4. Results of Applying the Procedure That Has Been Established

Figure 16 shows example time histories describing the physical quantities that are taken into account in the procedure presented above, i.e., the paths travelled by vehicle mass centers $y_{CA}(t)$ and $y_{CB}(t)$, tire sideslip angles $\alpha_A(t)$ and $\alpha_B(t)$, lateral accelerations $a_{yCA}(t)$ and $a_{yCB}(t)$, and trailer drawbar turning angle $\Delta\psi(t)$. To facilitate the interpretation of the graphs, curves representing the trajectory planning function $y_M(t)$ and the steering wheel angle $\delta_H(t)$ have also been plotted.

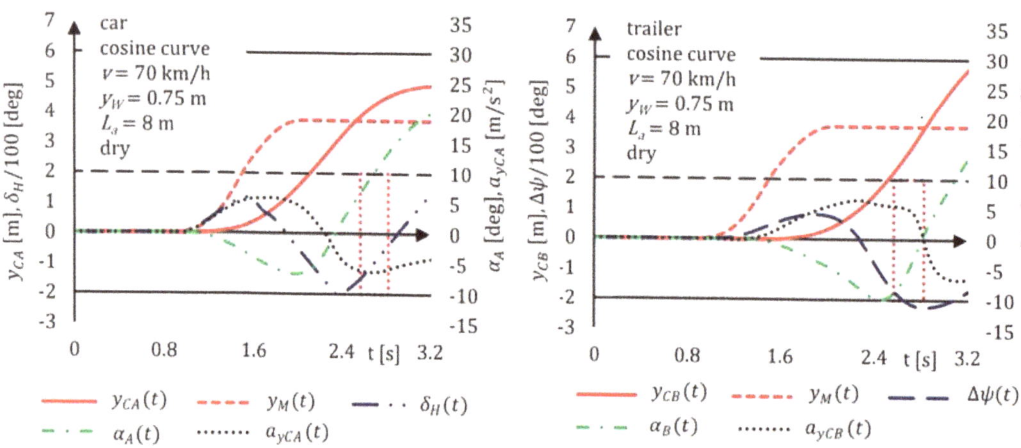

Figure 16. Curves analyzed in the procedure presented above (example obtained for the alternative with $v = 70$ km/h, dry road surface, and cosine method used to plan the trajectory).

To facilitate the inference, pursuant to the procedure presented, the simulation results have been brought together with respect to two points of view:

- as a set of characteristic or extreme values of the curves analyzed that represent the car and trailer behaviors (an example is shown in Table 1);
- as a synthetic presentation of the impact of L_a on characteristic values of some of the quantities under analysis (an example has been given in Figure 17).

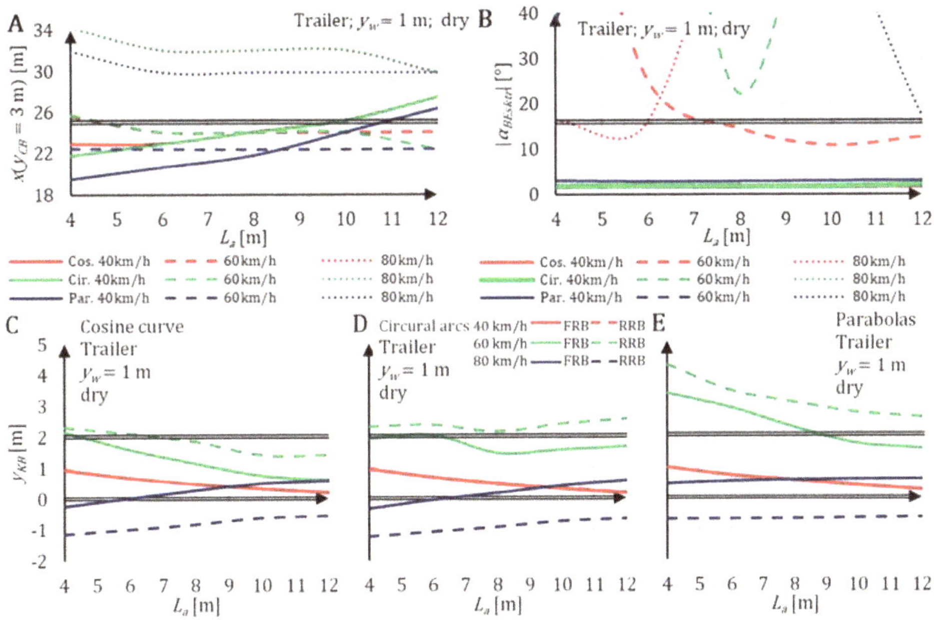

Figure 17. Summarized results of simulations of trailer's motion in a CT unit during an obstacle avoidance maneuver for three trajectory planning methods (explanation in the text above). Vehicle speed values: $v = 40$ km/h, 60 km/h, and 80 km/h; L_a values within a range of 4–12 m; road surface: dry.

Table 1 shows a fragment of a set of results obtained for five obstacle avoidance alternatives. Such summaries of values of the physical and logic quantities determined during individual simulations were used in the procedure described above.

Figure 17 presents an example of the impact of L_a on the results of simulation of the trailer (i.e., vehicle B) motion. In the graphs, the area of acceptable simulation results (according to the sub-criteria adopted) has been marked by double fine lines indicating the upper limit of the said area. A few curves have been plotted in the graphs that represent the functions taken into account in the procedure under consideration. In particular:

- Figure 17A shows the $x(y_{CB} = 3\ \text{m}) = f(L_a)$ curves, representing the distance travelled by the trailer for its lateral displacement $y_{CB}(x)$ to reach a value of 3 m. In the graph, there is a double fine line showing the distance of $x = 25$ m. The alternatives for which the $x(y_{CB} = 3\ \text{m})$ curves are situated below the double fine line help in the safe obstacle avoidance, i.e., they indicate the trajectory planning methods and the range of the L_a values that meet the Cb sub-criterion (see Table 2) regarding the stability of growth in $y_{CB}(x)$ during the obstacle avoidance manoeuvre.
- Figure 17B shows the $|\alpha_{BEkstr}(L_a)|$ curves, representing the absolute value of the tyre sideslip angle $\alpha_B(x)$. The double fine line shows the limit value of $|\alpha| < 16$ deg. The curves below the double fine line show the alternatives that meet the Cb sub-criterion, i.e., all the alternatives of the L_a values at $v = 40$ km/h; for $v = 60$ km/h and 80 km/h, however, the Cb sub-criterion is only met for a few L_a values and providing that the cosine method is used for planning the $y_M(x)$ trajectory.

- Figure 17C–E show the $y_{KB}(L_a)$ curves, representing the clearance between edge K and trailer corners (corners N_{FRB} and N_{RRB}) at the instant when the obstacle is passed by. The safe obstacle avoidance is only possible (i.e., the B criterion, see Table 2, is met) for the alternatives for which the y_{KB} values obtained fall between the double fine lines ($y_{KB} \in (0; 2\text{ m})$). Such a result was achieved e.g., for $v = 60$ km/h and $L_a \geq 7$ m in the cosine method, while in other methods, the N_{RRB} corner goes beyond the area of the lane available. For $v = 80$ km/h, the CT unit will hit the obstacle, regardless of the trajectory planning method used.

Table 2. Connection between the values given in individual columns of Table 1 and the procedure described above.

Columns of Table 1	5	6–7	8–10	11–13	14	15–16
Criteria and sub-criteria used in the procedure that are relevant to the values in specific columns	Cb	A; B	Cb	B	B	Ca

The results obtained for individual solution alternatives, exemplified in Table 1 and Figure 17, facilitated the use of the procedure presented. The procedure enabled the following decisions to be effectively made about the temporary solutions implemented in the control system for the distance x_0 of the CT unit approaching the obstacle:

- the trajectory planning method will be based on the cosine function;
- the optimum anticipation radius will be $L_a = 6$ m.

7. Analysis of the Motion of a CT Unit When Avoiding an Obstacle

At the next step, the CT unit's motion was analyzed with taking into account the temporary decisions made in Section 6. The cosine function was adopted as a basis for the method of planning the vehicle trajectory for the distance travelled by the CT unit when approaching the obstacle; the anticipation radius value was assumed as $L_a = 6$ m. For such assumptions, recommendations were worked out regarding the clearance margin value y_W necessary to plan a trajectory that would ensure safe avoidance of the obstacle. The area of searching for the y_W values is limited by the available space in the carriageway (R_{max} and y_0, see Figure 1). In the critical situation under analysis, the y_W value may be within a range of 0–2 m.

In consideration of the above, model tests were carried out for various values of clearance margin y_W and CT unit's speed v. The results of these tests made a basis for recommending the favorable y_W values.

7.1. Impact of Vehicle Speed on the Obstacle Avoidance Process

The vehicle speed is an effect of control system's decision made within following the general strategy of CT unit's motion. Figure 18 shows simulation results obtained for $v = 50$ km/h, 60 km/h, and 70 km/h.

The simulation was carried out while observing, in particular, the behavior of the CT unit driven on dry road surface to follow a trajectory $y_M(x)$ calculated for a clearance margin of y_W = 0.5 m. The applying of rising vehicle speed values as an input caused:

- increasing departure of trajectories $y_{CA}(x)$ and $y_{CB}(x)$ from trajectory $y_M(x)$, which can be seen in the animation (Figure 18A) and in Figure 18B;
- growing motorcar yaw ψ_A which helps to avoid a collision with the obstacle at $v = 50$ km/h (see the ψ_A curves in Figure 18A,C), but becomes hazardous at $v = 70$ km/h; similarly, the trailer yaw angle rose as well (see the ψ_B curve in Figure 18C);
- necessity to increase the steering wheel angle (Figure 18C); within the distance x_0, two extremums occurred in this angle and the span between them grew from $\Delta\delta_H = 134$ deg at $v = 50$ km/h to $\Delta\delta_H = 370.7$ deg at $v = 70$ km/h;

- growing span between the extreme values of the δ_H angle, which resulted in high lateral acceleration values, from a_{yEkstr} = 3.4–3.8 m/s^2 at v = 50 km/h to 4.7–6.7 m/s^2 at v = 60–70 km/h on dry road surface (Figure 18E); the acceleration values for the trailer exceeded those for the towing vehicle by 12–22%;
- rising extreme values of the tyre sideslip angles (for the rear axle wheels of the motorcar and the trailer axle wheels), from 3 deg at v = 50 km/h to 11 deg at v = 70 km/h (Figure 18D); the extreme values of the tire sideslip angle of the trailer wheels exceeded those of the motorcar's rear axle wheels by 7–30% (this trailer's predominance grew with rising vehicle speeds).

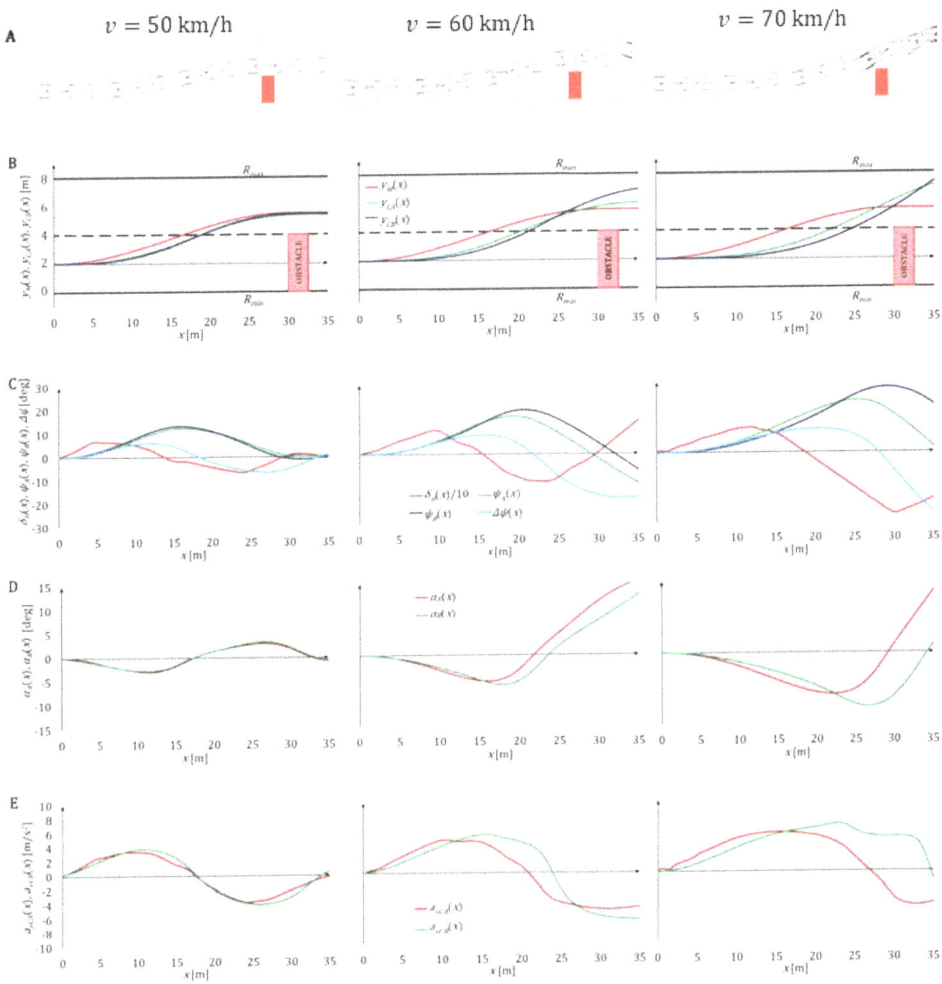

Figure 18. Obstacle avoidance simulation results obtained for the cosine method, L_a = 6 m, y_W = 0.5 m, dry road surface and different driving speeds (individual columns): (**A**) CT unit's motion animation; (**B**) path planning trajectory $y_M(x)$ and vehicles center of mass trajectory $y_{Cs}(x)$; (**C**) steering wheel angle $\delta_H(x)$, yaw angle $\psi_A(x)$, trailer drawbar turning angle $\Delta\psi(x)$; (**D**) tire sideslip angles for the rear axle of vehicles $\alpha_s(x)$; (**E**) lateral acceleration of vehicles center of mass $a_{yCs}(x)$.

The above shows that the higher vehicle speed, the more attention should be paid to the clearance margin values within y_W < 0.5 m when searching for the trajectory. If this is done, the trajectory curvature as well as the a_y values will be reduced. In consequence, the centrifugal force and the angle ψ_A of yaw of the car (and of the trailer, too) in the dangerous direction will decrease (Figure 18A). The risk of unstable trailer's motion resulting from high yaw angle values ψ_A and ψ_B has been confirmed by the research reported in [15]. Therefore, the lowering of the y_W values when planning the trajectory $y_M(x)$ may produce advantageous solutions.

7.2. Impact of the Clearance Margin y_W on the Safety of Obstacle Avoidance

The simulation results presented in Figures 19–21 have been obtained for clearance margin values y_W = 0–2.0 m and for the CT unit being driven on dry and wet road surface with a speed of v = 40–90 km/h. This has made it possible to formulate a more detailed recommendation for the selection of y_W.

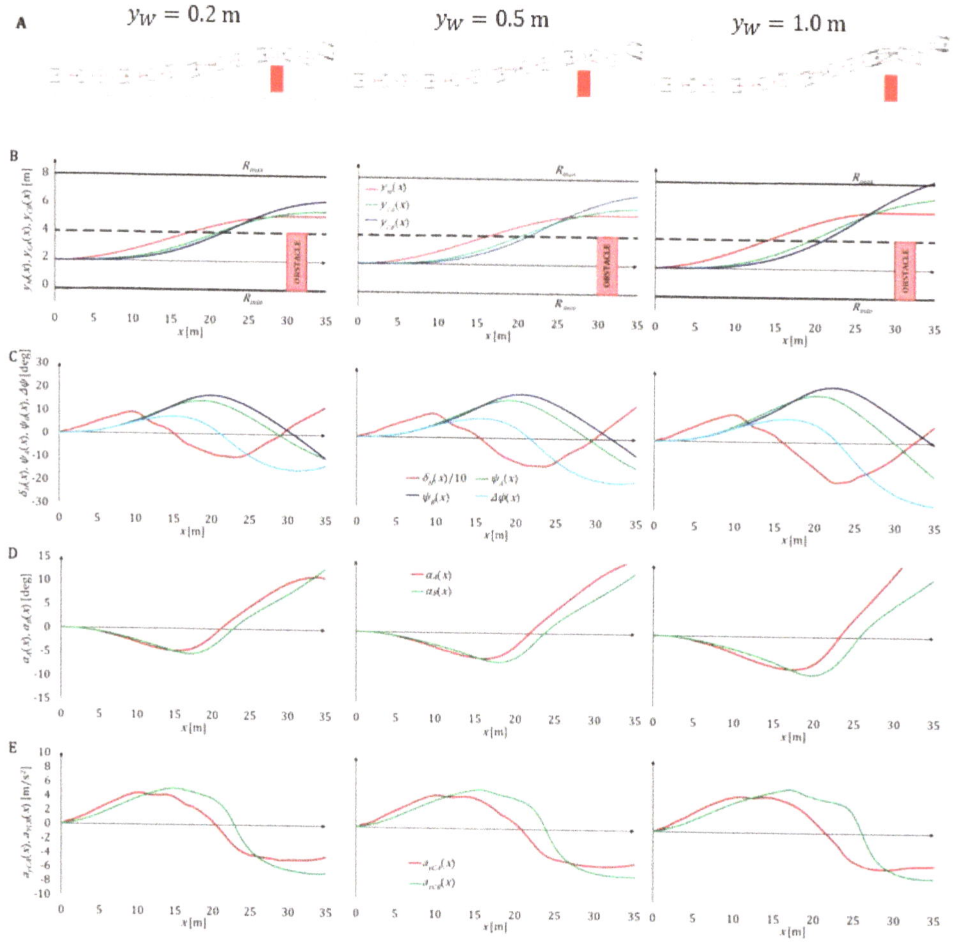

Figure 19. Curves representing selected physical quantities for the cosine trajectory planning method, L_a = 6 m, v = 60 km/h, dry road surface, and three clearance margin values y_W (individual columns): **(A)** CT unit's motion animation; **(B)** path planning trajectory $y_M(x)$ and vehicles center of mass trajectory $y_{Cs}(x)$; **(C)** steering wheel angle $\delta_H(x)$, yaw angle $\psi_A(x)$, trailer drawbar turning angle $\Delta\psi(x)$; **(D)** tire sideslip angles for the rear axle of vehicles $\alpha_s(x)$; **(E)** lateral acceleration of vehicles center of mass $a_{yCs}(x)$.

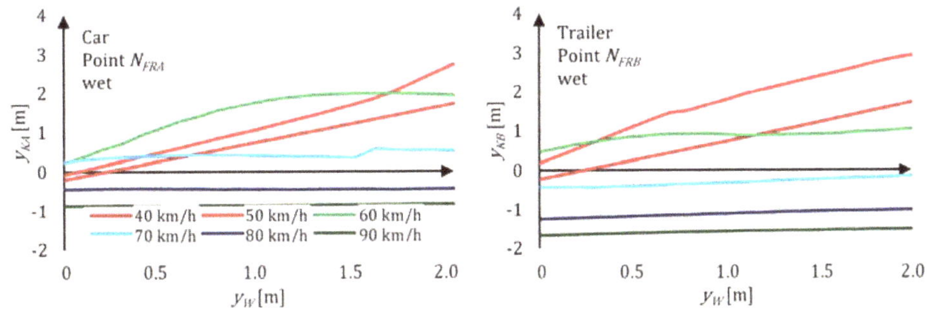

Figure 20. Distance y_K between the front corner of the car and trailer (N_{FRA}, N_{FRB}) and the obstacle when the latter was passed by wet road surface, y_W = 0.0–2.0 m.

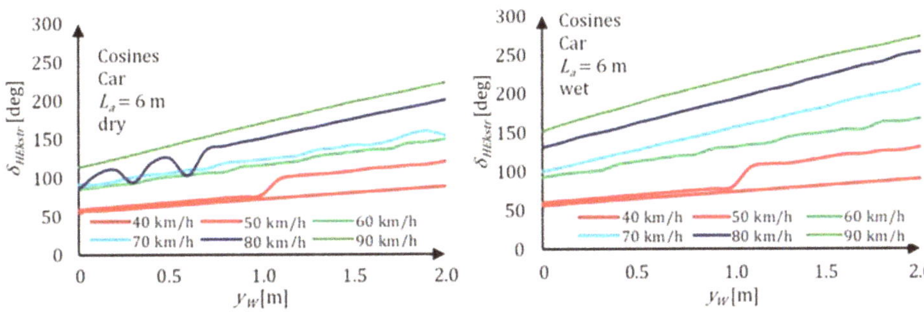

Figure 21. Extreme values of the steering wheel angle (the first extremum of the $\delta_H(x)$ curve) vs. the clearance margin y_W for vehicle speeds of v = 40–90 km/h.

Figure 19 presents the process of obstacle avoidance by a CT unit moving with a speed of v = 60 km/h, with the trajectory $y_M(x)$ having been calculated for y_W = 0.2 m, 0.5 m, and 1.0 m. The results of these simulations provide grounds for a statement that when the clearance margin value y_W was reduced from 1.0 m to 0.2 m, then:

- curvature of trajectory $y_M(x)$ and departure of trajectories $y_{CA}(x)$ and $y_{CB}(x)$ from trajectory $y_M(x)$ decreased by 38–40%, which can be seen in Figure 19B;
- clearances y_{KA} and y_{KB} decreased from 1.5 m to 0.3 m and from 1.9 m to 1.1 m, respectively (Figure 19A);
- extreme values of the steering wheel angle were reduced and the span $\Delta\delta_h$ between them, observed within the x_0 road section, dropped from 287.9 deg to 190.7 deg; trailer drawbar turning angle $\Delta\psi$ at the instant when the obstacle was being passed by dropped from 10.7 deg to 3.6 deg;
- lateral accelerations of the towing vehicle and the trailer, a_{yCA} and a_{yCB}, observed within the x_0 road section, declined from 5.0–6.1 m/s^2 to 4.4–5.1 m/s^2, respectively;
- motorcar tire sideslip angle α_A at the instant when the obstacle was being passed by declined from 13.8 deg to 10.0 deg.

It should be added here that when the clearance margin value was raised within the range $y_W > 0.5$ m in the vehicle driving conditions under analysis, then:

- at the instant when the obstacle was being passed by, the tire sideslip angle of the motorcar rear axle wheels rose from 10.0 deg to 13.8 deg, i.e., it reached values that made it difficult to control the vehicle movement (cf. the $F_y(\alpha)$ curve in Figure 8);

- the differences between the trajectories ($y_{CB}(x) - y_{CA}(x)$) and the yaw angles ($\psi_B(x) - \psi_A(x)$) of the trailer and the car increased as well, which may result in the instability of the CT unit's motion on the road section just beyond the obstacle.

The tests represented in Figure 19 have confirmed that the applying of low y_W values when planning the obstacle-avoiding trajectory would be well justified, but within a limited range of vehicle speeds (especially on wet road surface). This information was gained after the tests were extended to a vehicle speed range of v = 40–90 km/h.

In Figure 20, simulation results have been presented in the form of curves $y_K = f(y_W)$ plotted for the obstacle avoidance on wet road surface. When analyzing these results, it should be taken into account that the y_K values should be higher than zero, preferably within the range of 0–2 m, for the obstacle avoidance to be safe. The course of the green curve in Figure 20 confirms the conclusions drawn from the tests presented in Figure 19 for $v = 60$ km/h. For such a vehicle speed, the y_W values may be chosen from a wide range; previously, the recommendable value was specified as $y_W \cong 0.2$ m for dry road surface.

The angle $\delta_H(t)$ corresponds to a signal generated by the control system (Figure 3). Its values at the first extremum have been shown in Figure 21. During the initial part of the obstacle avoidance maneuver ($x \in (0; 0.5x_0 >$), they should steeply rise so that the CT unit would be able to follow the path with the curvature as planned. Such an effect may be obtained if a big distance margin y_W is adopted. With this objective in view, high clearance margin values should be used when planning the trajectory $y_M(x)$. However, it is not easy to make the CT unit avoid the obstacle this way at as low speeds as v > 60 km/h because of high values of lateral acceleration a_y, lateral force F_Q, and tire sideslip (Figures 18D and 19D), which increase the distances between the vehicle paths ($y_{CA}(x)$ and $y_{CB}(x)$) and the trajectory planned ($y_M(x)$) (Figure 21B for v = 70 km/h). This highlights one more of the dilemmas to be resolved when selecting the y_W values.

A synthetic summary of the simulation results presented in Figures 19–21 has made it possible to formulate some recommendations for the selection of y_W; simultaneously, it has highlighted the following dilemmas:

- It is recommendable to apply low y_W values because the trajectories thus planned do not require too high extreme values of the steering wheel angle and do not result in excessive values of lateral vehicle accelerations and tire sideslip angles; thanks to this, the vehicle path may run relatively close to $y_M(x)$.
- The application of high y_W values produces high $\delta_H(t)$ angle values even in the initial phase of the obstacle-avoiding process ($x \in (0; 0.5\,x_0\rangle$) and results in the planning of a trajectory that is characterized by big lateral displacements $y_{CA}(x)$ and $y_{CB}(x)$, especially at $v < 60$ km/h.
- High trajectory curvature at $v > 60$ km/h generates high values of lateral acceleration a_y and tire sideslip α, which may make it difficult for the CT unit to avoid the obstacle on the second part of the x_0 road section, i.e., where $x \in (0.5\,x_0; x_0\rangle$ (Figures 19 and 21).
- The control system generates higher extreme values δ_{HEkstr} on wet road surface than it does on dry road surface at identical y_W values. This causes a growth in the vehicle yaw angles ψ_A and ψ_B; high values of these angles make obstacle avoidance more difficult.
- On the wet road surface, the advantageous y_W values are lower than those recommendable for the dry road surface; on the other hand, the range from within they may be picked is narrower than that acceptable in the latter case.
- At $v = 80$–90 km/h, the y_W values under consideration do not offer a possibility of planning a trajectory that would ensure safe obstacle avoidance on wet road surface.

The very diverse impact of y_W on the obstacle avoidance by a CT unit will be made use of to build a set of clearance margin values recommendable for the planning of an obstacle-avoiding trajectory $y_M(x)$.

8. Selection of a Clearance Margin y_W for the Obstacle Avoidance Maneuver

The results of an assessment carried out according to the procedure prepared as described in Section 6 became a basis for selecting temporary parameters, i.e., a method of planning the trajectory $y_M(x)$ and determining the value of L_a for the critical situation under consideration. After the temporary parameters are determined as described above, a trajectory $y_M(x)$ is planned in the control system for a safe vehicle path to be obtained. The course of this path and the behavior of the vehicles moving along this path strongly depend on the dynamic characteristics of the CT unit, including the forces acting on the vehicles. In the critical situation under consideration, it is difficult to predict the values of such forces and the effects of their action (tire sideslip, skidding of wheels of individual axles, vehicles' yaw angles from the carriageway centerline, etc.). In such a situation, it is good to have a set of the y_W values that would facilitate the planning of a safe trajectory. The description provided in Section 2 and Figure 1 shows that the clearance margin in the critical situation under analysis should be within a range of $y_W \in \langle 0,\ 2\text{ m}\rangle$. Such a margin makes it possible to reserve a corridor for the CT unit's motion, wider than the CT unit width b. The said corridor is necessary because the vehicles move in positions yawed by angles ψ_A and ψ_B from the carriageway centreline when they are avoiding the obstacle. This yaw can be seen in Figures 14 and 15.

In the simulation tests carried out, various alternatives of the obstacle avoidance process were analyzed. To assess the alternatives, the distances y_{KA} and y_{KB} between the edge (corner) of the obstacle and the side of the car and the trailer when passing by the obstacle were taken as a basis. The test result is considered successful if, during the obstacle-avoiding phase of the maneuver, the CT unit safely passed by the obstacle and stayed within the lane planned also just beyond the obstacle, i.e., within the road section $x \in \langle x_0; x_0 + 10\text{ m}\rangle$ (cf. Table 1). When selecting the y_W (clearance margin) values for the trajectory planning, the following results were considered satisfactory:

- safe obstacle avoidance, i.e., none of the points within CT unit came into contact with the obstacle or left the lane planned;
- obstacle avoidance using the road shoulder.

If a collision took place between the CT unit and the obstacle contour or the CT unit found itself outside of the road (i.e., outside of the carriageway and the road shoulder) or overturned, such a test result was considered as a "collision with the obstacle" or maneuver failure. The test results have been presented in a synthetic form in Figures 22 and 23 and the clearance margin (y_W) values used to plan the safe trajectory $y_M(x)$ are specified in Table 3.

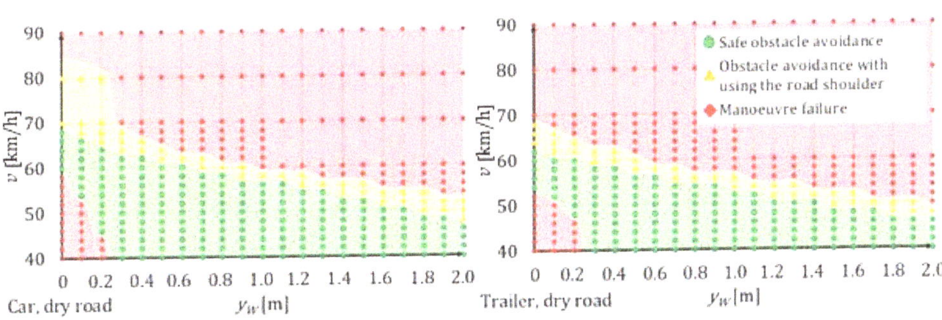

Figure 22. Area of choosing the y_W values when planning a vehicle trajectory for safe obstacle avoidance on dry road surface; motorcar and trailer.

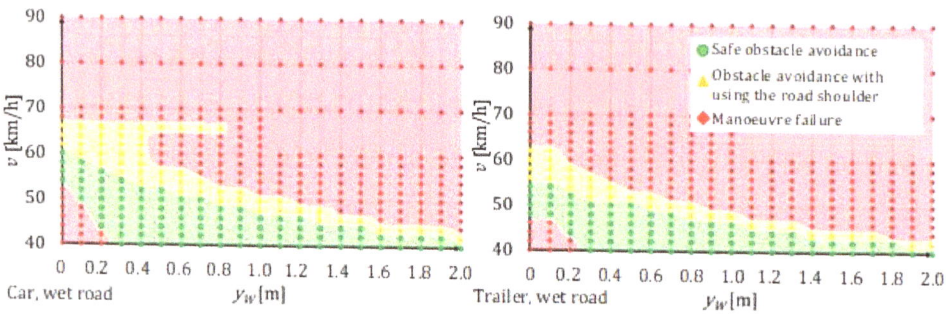

Figure 23. Area of choosing the y_W values when planning a vehicle trajectory for safe obstacle avoidance on wet road surface; motorcar and trailer.

Table 3. Possibilities of safe avoidance of the obstacle.

Road Surface	v [km/h]	y_W [m] for Planning a Trajectory that Would Ensure Safe Obstacle Avoidance		y_W [m] for Planning a Trajectory that Would Ensure Safe Obstacle Avoidance with Using the Road Shoulder	
		Motorcar as a Part of the CT Unit	CT Unit as a Whole	Motorcar as a Part of the CT Unit	CT Unit as a Whole
dry	40	0.3–2.0	0.3–2.0	0.3–2.0	0.3–2.0
	50	0.2–1.7	0.2–1.4	0.2–2.0	0.2–1.9
	60	0.0–0.6	0.0–0.2	0.0–0.9	0.0–0.5
	62	0.0–0.4	0.0	0.0–0.7	0.0–0.4
	66	0.0–0.1	–	0.0–0.4	0.0–0.1
	68	0.0	–	0.1–0.3	0.0
	70	–	–	0.0–0.2	–
	80	–	–	0.0–0.2	–
wet	40	0.3–2.0	0.3–2.0	0.3–2.0	0.3–2.0
	50	0.1–0.7	0.0–0.5	0.1–1.1	0.0–0.8
	60	0.0	–	0.0–0.4	0.0–0.1
	62	–	–	0.0–0.4	0.0–0.1
	66	–	–	0.0–0.8	–
	70	–	–	–	–

The area of acceptable selection of the y_W values, as plotted in Figure 22, shows that a motorcar in a CT unit moving on dry road surface would be able to avoid the obstacle even if travelling with a speed of up to $v = 68$ km/h, while the trailer towed by it would safely avoid the obstacle only if its speed did not exceed $v = 62$ km/h. If the road surface is wet (see Figure 23), a safe obstacle avoidance would only be possible if the CT unit's speed were limited to $v = 54$ km/h. Assuming the acceptability of using the road shoulder (up to 1 m wide) in the critical situation, the obstacle would be avoided if the vehicle speed were up to 68 km/h on dry road surface and up to 62 km/h on the wet road surface (see Table 3).

The tests described have shown the narrowness of the area of selection of the y_W values that would enable safe obstacle avoidance in the critical situation under consideration.

9. Summary

This study is dedicated to a critical (pre-accident or causing a collision hazard) road situation where immediate counteraction of the vehicle control system is required in a space that has been only partly defined. The hazards accompanying such a situation arise from the short time available for the reaction and from the complexity of the interactions taking place between the motorcar, trailer, controller, and road surface. The said interactions are strongly non-linear. In the study, an assumption has been made that the information received from the environment perception system will cause the settings of the control system of the CT unit to be re-adjusted as appropriate. The new settings will be introduced temporarily, for the time of passing by the obstacle, and their values will differ from those required at the stable vehicle drive before and after the obstacle avoidance maneuver.

In consideration of the above, the following was worked out:

- models of dynamics of a CT unit and of its control system;
- procedure of selection of the temporary system settings, i.e., a method of planning the trajectory $y_M(x)$ and the value of the anticipation radius L_a for the time of the critical situation;
- rudiments of selecting the value of the variable parameter, i.e., the clearance margin y_W.

Based on the above, tests were carried out to select the control system parameters (temporary and variable) that would be appropriate for the specific difficult road situation. The impact of the method of planning the trajectory $y_M(x)$ and the value of the anticipation radius L_a in the control system on the feasibility of safe avoidance of an obstacle having suddenly appeared has been examined. In consideration of high curvatures of the trajectories planned, high vehicle speeds, and dynamic action of the trailer, particular attention was paid to the impact of tire slip and skidding on vehicles' behavior when moving along the curvilinear path.

The research carried out has made it possible to formulate the following findings and conclusions for the CT unit's control system operating in the critical situation under analysis:

- The cosine method used to plan a trajectory $y_M(x)$ produced better performance of the obstacle avoidance maneuver in comparison with the other methods taken into account.
- The anticipation radius value at which the obstacle avoidance maneuver was most effective was $L_a = 6$ m.
- The area of selection of the y_W values for planning the vehicle trajectory for the obstacle avoidance time was narrow (Figures 22 and 23) and the narrower, the higher the vehicle speed was.
- The impact of growth in the driving speed on CT unit's behaviour during obstacle avoidance was not definite, because, e.g., small motorcar yaw angle values ψ_A helped to avoid a collision with the obstacle at $v = 50$ km/h and 60 km/h (see the ψ_A curves in Figure 18A,C) but became hazardous at $v = 70$ km/h.
- When a safe obstacle avoidance trajectory is planned for a speed exceeding 60 km/h, low clearance margin (ψ_A) values are required, i.e., the trajectory should be very close to the obstacle.

Results of the research being carried out support the current trends in the construction of vehicle control algorithms, where the definition of rigid operation rules is being abandoned in favor of immediate and ongoing shaping of the properties and areas of suitable solutions. This is of significant importance for rare critical situations. In such cases, the temporary data obtained from previous computer simulations and stored in the system controller are chiefly used.

Author Contributions: Conceptualization, L.P., P.S. and M.Z.; methodology, L.P. and M.Z.; software, P.S.; validation, T.P. and P.S.; formal analysis, L.P., M.Z., P.S.; investigation, L.P., M.Z. and P.S.; resources, T.P.; data curation, P.S.; writing—original draft preparation, L.P.; writing—review and editing, M.Z., P.S., M.G., T.P. and T.L.S.; visualization, M.Z., L.P. and P.S.; supervision, L.P.; project administration, M.Z.; funding acquisition, M.G. All authors have read and agreed to the published version of the manuscript.

Funding: This research received no external funding.

Conflicts of Interest: The authors declare no conflict of interest.

References

1. Katrakazas, C.; Quddus, M.; Chen, W.H.; Deka, L. Real-time motion planning methods for autonomous on-road driving: State-of-the-art and future research directions. *Transp. Res. Part C* **2015**, *60*, 416–442. [CrossRef]
2. Prochowski, L.; Ziubiński, M.; Szwajkowski, P.; Pusty, T.; Gidlewski, M. Experimental and Simulation Examination of the Impact of the Control Model on the Motion of A Motorcar with A Trailer in A Critical Situation. In Proceedings of the 15th International Conference Dynamical Systems—Theory and Applications DSTA, Łódź, Poland, 2–5 December 2019; pp. 409–422. [CrossRef]
3. Wu, B.; Zhu, X.; Shen, J.; Cang, X. Analysis of Steering Model for Emergency Lane Change Based on the China Naturalistic Driving Data. In Proceedings of the SAE World Congress Experience (WCX), Detroit, MI, USA, 4–6 April 2017. [CrossRef]
4. Jurecki, R.S.; Stańczyk, T.L. Driver model for the analysis of pre-accident situations. *Veh. Syst. Dyn.* **2009**, *47*, 589–612. [CrossRef]
5. Sun, C.; Zhang, X.; Xi, L.; Tian, Y. Design of a path-tracking steering controller for autonomous vehicles. *Energies* **2018**, *11*, 1451. [CrossRef]
6. Wang, P.; Gao, S.; Li, L.; Sun, B.; Cheng, S. Obstacle avoidance path planning design for autonomous driving vehicles based on an improved artificial potential field algorithm. *Energies* **2019**, *12*, 2342. [CrossRef]
7. Fényes, D.; Németh, B.; Gáspár, P. A Novel Data-Driven Modeling and Control Design Method for Autonomous Vehicles. *Energies* **2021**, *14*, 517. [CrossRef]
8. Moshchuk, N.; Chen, S.K.; Zagorski, C.; Chatterjee, A. Path planning for collision avoidance maneuver, in ASME 2013 International Mechanical Engineering Congress and Exposition. *Am. Soc. Mech. Eng. Digit. Collect.* **2013**, *56420*, V013T14A016. [CrossRef]
9. Korzeniowski, D.; Ślaski, G. Method of planning a reference trajectory of a single lane change manoeuver with Bezier curve. In *IOP Conference Series: Materials Science and Engineering*; IOP Publishing: Bristol, UK, 2016. [CrossRef]
10. Anderson, S.J.; Peters, S.C.; Pilutti, T.E.; Iagnemma, K. An optimal-control-based framework for trajectory planning, threat assessment, and semi-autonomous control of passenger vehicles in hazard avoidance scenarios. *Int. J. Veh. Auton. Syst.* **2010**, *8*, 190–216. [CrossRef]
11. Snider, J.M. *Automatic Steering Methods for Autonomous Automobile Path Tracking*; CMU-RITR-09-08; Robotics Institute: Pittsburgh, PA, USA, 2009.
12. Hu, X.; Chen, L.; Tang, B.; Cao, D.; He, D. Dynamic path planning for autonomous driving on various roads with avoidance of static and moving obstacles. *Mech. Syst. Signal Process.* **2018**, *100*, 482–500. [CrossRef]
13. Zanchetta, M.; Tavernini, D.; Sorniotti, A.; Gruber, P.; Lenzo, B.; Ferrara, A.; de Nijs, W. Trailer control through vehicle yaw moment control: Theoretical analysis and experimental assessment. *Mechatronics* **2019**, *64*, 102282. [CrossRef]
14. Hac, A.; Fulk, D.; Chen, H. Stability and control considerations of vehicle-trailer combination. *SAE Int. J. Passeng. Cars-Mech. Syst.* **2008**, *1*, 925–937. [CrossRef]
15. Plöchl, M.; Lugner, P. Passenger car and passenger car-trailer—Different tasks for the driver. *JSAE Rev.* **1999**, *20*, 543–548. [CrossRef]
16. Darling, J.; Tilley, D.; Gao, B. An experimental investigation of car-trailer high-speed stability. *Proc. Inst. Mech. Eng. Part D* **2009**, *223*, 471–484. [CrossRef]
17. Balal, E.; Cheu, R.L.; Sarkodie-Gyan, T. A binary decision model for discretionary lane changing move based on fuzzy inference system. *Transp. Res. Part C* **2016**, *67*, 47–61. [CrossRef]
18. Chen, G.; Zhang, W.G.; Zhang, X.N. Fuzzy neural control for unmanned robot applied to automotive test. *Ind. Robot* **2013**, *40*, 450–461. [CrossRef]
19. Ivanov, V. A review of fuzzy methods in automotive engineering applications. *Eur. Transp. Res. Rev.* **2015**, *7*, 29. [CrossRef]
20. Khalaji, A.K. PID-based target tracking control of a tractor-trailer mobile robot. *Proc. Inst. Mech. Eng. Part C* **2019**, *233*, 4776–4787. [CrossRef]
21. Yue, M.; Hou, X.; Gao, R.; Chen, J. Trajectory tracking control for tractor-trailer vehicles: A coordinated control approach. *Nonlinear Dyn.* **2018**, *91*, 1061–1074. [CrossRef]
22. Ni, J.; Hu, J. Dynamics control of autonomous vehicle at driving limits and experiment on an autonomous formula racing car. *Mech. Syst. Signal Process.* **2017**, *90*, 154–174. [CrossRef]
23. Taghavifar, H.; Rakheja, S. Path-tracking of autonomous vehicles using a novel adaptive robust exponential-like-sliding-mode fuzzy type-2 neural network controller. *Mech. Syst. Signal Process.* **2019**, *130*, 41–55. [CrossRef]
24. Gidlewski, M.; Jackowski, J.; Jemioł, L.; Żardecki, D. Sensitivity of a vehicle lane change control system to disturbances and measurement signal errors—Modeling and numerical investigations. *Mech. Syst. Signal Process.* **2021**, *147*, 107081. [CrossRef]

25. Mercorelli, P. Fuzzy Based Control of a Nonholonomic Car-Like Robot for Drive Assistant Systems. In Proceedings of the 2018 19th International Carpathian Control Conference (ICCC), Szilvasvarad, Hungary, 28–31 May 2018; pp. 434–439. [CrossRef]
26. Alshaer, B.J.; Darabseh, T.T.; Alhanouti, M.A. Path planning, modeling and simulation of an autonomous articulated heavy construction machine performing a loading cycle. *Appl. Math. Model.* **2013**, *37*, 5315–5325. [CrossRef]
27. Wang, H.W.; Yu, X.C.; Song, H.B.; Lu, Z.H.; Lloret, J.; You, F. A Global Optimal Path Planning and Controller Design Algorithm for Intelligent Vehicles. *Mob. Netw. Appl.* **2018**, *23*, 1165–1178. [CrossRef]
28. Road Traffic Office. *Road Traffic Accidents in Poland in 2019*; Polish Police Headquarters, Road Traffic Office: Warsaw, Poland, 2020.
29. Li, X.; Sun, Z.; Cao, D.; Liu, D.; He, H. Development of a new integrated local trajectory planning and tracking control framework for autonomous ground vehicles. *Mech. Syst. Signal Process.* **2017**, *87*, 118–137. [CrossRef]
30. Demonceaux, C.; Potelle, A.; Kachi-Akkouche, D. Obstacle detection in a road scene based on motion analysis. *IEEE Trans. Veh. Technol.* **2004**, *53*, 1649–1656. [CrossRef]
31. *PC-CRASH-A Simulation Program for Vehicle Accidents—Operating and Technical Manual, Version 10.0*; Dr. Steffan Datentechnik: Linz, Austria, 2013.
32. Steffan, H.; Moser, A. The trailer Simulation model of PC-Crash. *SAE Trans.* **1998**, *107*, 886–896. [CrossRef]
33. Rill, G. *Simulation von Kraftfahrzeugen*; Vieweg & Sohn Verlag GmbH: Braunschweig, Germany, 1994.
34. Hirschberg, W.; Rill, G.; Weinfurter, H. Tire model TMeasy. *Veh. Syst. Dyn.* **2007**, *45*, 101–119. [CrossRef]
35. Rill, G. TMeasy—A Handling Tire Model based on a three-dimensional slip approach. In Proceedings of the XXIII International Symposium on Dynamic of Vehicles on Roads and on Tracks (IAVSD 2013), Quingdao, China, 19–23 August 2013.
36. Hirschberg, W.; LČÁK, F.P.A.; Rill, G.; NÍK, J.Š.T.; Kintler, P. *TMeasy for Reliable Vehicle Dynamics Simulation*; SCIENTIFIC PROCEEDINGS 2009; Faculty of Mechanical Engineering, STU: Bratislava, Slovakia, 2009.
37. Rill, G. Vehicle dynamics. In *Lecture Notes*; Regensburg, Germany, 2009; Available online: http://pds7.egloos.com/pds/200801/19/01/Vehicle_Dynamics.pdf (accessed on 16 September 2020).
38. Stańczyk, T.L.; Strachowski, P. Assessment of the possibilities of determining the characteristics of a pneumatic tyre with the use of an original road test system. *Arch. Automot. Eng.* **2013**, *59*, 165–183. [CrossRef]
39. Luty, W. Tire transient properties in simulation of vehicle lateral dynamics in curvilinear motion, Prace Naukowe Politechniki Warszawskiej, Transport. *Oficyna Wydawnicza Politech. Warsz.* **2013**, *98*, 357–367.
40. *ISO 3888-1:1999, Passenger Cars—Test Track for A Severe Lane-Change Manoeuvre—Part 1: Double Lane-Change*; International Organization for Standardization: Geneva, Switzerland, 1999.
41. Saaty, T.L. *Fundamentals of Decision Making and Priority Theory with the Analytic Hierarchy Process*; RWS Publications: Pittsburgh, PA, USA, 2000.
42. Saaty, T.L.; Ozdemir, M.S. Why the magic number seven plus or minus two. *Math. Comput. Model.* **2003**, *38*, 233–244. [CrossRef]
43. Prochowski, L.; Pusty, T.; Gidlewski, M.; Jemioł, L. Experimental studies of the car-trailer system when passing by a suddenly appearing obstacle in the aspect of active safety of autonomous vehicles. In *IOP Conference Series: Materials Science and Engineering*; IOP Publishing: Bristol, UK, 2018; Volume 421, p. 032024. [CrossRef]

energies MDPI

Article

Assessment of the Influence of Road Infrastructure Parameters on the Behaviour of Drivers and Pedestrians in Pedestrian Crossing Areas

Marcin Budzynski [1,*], Anna Gobis [1], Lucyna Guminska [1], Lukasz Jelinski [1], Mariusz Kiec [2] and Piotr Tomczuk [3]

1 Faculty of Civil and Environmental Engineering, Gdansk University of Technology, 80-233 Gdansk, Poland; anna.gobis@pg.edu.pl (A.G.); lucyna.guminska@pg.edu.pl (L.G.); lukjelin@pg.edu.pl (L.J.)
2 Faculty of Civil Engineering, Cracow University of Technology, 31-155 Cracow, Poland; mkiec@pk.edu.pl
3 Faculty of Transport, Warsaw University of Technology, 00-662 Warsaw, Poland; piotr.tomczuk@pw.edu.pl
* Correspondence: mbudz@pg.edu.pl; Tel.: +48-604-460-466

Abstract: Pedestrians are participants and, most likely, fatalities in every third road traffic accident in Poland. Over 30% of all fatalities on Polish roads are pedestrians. Accidents with pedestrians are very often the result of various factors related to the infrastructure and behaviour of pedestrians and drivers. The objective of the work was to assess driver and pedestrian behaviour in pedestrian crossing areas. The research also served as a pilot study for similar work to be conducted across Poland, and constituted the basis for monitoring the behaviour of road users in the area of pedestrian crossings. Parameters which must be analysed were identified on the basis of field studies. Principles of selecting test sites were adopted, and measurement methods for pedestrian crossing areas are presented. The influence of the location of the selected test cross-section infrastructure parameters on the behaviour of road users in pedestrian crossing areas is demonstrated. The results of the study will be used as a basis for new solutions involving pedestrian crossing infrastructure designed to improve pedestrian safety. The results were also used in formulating new regulations for the design and maintenance of pedestrian crossings and recommendations for road safety auditors.

Keywords: road safety; pedestrian safety; pedestrian crossing; behaviour of drivers and pedestrians

Citation: Budzynski, M.; Gobis, A.; Guminska, L.; Jelinski, L.; Kiec, M.; Tomczuk, P. Assessment of the Influence of Road Infrastructure Parameters on the Behaviour of Drivers and Pedestrians in Pedestrian Crossing Areas. *Energies* **2021**, *14*, 3559. https://doi.org/10.3390/en14123559

Academic Editor: Mario Marchesoni

Received: 18 May 2021
Accepted: 11 June 2021
Published: 15 June 2021

1. Introduction

Pedestrians are among the groups of traffic participants who are very often deprecated by motorised participants, albeit a very important one in road traffic. Practically, every person able to move about within the road network is a pedestrian, with the length and aim of the trip taken on foot being the key. The distance to be covered by a traffic participant is the primary criterion which influences the choice of the type of trip [1]. Studies on transport behaviour carried out in Poland [2] make it possible to conclude that pedestrian trips constitute about 20% of the total trips in cities. In comparison, this share is ca. 35% in Barcelona and 31% in London [3].

Accidents involving pedestrians are very often the result of various factors related to the infrastructure and behaviour of pedestrians and drivers [4], and to the road infrastructure [5–7]. The factors which involve the behaviour of road infrastructure users include vehicle speeds and the degree to which they are exceeded, pedestrian speed, crossing the road illegally and how drivers behave towards pedestrians (giving way at pedestrian crossings) [8–10]. These factors will vary depending on traffic and pedestrian volumes and the pedestrian crossing location (built-up area, non-built-up area, road class and cross-section, intersections and midblock) [11–14]. Road infrastructure-related factors that have an effect on pedestrian safety include the cross-section [15] but also the width and length of pedestrian crossing, how far it is from an intersection, road section geometry (vertical and horizontal alignment), type and condition of road surface, presence of refuge islands and their geometric parameters, type of vertical and horizontal marking and its condition, signalised and not signalised crossing [16–19], distance to designated and illegal parking, presence of public transport stops, adequacy of drainage, facilities for the blind and visually impaired, pedestrian ramps, condition of pavements and lighting [20–23] and distances between pedestrian crossings [24]. Research on pedestrian crossings focuses on the interaction of drivers and pedestrians [5,25,26] using conflict analysis [27,28].

Pedestrian safety is among the most important challenges in implementing measures to improve road safety [29,30]. More than 1/5 of all fatalities in road accidents worldwide are pedestrians [31]. In several countries, these values are even higher and exceed 1/3 [32,33].

In 2008, in an effort to raise road safety levels across the EU to those of the best performing countries, the European Commission wrote and implemented Directive 2008/96/EC on road infrastructure safety management [34]. The European Commission's intention was to implement adequate tools to improve the safety of road infrastructure across the trans-European road network. Several years into the process, it was clear that the road safety actions were ineffective [35]. In 2019, Directive 2019/1936 of the European Parliament and of the Council was developed amending Directive 2008/96/EC on road infrastructure safety management [36], which extends its remit to roads co-financed from EU funds. In countries such as Poland, where pedestrians represent about 30% of all road deaths, implementing control measures to protect vulnerable road users should be of particular importance. These include road safety inspections (RSI) [37] and road safety audits (RSA). Through these measures, it is possible to significantly reduce the number of people killed or seriously injured in road accidents, including pedestrians [38]. Such work should be supported by scientific research on road user behaviour, which is a main subject in this paper. Pedestrian safety in Poland is at a very low level compared to the best European countries in this regard [39]. Every third fatality in a traffic accident in Poland is a pedestrian. In the years 2009–2018, approximately 94.5 thousand pedestrian accidents were registered in Poland, in which approximately 89.0 thousand pedestrians were injured, and 11.0 thousand pedestrians were killed (Figure 1). It should be emphasized that at that time, the number of pedestrian accidents and their victims decreased by over 40%. However, this is still too high a number. About 10% of all deaths in accidents in Poland were registered at pedestrian crossings (based on SEWIK, a police database).

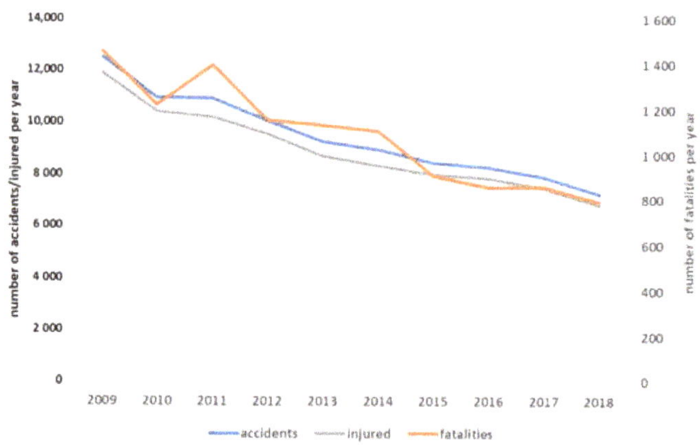

Figure 1. Pedestrian accident and victims in Poland in the years 2009–2018 (based on police's SEWIK data).

Ensuring pedestrian safety in the area of pedestrian crossings should be a priority for actions taken by road authorities. For this purpose, it is necessary to conduct systematic studies of pedestrian and driver behaviour. The lack of clear rules on when to give way to pedestrians gives rise to drivers' dangerous behaviour towards pedestrians, for example, driving the wrong way through pedestrian crossings and, on the other hand, dangerous behaviour of pedestrians themselves in traffic—careless running onto the roadway, crossing the road where it is illegal, etc. [40–42]. This problem also includes the lack of appropriate infrastructure for pedestrians moving across the roadway (refuge islands or appropriate marking) and along the roadway (sidewalks)—mainly on the outskirts of cities, segments of roads through built-up areas and segments of roads in non-built-up areas where pedestrian traffic may occur with significant intensity. Additionally, several existing pedestrian facilities do not meet traffic or safety standards (for example, their technical condition, location of pedestrian crossings in terms of visibility, public transport stops and gaps in footpaths). Dependencies between pedestrians and vehicles require relevant studies to ensure the safe movement of pedestrians.

Polish experience. Research in Poland analyses pedestrian behaviour by analysing digital footage from cameras, an approach also used in the MOBIS project. The objective of the project was to develop and test a method for assessing pedestrian crossing safety using an automatic analysis of video footage. The assessment was designed to identify conflicting movements of vehicles and pedestrians. The project helped us to test and optimise the method at different types of pedestrian crossings in urban areas: on single carriageways with and without a refuge island, dual carriageways and at traffic lights [43]. Studies of Poland's pedestrian crossings also analyse night-time luminance depending on road lighting [44,45]. A safety assessment of non-signalised pedestrian crossings showed that many drivers exceed the applicable speed limits [46]. Other studies indicate that there is a high risk to pedestrians on sections of transit roads in small towns [47] and cities [48].

International experience. A study was conducted in Australia designed primarily to test road user knowledge of regulations determining right-of-way in different situations [49]. Another research method implemented in Australia analysed child behaviour on major and local roads. The study paid special attention to where children stop, where they look around, how they analyse vehicle movement and how they cross the road [50]. In France, study was focused on analysing the behaviour and interrelations between pedestrians and drivers. The analysis of the results showed that when a pedestrian looks at an oncoming vehicle, the car is more than 10% more likely to stop [51].

Surveys conducted in the United Kingdom identified three main hazards to pedestrian safety: dual carriageway crossing, red light entry for drivers and reduced visibility by parked vehicles [52]. In Spain, cameras were deployed for automatic pedestrian detection at pedestrian crossings. In addition, a manual check was conducted of the gender, age, size of group or pedestrian baggage [53]. In Germany, road traffic behaviour was studied at signalised intersections [54]. Research in the Netherlands proved that pedestrians mostly struggle with excessive vehicle speeds, especially as regards people aged >65 [55]. A study in Sweden recorded road user behaviour using cameras [56,57]. The results of research in Belgium show that approximately 21% of the pedestrians violate traffic lights [58]. Research conducted in Finland concerned the speed of drivers and pedestrians, road traffic and the behaviour of road users in the area of pedestrian crossings [59]. Studies conducted in urban areas in Israel analysed the effects of pedestrian distraction at pedestrian crossings on pedestrian safety [60]. In Malaysia, driver behaviour was assessed at pedestrian crossings (accelerating and slowing) and pedestrian time lost [61]. A study in Estonia assessed the share of drivers running red lights, vehicle speed and estimated pedestrian and vehicle time lost [62]. Work conducted in China [63–65] studied pedestrian and driver behaviour at signalised pedestrian crossings. The country also conducted other research to measure vehicle speed and the distance between a stopped vehicle and a pedestrian [66]. Research [67,68] has shown that the greater the number of lanes in one direction, the greater the number of conflicts between drivers and pedestrians. Research in India [69,70] has indicated the problems that pedestrians have with the assessment of the possibility of crossing the road at crossings outside intersections. Research conducted in Slovakia concerned the use of mobile devices and headphones on pedestrian crossings [71].

The field research was also supported by simulation studies, during which the impact of different scenarios on the behaviour of drivers and pedestrians in the area of pedestrian crossings outside intersections was analysed [72–74].

Review of Polish and international research shows that field work must be conducted to observe pedestrian and driver behaviour at pedestrian crossings (as demonstrated further in the article). Real pedestrian crossing situations should be studied analysing crossing location, equipment, pedestrian traffic volume and pedestrian characteristics. In the case of drivers, observation is required to understand driver reaction on approaches to pedestrian crossings in relation to pedestrian behaviour. As a result, it will be possible to assess the effectiveness of existing solutions and improve pedestrian safety where it is most at risk, i.e., at pedestrian crossings [75]. In the majority of countries which study pedestrian behaviour and pedestrian–driver relations, this is not carried out systematically. Instead, the work analyses selected aspects only. It is critical for Poland to carry out such work due to the high levels of risk to pedestrians.

Taking into account the literature review, parameters were verified and selected, which helped to build a database that is the basis for analysing the behaviour of pedestrians and drivers and the pedestrian–driver relationship in pedestrian crossing areas:

- Pedestrian traffic parameters under research—field tests: pedestrian age, traffic volume, distance between the pedestrian and the edge of the roadway, how the pedestrian approaches the crossing, average speed of the pedestrian on the crossing, pedestrian delay before the crossing, pedestrian behaviour before entering the roadway (making sure that it is safe to enter), sudden entry onto the roadway (entry at red light) and pedestrians crossing in illegal locations in the vicinity of a designated crossing ($\pm$100 m).
- Vehicle parameters under research: speed of a vehicle approaching the crossing, speeding, entry at red light and traffic volume.

2. Materials and Methods

The main aim of this research was to assess the behaviour of pedestrians and drivers and the pedestrian–driver relationship at and in the area of pedestrian crossings. Such systematic studies are necessary due to the level of risk of pedestrian accidents. The objective was to make walking more attractive and improve pedestrian safety. To this end, field tests and survey studies were carried out in 70 test points in Poland's Pomorskie and Małopolskie regions, across an area that is diversified in terms of its development. The road cross-section and location of pedestrian crossings were also taken into consideration. The results of these studies make it possible to assess pedestrian safety at various types of crossings, associated with the street's cross-section.

2.1. Measurement Technique

The following instruments were used to record traffic parameters alongside pedestrian and driver behaviour:

- MioVision—a mobile camera which records and collects traffic data using a database platform. The camera was situated on a mast with a variable height, up to 7 m.
- MetroCount—a device which counts the traffic volume, including the structure and speed of vehicles, making it possible to collect data in the long term.

Depending on the test site, one or two video cameras were used to record images. The cameras were situated on a tripod at a distance of 10 m to 15 m behind a pedestrian crossing and located on the shoulder, lamppost or pavement along the road. The measuring equipment was located in a way which would not capture the attention of drivers or pedestrians. Driver behaviour was studied as cars approached the pedestrian crossing. The maximum range of image recording depended on road geometry and was between 40 m and 100 m. During the measurement, test cross-sections were designated in the field, every 10 m or 20 m, as a test basis for the assessment of average speed (Figure 2). The speed was calculated on the basis of the length of the measured segment and time travelled to the test cross-sections. Image analysis covered recording the time of consecutive events, i.e., the appearance of vehicles in the subsequent test cross-sections. Speed measurement included only the vehicles moving in free traffic, i.e., those which were not influenced by other vehicles in their vicinity. The following scenarios were selected in the analysis of vehicle speed on a pedestrian crossing:

- No pedestrian;
- With a pedestrian waiting before the crossing;
- With a pedestrian on the crossing.

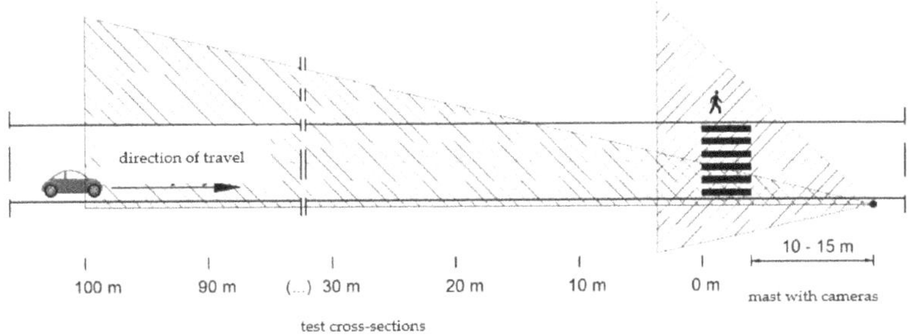

Figure 2. Diagram of field tests using video recording instruments.

For pedestrian traffic parameters, the camera's field of observation covered the entire pedestrian crossing under analysis and its access path. The analysis included the following:

- Average speed of the pedestrian crossing the street (counted between the edges of the carriageway);
- Pedestrian delay (waiting time for the road crossing);
- Distance to the vehicle when the pedestrian decides to enter the crossing and cross it.

Each test included pedestrians going across the crossing in both directions, with a division into age groups also included (<20 years—11.1%; 20–60 years—80.4%; >60 years—8.5%). In the case of drivers, age was not taken into account.

2.2. Selection of Test Sites

When selecting the test points, limitations were taken into account, making it possible to obtain a random sample of test points with a high uniformity of traffic conditions. This allowed a wide range of comparisons between test results. At each pedestrian crossing, the measurements were carried out for 6 h (10:00 a.m.–3:00 p.m.), in good weather conditions and in daylight. The research was conducted only on working days (Monday–Friday), in the months of April–June. In order to obtain such a set, test points were located as follows:

- Outside of horizontal and vertical curves with reduced visibility, which make it necessary to reduce speed compared to previous sections;
- In segments with a gradient of less than 4%;
- Outside of built-up areas in segments with comparable features in terms of both their function and the volume of pedestrian and vehicle traffic;
- In built-up areas in sections with comparable features in terms of the buildings and functions and the volume of pedestrian and vehicle traffic;
- In segments without local speed limits;
- In segments without speed cameras.

In order to compare vehicle speeds and pedestrian behaviour, selection criteria were designated for the test points. The following criteria were chosen:

- Type of area;
- Type of cross-section;
- Presence of traffic lights;

- Type of segment;
- Speed limit.

Field tests with pedestrians were carried out in three types of areas:

- Large and medium-sized cities (powiat/county level);
- Small towns;
- Rural areas.

In each of these areas, tests were performed in the following locations:

- In between intersections (midblock);
- At intersections that are not signalised;
- At signalised intersections.

For the locations selected above, detailed locations were specified, taking into account various cross-sections of the road segment:

- 1 × 2 (single carriageway, 1 lane in each direction);
- 1 × 4 (single carriageway, 2 lanes in each direction);
- 2 × 2 (dual carriageway, 2 lanes in each direction);
- 2 × 3 (dual carriageway, 3 lanes in each direction).

The current speed limit was identified for each of these test point's selection criteria and, for further analysis, 50 km/h (for built-up areas) and 70 km/h (for built-up areas with dual carriageways and for rural areas) were selected. Figure 3 presents a cumulative diagram of the test site.

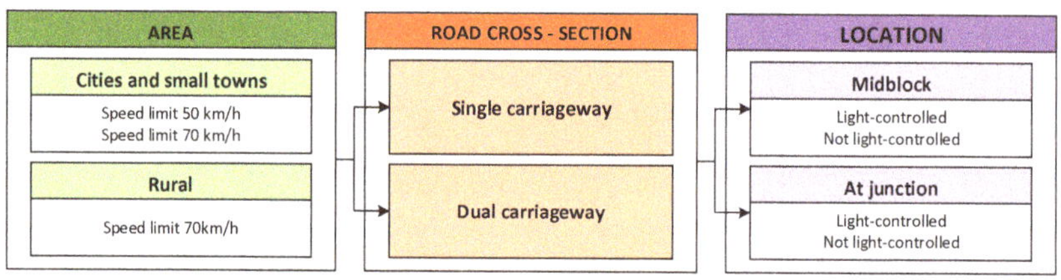

Figure 3. Diagram of test point selection.

2.3. Statistical Methods

Statistical methods were reduced to calculation descriptive statistics according to pedestrian and vehicle speed. Average speed, standard deviation and the max and min speed were calculated. Average speed and differences in speed for various locations of pedestrian crossings and the presence of pedestrians could be factors to assess the interaction between drivers and pedestrians. Therefore, in Section 3, results of the average speed comparison are presented. Additionally, to assess the statistical significance of speeds, a two-sample *t*-test for the difference of means was carried out. To assess speed distribution, the Shapiro–Wilk test was used to assess normal distribution.

3. Results

In order to compare vehicle speeds and pedestrian behaviour, cumulative analyses were produced for the test points. The following categories were chosen:

- Type of area;
- Type of cross-section;
- Presence of traffic lights;
- Type of segment.

In each of these categories, a speed limit was identified, currently in place at each given location. A division into 50 km/h and 70 km/h speed limits was adopted for cumulative comparisons. Based on the analysis of the collected data, conclusions were formulated on the behaviour of drivers in pedestrian crossing areas as per the criteria for locating test points. Descriptive statistics for each measurement sites as well as test cross-sections were calculated.

3.1. Test Points with 50 km/h Speed Limit

3.1.1. Type of Area

- Urban areas feature lower vehicle speeds on approach to the crossing than small towns, with no pedestrian by 5 km/h at a distance of 10 m from the crossing and 10 km/h at a distance of 50 m from the crossing.
- In both types of areas, similar vehicle speeds were recorded at a distance of 10 m from the crossing with a pedestrian waiting to cross.
- Urban areas feature higher vehicle speeds on approach to the crossing than small towns, with a pedestrian crossing the road by 5 km/h at a distance of 10 m from the crossing and 5 km/h at a distance of 50 m from the crossing.
- In both areas, a minor decrease in vehicle speed was recorded at a distance of 10 m from the crossing with a pedestrian waiting to cross compared to a situation with no pedestrian, and a considerable decrease in the case of pedestrians crossing the roadway (Figure 4).

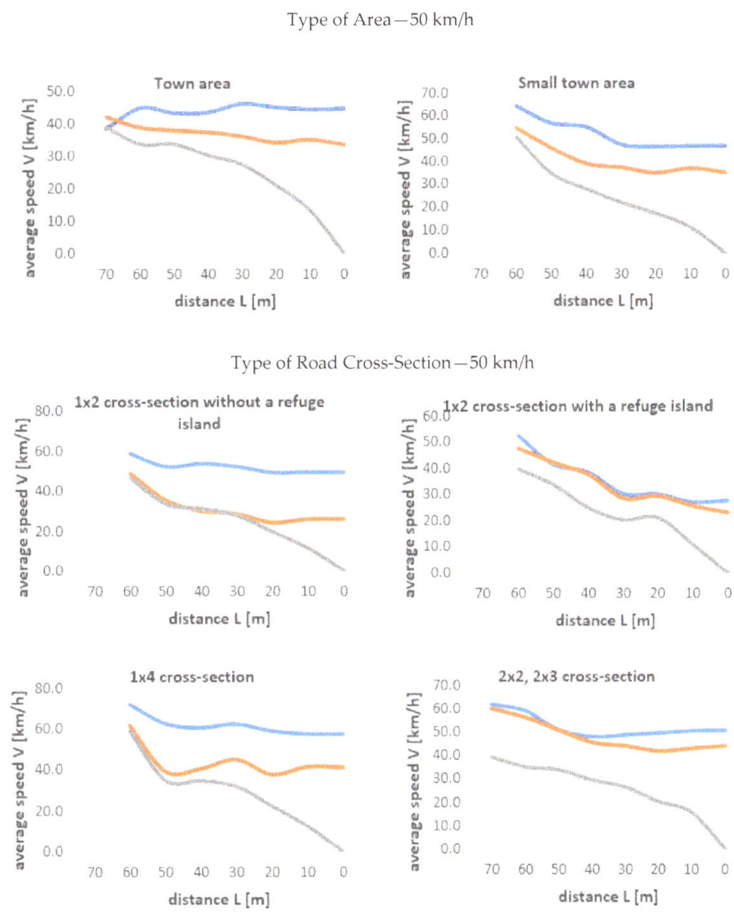

Figure 4. *Cont.*

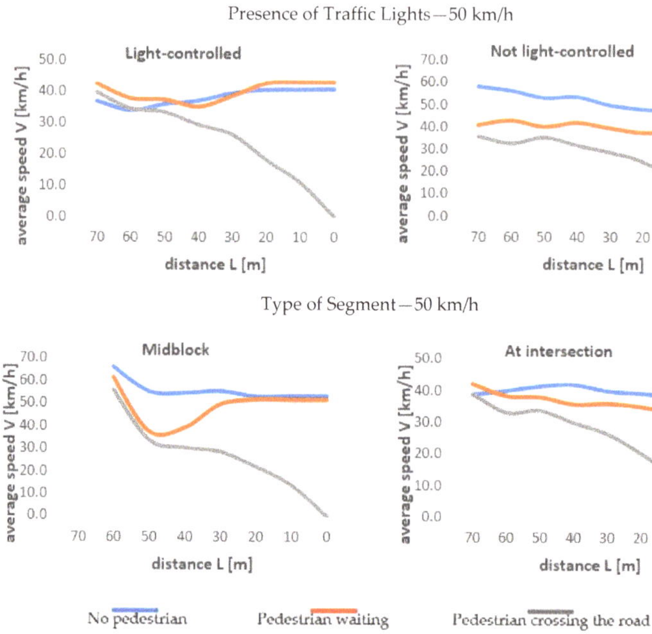

Figure 4. Vehicle speeds in pedestrian crossing areas—speed limit: 50 km/h.

3.1.2. Type of Road Cross-Section

- The lowest speed on approach to the crossing with no pedestrian, at a distance of 10 m from the crossing, was recorded for the 1 × 2 cross-section with a refuge island (30 km/h less than for the 1 × 4 cross-section; 23.5 km/h less than for dual carriageways).
- The lowest speed on approach to the crossing with a pedestrian waiting to cross was recorded for the 1 × 2 cross-section without a refuge island (16 km/h less than for the 1 × 4 and 1 × 2 cross-sections with a refuge island; 22 km/h less than for dual carriageways).
- The lowest speed on approach with a pedestrian crossing the roadway was recorded for the 1 × 2 cross-section with a refuge island, but the differences between the cross-sections were minor in this case (Figure 4).

For all types of cross-sections, vehicle speed with pedestrians waiting to cross was lower compared to situations when there was no pedestrian (by 16 km/h for 1 × 2, by 16 km/h for 1 × 4 and by 2 km/h for 2 × 2 and 2 × 3—Figure 4).

3.1.3. Presence of Traffic Lights

- With traffic lights, vehicle speeds on approach to the crossing were higher: by 7 km/h with no pedestrian, by 4 km/h when a pedestrian is waiting to cross and by 5.5 km/h when a pedestrian is crossing the roadway (at a distance of 10 m from the pedestrian crossing).
- With traffic lights, the speed with a pedestrian waiting to cross was higher by 1 km/h than when there was no pedestrian.
- Without traffic lights, vehicle speed on approach to the crossing with a pedestrian waiting to cross was 2 km/h lower than when there was no pedestrian.

It is noteworthy that the presence of traffic lights changes the speed on approach to the crossing to a minor degree, but light-controlled pedestrian crossings significantly increase pedestrian safety.

3.1.4. Type of Segment

- When there was an intersection, vehicle speeds at a distance of 10 m from the crossing were lower than in between intersections, by 0.5 km/h with no pedestrian, by 14 km/h with a pedestrian waiting and by 1 km/h with a pedestrian crossing the roadway, respectively.
- When a crossing was located in between intersections, the speed with a pedestrian waiting was 2 km/h lower than when there was no pedestrian, and when a crossing was at an intersection, the difference was 1 km/h (lower with a pedestrian waiting) (Figure 4).

A two-sample *t*-test for the difference of means to assess the statistical significance of differences in the average values of the observed speeds (Figure 5) was carried out. The impact of pedestrian presence, as well as the type of area, cross-section, segment and presence of traffic lights, was considered. Statistically insignificant results (p value > 0.05), which suggest no impact on driver behaviour, were obtained for the following:

- Light-controlled pedestrian crossings (all cases of pedestrian presence);
- No pedestrian vs. pedestrian waiting (dual carriageway roads and at an intersection);
- Pedestrian waiting vs. pedestrian crossing (the roads at small town area and two-lane roads);
- Pedestrian waiting—type of area and two-lane roads;
- Pedestrian crossing the road—type of area, cross-section and presence of traffic lights.

3.2. *Test Points with 70 km/h Speed Limit*

3.2.1. Type of Area

- A small town area had a lower vehicle speed on approach to the crossing than a rural area, with no pedestrian by 35 km/h at a distance of 10 m from the crossing and 18 km/h at a distance of 50 m from the crossing.
- A small town area had lower vehicle speeds on approach to the crossing than a rural area, with a pedestrian waiting by 25 km/h at a distance of 10 m from the crossing and 20 km/h at a distance of 50 m from the crossing.
- In a small town area, a higher speed was recorded with pedestrians waiting than when there was no pedestrian by 5 km/h; in a rural area with a pedestrian waiting, a lower speed was recorded with pedestrians waiting compared to a situation with no pedestrian by 1 km/h (this pertains to a distance of 10 m from the crossing) (Figure 6).

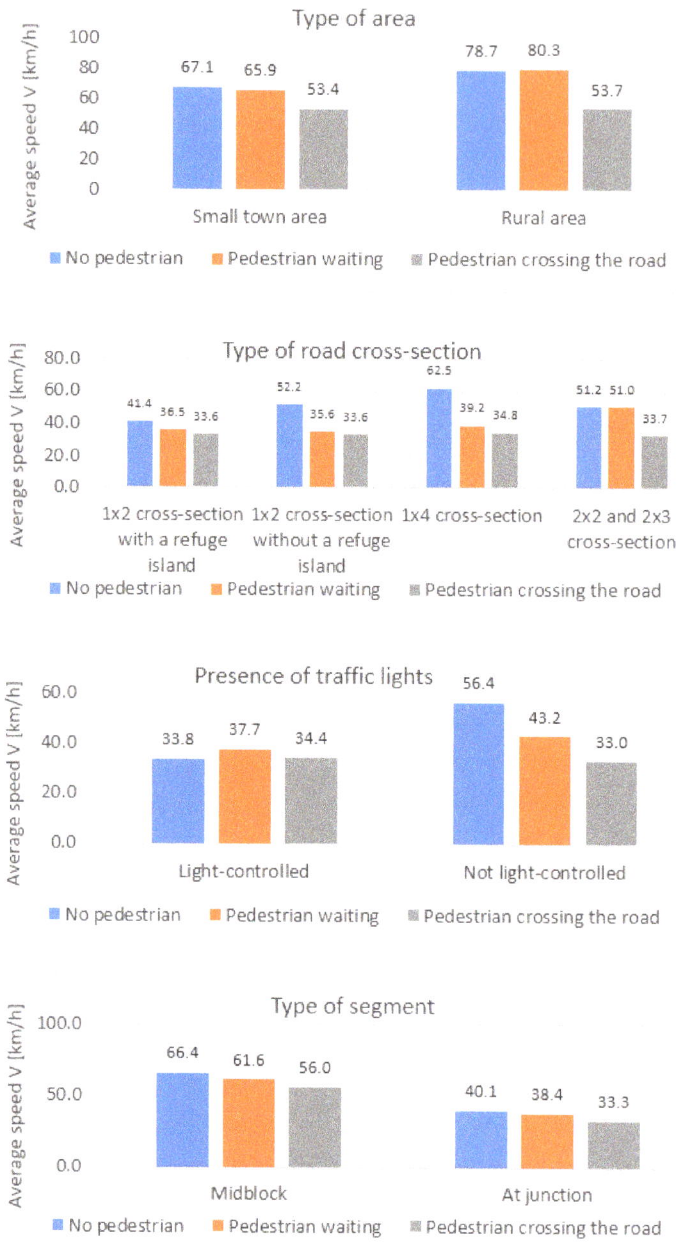

Figure 5. Comparison of speeds at pedestrian crossings in selected locations—speed limit of 50 km/h (measured at 50 m from the crossing).

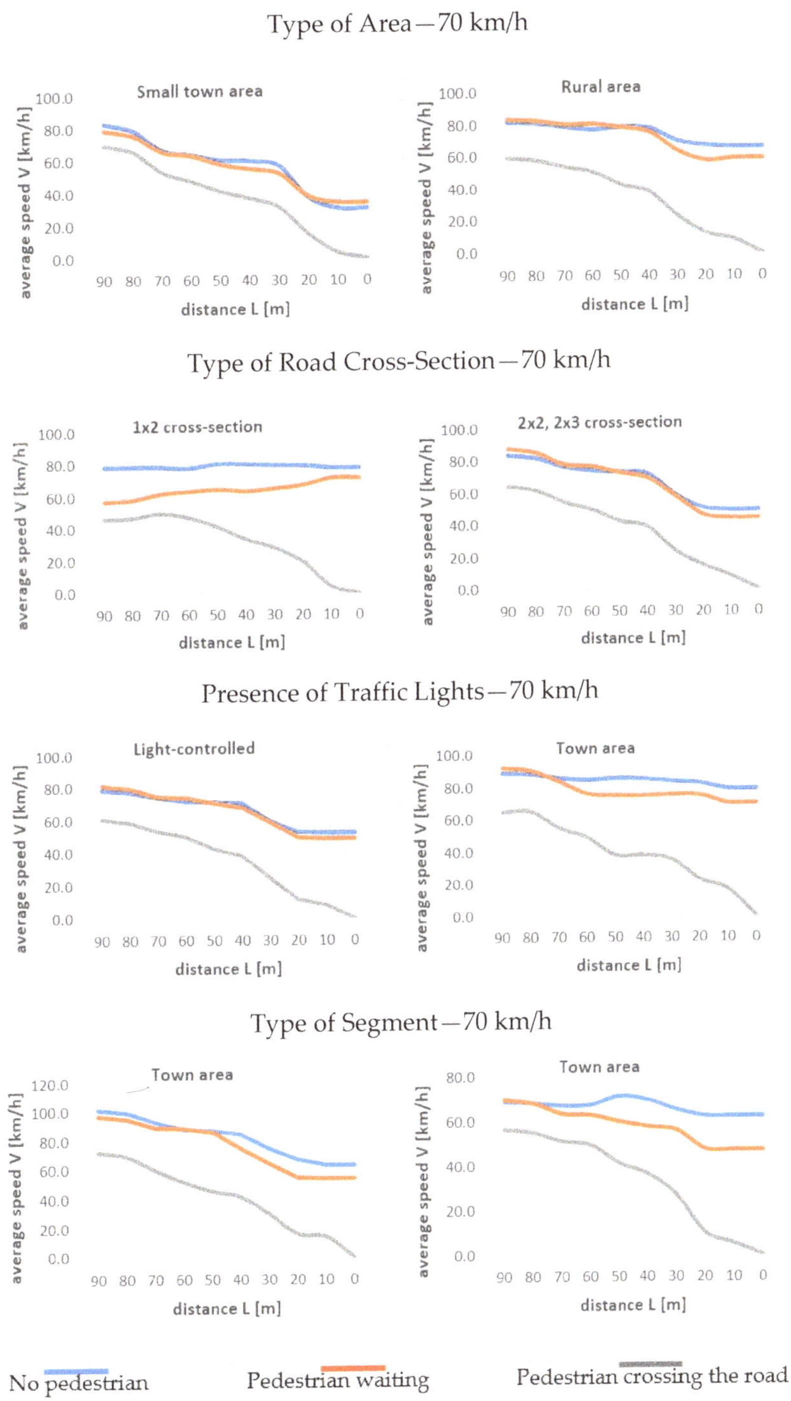

Figure 6. Vehicle speeds in pedestrian crossing areas—speed limit: 70 km/h.

3.2.2. Type of Road Cross-Section

- The lowest speed on approach to the crossing with no pedestrian, at a distance of 10 m from the crossing, was recorded for cross-sections 2 × 2 and 2 × 3 (29 km/h less than for cross-section 1 × 2).
- The lowest speed on approach to the crossing with a pedestrian waiting to cross was recorded for cross-sections 2 × 2 and 2 × 3 (27 km/h less than for cross-section 1 × 2).
- The lowest speed on approach to the crossing with a pedestrian crossing the roadway was recorded for cross-section 1 × 2, 4 km/h less than for cross-sections 2 × 2 and 2 × 3.
- For both types of cross-sections, a lower speed was recorded with a pedestrian waiting to cross compared to situations with no pedestrian (5–6 km/h) (Figure 6).

3.2.3. Presence of Traffic Lights

- With traffic lights, vehicle speeds on approach to the crossing were considerably lower, by 26 km/h with no pedestrian, by 20 km/h with a pedestrian waiting to cross and by 9 km/h with a pedestrian crossing the roadway (at a distance of 10 m from the pedestrian crossing).
- With traffic lights, the speed with a pedestrian waiting was 3 km/h lower than with no pedestrian.
- With no traffic lights, vehicle speed on approach to the crossing with a pedestrian waiting was 9 km/h lower than with no pedestrian (Figure 6).

It is noteworthy that the presence of traffic lights significantly changes the speed on approach to the crossing, and at the same time, light-controlled pedestrian crossings significantly improve pedestrian safety.

3.2.4. Type of Segment

- When there was an intersection, vehicle speeds at a distance of 10 m from the crossing were lower than in between intersections (with no pedestrian by 1 km/h and with a pedestrian waiting by 7 km/h), with a pedestrian crossing the roadway, lower speeds were recorded when there was an intersection—by 9 km/h.
- When a crossing was located in between intersections, the speed with a pedestrian waiting was 0.5 km/h lower than with no pedestrian and, in the case of a crossing at an intersection, this difference was 7 km/h (lower with a pedestrian waiting) (Figure 6).

A two-sample t-test for the difference of means to assess the statistical significance of differences in average values of the observed speeds (Figure 7) was carried out. The impact of pedestrian presence, as well as the type of area, cross-section, segment, and presence of traffic lights, were considered. Statistically insignificant results (p value > 0.05) were obtained for the following:

- No pedestrian vs. pedestrian waiting (type of area, dual carriageway roads, presence of traffic signals and type of segments)—drivers on roads with high speeds (70 km/h speed limit) were less willing to give priority to pedestrians.
- No pedestrian—type of cross-section had no impact on speed.
- Pedestrian crossing the road—behaviour of drivers was the same for both locations (type of area).

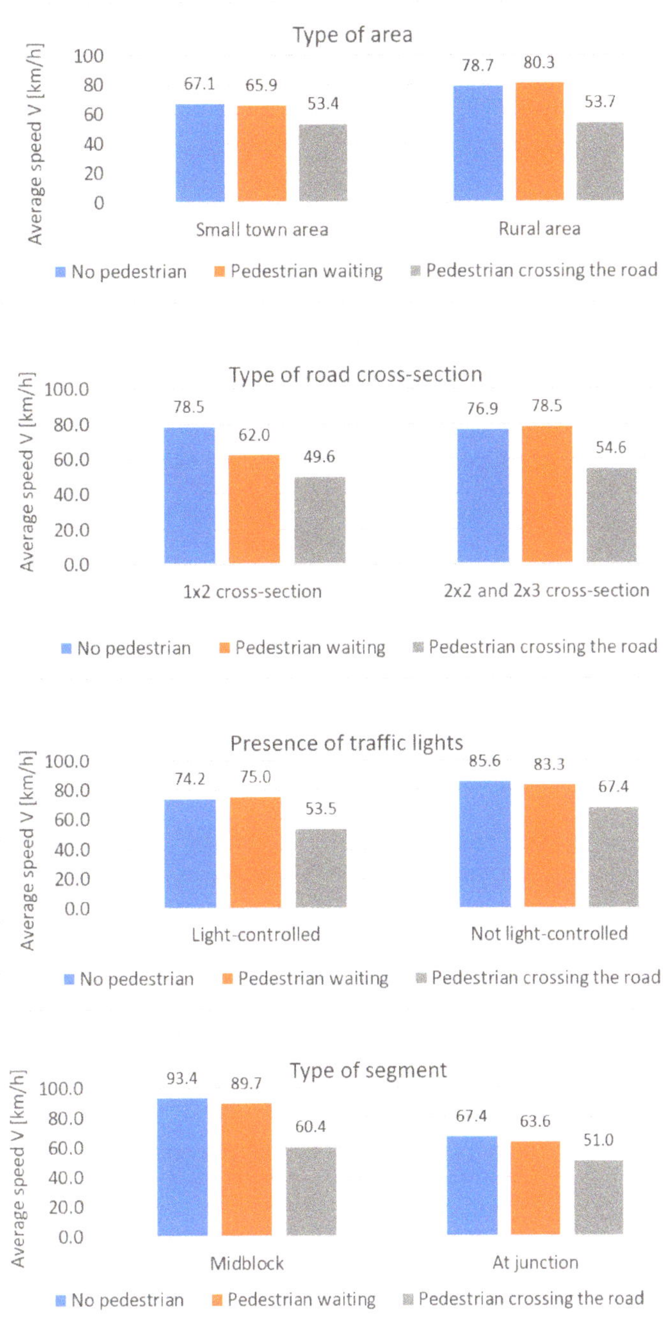

Figure 7. Comparison of speeds at pedestrian crossings in selected locations—speed limit of 70 km/h (measured at 70 m from the crossing).

3.3. Pedestrian Behaviour Studies

During the field tests, selected parameters and pedestrian behaviours were analysed—pedestrian speed on the crossing; time lost on crossings; the distance of the waiting pedestrian from the roadway; how the pedestrian approaches the crossing; making sure that it is safe before entering the roadway; dangerous behaviour—crossing at a red light and crossing the roadway outside of a pedestrian crossing. Based on the research, the following results were obtained (Table 1):

- Average speeds on crossings with a refuge island are lower than in cross-sections without a refuge island.
- There are no large differences in crossing speeds for various types of cross-section or area; the values are from 1.0 m/s to 1.6 m/s.
- There are big differences in time lost depending on the area—in a city area, a pedestrian requires on average 17 s to cross the roadway; in small town areas, it is 25 s; and in rural areas, 48 s.
- Out of the recorded dangerous pedestrian behaviours, the vast majority concern not making sure that it is safe to enter the roadway—84% (ca. 10% of all recorded pedestrians); crossing at a red light—8% (ca. 1% of all recorded pedestrians); and crossing in an unpermitted location—8% (ca. 1% of all recorded pedestrians).
- Depending on the type of area, the way a pedestrian approaches the crossing is mainly "normal" (without running up to the crossing, speeding up or slowing down)—97% in a city area, 89% in a small town area and 94% in a rural area.

3.4. Additional Research

Additional field tests were carried out in a highly urbanized city area, in three Polish cities: Gdynia, Gdansk and Warsaw. The tests were carried out in a reference period of very good weather conditions, including pedestrian crossings which are not light controlled. The study covered nine pedestrian crossings with diversified street cross-sections. Based on the analysis, it can be stated that there is a significant decrease in speed when a pedestrian is crossing the road, while no decrease in speed was recorded with a pedestrian waiting to cross. In certain cases, the value was higher than average (Table 2):

- The average speed at a distance of 70 m from the crossing was 39.6 km/h for all vehicles, 38.8 km/h with no pedestrians in the crossing area, 42.1 km/h with a pedestrian waiting to cross and 39.7 km/h with a pedestrian crossing the roadway.
- The average speed at a distance of 30 m from the crossing was 41.7 km/h for all vehicles, 45.9 km/h with no pedestrians in the crossing area, 41.2 km/h with a pedestrian waiting to cross and 27.7 km/h with a pedestrian crossing the roadway.
- The average speed at a distance of 0 m from the crossing was 36.5 km/h for all vehicles, 45.6 km/h with no pedestrians in the crossing area, 40.0 km/h with a pedestrian waiting to cross and 0 km/h with a pedestrian crossing the roadway.

The lowest average speed of vehicles approaching a pedestrian crossing with pedestrians crossing the roadway was recorded at roundabout-type intersections, and the highest when the crossing was located between intersections. This confirms that the location of a pedestrian crossing influences vehicle speed.

Table 1. Features of pedestrian traffic.

Group of Pedestrians	Average Pedestrian Speed [m/s]							Average Speed on All Pedestrian Crossings [m/s]	Average Time Loss [s]		Average Time Loss on All Pedestrian Crossings [s]	Distance of the Waiting Pedestrian from the Crossing [m]
	Crossing without a Refuge Island		Crossing with a Refuge Island									
	T1 (Whole Width of Road)	No. of People	T2 (1st Part of Road)	T3 (Island)	T4 (2nd Part of Road)	No. of People	Average Speed [m/s]		Crossing w/o a Refuge Island	Crossing with a Refuge Island		
Area												
City Area												
<20	1.39	231	1.70	1.44	1.64	164	1.60	1.53	7.4	6.9	7.2	0.94
20–60	1.27	1462	1.47	1.51	1.52	1543	1.50	1.44	18.4	7.7	12.9	0.92
>60	1.06	77	1.22	1.35	1.39	249	1.32	1.30	10.8	5.2	6.6	0.98
All	1.28	1770	1.46	1.48	1.52	1956	1.49	1.44	16.6	7.3	11.7	0.93
Small Town Area												
<20	1.17	30	1.42	1.11	1.59	60	1.37	1.34	10.4	29.1	22.8	1.46
20–60	1.29	41	1.60	1.16	1.49	234	1.42	1.41	36.8	27.5	28.9	1.18
>60	1.12	4	1.26	0.80	1.40	18	1.15	1.15	11.9	27.9	25.0	1.33
All	1.24	75	1.54	1.13	1.51	312	1.39	1.38	24.9	27.8	27.2	1.24
Rural Area												
<20	1.41	6	1.24	0.41	1.02	1	0.89	1.23	44.0	14.0	39.7	1.00
20–60	1.58	22	1.56	1.63	1.57	72	1.59	1.59	48.7	21.1	27.5	1.38
>60	-	0	-	-	-	0	-	-	-	-	-	-
All	1.54	28	1.55	1.62	1.52	73	1.56	1.56	47.7	21.0	28.4	1.36
Type of Road Cross-Section												
1 × 2 Cross-Section												
<20	1.38	223	1.63	1.32	1.63	185	1.52	1.48	6.9	10.3	8.5	1.21
20–60	1.22	661	1.37	1.46	1.52	984	1.45	1.41	20.2	6.1	11.7	0.97
>60	1.13	26	1.22	1.35	1.42	209	1.33	1.32	7.1	3.8	4.2	0.96
All	1.25	910	1.42	1.52	1.30	1378	1.41	1.38	16.6	6.3	10.4	1.00
1 × 4 Cross-Section												
<20	1.30	10	-	-	-	0	-	1.30	14.1	-	14.1	1.13
20–60	1.30	443	-	-	-	0	-	1.30	20.9	-	20.9	1.56
>60	1.01	41	-	-	-	0	-	1.01	13.8	-	13.8	1.41
All	1.28	494	-	-	-	0	-	1.28	20.1	-	20.1	1.53

Table 1. *Cont.*

Group of Pedestrians	Average Pedestrian Speed [m/s]							Average Speed on All Pedestrian Crossings [m/s]	Average Time Loss [s]		Average Time Loss on All Pedestrian Crossings [s]	Distance of the Waiting Pedestrian from the Crossing [m]
	Crossing without a Refuge Island		Crossing with a Refuge Island									
	T1 (Whole Width of Road)	No. of People	T2 (1st Part of Road)	T3 (Island)	T4 (2nd Part of Road)	No. of People	Average Speed [m/s]		Crossing w/o a Refuge Island	Crossing with a Refuge Island		
2 × 2 and 2 × 3 Cross-Section												
<20	1.32	34	1.61	1.50	1.63	40	1.58	1.52	18.1	24.3	21.5	0.64
20–60	1.33	421	1.62	1.48	1.52	865	1.54	1.51	16.2	16.1	16.1	0.81
>60	1.12	14	1.24	1.22	1.29	58	1.25	1.24	9.5	17.3	15.8	0.88
All	1.32	469	1.60	1.46	1.51	963	1.52	1.50	16.1	16.5	16.4	0.81
Presence of Traffic Lights												
Light-Controlled												
<20	1.44	66	1.47	1.64	1.49	49	1.54	1.51	22.7	36.9	28.7	0.85
20–60	1.26	865	1.60	1.59	1.49	689	1.56	1.47	28.1	22.3	25.5	1.02
>60	1.00	26	1.25	1.33	1.35	35	1.31	1.25	27.3	29.9	28.8	1.13
All	1.27	957	1.58	1.58	1.49	773	1.55	1.47	27.7	23.6	25.9	1.01
Not Light-Controlled												
<20	1.34	201	1.67	1.27	1.67	176	1.53	1.48	4.0	6.1	5.0	1.26
20–60	1.29	660	1.42	1.40	1.54	1160	1.45	1.43	7.8	3.9	5.3	0.82
>60	1.10	55	1.22	1.31	1.40	232	1.31	1.30	3.1	3.3	3.2	0.91
All	1.29	916	1.42	1.37	1.53	1568	1.44	1.42	6.7	4.0	5.0	0.89
Type of Segment												
Midblock												
<20	1.44	182	1.69	1.48	1.62	165	1.60	1.55	3.0	10.4	6.5	1.17
20–60	1.41	421	1.51	1.61	1.48	1007	1.53	1.52	7.5	13.4	11.7	0.92
>60	1.21	25	1.32	1.52	1.34	122	1.39	1.38	4.6	8.6	7.9	0.95
All	1.41	628	1.51	1.59	1.49	1294	1.53	1.51	6.0	12.6	10.4	0.95
At Intersection												
<20	1.22	85	1.46	0.98	1.64	60	1.36	1.32	20.6	19.5	20.1	0.82
20–60	1.22	1104	1.47	1.30	1.57	842	1.44	1.38	23.8	7.5	16.8	0.99
>60	1.00	56	1.15	1.15	1.43	145	1.24	1.21	13.7	5.2	7.6	1.04
All	1.21	1245	1.42	1.26	1.55	1047	1.41	1.35	23.1	7.9	16.2	0.98

Table 2. Average vehicle speed V [km/h] depending on the distance from the pedestrian crossing L [m].

Situation	Number of Cases	Vehicle Distance from the Crossing L [m]							
		0	10	20	30	40	50	60	70
No Pedestrian	2489	45.6	45.4	45.7	45.9	43.8	43.8	40.3	38.8
Pedestrian Waiting to Cross	920	40.0	40.0	37.7	41.2	37.5	38.3	38.9	42.1
Pedestrian Crossing the Roadway	708	0.0	13.6	20.3	27.7	30.3	33.9	33.8	39.0
All	4117	36.5	38.7	39.6	41.7	39.9	41.0	39.0	39.6

At the same time, observations were carried out on pedestrian crossings in selected test points, focused on assessing the behaviour of drivers with regard to pedestrians, on approach to the crossing. The results are presented in Figure 8. The tests were carried out for the 1 × 2 cross-section, a speed limit of 50 km/h, non-signalised crossing and no traffic calming features (islands separating traffic directions). The observations were carried out for vehicles moving in the lane adjacent to the sidewalk where the pedestrian was situated. The observations were assigned to four situations:

- A—the pedestrian is approaching the crossing (at a distance of ca. 34 m).
- B—the pedestrian is waiting to cross at the roadway edge.
- C—the pedestrian is entering the roadway (taking the first step onto the roadway).
- D—the pedestrian is in the lane opposite to that in which the vehicles being observed are moving.

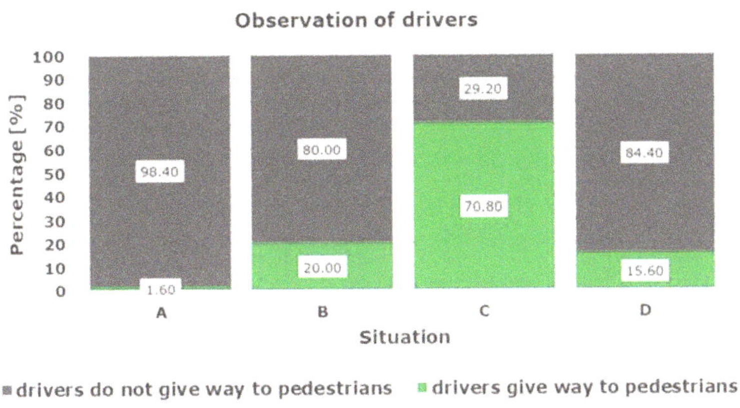

Figure 8. Results of observations of driver behaviour with regard to pedestrians in the crossing area.

The results indicate that only 1.6% of drivers stopped in order to give way to pedestrians approaching a crossing. The value of 20% of drivers giving way to pedestrians who were waiting to cross was not very large either, but it must be considered that, according to the regulations in place in Poland, the driver is not under an obligation to give way to pedestrians who are waiting to cross the roadway. Giving way to pedestrians in such situations is promoted as a positive way to behave in traffic. The result recorded in situation C is reason for serious concern: as many as 30% of drivers drive through a crossing, despite the fact that a pedestrian is already on the roadway. Such a behaviour of drivers leads to the negative statistics indicated in the first part of the article.

4. Conclusions

It follows from the review of Polish and international research performed to date that it is necessary to carry out studies on the behaviour of pedestrians and drivers in pedestrian crossing areas in order to plan, design and implement solutions to improve pedestrian safety.

In Poland, about 30% of all road fatalities are pedestrians. Records show that about 35% of all pedestrian fatalities occurred at a pedestrian crossing.

This study of driver and pedestrian behaviour in pedestrian crossing areas showed that vehicle speeds vary depending on the location—type of area, type of cross-section, section characteristics and pedestrian crossing control—and on the presence or absence of traffic lights. The study showed that speed limits are frequently exceeded near pedestrian crossings. This is why measures are necessary to improve pedestrian safety. They include pedestrian refuge islands and signalised crossings, especially for cross-sections such as 1×4, 2×2 or 2×3. Visibility at pedestrian crossings must be improved (vehicles parked, bus stops and road section geometry). Additional analyses of driver behaviour towards pedestrians waiting to cross, the level of conflicts and number of victims showed that the regulations must be revised. It is important to give priority to pedestrians waiting to cross the road, which has been implemented in many countries. This must be accompanied by educational campaigns, increased enforcement as regards driver behaviour in the initial period and ensuring that pedestrian crossing areas meet safety standards. The standards would include no designated pedestrian crossings if there are no traffic lights, a speed limit of more than 50 km/h and pedestrians having to cross the road at a second level (tunnel or footbridge), or with traffic lights, if there is more than one lane of traffic in the same direction.

More research on pedestrian safety is required and should be specifically aimed at identifying pedestrian crossing hazards, assessing the effectiveness of treatments and assessing pedestrian and driver behaviour on a regular basis by covering more parameters (e.g., different weather conditions or time of day).

Author Contributions: Conceptualization, M.B.; methodology, M.B., M.K. and P.T.; formal analysis, A.G., L.G. and L.J.; resources, A.G. and L.J.; data curation, A.G., L.J. and L.G.; writing—original draft preparation, M.B., L.J., A.G. and L.G.; writing—review and editing, M.B., M.K. and P.T.; project administration, M.B. All authors have read and agreed to the published version of the manuscript.

Funding: This research received no external funding.

Institutional Review Board Statement: Not applicable.

Informed Consent Statement: Not applicable.

Data Availability Statement: Not applicable.

Conflicts of Interest: The authors declare no conflict of interest.

References

1. Okraszewska, R.; Jamroz, K.; Michalski, L.; Zukowska, J.; Grzelec, K.; Birr, K. Analysing Ways to Achieve a New Urban Agenda-Based Sustainable Metropolitan Transport. *Sustainability* **2019**, *11*, 813. [CrossRef]
2. VIA VISTULA. *Gdańsk Traffic Research*; VIA VISTULA: Kraków, Poland, 2016.
3. Transport for London. Making London a walkable city. In *Mayor of London*; Transport for London: London, UK, 2004.
4. Martin, A. *Factors Influencing Pedestrian Safety: A Literature Review*; TRL Limited: Berk, UK, 2006.
5. Zhao, J.; Malenje, J.O.; Wu, J.; Ma, R. Modeling the interaction between vehicle yielding and pedestrian crossing behavior at unsignalized midblock crosswalks. *Transp. Res. Part F Traffic Psychol. Behav.* **2020**, *73*, 222–235. [CrossRef]
6. Fricker, J.D.; Zhang, Y. Modeling Pedestrian and Motorist Interaction at Semi-Controlled Crosswalks: The Effects of a Change from One-Way to Two-Way Street Operation. *Transp. Res. Rec. J. Transp. Res. Board* **2019**, *2673*, 433–446. [CrossRef]
7. Sheykhfard, A.; Haghighi, F. Assessment pedestrian crossing safety using vehicle-pedestrian interaction data through two different approaches: Fixed videography (FV) vs. In-Motion Videography (IMV). *Accid. Anal. Prev.* **2020**, *144*, 105661. [CrossRef]
8. Kadali, B.R.; Vedagiri, P. Evaluation of pedestrian crossing speed change patterns at unprotected mid-block crosswalks in India. *J. Traffic Transp. Eng. (Engl. Ed.)* **2020**, *7*, 832–842. [CrossRef]

9. Malin, F.; Luoma, J. Effects of speed display signs on driving speed at pedestrian crossings on collector streets. *Transp. Res. Part F Traffic Psychol. Behav.* **2020**, *74*, 433–438. [CrossRef]
10. Koh, P.; Wong, Y. Gap acceptance of violators at signalised pedestrian crossings. *Accid. Anal. Prev.* **2014**, *62*, 178–185. [CrossRef]
11. Stapleton, S.; Kirsch, T.; Gates, T.J.; Savolainen, P.T. Factors Affecting Driver Yielding Compliance at Uncontrolled Midblock Crosswalks on Low-Speed Roadways. *Transp. Res. Rec. J. Transp. Res. Board* **2017**, *2661*, 95–102. [CrossRef]
12. Stoker, P.; Garfinkel-Castro, A.; Khayesi, M.; Odero, W.; Mwangi, M.N.; Peden, M.; Ewing, R. Pedestrian Safety and the Built Environment. *J. Plan. Lit.* **2015**, *30*, 377–392. [CrossRef]
13. Quistberg, D.A.; Howard, E.J.; Ebel, B.E.; Moudon, A.V.; Saelens, B.; Hurvitz, P.M.; Curtin, J.E.; Rivara, F.P. Multilevel models for evaluating the risk of pedestrian–motor vehicle collisions at intersections and mid-blocks. *Accid. Anal. Prev.* **2015**, *84*, 99–111. [CrossRef]
14. Wang, W.; Guo, H.; Gao, Z.; Bubb, H. Individual differences of pedestrian behaviour in midblock crosswalk and intersection. *Int. J. Crashworthiness* **2011**, *16*, 1–9. [CrossRef]
15. Budzynski, M.; Guminska, L.; Jamroz, K.; Mackun, T.; Tomczuk, P. Effects of Road Infrastructure on Pedestrian Safety. *IOP Conf. Ser. Mater. Sci. Eng.* **2019**, *603*, 042052. [CrossRef]
16. Forde, A.; Daniel, J. Pedestrian walking speed at un-signalized midblock crosswalk and its impact on urban street segment performance. *J. Traffic Transp. Eng. (Engl. Ed.)* **2021**, *8*, 57–69. [CrossRef]
17. Sisiopiku, V.; Akin, D. Pedestrian behaviors at and perceptions towards various pedestrian facilities: An examination based on observation and survey data. *Transp. Res. Part F Traffic Psychol. Behav.* **2003**, *6*, 249–274. [CrossRef]
18. Shaaban, K.; Muley, D.; Mohammed, A. Analysis of illegal pedestrian crossing behavior on a major divided arterial road. *Transp. Res. Part F Traffic Psychol. Behav.* **2018**, *54*, 124–137. [CrossRef]
19. Bendak, S.; Alnaqbi, A.M.; Alzarooni, M.Y.; Aljanaahi, S.M.; Alsuwaidi, S.J. Factors affecting pedestrian behaviors at signalized crosswalks: An empirical study. *J. Saf. Res.* **2021**, *76*, 269–275. [CrossRef] [PubMed]
20. Jamroz, K.; Michalski, L.; Gaca, S. Pedestrian Safety. In *Handbook for Organizers Pedestrian Traffic*; National Council of Road Safety: Geneva, Switzerland, 2014.
21. Balasubramanian, V.; Bhardwaj, R. Pedestrians' perception and response towards vehicles during road-crossing at nighttime. *Accid. Anal. Prev.* **2018**, *110*, 128–135. [CrossRef]
22. Johansson, M.; Laureshyn, A.; Nilsson, M. Video Analysis of Pedestrian Movement (VAPM) under Different Lighting Conditions—Method Exploration. *Energies* **2020**, *13*, 4141. [CrossRef]
23. Jägerbrand, A.K. LED (Light-Emitting Diode) Road Lighting in Practice: An Evaluation of Compliance with Regulations and Improvements for Further Energy Savings. *Energies* **2016**, *9*, 357. [CrossRef]
24. Kruszyna, M. Evaluation of Distance Between Pedestrian Crossings by Students in One of the Polish Cities. *Arch. Civ. Eng.* **2013**, *59*, 547–559. [CrossRef]
25. Chen, C.; Zhang, G.; Yang, J.; Milton, J.C.; Alcántara, A. "Dely" An explanatory analysis of driver injury severity in rear-end crashes using a decision table/Naïve Bayes (DTNB) hybrid classifier. *Accid. Anal. Prev.* **2016**, *90*, 95–107. [CrossRef]
26. Hakkert, A.; Gitelman, V.; Ben-Shabat, E. An evaluation of crosswalk warning systems: Effects on pedestrian and vehicle behaviour. *Transp. Res. Part F Traffic Psychol. Behav.* **2002**, *5*, 275–292. [CrossRef]
27. Chaudhari, A.R.; Gore, N.; Arkatkar, S.; Joshi, G.; Pulugurtha, S.S. Deriving Pedestrian Risk Index by Vehicle Type and Road Geometry at Midblock Crosswalks under Heterogeneous Traffic Conditions. *J. Transp. Eng. Part A Syst.* **2020**, *146*, 04020123. [CrossRef]
28. Kadali, B.R.; Vedagiri, P. Proactive pedestrian safety evaluation at unprotected mid-block crosswalk locations under mixed traffic conditions. *Saf. Sci.* **2016**, *89*, 94–105. [CrossRef]
29. Campbell, B.; Zegeer, C.; Huang, H.; Cynecki, M. *A Review of Pedestrian Safety Research in the United States and Abroad*; Federal Highway Administration, U.S. Department of Transportation: Washington, DC, USA, 2004.
30. Rosén, E.; Stigson, H.; Sander, U. Literature review of pedestrian fatality risk as a function of car impact speed. *Accid. Anal. Prev.* **2011**, *43*, 25–33. [CrossRef] [PubMed]
31. WHO. *Pedestrian Safety*; World Health Organization: Geneva, Switzerland, 2013.
32. WHO. *Global Status Report on Road Safety*; World Health Organization: Geneva, Switzerland, 2018; Volume 151.
33. Jamroz, K.; Budzyński, M.; Romanowska, A.; Zukowska, J.; Oskarbski, J.; Kustra, W. Experiences and Challenges in Fatality Reduction on Polish Roads. *Sustainability* **2019**, *11*, 959. [CrossRef]
34. European Parliament and Council of the European Union. *Directive 2008/96/EC of the European Parliament and of the Council of 19 November 2008 on Road Infrastructure Safety Managemen 2008*; Publications Office of the European Union: Luxembourg, 2008.
35. Sitran, A.; Delhaye, E.; Uccelli, I. Directive 2008/96/EC On Road Infrastructure Safety Management: An Ex-post Assessment 5 years After its Adoption. *Transp. Res. Procedia* **2016**, *14*, 3312–3321. [CrossRef]
36. European Parliament and Council of the European Union. *Directive 2019/1936 of the European Parliament and of the Council of 23 October 2019 Amending Directive 2008/96/EC on Road Infrastructure Safety Management. 2019*; Publications Office of the European Union: Luxembourg, 2008.
37. Cafiso, S.; La Cava, G.; Montella, A. Safety Inspections as Supporting Tool for Safety Management of Low-Volume Roads. *Transp. Res. Rec. J. Transp. Res. Board* **2011**, *2203*, 116–125. [CrossRef]

38. Vaiana, R.; Perri, G.; Iuele, T.; Gallelli, V. A Comprehensive Approach Combining Regulatory Procedures and Accident Data Analysis for Road Safety Management Based on the European Directive 2019/1936/EC. *Safety* **2021**, *7*, 6. [CrossRef]
39. European Commission. *Traffic Safety Basic Facts on Pedestrians*; European Commission: Luxembourg, 2018.
40. Olszewski, P.; Szagała, P.; Wolański, M.; Zielińska, A. Pedestrian fatality risk in accidents at unsignalized zebra crosswalks in Poland. *Accid. Anal. Prev.* **2015**, *84*, 83–91. [CrossRef] [PubMed]
41. Olszewski, P.; Osińska, B.; Szagała, P.; Skoczyński, P.; Zielińska, A. Problems with Assessing Safety of Vulnerable Road Users Based on Traffic Accident Data. *Arch. Civ. Eng.* **2016**, *62*, 149–168. [CrossRef]
42. Jamroz, K.; Kustra, W.; Budzynski, M.; Zukowska, J. Pedestrian Protection, Speed Enforcement and Road Network Structure the key Action for Implementing Poland's Vision Zero. *Transp. Res. Procedia* **2016**, *14*, 3905–3914. [CrossRef]
43. Olszewski, P.; Czajewski, W.; Dąbkowski, P.; Kraśkiewicz, C.; Szagała, P. Assessment Of The Effectiveness Of Active Signage At Pedestrian Crossings. *Arch. Civ. Eng.* **2015**, *61*, 125–140. [CrossRef]
44. Tomczuk, P.; Chrzanowicz, M.; Mackun, T. Methodology for assessing the lighting of pedestrian crossings based on light intensity parameters. *MATEC Web Conf.* **2017**, *122*, 01008. [CrossRef]
45. Tomczuk, P.; Jamroz, K.; Mackun, T.; Chrzanowicz, M. Lighting requirements for pedestrian crossings–positive contrast. *MATEC Web Conf.* **2019**, *262*, 05015. [CrossRef]
46. Budzyński, M.; Jamroz, K.; Mackun, T. Pedestrian Safety in Road Traffic in Poland. *IOP Conf. Ser. Mater. Sci. Eng.* **2017**, *245*, 42064. [CrossRef]
47. Gaca, S.; Pogodzińska, S. Speed management as a measure to improve road safety on Polish regional roads. *Arch. Transp.* **2017**, *43*, 29–42. [CrossRef]
48. Bak, R.; Kiec, M. Influence of Midblock Pedestrian Crossings on Urban Street Capacity. *Transp. Res. Rec. J. Transp. Res. Board* **2012**, *2316*, 76–83. [CrossRef]
49. Hatfield, J.; Fernandes, R.; Job, R.S.; Smith, K. Misunderstanding of right-of-way rules at various pedestrian crossing types: Observational study and survey. *Accid. Anal. Prev.* **2007**, *39*, 833–842. [CrossRef] [PubMed]
50. Fildes, N.B.; Lee, J.S.; Kenny, D.; Foddy, W. *Survey of Older Road Users: Behavioural and Travel*; Monash University Accident Research Centre: Clayton, Australia, 1994.
51. Guéguen, N.; Meineri, S.; Eyssartier, C. A pedestrian's stare and drivers' stopping behavior: A field experiment at the pedestrian crossing. *Saf. Sci.* **2015**, *75*, 87–89. [CrossRef]
52. Evans, D.; Norman, P. Understanding pedestrians' road crossing decisions: An application of the theory of planned behaviour. *Health Educ. Res.* **1998**, *13*, 481–489. [CrossRef]
53. Alonso, I.P.; Fernández-Llorca, D.; Sotelo, M.A.; Bergasa, L.M.; De Toro, P.A.; Nuevo, J.; Ocana, M.; Garrido, M.A.G. Combination of Feature Extraction Methods for SVM Pedestrian Detection. *IEEE Trans. Intell. Transp. Syst.* **2007**, *8*, 292–307. [CrossRef]
54. Ni, Y. Pedestrian Safety at Urban Signalised Intersections. Ph.D. Thesis, Technische Universität Darmstadt, Darmstadt, Germany, 2010.
55. Hummel, T. Dutch Pedestrian Safety Research Review. *SWOV Inst. Road Saf. Res.* **1999**, 38.
56. Várhelyi, A. *Dynamic Speed Adaptation Based on Information Technology: A Theoretical Background*; Lund University: Lund, Sweden, 1996; Volume 166.
57. Johansson, C. Towards a Method to Improve Road Safety for Pedestrians and Cyclists, Especially in Child Pedestrian Environments. a Case Study in Boras, Sweden. Licentiate Thesis, University of Technology, Luleå, Sweden, 2001.
58. Diependaele, K. Non-compliance with pedestrian traffic lights in Belgian cities. *Transp. Res. Part F Traffic Psychol. Behav.* **2019**, *67*, 230–241. [CrossRef]
59. Pasanen, E. *Traffic Safety at Pedestrian Zebra Crossings Reports 7B/2007*; LINTU research programme: Helsinki, Finland, 2007.
60. Gitelman, V. National observational survey of pedestrian behaviour at crosswalks. In Proceedings of the International Conference on Safety and Mobility of Vulnerable Road Users: Pedestrians, Motorcyclists, and Bicyclists, Jerusalem, Israel, 30 May–2 June 2010.
61. Ibrahim, N.I.; Karim, M.R.; Kidwai, F.A. Motorists and pedestrian interaction at unsignalised pedestrian crossing. *East. Asia Soc. Transp. Stud.* **2005**, *5*, 120–125.
62. Nordh, U. Pedestrians ' safety and accessibility—A study of the situation at Pärnu mnt in central Tallinn. Master's Thesis, Lund University, Lund, Sweden, 2007.
63. Park, H.J.; Li, S.; Yu, W.; Yang, W.; Alhajyaseen, W.; Iryo-Asano, M. Pedestrian crossing behavior and compliance at signalized intersections. In Proceedings of the Road Safety on Five Continents (RS5C) 17th International Conference, Rio de Janeiro, Brazil, 17–18 May 2016.
64. Guo, H.; Gao, Z.; Yang, X.; Jiang, X. Modeling Pedestrian Violation Behavior at Signalized Crosswalks in China: A Hazards-Based Duration Approach. *Traffic Inj. Prev.* **2011**, *12*, 96–103. [CrossRef] [PubMed]
65. Ren, G.; Zhou, Z.; Wang, W.; Zhang, Y.; Wang, W. Crossing Behaviors of Pedestrians at Signalized Intersections: Observational Study and Survey in China. *Transp. Res. Rec. J. Transp. Res. Board* **2011**, *2264*, 65–73. [CrossRef]
66. Sun, R.; Zhuang, X.; Wu, C.; Zhao, G.; Zhang, K. The estimation of vehicle speed and stopping distance by pedestrians crossing streets in a naturalistic traffic environment. *Transp. Res. Part F Traffic Psychol. Behav.* **2015**, *30*, 97–106. [CrossRef]
67. Zhang, C.; Chen, F.; Wei, Y. Evaluation of pedestrian crossing behavior and safety at uncontrolled mid-block crosswalks with different numbers of lanes in China. *Accid. Anal. Prev.* **2019**, *123*, 263–273. [CrossRef] [PubMed]

68. Zhang, C.; Zhou, B.; Chen, G.; Chen, F. Quantitative analysis of pedestrian safety at uncontrolled multi-lane mid-block crosswalks in China. *Accid. Anal. Prev.* **2017**, *108*, 19–26. [CrossRef]
69. Pawar, D.S.; Patil, G.R. Critical gap estimation for pedestrians at uncontrolled mid-block crossings on high-speed arterials. *Saf. Sci.* **2016**, *86*, 295–303. [CrossRef]
70. Pawar, D.S.; Kumar, V.; Singh, N.; Patil, G.R. Analysis of dilemma zone for pedestrians at high-speed uncontrolled midblock crossing. *Transp. Res. Part C Emerg. Technol.* **2016**, *70*, 42–52. [CrossRef]
71. Mikusova, M.; Wachnicka, J.; Zukowska, J. Research on the Use of Mobile Devices and Headphones on Pedestrian Crossings—Pilot Case Study from Slovakia. *Safety* **2021**, *7*, 17. [CrossRef]
72. Wu, J.; Radwan, E.; Abou-Senna, H. Pedestrian-vehicle conflict analysis at signalized intersections using micro-simulation. In Proceedings of the Road Safety on Five Continents (RS5C) 17th International Conference, Rio de Janeiro, Brazil, 17–18 May 2016.
73. Lu, L.; Ren, G.; Wang, W.; Chan, C.-Y.; Wang, J. A cellular automaton simulation model for pedestrian and vehicle interaction behaviors at unsignalized mid-block crosswalks. *Accid. Anal. Prev.* **2016**, *95*, 425–437. [CrossRef] [PubMed]
74. Calvi, A.; D'Amico, F.; Ferrante, C.; Ciampoli, L.B. Effectiveness of augmented reality warnings on driving behaviour whilst approaching pedestrian crossings: A driving simulator study. *Accid. Anal. Prev.* **2020**, *147*, 105760. [CrossRef]
75. Soathong, A.; Wilson, D.; Ranjitkar, P.; Chowdhury, S. A Critical Review of Policies on Pedestrian Safety and a Case Study of New Zealand. *Sustainability* **2019**, *11*, 5274. [CrossRef]

energies

MDPI

Article

A Methodology for Evaluating Driving Styles in Various Road Conditions

Rafał S. Jurecki * and Tomasz L. Stańczyk

Faculty of Mechatronics and Mechanical Engineering, Kielce University of Technology, Avenue Tysiąclecia Państwa Polskiego 7, 25-314 Kielce, Poland; stanczyk@tu.kielce.pl
* Correspondence: rjurecki@tu.kielce.pl; Tel.: +48-41-342-4285

Abstract: For many institutions, it is important to evaluate a given driving technique as safe or unsafe based on measurable vehicle movement parameters. The paper constitutes a part of studies aimed at establishing a method of parameter-based evaluation of drivers in various road conditions, in other words, to create a so-called 'driver profile'. The tests were carried out on a 650 km route, on four varying road types. Longitudinal and lateral acceleration values are used to evaluate the driving style. An analysis is presented of the impact of the type and shape of road on acceleration values. The results demonstrate that the same driver, when driving the same vehicle on an expressway, an inter-urban road or in urban traffic, will move with various acceleration values. A detailed analysis of acceleration values and distributions was conducted. Interesting conclusions were drawn after excluding the so-called 'smooth driving' sections, by acceleration ranges of -0.5 to $0.5\ m/s^2$ from the analysis. This allowed for the evaluation of the structure of other longitudinal and lateral acceleration values. After this modification, the distributions showed specificity for the given road type, thereby allowing the road type used by the vehicle's driver to be recognized based solely on the distribution.

Keywords: road safety; driver behavior; driving profile; aggressive driving

Citation: Jurecki, R.S.; Stańczyk, T.L. A Methodology for Evaluating Driving Styles in Various Road Conditions. *Energies* **2021**, *14*, 3570. https://doi.org/10.3390/en14123570

Academic Editors: Stefano Bracco and Islam Safak Bayram

Received: 18 March 2021
Accepted: 11 June 2021
Published: 16 June 2021

1. Introduction

Road traffic safety (RTS) studies conducted in many countries have shown for many years that among the three main elements of the RTS system, i.e., human, vehicle, and environment, it is the human that contributes to the largest number of accidents. Therefore, the studies aimed at improving road safety concern mostly the human driver: his or her personality, behavior on the road, attitudes, emotional tension, knowledge, experience, and many other factors. The studies have a cognitive nature, and simply stated, it can be said in straightforward terms that they are devoted to identifying the reasons and circumstances of particular behaviors on the road.

From a practical viewpoint of, for example, people dealing in driver training, the police dealing in traffic congestion prevention, or insurance companies that determine vehicle insurance premium amounts, it is important to be able to evaluate driver skills in terms of safety. Motor vehicle operation has a widespread activity, which is why road traffic includes people who are driving very well and comply strictly with the road traffic regulations, but also people who devalue and ignore them. For the aforementioned persons and institutions, it is important to be able to evaluate driving techniques as safe or unsafe to various degrees, based on measurable vehicle movement parameters during vehicular operation rather than after a collision or accident. Therefore, due to the very high number of factors that may determine driver behavior on the road, the undertaking of such studies is both timely and of critical importance.

There is a very high number of papers published in the specialist literature in which the authors attempt to determine the driver's profile, his or her personality traits and predisposition (intellectual and psychomotor) with respect to vehicle operation. Many publications refer to specific traits, such as the drivers' physical or mental fitness as well as knowledge and skills, including attitudes and motivations. For cognitive purposes, but also due to the very high practical relevance, frequently discussed topics include the driver reaction time [1–5], studying the selection and execution of their preferred defensive maneuvers [6,7] as well as the driver's perception of the environment. The analysis covers vision after dark and sensitivity to glare. An important feature is the driver's hand-eye coordination, depending on various factors such as time of day, cognitive efficiency, and fatigue from a long day at work [8,9].

The aforementioned diversity of the papers results not only from the phenomenon's complexity and the large number of factors contributing to it, but also from the use of various research methods to study and analyze the vehicle driving process. Studies of the aforementioned driver features and behaviors can be conducted in various research environments: actual road or test sections and virtual-simulators and testing stations [10]. An increase in the elderly population, especially in well-developed countries, contributes to an increasing number of tests on drivers with various dysfunctions caused by mental diseases presenting at this age [11], or other diseases typical for this age, such as Alzheimer's or Parkinson's disease [12]. This is highly relevant during the development of new assistance systems dedicated to this specific group of drivers to limit their social exclusion constituting the effect of various dysfunctions related to such diseases.

A large problem in many countries is driving under the influence of alcohol or psychoactive substances (mainly drugs), which is why studies regarding their impact on driver behavior are conducted continuously in many centers [13,14]. In paper [15], the authors proposed a system aimed at analyzing driver behavior to be able to identify drivers who have consumed alcohol. The system, based on the use of a mobile phone and an acceleration sensor, enables evaluation of driver action by comparing the manner of executing typical maneuvers with the driver's activity patterns.

Many papers present studies aimed at identifying and sometimes evaluating the driving technique (style). Simple or more advanced tools to determine the driver's type based on certain indicators and evaluation of their mental inclination towards specific, negative behaviors are being developed. A very interesting and prospective trend is research aimed at recognizing the driver's intentions [16–18].

Simple measurable indicators are quite often used to evaluate the driver's actions [19,20]. These may include, e.g., vehicle position, vehicle speed [21], or longitudinal and lateral acceleration [22–24]. A radically different approach was provided in the paper [25], which presents a concept of continuous driver profiling by signals from the vehicle's built-in sensors via use of the machine learning method [26]. A set of 51 various data sets downloaded from the vehicle's CAN bus (acquired from 10 different controllers) was considered. The evaluation of the proposed method was conducted based on tests implemented by 10 drivers driving an identical vehicle on four different routes with a total length of 46 km. The drivers were driving the vehicle on, among others, an urban road, a motorway, and in a parking lot during working days at similar times.

The complexity of the developed driver driving method (driving style, technique, driver profile specification) evaluation algorithms depends largely on the main objective of the given studies. A simple method of evaluating the driver's driving method is presented in the paper [27]. The applied method, based on recording selected driving parameters on a smartphone with appropriate software installed, determines the driver's driving method based on indications of, among others, the acceleration sensor. According to the applied algorithm, the driver receives points when driving, which translates into the final rating: very bad (<750 points), bad, good, and very good (>990 points). The concept of a different, inexpensive and smart driver evaluation system aimed at identifying aggressive drivers is provided in the paper [28]. The system is based on data obtained from a three-axis accelerometer. Similar measurements are also discussed in other publications [29,30]. The paper [31] describes a system that allows for the identification of a driver's emotional excitation by measuring the driver's selected physiological parameters and the data on the vehicle's movement. The sensors establish wireless communication with the smartphone.

All data on the driver's behavior and driving method are displayed in real-time in the dedicated application (for the Android system).

The vehicle's speed and acceleration distributions measured at specific time intervals are the basis for evaluating the driving method of the driver described in the paper [16]. Such driving style analysis can be used to recognize the driver's mental and physiological condition and in particular to detect his or her drowsiness and fatigue.

Three-axis acceleration sensors constitute the foundation for a system aimed at determining the profile of the driver described in the paper [32]. The obtained acceleration values undergo a spectral analysis and are then used in conjunction with a suitable algorithm to enable recognition of an 'aggressive' driver and a 'safe' driver. The available literature also includes publications that include measurements on the positions of the controls, e.g., throttle or service brake [33,34]. In paper [35], Wakita et al. have proposed a method for identifying the driver's features based on signals recorded when driving behind a different vehicle. These signals included: the motion of the throttle; service brake; vehicle speed; and distance from the preceding vehicle. The tests were conducted in a driving simulator.

In paper [36], an instrumented vehicle-based experiment was designed to observe the drivers' action under various urban lane-changing scenarios. The 'in-vehicle' driver behavior data were obtained from the experiment and used to classify 40 drivers into four general groups according to the lane-changing maneuvers performed in an urban street environment.

The paper of Van Ly et al. [37], which profiled a driver enabling the identification of many dangerous maneuvers performed when driving, utilized the concept of using acceleration sensors that are already present in the vehicle (bus). The data was obtained from the CAN bus. The researchers focused on the 'observation' of three maneuvers: braking, turning, and acceleration. As result of the conducted analysis, it was pointed out that the evaluation of braking and acceleration gives better effects for drive identification than does an analysis of the acceleration process.

The method of comparing passenger surveys with acceleration measurements' results was used in the study presented in the publication [38]. The surveyed city bus passengers were tasked with evaluating, in a subjective manner, the vehicle's acceleration on a three-level scale, i.e., comfortable, normal and uncomfortable. For example, longitudinal accelerations deemed by the surveyed passengers as uncomfortable were in the range of 1.5–2.75 m/s^2. Sensors built into high-grade smartphones, such as accelerometer, gyroscope, magnetometer, GPS, and video cameras, were used in the study described in the paper [39]. It presents the concept of a system aimed at identifying typical and aggressive drivers based on data obtained by using the aforementioned sensors. Paper [40] features an analysis of driver behavior in urban conditions. It was demonstrated that the average speed and delay range were lower during peak traffic hours than at other times of bus operation. Interesting analyses concern the dependency between driving technique and driver gender. Men drove more aggressively and moved with higher acceleration values than did women. The difference was especially clear in the city. On most routes, drivers of both genders were driving at similar speeds. Driver behavior studies are often focused on attempts to determining specific parameters that would satisfactorily describe driver behaviors in selected road situations. Certain conventional indicators facilitating a broader driver behavior analysis are often designated. Examples of such indicators include, e.g., the headway or time to collision. Due to the universality of various devices that may directly or indirectly distract the drivers and simultaneously affect their actions, researchers are conducting tests aimed at determining the impact of the operation of various devices, e.g., radio, navigation system, or mobile phone, in various road situations [41,42].

In [43], large taxi floating car data (FCD) was used to empirically evaluate how traffic congestion-related negative moods, defined as state aggressiveness, affected drivers' speed choice. The results indicated that the speed model incorporating state aggressiveness could better predict the travel time than the traditional speed model that considers only the specific expected speed distribution. Publication [18] determines the criteria for evaluating driver behavior in three situations: when stationary; when driving, and when braking; whereas paper [44], to evaluate the drivers' driving styles, classifies vehicle exploitation in conditions during urban and extra-urban driving with simultaneous specification of driving in traffic and dynamic driving.

The presented paper constitutes a part of the studies aimed at establishing a method of parameter-based evaluation of driver behavior. This method assumes the continuous measurement of many parameters, some of which will be obtained from the vehicle CAN bus and includes utilization of various sensors in the vehicle's electronic control systems for specific assemblies (e.g., engine, suspension, brake system, etc.) as well as the active and passive safety systems. The method also projects the installation of additional sensors constituting special elements of measurement and recording modules-the so-called trackers. It is assumed that several variants of algorithms determining the so-called 'driver profile' will be developed. These profiles (algorithms) will vary in complexity based mainly on the number of parameters used. They will be developed hierarchically, starting with the simplest and proceeding to those of increasing complexity.

The short literature study presented above demonstrates that longitudinal and lateral acceleration values (as well as driving speed) are the parameters most often used to evaluate the driver's profile (driving style, technique, manner of driving). The authors also want to start their study by analyzing these acceleration values and verifying the factors (except habits and behaviors) that affect them the most. One of the most important factors contributing to the variety in the obtained parameters is, e.g., the type and performance of the driver's vehicle. High-capacity and high-torque engines installed in vehicles favor achieving high acceleration values. It can be expected that the values of acceleration, braking delays, and lateral acceleration in curvilinear motion in a high-class passenger vehicle with a high-capacity engine will be substantially higher (while ensuring the passengers' subjective feeling of comfort) than in the case of an urban transport bus. With respect to persons standing in such transport, even low longitudinal and lateral acceleration values can be experienced as excessive.

Another important factor that affects the acceleration values and variation is the road type and shape. When the same driver uses the same vehicle on different roads, he or she will achieve various acceleration values. Impactful will be such factors as, e.g., road grade, including the number of available traffic lanes; its profiling; straight section lengths; number and shape of corners; traffic organization, and traffic intensity. The topic of this paper is to study this specific factor.

2. Place of Conducted Measurements

The paper's development featured driving on test sections with a Ford Transit (Figure 1), generation VI station wagon (for 9 passengers), a curb weight of 2070 kg, an engine with a capacity of 2198 cm^3 and 92 kW, loaded with 320 kg.

Figure 1. Ford Transit test vehicle.

The test route was approximately 650 km in length, and the main movement parameters were measured when driving on it. Its selection was dictated by the fact that it could be divided into four sections varying in terms of the roadway number; traffic lanes number; traffic separation; profiling and finishing standard (e.g., motorway, expressway); straight sections length; intersection type (traditional and collision-free) and frequency; traffic lighting; roundabouts; and other elements used in road traffic engineering.

The vehicle's path was recorded in detail by the Globtrak™ system (Globtrak Polska sp. z o. o., Kielce, Poland), which allows, among others, for recording the GPS trail (Figure 2).

Figure 2. Course of the driven test route according to the Globtrak™ Systems.

3. Specification of Measurement Instrumentation

The testing featured utilization of specialist instrumentation including, among others, the following:

- The S-350 optoelectronic sensor from Corrsys Datron®, intended for measuring vehicle movement parameters: longitudinal and lateral speed as well as drift angle;
- TAA 3-axis linear acceleration sensor from Kistler®;
- TANS 3-axis linear and angular acceleration sensor from Kistler®;
- uEEP 12 Datron® data acquisition station along with a control tablet and the ARMS® software [45–47].

Due to the need to define the vehicle's movement parameters which were to be used in the specification of the driver's behavior, the data were recorded simultaneously with the frequency of 10 Hz.

4. Results

The testing featured a recording of the course of changes in the vehicle's driving parameters. An analysis of the driven route featured a distinction of four sections with a similar drive time and characterized by different driving parameter values. The distinguished sections correspond to the test route's road types.

The route's distinguished sections:

(a) Urban area (city);
(b) Non-urban area (single roadway with two traffic lanes);
(c) Two-roadway expressway;
(d) Motorway.

The breakdown of the obtained longitudinal acceleration values is presented in Figure 3 and Table 1 features their maximum and minimum values on particular road types. The positive values concern acceleration, while the negative values concern braking (deceleration).

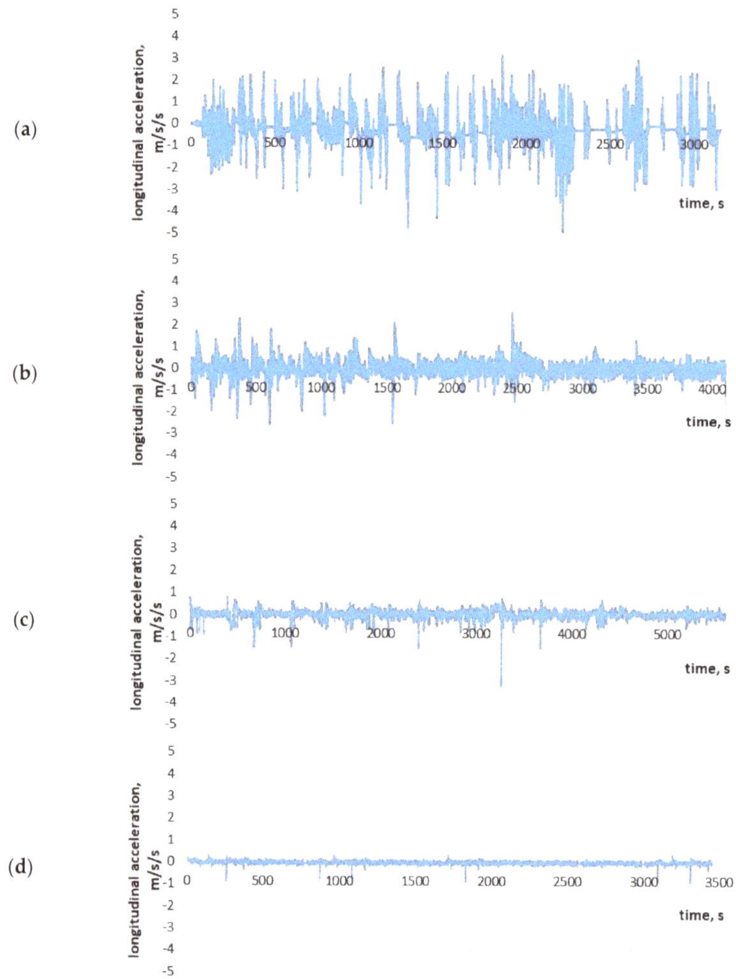

Figure 3. Values of longitudinal acceleration at specific sections on various road types: (**a**) urban area (city); (**b**) non-urban area (single roadway with two traffic lanes); (**c**) two-roadway expressway; (**d**) motorway.

Table 1. Longitudinal acceleration at specific road types.

Route	Maximum, m/s^2	Minimum, m/s^2
a	3.17	−4.88
b	2.61	−2.56
c	0.80	−1.59 (−3.20)
d	0.36	−0.89

The breakdown of the obtained lateral acceleration values is presented in Figure 4. Table 2 features their maximum and minimum values on particular road types.

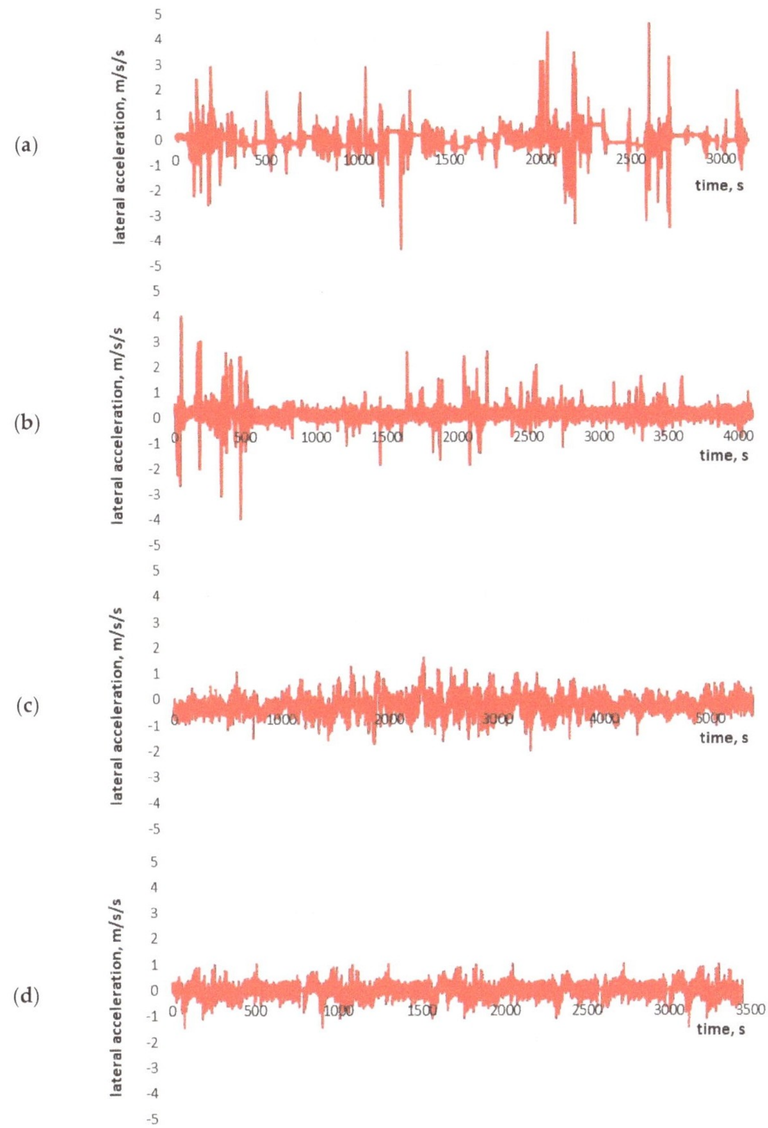

Figure 4. Values of lateral acceleration at specific sections on various road types: (**a**) urban area (city); (**b**) non-urban area (single roadway with two traffic lanes); (**c**) two-roadway expressway; (**d**) motorway.

Table 2. Lateral acceleration at specific road types.

Route	Maximum, m/s^2	Minimum, m/s^2
a	4.58	−4.33
b	3.97	−3.98
c	1.59	−1.98
d	1.01	−1.44

Figure 5 presents the ranges of longitudinal and lateral acceleration values in a different form that allows for the overall evaluation of the driving dynamics on particular road types.

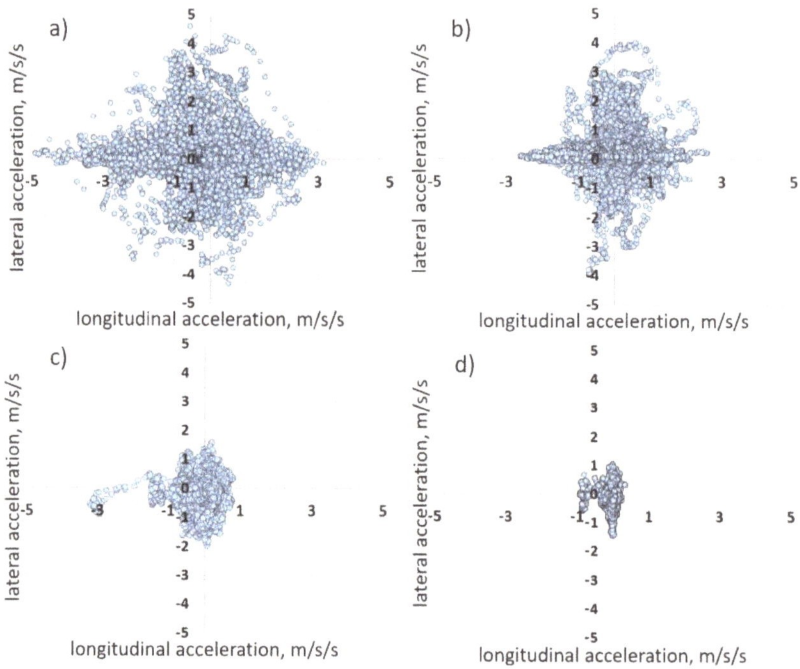

Figure 5. Comparison of the ranges of longitudinal and lateral acceleration values on various road types: (**a**) urban area (city); (**b**) non-urban area (single roadway with two traffic lanes); (**c**) two-roadway expressway; (**d**) motorway.

With this presentation, on the one hand it is possible to see the maximum longitudinal and lateral acceleration values, and on the other hand to see their diversity-filled blue fields in particular charts.

5. Analysis and Discussion

Figure 6 presents the frequency of occurrence of particular longitudinal acceleration values on the analyzed sections. Most of the driving time is dominated by acceleration values of 0.5 to -0.5 m/s^2. This interval was distinguished arbitrarily because it can be conventionally named as driving approximately at a constant speed or 'smooth driving'. These slight acceleration or delay values occur across the entire route. These are acceleration changes caused, e.g., by the topography (ascending or descending on a hill), or by the changing headwind or tailwind, and similar situations causing changes in driving speed. These can also include intentional driving speed corrections made by the driver to maintain the intended speed. Depending on the road type, the share of such acceleration values lies within a broader range of 61.9% to 99.5%, whereas the highest share of the 'smooth driving' occurs on the motorway, then on the expressway and non-urban area; the smallest share occurs in the urban area.

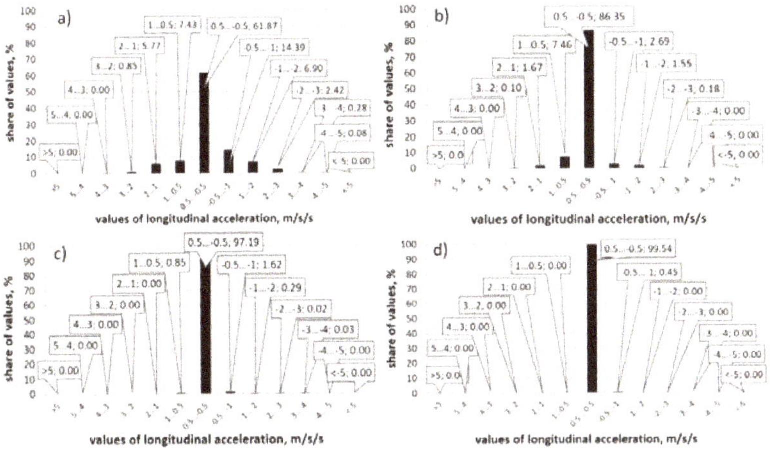

Figure 6. Share of longitudinal acceleration range occurrence on various road types: (**a**) urban area (city); (**b**) non-urban area (single roadway with two traffic lanes); (**c**) two-roadway expressway; (**d**) motorway.

When driving in the city, there is also a high variation in longitudinal acceleration values—the recorded acceleration values occur in eight separate ranges in the chart. The narrowing of these ranges occurs on the other road sections but is it most evident on the expressway and motorway.

As explained in the Introduction, the number of acceleration classes in the diagrams showing longitudinal and lateral acceleration was limited in accordance with the aim of the study (see the Acknowledgements). Increasing the number of classes, i.e., further dividing them, would be justified only if it helped understand the problem in more detail. Here, the purpose was driver risk assessment (driver profiling), which required a simple-to-use method of analysis to determine acceleration for different types of routes. Subdividing the classes of acceleration would make the interpretation and comparison of the results too complicated. A similar method of result presentation is used, e.g., in [36].

In the case of positive acceleration, mainly in urban traffic, it is possible to observe values of up to 3 m/s^2; however, their occurrence is sporadic. Negative values (delay during deceleration and braking) occur in the range of down to -4 m/s^2, wherein the lowest values were recorded sporadically only when driving in the city and on the expressway. What is interesting is that there were no cases when the delay values would exceed -4 m/s^2, i.e., intense braking, although the testing was conducted on a long route (650 km) on which situations requiring extensive braking can be expected. This was the case even though the driver was moving rather dynamically. This fact can be interpreted as confirmation of the driver's good skills, including anticipation of dangerous situations. A driver with lower skills would be more often 'forced' to brake extensively, most likely being surprised by the road situation. However, the revealed impact of the driver's experience and skills will be the topic of a separate study cycle. In the case of negative acceleration values, all charts are dominated by the range from -0.5 to -1 m/s^2.

Figure 7 shows the share of occurrence of various lateral acceleration value ranges when driving on different roads. Positive and negative values designate the acceleration's direction (left-right), hence the absolute values included. The figure also includes the distinguished range from -0.5 to 0.5 m/s^2. This interval can also be interpreted (in a fashion similar to the longitudinal acceleration values analysis) as the conventional 'smooth driving'. When driving on a straight road, drivers make mild corrections to the driving path to center the vehicle on the traffic lane if for any reason (operation of the radio, air conditioning, navigation system, etc.) they neared the left or right edge of the lane. This can also be caused by the varying crosswind affecting the vehicle. It is possible to specify other minor reasons inducing driving path correction, for example, the impact of pressure drop on the vehicle, caused by the presence of a large commercial vehicle on the side, etc.

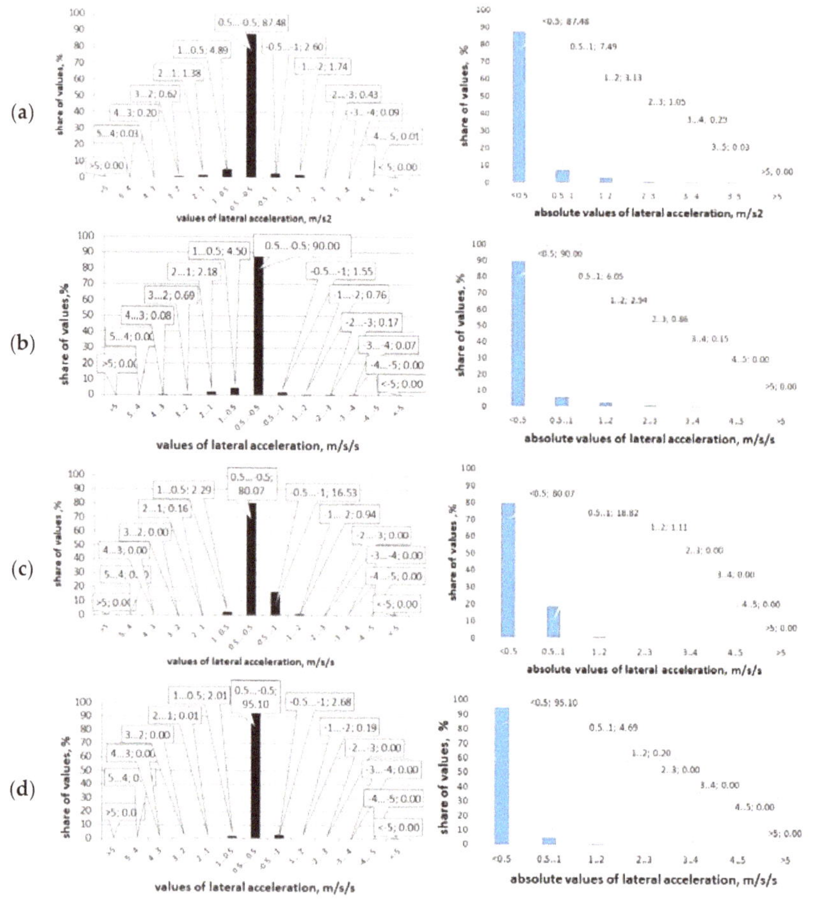

Figure 7. Share of occurrence of lateral acceleration ranges and their absolute values on various road types: (**a**) urban area (city); (**b**) non-urban area (single roadway with two traffic lanes); (**c**) two-roadway expressway; (**d**) motorway.

Values in the range of -0.5 to 0.5 m/s^2 were recorded for over 80% of the drive time, wherein the highest values were obtained when driving on the motorway (95%). It is worth noting that the shares of the lowest range for lateral acceleration values are within the substantially lowest interval of 80.4% to 95.1%. The interval length therefore amounts to 15.0%, whereas for longitudinal acceleration values (see Figure 6), the interval is twice as long and amounts to 99.5% − 61.9% = 37.6%.

The overwhelming prevalence of driving conventionally, referred to as 'smooth driving' or driving at approximately constant speed with its specific acceleration range of -0.5 to 0.5 m/s^2 makes the shares of driving at other acceleration values very small. However, in terms of the synthetic attempt of evaluating the driving techniques of different drivers, the cases of non-conventional or non-smooth driving are the more important. To take a look at these other cases, the sections of driving at acceleration values of -0.5 to 0.5 m/s^2 were excluded from further analysis. The modified waveforms were analyzed further. This allowed for the evaluation of the structure of other longitudinal and lateral acceleration values. The longitudinal acceleration values modified in the manner specified above are presented in Figure 8, and the lateral acceleration values are presented in Figure 9.

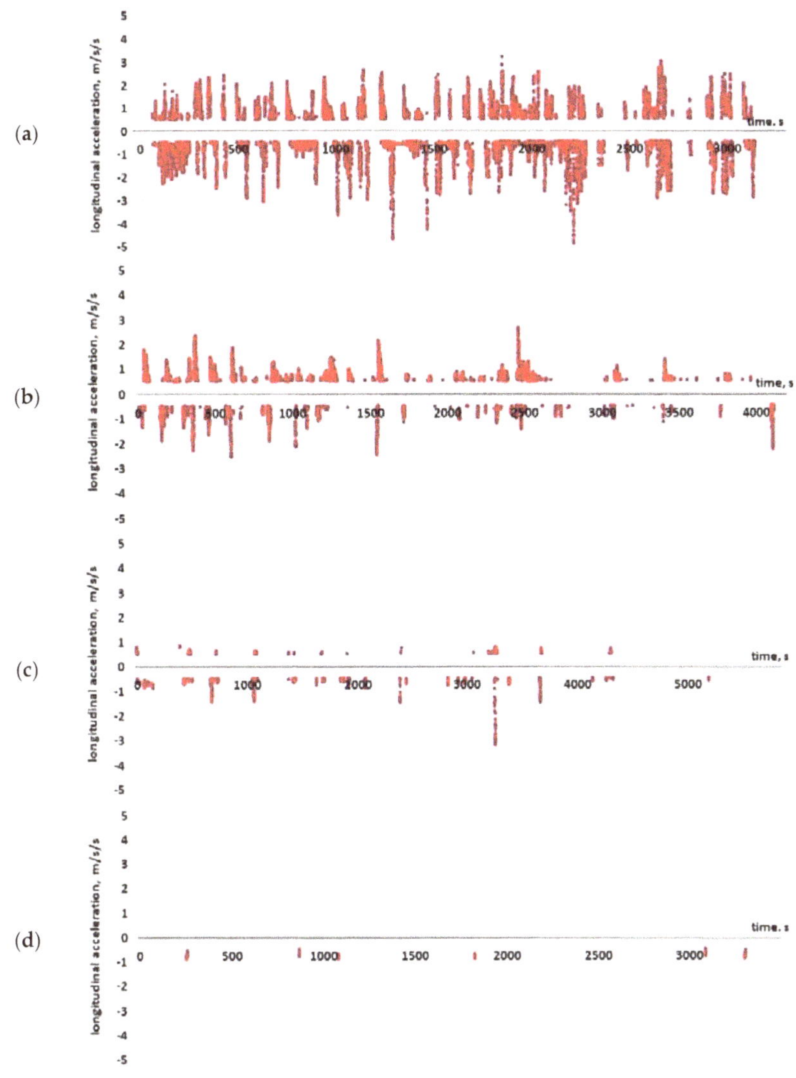

Figure 8. Longitudinal acceleration values excluding the 'constant speed' driving range on various road types: (**a**) urban area (city); (**b**) non-urban area (single roadway with two traffic lanes); (**c**) two-roadway expressway; (**d**) motorway.

(a) (b) (c) (d)

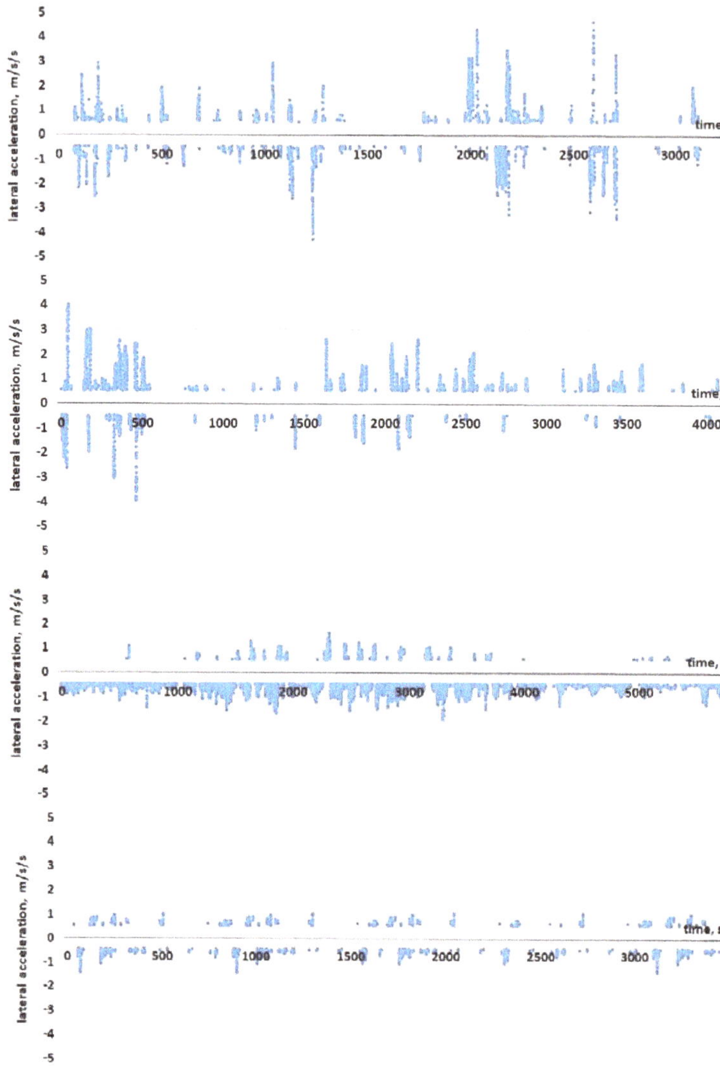

Figure 9. Lateral acceleration values excluding the 'constant speed' driving range on various road types: (**a**) urban area (city); (**b**) non-urban area (single roadway with two traffic lanes); (**c**) two-roadway expressway; (**d**) motorway.

Based on the charts presented in Figures 10 and 11, it is possible to evaluate the structure of the longitudinal and lateral acceleration values. In the case of the urban section driving, longitudinal acceleration in the range of -0.5 to -1 m/s^2, corresponding to mild braking is dominant, with a share of nearly 38%. The case is similar for expressway and motorway driving, wherein their shares amount to 58% and 100% accordingly. As for extra-urban route driving, slight acceleration in the range of 0.5 to 1 m/s^2, corresponding to smooth acceleration, has a share of 52.5%.

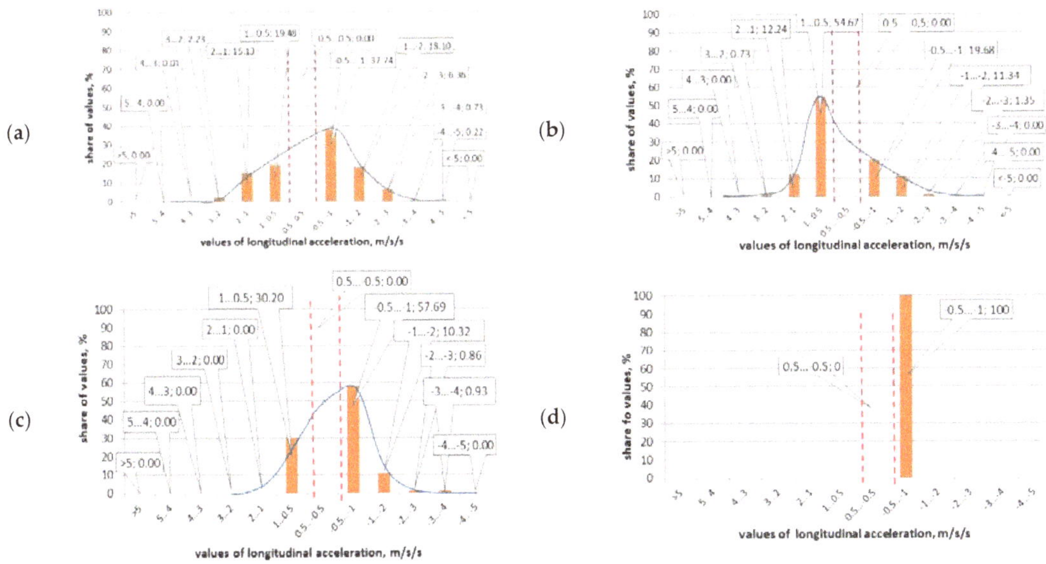

Figure 10. Longitudinal acceleration values excluding the 'constant speed' driving range on various road types: (**a**) urban area (city); (**b**) non-urban area (single roadway with two traffic lanes); (**c**) two-roadway expressway; (**d**) motorway.

The charts presented in Figure 10 demonstrate that the acceleration values are becoming more focused as the road standard increases: six diagram columns are visible in the chart for the urban road; for extra-urban roads—five; for the expressway—three; and only one for the motorway.

In the case of lateral acceleration values, Figure 11 presents distributions with consideration of the acceleration mark (turning right or left) and distributions of the accelerations' absolute values. The shares of driving at other acceleration values are visible only after such modifications. The first two charts demonstrate a high similarity, and only charts 3 and 4 differ from one another rather clearly. The high similarity of chart pairs 1 and 2 as well as 3 and 4, is visible only when the lateral accelerations' absolute values are used. Acceleration values in the range of 0.5 to 1.0 m/s^2 are dominant in all charts, whereas on the urban and extra-urban sections they slightly exceed 60% and reach 95–96% on the expressway and motorway. This indicates that on higher-grade roads, driving is substantially smoother.

The next two charts (on Figures 12 and 13) demonstrate acceleration distributions in the form of spline charts rather than bar graphs, wherein the curves of the given acceleration type were plotted on the common chart for all four of the studied road sections. Figure 12 shows that each of the longitudinal acceleration distribution curves has a different and individual shape. If the shapes would be repeated approximately for other vehicle types, then it is possible to state that the acceleration curves (modified in an aforementioned manner) can be used to recognize the road type on which the driving took place.

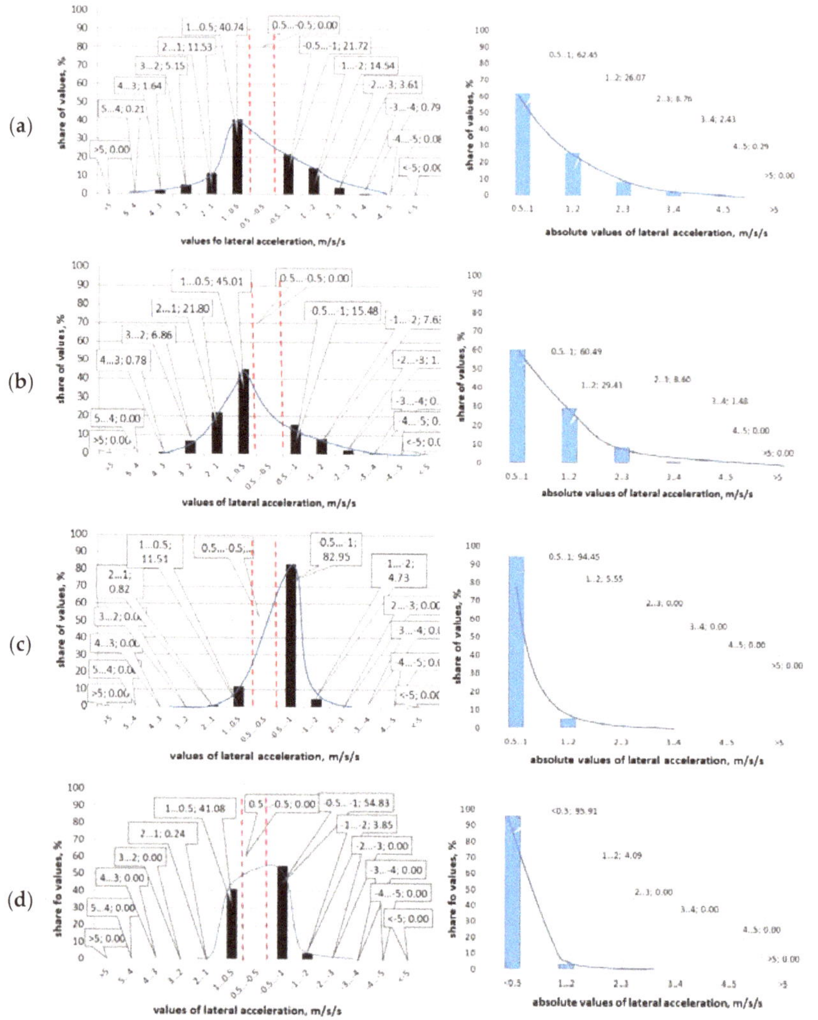

Figure 11. Lateral acceleration value ranges and their absolute values excluding the 'constant speed' driving range on various road types: (**a**) urban area (city); (**b**) non-urban area (single roadway with two traffic lanes); (**c**) two-roadway expressway; (**d**) motorway.

In the case of the lateral acceleration values presented in the left chart of Figure 13, their distribution curves also have individual shapes; however, it is more difficult to point out any regularities in the charts' shapes. However, a very interesting result is obtained when applying absolute value distribution curves-the right chart in Figure 13. It turns out that when charts 1 and 2, as well as 3 and 4, are paired, the charts are nearly identical. This means that in terms of the lateral dynamics, the vehicles' movement on each of the roads in the given pair is the same.

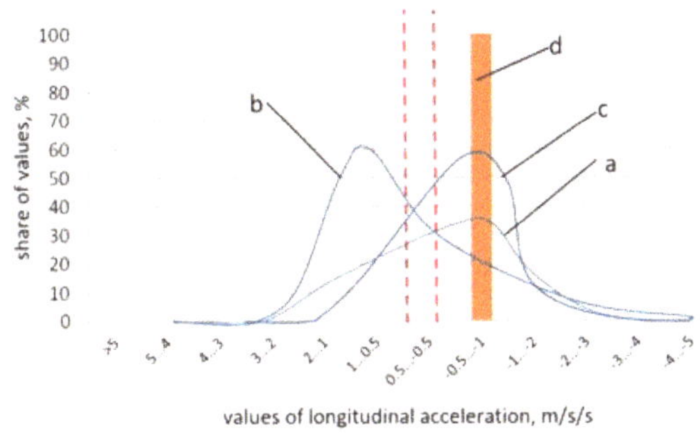

Figure 12. Breakdown of longitudinal acceleration curves on various road types: (a) urban area (city); (b) non-urban area (single roadway with two traffic lanes); (c) two-roadway expressway; (d) motorway.

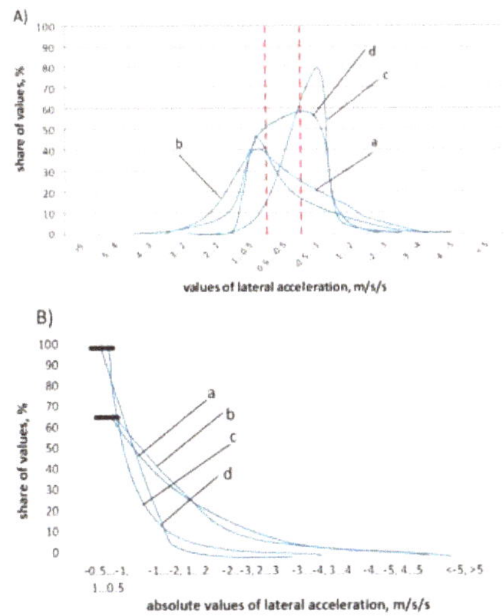

Figure 13. Breakdown of lateral acceleration values (**A**) and absolute values of lateral acceleration (**B**) on various road types: (a) urban area (city); (b) non-urban area (single roadway with two traffic lanes); (c) two-roadway expressway; (d) motorway.

6. Conclusions

In terms of the paper's main objective formulated in the final paragraph of the Introduction, i.e., the impact of the road type and shape on some of the basic movement parameters, i.e., longitudinal (positive and negative) and lateral acceleration values, the route selection must be evaluated as correct. It can be considered that the route included the four most-typical road types for many countries: urban area (city); single-roadway road with two traffic lanes in a non-urban area; expressway (with two roadways); and a motorway. In more detailed analyses, it is certainly possible to distinguish additional road types. However, in terms of achieving the paper's objective, the diversity of road type (and their features) in the presented set is sufficient to allow for achieving the attainment of the study's objective.

An analysis of the positive longitudinal acceleration values (vehicle acceleration) confirms the thesis about the strong impact of the road type on the occurring acceleration values. This conclusion can be drawn when observing the movement on particular roads, but this would be a qualitative conclusion. As result of the conducted testing, it is possible to formulate more specific quantitative conclusions. The charts' top parts presented in Figure 3 demonstrate that there is great diversity in the occurring values and in the variety of them. The urban area road features the highest values and the greatest variety of values. Both features clearly decline on other road types (the single-roadway extra-urban road, and the expressway) and are lowest on the motorway. It can be stated that on the other road types, the driving becomes 'smoother and easier' despite the fact that higher speeds are involved. Additional confirmation of this are the maximum values provided in Table 1, amounting respectively to: 3.17, 2.61, 0.8 and 0.36 m/s^2.

A similar regularity applies to negative acceleration, i.e., delays occurring at deceleration. The bottom parts of the charts in Figure 4 show the same gradation of roads in terms of the occurring values and their variety (if the value -3.2 m/s^2 line 3 of Table 2 is omitted, considering that according to Figure 3, it was an effect of the driver's one-time reaction to a dangerous situation, which can occur on any road).

In the case of the lateral acceleration values shown in Figure 4 and in Table 2, the road gradation in terms of the occurring values and their variety is the same as in the case of the longitudinal acceleration values.

The longitudinal acceleration distributions enable a more precise analysis of movement on particular road types. Each of the four charts features an overwhelming dominance of acceleration values in the range of 0.5 to -0.5 m/s^2. The share of these acceleration values amounted from 61.9% for the urban area road up to 99.5% for the motorway. In the analysis of the longitudinal acceleration values, the authors distinguished this interval and treated it conventionally, based on initial analyses, as driving at an approximately constant speed. When driving at a constant speed, there is some minor acceleration depending on the topography (e.g., driving up and down gentle slopes), gentle acceleration by several and up to over a dozen km/h to a new constant value, etc.

The lateral acceleration distributions also show an overwhelming dominance of acceleration values in the range of 0.5 to -0.5 m/s^2. The share of these acceleration values fluctuates from 80.1 to 95.1% on particular road sections. However, no gradation visible for longitudinal acceleration values has occurred in this case. Additionally, in this case, based on an analysis of the recorded vehicle movement parameters, the authors distinguished the aforementioned interval and deemed that the lateral acceleration values point to smooth cornering and smooth overtaking, bypassing, or traffic lane changing, which can also be deemed as 'smooth driving'.

The overwhelming dominance of driving conventionally referred to as 'smooth driving' on the longitudinal and lateral acceleration charts renders the shares of driving at other acceleration values very small. However, in terms of the synthetic attempt of evaluating the driving techniques of different drivers, these other cases are very important. To analyze them more precisely, the sections of driving at acceleration values of -0.5 to 0.5 m/s^2 were excluded from further analysis. This allowed for the evaluation of the structure of other longitudinal and lateral acceleration values.

The longitudinal and lateral acceleration values (in a time function) after the aforementioned modification were presented in Figures 8 and 9 accordingly. The charts presented in the figures more clearly show the differences in movement on particular road sections (especially for the longitudinal acceleration).

After excluding the conventional 'smooth driving' from the analysis, the longitudinal acceleration distributions demonstrate shares of driving at other (positive and negative) acceleration values. These charts clearly show that the acceleration value distributions become more focused along with increases in the road standard. A similar conclusion can be formulated for the analysis of lateral acceleration distributions. However, in this case, there is a similarity in the distribution's focus for roads 1 and 2, while distributions for roads 3 and 4 are much more focused (but similar to one another).

The analysis benefits most from the presentation of acceleration distributions not in the form of bar graphs, but the in the form of curves. The longitudinal acceleration distribution curve charts demonstrate that the acceleration distribution for the urban road is the least focused and, therefore, its modal value is the smallest. The curves for roads 2 (extra-urban) and 3 (expressway) are asymmetric but have a similar focus and similar modal values. The acceleration distribution for the motorway, after the aforementioned modification, is the most focused and features only a single column. The driving's monotonicity increases along with the increasing distribution focus. Figure 12 shows that each of the longitudinal acceleration distribution curves has a different and individual shape, therefore the road type on which the driving was taking place can be recognized based on the curve's shape.

In the case of the lateral acceleration values, their distribution curves also have individual shapes. However, interesting conclusions can be drawn when applying absolute value distribution curves. It turns out that when pairing the urban (1) and extra-urban road (2) as well as the expressway (3) and motorway (4), the curve charts are nearly identical. This means that in terms of the lateral dynamics, the vehicles' movement on each of the roads in the given pair is very similar.

Author Contributions: Conceptualization, R.S.J. and T.L.S.; methodology, R.S.J. and T.L.S.; software, R.S.J.; validation, R.S.J. and T.L.S.; formal analysis, R.S.J. and T.L.S.; investigation, R.S.J. and T.L.S.; resources, R.S.J. and T.L.S.; data curation, R.S.J. and T.L.S.; writing—original draft preparation, R.S.J. and T.L.S.; writing—review and editing, R.S.J. and T.L.S.; visualization, R.S.J.; supervision, T.L.S.; project administration, T.L.S.; funding acquisition, R.S.J. and T.L.S. All authors have read and agreed to the published version of the manuscript.

Funding: The research was funded by the National Centre for Research and Development, grant number POIR.04.01.04-00-0004/19-00.

Acknowledgments: The study was conducted as part of the research project titled 'Innovative system supporting the vehicle insurance risk evaluation dedicated to the UBI (Usage Based Insurance) no. POIR.04.01.04-00-0004/19-00 founded by the National Centre for Research and Development.

Conflicts of Interest: The authors declare no conflict of interest.

References

1. Bäumler, H. Reaction Time in Traffic. *Ferkehrsunfall Fahrz.* **2007**, *11*, 300–307.
2. Benderius, O.; Markkula, G.; Wolff, K.; Wahde, M. Driver behaviour in unexpected critical events and in repeated exposures—A comparison. *Eur. Transp. Res. Rev.* **2014**, *6*, 51–60. [CrossRef]
3. Green, M. How long does it take to stop? Methodological analysis of driver perception-brake times. *Transp. Hum. Factors* **2000**, *2*, 195–216. [CrossRef]
4. Jurecki, R.; Stańczyk, T.L. Analyzing driver response times for pedestrian intrusions in crash-imminent situations. In Proceedings of the 11th International Scientific and Technical Conference on Automotive Safety: Casta–Papiernicka, Častá, Slovakia, 18–20 April 2018; pp. 1–7. [CrossRef]
5. Jurecki, R.; Stańczyk, T.L. Driver model for the analysis of pre-accident situations. *Veh. Syst. Dyn.* **2009**, *47*, 589–612. [CrossRef]
6. Jurecki, R. Influence of the scenario complexity and the lighting conditions on the driver behaviour in a car-following situation. *Arch. Automot. Eng. Arch. Motoryz.* **2019**, *83*, 151–173. [CrossRef]
7. Kaysi, I.A.; Abbany, A.S. Modeling aggressive driver behavior at unsignalized intersections. *Accid. Anal. Prev.* **2007**, *39*, 671–678. [CrossRef] [PubMed]
8. Ahlström, C.; Anund, A.; Fors, C.; Åkerstedt, T. The effect of daylight versus darkness on driver sleepiness: A driving simulator study. *J. Sleep Res.* **2018**, 1–9. [CrossRef]
9. Di Milia, L.; Kecklund, G. The distribution of sleepiness, sleep and work hours during a long distance morning trip: A comparison between night–and non–night workers. *Accid. Anal. Prev.* **2013**, *53*, 17–22. [CrossRef] [PubMed]
10. Guzek, M. Simplex and complex reaction time of male drivers in various age—results of research with use of reflexometer. *Arch. Automot. Eng.* **2014**, *65*, 19–28.
11. Mitas, A.; Bugdol, M.; Ryguła, A. The psychophysiological conditionings of driver's work under the aspect of traffic safety. *Transp. Probl.* **2009**, *4*, 87–94.
12. Madeley, P.; Hulley, J.L.; Wildgust, H.; Mindham, R.H. Parkinson's disease and driving ability. *Neurol. Neurosurg. Psychiatry* **1990**, *53*, 580–582. [CrossRef] [PubMed]
13. Hindmarch, I. Psychomotor function and psychoactive drugs. *Br. J. Clin. Pharmacol.* **2004**, *58*, S720–S740. [CrossRef]
14. Ogden, J.D.; Moskowitz, H. Effects of Alcohol and Other Drugs on Driver Performance. *Traffic Inj. Prev.* **2004**, *5*, 185–198. [CrossRef] [PubMed]
15. Dai, J.; Teng, J.; Bai, X.; Shen, Z.; Xuan, D. Mobile phone based drunk driving detection. In Proceedings of the 2010 4th International Conference on Pervasive Computing Technologies for Healthcare, Munich, Germany, 22–25 March 2010; pp. 1–8. [CrossRef]

16. Rygula, A. Driving style identification method based on speed graph analysis. In Proceedings of the International Conference on Biometrics and Kansei Engineering, Cieszyn, Poland, 25–28 June 2009; pp. 76–79. [CrossRef]
17. Mierlo, J.; Maggetto, G.; Burgwal, E.; Gense, R. Driving style and traffic measures—Influence on vehicle emissions and fuel consumption. *Proc. Inst. Mech. Eng. Part D J. Automob. Eng.* **2004**, *218*, 43–50. [CrossRef]
18. Smoleń, P.; Starowicz, W. Koncepcja systemu oceniającego styl jazdy kierowcy w transporcie drogowym ładunków [Driving style evaluation system concept in cargo road transport]. *Transport Miejski i Regionalny.* **2018**, *8*, 18–23.
19. Merkisz, J.; Orszulak, B. Wstępna analiza rejestracji parametrów stylu jazdy kierowcy [Preliminary analysis of driving style parameters' recording]. *Logistyka Logist.* **2015**, *3*, 3210–3214.
20. Wang, J.; Lu, M.; Li, K. Characterization of longitudinal driving behavior by measurable parameters. *Transp. Res. Rec.* **2010**, *2185*, 15–23. [CrossRef]
21. Lajunen, T.; Karola, J.; Summala, H. Speed and Acceleration as Measures of Driving Style in Young Male Drivers. *Percept. Mot. Ski.* **1997**, *85*, 3–16. [CrossRef] [PubMed]
22. Af Wåhlberg, A.E. The relation of acceleration force to traffic accident frequency: A pilot study. *Transp. Res. Part F Traffic Psychol. Behav.* **2000**, *3*, 29–38. [CrossRef]
23. Merkisz, J.; Tarkowski, S. Czynniki dynamiczne i ich wpływ na subiektywne poczucie komfortu w autobusach miejskich [Dynamic sensors and their impact on subjective feeling of comfort in city buses]. *Postępy Nauki Techniki Sci. Technol. Adv.* **2012**, *14*, 169–178.
24. Reymond, G.; Kemeny, A.; Droulez, J.; Berthoz, A. Role of lateral acceleration in curve driving: Driver model and experiments on a real vehicle and a driving simulator. *Hum. Factors* **2001**, *43*, 483–495. [CrossRef]
25. Martinelli, F.; Mercaldo, F.; Orlando, A.; Nardone, V.; Santone, A.; Sangaiah, A.K. Human behavior characterization for driving style recognition in vehicle system. *Comput. Electr. Eng.* **2020**, *83*, 102504. [CrossRef]
26. Del Campo, I.; Asua, E.; Martínez, V.; Mata-Carballeira, Ó.; Echanobe, J. Driving style recognition based on ride comfort using a hybrid machine learning algorithm. In Proceedings of the 21st International Conference on Intelligent Transportation Systems (ITSC), Maui, HI, USA, 4–7 November 2018; pp. 3251–3258. [CrossRef]
27. Stoichkov, R. Android Smartphone Application for Driving Style Recognition. Project Thesis, Department of Electrical Engineering and Information Technology, Institute for Media Technology, Munich, Germany, 24 September 2013. Available online: http://www.eislab.fim.uni-passau.de/files/publications/students/Stoichkov-Projektarbeit.pdf (accessed on 12 November 2020).
28. Vaitkus, V.; Lengvenis, P.; Žylius, G. Driving style classification using long-term accelerometer information. In Proceedings of the 19th International Conference on Methods and Models in Automation and Robotics (MMAR), Międzyzdroje, Poland, 2–5 September 2014; pp. 641–644. [CrossRef]
29. Krotac, T.; Simlova, M. The Analysis of the Acceleration of the Vehicle for Assessing the Condition of the Driver'. In Proceedings of the Intelligent Vehicles Symposium, Alcalá de Henares, Spain, 3–7 June 2012; pp. 571–576.
30. Miyajima, C.; Ukai, H.; Naito, A.; Amata, H.; Kitaoka, N.; Takeda, K. Driver Risk Evaluation based on Acceleration, Deceleration, and Steering Behavior. In Proceedings of the IEEE International Conference on Acoustics, Speech, and Signal Processing, Prague, Czech Republic, 22–27 May 2011; pp. 1829–1832.
31. Vasey, E.; Ko, S.; Jeon, M. In-Vehicle Affect Detection System: Identification of Emotional Arousal by Monitoring the Driver and Driving Style. In Proceedings of the 10th International Conference on Automotive User Interfaces and Interactive Vehicular Applications, Toronto, ON, Canada, 23–25 September 2018; pp. 243–247. [CrossRef]
32. Zylius, G.; Vaitkus, V.; Lengvenis, P. Driving style analysis using spectral features of accelerometer signals. In Proceedings of the 9th International Conference on Computer Science & Education, Vancouver, BC, Canada, 22–24 August 2014; pp. 267–273.
33. Kudzia, K.; Stopa, S. Algorytm do określania ekonomiczności stylu jazdy na podstawie danych z systemów diagnostycznych samochodu [Algorithm for determining the driving style economy based on vehicle diagnostic systems' data]. *Logistyka Logist.* **2015**, *4*, 4320–4326.
34. Varhelyi, A.; Hjälmdahl, M.; Hyden, C.; Draskoczy, M. Effects of an active accelerator pedal on driver behaviour and traffic safety after long-term use in urban areas. *Accid. Anal. Prev.* **2004**, *36*, 729–737. [CrossRef] [PubMed]
35. Wakita, T.; Ozawa, K.; Miyajima, C.; Igarashi, K.; Katunobu, I.; Takeda, K.; Itakura, F. Driver identification using driving behavior signals. *IEICE Trans. Inf. Syst.* **2006**, *89*, 1188–1194. [CrossRef]
36. Sun, D.; Elefteriadou, L. Lane-changing behavior on urban streets: An "in-vehicle" field experiment-based study. *Comput. Aided Civ. Infrastruct. Eng.* **2012**, *27*, 525–542. [CrossRef]
37. Van Ly, M.; Martin, S.; Trivedi, M.M. Driver classification and driving style recognition using inertial sensors. In Proceedings of the IEEE Intelligent Vehicles Symposium (IV), Paris, France, 9–12 June 2013; pp. 1040–1045. [CrossRef]
38. Merkisz, J.; Pielecha, J.; Tarkowski, S. Pokładowe rejestratory parametrów ruchu i ich zastosowanie do oceny komfortu w autobusach miejskich [On-board movement parameter recorders and their use in the evaluation of comfort in city buses]. *Autobusy: Tech. Eksploat. Syst. Transp. Buses Technol. Exploit. Transp. Syst.* **2012**, *13*, 300–305.
39. Johnson, D.A.; Trivedi, M.M. Driving style recognition using a smartphone as a sensor platform. In Proceedings of the 14th International IEEE Conference on Intelligent Transportation Systems (ITSC), Washington, DC, USA, 5–7 October 2011; pp. 1609–1615. [CrossRef]
40. Ericsson, E. Variability in urban driving patterns. *Transp. Res. Part D Transp. Environ.* **2000**, *5*, 337–354. [CrossRef]

41. Amado, S.; Ulupınar, P. The effects of conversation on attention and peripheral detection: Is talking with a passenger and talking on the cell phone different? *Transp. Res. Part F Traffic Psychol. Behav.* **2005**, *8*, 383–395. [CrossRef]
42. Mohebbi, R.; Gray, R.; Tan, H.Z. Driver Reaction Time to Tactile and Auditory Rear–End Collision Warnings While Talking on a Cell Phone. *Hum. Factors* **2009**, *51*, 102–110. [CrossRef]
43. Huang, Y.; Sun, D.J.; Zhang, L.H. Effects of congestion on drivers' speed choice: Assessing the mediating role of state aggressiveness based on taxi floating car data. *Accid. Anal. Prev.* **2018**, *117*, 318–327. [CrossRef]
44. Kropiwnicki, J. Klasyfikacja warunków eksploatacji pojazdów z wykorzystaniem udziału czasu pracy silnika na biegu jałowym [Classification of vehicle exploitation conditions with the use of the engine idling time]. *Autobusy Tech. Eksploat. Syst. Transp. Buses Technol. Exploit. Transp. Syst.* **2011**, *12*, 204–209.
45. Jurecki, R.; Stańczyk, T.; Jaśkiewicz, M. Driver's reaction time in a simulated, complex road incident. *Transport* **2017**, *32*, 44–54. [CrossRef]
46. Jurecki, R.S.; Stańczyk, T.L. Driver reaction time to lateral entering pedestrian in a simulated crash traffic situation. *Transp. Res. Part F Traffic Psychol. Behav.* **2014**, *27*, 22–36. [CrossRef]
47. Szumska, E.M.; Jurecki, R. The Effect of Aggressive Driving on Vehicle Parameters. *Energies* **2020**, *13*, 6675. [CrossRef]

energies

MDPI

Article

Striated Tire Yaw Marks—Modeling and Validation

Wojciech Wach * and Jakub Zębala

Institute of Forensic Research in Kraków, 31-033 Kraków, Poland; jzebala@ies.gov.pl
* Correspondence: wwach@ies.gov.pl; Tel.: +48-12-61-85-763

Abstract: Tire yaw marks deposited on the road surface carry a lot of information of paramount importance for the analysis of vehicle accidents. They can be used: (a) in a macro-scale for establishing the vehicle's positions and orientation as well as an estimation of the vehicle's speed at the start of yawing; (b) in a micro-scale for inferring among others things the braking or acceleration status of the wheels from the topology of the striations forming the mark. A mathematical model of how the striations will appear has been developed. The model is universal, i.e., it applies to a tire moving along any trajectory with variable curvature, and it takes into account the forces and torques which are calculated by solving a system of non-linear equations of vehicle dynamics. It was validated in the program developed by the author, in which the vehicle is represented by a 36 degree of freedom multi-body system with the TMeasy tire model. The mark-creating model shows good compliance with experimental data. It gives a deep view of the nature of striated yaw marks' formation and can be applied in any program for the simulation of vehicle dynamics with any level of simplification.

Keywords: yaw marks striations; tire model; multibody dynamics; vehicle accident simulation

Citation: Wach, W.; Zębala, J. Striated Tire Yaw Marks—Modeling and Validation. *Energies* **2021**, *14*, 4309. https://doi.org/10.3390/en14144309

Academic Editor: Aldo Sorniotti

Received: 16 June 2021
Accepted: 15 July 2021
Published: 17 July 2021

1. Introduction

The rectified projection of tire yaw marks deposited on the road surface (often either from an orthophotomap, a point cloud or a total station survey) can be used in a macro-scale for:

- determination of subsequent vehicle positions and orientation during critical movement by matching the position of individual wheels to the corresponding marks;
- estimation of the vehicle's speed at the start of yawing.

For the latter—assuming the vehicle was moving on a horizontal and homogeneous surface along a curve of radius r—the critical speed formula (CSF) is most commonly used:

$$v = \sqrt{\mu \cdot g \cdot r}, \tag{1}$$

where: μ—tire-to-roadway coefficient of friction, r—radius of the mark, g = 9.8 m/s^2—gravitational acceleration.

Although this formula may seem simplistic, it has been shown repeatedly over the years that, under certain conditions, it can be considered as a quasi-empirical critical speed method (CSM). A broad overview of such conditions has been presented among others by Brach and Brach [1]. Sledge and Marshek [2] examined some refined forms of the CSF which account for the effects of, among others, vehicle weight distribution, slip angles, cornering stiffnesses and ABS. Cannon [3] has demonstrated that effective braking causes the CSF to overestimate the speed at the start of the yaw mark and that a 50 ft. chord appears to give acceptably low effective braking-related errors (<7%) for speeds of approximately 72 km/h (45 mph) and for speeds of approximately 97 km/h (60 mph) with light to moderate braking. Cliff et al. [4] have concluded that when using the peak coefficient of friction, both the CSF and simulation over-estimated the actual speed, whereas slide coefficient of friction under-estimated them. Braking tended to increase the results. Amirault and MacInnis [5] carried out a total of 29 tests at speeds of 80 to 95 km/h.

Bearing in mind non-braking and ABS braking tests, using 20 m chord measurements for the radius and the average braking coefficient of friction overestimated the measured speed by 4.1% ± 6.3% (±1σ), while using a center of gravity trajectory for the radius and the average braking coefficient of friction underestimated the measured speed by 2.0% ± 5.2% (±1σ).

Lambourn [6] conducted tests in which passenger cars were freely coasting, braking or under power when travelling in a curve at speed of 55 to 100 km/h, and proposed a procedure which makes it possible to limit the uncertainty of speed calculated from the CSF to ±10%. In [7] and [8] he concluded that his previously described CSM gives satisfactory results also in the event of light braking, heavy braking with ABS, acceleration and the operation of ESP. There was no sign of the cycling of the ABS. The yaw marks had the typical appearance, practically the same as in the case of low braking without ABS. Significantly less yaw or off-tracking were observed when compared with marks deposited with little or no braking. If the brakes were applied aggressively during the yaw, the amount of yaw would decrease. This feature—the reverse of usual yawing with the heavy non-ABS braking—is probably the result of the "select low" algorithm.

While most authors, notably Lambourn [6–8] and Amirault and MacInnis [5], limit their analysis to yawing with relatively low yaw rate, Cash and Crouch [9] derived a formula which accounts for a higher degree of vehicle yaw and any brake force (not only from driver input), resulting in a narrower error band than conventional CSMs. If the exact path and orientation of the collision vehicle are not readily apparent, their method allows the flexibility of considering ranges for the required values.

What distinguishes simulation methods is that they provide a deep view into the time histories of curvilinear movement parameters, including the critical speed, as well as identifying a set of data (e.g., steering angle, braking/accelerating level of particular wheels, yaw moment of inertia etc.) which enables a virtual vehicle to move along the actual tire marks.

The point of this article is the appearance of yaw marks in the micro-scale, that is from the perspective of the topology of striations forming the mark, which make it possible to infer the braking or acceleration status of the wheels (and sometimes even the steering angle of the front wheels). The yaw mark is essentially left by the entire tire footprint remaining in contact with the roadway (contact patch), but its blackness and distinctness depend on the local slip of the tread blocks relative to the road, and stress. The rule "the greater the stress, the more distinct the mark" applies both in the macro (when observing the entire yaw marks, the most pronounced are the marks of the external, loaded wheels) and in the micro scale (the most visible is the outer edge of a single yaw mark).

Yamazaki and Akasaka in their classic article [10] argue that deposition of striations is independent of the tread pattern and is caused by an in-plane bending moment that is transmitted from the roadway to the body of the tire via the tread contact patch during sharp cornering. They refer to this phenomenon as the bending buckling behavior of a steel-belted radial tire. Buckling occurs immediately in the contact patch in the presence of a large lateral reaction from the road and is specific to radial—not bias—tires. Yamazaki in [11] shows that sharp cornering turns on steel-belted radial tires often cause wavy wear along the periphery of the shoulder.

Beauchamp et al. in [12] summarize the literature concerning the yaw mark striations issue, analyze the differences in the mechanism in which striations are deposited, and discuss the relationship between tire mark striations and tire forces. They conclude that in the case of tires with pronounced shoulder blocks the striations are typically produced by these blocks whereas tires with very low pressure or without a tread pattern are more likely to deposit striations by buckling. In the absence of braking and acceleration the striation marks are parallel to the wheel rotation axis. When the brakes are applied aggressively, the striations will change to a direction more in line with the wheel trajectory but ABS prevents the tires from locking. Beauchamp et al. in [13] show examples where the striations reflect point loading of the tread shoulder blocks. In Figure 1 of their article they show a mark on which two stripes can be distinguished: lighter striations from the inside, being deposited by the tread, and darker striations from the outside, being deposited by the shoulder blocks. A scheme of formation of such a mark they demonstrate in Figure 3 of [13].

The authors of this work, in their research practice, have encountered all the types of yaw marks mentioned before—two examples are shown in Figure 1. In both cases the vehicle was yawing, but the tire mark (a) was deposited by a buckled, zero-pressure tire during a clockwise yaw, while the tire mark (b) was left by the leading shoulder blocks of the normal-pressure tire during a counter-clockwise yaw.

(a) (b)

Figure 1. Yaw marks deposited during a full scale vehicle yaw testing performed by Zębala et al. [14]: (**a**) resulting from in-plane buckling of a tire with zero pressure; (**b**) left by the tread shoulder blocks of a tire with normal pressure.

The article by Beauchamp et al. [13] does an excellent job of analyzing yaw mark striations from the viewpoint of its geometry and deriving equations for the calculation of longitudinal slip s_x using the striation marks angle θ and the slip angle α see analogous formulae (9) and (11) derived in this article). It was shown that the model offers insight into the braking actions of a driver at the time the tire marks were being deposited. The usefulness of such marks for accident analysis depends obviously on their quality and clarity, which affect the uncertainty in the measurement of the geometric parameters. Beauchamp et al. in [15] explore the sensitivity and uncertainty of the s_x equation. They prove that at $\alpha = 5^\circ$, the striations will change over 70 degrees between no braking and maximum braking, while at $\alpha = 85^\circ$, less than 2 degrees separate no braking and maximum braking. In the first case braking will likely be easy to distinguish; in the second, changes in striation angle from braking are unlikely to be detected.

Undoubtedly, all researchers who focus on point loading of the tread shoulder blocks as well as in-plane buckling are right in their specific areas, as in general, the appearance of a striated mark left during curvilinear motion depends on many factors, the main ones being:

- Camber angle of the wheel (which in turn depends on the suspension kinematics);
- Structure to the tire;

- Cornering, longitudinal and vertical stiffnesses of the tire (which depend among others things on tire pressure);
- Tire aspect ratio (sidewall height divided by tire width);
- Tire tread depth;
- Dynamic tire offset (pneumatic trail);
- Composition of rubber compound;
- Road surface properties; and
- Temperature in the contact patch.

However, regardless of whether the striations occur from buckling or tread blocks, their direction always follows the direction of the resultant tire velocity vector in the contact patch (called the wheel slip velocity and hereinafter referred to as $\mathbf{v}_{IO}$). The only difference lies in the pitch of the striations, which in the case of striations created by the tread shoulder blocks gives the opportunity to estimate the longitudinal slip s_x from the topology of the striations (according to the formula of Beauchamp et al. [13], see also Equations (9) to (12) in this article), while in the case of buckling does not give the same possibility because of lack of a buckling wave pitch (a momentary and unique period of the buckling wave).

An in-depth look at the mechanics of the yaw mark creation is very interesting not only as a mathematical problem, but first of all crucial from the angle of the uncertainty of vehicle accident analysis (see e.g., [16]).

2. Assumptions to the Model

For the development of a model of creating a striated tire yaw mark the following assumptions have been made:

1. To describe the tired wheel–road interaction, a semi-physical, non-linear tire model TMeasy was used [17,18]. All its features, such as longitudinal and lateral forces, aligning moment, other moments resulting from the wheel kinematics, first-order dynamics for longitudinal, lateral and torsional strain, and dynamic tire offset are taken into account.
2. The yaw mark creation model is universal, i.e., it applies to any curvilinear motion of a tire in lateral drift, when the wheel moves along any trajectory with variable curvature, and the wheel is subject to an unsteady force and moment as to the value and direction, calculated by solving the system of non-linear differential equations of vehicle dynamics.
3. As this area is in fact the dominant one, the yaw mark will be created by the tread elements of the tire shoulder forming the outer border (during curvilinear movement of the vehicle) of the patch in contact with the road. For the yaw mark curved to the left these will be the right patch border, and for the mark curved to the right the left patch border. As the striations being deposited by the internal tread elements of the patch are usually not very clear, they have been omitted; however, if necessary, it will be easy to extend the algorithm by following the model described.
4. It is possible to define any, including non-uniform, pitch of the tread shoulder blocks of the tire.
5. Semi-physical tire models of class TMeasy, Magic Formula [19], HSRI [20] etc. do not describe the tire in-plane buckling, therefore this phenomenon can instead be analyzed in two ways:
    (a) partially, i.e., as to the tire mark course and the striations direction only; it is then sufficient to enter any pitch of the tread shoulder blocks;
    (b) fully, i.e., as to the tire mark course, and striations direction and pitch, if the buckling pitch in the patch is known in some other way, which can be manually entered into the program.

3. Basic Terms Concerning Movement of a Wheel in a Bend

A concise, computer-friendly vector-matrix notation has been used. The italic small letters (e.g., v) mean scalars, bold small letters (e.g., $\mathbf{v}$)—vectors, and bold capital letters (e.g., **A**)—matrices. In the vector-matrix equations the Rill's subscript notation [18] has been used:

{} the symbol of a reference frame, e.g., the term {I} should be read "in the reference frame I";

{I} the ground-fixed inertial (global) reference frame with the origin at point I; the z_I axis is parallel to the gravitational acceleration vector **g** and points upward;

$\mathbf{r}_{IO,I}$ vector from point I to wheel center O; the subscript after the comma denotes the reference frame in which this vector is observed—here {I};

$\mathbf{v}_{IO,I}$ vector of absolute velocity of wheel center O, with respect to {I}.

A characteristic point of the tire-road patch Q, which is the origin of the Cartesian coordinate system with unit vectors on its axes $\mathbf{e}_x$, $\mathbf{e}_y$, $\mathbf{e}_n$ (shown in Figure 2a), is referred to as the contact point. In the TMeasy tire model, the system of forces acting on the wheel is reduced to an equivalent system of forces (F_x, F_y and F_z) and their moments (M_x, M_y and M_z), whose directions of action coincide with the directions of the unit vectors. The position of the rim center plane in relation to the road is determined by the position vector $\mathbf{r}_{IO}$ and the unit vector $\mathbf{e}_{ky}$, normal to this plane and defining the wheel rotation axis.

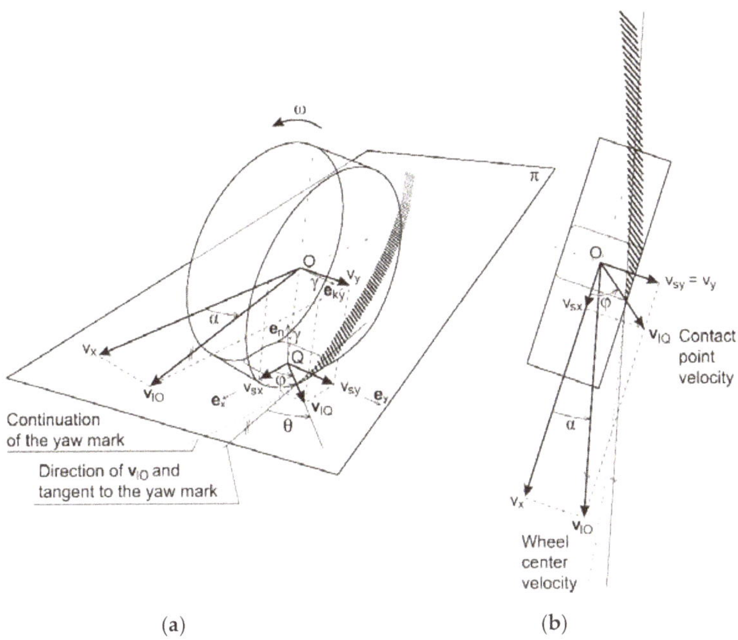

Figure 2. Velocities in the process of depositing striations by tire shoulder blocks during yaw: (**a**) perspective; (**b**) top view.

Camber γ as well as cornering cause the tire to deflect laterally and to offset the contact point Q against the rim center plane by the distance y_e (see Section 5.3). The road surface geometry in {I} defines the function:

$$z_s = z_s(x, y). \tag{2}$$

The current position of the contact point Q in the global reference frame {I} is given by the formula:

$$\mathbf{r}_{IQ,I} = \mathbf{r}_{IO,I} + \mathbf{r}_{OQ,I}\,, \tag{3}$$

where the vector $\mathbf{r}_{OQ,I}$ can be determined by the approach given in [18] or [21].

Figure 2a depicts a diagram of a wheel in lateral slip which rolls across the plane π with angular velocity ω. The following symbols have been adopted:

v_x and v_y—components of the wheel center velocity vector $\mathbf{v}_{IQ}$ parallel to π, the first of which lies on the direction of the longitudinal axis of the rim, and the other is perpendicular to it;
γ—camber angle;
α—wheel slip angle;
ω—wheel angular velocity about its spin axis given by the unit vector $\mathbf{e}_{ky}$;
v_{sx} and v_{sy}—components of the absolute velocity vector of the contact point $\mathbf{v}_{IQ}$ (wheel slip velocity); v_{sx} and v_{sy} are parallel to v_x and v_y, and read:

$$v_{sx} = v_x - r_d\omega, \tag{4}$$

$$v_{sy} = v_y, \tag{5}$$

r_d—dynamic tire radius;
φ—direction of the contact point velocity $\mathbf{v}_{IQ}$ against the longitudinal axis of the wheel (rim).

According to the ISO definition (see also Pacejka [19]), the longitudinal and lateral relative slips are:

$$s_x = -\frac{v_{sx}}{v_x} = -\frac{v_x - r_d\omega}{v_x}, \tag{6}$$

and

$$s_y = \tan\alpha = -\frac{v_{sy}}{v_x} \tag{7}$$

respectively. The vectors of the contact point velocity $\mathbf{v}_{IQ}$, the relative slip $\mathbf{s} = \left[s_x\ s_y\right]^T$ and the tangential force acting on the tire at the contact point $\mathbf{F} = \left[F_x\ F_y\right]^T$ have the same direction, and satisfy the relationship:

$$\tan\varphi = \frac{v_{sy}}{v_{sx}} = \frac{s_y}{s_x} = \frac{F_y}{F_x}, \tag{8}$$

wherein the components F_x and F_y are calculated according to the TMeasy (or any other) tire model.

It is easy to see that the relation (8) will also be fulfilled with other definitions of slips s_x and s_y, because they differ only in the denominator (e.g., at Rill $r_d|\omega|$ [18] or at Mitschke $\max\{r_d\omega, v_x\}$ [22]), which will disappear when inserted into the formula (8).

4. Wheel Velocities and Slips Versus Geometry of the Striated Tire Mark

Figure 2 shows, schematically, the process of making a striated yaw mark on the road surface with the tread shoulder blocks of a tire. The direction of the velocity vector $\mathbf{v}_{IQ}$ is also the direction of the contact point Q displacement relative to the road and, consequently, the direction of striations deposited on the road by the tire during yaw.

Dividing the formula (7) by (8) gives:

$$s_x = \frac{\tan\alpha}{\tan\varphi}, \tag{9}$$

and because $s_y = \tan\alpha$, hence

$$s_y = s_x \tan\varphi. \tag{10}$$

These relationships allow the wheel slips s_x and s_y to be calculated having only the striated yaw mark, as shown in Figure 3. Unlike the angle θ (representing the deviation of the striations direction from the tangent to the yaw mark, which is easy to measure on the mark), the slip and contact point velocity angles—α and φ respectively—are generally unknown because of the unknown orientation of the wheel relative to the mark. This difficulty applies in particular to the front wheel marks, as the steering angle of these wheels against the vehicle body is variable, and may even change over time, and are therefore impossible to determine by the yaw marks topology alone.

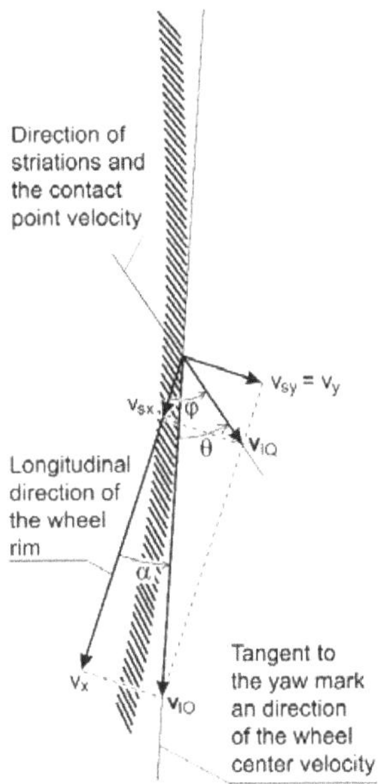

Figure 3. Determination of velocities and angles configuration from the striated yaw mark disclosed on the road.

That is why Beauchamp et al. [13], using simple geometric relationships, derived the following formula for the slip angle:

$$\alpha = \arcsin\frac{S \sin\theta}{T} - \theta\,, \tag{11}$$

where the distances S and T shown in Figure 4 mean:

S—the striation pitch measured along the yaw mark;
T—the pitch of the tread shoulder blocks measured on the tire circumference.

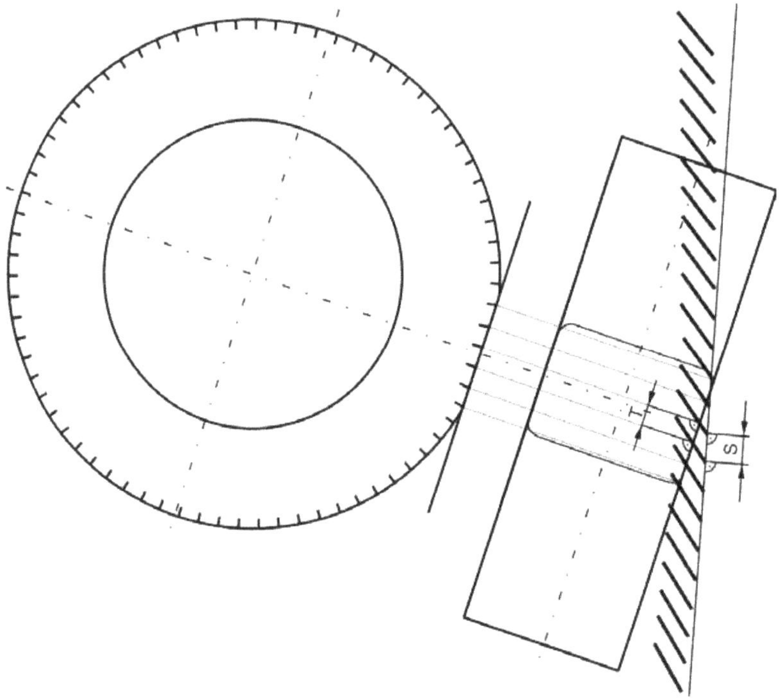

Figure 4. Measurement of distances S on the yaw mark and T on the tire shoulder.

From Figure 3 it follows that:

$$\varphi = \alpha + \theta. \quad (12)$$

To sum up, in order to calculate the wheel slippage at a selected point of the striated yaw mark one should:

- Measure the angle θ and distance S on the mark;
- Measure the distance T at the tire shoulder;
- Calculate the angles α and φ from the formulae (11) and (12) respectively (note: in the case of non-steered rear wheels, when, in addition, the marks of other wheels allow one to discover successive angular positions of the vehicle by matching the corresponding wheels, the orientation of the rear wheel relative to the yaw mark is known, therefore angles α and φ can be measured directly on the mark—see Figure 3);
- Calculate s_x and s_y from formulae (9) and (10).

The uncertainty of the results of such calculations related to non-uniform pitch of the striations S (as a consequence of the uneven tread pitch T aimed at reducing the noise generated by the tire) falls within the general uncertainty of this approach and, above all, the S and T measurement uncertainty.

5. Model of Deposition of the Yaw Mark Striations on the Road Surface

Figure 5a shows a diagram of tread shoulder blocks pitch, where:

n—number of blocks or grooves on the tread shoulder (consistently keeping to the chosen convention);

B_k, $k = 0, \ldots, n-1$—point indicating the k-th block (or groove);

d_k, $k = 0, \ldots, n-1$—distances between adjacent points B_k measured along an arc: with typical tires, without making a significant error, these can be measured in a straight line;

According to Figure 5a:

$$d_k = \begin{cases} \text{distance between the points } B_k \text{ and } B_{k+1} & \text{for } k = 0, \ldots, n-2 \\ \text{distance between the points } B_k \text{ and } B_0 & \text{for } k = n-1 \end{cases}$$

r_0—unloaded tire radius.

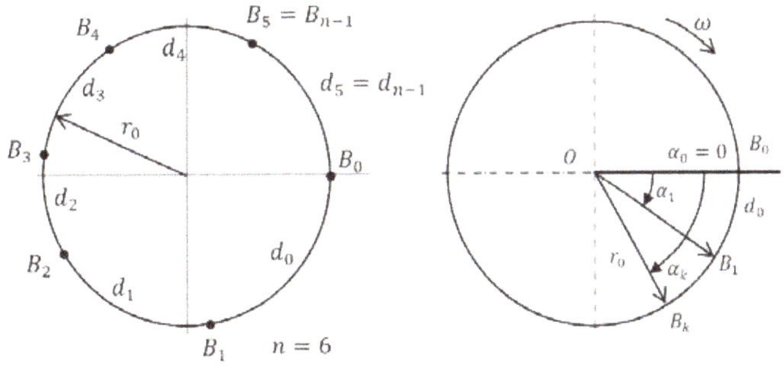

Figure 5. The diagram of tread shoulder blocks: (**a**) the symbols used to define the pitch of the tread blocks; (**b**) configuration at the instant $i = 0$, $t = 0$.

The values of parameters d_k and r_0 are constant throughout the simulation. In general, the pitch can be defined as non-uniform (d_k = constant), but in the simplest case, the program may suggest by default a uniform pitch according to the formula:

$$d_k = \frac{2\pi r_0}{n}, \ k = 0, \ldots, n-1. \tag{13}$$

Let's adopt additional symbols:

i—number of simulation step;
t—current simulation time;
h—simulation timestep.

The following calculation algorithm, repeated in each simulation step, can be proposed.

5.1. Determining the Position of Each Point B_k Relative to the Wheel-Fixed Reference Frame {O}

The origin of the wheel-fixed reference frame {O} is at the wheel center O, where the xz plane is the rim center plane and the y axis is the wheel rotation axis (with the unit vector $\mathbf{e}_{ky}$).

This can be done in the polar coordinate system shown in Figure 5b, where point B_k has coordinates (r_0, α_k). In each simulation step the angle α_k of each point is measured from the thick horizontal line, clockwise.

At the beginning of the simulation (i.e., $i = 0$, $t = 0$ s) point B_0 has the polar coordinates $(r_0, \alpha_0) = (r_0, 0)$ and the other points B_k have the coordinates (r_0, α_k), where:

$$\alpha_k = \alpha_k(t) = \alpha_{k-1} + \frac{d_k}{r_0} \text{ [rad]}, \ k = 1, \ldots, n. \tag{14}$$

This formula can be applied in the function Angular_position_of_block() shown in Appendix A (*Code 1*), in section *Code 1*.

5.2. Calculation of the Characteristic Dimensions of the Contact Patch

Assuming that the tire remains in full contact with the road over the entire tread width L_b, and the contact patch has a rectangular shape with length L_x and width L_y, these dimensions can be calculated using the approximation proposed by Rill in [21]:

$$L_x = L_x(t) \approx 2\sqrt{r_0 \Delta z} = 2\sqrt{r_0 \frac{F_z}{c_R}}\,, \tag{15}$$

$$L_y = L_y(t) \approx \frac{L_b}{\cos\gamma}\,, \tag{16}$$

where:

Δz is the total tire deflection given by the formula:

$$\Delta z = \Delta z(t) = r_0 - r_s = \frac{F_z}{c_R}, \tag{17}$$

$F_z = F_z(t)$—the normal reaction of the roadway to the tire at the contact point;
$c_R = \text{const}$—the radial stiffness of the tire;
$r_s = r_s(t)$—the loaded (static) tire radius;
$\gamma = \gamma(t)$—the tire camber angle, i.e., the inclination of the rim center plane against the roadway normal.

For example, for a tire 205/55 R16 at F_z = 4700 N and the pressure p = 2.5 bar, after adopting the data c_r = 265000 N / m and r_0 = 0.317 m, one gets $L_x \approx 0.15$ m.

5.3. Calculating the Geometry of Striations

In a single simulation step the coordinates of the points at which the shoulder blocks contact the road should be determined. They are shown in the global Cartesian coordinate system $\{I\}$.

In reality, the deposition of a yaw mark on the road surface depends on many local or temporary factors that are difficult to discover after the accident. As the main one is a sufficiently high lateral tire force, the following formula can be used to define the condition when the yaw mark should be created in the program:

$$F_y \geq \frac{p_{\%}}{100}\,\mu F_z\,, \tag{18}$$

where:

$F_y = F_y(t)$—current lateral force acting on the tire, calculated according to the TMeasy (or any other) tire model;
$F_z = F_z(t)$—current normal force to the roadway acting on the tire;
μ—tire-road friction coefficient; in general $\mu = \mu(x, y)$;
$p_{\%}$—percentage of the maximum horizontal force μF_z at which the yaw mark is to be made; by default $p_{\%} = 95\%$ can be adopted.

The length of the contact patch is limited by two boundary angles α_t' and α_t'' indicating the first and last point, respectively, of the tire circumference being in contact with the road (not to be confused with the point B_k indicating the tread element). They are illustrated in Figure 6 and given by the formulae:

$$\alpha_t' = \arccos\frac{L_x}{2r_0} = \arctan\frac{2r_s}{L_x} \tag{19}$$

and

$$\alpha_t'' = \frac{\pi}{2} + \arcsin\frac{L_x}{2r_0}. \tag{20}$$

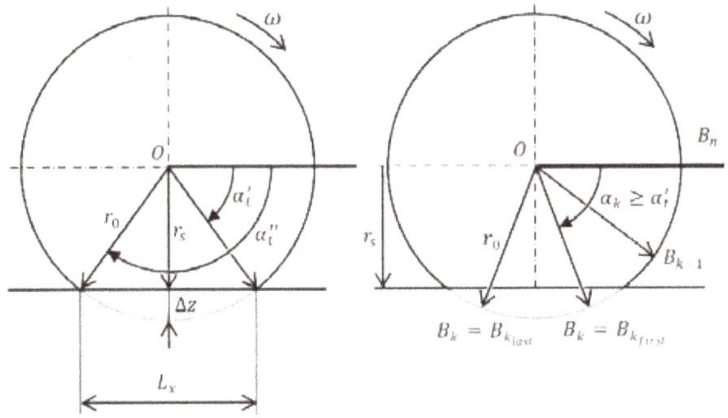

Figure 6. Configuration at the instant $i = 0$, $t = 0$ (note: the deflection of the tire is here exaggerated for the case of normal pressure).

In order to find the indexes k of all points B_k in contact with the road, to begin with the indexes of the first and last contact points have to be determined using the following short algorithms.

Determining the index of the first tread block in the contact patch k_{first}—see *Code 2* in Appendix A.

Determining the index of the last tread block in the contact patch k_{last}—see *Code 3* in Appendix A.

Now the coordinates of the points forming the striations of the yaw mark on the road in $\{I\}$ can be calculated. Figure 7 shows a simplified diagram of the lateral deflection of a tire while driving on a curvilinear track.

In fact, the deflection of the tire in the top view has a slightly more complex shape, but taking into account the assumptions made earlier, only the outer arc-shaped contour of the contact patch will be relevant (see Figure 4a,b in [10]). Thus, without making a significant error, it can be assumed that it is a fragment of a circle with a radius ρ determined by chord y_e and middle coordinate $2r_0$ from the formula:

$$\rho = \frac{1}{2}\left(|y_e| + \frac{r_0^2}{|y_e|}\right). \tag{21}$$

In Figure 7 the distances d_x and d_y lying on the road plane are indicated, which are distances from the tire-road contact point Q to the point B_k measured along and across the wheel, respectively. The first one is:

$$d_x = \frac{r_S}{\tan \alpha_k} \tag{22}$$

and the other

$$d_y = \frac{L_y}{2} + y_e - \delta. \tag{23}$$

Since

$$\delta = \sqrt{\rho^2 - d_x^2} - \rho + y_e\,, \tag{24}$$

hence the formula (23) takes the form:

$$d_y = \frac{L_y}{2} - \sqrt{\rho^2 - d_x^2} + \rho\,. \tag{25}$$

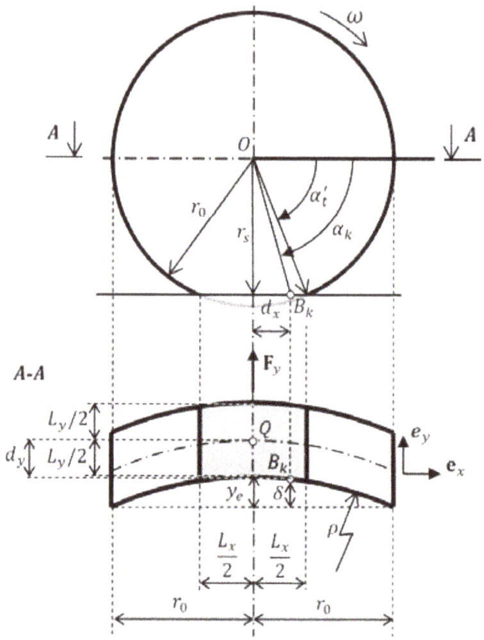

Figure 7. A simplified diagram of the tire lateral deflection while driving on a curvilinear track.

Finally, the position of the point B_k is:

$$\mathbf{r}_{IB_k,I} = \mathbf{r}_{IQ,I} + d_x\mathbf{e}_{x,I} - d_y\mathbf{e}_{y,I}\,, \tag{26}$$

where: $\mathbf{e}_{x,I}$, $\mathbf{e}_{y,I}$ are the unit vectors of the wheel in the point Q, shown in Figures 2a and 7.

The algorithm for calculation of the vector $\mathbf{r}_{IB_k,I}$ components is outlined in the section *Code 4* in Appendix A.

The next step of the simulation will be:

- Incrementation of the index indicating the simulation step $i = i + 1$;
- Time incrementation $t_i = t_{i-1} + h$;
- Calculation of the new angular position of the point B_0;

$$\alpha_{0_i} = \alpha_{0_{i-1}} + \omega_i h \text{ [rad]},$$

where ω_i is the current angular velocity of the wheel about its spin in [rad];

- Checking if the wheel has not already made a full rotation—See *Code 5* in Appendix A.
- Repeating the calculations from the formula (14)—strictly from the function Angular_position_of_block() (section *Code 1* in Appendix A)—to formula (26);
- Archiving the coordinates of the patch points drawing the striations.

A single striation is drawn by connecting the positions of the point B_{k_i} with lines in successive, adjacent steps i, as long as the condition (18) is satisfied. The striated yaw mark can be being drawn as the simulation proceeds or exported at the end as a drawing file (e.g., dxf).

6. Validation and Discussion

6.1. Stage 1. Measurement of Time Histories of Vehicle Motion Parameters (Actual Data)

The model was validated using the results of one of the full scale vehicle yaw tests performed as part of the Research Project No. VII/W-2014 of the Institute of Forensic

Research [14]. The test vehicle was a 2003 Volkswagen Passat 2.0 TD station wagon, with Firestone FireHawk 195/65R15 91T tires with their normal inflation pressure.

Experiments were performed in the summer, on a level, horizontal and dry asphalt road surface. In the test selected for validation, having established the speed of 53 km/h in a straight line the test driver steered the vehicle hard to the left causing the tires to break the grip on the roadway and to deposit striated yaw marks.

As a result of kinematic transformations of the raw measurement data the following parameters were obtained:

- A set of kinematic data fully describing the spatial movement of the vehicle, and, first of all, time histories of three components of the CG position vector and quasi-Euler body angles (transformations—see [23]);
- Time histories of the actual longitudinal s_x and lateral s_y slip of the front right wheel (transformations—see Appendix of [14]).

For example, assuming the notations as in Figure 8, the absolute velocity of the wheel center O, expressed in the local, rim fixed reference frame $\{O\}$, can be calculated from the vector equation:

$$\mathbf{v}_{IO,O} = \mathbf{A}_{OA}\mathbf{A}_{IA}^{\mathrm{T}}\mathbf{v}_{IO,I} = \mathbf{A}_{OA}\left[\mathbf{A}_{IA}^{\mathrm{T}}\mathbf{v}_{IA,I} - (\boldsymbol{\omega}_{IA,A} \times \mathbf{r}_{OA,A})\right], \quad (27)$$

where:

- $\mathbf{A}_{IA}$—rotation matrix from the local, body fixed reference frame $\{A\}$ to the global reference frame $\{I\}$;
- $\mathbf{A}_{OA}$—rotation matrix from from $\{A\}$ to $\{O\}$;
- $\mathbf{v}_{IO,I}$ —absolute velocity of the wheel center O in $\{I\}$;
- $\mathbf{v}_{IA,I}$—velocity of the GPS receiver mounting point A in $\{I\}$;
- $\boldsymbol{\omega}_{IA,A}$—angular velocity of the body in $\{A\}$;
- $\mathbf{r}_{OA,A}$—position vector from point O to A in $\{A\}$.

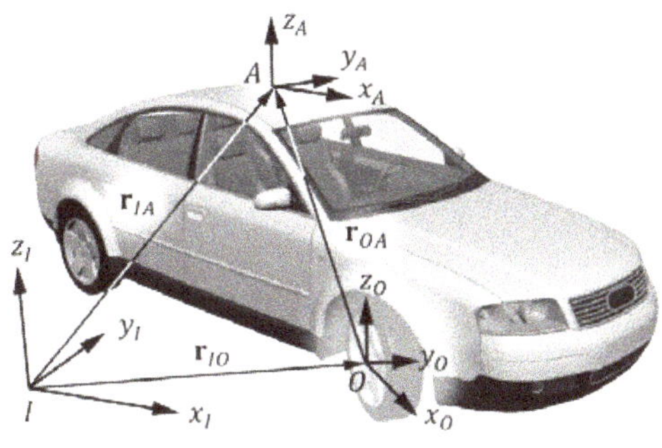

Figure 8. Coordinate systems: $\{I\}$—global, $\{A\}$—local vehicle-fixed (GPS receiver and/or IMU location), $\{O\}$—local rim fixed.

Thus, all parameters describing the movement of the vehicle (including all points, especially wheels) in the domains of time and space are known.

6.2. Stage 2. Vehicle Movement Simulation and Its Verification

The results of the measurements described in the Stage 1 were used to verify the yaw marks creation model, which is the very core of this article. It was implemented by the author-developed multibody dynamics simulation program Model.exe, briefly outlined in the Appendix of [24]. It was assumed that the vehicle is a multibody system with 36 degrees of freedom (DoF), composed of rigid bodies connected by geometric constraints, divided into the following partial-systems: basic—body with wheel suspensions, steering and drive. The equations of motion were generated using the Jourdain's principle. The tire forces were determined with the TMeasy model described in [17,18].

The vehicle motion was simulated, the model of which was parametrized according to the basic measurements data listed in Table 1. In simulation the measured time histories of the parameters shown in Figure 9 were adopted:

(a) steering wheel angle $\delta_H(t)$;
(b) external torques $M_{G1}(t) = M_{G2}(t)$ acting on the front wheels (left and right respectively) relative to their self-rotation axes.

Table 1. Technical data of the tested Volkswagen Passat

Parameter	Value	
Vehicle mass distribution on wheels:		
left front	465	kg
right front	475	kg
left rear	345	kg
right rear	350	kg
Distance of CG from front axle	1.148	m
CG height	0.55	m
Wheelbase	2.703	m
Steering system ratio	16:1	

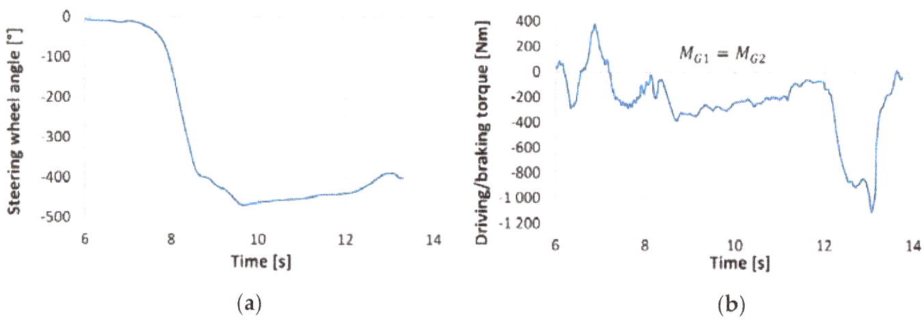

Figure 9. Inputs realized in simulation: (**a**) steering wheel angle; (**b**) torque acting on the front wheels.

The simulation results represented in Figure 10a,b of the time histories of various dynamic parameters, show good agreement with the actual data. In Figure 10c the simulated and actual positions and orientations of the vehicle have been compared. For as long as approximately 12 s they are almost identical, and some divergence of the CG paths occurs only at the very end of the movement. The differences may arise from the approximate torque waveform shown in Figure 9b, and partly from the simplified parametrization of the steering and drive systems and the neglected pavement unevenness.

In Figure 10a, a large discrepancy can be observed for the slip s_x of the front right wheel at the end of the simulation. The relative longitudinal slip s_x expressed by the formula (6) is the absolute slip (defined by the numerator) related to the speed v_x of the wheel center (denominator). Although the absolute slip within the entire time range 6–13 s is small, the low speed just before the vehicle stops (12–13 s) caused a sharp and disproportionate increase in the value of the relative slip s_x. In other words, the reason for the sudden increase in the measured s_x are low values of v_x and $r_d\omega$ together with inaccuracies in the independent measurement of this parameters.

Simulation, as a theoretical process, is free of measurement flaws, hence even small numbers are precise and correlated enough that dividing them gives reasonable results. That is why at the end of the simulation, at low speeds, a disturbing difference between the measured and simulated slips s_x occurred. This issue should not be overestimated.

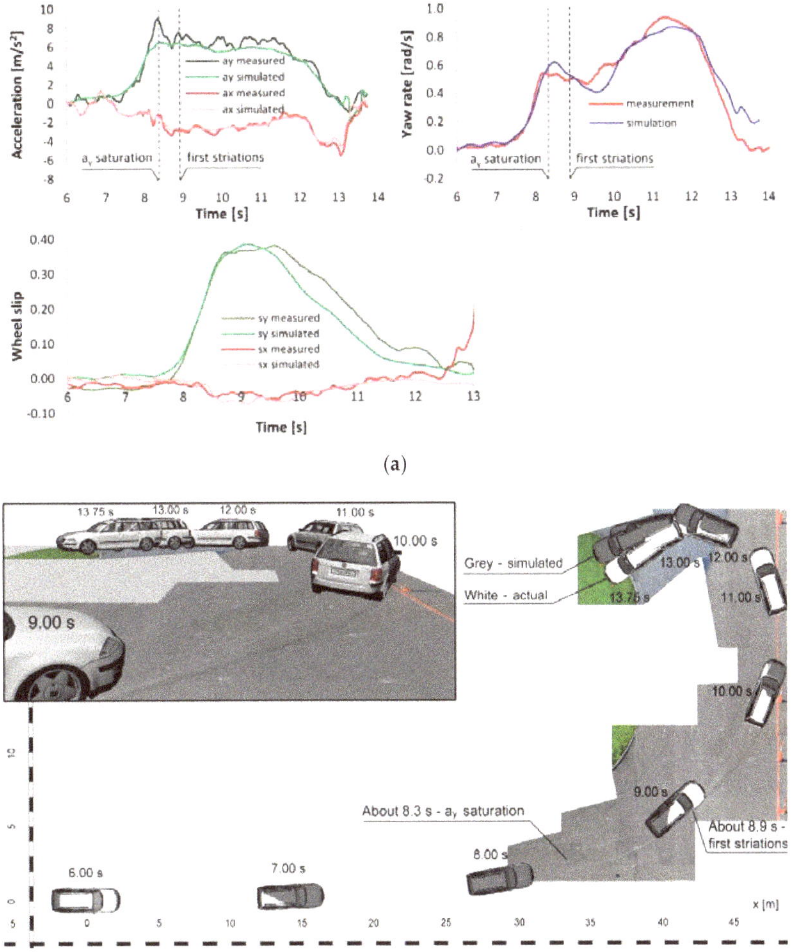

(a)

(b)

Figure 10. Comparison of the simulation and measurement results: (**a**) time histories of the longitudinal and lateral accelerations, yaw rate and slips; (**b**) vehicle positions and orientations in top view and in perspective.

6.3. Stage 3. Verification of the Striation Creation Model

In Figure 11 the striated yaw marks generated in the simulation have been superimposed on the actual ones (note that on the orthophotomap you can also see irrelevant marks from earlier runs). There is a strong convergence in both the paths of the marks and the direction of the striations, as shown in the close-ups of the time points $t = 8.9$ s, $t = 9.8$ s and $t = 10.9$ s. At $t = 8.9$ s, the virtual mark of the front right wheel is slightly shifted to the right of the actual mark, but this is due to fact that the authors refrained from fine-tuning the simulation indefinitely.

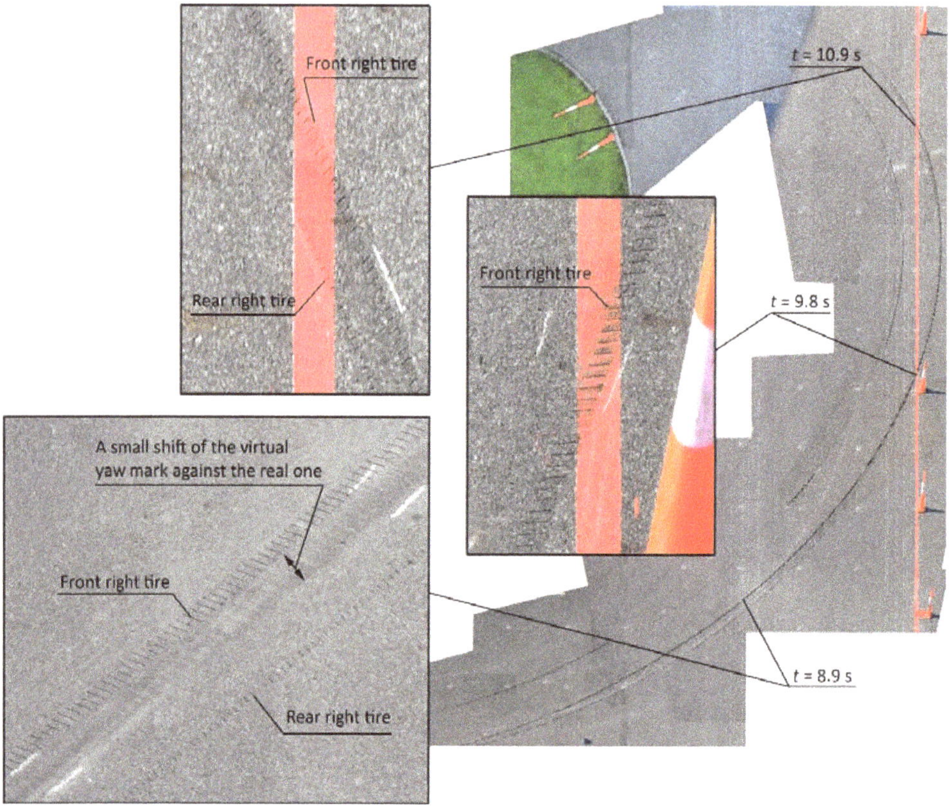

Figure 11. Comparison of striated yaw marks generated in simulation with actual ones (top view).

6.4. Example

Figure 12 shows yaw marks generated in the simulation of a severe step steer in a left turn maneuver resulting in breaking the adhesion of the tires on the roadway and the vehicle yawing. Basically, this maneuver is similar to that of Section 6, but what is especially interesting here are typical yaw mark characteristics: variation of the width depending on the vehicle orientation with respect to the CG velocity direction, change of the striation angle with respect to the tangent to the yaw mark (straightening), interweaving and fading.

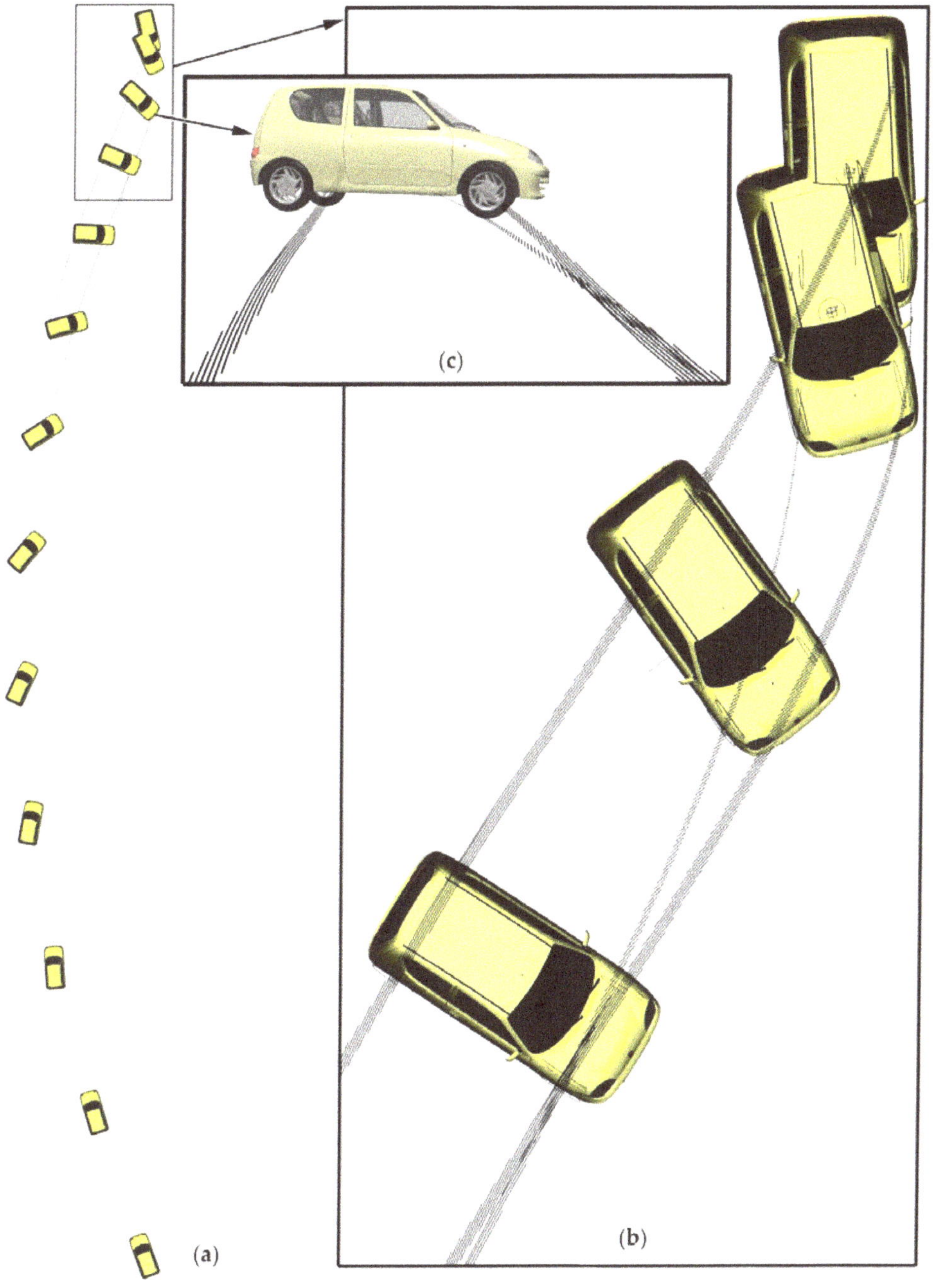

Figure 12. Yaw marks: (**a**) entire movement; (**b**) close-up of the end of yaw marks; (**c**) 3D view.

Remarks. The yaw mark creation model will still be valid on arbitrary 3D roads and typical road irregularities. Generally, bumps or potholes should result in gaps in the created yaw marks, but this problem has yet to be investigated.

In case of uncommon suspension systems and/or different tire types, e.g., forklift trucks [25,26] or other uncommon vehicles [27], the response of the unsprung masses is not expected to affect the striated yaw mark creation model effectiveness as long as the vehicle is moving along a non-deformable surface and the road wavelength is not larger than the tire circumference.

7. Conclusions

1. This work is, in some ways, a next step of the research of Beauchamp et al.—especially [13]. While they analyze striated yaw marks from the viewpoint of a reconstructionist who would like to know "what one can get from the marks uncovered at the accident scene" (deriving equations for calculation of longitudinal slip s_x as a function of the striation marks angle θ and the slip angle α), this work focuses on development of a model to create striated tire marks, which can be used in programs for simulation of vehicle accidents. While their formulas apply to planar mechanics and kinematics, this model takes into account the spatial vehicle multi-body dynamics and tire model, including the relative movement of the wheel, its position and orientation, tire deformation, suspension kinematics as well as forces and torques acting on the wheel. A mathematical model of striated tire yaw mark creation has been developed, intended for programs for the simulation of vehicle accidents. It implements the main features of the topology of such marks.
2. The application of the model is twofold. The first one is the simulation of the formation of striated yaw marks. The second one is to facilitate the understanding of the mechanics of the formation of such marks and the inference of the braking or acceleration state of the wheels based on the topology of the striations.
3. The mark creation model is universal, i.e., it applies to a tire moving along any trajectory with variable curvature, and it is subjected to any forces and torques calculated by solving a system of non-linear equations of vehicle dynamics.
4. The striated yaw mark is created by the tread shoulder blocks forming the outer border of the contact patch, as this area is actually dominant.
5. It is possible to define any, including non-uniform, pitch of the tread shoulder blocks.
6. The striated yaw mark creation model can be applied in programs for the simulation of vehicle dynamics with any degree of simplification and any tire model. In this paper, it was validated using the author-developed program Model.exe, in which the vehicle is represented by a 36 degree of freedom multi-body system with the TMeasy tire model.

Author Contributions: Conceptualization, W.W.; formal analysis, W.W.; methodology, W.W.; software W.W.; validation, W.W.; visualization, W.W; writing—original draft preparation, W.W.; resources, J.Z.; review and editing, W.W and J.Z. All authors have read and agreed to the published version of the manuscript.

Funding: This research received no external funding.

Institutional Review Board Statement: Not applicable.

Informed Consent Statement: Not applicable.

Data Availability Statement: Not applicable

Acknowledgments: This work was supported by the Institute of Forensic Research in Krakow—Project No. I/W/2017.

Conflicts of Interest: The authors declare no conflict of interest.

Appendix A

Code 1

```
// Start of simulation
int n; // Number of elements
```

```
int k; // Index
double t = 0; // Actual time of simulation
double α_0 = 0; // Initial angular position of element 0
void CTire::Angular_position_of_block()
{
   from k = 1 to n
   {
      α_k = α_{k-1} + d_k/r_0 [rad]; // Table α[k]; see equation (14)
      k = k + 1;
   }
}
```

Code 2

```
int q = n − 1; // index
from k = 0 to n − 1
{
   if (α_k ≥ α'_t ) & (α_q < α'_t )
   {
      k_first = k;
      exit;
   }
   else
   {
      q = k;
      k = k + 1;
   }
}
```

Code 3

```
int q = n − 1;
from k = 0 to n − 1
{
   if (α_k > α''_t )&(α_q ≤ α''_t )
   {
      k_last = q;
      exit;
   }
   else
   {
      q = k;
      k = k + 1;
   }
}
```

Code 4

```
from k = k_first to k_last
{
   // Determining the position of the tire-road contact r_IQ,I from equation (3):
   r_IQ_I();
   // Determining the point B_k position, along the tire longitudinal direction
   // (see Figure 6):
   if (tan α_k == π/2 )
      d_x = 0;
   else
```

$d_x = \frac{r_s}{\tan \alpha_k}$; // see (22)
$\rho = \frac{1}{2}\left(|y_e| + \frac{r_0^2}{|y_e|}\right)$; // see (21)
$d_y = \frac{L_y}{2} - \sqrt{\rho^2 - d_x^2} + \rho$; // see (25)
if $F_y > 0$
  $\mathbf{r}_{IB_k,I} = \mathbf{r}_{IQ,I} + d_x\mathbf{e}_{x,I} - d_y\mathbf{e}_{y,I}$; // left turn—the yaw mark is drawn
    // by the right tire shoulder; see (26)
else
  $\mathbf{r}_{IB_k,I} = \mathbf{r}_{IQ,I} + d_x\mathbf{e}_{x,I} + d_y\mathbf{e}_{y,I}$; // right turn—the yaw mark is drawn
    // by the left tire shoulder
}

Code 5
{
  $\alpha_{0_i} = \alpha_{0_i} + \Omega_i h$; // [rad]
  if $\alpha_{0_i} \geq 2\pi$
    $\alpha_{0_i} = \alpha_{0_i} - 2\pi$; // [rad]
  Angular_position_of_block();
};

References

1. Brach, R.M.; Brach, M. *Vehicle Accident Analysis and Reconstruction Methods*, 2nd ed.; SAE International: Pittsburgh, PA, USA, 2011; pp. 79–83.
2. Sledge, N.H., Jr.; Marshek, K.M. *Formulas for Estimating Vehicle Critical Speed from Yaw Marks—A review*; SAE Technical Paper 971147; SAE International: Pittsburgh, PA, USA, 1997. [CrossRef]
3. Cannon, J.W. *A Study of Errors in Yaw-Based Speed Estimates Due to Effective Braking*; SAE Technical Paper 2003-01-0888; SAE International: Pittsburgh, PA, USA, 2003. [CrossRef]
4. Cliff, W.E.; Lawrence, J.M.; Heinrichs, B.E.; Fricker, T.R. *Yaw Testing of an Instrumented Vehicle with and without Braking*; SAE Technical Paper 2004-01-1187; SAE International: Pittsburgh, PA, USA, 2004. [CrossRef]
5. Amirault, G.; MacInnis, S. *Variability of Yaw Calculations from Field Testing*; SAE Technical Paper 2009-01-0103; SAE International: Pittsburgh, PA, USA, 2009. [CrossRef]
6. Lambourn, R.F. The calculation of motor car speeds from curved tyre marks. *J. For. Sci. Soc.* **1989**, *29*, 371–386. [CrossRef]
7. Lambourn, R.F. *Braking and Cornering Effects with and without Anti-Lock Brakes*; SAE Technical Paper 940723; SAE International: Pittsburgh, PA, USA, 1994. [CrossRef]
8. Lambourn, R.F.; Jennings, P.; Knight, I. Critical speed yaw mark calculations with and without electronic stability control. In Proceedings of the 1st joint ITAI–EVU Conference (18th EVU and 9th ITAI Conference), Hinckley, UK, 25–27 September 2009; ITAI–EVU: Hinkley, UK, 2009; pp. 209–235.
9. Cash, S.; Crouch, M. Calculating vehicle speed from yaw mark analysis. In Proceedings of the 24th EVU Conference, Edinburgh, UK, 15–17 October 2015; EVU: Edinburgh, UK, 2015; pp. 185–206.
10. Yamazaki, S.; Akasaka, T. Buckling behavior in contact area of radial tire structure and skid marks left by tires. *JSAE Rev.* **1998**, *9*, 51–55.
11. Yamazaki, S. Friction coefficients of tires and skid marks left by tires. In Proceedings of the International Workshop on Traffic Accident Reconstruction, Tokyo, Japan, 12–13 November 1998; National Research Institute of Police Science: Chiba, Japan, 1998; pp. 41–51.
12. Beauchamp, G.; Pentecost, D.; Koch, D.; Rose, N. The relationship between tire mark striations and tire forces. *SAE Int. J. Trans. Saf.* **2016**, *4*, 134–150. [CrossRef]
13. Beauchamp, G.; Hessel, D.; Rose, N.A.; Fenton, S.J.; Voitel, T. Determining vehicle steering and braking from yaw mark striations. *SAE Int. J. Passeng. Cars Mech. Syst.* **2009**, *2*, 91–307. [CrossRef]
14. Zębala, J.; Wach, W.; Ciępka, P.; Janczur, R. *Determination of Critical Speed, Slip Angle and Longitudinal Wheel Slip Based on Yaw Marks Left by a Wheel with Zero Tire Pressure*; SAE Technical Paper No. 2016-01-1480; SAE International: Pittsburgh, PA, USA, 2016. [CrossRef]
15. Beauchamp, G.; Thornton, D.; Bortles, W.; Rose, N. Tire mark striations: Sensitivity and uncertainty analysis. *SAE Int. J. Trans. Saf.* **2016**, *4*, 121–127. [CrossRef]
16. Guzek, M.; Lozia, Z. Computing methods in the analysis of road accident reconstruction. *Arch. Comp. Meth. Eng.* **2020**. [CrossRef]
17. Hirschberg, W.; Rill, G.; Weinfurter, H. Tire model TMeasy. *Veh. Syst. Dyn.* **2007**, *45*, 101–119. [CrossRef]
18. Rill, G. *Simulation von Kraftfahrzeugen*; Vieweg & Sohn Verlag: Braunschweig/Wiesbaden, Germany, 1994.
19. Pacejka, H.B. *Tire and Vehicle Dynamics*, 2nd ed.; SAE: Pittsburgh, PA, USA, 2006.

20. Dugoff, H.; Fancher, P.S.; Segel, L. *An Analysis of Tire Traction Properties and Their Influence on Vehicle Dynamic Performance*; SAE Technical Paper No. 700377; SAE International: Pittsburgh, PA, USA, 1970. [CrossRef]
21. Rill, G. *Road Vehicle Dynamics: Fundamentals and Modeling*; CRC Press: New York, NY, USA, 2012.
22. Mitschke, M. *Dynamik der Kraftfahrzeuge, Band A: Antrieb und Bremsung*; Springer: Berlin/Heidelberg, Germany, 1995.
23. Wach, W. Reconstruction of Vehicle Kinematics by Transformations of Raw Measurement Data. In Proceedings of the XI International Scientific and Technical Conference Automotive Safety, Častá, Slovakia, 18–20 April 2018. [CrossRef]
24. Wach, W.; Struski, J. Rear wheels multi-link suspension synthesis with the application of a 'Virtual Mechanism'. In *SAE Special Publication SP-2019 Steering & Suspension Technology and Tire & Wheel Technology 2006*; SAE Technical Paper No. 2006-01-1376; SAE International: Pittsburgh, PA, USA, 2006. [CrossRef]
25. Martini, A.; Bonelli, G.P.; Rivola, A. Virtual testing of counterbalance forklift trucks: Implementation and experimental validation of a numerical multibody model. *Machines* **2020**, *8*, 26. [CrossRef]
26. Rebelle, J.; Mistrot, P.; Poirot, R. Development and validation of a numerical model for predicting forklift truck tip-over. *Veh. Syst. Dyn.* **2009**, *47*, 771–804. [CrossRef]
27. Odabaşı, V.; Maglio, S.; Martini, A.; Sorrentino, S. Static stress analysis of suspension systems for a solar-powered car. *FME Trans.* **2019**, *47*, 70–75. [CrossRef]

energies

Article

Measures of Functional Reliability of Two-Lane Highways

Krzysztof Ostrowski [1,*] and Marcin Budzynski [2]

[1] The Faculty of Civil Engineering, Cracow University of Technology, Warszawska 24 Street, 31-155 Cracow, Poland
[2] The Faculty of Civil and Environmental Engineering, Gdansk University of Technology, Narutowicza 11 Street, 80-233 Gdansk, Poland; mbudz@pg.edu.pl
* Correspondence: kostrowski@pk.edu.pl; Tel.: +48-604551175

Abstract: Rural two-lane highways are the most common road type both in Poland and globally. In terms of kilometres, their length is by far greater than that of motorways and expressways. They are roads of one carriageway for each direction, which makes the overtaking of slower vehicles possible only when there is a gap in the stream of traffic moving from the opposite direction. Motorways and express roads are dual carriageways that are expected to support high speed travel mainly over long distances. Express roads have somewhat lower technical parameters and a lower speed limit than motorways. Two-lane highways are used for both short- and long-distance travel. The paper presents selected studies conducted in Poland in 2016–2018 on rural two-lane highways and focuses on the context of the need for their reliability. The research was carried out on selected short and longer road sections located in various surroundings, grouped in terms of curvature change rate CCR, longitudinal slopes and cross-sections (width of lanes and shoulders). The studies of traffic volumes, travel time and travel speed, as well as traffic density, will be used to analyze traffic performance and identify measures of travel time reliability. The analyzed roads were characterized by good technical parameters and significant variability of traffic volume throughout the day, week and year. Some roads experience congestion, i.e., situations in which traffic volume Q is close to or above respective road capacity C. In order to determine the form of the suitable reliability measures, it will be important to determine the extent to which a road's geometric and traffic characteristics impact travel speed and time. The paper presents well-known reliability measures for dual carriageways and proposes new measures, along with an evaluation of their usefulness in the assessment of the functioning of two-lane highways.

Keywords: reliability measures; two-lane highways; travel speed; travel time; empirical research

Citation: Ostrowski, K.; Budzynski, M. Measures of Functional Reliability of Two-Lane Highways. *Energies* **2021**, *14*, 4577. https://doi.org/10.3390/en14154577

Academic Editor: Stefania Santini

Received: 31 May 2021
Accepted: 16 July 2021
Published: 28 July 2021

1. Introduction

Reliability is a major criterion for assessing selected elements of technical infrastructure such as transmission [1], information technology (IT) [2] or energy [3] infrastructure. The reliability of road infrastructure is also the subject of many studies, because of the role the parameter plays in traffic performance [4,5] and the safety of road users [6–8]. In the case of road safety, speed tests and testing of speed's impact on road safety measures are very important. In the tests [9] floating car data were used to achieve the goal. Ensuring the reliability of road infrastructure at a level acceptable to road users is a key aspect of planning and design decisions [10,11].

There are not many reliability analyses of two-way highways in scientific literature. Instead, researchers focus mainly on dual carriageways, i.e., motorways [12–14], expressways [15–18] or other dual carriageways [19,20], inter alia, analyzing the impact of Intelligent Transport System (ITS) solutions [21–23]. In simulation analyses and field research [24], the impact of selected parameters on the level of service (LOS) under heterogeneous traffic conditions for a two-lane highway was identified. The work [25] also analyzed (LOS) on the basis of estimation of passenger car unit values. The research [26] also pointed out the variability of traffic flow on individual road lanes. Some studies concern themselves with sections of motorways and expressways in urban areas [27–29]. Obtaining reliable data is an extremely important aspect of reliability studies. The work [30] presents various techniques for examining road traffic parameters. It compares the pneumatic tube detector method, video capturing method, moving observer method and the classic manual method. The studies [31] indicate the effectiveness of combining the moving observer method and digital image processing. The work [32] presents the effects of using stationary devices along the road to collect road traffic data. The research [33] provides an example of an effective use of video traffic monitoring. Modern techniques allow the use of Bluetooth technology to collect data on traffic parameters [34–36] and Lidar technology to collect data on road and its surroundings parameters [37–39].

Road traffic parameters depend on many factors, including the driver's psychophysiological characteristics, road and meteorological conditions [40]. A very important aspect influencing traffic conditions is constituted by road geometry, including the parameters of horizontal curves [41]. One factor related to driver behaviour is the distance between vehicles [42].

An example of research conducted on two-lane highways is provided by reliability analyses carried out on Poland's road network [43]. These studies were undertaken on higher standard roads managed by the General Directorate for National Roads and Motorways, with speed limits of 70–90 km/h, and a typical lane width of 3.5 m (with or without a hard shoulder). In Poland, these roads account for over 86% of all national roads, including motorways, expressways and accelerated main roads. According to the standards specified in the American method [44], these are first-class roads on which drivers expect travel speeds close to the speed limit. In Germany, a similar approach applies to roads with a similar function marked as EKLII and EKLIII [45].

The project [43] and work [14] also present studies on dual-carriageways, i.e., motorways, expressways and roads of lower technical class, on which speed limits in Poland are 140 km/h, 120 km/h and 100 km/h respectively. In Poland, motorways are roads of the highest technical standard, where traffic can be joined only through interchanges. In the case of expressways and other dual carriageways of the lower technical class, traffic can be joined through interchanges or through intersections (usually signalised).

The analyzed two-lane highways mainly support traffic functions typical of roads of higher technical classes, although they have a limited capacity (max. 3200 veh/h according to [44], approx. 2600 veh/h according to [45]) compared to high-speed roads (highways, expressways). Reliability, measured in terms of travel speed or time, is highly variable on the analyzed roads and depends on the time of day, day of the week or month. It also varies in the longer term (analysis by year). Therefore, it is necessary to identify the most important factors that influence their reliability levels and to indicate the best reference level for analyses conducted on two-lane highways.

The main aim of the analyses presented in the paper is to answer the research questions:

- Whether and to what extent selected traffic parameters impact the functional reliability measures of single carriageways and two-lane highways?
- Whether the measures and reference values for dual carriageways can be transplanted directly onto analyses of two-lane highways? An indirect aim pointing the directions of further research work revolves around answering the question of
- Whether the statistical parameters describing travel time variability are sufficient to analyze and assess the reliability of a road section in a probabilistic approach that takes into account the risk of the occurrence of road incidents happening during travel speeds exceeding the speed limit?

The answers will provide the foundation for an effective transformation of the existing road network, enabling the attainment of a standard of travel that will be acceptable to road users.

The paper is divided into several parts which include a review of literature providing a description of the reliability measures used, selected results of empirical studies conducted on Poland's two-lane highways, and reliability studies conducted on a selected road section where use is made of GPS data. At the end of the paper, conclusions are drawn and directions for further work are stated.

2. Materials and Methods

2.1. Reliability Measures

Travel time reliability depends on a benchmark and therefore has no fixed value. Its value is influenced by a number of factors of various origins [46,47], including traffic factors (traffic intensity, types of vehicles), geometry, road's location and type of surroundings, the knowledge of which is necessary in order to identify the reliability process, interactions between the variables and the correct interpretation of the results.

General factors influencing road reliability include:

- The presence of traffic control–traffic signals, including in particular incorrectly designed control parameters, rail-road crossings,
- Daily, weekly or seasonal fluctuations in traffic,
- Occasional events—various types of events making the traffic flow value different from the typical values of the flow on this road (religious, public holidays, days off, etc.),
- Road capacity—dependent on road geometry and a number of other factors e.g., the technical condition of road surface,
- Weather conditions, in particular snowfall, heavy or prolonged rainfall, fog,
- Road accidents and other road incidents blocking passing vehicles,
- Road works resulting in a taper of the road's cross-section, alternating traffic, temporary road blockage.

Considering the above division, it is possible to introduce a classification [48] that assigns the indicated factors to three different groups (Figure 1) on account of:

- Infrastructure, i.e., a road's geometry and its standard, including the road's curvature change rate CCR [49,50], longitudinal slope [51,52] and width [53,54] and traffic organization, including road works [55–57], temporary and permanent taper of the road's cross-section, and the presence of traffic lights [58,59],
- Road traffic, including traffic volume [60,61], its generic and directional structure [62,63] as well as road incidents [64,65],
- Road surroundings [66,67] and weather conditions [68–70].

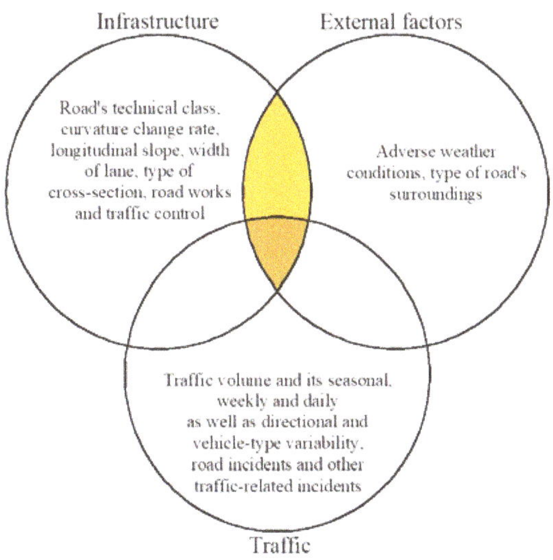

Figure 1. Interactions between sources of failure.

The diagram in Figure 1 shows interactions between the main sources of failure and the relationship between demand (traffic) and supply (infrastructure). The authors of paper [71] point out that the indicated interactions are the main determinants of road functionality. Both demand and supply vary over time, as both traffic and road capacity are influenced by various deterministic and random factors. Weather conditions, especially adverse ones, such as prolonged rainfall, snowfall, etc., have a significant impact on drivers' behaviour. The research [72–74] shows that road and intersection capacity decreases in such conditions by as much as 20%. The type of road surroundings resulting from the road's location often translates into the type of trips [75], which also determines drivers' behaviour. In cities and agglomerations where short trips, mostly related to commuting to work, shops, schools, and other facilities, predominate, the behaviour and expectations of vehicle drivers regarding traffic conditions and network reliability are completely different from such expectations outside cities, or during long-distance travel [O5] spread over a longer period of time. In the worst case scenario, all of these variables may affect travel time reliability. This situation occurs in the common sections of all three circles.

Over the past few decades, many studies have been conducted in the USA on existing roads to describe the reliability of travel times. In the research into and evaluation of reliability, generally available models were used, their modifications were created or completely new solutions were developed. Table 1 [76] shows an example of the application of reliability measures in practice, i.e., it lists selected US transport agencies and identifies indicators used by them to describe the functional reliability of roads.

Current methods of analysis [61] can be divided into:

- Statistical methods,
- Buffer time methods,
- Late travel indicators,
- Probabilistic methods,
- Skewness methods (treated as part of statistical methods in the paper).

Table 1. Reliability metrics used by selected US transport agencies.

Agency	Reliability Metrics Used
Georgia Regional Transportation Authority and Georgia DOT	Buffer Index
	Planning Time Index
Southern California Association of Governments	On-Time Index
	Buffer Index
Washington State DOT	95th Percentile Travel Time
National Transportation Operations Colation (NTOC)	Buffer Index
Maryland SHA	Travel Time Index
	Planning Time Index

Below, the authors present selected methods of analysis, including their advantages and disadvantages.

2.1.1. Statistical Methods

One of the oldest approaches to the description of travel time reliability used by Abishai Polus in 1979 [77] was based on a simple measure—standard deviation δ, showing the variability of the metric's value in relation to its mean value (Equation (1)). The author was one of the first to indicate the travel time variable as the best measure with which to describe a road's functional reliability.

$$\delta = \sqrt{\frac{1}{n-1} \cdot \sum_{i=1}^{n} (t_i - \bar{t})^2} \quad (1)$$

where:

n—number of travels,
t_i—i-travel time,
$\bar{t}$—average travel time.

The author defined a road's functional reliability using a measure of variability, i.e., travel time variance [78]. The higher the variance, the less reliable the road (Equation (2)).

$$R = \frac{1}{[Var(x)]^{\frac{1}{2}}} = \frac{1}{[E(x^2) - (E(x))^2]^{\frac{1}{2}}} \tag{2}$$

where: R—road's functional reliability, x—reliability measure (in this case–travel time), $Var(x)$—variance of reliability measure, $E(x)$—expected value of reliability measure.

The simplicity of the approach accounts for its advantage, while its low usefulness is a disadvantage because in most cases the empirical distributions describing the variability of travel time are not symmetrical and show considerable skewness. However, this did not prevent the development of these methods, and in subsequent editions, recommendations for reliability were developed based on the ranges of skewness as presented in publications [12,13,18]. The value of the standard deviation was used to build subsequent measures, such as the time window (Equation (3)) and the coefficient of variability (Equation (4)).

$$Time\ window = \bar{t} \pm \delta \tag{3}$$

$$Coefficient\ of\ variability = \frac{\delta}{\bar{t}} \cdot 100\% \tag{4}$$

where: $\bar{t}$— arithmetic mean of all travel times, δ—standard deviation of travel time.

The time window can have two values—one lower and another greater than the arithmetic mean by the value $\pm\delta$. The road user receives information about possible travel time discrepancies. In order to increase the scope of analyses, the δ may be multiplied. The travel time variation coefficient can be used to compare travel time variability between days, weeks, or road sections.

A measure that allows the comparison of traffic conditions between peak and off-peak times is the index of variation (Equation (5)).

$$Index\ of\ gualitative\ variation = \frac{V_{1,+95} - V_{1,-95}}{V_{0,+95} - V_{0,-95}} \tag{5}$$

where: $V_{1,+95}$, $V_{1,-95}$—upper and lower values between which there are 95% of travel times during the peak traffic period, $V_{0,+95}$, $V_{0,-95}$—upper and lower values between which 95% of travel times lie outside the peak hours.

The index of qualitative variation may apply to roads close to urban agglomerations, where increased traffic may occur because of urban traffic peaks. Due to the larger discrepancy between the peak and off-peak traffic, the value of the index is often greater than 1.0.

2.1.2. Buffer Time Methods

The term "buffer time" means extra time added to a specific activity in order to ensure no delays and an optimum time to reach the destination [79]. In the analysis of travel time reliability, it is the additional amount of time needed for the road user to reach their destination at their desired hour with a probability of 95%, taking the arithmetic mean of all travel times [18] (Equation (6)) as the expected travel time.

$$BT = t_{95} - \bar{t} \tag{6}$$

where: t_{95}—95th percentile of travel time, $\bar{t}$—average travel time.

The 95th percentile of travel time represents the worst possible traffic situation. This means that users may experience issues resulting in travel time delays in 1 out of 20 travels. Other percentiles can also be used depending on the needs of the research, e.g., 90th or 85th percentile [80].

The buffer time index (Equation (7)) is a derivative metric. It is calculated as the ratio of the buffer time to the arithmetic mean. This variable value exceeds 1.0. For example, a value of 1.7 means that in 95% of cases it takes 70% more time than the average travel time to reach the destination at the expected time [81]. Travel time reliability decreases as the buffer time index increases.

$$BTI = \frac{BT}{\bar{t}} \quad (7)$$

where: BT—buffer time, $\bar{t}$—average travel time.

Instead of average travel time, it is also possible to use median travel time. When the analyzed road section is divided into several sub-sections due to traffic volume or other factors differentiating individual road sub-sections, then the weighted average buffer time index can be calculated (Equation (8)):

$$BTI_i = \frac{\sum_{i=1}^{n}(BTI_i \cdot Q_i \cdot L_i)}{\sum_{i=1}^{n} Q_i \cdot L_i} \quad (8)$$

where: BTI_i—buffer time index for an i sub-section of the road, Q_i—traffic volume on i sub-section of the road, L_i—length of i sub-section of the road, n—number of sub-sections.

The BT, BTI and ABTI measures can be used to assess the reliability of city street networks and routes leading to daily destinations. These measures could be incorporated into GPS navigation systems to assist drivers in route planning. Other measures derived from the buffer time method are described below.

2.1.3. Planning Time

Planning time is the total time needed to reach the destination at the scheduled time with a 95% probability. The measure is easy to interpret and allows the user to plan the trip correctly. Comparing the 95th percentile from different hours shows the variability of road functional reliability over the course of the day. The measure is also important for the road manager, who, by using the PTI value, can classify the roads under his management in terms of reliability, as well as plan the reconstruction of sensitive sections of a road to improve travel time.

Another measure derived from PT is the planning time index (PTI), which informs how many times more time the road user needs to reach the destination relative to the travel time in free-flow conditions allowing the driver full freedom in choosing the speed (Equation (9)).

$$PTI = \frac{t_{95}}{t_o} \quad (9)$$

where: t_{95}—95th percentile of travel time, t_o—travel time in conditions perceived as free flow (Equation (10)).

$$t_o = \frac{3600}{v_f} \cdot L \quad (10)$$

where: v_f—travel time in free-flow conditions (determined e.g., based on [43–45]), L—length of road section.

PTI values exceed 0. Existing literature [82] allows us to find a categorization of *PTI* values in terms of service reliability level developed for practical use by road users, as well as road managers. It is recommended that the thresholds be adapted to the nature of the road (Table 2) [67].

Table 2. Thresholds values of *PTI* for individual levels of functional reliability.

Reliability Performance	*PTI* (–)
good	≤1.3
fair	1.3 ÷ 2.0
poor	>2.0

PTI is one of the variants of TTI, or travel time index, which can be calculated for any percentile (Equation (11)) .

$$TTI = \frac{t_x}{t_o} \tag{11}$$

where: t_x—any percentile of travel time.

Literature also contains other measures relating to situations of failure, i.e., the failure measure or the misery index, which show how many times more time is needed to make the longest trips in relation to the travel time in free-flow conditions (equation 12).

$$Failure\ rate = \frac{\bar{t}_{5\%}}{t_o} \tag{12}$$

where: $\bar{t}_{5\%}$—average travel time in the group of the longest trips, i.e., the 5th percentile; the so-called misery time.

All of the above-mentioned measures can be used separately or in combination (Figure 2) [46]. The latter option is more advisable and universal because it shows the size of various reliability measures and their comparison. The use of selected measures, and at best all measures, allows for a broader view of the issue of reliability and for finding appropriate solutions to improve traffic conditions.

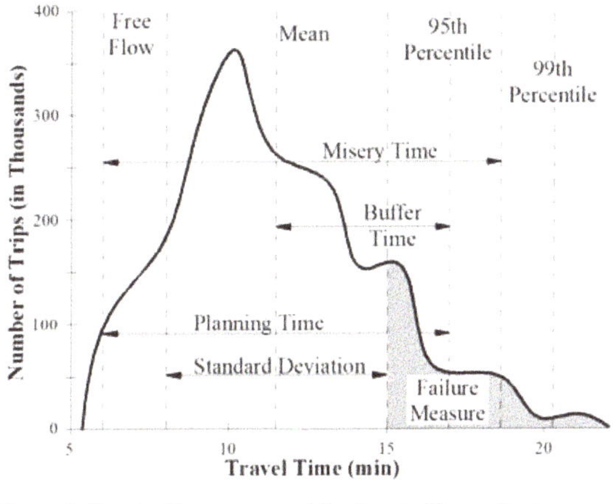

Figure 2. Chart with measures of the time buffer method.

In conducting research, it is important to select a suitable length of the road's section and research time to reflect the specific conditions on a given road and the objectives of the analyses. Time periods should be selected to reflect traffic conditions of a similar nature and intensity. It is recommended to conduct research and traffic observation on a continuous basis, using GPS devices and software. Morning and afternoon traffic peaks should not be combined in determining the 95th percentile [80].

Above the authors show selected more important measures used in the description of reliability of dual carriageways, where road features (road geometry) have a much smaller impact on driver behaviour, speed and travel time than on single carriageways, on which additionally vehicle overtaking occurs. The measures were subject to verification, analysis of variability and usefulness, and the possibility of adjusting their form to the analyses of two-lane highways. Traffic on single-carriageways is characterized by a significant variability of traffic intensity and speed over time and the presence of all types of vehicles on single lanes supporting traffic in two directions. Destinations are also more varied than on dual carriageways, which support mostly long-distance trips. Single-carriageways are used mainly in everyday travel, hence the assessment of their reliability is extremely important from the point of view of the average road user and road manager. Unfortunately, this type of road is omitted from the literature on reliability assessment. The next chapter will present selected results of travel speed studies conducted in the period 2016–2018 in Poland on two-lane highways [43], along with the indication of the extent to which statistically significant factors affect the travel speed and its variability relevant from the point of view of reliability analysis.

2.2. Travel Speed Research on Two-Lane Highways

This chapter presents selected results of research conducted as part of project [43] focused on the variability of travel speed and impact factors resulting from geometric, traffic and location features. The tests were carried out in favourable weather conditions, with very good visibility on selected road sections without intersections or other traffic disturbances.

When traffic volumes are not high, travel speed and thus travel time are significantly affected by the road's geometric features, such as road (CCR), radius and turning angle of the horizontal curve, longitudinal slopes, road width and type of cross-section, as well as traffic factors, including those related to the share of heavy vehicles in traffic. In view of the continuous changes in the car fleet, the roads' technical condition and geometry as well as drivers' behaviour, empirical research should be constantly updated, which was recently done and included in [43], resulting, among others, in a new regression model used to estimate travel speed on two-lane highways.

The tests were carried out in selected road cross-sections, at the beginning and end of the analysed section (speed and travel time were examined), and numerous trips behind the leader were made with continuous recording of data on the speed of moving vehicles (the leader).

The driving speed profile behind the leader was determined empirically using a GPS device. The influence of selected geometric features on vehicles' travel speed (free flow, no platoon traffic) is shown in Figure 3. The diagram shows that apart from the presence of horizontal curves, speed is also significantly affected by the length of the sections preceding the curves. It is assumed that a length of the section before the horizontal curve in excess of 400 m has no effect on the free-flow speed of vehicles. The chart shows speed limits for this road section due to the presence of horizontal curves and the values of superelevation on the curve implemented to ensure the safe passage of vehicles. Safe journeys should be made in accordance with the speed limit on the road. The chart shows that the speed limit is exceeded by approx. 10 km/h, which is informally allowed on Polish roads. From the point of view of the needs of road safety and the needs of road network reliability, preventive measures should be applied that would force drivers to drive within the speed limit, just like in other European countries which have a low number of deaths per 100,000 accidents. The impact of the presence of horizontal curves on travel speed and thus travel time is clearly visible.

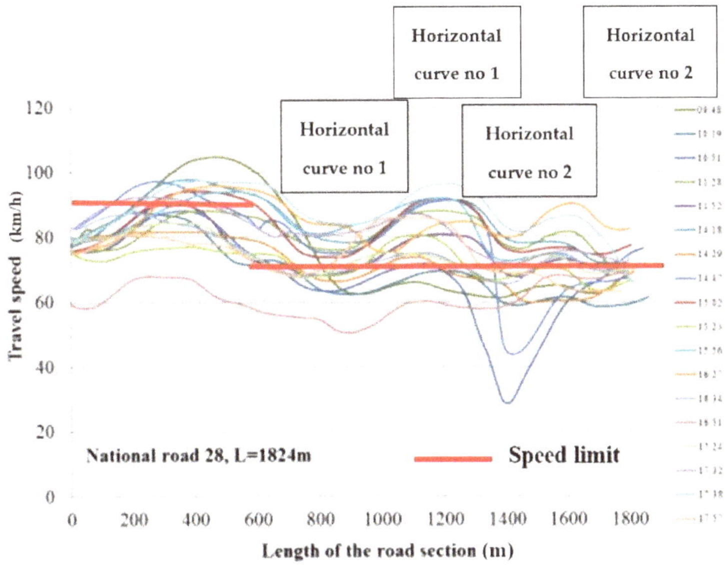

Figure 3. Selected drives behind the leader showing the impact of horizontal curves located close to each other on the variability of vehicles' free-flow speed.

Using the empirically collected data from 96 sections of two-lane highways with a length of 1 km to 5 km [43], relating to features of traffic, road and surroundings (Table 3), relationships were created showing the variability of travel speed, and thus indirectly also travel time. Selected interactions of travel speed and traffic volume, CCR and the interaction of the road's longitudinal slope and the share of heavy vehicles on two-lane highways are presented below. The significant dispersion of points arises from an aggregation of data from all research sites (96) and directions of traffic (192) and the differences between them.

Table 3. Technical parameters of two-lane highways covered by the RID-I−50 Project.

Road Design Parameter	Range of Parameters
Design speed S_d (km/h)	40–100, road serpentine 15–30
Speed limit S_{SL} (km/h)	40–90
Technical class of road	Z–S
Horizontal curve radius R_H (m)	30–3200
Vertical curve radius R_V (m)	not specified
Type of cross-section	1 × 2; 2 + 1
Lane width s (m)	3.0–3.5
Hard shoulder width s_{up} (m)	0–1.5
Average weighted longitudinal slope i (m)	0.1–9.0
Length of measured section L (m)	400–3900
Curvature change rate CCR (g/km)	0–630
Percentage of sections where overtaking is possible p_w (%)	0–100
Access—point density A_p (A_p/km)	0–42

Figure 4 shows a slight influence of traffic volume on the speed of vehicles and additionally illustrates the number of overtakes typical of two-lane highways. The number of overtakes depends on the traffic volume across the road's cross-section, traffic's directional structure and the number of slow-moving vehicles. Road's CCR undoubtedly has an influence on the number of overtakes. The greater the CCR, the smaller the number of overtakes. High values of CCR often coexist with high values of longitudinal slopes.

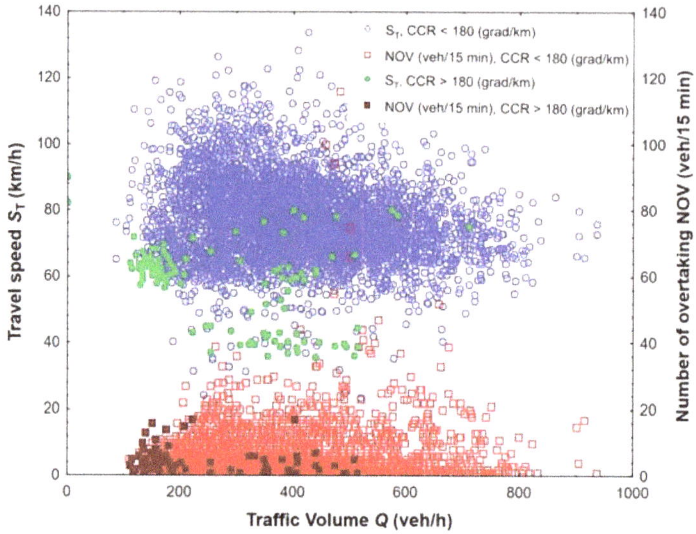

Figure 4. Interaction between travel speed, the number of overtakes and traffic volume throughout a road's cross-section, with the traffic density below 30 veh/km. Curvature change rate CCR divided into two groups–below and above 180 grad/km. The sample size is 12,565 cases.

With the traffic volume in a road's cross-section exceeding 800 veh/h, the number of overtakes is much smaller. This is due to traffic characteristics as well as road characteristics. In these circumstances, drivers will not overtake for safety reasons and follow a slow-moving vehicle until it can be overtaken. The more tortuous the road, the fewer overtaking manoeuvres.

The impact of a road's CCR on a vehicle's travel speed is very significant (Figure 5). Apart from a reduction in the median and average values of speed, the range of vehicle speed variability also decreases as the road CCR increases. The speed dispersion is much smaller on roads with greater rather than smaller CCR. The greater the road's curvature change rate CCR, the smaller the standard deviation of travel time and the greater the kurtosis. The chart also shows that the types of speed distributions in the distinguished ranges of CCR for the entire data set, including for low and high traffic volume, deviate from the normal distribution. High dispersion of speed at low curvature change rate CCR is due to a road's features and the possibility of overtaking slow-moving vehicles. At high CCR, a significant share of platoon traffic is manifest leading to smaller speed dispersion.

Figure 6 shows the combined impact of a road's longitudinal slope and the share of heavy vehicles in traffic on vehicles' travel speed. The statistical analyses carried out in [43] show that a road's longitudinal slope alone does not have as strong an impact on speed as a variable of the product of longitudinal slope and the share of heavy vehicles in traffic. A decrease in the impact of longitudinal slope in relation to previous studies [83] results from the improvement of the vehicle fleet in Poland. Nowadays, the capabilities of vehicle engines usually guarantee a smooth drive on a road with a variable longitudinal slope. Only heavy goods vehicles slow down when driving up or downhill, thus limiting the travel speed of other road users.

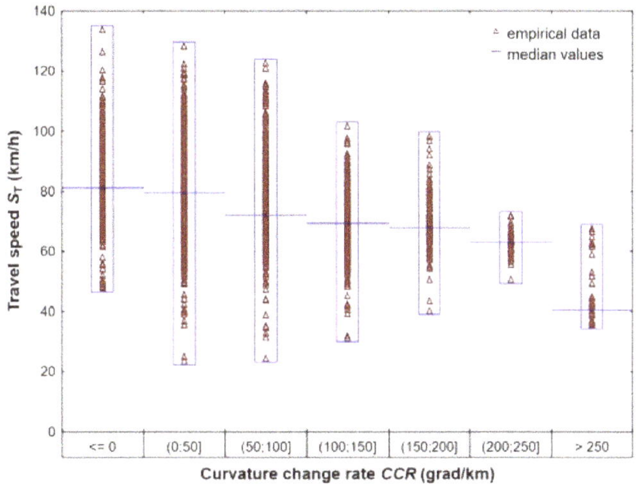

Figure 5. Interaction of travel speed and a road's curvature change rate CCR.

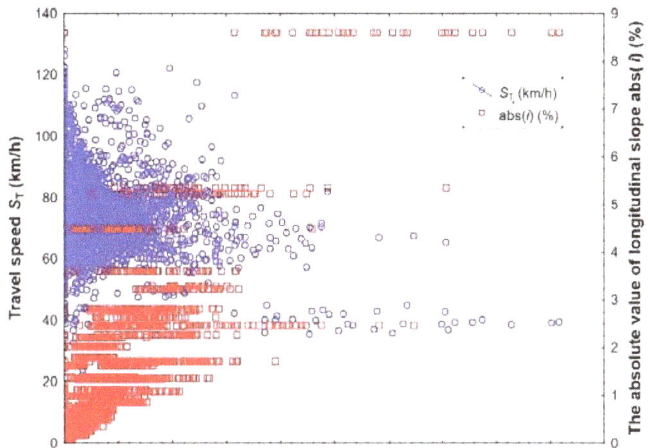

Figure 6. Interaction between travel speed and the combined effect of slope and the share of heavy vehicles.

Analysis of changes in a road's geometric features along the analyzed road sections shows that their impact on drivers' behaviour is varied and significant. When analyzing speed variability and thus travel times over a longer section (over 5 km), such a road section should be divided into sections that are homogeneous in terms of the described geometric and location features. The structure of traffic volume does not usually change significantly along the length of the road section and over the analyzed period of time. The division of the section into shorter sections allows you to analyze changes in travel speed and times and to eliminate places limiting travel speed and time e.g., by rebuilding the section.

The research shows that a road's geometric parameters and the share of heavy vehicles on roads with a significant longitudinal slope have a significant impact on the travel speed and thus travel time, including in free flow conditions. Therefore, these factors have a significant impact on the reliability of two-lane highways and should be included in reliability measures.

3. Results and Discussion

3.1. Functional Reliability of Two-Lane Highways

The functioning of two-lane highways differs significantly from that of dual carriageways in terms of the number and nature of factors affecting the speed of vehicle movement. In the case of two-lane highways, it is the traffic characteristics that mainly determine vehicle speed. Road features do matter but play a secondary role due to a need to adjust them to the high technical class of the road, its function and expected high travel speed in the early stages of the road's design. In the case of two-lane highways, the functional assessment of the road is significantly affected not merely by traffic characteristics, but by the geometric characteristics of the road and of its surroundings as well. Current computational methods mainly aim to assess the quality of traffic performance through analysis of measures of conditions including the average travel speed and the percentage of driving time in platoons [44], traffic density [43,45] and other indirect metrics such as the degree of capacity utilization or the reserve capacity. These measures are mainly used by road managers for decision-making, design or operational purposes.

Reliability measures related mainly to travel time (described in Section 2) may be used by road managers and additionally by road users to plan travel time or make up-to-date decisions on route selection if the data are provided on an ongoing basis e.g., via navigation systems. Road managers may also use reliability measures to analyze the functioning of the road in order to decide on the reconstruction of road sections in order to improve the road's standard and speed and shorten travel time.

3.2. Scenario Analysis and Reliability Measures of Two-Lane Highways

Analyses of the reliability of road functioning are based on the decisions of road users who drive along the road and generate travel times adequate to the traffic situation. Road managers can analyse such trips and on this basis decide on the type of measures that can be put in place to reduce travel times and improve road reliability and traffic safety. Such analyses can be conducted for the entire road or for selected sensitive road sections.

Time-related measures of reliability (Section 2) are closely related to road speed, which is limited by law, or locally by road signs. In the case of two-lane highways, the location of road signs often results from the characteristics of the road and the need to ensure safe passage. Free flow travel at speeds close to the permitted speed limit usually means a high level of functional reliability for people who are aware of the route they are taking. There is some synergy between the local speed limits set by road managers in hazardous locations and travel time reliability related to this speed.

The research conducted in [43] shows that roads with lower technical parameters have a lower capacity value, therefore, increased traffic relatively quickly leads to an unfavourable level of service and a significant increase in travel time as seen in Figure 7. The green colour representing a smooth drive does not change on the access sections of the serpentine at either of the analyzed times, and the colour of the highly winding sections does change, indicating a low travel speed in increased traffic [84].

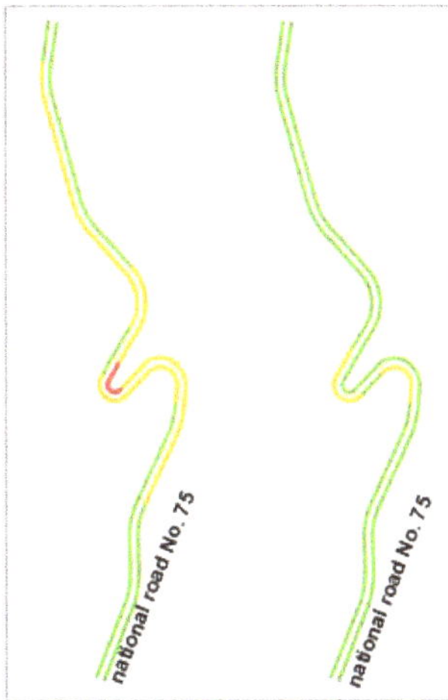

Figure 7. Change in the travel speed caused by a road's geometrical features and traffic features. The left panel shows the situation at 2:30 p.m. in conditions of a higher traffic volume, and the right one at 8:30 p.m. when traffic volume is low.

A justified and correct location of speed limit signs reflecting the local technical parameters of the road and traffic safety considerations should correspond well with a speed reflecting the 85th percentile and indirectly with the possible capacity of a given road section, as traffic volume reaches the optimum value. Thus, it can be noted that the travel time determined by the speed limit on the road is the appropriate travel time reference value in operational reliability analyses. If a road section consists of several subsections with different speed limits, the average weighted speed and travel time should be determined. The above studies show three possible scenarios of functional reliability analysis in relation to two-lane highways:

Scenario 1: The road is functioning reliably. Travel time is similar to the travel time at the speed limit (85th percentile) or possibly higher. Traffic volumes Q in both directions are much lower than road capacity C ($V/C < 0.5$). Overtaking is mainly determined by the road's geometry, i.e., its CCR and longitudinal slopes, etc. Drivers' acceptance level of traffic performance is high.

Scenario 2: Traffic volumes in both directions are much higher than in scenario 1 (V/C ranges from 0.5 to 0.9), and overtaking is limited both by the road's geometry and by higher traffic volumes of the analyzed direction and the opposite one. In drivers' opinion, the road may be functioning reliably and unreliably. Variability of the traffic volume affects traffic performance, and that's why drivers' acceptance level of traffic performance may differ.

Scenario 3: The road functions unreliably. Travel time is long and drivers, who travel at a much lower speed than the permitted speed, find it unacceptable. Traffic volume is close to or exceeds road capacity ($V/C > 0.9$). Overtaking is not possible due to high levels of traffic volume. Geometric factors play a minor role. Drivers' acceptance level of traffic conditions is very low.

The use of the limit speed as a reference value for determining reliability measures does not disqualify the use of other types of speed and thus travel times (statistical parameters). However, from the point of view of road and traffic characteristics, traffic safety needs, and the needs of practical application, this value is the most appropriate one for two-lane highways. Having analyzed the reliability measures used (Section 2), the planning time index (Equation (13)) and the travel time index (Equation (14)) are the most suitable metrics for two-lane highways, assuming that the base travel time refers to the permissible speed on the road or other local limits when a road section is divided into shorter subsections (in which case it is advisable to determine the weighted average PTI_{SL} and TTI_{SL}).

The thresholds values of PTI_{SL} and TTI_{SL} for individual levels of functional reliability are presented in Table 2. The indicated division of the section corresponds to the three analyzed reliability scenarios described in this chapter.

$$PTI_{SL} = \frac{t_{95}}{t_{SL}} \tag{13}$$

$$TTI_{SL} = \frac{t_x}{t_{SL}} \tag{14}$$

where: t_{95}—95th percentile of travel time, t_x—any percentile of travel time, t_{SL}—travel time determined by speed limits on the road.

Additionally, other statistical measures cited in Section 2 can be used to describe travel time variability. When the road is located near a large agglomeration and the influence of urban activity on traffic variability is noticeable, the buffer time index referring to the average value may also be used, as long as it is calculated separately for individual sub-sections of the road's section with different speed limits (Equations (7) and (8)).

3.3. Sample Reliability Analysis for a Selected Road Section

A section of a two-way highway marked as national road No. 47 between Rabka Zdrój and Klikuszowa, Poland was selected for analysis and was subsequently divided into two subsections. The first subsection, 1459 m long, starts outside the section with a 2 × 2 cross-section, and the second, 750 m long, consists of straight subsections of the road and a horizontal curve as denoted with the red arrows in Figure 8. The speed limit along the first subsection is 90 km/h, and within the second section, the speed limit of 60 km/h was introduced due to the presence of the horizontal curve having a large angle and a small radius. The speed limit of 60 km/h ensures safe vehicle driving on the road curve reflecting its technical parameters, including the superelevation used. Driving at a higher speed increases the risk of the vehicle skidding off the road and thus the occurrence of a traffic blockage, which would adversely affect the functional reliability of the analyzed road section.

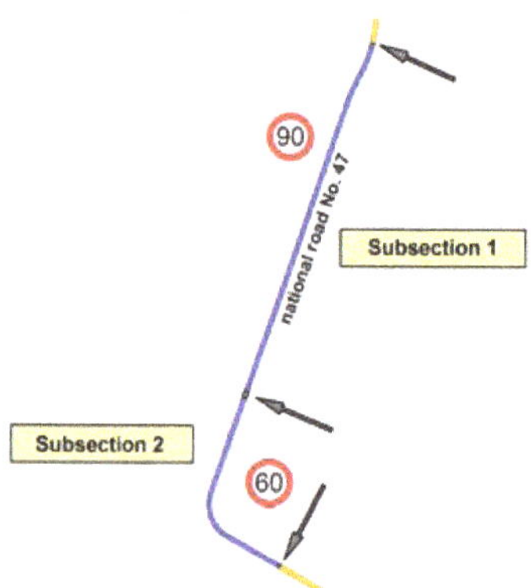

Figure 8. Analyzed section of the road divided into two subsections with speed limits of 90 km/h and 60 km/h.

The road traffic parameters were measured on the selected section of the road (Figure 8) around the clock, for 12 consecutive days of the week, i.e., from 13:15 on 13 May 2021 until 13:10 on 25 May 2021 using the GPS technology. The travel times of all vehicles (without a break-down into light and heavy vehicles) were obtained automatically from Google-associated paid applications for one direction, i.e., Kraków–Zakopane. The data were collected at 15-min intervals for which the average value of travel time was determined, with a break-down for each of the analyzed subsections (Figure 8) The introduction of a division of time into 15-min intervals results from the classic and practical approach to the study of traffic volumes and traffic conditions [O2]. The HCM method [O6] also indicates analyzing changes in traffic conditions in consecutive 15-min intervals as one of the possible analysis periods.

Observation of the traffic volume on the analyzed section of the national road reveals considerable variability of traffic during the 24 h time period, in relation to work-related, recreational and tourist traffic. The traffic volume on the national road No. 47 on weekend days often exceeds road capacity and contributes to congestion. Table 4 summarizes the statistical parameters describing the variability of travel time in consecutive 15-min intervals of the analyzed 12-day period, with a break-down into two analyzed road subsections. Additionally, the table specifies the unit travel time t_{100} (s/100 m) relating to two subsections of the road having the same length of 100 m. The summary aims to show the impact of different road features on travel time. Eliminating the impact of the different lengths of individual road subsections on travel time will allow comparison of the statistical parameters describing the variability of the unit travel time on a straight part of the road (Section 1) and on a horizontal curve (Section 2), with different speed limits in place on the road (Figure 9) and under the same traffic and weather conditions.

Table 4. Statistical parameters describing travel time variability determined for subsequent 15 min. time intervals.

Variable	Subsection	Sample Size	Average	Median	Min	Max	Percentile 5%	Percentile 95%	Standard Deviation	Coefficient of Variation
t_{100} (s/100 m)	1	1090	4.7	4.7	4.0	11.0	4.2	5.6	0.61	12.9
t (s)	1	1090	69.2	68.0	58.0	160.0	62.0	81.0	8.96	12.9
t_{100} (s/100 m)	2	1090	6.0	5.5	4.4	35.0	4.8	7.9	2.1	35.2
t (s)	2	1090	44.8	41.0	33.0	264.0	36.0	59.0	15.8	35.2

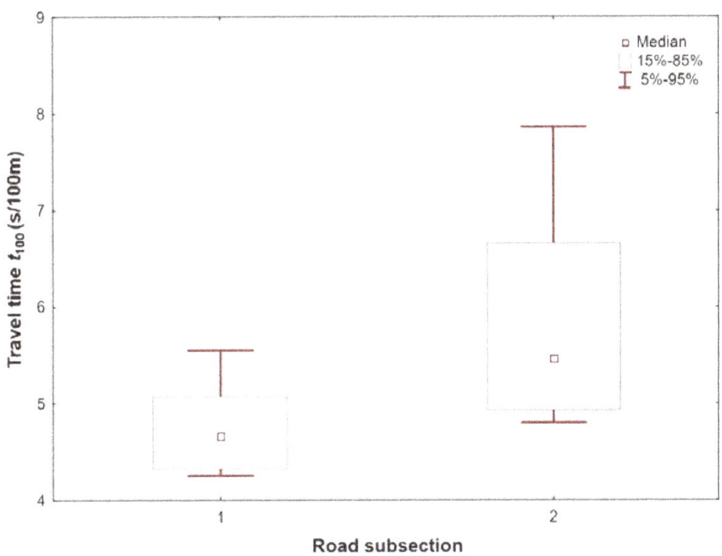

Figure 9. Comparison of 50, 85 and 95% quantile of unit travel time t_{100} determined for a 100 m length of subsections 1 and 2.

The analyses revealed a much greater value and variability of the unit travel time t_{100} along the horizontal curve (Section 2) than along the straight stretch of the road (Section 1).

They showed a significant impact of the presence of the horizontal curve, markings limiting the speed to 60 km/h and road safety devices (roadside barriers) on the speed of vehicles, and thus on travel time. The impact of the presence of horizontal curves on travel speed will be the greater, the greater the number of horizontal curves on the analyzed section (Figure 5), and when the horizontal curves have a small radius and a superelevation along the curve which are not adjusted to the speed limit on the road. The situation is made worse by the road's design featuring large turning angles and small curve radii at the same time. Then the parameters of the horizontal curve do not allow for driving along the curve at a speed of 90 km/h and there appears a need to introduce a speed limit necessitated by the requirements of traffic safety (Figures 3 and 8). In most cases, drivers adjusted their speed (85% quantile) to the applicable speed limit on the road, accepting the reasonableness of vertical markings ensuring safe driving on the curve, and thus accepting the increased travel times resulting from its presence.

Table 5 presents the results of planning time index analyses conducted in a classic way (Equation (9)) and in the way proposed in the paper (Equation (13)), i.e., relating to the travel time in accordance with the speed limit enforced on the road. The travel time t_0, in conditions considered to allow free flow was determined based on the American recommendations [44]. Method [44] states that the initial speed of vehicles in free flow traffic can be determined by increasing the speed limit on a two-way highway by 15 km/h.

Table 5. Summary of calculation data and analysis results regarding *PTI* and PTI_{SL} values.

Subsection	t_{95} (s)	t_0 (s)	*PTI* (Equation (9))	t_{SL} (s)	PTI_{SL} (Equation (13))
1	81.0	50.0	1.62	58.4	1.39
2	59.0	36.0	1.64	45.0	1.31

Comparing the results of the analyses, it can be concluded that the classical analysis performed on the basis of Equation (9) yields much worse results than in the case relating to travel time in accordance with the speed limit. In both cases, in accordance with the criteria in Table 2, the same fair level of functional reliability was obtained. The classical approach refers to much higher travel speeds typical of free flowing traffic. An analysis performed in this way results from vehicle drivers' expectations which are overestimated and inconsistent with applicable regulations, expectations which ignore road safety. It is worth noting that any increase in vehicle speed above the applicable speed limit increases the risk of road incidents. In the analyzed case, the risk will be much greater on a horizontal curve than on a straight stretch of the road (risk of the vehicle skidding off the road at a higher speed).

When assessing the road's functional reliability in a dependable manner, apart from the impact of traffic variability, one should also take into account the possibility of traffic jams resulting from road incidents. When these occur, they block the possibility of driving on a two-way highway, and their high frequency, e.g., due to increased travel speed, reduces the level of the road's functional reliability and the reliability of the analyses. The number of road incidents strongly corresponds to the traffic volume on the road and therefore it is important to maintain vehicle speed below the speed limit by applying appropriate preventive (fines) and engineering measures (speed limit signs, speed cameras, sectional speed measurements, etc.). By showing a slightly worse functional reliability of the road in the classic approach (referring to the free flow speed), it is possible to incorrectly assess the functional state of the road and incorrectly plan financial outlays. Therefore, the conducted analyses indicate a very strong correlation between functional reliability measures and road safety levels, defined, inter alia, in terms of geometric parameters and applicable speed limits.

4. Conclusions

The research carried out on two-way highways allowed to achieve the goal of the analysis and provided an answer to the research question. The geometrical differentiation of two-lane highways has a significant impact on drivers' behaviour and the speeds they achieve, and, consequently, on travel times. On the analyzed roads, especially when there are heavy vehicles in traffic, one can often observe platoon driving with vehicles following the leader. Its duration depends not only on the volume of heavy vehicle traffic and traffic from the opposite direction but also on the geometric features of the road, such as CCR, longitudinal slope, the width of the road and the type of shoulder. Travel time may also be occasionally inflated by adverse weather conditions that adversely affect drivers' behaviour and road capacity, as well as other random incidents on the road often occurring due to increased travel speeds. On the basis of Polish research, the influence of these factors on travel speed and thus on travel time as well as a road's functional reliability was demonstrated. Analyses of the currently used reliability measures conducted mainly for dual carriageways indicate the need for an appropriate reference level that will reflect the road type and its standard. The reference parameter should combine the appropriate road and traffic characteristics from the point of view of describing reliability, functionality and traffic safety. The paper proposes that permissible speed is the best reference level for two-lane highways. The proposed parameter is different from that recommended for dual carriageways of high technical parameters. The speed limit on single carriageways changes rather frequently along the road and depends on the road's characteristics, indicating to drivers the safe and reliable travel speed. Reliability analyses require therefore that a road section be divided into shorter sub-sections. The influence of traffic characteristics (traffic volume and structure) differentiates travel time in relation to the time resulting from the speed limit, and its comparison with the marginal values allows classifying individual sub-sections and the entire section's functional reliability.

Current reliability analyses barely take into consideration the need to relate the reference level (arising e.g., from the free flow speed or various quantiles of travel speed) to the needs of road safety (resulting e.g., from the applied speed limits on the road or from risk analyses of road incidents). Often it is difficult and complicated to determine the value of free flow speed and, consequently, such determination may not correspond to the actual conditions. Pertinent literature does not assess the value of the difference between travel time arising, e.g., from the 85% quantile and the travel time resulting from the speed limits enforced on the road. The size of the analyzed difference may be related to the indicators concerning the number of road incidents and the risk of their occurrence. Further analyses and studies in this field are necessary, not only for single carriageways but also for dual carriageways. The paper shows gaps in research in the field of continuous traffic tests and reliability analyses into two-lane highways, and indicates the need to develop and quantify reliability measures, thanks to which actions can be taken to improve traffic flow on the analysed roads. The use in the presented analyses of a tool for continuous measurement of travel time using GPS devices installed in vehicles and mobile phones allowed for analysis of the variability of travel time on the analyzed road section over a long period of time, and thus allowed indicating the impact of selected road and traffic factors and the need to incorporate traffic safety recommendations into reliability analyses. In the development of methods of road reliability analysis, it is necessary to develop tools for the continuous collection and analysis of traffic data.

Author Contributions: Conceptualization, K.O.; methodology, K.O. and M.B.; formal analysis, K.O.; resources, K.O.; data curation, K.O.; writing—original draft preparation, K.O. and M.B.; writing—review and editing, K.O. and M.B.; project administration, K.O. and M.B. All authors have read and agreed to the published version of the manuscript.

Funding: This research received no external funding.

Institutional Review Board Statement: Not applicable.

Informed Consent Statement: Not applicable.

Data Availability Statement: Not applicable.

Conflicts of Interest: The authors declare no conflict of interest.

References

1. Hou, X.; Wang, Y.; Zhang, P.; Qin, G.; Hou, W. Qin Non-Probabilistic Time-Varying Reliability-Based Analysis of Corroded Pipelines Considering the Interaction of Multiple Uncertainty Variables. *Energies* **2019**, *12*, 1965. [CrossRef]
2. Teslyuk, V.; Sydor, A.; Karovič, V.; Pavliuk, O.; Kazymyra, I. Modelling Reliability Characteristics of Technical Equipment of Local Area Computer Networks. *Electronics* **2021**, *10*, 955. [CrossRef]
3. Coudray, R.; Mattei, J. System reliability: An example of nuclear reactor system analysis. *Reliab. Eng.* **1984**, *7*, 89–121. [CrossRef]
4. Petkevičius, K.; Maskeliūnaitė, L.; Sivilevičius, H. Determining travel conditions on motorways for automobile trasport based on the case study for Lithuanian highways. *Transport* **2019**, *34*, 89–102. [CrossRef]
5. Nicholson, A. Travel time reliability benefits: Allowing for correlation. *Res. Transp. Econ.* **2015**, *49*, 14–21. [CrossRef]
6. Ji, X.-F.; Wu, Y.-X.; Yuan, H.-Z.; Yang, W.-C.; Hu, C.-Y.; Lu, H. Influences of Traffic Flow Characteristics on Accident Severity on Secondary Roads. *Zhongguo Gonglu Xuebao/China J. Highw. Transp.* **2020**, *33*, 135–145.
7. Sun, J.; Sun, J. Proactive assessment of real-time traffic flow accident risk on urban expressway. *Tongji Daxue Xuebao/J. Tongji Univ.* **2014**, *42*, 873–879.
8. Kustra, W.; Żukowska, J.; Budzyński, M.; Jamroz, K. Injury Prediction Models for Onshore Road Network Development. *Pol. Marit. Res.* **2019**, *26*, 93–103. [CrossRef]
9. Ambros, J.; Jurewicz, C.; Chevalier, A.; Valentová, V. Speed-Related Surrogate Measures of Road Safety Based on Floating Car Data. *Inventive Comput. Inf. Technol.* **2021**, *207*, 129–144. [CrossRef]
10. Lyman, K.; Bertini, R.L. Using Travel Time Reliability Measures to Improve Regional Transportation Planning and Operations. *Transp. Res. Rec. J. Transp. Res. Board* **2008**, *2046*, 1–10. [CrossRef]
11. Meyer, M.D. *Transportation Planning Handbook: Institute of Transportation Engineers*; John Wiley & Sons, Inc.: Hoboken, NJ, USA, 2016.
12. Brilon, W.; Geistefeldt, J.; Zurlinden, H. Implementing the Concept of Reliability for Highway Capacity Analysis. *Transp. Res. Rec. J. Transp. Res. Board* **2007**, *2027*, 1–8. [CrossRef]
13. Bhouri, N.; Aron, M.; Scemama, G. Travel time reliability with and without the dynamic use of hard shoulder: Field assessment from a French motorway. *J. Traffic Transp. Eng. Engl. Ed.* **2016**, *3*, 520–530. [CrossRef]
14. Olszewski, P.; Dybicz, T.; Jamroz, K.; Kustra, W.; Romanowska, A. Assessing Highway Travel Time Reliability using Probe Vehicle Data. *Transp. Res. Rec. J. Transp. Res. Board* **2018**, *2672*, 118–130. [CrossRef]
15. Romanowska, A.; Jamroz, K.; Olszewski, P. Review of methods for assessing traffic conditions on basic motorway and expressway sections. *Arch. Transp.* **2019**, *52*, 7–25. [CrossRef]
16. Mehran, B.; Nakamura, H. Implementing Travel Time Reliability for Evaluation of Congestion Relief Schemes on Expressways. *Transp. Res. Rec. J. Transp. Res. Board* **2009**, *2124*, 137–147. [CrossRef]
17. Farrag, S.G.; El-Hansali, M.Y.; Yasar, A.-U.-H.; Shakshuki, E.M.; Malik, H. A microsimulation-based analysis for driving behaviour modelling on a congested expressway. *J. Ambient. Intell. Humaniz. Comput.* **2020**, *11*, 5857–5874. [CrossRef]
18. Adachi, T.; Ishida, T.; Yaginuma, H.; Asakura, Y. Empirical analysis for the effects of a new expressway section on travel time reliability and drivers route choice. In Proceedings of the 5th International Symposium on Transportation Network Reliability, Hong Kong, 18–19 December 2012.
19. Tay, A.C.; Lee, H.F.; Zheng, Y. Traffic Assessment for Completion of Road Upgrading from Secondary Road to Dual Carriageway. *IOP Conf. Ser. Mater. Sci. Eng.* **2019**, *495*, 012098. [CrossRef]
20. Varsha, V.; Pandey, G.H.; Rao, K.R.; Bindhu, B. Determination of Sample Size for Speed Measurement on Urban Arterials. *Transp. Res. Procedia* **2016**, *17*, 384–390. [CrossRef]
21. Oskarbski, J.; Kamiński, T.; Kyamakya, K.; Chedjou, J.C.; Żarski, K.; Pędzierska, M. Assessment of the Speed Management Impact on Road Traffic Safety on the Sections of Motorways and Expressways Using Simulation Methods. *Sensors* **2020**, *20*, 5057. [CrossRef]
22. Cafiso, S.; D'Agostino, C.; Kiec, M.; Pogodzinska, S. Application of an Intelligent Transportation System in a Travel Time Information System: Safety Assessment and Management. *Transp. Res. Rec. J. Transp. Res. Board* **2017**, *2635*, 46–54. [CrossRef]
23. Kattan, L.; de Barros, A.G.; Saleemi, H. Travel behavior changes and responses to advanced traveler information in prolonged and large-scale network disruptions: A case study of west LRT line construction in the city of Calgary. *Transp. Res. Part F Traffic Psychol. Behav.* **2013**, *21*, 90–102. [CrossRef]
24. Vivek. A new model to evaluate percent-time-spent-following on two-lane highways. *Lect. Notes Civ. Eng.* **2020**, *45*, 377–387.
25. Raj, P.; Sivagnanasundaram, K.; Asaithambi, G.; Shankar, A.U.R. Review of Methods for Estimation of Passenger Car Unit Values of Vehicles. *J. Transp. Eng. Part A Syst.* **2019**, *145*, 04019019. [CrossRef]
26. Pompigna, A.; Rupi, F. Lane-Distribution Models and Related Effects on the Capacity for a Three-Lane Freeway Section: Case Study in Italy. *J. Transp. Eng. Part A Syst.* **2017**, *143*, 05017010. [CrossRef]

27. Harwood, D.W.; Bauer, K.M.; Potts, I.B. Development of Relationships between Safety and Congestion for Urban Freeways. *Transp. Res. Rec. J. Transp. Res. Board* **2013**, *2398*, 28–36. [CrossRef]
28. Biswas, S.; Chakraborty, S.; Chandra, S.; Ghosh, I. Kriging-Based Approach for Estimation of Vehicular Speed and Passenger Car Units on an Urban Arterial. *J. Transp. Eng. Part A Syst.* **2017**, *143*, 04016013. [CrossRef]
29. Ma, Y.; Gu, X.; Lee, J.; Xiang, Q. Investigating the Affecting Factors of Speed Dispersion for Suburban Arterial Highways in Nanjing, China. *J. Adv. Transp.* **2019**, *2019*, 1–11. [CrossRef]
30. Lowe, W.U.A.; Mendis, H.S.A.; Sathyaprasad, I.M.S. A Comparative Study of Speed and Flow Measurements Methods as Applied to Four Lane Dual Carriageway Roads. *Proc. EECE 2020* **2019**, *44*, 211–223. [CrossRef]
31. Guerrieri, M.; Parla, G.; Mauro, R. Traffic Flow Variables Estimation: An Automated Procedure Based on Moving Observer Method. Potential Application for Autonomous Vehicles. *Transp. Telecommun. J.* **2019**, *20*, 205–214. [CrossRef]
32. Olia, A.; AbdelGawad, H.; Abdulhai, B.; Razavi, S.N. Optimizing the Number and Locations of Freeway Roadside Equipment Units for Travel Time Estimation in a Connected Vehicle Environment. *J. Intell. Transp. Syst.* **2017**, *21*, 296–309. [CrossRef]
33. Pamula, T. Road Traffic Conditions Classification Based on Multilevel Filtering of Image Content Using Convolutional Neural Networks. *IEEE Intell. Transp. Syst. Mag.* **2018**, *10*, 11–21. [CrossRef]
34. Gong, Y.; Abdel-Aty, M.; Park, J. Evaluation and augmentation of traffic data including Bluetooth detection system on arterials. *J. Intell. Transp. Syst.* **2019**, 1–13. [CrossRef]
35. Hoseinzadeh, N.; Liu, Y.; Han, L.D.; Brakewood, C.; Mohammadnazar, A. Quality of location-based crowdsourced speed data on surface streets: A case study of Waze and Bluetooth speed data in Sevierville, TN. *Comput. Environ. Urban Syst.* **2020**, *83*, 101518. [CrossRef]
36. Villiers, C.; Nguyen, L.D.; Zalewski, J. Evaluation of traffic management strategies for special events using probe data. *Transp. Res. Interdiscip. Perspect.* **2019**, *2*, 100052. [CrossRef]
37. Hu, H.; Lu, Z.; Wang, Q.; Zheng, C. End-to-End Automated Lane-Change Maneuvering Considering Driving Style Using a Deep Deterministic Policy Gradient Algorithm. *Sensors* **2020**, *20*, 5443. [CrossRef] [PubMed]
38. Ural, S.; Shan, J.; Romero, M.A.; Tarko, A. Road and roadside feature extraction using imagery and lidar data for transportation operation. *ISPRS Ann. Photogramm. Remote. Sens. Spat. Inf. Sci.* **2015**, *II-3/W4*, 239–246. [CrossRef]
39. Soni, R.; Vasudevan, V.; Dutta, B. Analysis of overtaking patterns of Indian drivers with data collected using a LiDAR. *Transp. Res. Part F Traffic Psychol. Behav.* **2020**, *74*, 139–150. [CrossRef]
40. Baskov, V.; Ignatov, A.; Gamayunov, P.; Igitov, S. Influence of elements of the "driver-car-road-environment" system on emergence of the transport jam. *ARPN J. Eng. Appl. Sci.* **2019**, *14*, 1093–1099.
41. Ali, E.K.; Hashim, I.H.; Shwaly, S.A.; Zidan, Z.M.; El-Badawy, S.M. Risk assessment of horizontal curves using reliability analysis based on Google traffic data. *Innov. Infrastruct. Solut.* **2021**, *6*, 1–13. [CrossRef]
42. Liu, T. Selpi Comparison of Car-Following Behavior in Terms of Safety Indicators between China and Sweden. *IEEE Trans. Intell. Transp. Syst.* **2020**, *21*, 3696–3705. [CrossRef]
43. Ostrowski, K.; Dybicz, T.; Kustra, W.; Olszewski, P.; Chodur, J.; Jamroz, K. Modern methods of calculating the road capacity and assessment of traffic conditions of roads outside municipal agglomerations, including express roads. In *Project RID-I-50. GDDKiA and NCBiR*; Cracow, Warsaw; Gdansk, Poland; 2019; Available online: https://nauka.pk.edu.pl/index.php?option=com_content&task=view&id=32&Itemid=39&pokaz=projekt&id_pr=1196 (accessed on 8 June 2021).
44. Transportation Research Record. TRB Highway Capacity Manual. In *A Guide for Multimodal Mobility Analysis*, 6th ed.; The National Academies of Sciences, Engineering, and Medicine: Washington, DC, USA, 2016.
45. Baier, M.W.; Brilon, G.; Hartkopf, K.; Lemke, R.; Maier, M.S. *HBS2015 Handbuch für die Bemessung von Straßenverkehrsanlagen, Forschungsgesellschaft für Straßen und Verkehrswesen*; FGSV: Köln, Germany, 2015.
46. HNTB Corporation; Strategic Highway Research Program Reliability Focus Area. Transportation Research Board Analytical Procedures for Determining the Impacts of Reliability Mitigation Strategies. *Anal. Proced. Determ. Impacts Reliab. Mitig. Strateg.* **2012**. [CrossRef]
47. Associates, I.K. Strategic Highway Research Program Reliability Focus Area; Transportation Research Board Evaluating Alternative Operations Strategies to Improve Travel Time Reliability. *Eval. Altern. Oper. Strateg. Improv. Travel Time Reliab.* **2013**. [CrossRef]
48. OECD. *Improving Rellabillty on Surface Transport Networks*; OECD: Paris, France, 2010.
49. Lin, Y.; Niu, J. Effect of Curvature Change Rate of Highway Horizontal Curve in the Path of a Vehicle. *ICTIS 2011* **2011**, 904–912. [CrossRef]
50. Sil, G.; Nama, S.; Maji, A.; Maurya, A.K. Effect of horizontal curve geometry on vehicle speed distribution: A four-lane divided highway study. *Transp. Lett.* **2020**, *12*, 713–722. [CrossRef]
51. Russo, F.; Biancardo, S.A.; Busiello, M. Operating speed as a key factor in studying the driver behaviour in a rural context. *Transport* **2016**, *31*, 260–270. [CrossRef]
52. Vorobjovas, V. Assurance of the Function of Low-Volume Roads for the Improvement of Driving Conditions. *Balt. J. Road Bridg. Eng.* **2011**, *6*, 67–75. [CrossRef]
53. Jensen, S.U. Car Drivers' Experienced Level of Service on Freeways. *Transp. Res. Rec. J. Transp. Res. Board* **2017**, *2615*, 132–139. [CrossRef]
54. Pinna, F. Free flow speed and drivers behavior in Italian arterial roads. *Adv. Transp. Stud.* **2020**, *50*, 31–48.

55. Haseman, R.J.; Wasson, J.S.; Bullock, D. Real-Time Measurement of Travel Time Delay in Work Zones and Evaluation Metrics Using Bluetooth Probe Tracking. *Transp. Res. Rec. J. Transp. Res. Board* **2010**, *2169*, 40–53. [CrossRef]
56. Bharadwaj, N.; Edara, P.; Sun, C.; Brown, H.; Chang, Y. Traffic Flow Modeling of Diverse Work Zone Activities. *Transp. Res. Rec. J. Transp. Res. Board* **2018**, *2672*, 23–34. [CrossRef]
57. Kianfar, J.; Abdoli, S. Deterministic and Stochastic Capacity in Work Zones: Findings from a Long-Term Work Zone. *J. Transp. Eng. Part A Syst.* **2021**, *147*, 04020141. [CrossRef]
58. Jeong, J.P.; Kim, J.; Hwang, T.; Xu, F.; Guo, S.; Gu, Y.J.; Cao, Q.; Liu, M.; He, T. TPD: Travel Prediction-based Data Forwarding for light-traffic vehicular networks. *Comput. Netw.* **2015**, *93*, 166–182. [CrossRef]
59. Khattak, Z.H.; Magalotti, M.J.; Fontaine, M.D. Operational performance evaluation of adaptive traffic control systems: A Bayesian modeling approach using real-world GPS and private sector PROBE data. *J. Intell. Transp. Syst.* **2019**, *24*, 156–170. [CrossRef]
60. Gu, Y.; Wang, Y.; Dong, S. Public Traffic Congestion Estimation Using an Artificial Neural Network. *ISPRS Int. J. Geo-Inf.* **2020**, *9*, 152. [CrossRef]
61. Geistefeldt, J.; Shojaat, S. Comparison of Stochastic Estimates of Capacity and Critical Density for U.S. and German Freeways. *Transp. Res. Rec. J. Transp. Res. Board* **2019**, *2673*, 388–396. [CrossRef]
62. Figliozzi, M.A.; Wheeler, N.; Albright, E.; Walker, L.; Sarkar, S.; Rice, D. Algorithms for Studying the Impact of Travel Time Reliability along Multisegment Trucking Freight Corridors. *Transp. Res. Rec. J. Transp. Res. Board* **2011**, *2224*, 26–34. [CrossRef]
63. Samandar, M.S.; Williams, B.; Ahmed, I. Weigh Station Impact on Truck Travel Time Reliability: Results and Findings from a Field Study and a Simulation Experiment. *Transp. Res. Rec. J. Transp. Res. Board* **2018**, *2672*, 120–129. [CrossRef]
64. Hojati, A.T.; Ferreira, L.; Washington, S.; Charles, P.; Shobeirinejad, A. Reprint of: Modelling the impact of traffic incidents on travel time reliability. *Transp. Res. Part C Emerg. Technol.* **2016**, *70*, 86–97. [CrossRef]
65. Collins, A.; Foytik, P.; Frydenlund, E.; Robinson, R.M.; Jordan, C.A. Generic Incident Model for Investigating Traffic Incident Impacts on Evacuation Times in Large-Scale Emergencies. *Transp. Res. Rec. J. Transp. Res. Board* **2014**, *2459*, 11–17. [CrossRef]
66. Tay, R.; Churchill, A.; De Barros, A.G. Effects of roadside memorials on traffic flow. *Accid. Anal. Prev.* **2011**, *43*, 483–486. [CrossRef]
67. Hazim, N.; Shbeeb, L.; Abu Salem, Z. Impact of Roadside Fixed Objects in Traffic Conditions. *Eng. Technol. Appl. Sci. Res.* **2020**, *10*, 5428–5433. [CrossRef]
68. Zhao, L.; Chien, S.I.-J. Analysis of Weather Impact on Travel Speed and Travel Time Reliability. *CICTP 2012* **2012**, 1145–1155. [CrossRef]
69. Kim, J.; Mahmassani, H.S.; Dong, J. Likelihood and duration of flow breakdown: Modeling the effect of weather. *Transp. Res. Rec.* **2010**, *2188*, 19–28. [CrossRef]
70. Thakuriah, P.; Tilahun, N. Incorporating Weather Information into Real-Time Speed Estimates: Comparison of Alternative Models. *J. Transp. Eng.* **2013**, *139*, 379–389. [CrossRef]
71. Sumalee, A.; Sumalee, A.; Watling, D.; Lecturer, S. Travel time reliability in a network with dependent link modes and partial driver response. *J. East Asia Soc. Transp. Stud.* **2003**, *5*, 1687–1701.
72. Chodur, J.; Ostrowski, K.; Tracz, M. Variability of Capacity and Traffic Performance at Urban and Rural Signalised Intersections. *Transp. Res. Procedia* **2016**, *15*, 87–99. [CrossRef]
73. Chodur, J.; Ostrowski, K.; Tracz, M. Impact of Saturation Flow Changes on Performance of Traffic Lanes at Signalised Intersections. *Procedia Soc. Behav. Sci.* **2011**, *16*, 600–611. [CrossRef]
74. Ostrowski, K.; Tracz, M. Availability and reliability of a signalised lane. *Transp. B Transp. Dyn.* **2018**, *7*, 1044–1061. [CrossRef]
75. Ostrowski, K.; Chodur, J. The impact of selected road and traffic features on travel speed on two lane highway. *Road Rail Infrastruct. V* **2018**, *5*, 1129–1135. [CrossRef]
76. Haghani, A.; Zhang, Y.; Hamedi, M. *Impact of Data Source on Travel Time Reliability Assessment*; Grant DTRT12-G-UTC03; Mid-Atlantic Universities Transportation Center: University Park, PA, USA, 2014.
77. Polus, A. A study of travel time and reliability on arterial routes. *Transportation* **1979**, *8*, 141–151. [CrossRef]
78. Polus, A.; Shofer, J.L. Analytical Study of Freeway Reliability. *Transp. Eng. J. ASCE* **1976**, *102*, 857–870. [CrossRef]
79. Available online: https://www.velaction.com/buffer-time/ (accessed on 6 May 2021).
80. FHWA Final Report FHWA-HOP-05-018. Monitoring urban freeways in 2003. In *Current Conditions and Trends From Archived Operations Data*; Federal Highway Administration Office of Operations: Washington, DA, USA, 2004.
81. US Department of Transportation Urban Congestion Reports-Operations Performance Measurement-FHWA Office of Operations. Available online: https://ops.fhwa.dot.gov/perf_measurement/ucr/ (accessed on 6 May 2021).
82. Transportation Research Board. TRB S2-L05-RR-2. In *Guide to Incorporating Reliability Performance Measures into the Transportation Planning and Programming Processes*; The National Academies of Sciences, Engineering, and Medicine: Washington, DC, USA, 2014.
83. Jamroz, K.R.; Krystek, A.; Cielecki, J.; Kempa, T.; Sandecki, J.; Sosin; Szczuraszek, T.; Zieliński, T.; Hoppe, L.M.; Michalski, L.; et al. *Polish Guidelines on Rural Roads Capacity Estimation*; Gdansk University of Technology: Gdańsk; Poland, 1986.
84. Available online: https://www.google.pl/maps/dir///@49.7224264,20.6424265,15z/data=!4m2!4m1!3e0!5m1!1e1 (accessed on 6 May 2021).

energies

MDPI

Article

A Comparison of Lane Marking Detection Quality and View Range between Daytime and Night-Time Conditions by Machine Vision

Darko Babić [1], Dario Babić [1,*], Mario Fiolić [1], Arno Eichberger [2] and Zoltan Ferenc Magosi [2]

[1] Faculty of Transport and Traffic Sciences, University of Zagreb, Vukelićeva 4, 10000 Zagreb, Croatia; darko.babic@fpz.unizg.hr (D.B.); mario.fiolic@fpz.unizg.hr (M.F.)

[2] Institute of Automotive Engineering, Graz University of Technology, Inffeldgasse 11/II, A-8010 Graz, Austria; arno.eichberger@tugraz.at (A.E.); zoltan.magosi@tugraz.at (Z.F.M.)

* Correspondence: dario.babic@fpz.unizg.hr

Abstract: Lateral support systems in vehicles have a high potential for reduction of lane departure crashes. To profit from their full potential, such systems should function properly in adverse conditions. Literature indicates that their accuracy varies between day and night-time. However, detailed quantifications of the systems' performance in these conditions are rare. The aim of this study is to investigate the differences in detection quality and view range of Mobileye 630 in dry daytime and night-time conditions. On-road tests on four rural road sections in Croatia were conducted. Wilcoxon signed-rank test was used to test the difference between the number of quality rankings while absolute average, average difference and standard deviation were used to analyse the view range. Also, a paired samples t-test was used to test the difference between conditions for each line on each road. The overall results confirm that a significant difference in lane detection quality view range exists between tested conditions. "Medium" and "high" detection confidence (quality level 3 and 2), increased by 5% and 8% during night-time compared to daytime while level 0 ("nothing detected") decreased by 12%. The view range increased (almost 16% for middle line) during daytime compared to night-time. The findings of this study expand the existing knowledge and are valuable for research and development of machine-vision systems but also for road authorities to optimize the markings' quality performance.

Keywords: ADAS; lateral support systems; lane detection; automated driving; visibility; lane keeping systems

Citation: Babić, D.; Babić, D.; Fiolić, M.; Eichberger, A.; Magosi, Z.F. A Comparison of Lane Marking Detection Quality and View Range between Daytime and Night-Time Conditions by Machine Vision. *Energies* **2021**, *14*, 4666. https://doi.org/10.3390/en14154666

Academic Editors: Guzek Marek, Rafał Jurecki, Wojciech Wach and Thanikanti Sudhakar Babu

Received: 10 June 2021
Accepted: 30 July 2021
Published: 1 August 2021

1. Introduction

Lane departure crashes are one of the most common types of road accidents. In the US alone, 51% of all fatal accidents are caused by lane departure, i.e., a vehicle crossing the edge or the centre line [1]. There is a variety of contributing factors involving the driver, the vehicle, the road and its surroundings (the environment) but it is proven that rural roads with low traffic volume and density are more likely to contribute to road departure accidents [2]. This is mainly due to higher travelling speed, distracted driving and/or fatigue. Different safety measures have shown positive results in decreasing lane departure accidents [3–5], yet the overall problem still exists.

A potentially promising solution to the aforementioned problem lies in automated driving and Advanced Driver Assistance Systems (ADAS), which perceive the static and the dynamic content of the environment around the vehicle and thus assist the human driver in driving. An important task during environment perception is lane detection needed for Lateral Support Systems (LSS), which comprise of lane departure warning and/or lane keeping assistance. The main purpose of LSS is to prevent road accidents caused by road departure or entrance in the lane of other vehicles. Due to high fatality rate in such accidents, the use of different LSS could significantly improve the overall road safety [6–9].

In literature, two main technologies for LSS are reported: LIDARs (Light Detection and Ranging) and vision-based cameras. The main advantage of LIDAR is the fact that it uses an active light source and thus does not rely on the variabilities associated with external/natural lighting as do regular vision-based cameras [10]. However, LIDAR technology is usually used for adaptive cruise control, while lane detection is only partially feasible and mainly used in expensive demonstrator vehicles for highly automated driving [11]. Therefore, current market-ready systems use passive vision-based cameras and image processing to collect and analyse the data from roads [12]. In general, camera-based lane detection starts from image pre-processing which includes different corrections of the collected image (such as exposure correction and shadow removal) and feature extraction. This is then followed by feature detection and model fitting, and then time integration to keep temporal and position consistency [13].

The proper function of LSS depends on several factors [11,14,15]: the quality of the camera (focal distance and camera velocity), condition, colour, width and visibility of lane markings (daytime visibility, night-time visibility—retroreflection and contrast to pavement), lane marking configuration (full/dashed, length of dashed lines), driving speed, weather conditions, general visibility of the environment, sun direction, pavement characteristics (type, condition and texture), geometry of the road, type of road edge (structured/unstructured) and combinations of the above factors.

Several studies have investigated how lane markings' characteristics affect the accuracy of lane detection under different conditions. One of the first such studies was conducted in Sweden in 2010 with the aim of testing various types of lane markings (flat/profiled, new/existing) under different weather and lighting conditions [16]. The study concluded that in dry daytime conditions, the luminance coefficient must be at least 5 mcd/lx/m^2 higher than the road surface and that it should be at least 85 mcd/lx/m^2. Furthermore, the study also found that roads wider than seven meters need to have a centre line in order for LSS to become active. Finally, the study highlighted the importance of increased visibility of lane markings in wet and rainy conditions. In 2016, a research was conducted with the aim of identifying the effects of lane markings' characteristics (width, colour and retroreflectivity) on the performance of a machine-vision system [17]. The study concluded that the view range of the investigated machine-vision (Mobileye) is between 6–18 m in front of the vehicle and that, at night-time, the retroreflectivity of lane markings affected the reading quality. Namely, lane markings with higher retroreflectivity increased the reading level and confidence. Also, wider lane markings (15 cm width) were read better when compared with narrower markings (10 cm width), regardless of stripe colour. Similar results were obtained in a 2017 study [18]. The results indicated that the machine-vision (Mobileye) detection of lane markings generally increased with the increase of retroreflection and contrast ratio. However, the authors highlighted that factors such as light bloom from a low-angled sunlight or visual occlusion from rain, snow, or fog may also influence the detection and readability of machine-vision. Furthermore, such systems generally detect markings with the minimal retroreflectivity of 100 mcd/lx/m^2 but do not necessarily provide the strongest detection. An extensive study, which consisted of interviews of stakeholders and on-road and off-road testing, was conducted in Australia in order to determine the implications of road markings for machine vision [11]. Testing scenarios included several test cases which included the impact of different road markings' characteristics (daytime dry luminance coefficient—Qd, daytime dry contrast ratio, day wet contrast ratio, night dry retroreflectivity—R_L, night dry contrast ratio, night wet (recovery) retroreflectivity—R_L, night wet contrast ratio width, marking width), different complex situations (such as road markings' perceptual measures), non-marked edge line, road curvation etc. The authors used several vehicles and a Mobileye camera to test lane detection depending on different scenarios. Based on data analysis, it was concluded that machine-vision detection of solid lines is "better" when compared to dashed lines with same characteristics (equal width, brightness and maintenance). Weather conditions also impacted machine-vision readings differently. Namely, minimal ambient lighting (such as streetlights or low-angled sunlight) may improve the contrast ratio due to reduced specular diffusion and, thus improve line detection. On the other hand, with excessive ambient lighting, machine-vision systems can suffer from 'light bloom'. Furthermore, the contrast ratio for night-time visibility of between 5-to-1 and 10-to-1 between lane markings and the surrounding substrate is needed for proper functioning of the machine-vision system. Also, lane detection during the day was generally less effective than at night-time due to the complexity of visual clutter evident during daylight hours and the fact that retroreflective properties of well-maintained lane markings provide greater contrast during night-time. In addition, the study found that several other factors, such as driving speed, marking width, maintenance practices etc. influence the accuracy of machine-vision.

Most recently, a researcher at the Department of Civil Engineering and Architecture at the University of Catania used Automatic Road Analyzer (ARAN) and Mobileye 6.0 system to investigate how different road factors (road characteristics and conditions) impact the performance of the LSS system [19]. The ARAN was used to obtain measures of road geometric characteristics (cross section, gradients, horizontal and vertical alignment) which were then synchronised with the Mobileye. In addition, the luminance coefficient of the lane marking in diffuse lighting conditions (Qd) was detected with a portable retroreflectometer. Based on the data analysis using a Decision Tree Method, authors concluded that when daytime visibility (Qd) of road markings is lower than 153 mcd/lx/m^2, the probability of LSS failing rises to 11.4% for the calibration sample and 14.35% for the validation sample. Also, curved road sections (with R < 141 m) showed a higher percentage of faults than the average 3% in the test conditions. On the other hand, the average driving speed did not result in any significant changes in LSS accuracy. Overall, the results suggest that a Qd higher than 153 mcd/lx/m^2 improves the detection of lane markings using a Mobileye lane detection system.

Based on literature findings, one can conclude that the function of LSS is influenced by a number of factors and their interaction. However, it is still largely unknown to what extent each factor influences LSS, precisely due to their high number and their interaction (one factor may significantly influence another one in certain conditions). The aforementioned can be seen from the example of "visibility factor", i.e., lane detection between daytime and night-time. Several studies indicated that the accuracy of lane detection varies between day and night-time, however it is still largely unknown to what extent these differences go. For this reason, the aim of this study is to conduct on-road tests in order to determine and compare the detection quality and view range of the machine-vision system during dry daytime and night-time conditions. Based on the aforementioned aim and literature review, the hypotheses of the study are as follows:

- A statistically significant difference of detection quality and view range of lane detection system exists between daytime and night-time conditions;
- Detection quality of lane markings is "better" during night-time compared to daytime;
- View range of the machine-vision system will be longer during daytime compared to night-time.

The results of the study are important for two reasons. First, they provide a valuable input to researchers and developers regarding the "real-world" functioning of machine-vision lane detection. The insight into the variations of lane detection accuracy in different visibility conditions may help further development of such systems. Second, the results may help road authorities in optimizing the quality performance of road markings. Namely, it is still not entirely clear what the minimal quality requirements for road markings are in order to provide adequate accuracy of machine-vision. However, knowing which visibility conditions are more problematic for machine-vision may help road authorities in prioritizing maintenance activities as well as defining minimal visibility levels of road markings.

The manuscript is structured in five main sections. Section 1 presents the research problem and literature findings related to the impact of road markings quality on LSS.

Section 2 describes the research equipment, testing procedure, road sections chosen for the purpose of this research and data analysis. The results of on-road tests are divided into two subsections presented in Section 3: Quality of lane markings' detection and view range of lane markings. In Section 4 the obtained results are discussed and compared to findings from previous studies, including limitations and suggestions for future studies. The last section (Section 5) presents the conclusion of the study and provides potential practical application for the obtained results.

2. Materials and Methods

2.1. Apparatus

The data related to lane detection was recorded using Mobileye 630 system implemented in the testing vehicle (BMW640i) of the Institute of Automotive Engineering, Graz University of Technology (Figure 1). The system is developed by a leading supplier of camera systems for ADAS and it was previously used in several studies [11,14,17,19]. The Mobileye 630 system uses a digital camera with the 38 degrees horizontal and 28 degrees vertical field of view located behind the front windshield inside the vehicle. Using image processing chips, the camera enables high-performance real-time image processing (15 frames per second) of different objects on roads such as lane markings, pedestrians etc. The system, among others, conforms to the Directive 72/245/EEC for electronic equipment which can be built in road vehicles and enables extraction of recorded data for further processing. For the purpose of this study, we recorded the data related to the type of detected longitudinal marking (continuous or dashed), approximate marking width, view range and the quality of the marking both for middle and edge lines. The vehicle was also equipped with a precise measurement system to record the vehicle's trajectory by a combination of GPS localization (Novatel OEM-6–RT2 receiver) and inertial measurement unit (GENESYS ADMA G-III).

Figure 1. Testing vehicle with Mobileye 360 implemented behind the front windshield inside the vehicle.

2.2. Test Road Sections and Procedure

The study was conducted on four rural road sections in Croatia in total length of 120.8 km. The roads were two-way roads with 3.5 m lane width and low traffic volumes. Three road sections were marked with the middle line and partially with edge lines while the fourth road had only the middle line. All markings were white, 15 cm wide and made from solventborne paint (Type I). The main characteristics of the road sections are presented in Table 1.

Table 1. Characteristics of test road sections.

#	Road Lenght (km)	Markings Width (cm)	Length of the Middle Line (km)	Age of the Marking	Length of the Edge Line (km)
1	32.21	15	Solid: 20.61 Dashed: 11.60	Middle: <6 months Edge: >1 year	Solid: 11.73
2	20.53	15	Solid: 14.68 Dashed: 5.85	Middle: <6 months Edge: >1 year	Solid: 11.30
3	38.05	15	Solid: 15.00 Dashed: 23.05	Middle: <6 months Edge: >1 year	Solid: 22.27
4	30.01	15	Solid: 30.08	Middle: >1 year	-

The roads were selected based on the fact that they are rural with low traffic volumes. Rural roads with low traffic volumes have higher risk of road departure accidents [2] and thus the importance of LSS is increased. Also, roads were selected since they have the recommended width of lane markings for LSS [20–22] and on the majority of their length, there is no road lighting. Road lighting was present only on short sections located in populated areas. However, these sections were excluded from the analysis (see Section 2.3).

Each road was "measured" twice: once during daytime and once during night-time. The measurements for night-time conditions were conducted on 21 September 2020 and on 22 September 2020 for daytime conditions. The measurements were done between 10:15 h and 13:00 h during daytime and between 19:20 h and 22:20 h during night-time. In both conditions, the weather was dry and the sky clear. The driving speed was in accordance with the speed limit and differed between 60 km/h and 80 km/h. In this way, we tried to control the impact of speed on the detection of markings since literature suggests that speed has a varying impact on machine-vision (some improve at higher speeds, some degrade) [11].

2.3. Data Analysis

Raw data from the Mobileye device was extracted using Control Area Network (CAN) bus interface separately for each road. As stated in Section 2.1, the data was recorded with Mobileye, and two main variables were analysed: view range and the detection quality of the markings. View range was determined in meters (maximal value 80 m) while the quality level was ranked on the scale from 0 to 3, where 0 equalled "nothing detected", 1 presented "low detection confidence", 2 "medium detection confidence" and 3 "high detection confidence". The thresholds for each quality level are not known to the authors since this information is a "know-how" of the manufacturer. The sampling rate of the camera was set at 100 Hz and GPS coordinates were recorded for each sample.

Since smaller parts of the roads are passing through settlements with road lighting, we have excluded those sections from the analysis in order to eliminate the impact of such environmental lighting on Mobileye during night-time. Also, since edge markings were not located on the whole length of the analysed roads, sections without edge lines are also excluded from the analysis.

The results of the "measurements" between day and night conditions were correlated based on the GPS coordinates. Two analyses were conducted: (1) quality of lane marking detection; and (2) view range. With respect to (1), the number of samples per each quality level was calculated for each road. Wilcoxon signed-rank test was used to test the difference in the number of quality rankings on each road during daytime and night-time. The Wilcoxon signed-rank test is a non-parametric equivalent of the dependent t-test, meaning that it is based on the differences between scores in the two comparing conditions. Once the differences are calculated, they are ranked and the sign of the difference (positive or negative) is assigned to the rank [23]. To calculate the significance of the test statistic (T), the mean ($\overline{T}$) and the standard error ($SE_{\overline{T}}$) are used as shown in the following equation:

$$\overline{T} = \sqrt{n(n+1)4} \quad (1)$$

$$SE_{\overline{T}} = \sqrt{\frac{n(n+1)(2n+1)}{24}} \tag{2}$$

Based on the test statistic, the mean of the test statistic and the standard error, z-score is calculated using the Equation (3):

$$z = \frac{T - \frac{n(n+1)}{4}}{\sqrt{\frac{n(n+1)(2n+1)}{24}}} \tag{3}$$

The view range analysis included the calculation of the absolute averages for each road, line and visibility conditions as well as the average difference and standard deviation when daytime reading quality was higher compared to night-time and vice versa for both markings (middle and edge). Finally, a paired samples t-test was used to test the difference between night-time and daytime conditions for each line on each road. In general, a paired samples t-test compares the means of two measurements taken from the same individual object, or related units. These "paired" measurements can represent a measurement taken at two different times, a measurement taken under two different conditions or measurements taken from two halves or sides of a subject or experimental unit. The test compares the mean difference between samples ($\overline{D}$) to the difference that is expected to be found between population means (μ_D), and then takes into account the standard error of the differences ($S_D/\sqrt{N}$), as shown in the following equation:

$$t = \frac{\overline{D} - \mu_D}{S_D/\sqrt{N}} \tag{4}$$

If the null hypothesis is true, then it is expected that there is no difference between the population means ($\mu_D = 0$). The purpose of the test is to determine whether there is statistical evidence that the mean difference between paired observations is significantly different from zero [23].

In both tests (Wilcoxon signed-rank and paired samples t-test), the significant level was set at 0.05. IBM SPSS 26 was used for statistical analysis.

3. Results

As described in Section 2.3, raw data for each road was purified after extraction in order to exclude road sections without edge markings as well as parts with road lighting. The final data used in the analysis included 2,649,472 samples and for each sample the quality of the detection and the view range were recorded. The total number of samples per each quality level, line type (middle/edge) and visibility conditions (night-time/daytime) is presented in Table 2.

Table 2. Number of analysed samples.

Quality Level	Night-Time		Daytime		Total
	Middle	Edge	Middle	Edge	
0	136,547	269,928	124,530	330,601	861,606
1	21,912	7968	22,029	7767	59,676
2	76,333	111,206	103,074	69,030	359,643
3	523,046	177,796	508,205	159,500	1,368,547
Total	757,838	566,898	757,838	566,898	2,649,472

Further analysis is divided into two subsections: Quality of lane markings' detection and view range of lane markings, and presented in the following chapters.

3.1. Quality of Lane Markings' Detection

During night-time, 69% of all readings of the middle line were ranked as level 3, i.e., "high detection confidence". Around 18% of middle markings were not detected (level 0), almost 2% had "low detection confidence" (level 1) while 10% had "medium detection confidence" (level 2). On the other hand, night-time readings for the edge line are as follows: 31% level 3, 20% level 2, 1% level 1 and 48% level 0.

The overall results for the middle line show that the number of samples classified as level 3 and 0 differed around 2% between daytime and night-time with lower number of samples recorded during daytime. On the other hand, the number of samples classified as level 2 during daytime increased by reaching 3.5% difference compared to night-time, while level 1 varied slightly between the two conditions. A similar result was found for the edge marking. During daytime, the number of level 3 and 2 samples decreased compared to night-time, with differences of around 3% and 7% respectively. The number of level 1 samples increased during daytime and differed by 10.7% to the night-time, while level 1 stayed approximately the same.

The afore-presented results are shown in Figure 2.

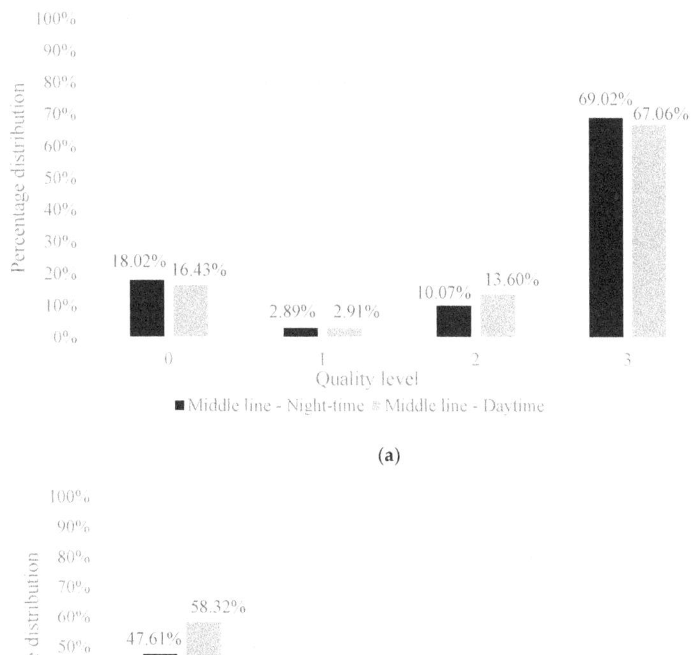

(a)

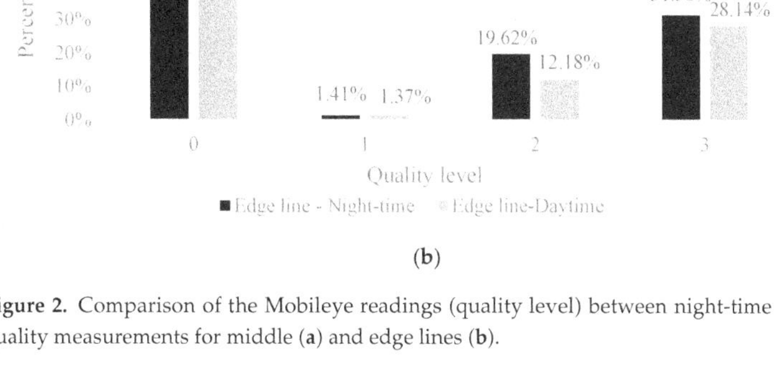

(b)

Figure 2. Comparison of the Mobileye readings (quality level) between night-time and daytime quality measurements for middle (**a**) and edge lines (**b**).

Overall, when analysing all roads and lines together, on average 50% of daytime readings were classified as level 3, 13% as level 2, 2.2% as level 1 and 34% as level 0. On the other hand, those averages changed to some extent during night-time as visible from the Figure 3. Namely, the average number of level 3 readings decreased by almost 5% in daytime compared to night-time conditions. It is similar for level 2 where the decrease is 8% while the number of level 1 readings decreased slightly (0.28%). On the other hand, the number of level 0 readings in daytime increased by 12% compared to night-time.

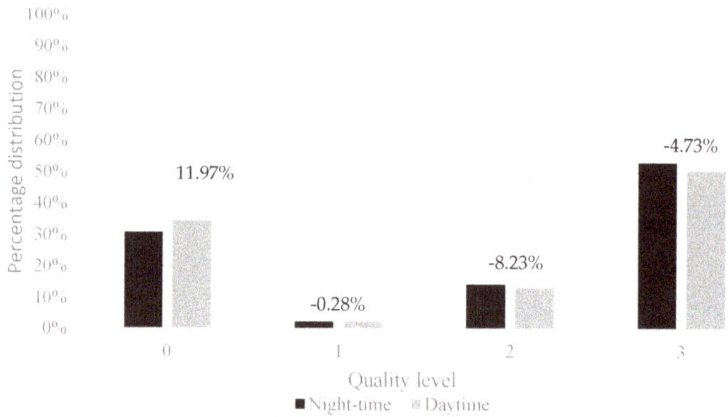

Figure 3. Number of overall reading qualities during daytime and night-time as well as their differences (day–night).

When looking at each road separately, the results show that on all the roads the percentage of level 3 readings is higher during night-time when compared to daytime, reaching the maximum difference on road four—3.52%. On two roads (one and four), the percentage of level 2 daytime readings is higher compared to night-time and lower on other two roads (two and three). The percentages of level 1 readings remained approximately the same in both conditions while the number of level 0 readings is higher during daytime on three roads (one, two and three) with the maximum of 6.86%. The results for each road are presented in Table 3.

Table 3. Frequencies and percentages of samples by each marking quality rank for middle and edge lines during night-time.

Road	Marking Quality	Daytime	Night-Time	Daytime–Night-Time
1	0	32.61%	31.38%	1.24%
	1	2.70%	2.99%	−0.29%
	2	19.60%	17.70%	1.89%
	3	45.10%	47.93%	−2.84%
2	0	46.48%	39.62%	6.86%
	1	3.18%	2.89%	0.29%
	2	6.98%	10.92%	−3.94%
	3	43.36%	46.57%	−3.21%
3	0	28.59%	22.98%	5.60%
	1	1.50%	1.43%	0.07%
	2	11.31%	15.57%	−4.26%
	3	58.61%	60.02%	−1.41%
4	0	53.25%	47.07%	−0.42%
	1	1.48%	1.64%	0.00%
	2	16.04%	13.67%	3.94%
	3	29.23%	37.62%	−3.52%

In order to test the statistical difference between the detection quality of lane markings during daytime compared to night-time, a Wilcoxon signed-rank test was used. Table 4 presents the results of the aforementioned tests, separately for middle and edge lines on all four roads. Although the majority of samples had "equal readings" in both conditions (day = night), it can be concluded that for all four roads a statistically significant difference in the number of samples between visibility conditions still exists ($p < 0.05$). Roads 1 and 4 have more samples which had higher readings during night-time compared to daytime, while it is the opposite on roads 2 and 3.

Table 4. Results of Wilcoxon signed-rank test for middle and edge lines on analysed roads.

Road	Comparison	Middle Line			Edge Line		
		N	Z	Asymp. Sig.(2-Tailed)	N	Z	Asymp. Sig.(2-Tailed)
1	Day < Night	59.899	−4.62	<0.05	45.890	−20.532	<0.05
	Day > Night	40.474			48.913		
	Day = Night	98.369			103.939		
2	Day < Night	20.177	−35.144	<0.05	36.485	−109.836	<0.05
	Day > Night	32.018			13.910		
	Day = Night	79.705			81.505		
3	Day < Night	38.225	−14.26	<0.05	85.202	−76.954	<0.05
	Day > Night	40.965			62.247		
	Day = Night	157.066			88.807		
4	Day < Night	47.808	−8.297	<0.05	-	-	-
	Day > Night	45.717			-		
	Day = Night	97.415			-		

3.2. *View Range of Lane Markings*

Overall, the absolute average of Mobileye's view range for the middle line during night-time was 34.07 ± 22.23 m. During daytime, the view range increased and averaged 39.42 ± 25.36 m. On the other hand, the range for the edge lines was lower compared to the middle lines in both daytime and night-time conditions, averaging 17.69 ± 24.38 m and 17.01 ± 20.48 m, respectively. Absolute averages of the view range for each road, line and visibility condition is presented in Table 5.

Furthermore, we calculated the average difference and the standard deviation of the view range when daytime reading quality was higher compared to night-time and vice versa for both markings (middle and edge). On average, the difference in the view range for the middle line when daytime had higher reading quality was around 29 m with the standard deviation of around 21 m. On the other hand, when reading quality was higher during night-time, a slight decrease of the view range difference and standard deviation was recorded. There is a similar trend for the edge line as well but the difference in view range for "Daytime > Night-time" is almost 40 m and "Night-time > Daytime" around 27 m as shown in Table 6.

Table 5. Absolute averages of the view range for each road, line and visibility condition.

Line/Condition	Mean	Std.	95% Confidence Interval	
			Lower Bound	Upper Bound
		Road 1		
Middle line/Night-time	33.16	21.89	33.07	33.26
Edge line/Night-time	15.67	20.54	15.58	15.76
Middle line/Daytime	38.93	24.66	38.82	39.04
Edge line/Daytime	21.43	27.25	21.31	21.55
		Road 2		
Middle line/Night-time	33.83	20.58	33.72	33.94
Edge line/Night-time	9.58	15.72	9.49	9.66
Middle line/Daytime	41.57	24.64	41.44	41.71
Edge line/Daytime	6.09	16.77	6.00	6.18
		Road 3		
Middle line/Night-time	42.68	22.82	42.58	42.77
Edge line/Night-time	25.79	25.19	25.69	25.89
Middle line/Daytime	46.43	24.17	46.34	46.53
Edge line/Daytime	25.53	29.11	25.42	25.65
		Road 4		
Middle line/Night-time	26.60	23.64	26.49	26.71
Middle line/Daytime	30.72	27.96	30.59	30.84

Table 6. Average differences in view range of middle and edge lines for each road and condition.

Road	Middle Line				Edge Line			
	Daytime > Night		Daytime < Night		Daytime > Night		Daytime < Night	
	Average Diff. (m)	Std.	Average Diff. (m)	Std.	Average Diff. (m)	Std.	Average Diff. (m)	Std.
1	26.43	18.36	22.97	17.30	37.40	21.03	21.31	17.51
2	31.39	25.90	26.26	24.00	38.80	20.93	29.30	15.12
3	27.68	21.03	24.95	18.03	40.71	25.31	32.74	22.47
4	30.67	19.77	30.00	18.39	-	-	-	-
Average	29.04	21.26	26.04	19.43	38.97	22.42	27.78	18.36

In addition, a paired samples t-test was used to test whether a significant difference between night-time and daytime conditions for each line on each road exists. In total 7 pairs were made as shown in Table 7. The results of the t-test show that the view range between all pairs is statistically different ($p < 0.05$), i.e., that Mobileye view range differed for all test cases between daytime and night-time conditions.

Table 7. Results of paired samples *t*-test.

	Road/Pairs	Paired Differences					*t*	*P* (2-Tailed)
		Mean	Std.	Std. Error Mean	95% Confidence Interval of the Difference			
					Lower	Upper		
1	Middle line-Night-time vs. Middle line-Daytime	−5.76	29.51	0.06	−5.89	−5.63	−87.09	<0.05
2	Edge line-Night-time vs. Edge line-Daytime	−5.76	27.11	0.06	−5.88	−5.64	−94.85	<0.05
3	Middle line-Night-time vs. Middle line-Daytime	−7.74	30.22	0.08	−7.90	−7.57	−93.02	<0.05
4	Edge line-Night-time vs. Edge line-Daytime	3.49	22.76	0.06	3.36	3.61	55.67	<0.05
5	Middle line-Night-time vs. Middle line-Daytime	−3.75	32.58	0.06	−3.88	−3.62	−56.04	<0.05
6	Edge line-Night-time vs. Edge line-Daytime	0.25	35.78	0.07	0.11	0.40	3.49	<0.05
7	Middle line-Night-time vs. Middle line-Daytime	−4.12	32.50	0.07	−4.26	−3.97	−55.39	<0.05

4. Discussion

Although several studies [11,14,15] investigated how different factors affect lane detection and thus proper functioning of LSS, gaps in literature still exist. Mainly, these gaps are related to determining the adequate levels of lane markings' visibility in different conditions. For this reason, an on-road test was conducted to determine and compare the detection quality and view range of machine-vision system during daytime and night-time.

The overall results show that the number of each quality level (0–3) as output from the camera differed between visibility conditions. Namely, the average number of level 3 and 2 readings decreased by 5% and 8% in daytime compared to night-time conditions. The share of level 1 remained approximately the same (slight decrease—0.28%) while the number of level 0 readings in daytime increased by 12% compared to night-time indicating potential failures in lane detection during daytime. Although the conducted tests show that a significant difference of quality readings between visibility conditions exists, it has to be noted that differences are relatively small and may not influence the functioning of the lane detection system. The main concern from the practical point is related to the 12% increase of level 0 readings during daytime compared to night-time. Since level 1 readings in both conditions stayed approximately the same, it is reasonable to conclude that in some cases markings were not detected by Mobileye (level 0) during daytime while during night-time they were detected with medium (level 2) or high detection confidence (level 3). Since the driving speed, the sun direction in daytime conditions and the impact of road lightning in night-time conditions were controlled, the potential reason for such results is related to the difference in visual complexity between the two conditions (day vs night-time). Due to the complexity of visual clutter during daytime, the contrast ratio between the marking and the road may differ and decrease thus affecting the proper functioning of lane detection. This is also suggested by previous studies in which it was found that the optimum contrast between marking and road surface should be around 3:1 during daytime [11,19]. On the other hand, the aforementioned contrast ratio during night-time is generally much higher due to the fact that the surrounding environment is dark and the visibility of lane markings is achieved with the use of retroreflective materials, i.e., glass beads which return the incoming light ray from the vehicle headlights back to the source (driver) [24]. In addition, the number of samples classified as level 0 detection quality was much higher for edge markings compared to middle markings, both during daytime and nigh-time. This suggests that the quality of the marking plays an important role. On three roads the middle line was relatively new (<6 months,) and on one road it was older than one year. On the other hand, edge markings on all three roads were older than one year. Several studies proved that age affects visibility properties (daytime and night-time visibility) of lane markings, especially for paint markings [25–28]. Since the service life of paint markings is usually around one year, it is reasonable to conclude that the visibility properties of edge lines in this study were lower compared to newer middle lines.

Although the detection quality slightly decreased in daytime compared to night-time, the view range increased, however differently for middle and edge lane marking: the absolute average and the standard deviation of the view range for the middle line during daytime was 39.42 $\pm$ 25.36 m compared to 34.07 $\pm$ 22.23 m in night-time conditions. The range for edge lines was shorter compared to middle lines in both daytime and night-time conditions averaging 17.69 $\pm$ 24.38 m and 17.01 $\pm$ 20.48 m, respectively. Such results are lower compared to the study [17]. However, this is mainly due to the different methodology. The aforementioned study tested lane detection in a static environment with consistent parameters (new markings, lighting, road surface retroreflectivity, etc.) and used older version of Mobileye (560). Overall, form the practical point of view, the differences of the results between visibility conditions are, as in the case of detection quality, relatively small. However, there is a much higher difference between the view range of middle and edge markings. This may be due to the markings' age. As described in previous paragraph, the age of the marking affects its visibility properties (daytime and night-time visibility). Since edge lines in this study were older compared to the middle lines, it is reasonable to conclude that their visibility properties were lower compared to newer middle lines. Furthermore, relatively high standard deviations of the view range (around 20 m) for all markings suggest that, beside the markings age and its visibility, road geometry plays an important role in lane marking detection. Since Mobileye camera is fixed to the vehicle windshield, it is not moving like human drivers' head and eyes would, meaning that it "looks" at a relatively fixed area at the more or less the same angle, i.e. the viewing area and angle are changed only with the change of the vehicles movement. This indicates that with the change of road geometry such as curves, dips, slopes etc., middle and edge lane markings will be "seen" by the "fixed" machine-vision at different distances. The aforementioned findings further support the findings from literature [11,14,15].

Although this study provided valuable results, there are several limitations. First, lane markings' visibility properties were not taken into account due to the fact that the main objective of the study is to compare the detection quality and view range of a machine-vision system during daytime and night-time and to analyse how these values change depending on the visibility conditions alone. Further research is needed to determine the relationship between daytime and night-time visibility on the detection quality and view range of LSSs. Furthermore, we did not evaluate how different road geometry and lane markings' configurations (dashed vs solid) affect the detection of markings. Existing literature indicates that both road geometry and configuration of lane markings are potential influencing factors [11]. However, due to the lack of data related to the road geometry (curves, dips, slopes etc.) and exact location of each type of dashed lines, such analysis was not conducted. Also, the daytime measurements were conducted between 10:15 h and 13:00 h which may have, to some extent, affected the results since the sun direction is one of the factors that affects lane markings' detection by machine-vision systems [11,14,15]. However, all measurements were conducted on the same day under the clear sky and on roads whose direction is not in the direction of the sun (north-south direction) so the effect of the aforementioned factor is limited and negligible. The effect of driving speed was also not analysed, although the literature suggests that vehicle speed has a varying impact on machine-vision systems—some improve at higher speeds and some degrade [11]. Since the tests in this study were conducted on rural roads, the driving speed was in accordance with the speed limit (between 60 km/h and 80 km/h) and thus the impact of speed was controlled. Lastly, only one machine-vision system (Mobileye 630) was tested and overall results may not be applicable to other such systems.

Since the differences between lane markings' detection found in this study are relatively small, more studies are needed to further validate here presented results. Besides determining the relationship between daytime and night-time visibility on the detection quality and view range of LSS, future studies should focus on detailed investigation of the influence of other factors on the LSS such as weather conditions, road geometry (curves, dips, slopes etc.), markings' characteristics (colour, width, configuration), driving speed, road surface characteristics (type, condition and texture) etc., as well as combinations of these factors. Therefore, we propose that future research combines the field and on-road tests in order to further expand the existing knowledge and thus improve the overall quality of machine-vision systems as well as to determine adequate properties of lane markings.

5. Conclusions

The results of this study indicate that lane marking detection quality and view range by machine-vision system differs to some extent between dry daytime and night-time conditions. The field test results show that detection quality of lane markings is "better" during night-time compared to daytime. However, it has to be noted that the aforementioned differences are relatively small and may not critically influence the functioning of the lane detection system. Nevertheless, the results presented here support previous findings and provide further proof that visibility conditions play an important role in lane detection and that, during daytime, the complexity of visual clutter decreases the contrast ratio between the marking and the road surface and thus affects the detection quality and view range of machine-vision. In addition, the results suggest that other factors such as road geometry, markings' age and quality are also influential and should be further evaluated.

Overall, the findings of this study provide a quantization of the effect of surrounding visibility on lane detection and thus contribute to expanding the existing knowledge regarding lane detection by machine-vision. Here presented methodology and results may be useful for researchers in designing and evaluating similar studies. Furthermore, the results may be useful to road authorities. Although, the study did not evaluate how different visibilities of road markings affect their detection by machine-vision, the results support previous findings which indicate that the detection of lane markings is much more problematic for machine-vision during daytime compared to night-time. This finding is useful for road authorities and may help them in prioritizing and optimizing road marking maintenance activities. Depending on the weather and traffic conditions and general road characteristics (type of the road, width, general road geometry, surface condition etc.), road authorities should adopt different maintenance policies to ensure proper and timely maintenance and thus adequate quality of road markings needed for both for human drivers as well as machine-vision systems. Finally, the presented findings may help the developers of machine-vision systems in detecting critical situations and conditions which negatively affect lane marking detection.

Author Contributions: Conceptualization, D.B. (Darko Babić), D.B. (Dario Babić) and M.F.; methodology, D.B. (Darko Babić), D.B. (Dario Babić) and M.F.; formal analysis, D.B. (Dario Babić) and M.F.; investigation, D.B. (Darko Babić), D.B. (Dario Babić), M.F., A.E. and Z.F.M.; resources, D.B. (Darko Babić) and A.E.; data curation, D.B. (Dario Babić) and Z.F.M.; writing—original draft preparation, D.B. (Dario Babić), M.F., A.E.; visualization, D.B. (Dario Babić); supervision, D.B. (Darko Babić); project administration, D.B. (Darko Babić); funding acquisition, D.B. (Darko Babić). All authors have read and agreed to the published version of the manuscript.

Funding: This research was a part of the project entitled "Establishing a Methodology for Testing and Evaluating the ADAS Systems" funded by University of Zagreb (Potpore za temeljno financiranje znanstvene i umjetničke djelatnosti Sveučilišta u Zagrebu u ak. god. 2019/2020).

Institutional Review Board Statement: Not applicable.

Informed Consent Statement: Not applicable.

Data Availability Statement: The data used to support the findings of this study are available from the corresponding author upon request.

Conflicts of Interest: The authors declare no conflict of interest.

References

1. Federal Highway Administration. Roadway Departure Safety. 2020. Available online: https://safety.fhwa.dot.gov/roadway_dept/ (accessed on 20 December 2020).
2. Xiong, X.; Chen, L.; Liang, J. Analysis of roadway traffic accidents based on rough sets and Bayesian networks. *Promet Traffic Transp.* **2018**, *30*, 71–81. [CrossRef]
3. National Cooperative Highway Research Program. Best Practices in Lane-Departure Avoidance and Traffic Calming. Project 20-68A, USA. 2011. Available online: http://onlinepubs.trb.org/onlinepubs/nchrp/docs/NCHRP20-68A_09-03.pdf (accessed on 11 December 2020).
4. American Association of State Highway and Transportation Officials. Driving down Lane-Departure Crashes. Report, USA. 2008. Available online: http://www.virginiadot.org/business/resources/LocDes/Lane_Departures_PLD-1.pdf (accessed on 11 December 2020).
5. Babić, D.; Fiolić, M.; Babić, D.; Timothy, G. Road markings and their impact on driver behaviour and road safety: A systematic review of current findings. *J. Adv. Transp.* **2020**, 7843743. [CrossRef]
6. Eichberger, A.; Rohm, R.; Hirschberg, W.; Tomasch, E.; Steffan, H. RCS-TUG Study: Benefit potential investigation of traffic safety systems with respect to different vehicle categories. In Proceedings of the 22nd International Conference on the Enhanced Safety of Vehicles (ESV), Washington, DC, USA, 13–16 June 2011.
7. Sternlund, S. The safety potential of lane departure warning systems—A descriptive real-world study of fatal lane departure passenger car crashes in Sweden. *Traffic Injury Prev.* **2017**, *18*, 18–23. [CrossRef] [PubMed]
8. Sternlund, S.; Strandroth, J.; Rizzi, M.; Lie, A.; Tingvall, C. The effectiveness of lane departure warning systems—A reduction in real-world passenger car injury crashes. *Traffic Injury Prev.* **2017**, *18*, 225–229. [CrossRef] [PubMed]
9. Penmetsa, P.; Hudnall, M.; Nambisan, S. Potential safety benefits of lane departure prevention technology. *IATSS Res.* **2019**, *43*, 21–26. [CrossRef]
10. Hata, A.Y.; Wolf, D. Road marking detection using LIDAR reflective intensity data and its application to vehicle localization. In Proceedings of the 17th International IEEE Conference on Intelligent Transportation Systems (ITSC), Qingdao, China, 8–11 October 2014. [CrossRef]
11. Austroads. Implications of Pavement Markings for Machine Vision. Research Report AP-R633-20. 2020. Available online: https://austroads.com.au/publications/connected-and-automated-vehicles/ap-r633-20 (accessed on 20 December 2020).
12. Nguyen, V.; Kim, H.; Jun, S.; Boo, K. A study on real-time detection method of lane and vehicle for lane change assistant system using vision system on highway. *Eng. Sci. Technol.* **2018**, *21*, 822–833. [CrossRef]
13. Hillel, A.B.; Lerner, R.; Levi, D.; Raz, G. Recent progress in road and lane detection: A survey. *Mach. Vis. Appl.* **2014**, *25*, 727–745. [CrossRef]
14. Texas A&M Transportation Institute. Road Markings for Machine Vision, NCHRP Project 20-102(6). 2016. Available online: http://sp.scote.transportation.org/Documents/2016%20SCOTE%20Meeting/Tuesday%207JUN17/NCHRP%2020-102(6)%20update.pdf (accessed on 20 December 2020).
15. Mistry, V.H.; Makwana, R. Survey: Vision based road detection techniques. *Comput. Sci. Inf. Technol.* **2014**, *5*, 4741–4747.
16. Lundkvist, S.O.; Fors, C. *Lane Departure Warning System—LDW*; VTI: Linköping, Sweden, 2010.
17. Potters Industry and Mobileye. Pavement Markings Guiding Autonomous Vehicles—A Real World Study. 2016. Available online: https://higherlogicdownload.s3.amazonaws.com/AUVSI/14c12c18-fde1-4c1d-8548-035ad166c766/UploadedImages/documents/Breakouts/20-2%20Physical%20Infrastructure.pdf (accessed on 20 December 2020).
18. Carlson, P.J.; Poorsartep, M. Enhancing the roadway physical infrastructure for advanced vehicle technologies: A case study in pavement markings for machine vision and a road map towards a better understanding. In Proceedings of the 96th Annual Meeting Transportation Research Board, Washington, DC, USA, 8–12 January 2017.
19. Pappalardo, G.; Cafiso, S.; Di Graziano, D.; Severino, A. Decision tree method to analyze the performance of lane support systems. *Sustainability* **2021**, *13*, 846. [CrossRef]
20. European Union Road Federation. *Marking the Way towards a Safer Future*; ERF: Brussels, Belgium, 2013.
21. American Traffic Safety Services Association. ATSSA Policy on Road Markings for Machine Vision Systems. 2019. Available online: https://www.reflective-systems.com/wp-content/uploads/2019/04/Policy-re-Road-Markings-for-Machine-Vision-Systems.pdf (accessed on 20 December 2020).
22. EuroRAP. Roads That Cars Can Read. 2011. Available online: http://www.eurorap.org/wp-content/uploads/2015/04/20110629-Roads-That-Cars-Can-Read-June-2011.pdf (accessed on 20 December 2020).
23. Howell, C.D. *Statistical Methods for Psychology*, 7th ed.; Cengage Learning: Belmont, CA, USA, 2010.
24. Babić, D.; Ščukanec, A.; Babić, D. Determining the correlation between daytime and night-time road markings visibility. *Balt. J. Road Bridge Eng.* **2016**, *11*, 283–290. [CrossRef]
25. Babić, D.; Ščukanec, A.; Babić, D.; Fiolić, M. Model for predicting road markings service life. *Balt. J. Road Bridge Eng.* **2019**, *14*, 341–359. [CrossRef]

26. Sitzabee, W.E.; Hummer, J.E.; Rasdorf, W. Pavement Marking degradation modelling and analysis. *J. Infrastruct. Syst.* **2009**, *15*, 190–199. [CrossRef]
27. Hummer, J.E.; Rasdorf, W.; Zhang, G. Linear mixed-effects models for paint pavement-marking retroreflectivity data. *J. Transp. Eng.* **2011**, *137*, 705–716. [CrossRef]
28. Mull, D.M.; Sitzabee, W.E. Paint pavement marking performance prediction model. *J. Transp. Eng.* **2011**, *138*, 618–624. [CrossRef]

energies

Article

Parking Pose Generation for Autonomous Freight Collection by Pallet Handling Car-like Robot

Khayyam Masood [1,2,*,†], David Pérez Morales [2], Vincent Fremont [2], Matteo Zoppi [1] and Rezia Molfino [1]

1 PMAR Robotics, DIME, University of Genova, 16126 Genova, Italy; zoppi@dimec.unige.it (M.Z.); molfino@dimec.unige.it (R.M.)
2 LS2N, CNRS, Ecole Centrale de Nantes, 1 Rue de la Noë, 44300 Nantes, France; david.perezmorales@ls2n.fr (D.P.M.); vincent.fremont@ec-nantes.fr (V.F.)
* Correspondence: khayyam.masood@edu.unige.it; Tel.: +39-3205304054
† Current address: Room 213, LS2N, CNRS, Ecole Centrale de Nantes, 1 Rue de la Noë, 44300 Nantes, France.

Abstract: This paper focuses on autonomous navigation for an electric freight vehicle designed to collect freight autonomously using pallet handling robots installed in the vehicle. Apart from autonomous vehicle navigation, the primary hurdle for vehicle autonomy is the autonomous collection of freight irrespective of freight orientation/location. This research focuses on generating parking pose for the vehicle irrespective of the orientation of freight for its autonomous collection. Freight orientation is calculated by capturing the freight through onboard sensors. Afterward, this information creates a parking pose using mathematical equations and knowledge of the vehicle and freight collection limitations. Separate parking spots are generated for separate loading bays of the vehicle depending on the availability of the loading bay. Finally, results are captured and verified for different orientations of freight to conclude the research.

Keywords: parking generation; freight handling; autonomous vehicles; driver-less parking

Citation: Masood, K.; Morales, D.P.; Fremont, V.; Zoppi, M.; Molfino, R. Parking Pose Generation for Autonomous Freight Collection by Pallet Handling Car-like Robot. *Energies* **2021**, *14*, 4677. https://doi.org/10.3390/en14154677

Academic Editor: Guzek Marek

Received: 5 July 2021
Accepted: 30 July 2021
Published: 1 August 2021

1. Introduction

Vehicle FURBOT (Freight Urban RoBOTic vehicle) is a complete drive-by-wire electric freight vehicle designed to operate in last-mile delivery operations in an urban environment. The vehicle was part of the European Green Vehicles Initiative (EGVI) funded under the umbrella of the FP7-Transport European project ending in December 2015 [1,2]. The vehicle was completed as an entire drive-by-wire vehicle with the possibility of being upgraded to an autonomous vehicle. Currently, the project is being developed to convert it from a drive-by-wire to a completely autonomous vehicle. Thus, the necessity of autonomous freight collection is generated.

For autonomous vehicle navigation, a mathematical model has been built for performance evaluation of the vehicle [3] which is later used for developing sensor-based strategies for obstacle avoidance [4]. Further automation is achieved by controlling the pallet handling robot using hydraulic pressure-flow rate control [5]. Strategies are built to convert the vehicle from drive-by-wire to a completely autonomous vehicle [6]. Additionally, work on perception and control strategies for autonomous docking of the freight has also been previously studied [7]. This work is the next step in attaining the automation of the vehicle for it to auto-load its freight.

FURBOT has to take part in SHOW (SHared automation Operating models for Worldwide adoption) project where it is required to deliver freight autonomously to customers across an urban area. Currently, the vehicle is not equipped with proper algorithms for the autonomous collection of freight. However, the vehicle will be equipped with autonomous navigation software, which has its own autonomous parking algorithms [8,9]. The autonomous freight collection requires the vehicle to park next to the freight, such that it can collect freight autonomously. For that, a need for generation of correct parking pose is required so the already available parking algorithms can maneuver the vehicle to the correct parking pose. Thus, the need for our work.

The role of autonomous vehicles in an urban environment is still being studied, and there is a gap in understanding the role of urban autonomous vehicles in smart urban mobility [10]. Autonomous freight delivery and their adoption in society are in preliminary stages, and surveys are being conducted for their acceptability in the society [11]. Further studies are being conducted to understand the role of autonomous vehicles in logistics [12]. Still, the concept of using manned deliveries for last-mile delivery through autonomous vehicles is being studied [13]. Privately-owned autonomous vehicles not designed for freight delivery are being looked into, and their potential is being discussed [14]. Currently, autonomous freight delivery through a designated autonomous freight vehicle is still a new subject, and the latest research proves that the solution for autonomous freight delivery is yet to be explored.

Freight delivery in an urban environment comes with many predefined constraints, e.g., loading constraints, vehicle routing problems [15] or compartment and order fulfillment related issues [16]. Many types of loading constraints and vehicle routing problems are discussed in [15,17]. Vehicle compartment issues are further divided into the flexibility of compartment sizes, assignment of product types, and share-ability of compartments. In contrast, order fulfillment issues comprise the mode of demand fulfillment and the total number of visits per customer [16]. However, all these logistic issues do not deal with how the freight needs to be loaded in the vehicle, especially dealing with the freight loading autonomously.

The generation of a parking spot for autonomous vehicles has also been studied in-depth in the last decade. Many next-generation vehicles are already equipped with autonomous parking [18]. New research in parking solutions varies in finding niche problems within the vehicle parking solution domain, including vision-based indoor parking [19], parking of fleet of vehicles [20], collision-avoidance-based parking [21] and sensor (Lidar)-based parking solutions [22]. However, it is challenging to find parking solutions for freight vehicles, especially for freight loading. The closest work on autonomous parking for articulate vehicles is studied in [23], but it is still insufficient in resolving freight-based parking solutions. As FURBOT is a designated freight handling vehicle, the parking solution for autonomous loading of freight needs to be resolved.

Path generation and detecting correct parking pose for the vehicle are interesting subjects, especially in autonomous vehicles as they need to park by themselves. Usual work on vehicle parking pose estimation comes from parking pose marking recognition [24] or using a camera to detect these markings [25]. Apart from parking pose recognition from sensor feedback, estimation of parking pose requires solving geometrical equations for path generation or calculating current parking pose [26,27]. This research also exploits geometrical equations after freight detection to generate a parking pose that can solve the autonomous freight collection problem.

Electric freight vehicle poses a unique challenge for the creation of parking pose concerning freight for its collection. As it is not possible to make the freight perfectly align with any known orientation before loading the freight. Furthermore, there is a hassle for aligning freight to an orientation each time before loading the freight, even if a solution exists. It is much easier to consider autonomously loading the freight through aligning the vehicle w.r.t the freight. However, previous research in this domain is not available, i.e., the generation of parking pose for a freight handling vehicle for freight collection. Furthermore, the vehicle FURBOT is unique because it loads the freight sideways within itself through its loading bays, unlike typical forklift trucks.

This research is a step towards an autonomous collection of freight for the vehicle, especially the part of creating autonomous parking pose w.r.t freight keeping in line the constraints posed by vehicle FURBOT. Currently, there is no alternative solution for collecting freight autonomously for such vehicles. The ability to create a parking pose w.r.t an inanimate object, which is subject to change its orientation and position, is not previously studied and discussed, thus the need of this research. The highlight of this research is creating a parking pose for such a vehicle, keeping in mind the constraints posed by the freight and the vehicle. Finally, testing the results to validate the proposed mathematical modeling for the solution to the unique problem.

In the next section, we discuss the problem and approach to the parallel freight parking issue, discussing the issues concerning freight detection, how the vehicle should be oriented for collection of the freight, and how parking spot should be defined w.r.t freight. Following, in Section 3, the software and control architecture of the vehicle is discussed, focusing on desired parking pose for the collection of freight. In Section 4, a mathematical model for vehicle parking pose is discussed. This section also covers the vehicle kinematic model and mathematical equations and notations for creating the parking pose for each loading bay of the vehicle. Section 5 elaborates on the desired results achieved of the parking pose for each loading bay of the vehicle, and the results are further verified for different freight poses, and the parking pose is validated as per requirement. Finally, in Section 6, the conclusions of the research are discussed.

2. Problem and Approach

The considered vehicle has to park, keeping the freight on the vehicle's right-hand side for loading the freight through the designed loading bays. There are three core steps in defining the parking spot for the vehicle for the autonomous collection of freight. The critical issues in defining the parking spot are freight detection, the vehicle's orientation, and defining parking spot reference to freight. These issues are further elaborated in their sub-sections below.

2.1. Freight Detection

The first step in defining the parking spot for the vehicle for loading freight autonomously is the detection of freight location. The freight vehicle is equipped with 3D LIDAR, which can identify and distinguish freight from the environment as the freight has a unique shape, size and color from the usual environment. Image feedback from an on-board camera is also used to distinguish freight from the environment. The design and dimensions of the vehicle freight box are given in Figure 1 (dimensions in mm), which also show the freight edges (highlighted in red) that need to be captured for correctly identifying the parking spot next to freight. Further dimensions for the freight box are available in [28].

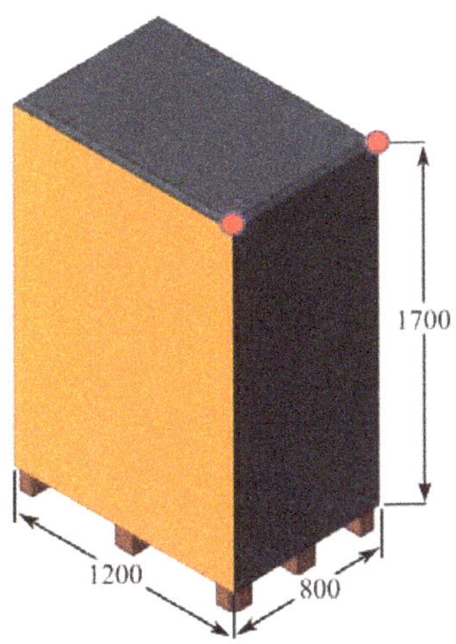

Figure 1. The design of FURBOT freight box.

The vehicle can load two freight boxes simultaneously, and because of the dimensions of the freight, it loads the freight from the depth edge of the freight (measuring 800 mm, as seen in Figure 1), as shown in Figure 2.

Figure 2. Loading of freight.

2.2. Orientation of Vehicle

The next consideration for loading is to park the vehicle next to the freight in such a manner that the vehicle's right side is facing the freight, as shown in Figure 3, since the loading mechanism can only operate from the right-hand side of the vehicle. The maximum tolerance for collection of freight between the freight and vehicle is 300 mm [29]. This limitation, along with the width of the vehicle, defines our parking spot's total width.

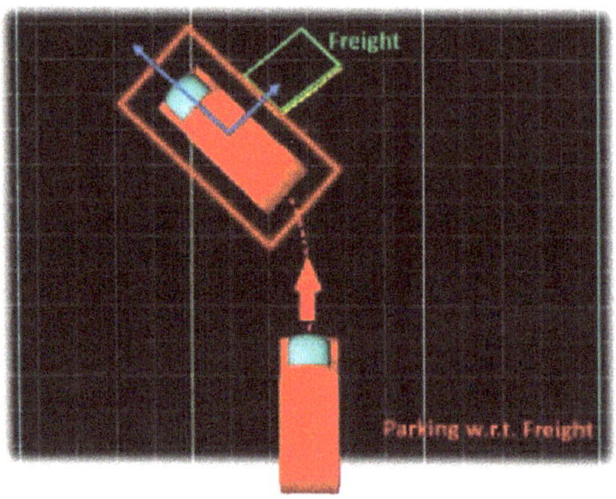

Figure 3. Parking w.r.t freight.

Furthermore, the vehicle is equipped with two separate loading bays, as shown in Figure 4. The freight has to be loaded in the vacant loading bay, i.e., if bay 1 is occupied, then load the freight in bay 2. As the loading bays are not centrally aligned with the freight, the vehicle has to adjust its parking spot accordingly so that the freight can be loaded in the available loading bay. Furthermore, the freight also needs to be aligned with the center of the forklift of the loading bays (as shown in Figure 4) for the forklifts to be inserted into the freight pallet.

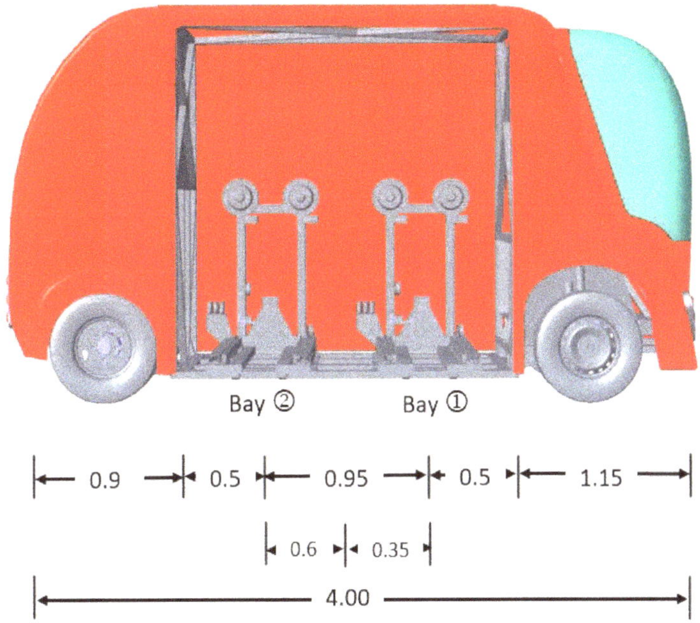

Figure 4. Vehicle dimensions w.r.t loading bays.

2.3. Defining Parking Spot Reference to Freight

The vehicle dimensions are given in Table 1. The width of the vehicle plus the tolerance to the freight defines the width of the parking space. To maintain a reasonable amount of parking maneuvers ($\leq$3 for typical cases) [30], the length of the parking space is chosen to be 2 m more than the length of the vehicle.

Table 1. Vehicle dimensions in meters.

Vehicle Dimensions (m)	Length	Width	Height
	4	1.5	2

Taking the loading of the freight and vehicle dimensions into consideration, the overall parking spot defined for the vehicle is a box of 2.1 $\times$ 6.0 m, as shown in Figure 3.

3. Software and Control Architecture

To speed up the development and integration process with the real experimental vehicle, Robot Operating System (ROS)-based autonomous vehicle's software architecture (called ICARS (Software being developed at LS2N (Laboratoire des Sciences du Numérique de Nantes) www.ls2n.fr (accessed on 15 July 2021)) will be exploited (Figure 5). Considering the task to be accomplished, a particular interest is placed on the Multi-Sensor-Based Predictive Controller (MSBPC) used for parking [31]. As the name of the approach suggests, the parking controller is based on a combination of Model Predictive Control (MPC) and Multi-Sensor-Based Control in order to perform safe (collision-free) parking operations that rely solely on locally perceived sensor features signals and without needing to explicitly plan any path. Moreover, since the technique exploits locally perceived sensor features at each time instant, no localization system is inherently required. The MSBPC approach is now recalled.

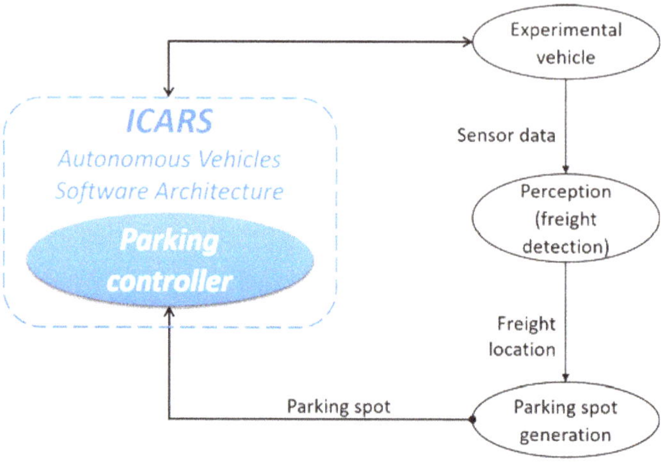

Figure 5. FURBOT software architecture.

An internal-model-control (IMC) structure [32] is used as a basis for formalizing the MSBPC approach (Figure 6). The robotized vehicle and perception system compose the System block. The input to this block is the control variable $\mathbf{v}_r = [v, \delta]^T$, where the longitudinal velocity is denoted by v and the steering angle by δ, while its output is the current value of the sensor features (i.e., corners of the parking spot). The reference signal $\mathbf{s}^*$ is the desired value of the output $\mathbf{s}$. General discrepancies between the current sensor features and the values that were predicted from the model (e.g., modeling errors and disturbances) are represented by the error signal ϵ:

$$\epsilon(n) = \mathbf{s}(n) - \mathbf{s}_{\mathrm{mp}}(n) \tag{1}$$

where n denotes the current time.

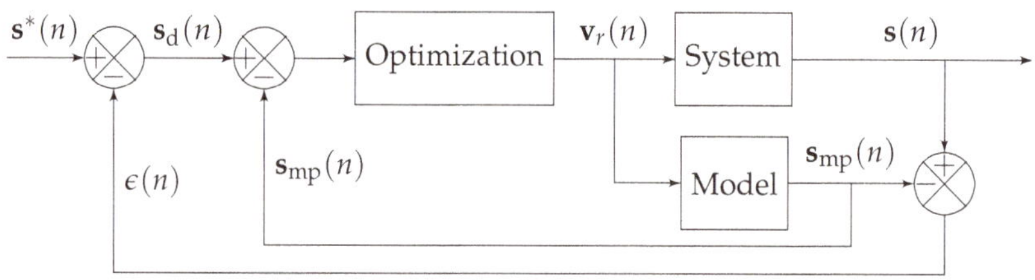

Figure 6. Control structure [31].

The difference between the desired value $\mathbf{s}_{\mathrm{d}}$ and the predicted model output $\mathbf{s}_{\mathrm{mp}}$ is minimized by an optimization algorithm. Following Figure 6:

$$\mathbf{s}_{\mathrm{d}}(n) = \mathbf{s}^*(n) - \epsilon(n) = \mathbf{s}^*(n) - (\mathbf{s}(n) - \mathbf{s}_{\mathrm{mp}}(n)), \tag{2}$$

from where one can deduce

$$\mathbf{s}_{\mathrm{d}}(n) - \mathbf{s}_{\mathrm{mp}}(n) = \mathbf{s}^*(n) - \mathbf{s}(n), \tag{3}$$

thus, to track $\mathbf{s}^*$ by $\mathbf{s}$ is equivalent to track $\mathbf{s}_{\mathrm{d}}$ by $\mathbf{s}_{\mathrm{mp}}$.

The interaction model described in [31] is used in order to predict the evolution of the sensor features $\mathbf{s}_{\mathrm{mp}}$ over a finite horizon N_p. The cost function is to be minimized with respect to a control sequence $\tilde{\mathbf{v}}_r$ over N_p and depends mainly on the difference between $\mathbf{s}_{\mathrm{d}}$

and $\mathbf{s}_{mp}$. As with any classical Model Predictive Control technique, only the first element $\mathbf{v}_r(n)$ of the optimal control sequence is applied to the system at each iteration.

The controller (implemented in C++) runs online at 10 Hz using the solver NLopt [33] with a Sequential Least Squares Programming (SLSQP) algorithm [34]. Furthermore, it is assumed that both the vehicle's longitudinal velocity and steering angle are controllable; thus, lower-level controllers that directly interface with the actuators are out of the scope of the parking approach.

The simulation inputs are four points generated based on a parked pose and the type of parking spot. This information is then transferred as a topic to the control node, which, in turn, generates a control command as a topic as well. The complete software is constructed in a manner that it is agnostic; it is working in a simulation environment or in a real vehicle.

Therefore, with the experimental vehicle being interfaced with ICARS, two additional ROS nodes would have to be developed: A perception one to extract the freight's pose from sensory data and another one to generate a parking spot next to the freight. Once the parking spot has been successfully generated from the sensory data, the four corners that define it would have to be sent to the ICARS parking controller to park the vehicle in the desired pose to pick up the freight.

4. Mathematical Modeling and Notation

4.1. Vehicle Kinematic Model

The vehicle kinematic model for a rear-wheel driven vehicle as taken from [8] is presented by Equation (4). The freight vehicle also follows the same kinematic model for its drive.

$$\begin{bmatrix} \dot{x} \\ \dot{y} \\ \dot{\theta} \\ \dot{\phi} \end{bmatrix} = \begin{bmatrix} \cos\theta \\ \sin\theta \\ \tan\phi/l_{wb} \\ 0 \end{bmatrix} v + \begin{bmatrix} 0 \\ 0 \\ 0 \\ 1 \end{bmatrix} \dot{\phi} \tag{4}$$

where v and $\dot{\phi}$ are longitudinal and steering velocities. Since parking maneuvers are performed at low speed, one can consider the kinematic model as accurate enough. Further notations and elaboration of the model as given by Equation (4) are represented in Figure 7a. The vehicle used for parking evaluation is further represented in Figure 7b. There is a designated space for the vehicle operator, as shown in the figure. The battery pack and electronics bay holds the complete electronics for the operation of the vehicle.

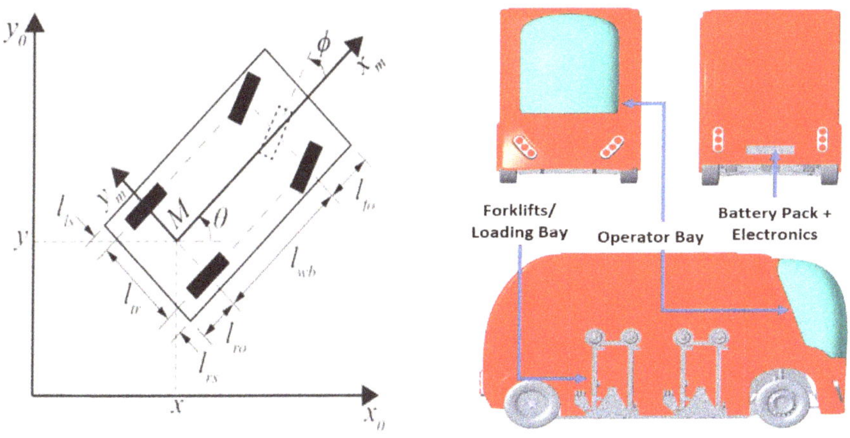

(a) (b)

Figure 7. (**a**) Vehicle kinematic model. (**b**) FURBOT vehicle.

4.2. Points Acquisition for Parking Spot

For correctly identifying parking spots for the vehicle, the acquisition of the freight corners (highlighted by red), as shown in Figure 1, is required through sensor feedback. From there, using these points, the freight center point is calculated using Equation (5).

$$x_{fc} = \frac{(x_{f1} + x_{f2})}{2}, y_{fc} = \frac{(y_{f1} + y_{f2})}{2} \tag{5}$$

Furthermore, knowledge from these two corner points is also used to calculate the inclination angle θ_f of the freight using the Pythagoras theorem (Equation (6)).

$$\theta_f = \tan^{-1}(\frac{y_{f2} - y_{f1}}{x_{f2} - x_{f1}}) \tag{6}$$

Using the width of the parking spot d_w and information of the freight center point, the vehicle center point is calculated from the freight using Equation (7).

$$x_{vc} = x_{fc} + \frac{d_w}{2} \cos(\theta_f + \frac{3\pi}{2}), y_{vc} = y_{fc} + \frac{d_w}{2} \sin(\theta_f + \frac{3\pi}{2}) \tag{7}$$

The above-mentioned points are further explained in Figure 8. Using these points, freight collection parking spots are calculated. This is further explained in the next subsection.

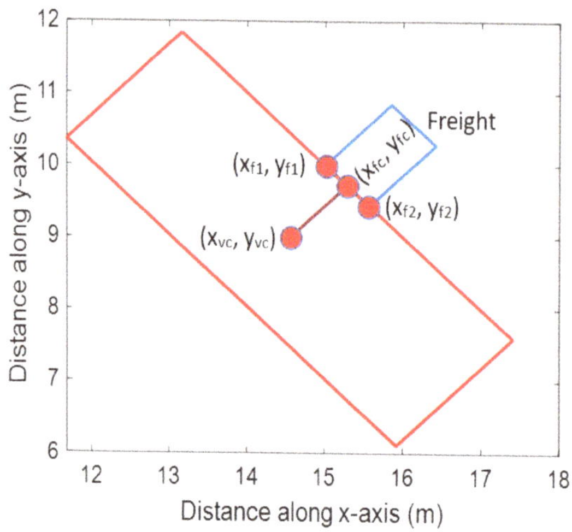

Figure 8. Point definition for parking spot.

4.3. Parking Area with Respect to Loading Bays

As the vehicle is equipped with two loading bays, the parking spot w.r.t each loading bay is different. To load the cargo in the front-loading bay, the vehicle needs to park a certain distance behind its center point, as shown in Figure 4. If the first loading bay is occupied by the previously loaded freight, then the vehicle needs to park a little ahead of its center point by a distance so that the forks are perfectly aligned with the freight to load the freight in the second loading bay. The center point of the parking spot thus varies depending upon the availability of the loading bay. This is further explained in Equation (8).

$$x_{pc} = \begin{cases} x_{vc} + d_{b1} \cos\theta_f, \\ x_{vc} - d_{b2} \cos\theta_f, \end{cases}, y_{pc} = \begin{cases} y_{vc} + d_{b1} \sin\theta_f, & \text{if bay1} = 1 \\ y_{vc} - d_{b2} \sin\theta_f, & \text{otherwise} \end{cases} \tag{8}$$

where d_{b1} and d_{b2} are the positive distances of loading bay 1 and 2 from vehicle center (measurement shown in Figure 4), and (x_{vc}, y_{vc}) is the center-point of parking spot. The condition $bay1 = 1$ denotes the condition of availability of loading bay 1. If bay 1 is available to load the freight, then the first condition applies or else the parking spot is generated w.r.t second loading bay.

After defining the center of the parking spot, the four corners of the parking spot can be defined through Equations (9)–(12) using the knowledge of parking width (d_w), parking length (d_l), angle of the freight (θ_f) and parking center-point (x_{pc}, y_{pc}). The collection of the first corner of the parking pose requires the knowledge of parking pose width + length. The derivation of equations is based on trigonometric relations and point transformation from the known parking center-point to the first edge of the parking pose. Since all four lines of the parking pose are parallel or perpendicular to the freight, the solution is formed by adding the respective angle $(90^\circ, 180^\circ, 270^\circ)$ to the angle of the freight θ_f for calculation of next point of the parking pose.

$$\begin{aligned} x_{p1} &= x_{pc} + \tfrac{d_w}{2}\cos(\theta_f + \tfrac{\pi}{2}) + \tfrac{d_l}{2}\cos(\theta_f + \pi), \\ y_{p1} &= y_{pc} + \tfrac{d_w}{2}\sin(\theta_f + \tfrac{\pi}{2}) + \tfrac{d_l}{2}\sin(\theta_f + \pi) \end{aligned} \tag{9}$$

$$x_{p2} = x_{p1} + d_w \cos(\theta_f + \frac{3\pi}{2}), y_{p2} = y_{p1} + d_w \sin(\theta_f + \frac{3\pi}{2}) \tag{10}$$

$$x_{p3} = x_{p2} + d_l \cos(\theta_f), y_{p3} = y_{p2} + d_l \sin(\theta_f) \tag{11}$$

$$x_{p4} = x_{p3} + d_w \cos(\theta_f + \frac{\pi}{2}), y_{p4} = y_{p3} + d_w \sin(\theta_f + \frac{\pi}{2}) \tag{12}$$

The four corners of the parking pose $(x_{p1}, y_{p1}), (x_{p2}, y_{p2}), (x_{p3}, y_{p3})$ and (x_{p4}, y_{p4}) are the consequent four corners of the parking pose. These points are further explained later.

The summary of the whole pose generation solution is as follows. The vehicle is required to detect freight and acquire the corner points of the freight from the side where it can be loaded. Afterwards, Equation (5)–(12) are solved to get the correct pose for loading freight into the vehicle. The complete process of correct pose generation is further summarized in the flowchart presented in Figure 9.

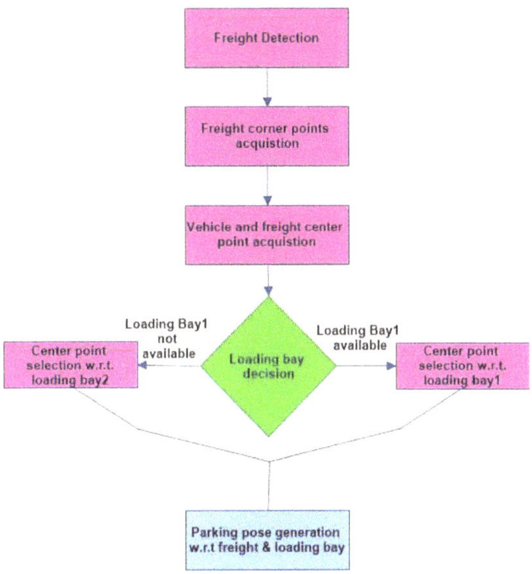

Figure 9. Flowchart for parking pose generation.

5. Results

By using the mathematical equations and notations discussed in the previous section, the node for the vehicle parking spot was run and tested to verify if the parking spot is generated according to the loading bays. The results are further discussed in the sections below for further clarity.

5.1. Parking Spot Definitions w.r.t Loading Bays

The results are first validated for generating the respective parking spot for front and rear loading bays. Figure 10 shows the output of parking pose generated for randomly positioned freight at a 45 degree angle. The blue-colored text represents the points for the parking spot definition concerning the front loading bay, whereas the red-colored text represents the parking spot concerning the rear loading bay. The red line from the parking spot center towards freight represents the right-hand side of the vehicle, thus specifying the heading of the vehicle.

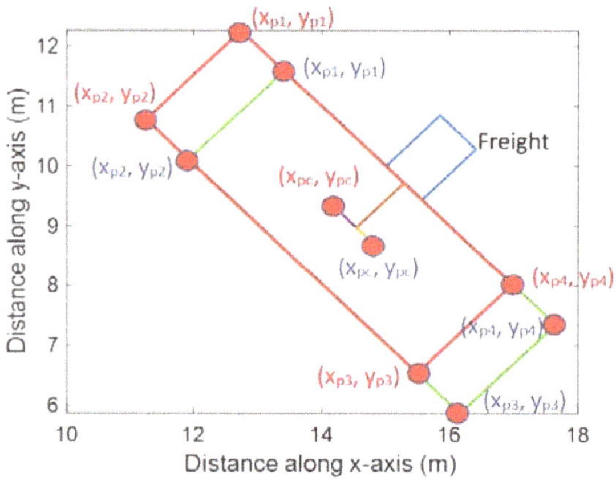

Figure 10. Parking spot with respect to loading bays.

From Figure 10, we can observe that the vehicle parks behind the center-point of the vehicle for loading freight in the first loading bay, whereas it parks a little ahead of its center-point when it is loading freight in the second loading bay. Furthermore, points 1 and 2 show the front end of the parking pose, and points 3 and 4 show the rear end of the pose generated. These results are also in coherence with the physical vehicle anatomy and the mathematical methodology proposed, henceforth validating our results.

5.2. Results for Different Freight Placement

As it is not possible to make sure that the orientation of freight is perfectly aligned with any pre-determined reference, the parking spot for the vehicle must be generated automatically, keeping in view the orientation of the freight. This conditional issue has been taken into consideration prior in the mathematical modeling of the parking spot. The code generated is validated for different orientations of the freight (Figure 11) to verify that the correct parking spot is generated irrespective of the orientation of the freight.

The results achieved, as shown in Figure 11, show four different orientations of freight (225, 0, 90 and 135 degrees, respectively) and the respective parking spot generated for each loading bay. The results show that irrespective of the orientation of the freight, the parking spot is generated accordingly for the vehicle for collecting the freight for the available loading bay.

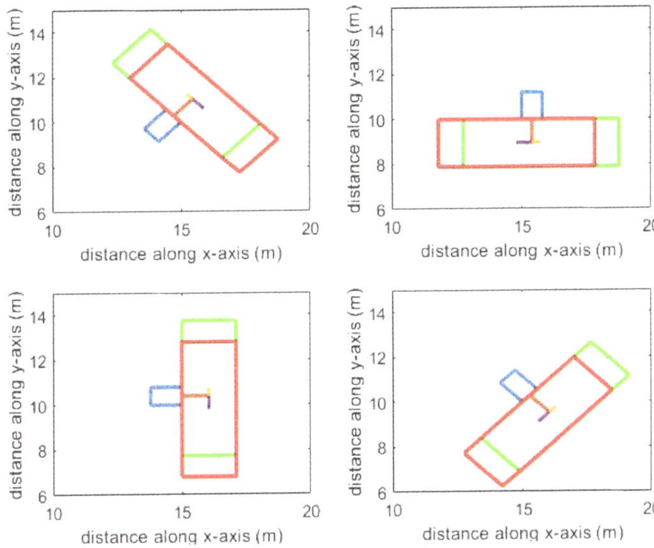

Figure 11. Parking spot relative to oriented loading bays.

Furthermore, the results presented also validate our purpose of generating parking pose autonomously irrespective of the orientation of the freight. The results also show that the vehicle will always park against the freight keeping the freight on the right-hand side of the vehicle so that it can be loaded in the available loading bay, i.e., the heading of the vehicle and loading bay location is always kept into consideration during the autonomous generation of a parking pose. The consideration of the freight w.r.t vehicle is shown by the red line generated from the center of the pose towards the freight (as shown in the figure with the red line). This allows the vehicle to know the direction of the parking pose and the heading angle for the vehicle. The parking pose points generated with the help of Equations (9)–(12) lets the vehicle know the heading of the vehicle as well. Point 1 and 2 depict the points where the front of the vehicle should face, and point 3 and 4 depict the points for the rear end of the vehicle.

5.3. Vehicle Parking in ROS Environment

As discussed earlier in Section 3, ROS environment-based software architecture ICARS is exploited to speed up the actual experiments. Using the already developed parking schemes within the software architecture [8,9,31], we can park the vehicle in the designated parking pose. The already built parking algorithms are sufficient for maneuvering the vehicle within the parking pose generated autonomously for the freight collection. Figure 12 shows the time-wise parking maneuver for vehicle FURBOT in the parking pose generated for the freight. Due to the constraint environment, the vehicle cannot park after detecting the freight because it needs to fulfill the requirement to park the vehicle, keeping the right side of the vehicle towards the freight. Thus, the vehicle has to move ahead of the freight and park while reversing, as shown in Figure 12. These results are generated in ICARS software architecture with input from the simulated sensors mounted on the vehicle.

In Figure 12, the orange box depicts the freight placement and location. The green rectangle represents the parking pose generated w.r.t freight. The red trail left behind the vehicle denotes the performed parking maneuver, and the red lines in front or at the back of the vehicle depict the direction and steering angle of the vehicle.

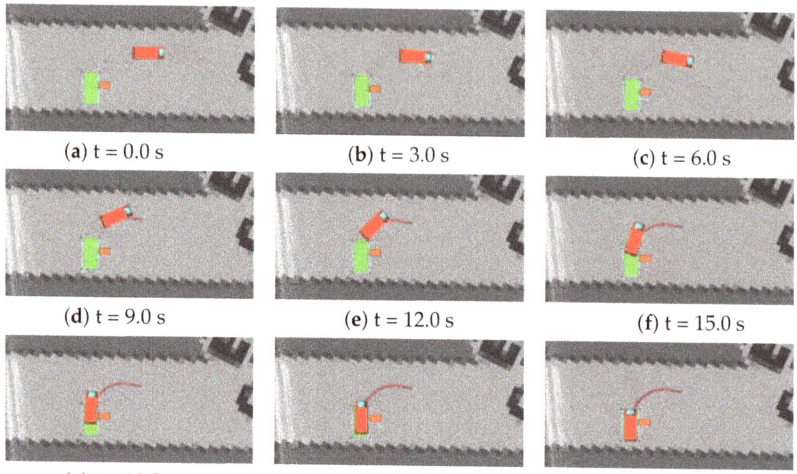

(a) t = 0.0 s (b) t = 3.0 s (c) t = 6.0 s
(d) t = 9.0 s (e) t = 12.0 s (f) t = 15.0 s
(g) t = 18.0 s (h) t = 21.0 s (i) t = 24.0 s

Figure 12. Vehicle parking w.r.t freight at different time intervals.

The virtual experiment is performed in the already installed maps of ICARS software architecture. The environment is constrained by natural buildings surrounding the area. Furthermore, the parking algorithm can detect and avoid obstacles, e.g., pedestrians, in case the vehicle can detect them [31]. The results validated in the ICARS environment prove that the methodology behind generating such parking pose for autonomous freight collection is valid and resolves the issue for autonomous parking pose generation w.r.t freight.

6. Conclusions and Perspectives

Research perspectives for this research were to highlight the gap between currently available research and how to resolve autonomous parking for freight collection. For addressing this issue, collecting the orientation and freight location are key values for the proposed algorithm. Using these key values, we can identify the parking pose for the vehicle for it to load the freight autonomously.

Major findings of this research discuss and resolve the issue related to parking the vehicle next to the freight for loading it into the vehicle. The hurdle of generating a parking spot that could lead to the successful loading of freight is looked into. Control architecture and vehicle dynamics were previously built and checked against different parking conditions, i.e., perpendicular, parallel and angled parking. This enabled the research to focus on the parking pose of the vehicle for freight collection. The solution required an algorithm that can define parking pose for our unique problem of loading the freight autonomously. With this research, this issue is resolved.

Concerning the correct pose for autonomous loading in respective bays of the vehicle, detection of freight from the surroundings is required. Once localization of freight is attained, then the proposed mathematical model suffices to generate the correct parking pose for the vehicle. However, the detection of freight and its localization is out of the scope of this research and will be looked into our future work. The parking pose length and width are defined, keeping minimal maneuvers required and distance to freight into consideration.

Considerations are kept for loading the freight in the correct loading bay. Separate parking spots are generated depending upon which loading bay is to be used for loading the freight into the vehicle. Simulating the freight with different orientations and generating the parking spot accordingly is also verified. The mathematical model verifies that if the freight is correctly identified from the environment, the parking spot generated will be correct irrespective of the orientation or location of the freight.

The main distinctive factor for the vehicle while demonstrating level 4 autonomy is the autonomous collection of freight which sets an electric freight vehicle apart from conventional vehicles. With the help of a previously designed parking controller, the identification of the freight through sensor's feedback and the definition of parking pose through this research, the vehicle can now align itself with the freight for autonomous collection of freight, increasing the level of autonomy of the vehicle.

This research will help to create an alternative approach for parking pose generation, especially in reference to an inanimate object. Researchers working in developing autonomous freight handling issues can directly benefit from this research. This research can further be used in developing parking pose algorithms where it might be necessary to park against an inanimate object, e.g., a bus stop or a door to a facility/delivery location. Currently, there is an unavailability of previous research in this field and solutions, where being studied might require continuous input from sensors in order to achieve the eventual goal where sensor output has to achieve an eventual goal to find a solution. This solution requires comparatively minimal input from sensors (for identification and localization); once attained, it does not require keeping on checking with sensors' feedback.

Future work involves the use of correct sensors (3D-Lidar in particular) for extracting information from the freight from the environment. This includes extracting data through 3D-Lidar's point-cloud data and reconstructing the freight in a virtual environment. Afterwards, the information from point-cloud data will extract the orientation and freight corner points. This perception module will help us extract the freight pose, which will later be used by the parking pose module for the autonomous generation of a parking spot regarding the freight.

Author Contributions: Conceptualization, K.M., V.F. and M.Z.; methodology, K.M., D.P.M. and V.F.; validation, M.Z. and V.F.; formal analysis, K.M. and D.P.M.; investigation, K.M. and V.F.; resources, M.Z. and V.F.; data curation, K.M.; writing—original draft preparation, K.M., D.P.M. and V.F.; writing—review and editing, K.M., D.P.M. and V.F.; visualization, K.M., D.P.M. and V.F.; supervision, R.M., V.F. and M.Z.; project administration, R.M., V.F. and M.Z.; and funding acquisition, R.M., V.F. and M.Z. All authors have read and agreed to the published version of the manuscript.

Funding: This research received no external funding.

Institutional Review Board Statement: Not applicable.

Informed Consent Statement: Not applicable.

Acknowledgments: This paper was supported by the European Union's Horizon 2020 research and innovation program under grant agreement No. 875530, project SHOW (SHared automation Operating models for Worldwide adoption).

Conflicts of Interest: The authors declare no conflict of interest. The funders had no role in the design of the study; in the collection, analyses, or interpretation of data; in the writing of the manuscript, or in the decision to publish the results.

Abbreviations

The following abbreviations are used in this manuscript:

FURBOT	Freight Urban Robotic Vehicle
SHOW	SHared automation Operating models for Worldwide adoption

References

1. Freight Urban RoBOTic Vehicle. 8 August 2016. Available online: https://cordis.europa.eu/project/id/285055 (accessed on 19 March 2021).
2. FURBOT, EGVI. 9 January 2018. Available online: https://egvi.eu/research-project/furbot/ (accessed on 19 March 2021).

3. Masood, K.; Zoppi, M.; Molfino, R. Mathematical Modelling for Performance Evaluation Using Velocity Control for Semi-autonomous Vehicle. In Proceedings of the 15th International Conference on Soft Computing Models in Industrial and Environmental Applications (SOCO 2020), Burgos, Spain, 16–18 September 2020; Herrero, Á., Cambra, C., Urda, D., Sedano, J., Quintián, H., Corchado, E., Eds.; Advances in Intelligent Systems and Computing; Springer: Cham, Switzerland, 2021; Volume 1268. [CrossRef]
4. Masood, K.; Molfino, R.; Zoppi, M. Simulated Sensor Based Strategies for Obstacle Avoidance Using Velocity Profiling for Autonomous Vehicle FURBOT. *Electronics* **2020**, *9*, 883. [CrossRef]
5. Masood, K.; Dauptain, X.; Zoppi, M.; Molfino, R. Hydraulic Pressure-Flow Rate Control of a Pallet Handling Robot for an Autonomous Freight Delivery Vehicle. *Electronics* **2020**, *9*, 1370. [CrossRef]
6. Masood, K.; Zoppi, M.; Fremont, V.; Molfino, R.M. From Drive-By-Wire to Autonomous Vehicle: Urban Freight Vehicle Perspectives. *Sustainability* **2021**, *13*, 1169. [CrossRef]
7. Clarembaux, L.G.; Pérez, J.; Gonzalez, D.; Nashashibi, F. Perception and Control Strategies for Autonomous Docking for Electric Freight Vehicles. *Transp. Res. Procedia* **2016**, *14*, 1516–1522. [CrossRef]
8. Morales, D.P.; Kermorgant, O.; Quijada, S.D.; Martinet, P. Laser-Based Control Law for Autonomous Parallel and Perpendicular Parking. In Proceedings of the 2018 Second IEEE International Conference on Robotic Computing (IRC), Laguna Hills, CA, USA, 31 January–2 February 2018. [CrossRef]
9. Pérez-Morales, D.; Kermorgant, O.; Domínguez-Quijada, S.; Martinet, P. Autonomous Perpendicular And Parallel Parking Using Multi-Sensor Based Control. In Proceedings of the 9th Workshop on Planning, Perception and Navigation for Intelligent Vehicles at IEEE/RSJ International Conference on Intelligent Robots and Systems, Vancouver, CO, Canada, 24–28 September 2017.
10. Golbabaei, F.; Yigitcanlar, T.; Bunker, J. The role of shared autonomous vehicle systems in delivering smart urban mobility: A systematic review of the literature. *Int. J. Sustain. Transp.* **2020**, 1–18. [CrossRef]
11. Amiri, A.M.; Ferguson, M.R.; Razavi, S. Adoption patterns of autonomous technologies in Logistics: evidence for Niagara Region. *Transp. Lett.* **2021**, 1–12. [CrossRef]
12. Willems, L. Understanding the Impacts of Autonomous Vehicles in Logistics. *Digit. Transform. Logist.* **2021**, 113–127. [CrossRef]
13. Reed, S.; Campbell, A.M.; Thomas, B.W. The Value of Autonomous Vehicles for Last-Mile Deliveries in Urban Environments. *Manag. Sci.* **2021**. [CrossRef]
14. Schlenther, T.; Martins-Turner, K.; Bischoff, J.F.; Nagel, K. Potential of Private Autonomous Vehicles for Parcel Delivery. *Transp. Res. Rec. J. Transp. Res. Board* **2020**, *2674*, 520–531. [CrossRef]
15. Pollaris, H.; Braekers, K.; Caris, A.; Janssens, G.K.; Limbourg, S. Vehicle routing problems with loading constraints: State-of-the-art and future directions. *OR Spectr.* **2014**, *37*, 297–330. [CrossRef]
16. Ostermeier, M.; Henke, T.; Hübner, A.; Wäscher, G. Multi-compartment vehicle routing problems: State-of-the-art, modeling framework and future directions. *Eur. J. Oper. Res.* **2021**, *292*, 799–817. [CrossRef]
17. Bortfeldt, A.; Homberger, J. Packing first, routing second—A heuristic for the vehicle routing and loading problem. *Comput. Oper. Res.* **2013**, *40*, 873–885. [CrossRef]
18. Research, H.A. Self-Parking Cars: Quick Guide, Car and Driver. 30 November 2020. Available online: https://www.caranddriver.com/research/a31995350/self-parking-cars-quick-guide/ (accessed on 20 March 2021). [CrossRef]
19. Kang, D.H.; Kang, C.M.; Kim, J.-S.; Kim, S.; Kim, W.-Y.; Lee, S.-H.; Chung, C.C. Vision-based autonomous indoor valet parking system. In Proceedings of the 2017 17th International Conference on Control, Automation and Systems (ICCAS), Jeju, Korea, 18–21 October 2017.
20. Shen, X.; Zhang, X.; Borrelli, F. Autonomous Parking of Vehicle Fleet in Tight Environments. In Proceedings of the 2020 American Control Conference (ACC), Denver, CO, USA, 1–3 July 2020.
21. Zhang, X.; Liniger, A.; Sakai, A.; Borrelli, F. Autonomous Parking Using Optimization-Based Collision Avoidance. In Proceedings of the 2018 IEEE Conference on Decision and Control (CDC), Miami, FL, USA, 17–19 December 2018.
22. Lee, B.; Wei, Y.; Guo, I.Y. Automatic parking of self-driving car based on Lidar. In Proceedings of the ISPRS-International Archives of the Photogrammetry, Remote Sensing and Spatial Information Sciences, Wuhan, China, 18–22 September 2017; Volume XLII-2/W7, pp. 241–246.
23. Kusumakar, R. Autonomous Parking for Articulated Vehicles. Master's Thesis, HAN University of Applied Science, Arnhem, The Netherlands, 2017.
24. Jung, H.G.; Kim, D.S.; Yoon, P.J.; Kim, J. Parking Slot Markings Recognition for Automatic Parking Assist System. In Proceedings of the 2006 IEEE Intelligent Vehicles Symposium, Meguro-Ku, Japan, 13–15 June 2006.
25. Li, S.; Hai, Y. Estimating camera pose from H-pattern of parking lot. In Proceedings of the 2010 IEEE International Conference on Robotics and Automation, Anchorage, Alaska, 3–8 May 2010.
26. Du, X.; Tan, K.K. Autonomous Reverse Parking System Based on Robust Path Generation and Improved Sliding Mode Control. *IEEE Trans. Intell. Transp. Syst.* **2015**, *16*, 1225–1237.
27. Vorobieva, H.; Glaser, S.; Minoiu-Enache, N.; Mammar, S. Automatic parallel parking with geometric continuous-curvature path planning. In Proceedings of the 2014 IEEE Intelligent Vehicles Symposium Proceedings, Dearborn, MI, USA, 8–11 June 2014. [CrossRef]

28. Cepolina, E.M.; Farina, A. An optimization methodology for the consolidation of urban freight boxes. In Proceedings of the 15th International Conference on Harbor, Maritime and Multimodal Logistics Modeling and Simulation, Athens, Greece, 23–25 September 2013; pp. 46–52.
29. Muscolo, G.G.; Leonardo, L.M.; Pietronave, G.; Zoppi, M.; Molfino, R. Industrial solutions for loading/unloading goods on a full electrical freight urban robotic vehicle. *Int. J. Veh. Syst. Model. Test.* **2015**, *10*, 366.
30. Vorobieva, H.; Glaser, S.; Minoiu-Enache, N.; Mammar, S. Automatic parallel parking in tiny spots: Path planning and control. *IEEE Trans. Intell. Transp. Syst.* **2015**, *16*, 396–410. [CrossRef]
31. Pérez-Morales, D.; Kermorgant, O.; Domínguez-Quijada, S.; Martinet, P. Multi-Sensor-Based Predictive Control for Autonomous Parking in Presence of Pedestrians. In Proceedings of the ICARCV 2020 16th International Conference on Control, Automation, Robotics and Vision, Shenzhen, China, 13–15 December 2020. [CrossRef]
32. Morari, M.; Zafiriou, E. *Robust Process Control*; Prentice Hall: Englewood Cliffs, NJ, USA, 1989.
33. Johnson, S.G. The NLopt Nonlinear-Optimization Package. Available online: http://ab-initio.mit.edu/nlopt (accessed on 1 July 2021).
34. Kraft, D. *A Software Package for Sequential Quadratic Programming*; Wiss. Berichtswesen d. DFVLR: Koeln, Germany, 1988; Volume 88.

energies

MDPI

Article

The Future of Autonomous Vehicles in the Opinion of Automotive Market Users

Monika Stoma [1], Agnieszka Dudziak [1,*], Jacek Caban [2] and Paweł Droździel [2]

[1] Faculty of Production Engineering, University of Life Sciences in Lublin, Gleboka Street 28, 20-612 Lublin, Poland; monika.stoma@up.lublin.pl
[2] Faculty of Mechanical Engineering, Lublin University of Technology, Nadbystrzycka Street 36, 20-618 Lublin, Poland; j.caban@pollub.pl (J.C.); p.drozdziel@pollub.pl (P.D.)
* Correspondence: agnieszka.dudziak@up.lublin.pl

Abstract: Contemporary trends are focused on the development of the so-called smart, connected and multimedia cars as well as actions in the field of driving autonomy, and these trends may lead to changes in the structure of the industry through the emergence and growth of the importance of new entities. The article presents the concept of autonomous vehicles (AVs) and the way it is perceived by users of traditional cars. Surveys were carried out in various age groups on the possibilities of developing AVs in Poland. The group of respondents were inhabitants of a rural area, small towns and cities with over 300,000 inhabitants. Based on our own research, it can be concluded that, due to many different factors, including costs, legal regulations and conviction, among others, AVs will not appear so soon in common use on Polish roads. The results of the research showed that the majority of respondents consider hybrid vehicles (HVs) and then electric vehicles (EVs) to be the dominant type of vehicles in the near future in Poland, at the same time pointing at the long process of adopting AV technology.

Keywords: autonomous level; autonomous technology; vehicles; road traffic; survey study

Citation: Stoma, M.; Dudziak, A.; Caban, J.; Droździel, P. The Future of Autonomous Vehicles in the Opinion of Automotive Market Users. *Energies* **2021**, *14*, 4777. https://doi.org/10.3390/en14164777

Academic Editors: Guzek Marek and Yair Wiseman

Received: 30 June 2021
Accepted: 4 August 2021
Published: 6 August 2021

1. Introduction

The role of the automotive industry in industrialization is huge. This industry is one of the most important for the Polish economy. It is also responsible for a significant number of jobs, as well as many different investments and revenues for the public finance sector. Importantly, this industry directly or indirectly affects other sectors of the economy. Three technological megatrends of the fourth industrial revolution (Industry 4.0) are indicated, i.e., 1. Communication, 2. artificial intelligence and 3. flexible automation. The technology of the fully autonomous vehicle (AV) is best adapted to these megatrends, especially in the case of connected and autonomous vehicles (CAVs), which enable communication between vehicles, infrastructure and other road users (so-called V2X connectivity) [1,2]. Automated driving is a major trend in the automotive industry as it promises to increase road safety and driver comfort [3].

One of the key elements in the development of AV technology is the progress related to the conversion of vehicle drive systems to electric drives. Currently, more and more vehicle manufacturers offer hybrid or electric vehicles. This will certainly affect the development of the industry related to electric driving and increase the level of employment in these sectors of the economy. Over the past decade, significant progress has been made in the field of automated driving systems (ADS), and AV technology is gaining more and more attention from vehicle manufacturers, technology companies, decision makers and the general public. Recent dynamic changes in vehicle technology like advanced driver assistance systems (ADAS) (such as, e.g., automatic braking, automatic cruise control, intelligent speed assistance, line maintenance assistance systems, etc.) bring us closer to increasingly autonomous and independent vehicles, which will use these solutions.

This development is supplemented by the parallel development of connectedness and communication in vehicles. Therefore, taking into account current dynamics and progress, it can be expected that these systems will continue to develop, and the technology of automated driving will lead to a change in the paradigm in transport systems in terms of comfort of use, choice of mode and business models [4].

Autonomous vehicles include various vehicles, including passenger cars, trucks and drones, which are based on artificial intelligence with varying degrees of human participation. By processing and analyzing billions of data from a number of sensors, cameras and radar systems every second, AVs can effectively "see" the road and respond to changing conditions or overcome obstacles.

The ability of autonomous vehicles to operate without human intervention depends on their level of technological sophistication, in accordance with the current six-degree autonomy scale proposed by the International Society of Automotive Engineers (SAE) [5,6] (see Figure 1) from level 0 (without automation) to level 5 (full unlimited automation); levels 1 to 3 are considered "semi-autonomous":

- Level 0: the driver performs all tasks related to driving and there is no automation.
- Level 1 (Driver's assistant): It is implemented by systems that automate a specific element of driving, and the driver is obliged to keep their hands on the steering wheel and watch the traffic on the road. So, the driver controls most driving functions, but under certain conditions the vehicle may be able to adjust the cruise control speed or stay on the road lane.
- Level 2 (partial automation): Corresponds to semi-autonomous driving—in the case of traffic jams on the road, the vehicle can autonomously take over driving, steering and braking. The car can both accelerate/decelerate and perform basic steering functions. The driver is still responsible for steering the navigation (e.g., exit from the highway, change of lane or turn onto a new street).
- Level 3 (conditional automation): on-board systems are already able to take over all driving functions, but only in certain cases; however, the driver must be alert at all times and ready to take over—their car can therefore monitor the driving environment and accelerate, turn or brake, but still awaits human intervention upon notification.
- Level 4 (high automation): fully autonomous driving, vehicles communicate with each other and inform each other about, e.g., change of lane, and the driver does not have to constantly observe the surrounding traffic on the road; the car can control all aspects of driving and operate without human intervention, but only under certain conditions.
- Level 5 (full automation): the car is fully autonomous in all driving conditions and does not require human intervention—the technology system can perform all driving tasks in all circumstances, and the passengers are only passive passengers and never have to participate in driving and perform any driving tasks.

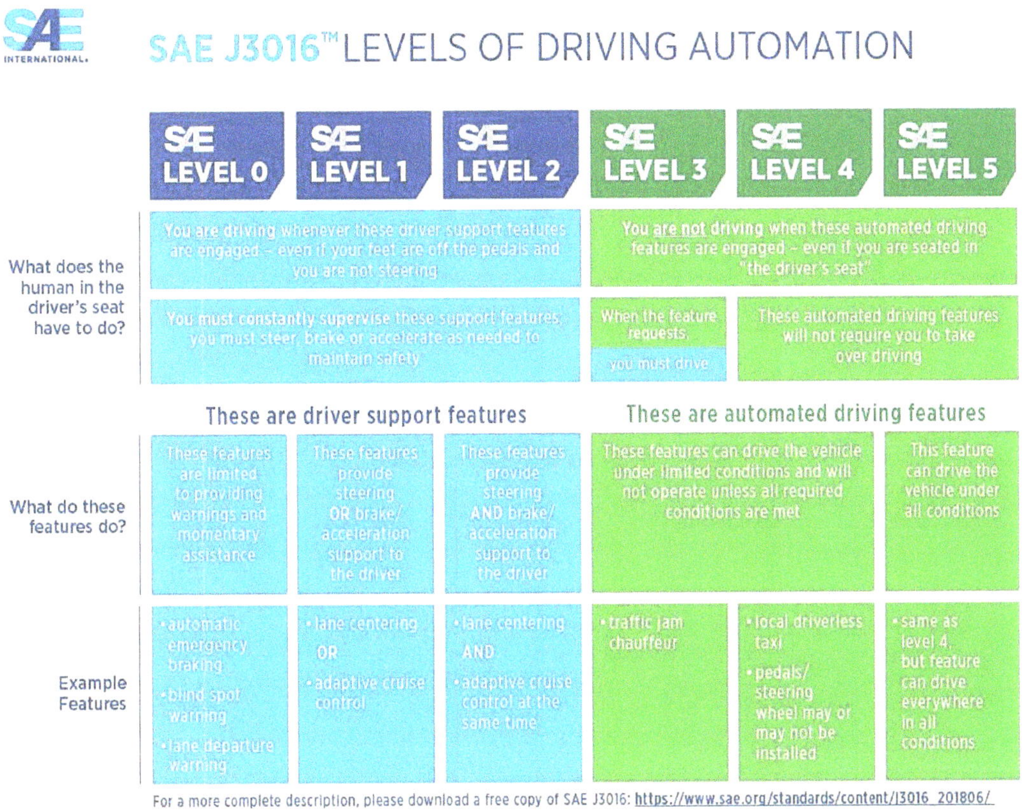

Figure 1. Level of driving automation proposed by SAE [7].

The latest car communication is designed to allow continuous, reliable and fast interaction between moving vehicles. They are usually divided into four use cases: vehicles to other vehicles (V2V), vehicles to the road-side infrastructure (V2I), vehicles to pedestrians (V2P), vehicles to devices (V2D) and vehicles to the cellular network (V2N). Together, these use cases are known as V2X—vehicles to everything [8]. It is anticipated that the wide-scale application of V2X technology can greatly improve transport safety and particularly significantly reduce vehicle collisions, especially in light vehicle accidents, by improving situational awareness. There are two types of V2X communication technology depending on the underlying technology being used: WLAN-based and cellular-based (C-V2X) technologies based on LTE [9–13]. Technology based on Wi-Fi is based on the standard IEEE802.11p for vehicular communication. It is also known as ITS-G5—Wireless short range to Intelligent Transport System, or DSRC—Dedicated Short-Range Communications (American or European protocol, respectively) [12]. C-V2X uses 3GPP—The Third Generation Partnership Project, 4G—The fourth generation, LTE—long-term evolution, or 5G—the fifth generation new radio (NR) connectivity to transmit and receive signals [13]. It uses two complementary transmission modes. The first is V2V, V2I and V2P. In this mode, C-V2X works independently of the cellular networks and it uses a PC5 interface for communication. The second mode is cellular network communications, in which C-V2X employs the mobile telephony network to enable vehicles to receive information about road and traffic conditions in the area. It uses LTE–Uu interface for communication, particularly for V2N [12,13].

AVs, despite the fact that they are still at an early stage of development and implementation, could mean a huge revolution not only in transport and the automotive industry, generating significant benefits in the long run in terms of transport accessibility, safety, traffic flow, emissions, fuel consumption and comfort. Many scientists believe that the large-scale deployment of AVs will bring about transformational changes in mobility and accessibility, travel patterns, safety and security, energy efficiency, emissions, employment, data availability, management and business models [14–16]. In the publications [17–20], the authors note the benefits for logistics and technical tasks in the workplace. Another aspect of the use of AVs is the management of urban space and the change in the demand for parking spaces depending on the change of ownership and shared use of autonomous vehicles (SAVs) [21,22]. SAVs represent an emerging alternative for driverless and no-demand transport [23], offering a compromise between private ownership and public transport [24]. Advanced technologies and systems used in this type of vehicles will also have a significant impact on other sectors of the modern market, such as trade, logistics, construction, insurance and related industries. A widespread transition to autonomous and electric vehicles would also change our daily lives. GM CEO Mary Barra refers to this future as "zero accidents, zero emissions and zero fatalities" [25]. In the literature on the subject, one can find many scientific papers on the vision zero perspective in various European countries [26–29]. These are, however, mainly papers about the "vision zero" concept as such, and there is only a limited number of empirical studies available. The European Union set out a "vision zero" target of reducing the number of fatal road accidents to almost zero by 2050. However, all these potential economic, ecological and social benefits will not be achieved until AVs are accepted and used by the majority of society [30,31]. Another important issue often raised in various studies is the regulation of legal issues [32–34], primarily concerning liability and security [35].

Another very important aspect related to AVs is the concept of smart cities, which practically focuses on the transformation of cities based on sustainable development. Smart city-related guidelines supply EU countries with general ideas on handling and controlling social, economic and technological change. Technologies for transport face the great challenges by globalization, re-urbanization and the change of social mobility behavior [36]. Passenger transportation is an indispensable and elementary service, in addition, there is a large share of freight traffic in urban areas. For their current problems there are several answers, one of them being smart cities' sub-systems, or smart mobility [37,38]. Smart mobility can be divided into two segments: (1) innovative solutions and (2) development of current services. Innovative solutions are not present in every urban transportation system; however, they play a main role in smart mobility-oriented development [39]. AVs and EVs are tools on the vehicle side. Mobility as a service (MaaS) is a new concept [39], with which both demand-driven service planning and the personalization of services are possible.

One of the important and highly researched effects of AVs is their effect on urban space usage and parking tendencies. Parking services are also moving to automated solutions; P+R parking lots and connectivity with public transportation networks are the most important issues [36]. One of the latest research directions is urban space saving by normalizing parking issues. Even a third of the total traffic time during peak traffic periods is related to finding parking spaces in congested urban areas. The appropriate and adequate information about the number of free places, their location, etc., can reduce even by 30% the traffic volume in some cases [40]. The parking management has to be integrated in the traffic management and parking-related measures have to be adjusted to traffic management measures [40]. AVs also cope well with this problem thanks to V2V and V2I communication.

Due to the fact that AVs are currently undergoing various tests and have not yet been introduced to the market on a large scale, few people have been in contact with this technology so far. However, the concept itself raises a lot of controversy and doubt, and it seems that it still remains in the field of innovative solutions. As already mentioned, after analyzing the available literature, it was found that there are still few results of research carried out in this area, allowing for the assessment of the knowledge of the AV concept, its popularity, possibilities of implementation, as well as barriers inhibiting or even preventing its dissemination. This is particularly true for less urbanized, industrialized or typically agricultural areas, where smaller towns and rural areas predominate, and their inhabitants are often not up-to-date with modern technologies. Taking the above into account, the aim of the article is an attempt to partially fill the literature gaps, and hence to examine and assess the attitudes and perception of potential AV users and current users of traditional cars, of the concept of this type of vehicle, as well as to assess their development prospects, mainly through the prism of barriers and factors inhibiting the introduction of the analyzed solution on Polish roads.

2. Materials and Methods

In order to assess the prospects for the development of AVs in the assessment of their potential users, we conducted our own research in the group of adults over 18 years old, which resulted from the possibility of having a driving license. The study was carried out in two stages: a pilot study and a proper study. In the first stage a questionnaire, which was especially developed for the purpose of this study, was tested on a group of 10 randomly selected people. Based on the feedback from the pilot test, in the next stage of the research, the questionnaire was improved and the authors constructed the final version of the measurement tool used in the main research—a structured proprietary survey questionnaire.

The questions in the survey were mainly closed; they were developed unambiguously so that they did not require supplementary comments, and the respondents were asked to select one correct answer from several available options. Several of the questions were multiple choice questions, also with the option of giving your own answer, which was, however, always indicated in the questionnaire. The questionnaire, except for the metric questions, enabling the socio-demographic characteristics of the respondents due to various grouping variables, contained questions mainly about the following:

- Knowledge of autonomous vehicle technology;
- Attitude to this technology;
- Barriers and challenges resulting from the introduction of this technology.

The purpose of the survey was explained to the respondents, and the confidentiality of results was emphasized.

The way to collect information was to conduct a directed interview in four age groups, taking into account the sex of respondents, their place of residence and other grouping variables, mainly concerning driver status (as shown in Table 1). The random sampling method was used—simple random selection (without returning). Due to the fact that data were obtained by taking population samples, they can be considered as descriptive research based on the method of data collection, and more specifically—a survey. The number of the examined group was 579 people.

Table 1. Socio-demographic profile of the population surveyed.

Socio-Demographic Profile	Number of Respondents	Percentage Share [%]
Total	579	100.0
Gender:		
Female	219	37.8
Male	360	62.2
Age:		
19–25 years old	276	47.7
26–40 years old	140	24.2
41–60 years old	138	23.8
60 years and more	25	4.3

Table 1. *Cont.*

Socio-Demographic Profile	Number of Respondents	Percentage Share [%]
Place of residence:		
rural area	198	34.2
city to 100,000 residents	118	20.4
100,000–300,000 residents	58	10.0
city with more than 300,000 residents	203	35.4
Driving license:		
yes	506	87.4
no	73	12.6

The collected data were presented using graphs and then subjected to basic statistical analysis adequate to the nature of the variables.

3. Results and Discussion

To achieve the goal set in the study, the results obtained based on questionnaires were analyzed and presented in descriptive and graphic form. The majority of respondents were men, young people (18–25 years old), with secondary and higher education and living mainly in the countryside or in large cities (over 300,000 inhabitants). The detailed socio-demographic characteristics of the respondents are presented in Table 1.

Polish society is becoming more and more mobile, which can be seen, among other factors, by increasing car sales. The automotive industry is preparing for a revolution on many levels. The technical revolution concerns mainly the issue of driving—electrification is approaching fast. The industry's social responsibility requires investing in new, environmentally friendly driving options. This is a very serious challenge for manufacturers who have to prepare the entire network of suppliers for the new assortment, as well as the service network to handle it [41].

The concept of an AV is also part of the automotive market development strategy, but its perception by current drivers in Poland seems to be at a highly differentiated level.

3.1. *Autonomy Levels in the Automotive Market*

In recent years, various surveys of public opinion and user acceptance regarding the perception and adoption of ADS have been conducted worldwide [32,42,43]. Begg [44] developed a survey on the likelihood of AV adoption aimed at UK transport experts to establish their perceptions of whether and when interviewees expect AVs to become a reality. In this survey, 28% of respondents said vehicles with level 3 autonomous driving technology will be available on UK public roads by 2040, and almost 25% said that the implementation of AVs would improve road transport safety.

In turn, Kyriakidis et al. [45] conducted a public opinion poll on automated driving among 4886 respondents in 109 countries. In this survey, respondents indicated that fully automated driving (level 5) would be easier than manual driving, while partially automated driving (level 3) was perceived as more difficult. Concerns focused on the hacking and misuse of software, legal issues and security. In addition, 20% of respondents said they would be willing to pay USD 7000 more for a fully level 5 AV, and nearly 70% said AVs could gain around 50% of the market share by 2050.

Another area of research is the analysis of the service side provided by AVs. Automated and autonomous freight transport will expectedly consist of high capacity (e.g., trucks) and small capacity (e.g., vans) vehicles [46]. The small capacity vehicles will be mainly applied in the urban environment for many transportation tasks. Reference [47] presents research on the readiness to entrust one's safety and health while traveling in an autonomous ambulance. As shown, the respondents approach this solution with great caution and uncertainty. As mentioned in Reference [48], the gradual introduction of autonomous ambulance technology through the trial transportation of patients with noncritical injuries may enhance users' trust and lead to the spread of this technology. Currently, in the area of services, innovative solutions are also used in the field of city logistics: EVs, electric cargo bikes (E-CB), new techniques for modeling and controlling traffic [49–51] and AVs are available. Demand-based ICT applications (hardware and software) are also spreading.

The research carried out by the authors of this work showed different tendencies in the perception of the concept of AVs by the respondents (not only active drivers) depending on the grouping variable adopted (age, gender, place of residence and driver's experience). As can be seen from the data presented in Table 1, the majority of the respondents were men (62.2%). The most numerous age group consisted of young respondents aged 19–25—they constituted 47.7% of all respondents. Most people, 35.4%, declared that they came from a city with more than 300,000 inhabitants, and the vast majority of them constituted a group of self-declared drivers, i.e., people holding a driving license, at 87.4%.

The conducted research shows that the belief of respondents regarding the level of autonomy is clearly the highest in relation to level 1. The higher the level of autonomy, the more skeptical both women and men are regarding this solution. The importance of the different types of autonomy is presented in Table 2 and Figure 2.

Table 2. The importance of different types of AVs by respondents based on gender.

Autonomy Level	Woman	Man	Total
Level 1	69	114	183
Level 2	64	104	168
Level 3	61	101	162
Level 4	18	30	48
Level 5	7	11	18
Generally	219	360	579

Summary: Calculating the cardinality; the number of marked cells > 10 Chi^2 Pearson: 17.4178, df = 4, $p = 0.001603$.

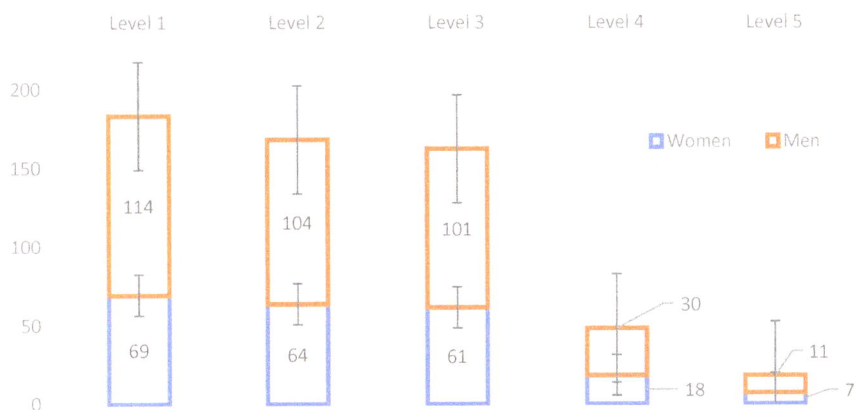

Figure 2. Perception of the levels of autonomy for the AV market among respondents by gender.

In the case of autonomy levels 1, 2 and 3, the significance of the survey is quite high in the opinion of the respondents, because they assess the concept as the most likely form for cars on the market. It should be remembered that these levels apply to standard vehicles supported by autonomous solutions (SVs+).

Respondents positively evaluated the concept of AVs up to level 3, while levels 4 and 5 were rated quite poorly. These are already quite innovative solutions, and in the opinion of the respondents, not very realistic for the automotive market in the near future.

The conducted research also showed that the respondents know the concept of AVs on average—only 30.4% of them have heard and know the concept; while more than half (58.2%) declared that they had heard something about it, and 11.4% of the respondents did not know what the idea was about.

On the other hand, due to the popularity of the progressive introduction of autonomous vehicle solutions on the market, drivers with five years seniority believe that AVs will become popular in more than 20 years. Drivers with 5–15 years of experience, as in the case of drivers who have a longer experience, i.e., 15–30 years as a driver, decided that it would take place in the period of 10–20 years. On the other hand, drivers with the longest experience, i.e., over 30 years, similarly to the youngest ones, said that, in their opinion, the development of AVs would take place in over 20 years (Figure 3).

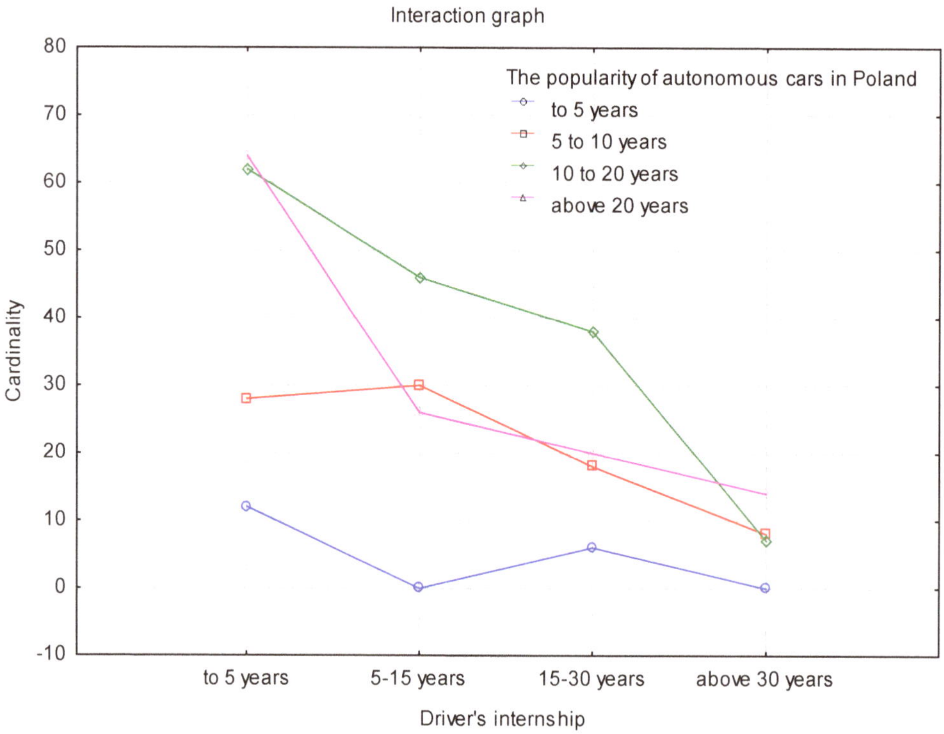

Figure 3. The popularity of AVs in Poland in the opinion of the respondents due to the driver's experience.

Among the different groups of respondents by age, the concept of the development of the AV market with regard to the level of autonomy was highly diversified (Figure 4a,b). In the 19–25 age group, women were rather unequivocally focused on development over 10–20 years for level 2 of autonomy, while in the same age group, men were not so clearly determined; the majority of indications concerned level 3 (development in the perspective of 10–20 years), the remaining answers were similarly divided into level 1 (development in the perspective of more than 20 years) and level 2 (period of 10–20 years). It should be added that according to women in this age group, the introduction of AVs at level 4 is unlikely, and at level 5, even impossible. In turn, men in this group had a more positive attitude to higher levels of autonomy than women, especially in levels 3 and 4.

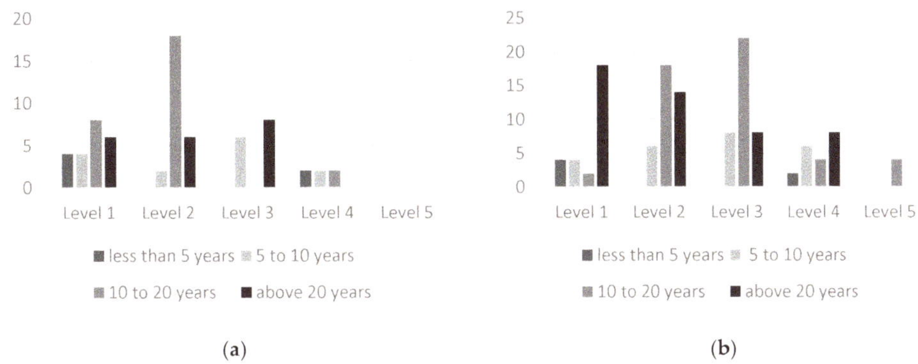

Figure 4. Perception of the levels of autonomy for the AV market among respondents by gender in the 19–25 age group: (**a**) female; (**b**) male.

The 26–40 age group (Figure 5a,b) indicated the following trends in the perception of levels of autonomy: for women—the most likely spread of AVs was at level 3 (development over the 10–20 years period), while—similarly to the perspective in the previously analyzed age group—the higher levels of autonomy (i.e., levels 4 and 5) were not taken into account at all. It should be added that a large group of women in this age group believe that the introduction of AVs, starting from level 1, will only be possible in 20 years' time at the earliest.

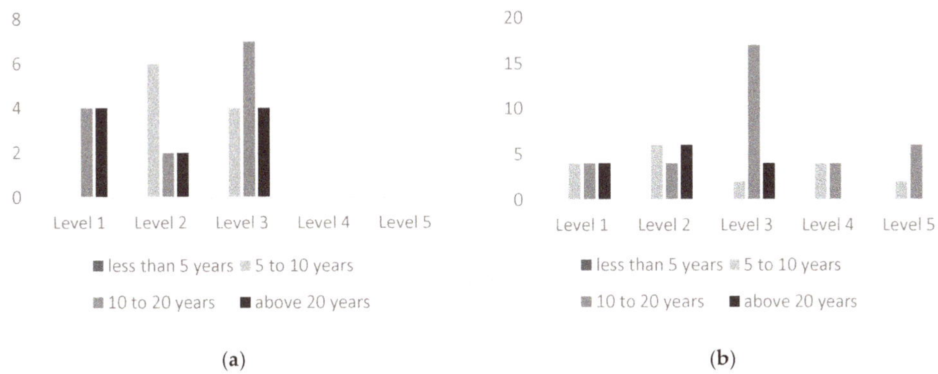

Figure 5. Perception of the levels of autonomy for the AV market among respondents by gender in the 26–40 age group: (**a**) female; (**b**) male.

On the other hand, men also mainly see sense in level 3 of autonomy in the period of 10–20 years; however, they are more optimistic than women in this age range in regard to higher levels of vehicle autonomy—indicating a period of 10 to 20 years for perceived development (even for level 5).

The opinions of the respondents in the 41–60 age group were also varied (see Figure 6a,b). Women mainly indicated the 5–10-year development perspective for level 3, while men were less determined, and equally indicated level 1 (over 20 years) and level 3 (development over 10–20 years).

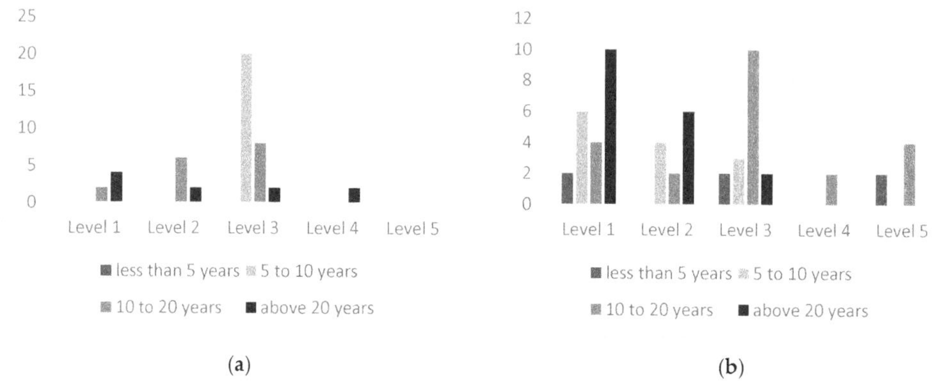

Figure 6. Perception of the levels of autonomy for the AV market among respondents by gender in the 41–60 age group: (**a**) female; (**b**) male.

The smallest age group participating in the study—only 4.3% of the respondents—were elderly people over 60 (Figure 7a,b). In this age group, among women, the indications for level 2 of autonomy (in the period of 10–20 years) were dominant, and levels 3, 4 and 5 were not taken into account at all. On the other hand, men indicated the development perspective for AVs for a period of more than 20 years for level 1, i.e., a small support for a car, which today, among modern cars on the road, perform various functions available on the automotive market (cruise control, reversing or parking assistant, etc.). Thus, levels 4 and 5 were omitted as a potential future trend.

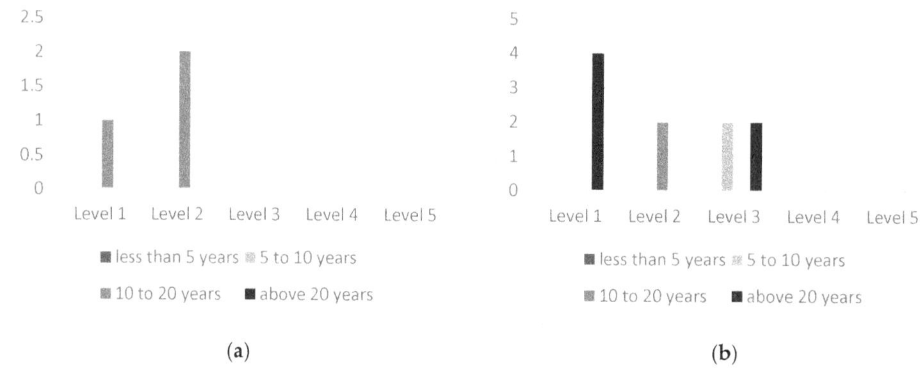

Figure 7. Perception of the levels of autonomy for the AV market among respondents by gender in the age group above 60 years: (**a**) female; (**b**) male.

In further stages of the interpretation of the obtained research results, reference was made to replacing traditional cars with different types of vehicles: HVs, EVs, standard vehicles, but supported by selected functions of autonomous vehicles and, of course, AVs. The analysis shows that among people who know the concept, as many as 68.4% say that AVs will gradually replace the traditional ones. On the other hand, in terms of the automotive market, according to the respondents, HVs will be the most popular in Poland in the coming years (39.4% of respondents' indications) as well as standard vehicles (SVs)—21.2%. It should be added that a similar number of responses were obtained for SVs with autonomous power steering and EVs—18.7% and 18%, respectively. AVs were indicated by only 2.8% of the respondents.

In order to better present the obtained results, in Figures 8–11, in relation to the above analysis, basic descriptive statistics (location measures) are presented including the median, quartiles, minimum and maximum for two variables; the ranges of the variable "place of residence of respondents" are marked on the horizontal axis, while on the vertical axis, individual categories for the variable "vehicle type which, in the opinion of respondents, will be the most popular in the near future" are marked. The survey lists different types of vehicles in terms of their propulsion: standard vehicle—SV, hybrid vehicle—HV, electric vehicle—EV, standard vehicle with autonomous support—SV+ and autonomous vehicle—AV. An autonomous vehicle can be propelled in various ways (SV—petrol or diesel, EV or HV), but in this questionnaire, respondents did not distinguish between the driving method according to its propulsion. Thus, vehicles (SV, EV, HV and SV+) are considered as human-driven vehicles (traditional vehicles), and an AV is to be understood as a self-steering vehicle without human intervention.

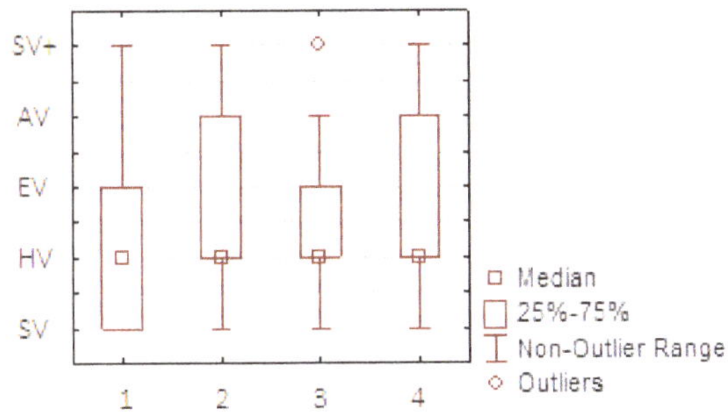

Figure 8. Perception of the popularity of particular types of vehicles among the respondents according to the place of residence: 1—city with more than 300,000 residents; 2—city to 10,000 residents; 3—100,000–300,000 residents; 4—rural areas.

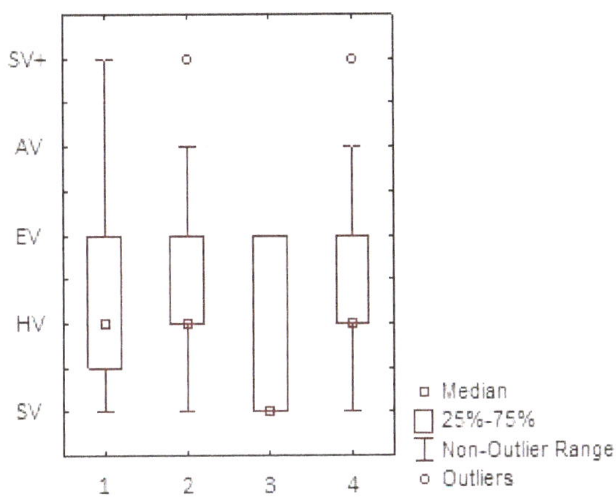

Figure 9. Perception of the popularity of particular types of vehicles among the respondents in terms of the respondent's age: 1—41–60 years; 2—26–40 years; 3—60 years and more; 4—19–25 years.

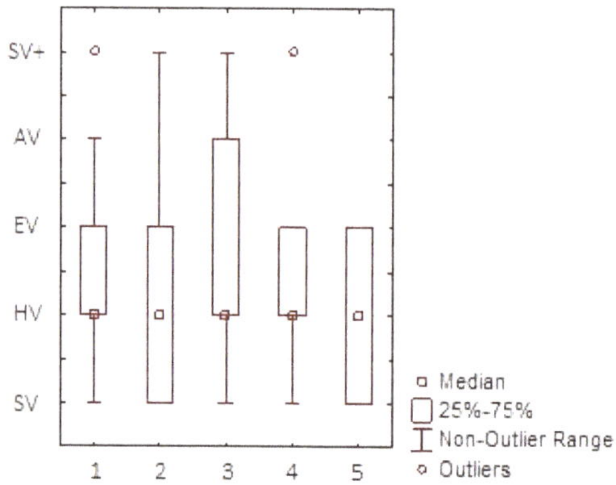

Figure 10. Perception of the popularity of particular types of vehicles among the respondents due to their driving license status: 1—more than 10 years; 2—6 to 10 years; 3—1 to 5 years; 4—none; 5—less than 1 year.

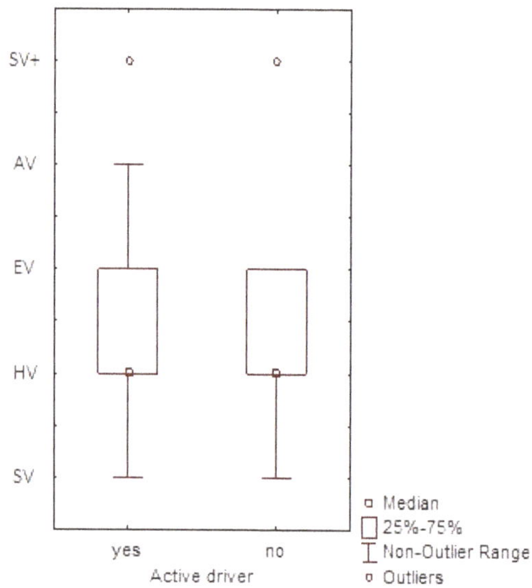

Figure 11. Perception of the popularity of particular types of vehicles among the respondents due to their driver's status.

As shown in Figure 8, over 50% of respondents were living in large cities with more than 300,000 inhabitants, and they mainly indicated HVs, but also SVs and EVs. People living in cities from 100,000 to 300,000 inhabitants—mainly indicated HVs and EVs. On the other hand, people living in smaller cities and rural areas, in addition to HVs, also indicated EVs.

As shown in the data presented in Figure 9, the majority of respondents aged 41–60 indicated mainly HVs, but also SVs+. People aged 26–40 indicated perceptions similar to people from the previous age group, but adding EVs. People in the age group over 60, however, most often indicated the popularity of SVs, also considering EVs and HVs. The respondents from the youngest age group indicated HVs as the most popular in the near future and also mentioned SVs+.

As can be seen from the data presented in Figure 10, the impact of the status in relation to the driving license on the popularity of particular types of vehicles in the near future was marked by the popularity of HVs by all people. For respondents who had a driving license for more than 10 years, they also showed EVs and SVs+. In the case of respondents who had a driving license for 6 to 10 years, they indicated SVs and SVs+. The respondents who had a driving license in the range of 1–5 years showed HVs, EVs and AVs. Among the respondents without a driving license, they indicated the popularity of HVs and EVs. In turn, respondents who had a driving license for less than a year indicated HVs, EVs and SVs.

Among active drivers (Figure 11), the respondents selected HVs and EVs as well as SVs and AVs. In the second group, the respondents selected HVs, EVs and SVs. The above analysis shows that most respondents predicted the greatest popularity in the near future for HVs, followed by EVs. At the same time, a significant proportion of respondents indicate SVs. Based on these studies, it can be concluded that relatively few people are convinced about the technology of AVs in Poland in the near future, and these are respondents who are active drivers.

3.2. Barriers and Challenges in the Field of Autonomous Vehicles

There are also barriers and factors that slow down the large-scale introduction of AVs. Some of them result from the mentality of the society—attitudes, preferences and approaches to using this type of vehicle. It was indicated that a large part of potential users do not feel safe in a vehicle without a driver. Moreover, they are not prepared to give up the freedom and independence that comes with owning and driving their own vehicle. On the other hand, potential users are concerned about the need to control and react quickly in emergency situations, such as a failure of the automation or the vehicle exceeding its functional limits [45].

It is also likely that the infrastructure and the applicable legal regulations (regulating the principles of operation of AVs or insurance regulations) will block the widespread introduction of self-driving cars. It should be remembered that AVs must learn to drive not only in predictable conditions, but also in imperfect and dynamic conditions, in which both human behavior, weather and various other obstacles may create difficult situations on the road. In addition, attention is drawn to the high cost of acquiring such a vehicle due to the complexity of software and cybersecurity, extensive hardware requirements for video systems and the challenges of large-scale network management.

As argued by Shariff et al. [52], the greatest obstacles to mass AV adaptation may be psychological, not technological. If AVs are not widely accepted by the society, road safety cannot be improved and the anticipated benefits for society and the environment cannot be achieved [53].

Adapting AVs to public roads is therefore associated with many challenges. Such vehicles will have to be more integrated with national intelligent transport infrastructures and systems, such as satellite navigation systems in vehicles, traffic signal control systems, information about parking lots, weather forecasts, de-icing bridges, automatic number plate recognition systems or speed cameras for monitoring related applications.

Research carried out by Silberg et al. on technology acceptance, adoption and application of automation technology in vehicles has gained great value in research in the field of transport [54]. They conducted a survey addressed to focus groups in California, New Jersey and Illinois in the USA, asking for their opinion on AVs. It was found that respondents would be more interested in adopting AVs if they received incentives such as autonomous vehicle lanes. In addition, people over 60 and people aged 18 to 25 showed the highest readiness for this type of solution. In contrast, Schoettle and Sivak [55] surveyed public opinion on autonomous and AVs among 1533 respondents in the USA, Great Britain and Australia. The survey found that most respondents were interested in having a fully self-contained vehicle technology, but most respondents said they would not be willing to pay extra for this technology. Respondents in the USA expressed greater concern than those in the UK or Australia regarding data privacy, interactions with self-driving vehicles, learning to use vehicles and operating the vehicle in bad weather.

On the other hand, Underwood [56] examined the opinion of 217 transport experts on automated vehicles. Respondents identified legal accountability and regulation as the most difficult barriers to deploying fully automated vehicles, and social and consumer acceptance was considered the least difficult. Moreover, other recent opinion polls have shown that the general public shows some resistance or a neutral attitude to AV technology [24,45].

The own research conducted by the authors of this study also revealed various barriers that may be key in the perception of the development of the concept of AVs. Further analysis was carried out in the breakdown of respondents by age and gender.

Women aged 19–25 see the greatest barrier in the costs of purchasing such a vehicle (21.9% of responses) and safety issues (17.8%). Women aged 26–40 point to the price (15.8%) and the question of liability in dispute (15.8%), while women aged 41–60 are concerned with the price (26.7%) and safety 14.8%). On the other hand, women over 60, similar to the youngest ones, pay more attention to the price (22.6%) and safety issues (16.1%). The results of this analysis are shown in Figure 12.

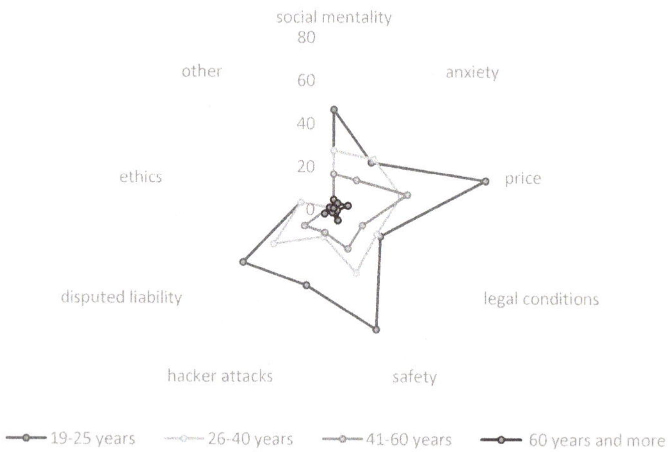

Figure 12. Barriers limiting the positive perception of the concept of AV development in the opinion of women.

In the case of barriers limiting the perception of AV in the opinion of men (Figure 13), it can be noticed that respondents aged 19–25 most often indicated the price and the contentious responsibility (20.3% and 15.9%, respectively). Men in the 26–40 age group see the main barriers in the development of AVs in terms of safety (17.7%) and the cost of buying a vehicle (15.7%), while in the age group of 41–60, the issue that is a more serious barrier concerns the price (21.6%) and potential hacking attacks (17.1%), which other groups of respondents did not pay attention to. On the other hand, respondents in the age group over 60 indicated the price as the main barrier—as many as 50% of respondents in this group expressed their opinion on it, while 25% of respondents were afraid of this solution at all.

From the obtained results, it can be drawn unequivocally that the main barriers are costs, safety issues and disputable liability between road users, in the case of a potential AV driving. Similar conclusions were presented in References [55,57]. In the case of high operating costs of such a system, these concerns are also confirmed by the results of studies presented by Brown et al. [58] and Casley et al. [59] or Shabanpour et al. [60].

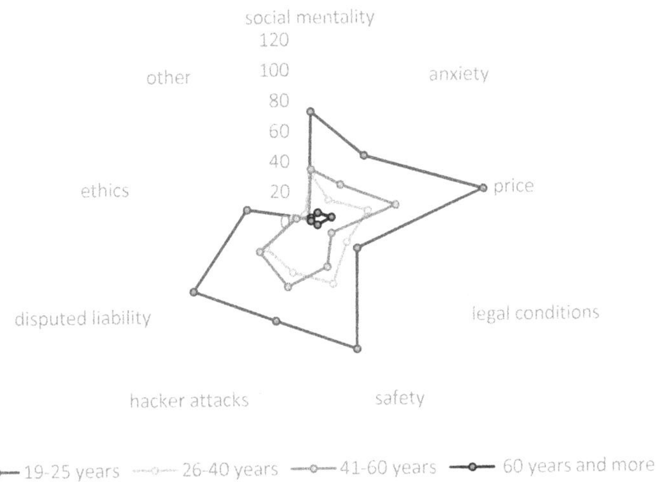

Figure 13. Barriers limiting the positive perception of the concept of AV development in the opinion of men.

3.3. *The Future of Autonomous Vehicles in Poland Based on Research Results*

Based on the analysis of the literature and our own research, it can be indicated that, in most cases, the widespread use of AVs in road networks would lead to the improvement of road safety in the form of fewer road accidents, as well as savings in energy consumption by vehicles and increased efficiency. Most studies also agreed on potential obstacles to AV adoption, such as legal liability and ethical issues, privacy concerns, cybersecurity and hacking issues, as well as the high cost of vehicles and related technologies.

Research has shown that, in Poland, the process of replacing traditional human-steering vehicles with AVs will be gradual and slow. This process will depend mainly on the adaptation of society to this type of innovation and changes, as well as the wealth of the society. As shown in Reference [8], the introduction of this type of vehicle on a large scale—in the most realistic variant—will take place in the next 20–25 years.

According to the research, most respondents believe that the dominant type of vehicles in Poland in the near future will be traditional human-steering HVs and EVs. The re-search also showed an attachment to SVs (with combustion engines), especially of respondents in the last age group—over 60 years of age.

The most important premises resulting from the analysis of the test results are as follows:

- Respondents positively evaluated the concept of AVs up to level 3 (SVs+), while levels 4 and 5 were rated quite poorly. Moreover, in each age group, there was a greater awareness of the introduction of vehicle autonomy levels among men than among women, who did not show level 5, and level 4 was shown only sporadically.
- Only about 30% of respondents knew AV technology and almost 60% "heard something about this technology". This shows how low the awareness is about innovative solutions in the road transport sector among the respondents.
- The average time of AV introduction in Poland was estimated at 10–20 years and over 20 years.
- Due to the place of residence, the most popular vehicles in the future were indicated by the following: HVs, EVs, AVs and SVs, followed by SVs+.
- Due to the age of the respondents, the most popular vehicles in the future were indicated as HVs, EVs and SVs, followed by AVs and SVs+.

- Due to the length of driving license, the most popular vehicles in the future were indicated by the following, successively: HVs, EVs, SVs and AVs, with SVs+ last.
- In the group of active drivers, the respondents indicated the popularity of HVs, EVs and SVs, and then AVs. On the other hand, HVs, EVs and SVs were indicated as the most popular among inactive drivers.
- Among the main barriers for the introduction of AVs in Poland, the respondents indicated the following: price, safety, disputed liability, social mentality and hacker attacks.

The analysis of the research results shows that the respondents showed great attachment to traditional self-driving, at the same time pointing to the popularity of hybrid, electric and internal combustion engine vehicles. This shows that technology of AVs in Poland is more distant than, for example, in the USA or other, more developed EU countries as well as in the UK. That is why it is so important to promote the benefits of AVs. Activities in the mental sphere are very important—there needs to be promotion of the AV concept and related technologies such as the Smart City, car sharing and other innovative solutions in road transport in urban areas.

4. Conclusions

The article presents our own research on the future of autonomous vehicle technology, obtained as a result of surveys. It summarizes the benefits and barriers to the widespread use of AVs that were seen as the most important in each study. As indicated in the literature in most cases, the widespread use of AVs on road networks would lead to fewer road accidents, increased fuel savings and increased productivity. Most studies also agreed on potential obstacles to AV adoption, such as legal liability, ethical issues, privacy concerns, cybersecurity and hacking issues, as well as the high cost of vehicles and related technologies.

Therefore, it seems that the replacement of traditional cars with AVs will take place gradually and will depend mainly on the adaptation of society to this type of innovation and related changes. Hence, experts and researchers predict that the introduction of AVs on a large scale will take place in the next 20 to 25 years [8]. The degree of advancement of this process, its scale and form will, of course, be different in individual countries—in smaller, well-developed countries with modern infrastructure, such as the Netherlands, AVs may already become widespread in the years 2030–2040. However, in the case of larger and less-developed countries, this may not happen until 2040–2050 or even later. Level 1 and level 2 of autonomy are expected to be introduced extensively, i.e., autonomous functions in commercial vehicles (comprehensive lane assistance and automatic braking and parking functions, adaptive cruise control, blind spot monitoring, etc.), which were previously only available in luxury cars. By contrast, true automation, from level 3 onwards, may prove very limited in most national economies due to insurmountable problems with driver inattention.

Summing up, it can be said that the process of adopting AV technology in Poland will take a relatively long time. Most of the respondents indicated about a 20-year period of adaptation of higher levels of vehicle autonomy.

Author Contributions: Conceptualization, M.S., A.D. and J.C.; methodology, M.S. and A.D.; software, A.D.; validation, M.S., J.C. and P.D.; formal analysis, A.D., J.C. and P.D.; investigation, M.S., A.D. and J.C.; resources, A.D., M.S. and J.C.; data curation, A.D.; writing—original draft preparation, M.S., A.D., J.C. and P.D.; writing—review and editing, M.S., A.D., J.C. and P.D.; visualization, A.D. and P.D.; supervision, M.S. and P.D.; project administration, A.D. All authors have read and agreed to the published version of the manuscript.

Funding: Funded from the 'Excellent science' program of the Ministry of Science and Higher Education as a part of the contract no. DNK/SP/465641/2020 "The role of the agricultural engineering and environmental engineering in the sustainable agriculture development".

Institutional Review Board Statement: Not applicable.

Informed Consent Statement: Not applicable.

Data Availability Statement: Not applicable.

Conflicts of Interest: The authors declare no conflict of interest.

References

1. Nikitas, A.; Kougias, I.; Alyavina, E.; Tchouamou, E.N. How can autonomous and connected vehicles, electromobility, brt, hyperloop, shared use mobility and mobility-as-a-service shape transport futures for the context of smart cities? *Urban Sci.* **2017**, *1*, 36. [CrossRef]
2. Gavanas, N. Autonomous Road Vehicles: Challenges for Urban Planning in European Cities. *Urban Sci.* **2019**, *3*, 61. [CrossRef]
3. Riedmaier, S.; Schneider, D.; Watzenig, D.; Diermeyer, F.; Schick, B. Model validation and scenario selection for virtual-based homologation of automated vehicles. *Appl. Sci.* **2021**, *11*, 35. [CrossRef]
4. Chan, C.Y. Advancements, prospects, and impacts of automated driving systems. *Int. J. Transp. Sci. Technol.* **2017**, *6*, 208–216. [CrossRef]
5. SAE International. *SAE J3016_201806, Taxonomy and Definitions for Terms Related to Driving Automation Systems for On-Road Motor Vehicles*; SAE International: Washington, DC, USA, 2016.
6. Panagiotopoulos, I.; Dimitrakopoulos, G. An empirical investigation on consumers' intentions towards autonomous driving. *Transp. Res. Part C* **2018**, *95*, 773–784. [CrossRef]
7. SAE Standards News: J3016 Automated-Driving Graphic Update. Available online: https://www.sae.org/news/2019/01/sae-updates-j3016-automated-driving-graphic (accessed on 19 July 2021).
8. Casey, J.P. The Future of Autonomous Vehicles. Available online: https://www.roadtraffic-technology.com/comment/future-autonomous-vehicles/ (accessed on 19 May 2021).
9. Asselin-Miller, N.; Biedka, M.; Gibson, G.; Kirsch, F.; Hill, N.; White, B.; Uddin, K. *Study on the Deployment of C-ITS in Europe: Final Report*; European Commission: Brussels, Belgium, 2016.
10. 5GAA. *C-ITS Vehicle to Infrastructure Service: How C-V2X Technology Completely Changes the Cost Equation for Road Operators*; White Paper; 5G Automotive Association: Munich, Germany, 2018.
11. Analysis Mason Limited, 5GAA. *Socio-Economics Benefits of Cellular V2X*; 5G Automotive Association: Munich, Germany, 2017.
12. Martínez, A.; Cañibano, E.; Romo, J. Analysis of Low Cost Communication Technologies for V2I Applications. *Appl. Sci.* **2020**, *10*, 1249. [CrossRef]
13. Miao, L.; Virtusio, J.J.; Hua, K.-L. PC5-Based Cellular-V2X Evolution and Deployment. *Sensors* **2021**, *21*, 843. [CrossRef]
14. Bartuska, L.; Labudzki, R. Research of basic issues of autonomous mobility. LOGI 2019-Horizons of Autonomous Mobility in Europe. *Transp. Res. Procedia* **2020**, *44*, 356–360. [CrossRef]
15. Charness, N.; Yoon, J.S.; Souders, D.; Stothart, C.; Yehnert, C. Predictors of attitudes toward autonomous vehicles: The roles of age, gender, prior knowledge, and personality. *Front. Psychol.* **2018**, *9*, 2589. [CrossRef]
16. Das, S.; Sekar, A.; Chen, R.; Kim, H.C.; Wallington, T.J.; William, E. Impacts of autonomous vehicles on consumers time-use patterns. *Challenges* **2017**, *8*, 32. [CrossRef]
17. Nieoczym, A.; Tarkowski, S. The modeling of the assembly line with a technological Automated Guided Vehicle (AGV). *Logforum* **2011**, *7*, 35–42.
18. Papa, E.; Ferreire, A. Sustainable accessibility and the implementation of automated vehicles: Identifying critical decisions. *Urban Sci.* **2018**, *2*, 5. [CrossRef]
19. Pettigrew, S.; Fritschi, L.; Norman, R. The potential implications of autonomous vehicles in and around the workplace. *Int. J. Environ. Res. Public Health* **2018**, *15*, 1876. [CrossRef] [PubMed]
20. Stopka, O. Modeling the Delivery Routes Carried out by Automated Guided Vehicles when Using the Specific Mathematical Optimization Method. *Open Eng.* **2020**, *10*, 166–174. [CrossRef]
21. Siqueira Silva, D.; Csiszár, C.; Földes, D. Autonomous vehicles and urban space management. *Sci. J. Sil. Univ. Technol. Ser. Transp.* **2021**, *110*, 169–181. [CrossRef]
22. Tian, Z.; Feng, T.; Tommermans, H.J.P.; Yao, B. Using autonomous vehicles or shared cars? Results of a stated choice experiment. *Transp. Res. Part C Emerg. Technol.* **2021**, *128*, 103117. [CrossRef]
23. Fagnant, D.J.; Kockelman, K.M. Dynamic ride-sharing and fleet sizing for a system of shared autonomous vehicles in Austin, Texas. *Transportation* **2018**, *45*, 143–158. [CrossRef]
24. Haboucha, C.J.; Ishaq, R.; Shiftan, Y. User preferences regarding autonomous vehicles. *Transp. Res. Part C Emerg. Technol.* **2017**, *78*, 37–49. [CrossRef]
25. Self-Driving Cars will Profoundly Change the Way People Live. The Economist. 1 March 2018. Available online: https://www.economist.com (accessed on 4 August 2021).
26. Jamroz, K.; Kustra, W.; Budzynski, M.; Zukowska, J. Pedestrian protection, speed enforcement and road network structure the key action for implementing Poland's Vision Zero. *Transp. Res. Procedia* **2016**, *14*, 3905–3914. [CrossRef]
27. Kristianssen, A.C.; Andersson, R.; Belin, M.A.; Nilsen, P. Swedish Vision Zero policies for safety–A comparative policy content analysis. *Saf. Sci.* **2018**, *103*, 260–269. [CrossRef]

28. Rosencrantz, H.; Edvardsson, K.; Hansson, S.O. Vision Zero–Is it irrational? *Transp. Res. Part A Policy Pract.* **2007**, *41*, 559–567. [CrossRef]
29. Zwetsloot, G.I.J.M.; Kines, P.; Ruotsala, R.; Drupsteen, L.; Merivirta, M.L.; Bezemer, R.A. The importance of commitment, communication, culture and learning for the implementation of the Zero Accident Vision in 27 companies in Europe. *Saf. Sci.* **2017**, *96*, 22–32. [CrossRef]
30. Xu, Z.; Zhang, K.; Min, H.; Wang, Z.; Zhao, X.; Liu, P. What drives people to accept automated vehicles? Findings from a field experiment. *Transp. Res. Part C* **2018**, *95*, 320–334. [CrossRef]
31. Payre, W.; Cestac, J.; Delhomme, P. Intention to use a fully automated car: Attitudes and a priori acceptability. *Transp. Res. Part F* **2014**, *27*, 252–263. [CrossRef]
32. Gkartzonikas, C.; Gkritza, K. What have we learned? A review of stated preference and choice studies on autonomous vehicles. *Transp. Res. Part C Emerg. Technol.* **2019**, *98*, 323–337. [CrossRef]
33. Poliak, M.; Mrnikova, M.; Simurkova, P.; Medvid, P.; Poliakova, A.; Hernandez, S. Social law in road transport like tool safety road transport. In Proceedings of the 2018 XI International Science-Technical Conference Automotive Safety, Casta, Slovakia, 18–20 April 2018; IEEE: Piscataway, NJ, USA. [CrossRef]
34. Törő, O.; Bécsi, T.; Aradi, S. Design of lane keeping algorithm of autonomous vehicle. *Period. Polytech. Transp. Eng.* **2016**, *44*, 60–68. [CrossRef]
35. Moravcik, L. Typové schválenie autonómnych (samojazdiacich) vozidiel, Type approval of autonomous (self-driving) vehicles. *Perner's Contacts* **2020**, *15*, 2. [CrossRef]
36. Nagy, S.; Csiszár, C. The quality of smart mobility: A systematic review. *Sci. J. Sil. Univ. Technol. Ser. Transp.* **2020**, *109*, 117–127.
37. Škultéty, F.; Beňová, D.; Gnap, J. City logistics as an imperative smart city mechanism: Scrutiny of clustered EU27 capitals. *Sustainability* **2021**, *13*, 3641. [CrossRef]
38. Del Vecchio, P.; Secundo, G.; Maruccia, Y.; Passiante, G. A system dynamic approach for the smart mobility of people. Implications in the age of big data. *Technol. Forecast. Soc. Chang.* **2019**, *149*, 119771. [CrossRef]
39. Jittrapirom, P.; Caiati, V.; Feneri, A.M.; Ebrahimigharehbaghi, S.; Gonzalez, M.J.A.; Narayan, J. Mobility as a service: A critical review of definitions, assessments of schemes and key challenges. *Urban Plan.* **2017**, *2*, 13–25. [CrossRef]
40. Sandor, Z.P.; Csiszar, C. Role of integrated parking information system in traffic management. *Period. Polytech. Civil Eng.* **2015**, *59*, 327–336. [CrossRef]
41. Automotive Industry Report 2018/2019. Available online: https://www.pzpm.org.pl/Publikacje/Raporty/Raport-branzy-motoryzacyjnej-2018-2019 (accessed on 4 August 2021).
42. Shin, J.; Bhat, C.R.; You, D.; Garikapati, V.M.; Pendyala, R.M. Consumer preferences and willingness to pay for advanced vehicle technology options and fuel types. *Transp. Res. Part C* **2015**, *60*, 511–524. [CrossRef]
43. Daziano, R.A.; Sarrias, M.; Leard, B. Are consumers willing to pay to let cars drive for them? Analyzing response to autonomous vehicles. *Transp. Res. Part C* **2017**, *78*, 150–164. [CrossRef]
44. Begg, D. *A 2050 Vision for London: What are the Implications of Driverless Transport*; Transport Times: London, UK, 2014.
45. Kyriakidis, M.; Happee, R.; De Winter, J.C.F. Public opinion on automated driving: Results of an international questionnaire among 5000 respondents. *Transp. Res. Part F* **2015**, *32*, 127–140. [CrossRef]
46. Csiszár, C.; Földes, D. System model for autonomous road freight transportation. *Promet Traffic Transp.* **2018**, *30*, 93–103. [CrossRef]
47. Zarkeshev, A.; Csiszár, C. Are people ready to entrust their safety to an autonomous ambulance as an alternative and more sustainable transportation mode? *Sustainability* **2019**, *11*, 5595. [CrossRef]
48. Winter, S.R.; Keebler, J.R.; Rice, S.; Mehta, R.; Baugh, B.S. Patient perceptions on the use of driverless ambulances: An affective perspective. *Transp. Res. Part F Traffic Psychol. Behav.* **2018**, *58*, 431–444. [CrossRef]
49. Balm, S.; Browne, M.; Leonardi, J.; Quak, H. Developing and evaluation framework for innovative urban and interurban freight transport solutions. *Procedia Soc. Behav. Sci.* **2014**, *125*, 386–397. [CrossRef]
50. Schliwa, G.; Armitage, R.; Aziz, S.; Evans, J.; Rhoades, J. Sustainable city logistics–making cargo cycles viable for urban freight transport. *Transp. Bus. Manag.* **2015**, *15*, 50–57. [CrossRef]
51. Tettamanti, T.; Varga, I.; Szalay, Z. Impacts of autonomous cars from a traffic engineering perspective. *Period. Polytech. Transp. Eng.* **2016**, *44*, 244–250. [CrossRef]
52. Shariff, A.; Bonnefon, J.-F.; Rahwan, I. Psychological roadblocks to the adoption of self-driving vehicles. *Nat. Hum. Behav.* **2017**, *1*, 694–696. [CrossRef] [PubMed]
53. Noy, I.Y.; Shinar, D.; Horrey, W.J. Automated driving: Safety blind spots. *Saf. Sci.* **2018**, *102*, 68–78. [CrossRef]
54. Silberg, G.; Manassa, M.; Everhart, K.; Subramanian, D.; Corley, M.; Fraser, H.; Sinha, V. Self-driving Cars: Are we ready? *Tech. Rep. KPMG* **2013**, 1–36. Available online: https://assets.kpmg/content/dam/kpmg/pdf/2013/10/self-driving-cars-are-we-ready.pdf (accessed on 4 August 2021).
55. Schoettle, B.; Sivak, M. *A Survey of Public Opinion about Autonomous and Self-driving Vehicles in the U.S., the U.K., and Australia*; Transportation Research Institute, University of Michigan: Ann Arbor, MI, USA, 2014.
56. Underwood, S.E. Automated vehicles forecast vehicle symposium opinion survey. In Proceedings of the Automated Vehicles Symposium, San Francisco, CA, USA, 15–17 July 2014.
57. König, M.; Neumayr, L. Users' resistance towards radical innovations: The case of the self-driving car. *Transp. Res. Part F Traffic Psychol. Behav.* **2017**, *44* (Suppl. C), 42–52. [CrossRef]

58. Brown, B.; Drew, M.; Erenguc, C.; Hasegawa, M.; Hill, R.; Schmith, R.; Ganula, B. Global Automotive Consumer Study: The Changing Nature of Mobility–Exploring Consumer Preferences in Key Markets Around The World. Technical Report, Deloitte. 2014. Available online: https://www2.deloitte.com/content/dam/Deloitte/global/Documents/Manufacturing/gx-mfg-geny-automotive-consumer.pdf (accessed on 19 May 2021).
59. Casley, S.V.; Jardim, A.S.; Quartulli, A.M. A Study of Public Acceptance of Autonomous Cars, Interactive Qualifying Project, Worcester Polytechnic Institute. 2013. Available online: https://web.wpi.edu/Pubs/E-project/Available/E-project-043013-155601/unrestricted/A_Study_of_Public_Acceptance_of_Autonomous_Cars.pdf (accessed on 21 May 2021).
60. Shabanpour, R.; Golshani, N.; Shamshiripour, A.; Mohammadian, A.K. Eliciting preferences for adoption of fully automated vehicles using best-worst analysis. *Transp. Res. Part C Emerg. Technol.* **2018**, *93*, 463–478. [CrossRef]

energies

MDPI

Article

Parameters Influencing on Electric Vehicle Range

Emilia M. Szumska and Rafał S. Jurecki *

Department of Automotive Engineering and Transport, Faculty of Mechatronics and Mechanical Engineering, Kielce University of Technology, Ave. Tysiąclecia Państwa Polskiego 7, 25-314 Kielce, Poland; eszumska@tu.kielce.pl
* Correspondence: rjurecki@tu.kielce.pl

Abstract: There is a range of anxiety-related phenomena among users and potential buyers of electric vehicles. Chief among them is the fear of the vehicle stopping and its users getting "stuck" before reaching their designated destination. The limited range of an electric vehicle makes EV users worry that the battery will drain while driving and the vehicle will stall on the road. It is therefore important to know the factors that could further reduce the range during daily vehicle operation. The purpose of this study was to determine the effect of selected parameters on a battery's depth of discharge (DOD). In a simulation study of an electric vehicle, the effects of the driving cycle, ambient temperature, load, and initial state of charge of the accumulator on the energy consumption pattern and a battery's depth of discharge (DOD) were analyzed. The simulation results confirmed that the route taken has the highest impact on energy consumption. The presented results show how significantly the operating conditions of an electric vehicle affect the energy life. This translates into an electric vehicle's range.

Keywords: electric vehicle; electric vehicle range; depth of discharge

Citation: Szumska, E.M.; Jurecki, R.S. Parameters Influencing on Electric Vehicle Range. *Energies* **2021**, *14*, 4821. https://doi.org/10.3390/en14164821

Academic Editor: Adolfo Dannier

Received: 29 June 2021
Accepted: 4 August 2021
Published: 7 August 2021

1. Introduction

Electric vehicles are now widely available. They can be found in virtually every vehicle category, from small passenger cars to SUVs, and sports cars to light duty vehicles. EVs are easy to find in the offerings of most car companies. In recent years, corporations have emerged that specialize in producing electric vehicles only, such as Tesla. In the last quarter of 2020, electric vehicles accounted for 16.5% of all newly registered passenger cars in the EU. Overall, in 2020, the share of newly registered electric vehicles was 10.5% of all new vehicles in the EU [1]. However, when looking at the number of all passenger cars in use in the EU, EVs only represent 0.2% [2]. The highest share of electric vehicles is in Norway, at 56%. The largest number of electric vehicles is in China, at 2.4 million [3,4]. Car sharing companies have sprung up in many major cities, offering only electric vehicles. According to the results of the analyses presented in [5,6], the number of people using this type of service is growing. The number of electric taxis is also gradually increasing. Many taxi companies have decided to purchase electric or hybrid vehicles [7–11]. An example for many cities can be found in Amsterdam (The Netherlands), where a taxi company has decided to replace its entire fleet of vehicles with electric ones by 2025 [12]. The largest number of e-taxis is in Beijing, China [13,14].

Compared to vehicles with a combustion engine, electric vehicles have many advantages. The main and undeniable one is the zero emission of harmful compounds at the place of their use. Electric vehicles also emit less noise while driving. In the low-speed range up to 50 km/h, it can be up to 7 dB(A) depending on the vehicle speed [15–18]. The noise generated by an electric vehicle in city traffic reaches 40–60 dB(A). For comparison, a passenger car powered by an internal combustion engine emits an average of 60–70 dB(A) in city traffic.

Another advantage is the much lower cost of ownership of electric vehicles compared to vehicles with internal combustion engines. As shown by the results of TCO analyses presented in [19–22], the running costs of electric vehicles are much lower than those of internal combustion vehicles.

One of the advantages of EVs is the much higher efficiency of the electric power train compared to conventional drives equipped with internal combustion engines. It is estimated that the efficiency of the electric power train is 90–98% [23–25]. In vehicles with a conventional power train, where the energy source is the internal combustion engine, the energy from fuel combustion is thermal energy, which is then converted into mechanical energy. It is estimated that the efficiency of energy transfer from the source (internal combustion engine) to the driving wheels in conventional propulsion is 42–45% [26–28]. In addition, the torque generated by the electric motor is available almost from zero rotational speed. This allows the parameters of the electric motor driving the vehicle to be selected in such a way as to ensure that the driving speed can be changed within the required range without the use of an additional transmission. This reduces the mechanical losses of the drive system. When stationary, e.g., in traffic, electric motors do not run, do not require sustained operation, and therefore do not generate losses associated with motor idling [29,30]. When operating in generator mode, the electric motor can recover some of its kinetic energy during braking. This energy is used to additionally recharge the battery. In vehicles powered solely by internal combustion engines, during braking, the energy is dissipated into the environment in the form of heat and thus irretrievably lost.

Despite their many significant advantages, electric vehicles also have disadvantages. Undoubtedly, one of them is the cost of buying an EV. The price of a small electric car is often close to that of a premium class conventional car. To prevent this, many states offer subsidies or other incentives to willing buyers, such as free parking in city centers, free entry into cities, or the ability to drive using bus lanes.

Another disadvantage of electric vehicles is the long battery charging time. Depending on the charging method, it is several hours. The availability of charging stations and points is limited in many countries. In small towns especially, charging infrastructure is still not sufficiently developed. In many countries this problem also applies to the availability of charging stations along highways and freeways.

The disadvantage of most electric vehicles is also the short range on a single charge. Depending on battery capacity and operating conditions, electric vehicles can cover up to 400 km on a single charge (e.g., Nissan Leaf, Renault Zoe, BMW i3 120 Ah), while vehicles equipped with internal combustion engines, depending on fuel tank capacity, can cover a much longer distance on a single fill-up.

The issue of short range is one of the barriers to the spread of electric vehicles. The phenomenon of range anxiety has been reported among EV users and those considering their purchases. Although the on-board computer displays the estimated range of the vehicle, there are situations on the road that cannot be predicted. Users or potential buyers are concerned that the battery will run down before the end of the journey or that there is no charging point in the area.

When reviewing the literature on the topic of electric vehicle range, many papers include research and analysis of the impact of factors that limit the range of electric vehicles. These include operating conditions, driver's driving style, ambient temperature, vehicle load, etc. Many works are devoted to studying the energy efficiency of EVs in a specific driving cycle. Ongoing research is also aimed at adapting the control strategy for energy flow during braking between mechanical brakes and the energy recovery system. This makes it possible to extract as much energy as possible during braking.

The purpose of this paper is to present the factors and problems associated with electric vehicle coverage and to review previous research. The second part of the paper presents the methodology of a simulation study of an electric vehicle, in which the effects of driving cycle, energy level in the accumulator, ambient temperature, and load on the energy consumption of an electric vehicle were analyzed. This paper is a prelude to an experimental study aimed at investigating and understanding the factors affecting the range of an electric vehicle under real-world conditions.

2. Range Anxiety

Range anxiety is the travel anxiety and stress experienced by the driver of an electric vehicle caused by decreasing or low energy in the battery. It is the fear of the vehicle stopping and its users getting "stuck" before reaching their designated destination [31]. What distinguishes the concern about the range of electric vehicles from the classic problems associated with the use of vehicles powered by conventional fuels is the fear of not being able to find a charging station in time. Using vehicles with internal combustion engines, finding a refueling station during a long journey is not a problem.

For electric vehicle users, there is a concern that when traveling to areas of the city or regions unfamiliar to them, drivers will not encounter available EV chargers along their route. What is important is that many EV users have no experience using a public EV charger. In fact, most EV owners charge their vehicles primarily at home or at their place of work.

The authors of the paper [32] noted that range anxiety is expressed on four levels:

- Cognitive level: manifested by negative beliefs related to the range, e.g., fear of running out of energy and not being able to reach the destination.
- Emotional level: causing, e.g., anxiety, nervousness, or fear in the situation.
- Behavioral level: expressed by a change in a driver's behavior that reduces immediate anxiety by increasing feelings of safety and control. Demonstrated, for example, by certain actions such as changing driving style to save energy or frequently checking the on-board computer readouts (for range, energy level, position), or even aggressive behavior manifested, for example, by tapping fingers on the steering wheel, shouting, honking, or aggressive gestures.
- Physiological level: manifested in the body's response to the situation, e.g., increased blood pressure, heart rate, heart rate variability, cortisol level, pupil diameter, respiratory rate, etc.

Based on the results of the survey presented in [33], the most frequent behaviors of electric vehicle drivers experiencing range anxiety were limiting the speed, changing the driving style, and searching for a nearby charging point. The actual behavior change represents the instability of the driver's behavior under stress and is intended to reduce the risk of getting stuck when traveling. To conserve energy and increase range, drivers choose to reduce speed.

In light of the frequent discussion of this psychological phenomenon, many studies have sought an empirical understanding of electric vehicle drivers' range anxiety. The results confirm that visualizing the energy level in the accumulator contributes to anxiety in some drivers. Electric vehicle drivers feel uncomfortable when the battery charge drops below a comfortable threshold subjectively accepted by the vehicle user. The paper [34] presents the results of an experiment in which participants were asked to drive a 19-mile distance in an electric vehicle. One half of the participants drove a vehicle with a low energy level in the accumulator (30%) and the other with a fully charged battery. The authors conclude that low initial SOC has a significant effect on range anxiety, trust in the vehicle, and driving behavior.

Based on the experimental results, the authors of the paper [35] noted that range anxiety increases when the driver observes a decreasing energy level in the battery while driving. This situation makes it impossible for the driver to predict how far he can still travel with the remaining energy level. In the paper [36], it was shown that the anxiety of electric vehicle drivers increases when the energy level in the battery drops below 50% of its capacity. Based on the surveys presented in [37], a significant number of respondents stated that charging should be performed at 30% energy level in the battery.

Referring to the literature, reducing perceived anxiety in electric vehicle users can be achieved in two ways. The first way is to increase the nominal range, e.g., by using batteries with higher energy capacity or replacing the electric vehicle with REV (extended-range electric vehicles) or plug-in hybrids (HEV plug-in). The second way is to develop the charging infrastructure by increasing the number of fast charging stations so that they are as easily accessible as gas stations.

Many researchers conduct intensive research on information systems for electric vehicle drivers. The have developed applications and systems that warn of the need to charge the battery, inform drivers about the location of the nearest charging station, or allow drivers to determine whether the battery will need to be recharged on the planned route [38–41]. In a survey presented in [37], participants stated that the distance between charging stations should be on average 0.12 km less than the distance between traditional gas stations. According to the respondents, the preferred distance between charging points should be 5 km. A lot of work has gone into the proper siting of charging stations so that they are easily accessible to electric vehicle users. To develop a network of charging stations, the researchers used methods such as linear programming [42,43], genetic algorithm [44,45], location approach [46], and fuzzy neural network [47].

To reduce the charging time for electric vehicles, the authors of [48,49] suggested that in addition to charging stations for electric vehicle users, battery swap points should be widely available where users can swap a discharged battery for a fully charged one. This solution would eliminate the long battery charging time.

Technology that relies on inductive coils embedded in the road lanes would allow the battery to be charged while driving. Installed on highways, this solution would allow electric vehicles to be more mobile and help alleviate range concerns. However, this is a solution of the future, as weight deployment of this technology seems difficult and very expensive.

3. Factors Affecting the Range of an Electric Vehicle

The range of an electric vehicle is understood as the distance the vehicle can travel before the battery has to be recharged. The range of the car depends on the amount of energy in the batteries and the unit consumption. Higher energy consumption due to additional factors shortens the range a vehicle can cover on a single battery charge. The range of an electric vehicle depends on a large number of factors.

According to [50], the range of electric vehicles generally depends on three main classes of factors: vehicle design, driver's driving style and use conditions. Some parameters are invariable, e.g., vehicle type, gearbox type, number of seats, weight of electric drive, weight and type of battery, road infrastructure, and availability of battery charging infrastructure.

Other parameters such as the battery's state of charge (SOC), state of health (SOH), driver's behavior, traffic volume, weather factors, etc. are subject to change. The EV range reported by manufacturers is directly related to a linear estimate of the maximum range that can be achieved by an electric vehicle, based on a real-time estimate of the battery's state of charge [51].

In [52], it was shown that the weight of the vehicle and the energy capacity of the batteries have a major influence on the range of an electric vehicle. When an electric vehicle has a high-capacity battery pack and is used for short distances, it is less efficient than a vehicle with a low-capacity battery. The results of the experiment presented in [53] show that the estimated average range of an electric vehicle without additional load during acceleration to 100 km/h over a distance of 5 km was 97.01 km, and an additional load of 270 kg reduced the range by 8.2%. Driving a car with an extra load also reduces the average speed. Based on the results of the study presented in [54], it was estimated that the actual energy consumption of electric vehicles increases by 60% and 40%, respectively, with each doubling of the vehicle weight, but only by 5% with each doubling of the rated power of the electric motor.

Iclodean et al.l. in work [55] present simulation studies of an electric vehicle equipped with batteries of the same energy capacity based on different battery technologies: Li-Ion, Na-NiCl2, Ni-MH, Li-S. Simulation studies were conducted based on an actual velocity profile. The results show that among all the battery technologies tested, equipping the vehicle with Na-NiCl2 sodium-nickel batteries had the lowest average energy consumption. On the other hand, the highest average energy consumption was recorded when the vehicle was equipped with Li-S lithium-sulphur batteries. The authors of [53] conclude that an increase in the capacity of the batteries of an electric vehicle may deteriorate the vehicle acceleration parameters, increase its operating costs, and reduce the cargo space when batteries of the same technology are used.

Electric vehicles have the ability to recover some of the energy in the braking process. The braking system converts kinetic energy during braking into electrical energy via the electric motor, which then acts as a power generator. In this way, part of the energy that is converted into heat during braking in a conventional vehicle, in this case in the form of electricity, can be recovered and stored in batteries and then used to propel the vehicle. The proportions of power sharing between the braking torque for energy recovery and the friction torque in the braking system are determined by data exchange between the controllers of the power train and the braking system. During deceleration of the vehicle, energy recovery can only be performed to a certain limit of vehicle speed proportional to the rotational speed of the electric motor. If the electric motor reaches the torque limit, the torque must be continuously reduced by mechanical elements as the rotational speed is reduced. The braking system of an electric vehicle should be so configured that some energy is recovered by the electric motor when the accelerator pedal is released.

There are two main strategies for sharing braking energy between a regenerative braking system and a mechanical braking system. The parallel braking strategy involves the simultaneous operation of the energy recovery system and the mechanical brakes. This is a relatively simple braking strategy but provides the lowest energy recovery. The series strategy envisages that the energy recovery system is activated first, and the mechanical brakes are applied only when the maximum level of recoverable electrical energy related to the kinetic energy of the vehicle is exceeded. Many researchers have undertaken to improve the performance of the energy recovery system by developing strategies for distributing the braking force between the mechanical brakes and the regenerative braking system. Examples can be found in [56–58], among others. The results presented in [59] show a significant effect of speed, deceleration, and vehicle weight on the value of energy recovered during braking. Road test results presented in [60] showed that under urban traffic conditions, regenerative braking covers almost 23% of the driving time, with the level of regeneration highly dependent on the driver's driving technique.

Driving conditions and the velocity profile used also affect the level of energy consumption and thus the range of an electric vehicle. Each route is characterized by certain parameters, including length, infrastructure, speed limits, and topography. On urban routes, the time of day is additionally important due to varying traffic volumes and congestion. The results presented in [61–64], among others, show that the type of route covered can have a significant impact on the level of energy needed to complete it. In [65], it was shown that driving on a road with varying road gradients significantly affects the changes in instantaneous power consumed by the vehicle and SOC parameters, and this translates into vehicle range.

The energy values obtained from the representative driving cycles used in the approval tests do not reflect the energy level obtained in real driving conditions. Based on the results presented in [66], the energy values obtained in the standardized cycles (FTP72, FTP75, JC 08, Japan10, NRDC, ECE-15) compared to the energy consumption in real driving conditions differ by 9.65–21.17% compared to the energy consumption in real driving conditions. To accurately estimate the energy consumption and driving range of electric vehicles under real-world road conditions, it is best to construct representative driving cycles typical of a city or region.

The ambient temperature also indirectly affects the range of an electric vehicle. This is due to the need to power additional equipment: at high temperatures—air conditioning, at low temperatures—the cabin ventilation and heating system. In the results of operational tests of electric drive vehicles with different battery energy capacity presented in [67–69], with respect to a temperature of 20 °C, driving with air conditioning turned on causes an increase in energy consumption. The results presented in the paper [70] show that using full cabin heating in winter conditions at −20 °C can reduce the range by up to 60%.

In addition to the parameters mentioned above, the range of the electric vehicle is also influenced by the driving style, infrastructure design, traffic intensity, and weather.

Driving style reflects the way the driver habitually drives. The way the vehicle is driven has a significant impact on the energy consumption of a vehicle with both electric and conventional drive systems. In the case of electric vehicles, energy consumption is associated with range. Many studies have confirmed that aggressive driving styles record higher values of energy consumption than drivers driving calmly. Aggressive driving reflects abrupt road maneuvers, rapid acceleration, and hard braking. Large oscillations in the driving speed have a negative impact on the energy consumption of electric vehicles. The greater the oscillation, the greater the energy consumption. As shown in [71], aggressive driving with high-speed oscillations may contribute to an increase in energy consumption by as much as 53%. On the basis of research on energy consumption in various traffic conditions during rush hours, the authors of the study [72] found that aggressive drivers cause on average 43% higher energy consumption compared to non-aggressive drivers. In addition, in lower traffic intensity, aggressive drivers cause 41.5% higher average energy consumption than calm drivers. It is also interesting that longer use of an electric vehicle contributes to a change in driving style. The research results presented in [73] show that after six months of use of an EV, as many as 73% of electric vehicle drivers changed their driving style. The use of an electric car made drivers change their driving style in terms of lower speed, less aggression, and more economical driving.

The type of road, road conditions, and elements of road infrastructure also determine the parameters of the vehicle's motion [74,75], and thus affect energy consumption. Curved road sections, speed limits, traffic-calming measures, and the arrangement of intersections with traffic lights all influence energy consumption through acceleration forces caused by numerous stops and accelerations. The results presented in [76] show that the average energy consumption in heavy traffic conditions is 15.6% higher than in smooth driving conditions.

Weather conditions also affect energy consumption. In the simulation results presented above it was shown that the ambient temperature has a great impact on the energy consumption of electric vehicle. It was shown in [71] that wind influences energy consumption in different ways for different driving speeds. Based on simulation studies, it was found that, with a strong headwind of 100 km/h, the energy consumption was nearly three times higher than in no wind conditions. The authors of the work [77] showed that electricity consumption is much higher in winter. Based on the research, it has been shown that the energy consumption of an electric vehicle increases by about 34% in relation to summer conditions. This means that, for countries where the winter season is relatively long, this translates into a shorter range of the electric vehicle in that period.

4. Range Estimation Models in Electric Vehicles

Many researchers have attempted to develop methods to predict range with the remaining energy level in the battery. Generally, to estimate the range, the energy control system relies on information regarding the current energy level of the battery, the energy value of a fully charged battery, and the energy consumption of additional loads, such as air conditioning. There are two main methods of range estimation for electric vehicles: history-based estimation and model-based estimation.

Range estimation methods may be based on historical energy consumption data calculated from a vehicle's dynamic parameters acquired during a journey or from the last journey. These data are used to predict vehicle range based on the energy remaining in the battery. The predicted energy consumption is estimated as the average energy consumption of the last kilometers driven. This method does not take into account the influence of actual driving style, or road and environmental conditions. Estimation based on historical data has limited accuracy but has the advantage of not including a model. The method of estimating range based on historical data is used in current commercially available electric vehicles. In the case of Tesla's Model 3, the mobile application displays two types of range, either an average range based on last-mile energy consumption data of 10, 25, or 50 km, or one determined by temporary range, taking into account the last few kilometers driven. In the case of the Renault Zoe, the remaining range is calculated based on the average consumption of the last 200 km. In the case of the Nissan Leaf, the range is estimated based on the average energy consumption of previously completed trips [78,79].

Many researchers have attempted to improve range estimation strategies based on historical data by inputting planned route data or current weather conditions. In [80], the predicted future power demand of an electric vehicle is determined based on the energy consumption history, acceleration and speed, and road information from a previously downloaded map. In [81], a method for estimating SOC and range is proposed considering the location-dependent operating conditions and power transmission efficiency of the power train. The electric vehicle range estimation model developed in [82] is a hybrid structure system combining a physical equation-based model with empirical data of the future driving profile and the vehicle's historical dynamic parameters.

In model-based estimation, a mathematical model is developed in which some dependent variables are expressed as functions of the independent variables. Model-based range estimation was developed to improve accuracy by estimating future energy consumption and thus range after predicting the future travel velocity profile. This approach is broken down into two steps. The first step is to predict the future velocity profile considering the route information and speed limits from GPS data. In the second step, the dynamic parameters of the vehicle and driving style are taken into account. In this method, the range is calculated while driving and may change [83]. Model-based range estimation has developed methods such as:

- Contour positioning system (CPS): Based on the current geographical location, the route parameters are taken to calculate the required force and torque for the route, and then the energy to be released from the battery is estimated [84,85].
- Dynamic range estimator (DRE): In addition to the route parameters, the driver's behavior is taken into account for estimating the energy consumption. This is represented by the acceleration value and the efficiency of the driver, and the use of additional electric auxiliary equipment (e.g., air-conditioning) [83,86,87].
- Big data analysis. This includes a mathematical model that involves a complex process of examining large data sets to discover certain patterns, correlations, or trends that are used to estimate some assumed variable. In estimating the range of an electric vehicle using this method, three steps are distinguished: (1) determining the route profile based on GPS data; (2) estimating the battery life considering the relationship of internal resistance and SOH; (3) estimating the energy consumption based on the selected route and the energy capacity of the battery, and calculating the range based on these values [88].
- Probabilistic estimation maps: Energy consumption is predicted as a random variable, and the probability of reaching the destination is obtained by estimating the energy required for the trip based on route parameters and the driver's behavior [89,90].

Less computational effort is required in estimating electric vehicle range using mathematical models than physical models, but they are less accurate because they are based on statistical and probability methods.

5. Methodology of Simulation

5.1. Vehicle

The electric vehicle model was created in AVL Cruise. The vehicle was equipped with an electric motor with a maximum power of 102 kW at 3000 rpm rotational speed and a 45 kWh battery with a maximum power of 160 kW. The powertrain has a single-ratio transmission, with a gear ratio of 6.058. Table 1 presents the vehicle's technical parameters.

Table 1. Vehicle parameters.

Type of Parameters	Parameters, Units	Value
Vehicle	curb weight, kg	1700
	gross weight, kg	1980
	frontal area, m^2	1.97
	wheel base, m	2.77
	height of gravity center, m	0.5
	drag coefficient	0.36
Electric motor (asynchronous motor)	maximum speed, rpm	3000
	maximum power, kW	102
Li-Ion battery	energy capacity, kWh	45
	maximum power, kW	160
	maximum charge, Ah	60
	maximum voltage, V	320
	number of cell per row	120
	number of cell rows	5

The electric vehicle model has regenerative braking capability, based on the velocity profile. The parallel regenerative braking strategy consists of a simultaneous operation of the electric and mechanical brakes. The energy recovered is limited by the current limitation of the electric motor and battery. The energy control system calculates the electric motor load signal in both traction and recuperation modes.

In the vehicle, the components of the electric power train are cooled by a coolant. The piping system is distributed to take heat away from the main components. The heat generated by the electric machine is transferred to cooling channels symmetrically distributed around the circumference of the stator. There are plates between battery cells, made up of channels, through which coolant flows to protect the battery pack from overheating. The simulations did not take into account the impact of additional energy consumers, such as air conditioning, audio system, heated windows, and mirrors, on the energy consumption of an electric vehicle.

5.2. Driving Cycles

A driving cycle is a velocity profile recorded as a function of time or distance that reflects the conditions of a vehicle on the streets of a particular city, region, or route. It is a sequence of repeating modules, containing phases of acceleration, constant driving speed, braking, and stopping. The following factors have an important influence on the shape of the cycle:

- Infrastructure, e.g., location and position of traffic lights, types of intersection, location of bus stops, road type (urban, suburban, highway).
- Vertical profile of the route.
- Traffic volume.
- Dynamic properties of the car.

Driving cycles standardized by legislators are used in approval tests for passenger cars and light-duty trucks. The laboratory tests of vehicles are carried out on a chassis dynamometer and reflect urban road conditions. As a result of the experiment, the emission of toxic compounds and fuel consumption in the velocity profile are obtained [91–94].

Until recently, the NEDC (New European Driving Cycle) was in force in the EU. NEDC has recently been replaced by the WLTC (Worldwide Harmonized Light Vehicles Test Cycle). The WLTC was developed based on velocity profiles of actual trips under traffic conditions characteristic of cities in Europe, Japan, and India [95]. In the United States, the FTP-75 (Federal Test Procedure) is used for approval testing of passenger cars and light trucks. This cycle includes four tests: FTP-75 which mimics inner city traffic conditions, HWFET which reflects highway driving conditions, SFTP US06 which imitates aggressive driving style, and the optional SFTP SC03 test which allows researchers to examine the impact of the air conditioning system on engine performance and emissions.

In many works and projects, driving cycles developed for a specific city or region can be found, called real-world driving cycles. These cycles are created from recorded velocity profiles while driving. Many papers have been devoted to the methodology of driving cycle development. Examples of a driving cycle developed exclusively for electric vehicles can be found in [96,97], among others. Vehicle tests conducted on the basis of a velocity profile reflecting the traffic conditions of a given city make it possible to estimate more precisely the values of energy and fuel consumption, as well as the level of emission of harmful exhaust compounds for specific traffic conditions.

Two NEDC or EPA (Environmental Protection Agency) tests are typically used to determine the range of an electric vehicle during testing. The range determined acc. to EPA is 30% shorter than the range determined using NEDC. It is worth noting that air conditioning, audio system, heated windows, windshield wipers, and other electrical devices that affect energy consumption are not run in the vehicles during testing.

In this paper, three cycles were used for simulation studies: The NEDC and WLTC and a cycle containing a velocity profile reflecting highway driving. The NEDC and WLTC cycles cover urban and extra-urban driving. The parameters of the analyzed cycles are presented in Table 2.

Table 2. Comparison of test cycles.

Parameters	NEDC	WLTC	Highway Cycle
duration, s	1180	1800	1200
distance, km	11	23.25	35
average velocity, km/h	34	46	104
maximum velocity, km/h	120	131	131
average acceleration, m/s^2	0.53	0.53	1.34
maximum acceleration, m/s^2	1.06	1.67	1.58
average deceleration, m/s^2	−0.75	−0.58	−1.41
maximum deceleration, m/s^2	−1.39	−1.50	−1.49

The NEDC cycle takes 1180 s and measures 11 km. In the part reflecting urban driving, the cycle is formed by modules consisting of stopping, accelerating, driving at constant speed, and braking. In this section, the vehicle accelerates to a maximum velocity of 50 km/h. The urban part accounts for about 66% of the cycle. In the test stage reflecting extra-urban driving, there are no stops, and the vehicle accelerates to a maximum velocity of 120 km/h. The average cycle velocity is 34 km/h. The proportion of stops in the cycle is 23%. The velocity profile for the NEDC cycle is shown in Figure 1.

The WLTC (Figure 2) driving cycle is divided into four parts, corresponding to different driving velocities: low, medium, high, and very high. It is more dynamic than NEDC and has a velocity profile that is more similar to the actual driving cycle. The length of the test is 23.25 km. The maximum velocity of the cycle is 131 km/h and the average velocity is 46.5 km/h. The share of stopping in the whole cycle is 13%. The urban driving portion of the cycle accounts for 52%. The Highway Cycle takes 1200 s and measures 35 km. The average cycle speed is 104 km. The vehicle accelerates to a maximum of 131 km/h. The cycle velocity profile is shown in Figure 3.

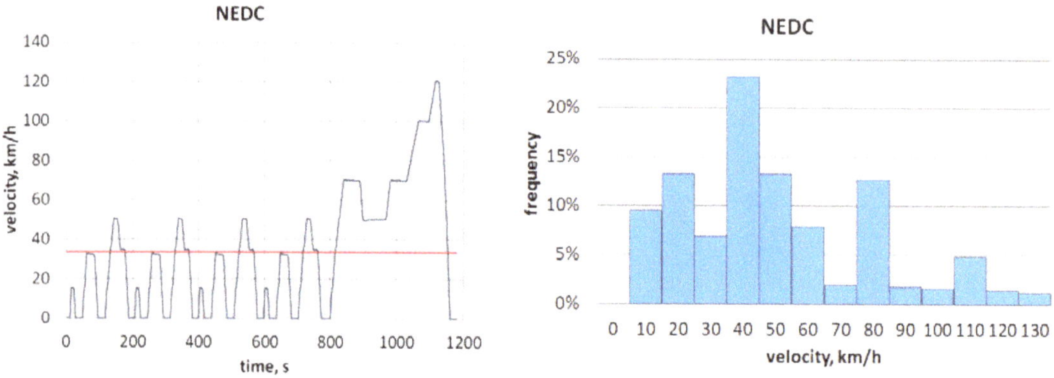

(**a**) (**b**)

Figure 1. NEDC cycle; (**a**) velocity profile (average velocity indicated by red line), (**b**) velocity distribution over the cycle.

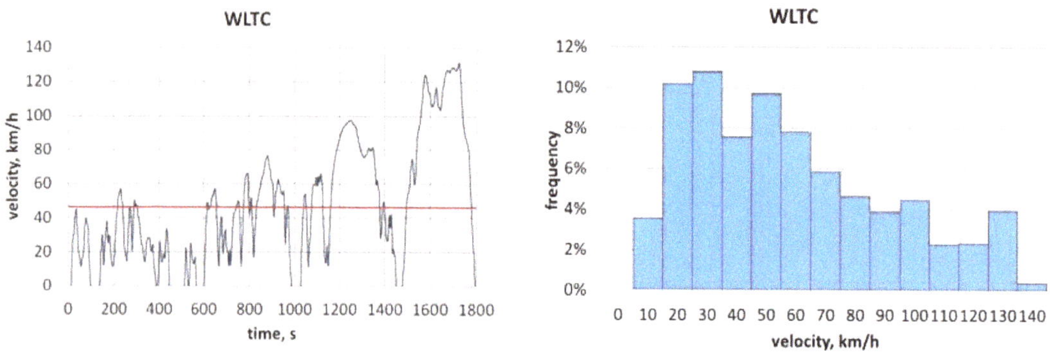

(**a**) (**b**)

Figure 2. WLTC cycle; (**a**) velocity profile (mean velocity indicated by red line), (**b**) velocity distribution over the cycle.

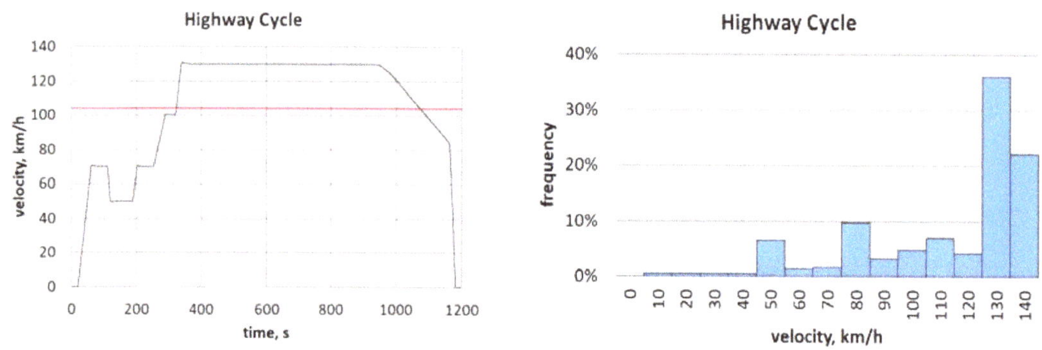

(**a**) (**b**)

Figure 3. Highway Cycle; (**a**) velocity profile (mean velocity indicated by red line), (**b**) velocity distribution over the cycle.

5.3. AVL Cruise Vehicle Simulation Program

AVL Cruise software (2020, AVL—Advanced Simulation Technologies, Graz, Austria) was used in the simulation studies. The program allows the modeling and simulation of vehicles equipped with different types of power trains. The vehicle model available in the program is a system built of interconnected subsystems, namely vehicle components and drive components. The program allows researchers to:

- obtain a reliable and accurate fuel consumption forecast of the developed vehicle;
- analyze energy flow, power distribution, and losses in the power train, from the power source to the wheels;
- find the optimum balance between fuel consumption, emissions, and the drive's traction properties;
- analyze torsional vibrations of flexible chassis (under dynamic load);
- analyze temperature distribution in power train components.

When creating a vehicle model in AVL Cruise, the user has the opportunity to use an extensive library of real vehicles and power train components with their parameters and characteristics. It is possible to create one's own vehicle or power train model in AVL Cruise M. The developed vehicle can then be entered into AVL Cruise, where a simulation is performed. During the simulation, it is possible to select additional equipment in the vehicle (such as air conditioning or power steering), the influence of external conditions, i.e., wind force, ambient or road temperature, and to generate a random driving cycle, which may include driving in urban, suburban, or highway conditions.

Simulation calculations are performed using a mixed, backward/forward procedure. This enables effective and more efficient estimation of the influence of input parameter values on the efficiency of power train components. All possible combinations of variables are performed in the calculation. As a result of the simulation, the user is provided with values for fuel consumption, emissions and vehicle efficiency, maximum climbing ability and acceleration times, as well as graphs and performance characteristics of drive components.

Simulation studies enable quick and easy analysis of the influence of selected factors on the efficiency of the vehicle and drive components.

6. Results

In the simulation studies conducted, an electric passenger car was analyzed under various driving conditions. During the simulation, the vehicle was tested:

- in three driving cycles with different traffic dynamics (NEDC, WLTC, highway cycle);
- at different initial battery charge levels (40%, 50%, 60%, 70%, 80%, 90%, 100%);
- at different loads (50, 100, 150, 200, 250 kg);
- at different ambient temperatures (−20 °C, −10 °C, 0 °C, 10 °C, 20 °C, 40 °C).

During the analyses, the influence of the factors listed above on the energy level in the accumulator (SOC; state of charge), the degree of accumulator discharge (DOD; depth of discharge), total energy consumption, and average energy consumption were investigated.

In the simulation studies conducted, three driving cycles were analyzed. The NEDC and WLTC cycles cover urban and suburban driving, and the highway cycle reflects highway driving. The energy consumption of the electric vehicle during the analyzed cycles is shown in Figure 4. Simulation studies were conducted with a constant ambient temperature of 20 °C and a constant load of 50 kg. The impact of the driving cycle has a significant impact on energy consumption. It is noticeable that the higher the speed of travel, the greater the amount of energy required to maintain it. As the simulation results show, the highest average energy consumption was recorded in the cycle reflecting highway driving. The energy expended to cover a 35 km stretch on a highway is about 8.6 kWh. In cycles combining urban and extra-urban driving, energy consumption is significantly lower. Energy consumption in the NEDC cycle was approximately 2.2 kWh. In the WLTC cycle, characterized by a more dynamic velocity profile, the energy consumption was about 4.8 kWh. It is worth noting that the WLTC cycle is about twice as long as the NEDC cycle.

Converting this energy into 1 km of distance traveled, it can be seen that the tests carried out yielded: 0.2, 0.21, and 0.24 kWh/km, respectively. The average energy consumption per km is shown in Figure 5.

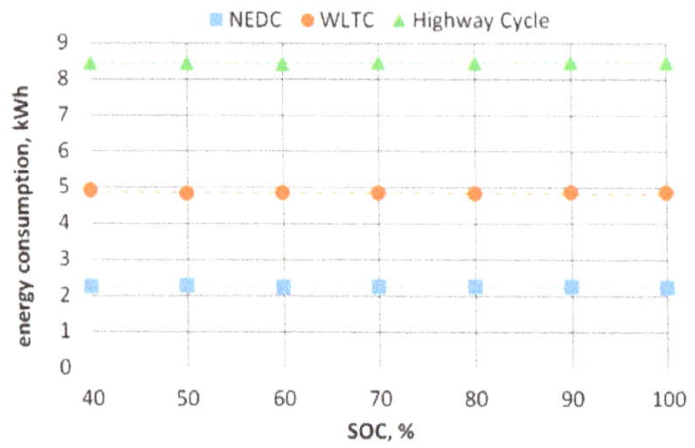

Figure 4. Energy consumption of the analyzed driving cycles at different SOC.

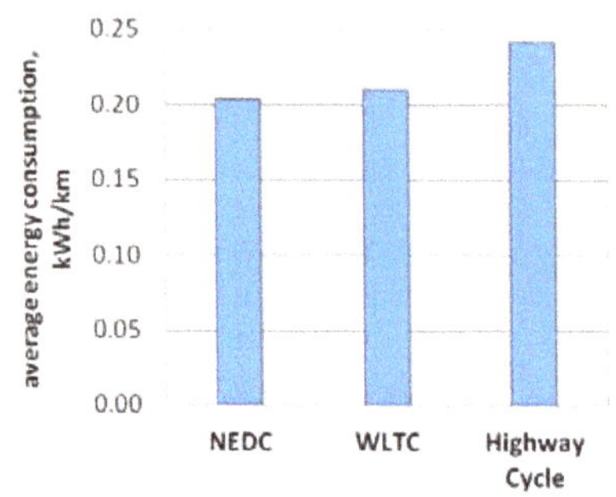

Figure 5. Average energy consumption of the analyzed driving cycles.

The values of the average energy consumption in the analyzed cycles are similar. The lowest average energy consumption was recorded on the NEDC cycle. The highest average energy consumption was recorded by the electric vehicle in a cycle reflecting highway driving. It is worth noting that the differences in the values of the average energy consumption in these two cycles are not high, as they are only 16%. The course of the battery's state of charge for different initial state of charge values in the selected cycles is shown in Figure 6.

As can be seen from the courses of the state of charge (SOC) for the NEDC and WLTC cycles, there were no sudden drops in energy during city driving. It can be seen that energy consumption increases, especially in the part of the cycle reflecting non-urban driving. This is when the vehicle is accelerated to higher velocities, resulting in a higher demand for energy, causing the energy level in the accumulator to drop. In order to determine the effect of the implemented cycle at different levels of state of charge (SOC), the depth of discharge value was used. Its value during the analyzed driving cycles is shown in Figure 7.

(a)

(b)

(c)

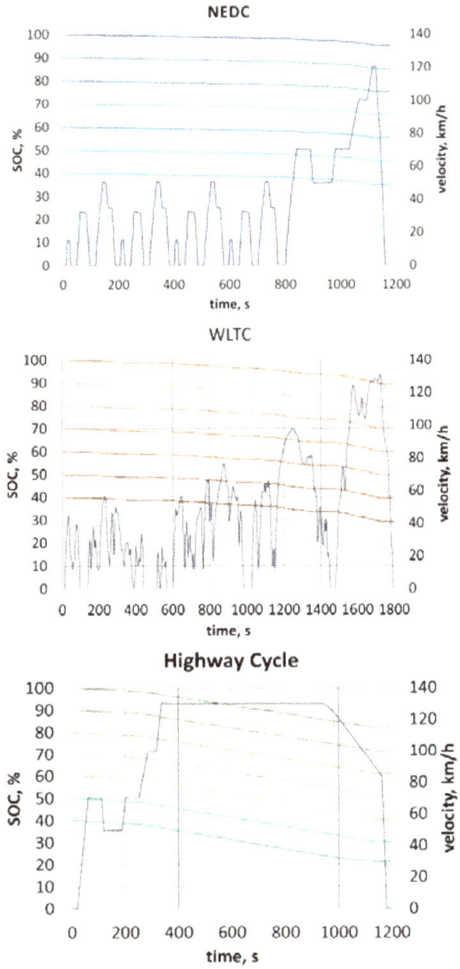

Figure 6. State of charge (SOC) under analyzed driving cycles; (**a**) NEDC, (**b**) WLTC, (**c**) Highway Cycle.

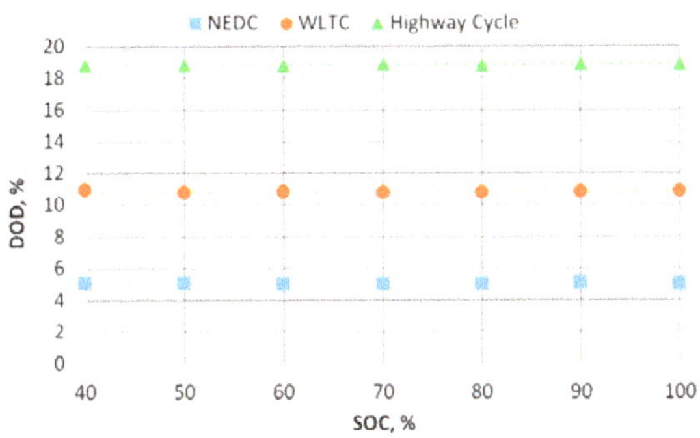

Figure 7. Depth of discharge (DOD) under the cycle.

Depending on the initial state of charge of the battery, the drop in energy level after completing the NEDC cycle is 4.94–5.02% and in the WLTC cycle it is 10.72–10.91%. The depth of discharge of the battery after a cycle reflecting highway driving is 18.74–18.85%. Thus, it can be confirmed that, regardless of the initial state of charge of the battery, its level of discharge is dependent on how the route is taken. Figure 8 shows the battery current waveform for the analyzed driving cycles at initial state of charge of 60% and 100%.

(a) (b) (c)

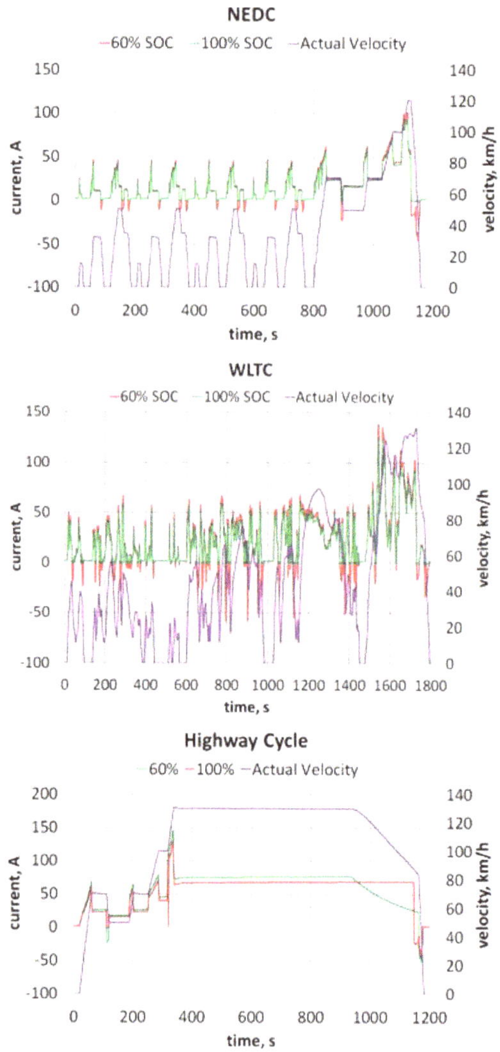

Figure 8. Battery current in analyzed driving cycles; (**a**) NEDC, (**b**) WLTC, (**c**) Highway Cycle.

The current waveform directly depends on the speed changes implemented. During acceleration a current of 150 A is drawn, and it is possible to charge with a current of more than 50 A during braking. Electric vehicles have the possibility to recover part of the energy during braking. However, this is determined by the energy level in the accumulator and the power of the electric machine in generator mode. The battery management system (BMS) decides how much energy can be directed to the accumulator. As presented in Figure 9, a fully charged battery (SOC 100%) is not able to store any more recovered energy.

Only when the energy level is below 80% can the battery be recharged. If the energy level is, for example, 60%, then the accumulator can accept much more recuperated energy.

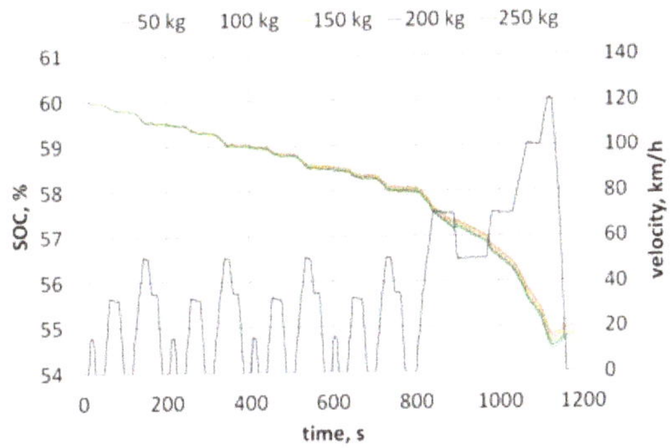

Figure 9. State of charge (SOC) under the vehicle load.

The efficiency of the electric power train is affected by the weight of the vehicle. The higher the load, the more energy the vehicle needs to overcome the resistance to motion. Figure 8 shows the level of discharge of a battery pack during the NEDC cycle at different vehicle load levels, and Figure 10 shows the depth of discharge of the battery. The initial state of charge of the accumulator was 60%, the ambient temperature was 20 °C, and the vehicle load was 50 kg.

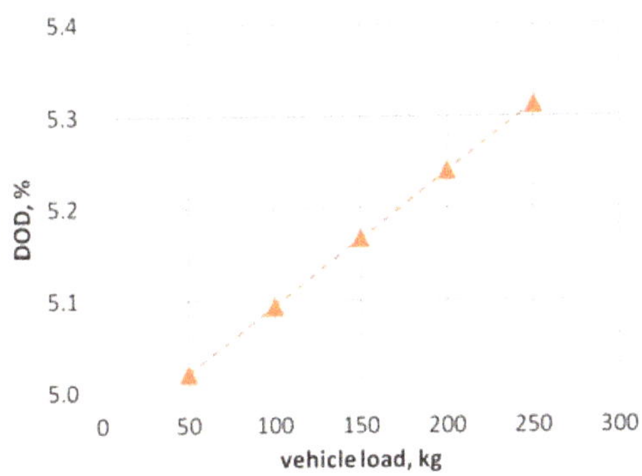

Figure 10. Depth of discharge (DOD) under the vehicle weight.

From the simulation results, it can be concluded that the load and weight of the vehicle significantly affect the energy consumption. With a vehicle load of 250 kg, energy consumption is 7% higher than with a load of 50 kg. The higher the load, the higher the energy consumption. By considering the level of depth of discharge of the accumulator at the analyzed loads, it can be concluded that this is a linear relationship.

The battery state of charge at the temperatures considered is shown in Figure 11.

An important issue related to the energy consumption of an electric vehicle is also the influence of the ambient temperature. In this study, the effect of ambient temperature on the energy consumption of the NEDC cycle was investigated at the following ambient temperatures: −20 °C, −10 °C, 0 °C, 10 °C, 20 °C, 40 °C. The initial battery's state of charge was 60%. In the study, it was assumed that the set ambient temperature is also the temperature of the battery.

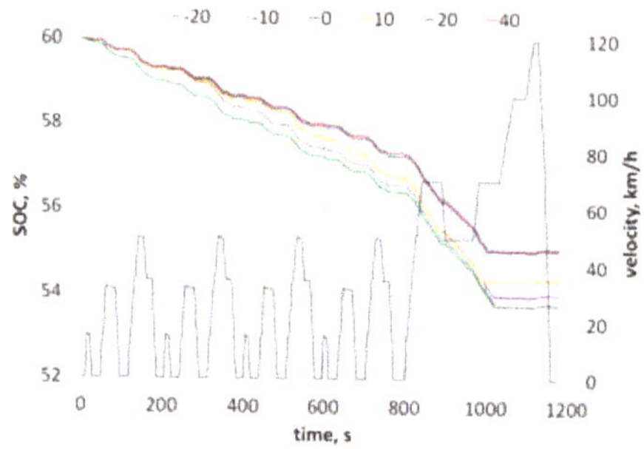

Figure 11. Coolant temperature in battery during NEDC.

From the simulation results shown in the graphs, it is clear that temperature has a significant impact on the energy consumption level. In the case of a temperature of 40 °C, the battery's level of discharge after performing the NEDC cycle is similar to the course at 20 °C. The energy level of the battery at temperatures below 0 °C is 24% lower compared to the test results at 20 °C. At a temperature of 10 °C, these differences are 13%. The depth of discharge of the battery at selected temperatures on the NEDC cycle is shown in Figure 12.

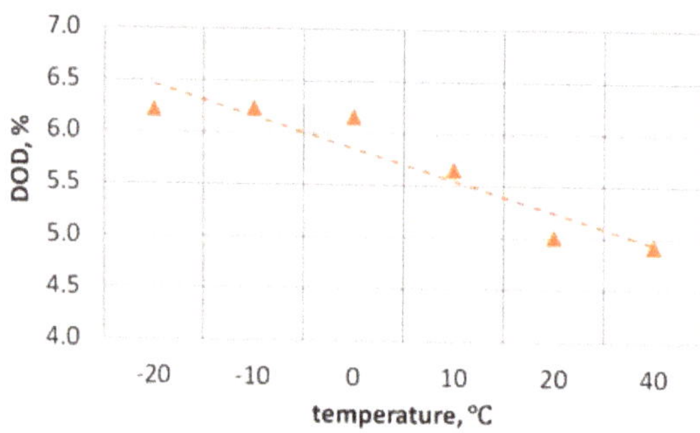

Figure 12. Depth of discharge (DOD) under different ambient temperature in NEDC.

The depth of discharge of the battery assumes the highest values at negative temperatures. For temperatures above 20 °C, the DOD takes comparable values. As previously mentioned, in an electric vehicle model, the drive components are cooled using a cooling system where the coolant is liquid. The coolant temperature during the NEDC cycle at a given ambient temperature is shown in Figure 13.

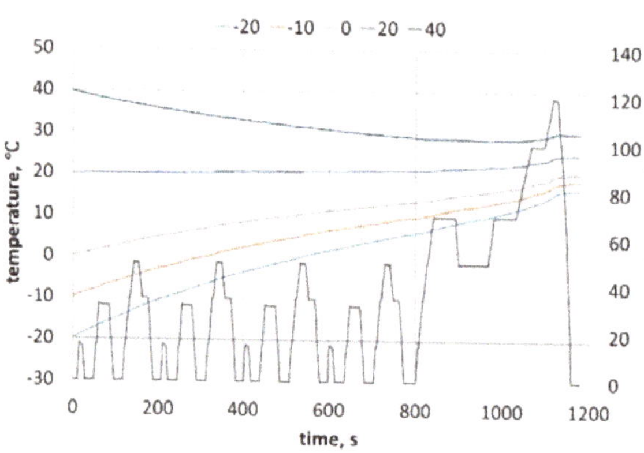

Figure 13. Battery coolant temperature during NEDC.

The battery thermal management system (BTMS) tries to keep the battery temperature at a certain level. In the case of high ambient temperature, the BTMS is designed to prevent overheating. Hence, as seen in Figure 12, the temperature of the liquid decreases and thus cools the batteries. When the ambient temperature is below 0 °C, the task of the system is to heat the battery. This can be observed in the figure. The temperature of the coolant in the system increases.

7. Discussion

The aim of the analyses undertaken was to determine how various factors affect the depth of discharge. Drivers' awareness of how a car's battery discharges, what actions to avoid while driving, or what actions are desirable can significantly offset concerns about electric vehicle use.

Thus, this paper analyzes the effects of distance, the initial state of charge, load, and external temperature.

Of the factors analyzed, the battery energy level is most affected by the driving cycle. On the basis of the presented results concerning the level of depth of discharge of the accumulator, it can be concluded that the length, the shape of the road, the route, and thus the driving dynamics significantly affect energy consumption. This, in turn, can translate into vehicle range. In this paper, three driving cycles characterized by different driving dynamics were analyzed. The NEDC cycle is the shortest of the cycles analyzed. It is 11 km long and 66% of the route reflects urban conditions. Average energy consumption on the NEDC cycle is 0.20 kWh/km and DOD is 4.95%. Another cycle, WLTC, has a more dynamic velocity profile and higher average speed than the NEDC cycle. The WLTC cycle measures 23 km, and with 52% urban driving represents the routes of the cycle. Average energy consumption for this cycle was 0.21 kWh/km (DOD is 10.75%). In a cycle reflecting highway driving, average energy consumption was 0.24 kWh/km. This cycle measures 35 km with an average speed of 104 km/h. In this cycle, DOD was 18.8%. The discrepancy of the depth of discharge, as shown in the simulations, is quite significant. Table 3 shows the values of energy consumption in the analyzed cycles with a load of 50 kg and an initial state of charge of the accumulator equal to 60%. The ambient temperature was 20 °C.

Table 3. Energy consumption of the analyzed cycles.

Cycle	Energy Consumption		DOD, %
	Total, kWh	Average, kWh/km	
NEDC	2.23	0.20	4.95
WLTC	4.74	0.21	10.75
Highway Cycle	8.43	0.24	18.74

Table 4 shows the energy consumption values and the depth of discharge for the NEDC cycle with an accumulator's initial state of charge of 60% and a load of 50 kg.

Table 4. Energy consumption and depth of discharge at different temperatures.

Temperature, °C	−20	−10	0	10	20	40
total energy consumption, kWh	2.93	2.81	2.64	2.60	2.23	2.27
average energy consumption, kWh/km	0.27	0.26	0.24	0.24	0.20	0.21
differences in average energy consumption, %	23.89	20.64	15.53	14.23	–	1.76
DOD, %	6.23	6.22	6.15	5.65	4.95	4.92

The ambient temperature also has a significant impact on energy consumption. In the analyzed driving cycle at temperatures below 0 °C, average energy consumption can increase by up to 24% compared to an ambient temperature of 20 °C. At negative ambient temperatures, the DOD was 6.2%. Average energy consumption is less affected by high ambient temperatures. The energy consumption values at 40 °C are similar to the energy consumption values at 20 °C.

Table 5 shows the energy consumption values and discharge rate of the battery in the NEDC cycle with an initial state of charge of the accumulator of 60% and an ambient temperature of 20 °C.

Table 5. Energy consumption and depth of discharge under different loads.

Load, kg	50	100	150	200	250
total energy consumption, kWh	2.23	2.26	2.29	2.36	2.39
average energy consumption, kWh/km	0.20	0.21	0.21	0.22	0.22
differences in average energy consumption, %	–	1.33	2.62	5.51	6.69
DOD, %	4.95	5.02	5.09	5.17	5.24

To a lesser extent, DOD levels are affected by vehicle load. The average energy consumption of an electric vehicle with a load of 250 kg is 7% higher than when the vehicle load is 50 kg. With a maximum load of 250 kg, the discharge rate of the accumulator is 5.3%.

Table 6 shows the energy consumption values and discharge rate of the battery in the NEDC cycle at a different state of charge of the accumulator with a load of 50 kg and an ambient temperature of 20 °C.

Table 6. Energy consumption and depth of discharge at a different initial SOC.

SOC, %	40	50	60	70	80	90	100
total energy consumption, kWh	2.26	2.25	2.23	2.24	2.24	2.25	2.25
average energy consumption, kWh/km	0.21	0.21	0.20	0.20	0.20	0.21	0.21
differences in average energy consumption, %	−1.35	−0.90	–	−0.45	−0.45	−0.90	−0.90
DOD, %	5.03	5.01	4.95	4.96	4.98	5.00	5.01

Power consumption is slightly affected by the initial state of charge of the battery. The differences in the average energy expended per driving cycle with different initial battery state of charge are about 1%.

A summary of the effects of the factors in this article on the range of depth of discharge is shown in Figure 14.

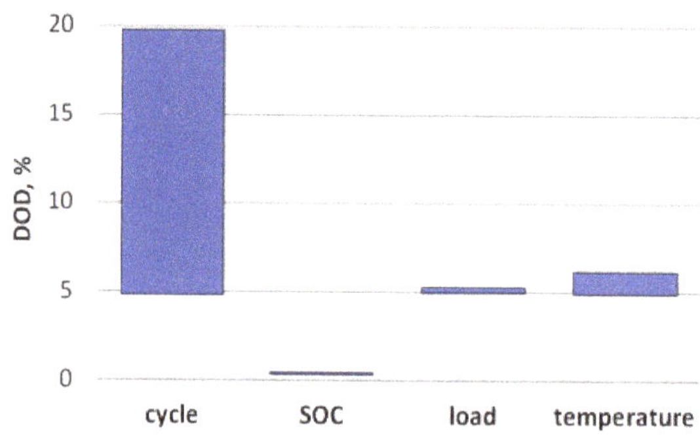

Figure 14. Influence of selected parameters on battery energy level.

8. Conclusions

Recently, range anxiety has been observed among electric vehicle users. The limited range of an electric vehicle makes EV users worry that the battery will drain while driving and the vehicle will stall on the road. The results of the research presented in the papers cited in Section 2 show that range anxiety affects driving behavior at the emotional, psychological, behavioral, and physiological levels. Seeing low energy levels makes drivers feel anxious, and under stress, they may change their driving style. Drivers mostly choose to reduce speed to save energy and thus increase range. Moreover, the change in driving style includes behaviors that are considered dangerous and aggressive, such as tapping your fingers on the steering wheel, yelling, honking, or aggressive gestures to other road users.

So, it is worth exploring the factors that further reduce range during daily electric vehicle operation that EV users are often unaware of. The range of an electric vehicle is affected by many factors, primarily the route taken, the load on the vehicle, and the ambient temperature. The purpose of this study was to determine the effect of selected parameters on a battery's depth of discharge (DOD). In the simulation study of an electric vehicle, the effects of the driving cycle, ambient temperature, load, and the initial state of charge of the accumulator on the energy consumption pattern and a battery's depth of discharge (DOD) were analyzed. The presented results show how significantly the operating conditions of an electric vehicle affect the energy life. This translates into the distance the vehicle can travel on a single charge.

Simulation results confirmed that the route significantly affects energy consumption. Urban driving involves frequent acceleration and braking phases which can contribute to a significant increase in energy consumption, but also provide the possibility of energy recuperation during braking. Higher energy consumption is observed at high speeds, i.e., when driving on a highway or in extra-urban driving conditions, at a fixed speed, without the possibility of energy recovery. In the NEDC cycle, urban driving accounts for 66%. The average energy consumption in NEDC was 2.20 kWh (DOD—4.95%). In the WLTC, the urban driving portion of the cycle accounts for 52%. Average energy consumption was 0.21 kWh (DOD was 10.75%). In a cycle representing highway driving, average energy consumption was 0.24 kWh (DOD was 18.8%). The differences in the values of the average energy consumption between the NEDC cycle and the cycle reflecting driving on the highway is 16%. Therefore, it can be concluded that as the average speed of the driving cycle increases, the energy consumption also increases and therefore the range decreases.

The ambient temperature has also an impact on the energy consumption of an electric vehicle. Simulation studies have shown that the highest energy consumption occurs during negative ambient temperature. At temperatures below 0 °C, average energy consumption can increase by up to 24% compared to an ambient temperature of 20 °C. On the other hand, energy consumption is less affected by high ambient temperatures. Based on simulation results, when the ambient temperature was 40 °C, the average energy consumption was 2% higher than at an ambient temperature of 20 °C. Therefore, it can be pointed out that the anxiety of drivers that the range of an electric vehicle is significantly reduced at negative ambient temperatures can be confirmed.

Based on the simulation results, the load affected on energy consumption of EV. Driving with a maximum load can result in an increase in average energy consumption of 7% compared to driving with a load of 50 kg. It can be observed that the energy consumption increases in a linear manner with the load.

The range of an electric vehicle depends on many factors. Based on the results of the simulation research, it can be concluded that the users of electric vehicles should take additional factors into account when estimating the vehicle range and planning the route. Such factors should mainly include the type of route, ambient temperature, and vehicle load, as these factors can affect energy consumption and thus reduce the range. The authors are aware that, apart from the analyzed parameters, there are many other factors influencing the range of an electric vehicle. More research is needed to deepen our understanding of the parameters that may limit the range of an EV. In further work, the authors want to investigate the impact of driving style, traffic intensity, and infrastructure elements on the operational parameters of an electric vehicle.

Author Contributions: Conceptualization, E.M.S. and R.S.J.; methodology, E.M.S.; software, E.M.S.; validation, E.M.S.; formal analysis, E.M.S. and R.S.J.; investigation, E.M.S.; resources, E.M.S.; data curation, E.M.S.; writing—original draft preparation, E.M.S. and R.S.J.; writing—review and editing, E.M.S. and R.S.J.; visualization, E.M.S. and R.S.J.; supervision, E.M.S. and R.S.J.; project administration, E.M.S.; funding acquisition, R.S.J. Both authors have read and agreed to the published version of the manuscript.

Funding: This research received no external funding.

Institutional Review Board Statement: Not applicable.

Informed Consent Statement: Not applicable.

Data Availability Statement: Not applicable.

Conflicts of Interest: The authors declare no conflict of interest.

References

1. New Passenger Car Registrations in the EU by Alternative Fuel Type, European Automobile Manufacturers' Association (ACEA). Available online: https://www.acea.be/press-releases/article/fuel-types-of-new-cars-electric-10.5-hybrid-11.9-petrol-47.5-market-share-f (accessed on 20 March 2021).
2. *Vehicle in Use*; ACEA Report; European Automobile Manufacturers' Association (ACEA): Brussels, Belgium, 2021.
3. *Global EV Outlook 2020 Entering the Decade of Electric Drive?* International Energy Agency (IEA): Paris, France, 2020.
4. The Global Electric Vehicle Market in 2021: Statistics & Forecasts. Available online: https://www.virta.global/global-electric-vehicle-market (accessed on 20 March 2021).
5. Turoń, K.; Sierpiński, G. Car-Sharing in Urban Transport Systems—Overview of Europe and Asia. In *Challenges of Urban. Mobility, Transport. Companies and Systems*; (Springer Proceedings in Business and Economics); Springer: Berlin/Heidelberg, Germany, 2019; pp. 89–99. [CrossRef]
6. Münzel, K.; Boon, W.; Frenken, K.; Blomme, J.; van der Linden, D. Explaining carsharing supply across Western European cities. *Int. J. Sustain. Transp.* **2019**, *14*, 243–254. [CrossRef]
7. Ramos, É.M.S.; Bergstad, C.J.; Chicco, A.; Diana, M. Mobility styles and car sharing use in Europe: Attitudes, behaviours, motives and sustainability. *Eur. Transp. Res. Rev.* **2020**, *12*, 13. [CrossRef]
8. Study on Passenger Transport by Taxi, Hire Car with Driver and Ridesharing in the EU, European Commission Directorate-General for Mobility and Transport. 2016. Available online: https://ec.europa.eu/transport/modes/road/studies/study-passenger-transport-taxi-hire-car-driver-and-ridesharing-eu_cs (accessed on 21 March 2021).
9. Mathieu, L.; Poliscanova, J. *Mission (Almost) Accomplished. Carmakers' Race to Meet the 2020/21 CO_2 Targets and the UE Electric Cars Market*; Transport & Environment: Brussels, Belgium, 2020.
10. Merkisz-Guranowska, A.; Maciejewski, M. The Implementation of the Electric Taxi Fleet in the City of Poznan, Poland. *WIT Trans. Built Environ.* **2015**, *146*, 243–254. [CrossRef]
11. Moniot, M.; Rames, C.; Burrell, E. Feasibility Analysis of Taxi Fleet Electrification using 4.9 Million Miles of Real-World Driving Data. *SAE Tech. Pap.* **2019**, 1–10. [CrossRef]
12. Tamis, M.; van den Hoed, R. Moving a Taxi Sector to Become Electric: Characterizing Taxi Drivers Interested in Purchasing a Full Electric Vehicle. *World Electr. Veh. J.* **2020**, *11*, 20. [CrossRef]
13. Wang, G.; Zhang, Y.; Fang, Z.; Wang, S.; Zhang, F.; Zhang, D. FairCharge: A Data-Driven Fairness-Aware Charging Recommendation System for Large-Scale Electric Taxi Fleets. *Proc. ACM Interact. Mob. Wearable Ubiquitous Technol.* **2020**, *4*, 25. [CrossRef]
14. Hall, D.; Moultak, M.; Lutsey, N. Electric Vehicle Capitals of the World: Demonstrating the Path to Electric Drive. International Council on Clean Transportation. 2017. Available online: https://theicct.org/publications/EV-capitals-of-the-world (accessed on 21 March 2021).
15. Maffei, L.; Masullo, M. Electric Vehicles and Urban Noise Control Policies. *Arch. Acoust.* **2014**, *39*, 333–341. [CrossRef]
16. Pallas, M.-A.; Kennedy, J.; Chatagnon, R.; Berengier, M.; Czuka, M.; Conter, M.; Muirhead, M. Towards a model for electric vehicle noise emission in the European prediction method CNOSSOS-EU. *Appl. Acoust.* **2016**, *113*, 89–101. [CrossRef]
17. Łebkowski, A. Electric Vehicles—The Sound Of Silence. *Masz. Elektr. Zesz. Probl. Nr 1* **2016**, *109*, 155–159.
18. Bekker, A. Influences of Electric Propulsion on Vehicle Vibro-acoustics. *R D J. South. Afr. Inst. Mech. Eng.* **2014**, *30*, 47–54.
19. Szumska, E.; Jurecki, R.; Pawełczyk, M. Assessment of Total Costs of Ownership for Midsize Passenger Cars with Conventional and Alternative Drive Trains. *Commun. Sci. Lett. Univ. Zilina* **2019**, *21*, 21–27. [CrossRef]
20. Hagman, J.; Ritzén, S.; Janhager Stier, J.; Susilo, Y. Total cost of ownership and its potential implications for battery electric vehicle diffusion. *Res. Transp. Bus. Manag.* **2016**, *18*, 11–17. [CrossRef]
21. Kumar, P.; Chakrabarty, S. Total Cost of Ownership Analysis of the Impact of Vehicle Usage on the Economic Viability of Electric Vehicles in India. *Transp. Res. Rec.* **2020**, *2674*, 563–572. [CrossRef]
22. Letmathe, P.; Suares, M. A consumer-oriented total cost of ownership model for different vehicle types in Germany. *Transp. Res. Part D Transp. Environ.* **2017**, *57*, 314–335. [CrossRef]

23. Stevic, Z.; Radovanovic, I. Energy Efficiency of Electric Vehicles, New Generation of Electric Vehicles. In *New Generation of Electric Vehicles*; Stevic, Z., Ed.; Intech-Open: Belgrade, Serbia, 2012; pp. 93–134. [CrossRef]
24. Mirchevski, S. Energy Efficiency in Electric Drives. *Electronics* **2012**, *16*, 46–49. [CrossRef]
25. Gu, J.; Ouyang, M.; Lu, D.; Li, J.; Lu, L. Energy efficiency optimization of electric vehicle driven by in-wheel motors. *Int. J. Automot. Technol.* **2013**, *14*, 763–772. [CrossRef]
26. Kobayashi, S.; Plotkin, S.; Ribeiro, S.K. Energy efficiency technologies for road vehicles. *Energy Effic.* **2009**, *2*, 125–137. [CrossRef]
27. Han, J.; Vahidi, A.; Sciarretta, A. Fundamentals of energy efficient driving for combustion engine and electric vehicles: An optimal control perspective. *Automatica* **2019**, *103*, 558–572. [CrossRef]
28. Albatayneh, A.; Assaf, M.; Alterman, D.; Jaradat, M. Comparison of the Overall Energy Efficiency for Internal Combustion Engine Vehicles and Electric Vehicles. *Environ. Clim. Technol.* **2020**, *24*, 669–680. [CrossRef]
29. Ehsani, M.; Gao, Y.; Emadi, A. *Modern Electric, Hybrid Electric, and Fuel Cell Vehicles—Fundamentals, Theory, and Design*, 2nd ed.; CRC Press: Boca Raton, FL, USA, 2017.
30. Liu, H.; Chen, X.; Wang, X. Overview and Prospects on Distributed Drive Electric Vehicles and Its Energy Saving Strategy. *Przegląd Elektrotechniczny Electr. Rev.* **2012**, *88*, 122–125.
31. Wang, J.; Besselink, I.; Nijmeijer, H. Electric Vehicle Energy Consumption Modelling and Prediction Based on Road Information. *World Electr. Veh. J.* **2015**, *7*, 447–458. [CrossRef]
32. Musabini, A.; Nguyen, K.; Rouyer, R.; Lilis, Y. Influence of Adaptive Human–Machine Interface on Electric-Vehicle Range-Anxiety Mitigation. *Multimodal Technol. Interact.* **2020**, *4*, 4. [CrossRef]
33. Yuan, Q.; Hao, W.; Su, H.; Bing, G.; Gui, X.; Safikhani, A. Investigation on Range Anxiety and Safety Buffer of Battery Electric Vehicle Drivers. *J. Adv. Transp.* **2018**, *2018*, 8301209. [CrossRef]
34. Jung, M.F.; Sirkin, D.; Gür, T.M.; Steinert, M. Displayed Uncertainty Improves Driving Experience and Behavior: The Case of Range Anxiety in an Electric Car. In Proceedings of the 33rd Annual ACM Conference on Human Factors in Computing Systems (CHI '15), Seoul, Korea, 18–23 April 2015; pp. 2201–2210. [CrossRef]
35. Graham-Rowe, E.; Gardner, B.; Abraham, C.; Skippon, S.; Dittmar, H.; Hutchins, R.; Stannard, J. Mainstream consumers driving plug-in battery-electric and plug-in hybrid electric cars: A qualitative analysis of responses and evaluations. *Transp. Res. Part A Policy Pract.* **2012**, *46*, 140–153. [CrossRef]
36. Valentine-Urbschat, M.; Bernhart, W. *Powertrain 2020—The Future Drives Electric*; Roland Berger Strategy Consultants: Munich, Germany, 2009.
37. Pevec, P.; Babic, J.; Carvalho, A.; Ghiassi-Farrokhfal, Y.; Ketter, W.; Podobnik, W. Electric Vehicle Range Anxiety: An Obstacle for the Personal Transportation (R)evolution? In Proceedings of the 4th International Conference on Smart and Sustainable Technologies (SpliTech), Split, Croatia, 18–21 June 2019; pp. 1–8. [CrossRef]
38. Rauh, N.; Franke, T.; Krems, J.F. Understanding the Impact of Electric Vehicle Driving Experience on Range Anxiety. *Hum. Factors* **2014**, *57*, 177–187. [CrossRef] [PubMed]
39. El-Fedany, I.; Kiouach, D.; Alaoui, R. Application Design Aiming to Minimize Drivers' Trip Duration through Intermediate Charging at Public Station Deployed in Smart Cities. *World Electr. Veh. J.* **2019**, *10*, 67. [CrossRef]
40. Tannahill, V.R.; Sutanto, D.; Muttaqi, K.M.; Masrur, M.A. Future vision for reduction of range anxiety by using an improved state of charge estimation algorithm for electric vehicle batteries implemented with low-cost microcontrollers. *IET Electr. Syst. Transp.* **2015**, *5*, 24–32. [CrossRef]
41. Chen, R.; Liu, X.; Miao, L.; Yang, P. Electric Vehicle Tour Planning Considering Range Anxiety. *Sustainability* **2020**, *12*, 3685. [CrossRef]
42. Guo, F.; Yang, J.; Lu, J. The battery charging station location problem: Impact of users' range anxiety and distance convenience. *Transp. Res. Part E Logist. Transp. Rev.* **2018**, *114*, 1–18. [CrossRef]
43. Cui, S.; Zhao, H.; Wen, H.; Zhang, C. Locating Multiple Size and Multiple Type of Charging Station for Battery Electricity Vehicles. *Sustainability* **2018**, *10*, 3267. [CrossRef]
44. Liu, G.; Kang, L.; Luan, Z.; Qiu, J.; Zheng, F. Charging Station and Power Network Planning for Integrated Electric Vehicles (EVs). *Energies* **2019**, *12*, 2595. [CrossRef]
45. Akbari, M.; Brenna, M.; Longo, M. Optimal Locating of Electric Vehicle Charging Stations by Application of Genetic Algorithm. *Sustainability* **2018**, *10*, 1076. [CrossRef]
46. Pagany, R.; Marquardt, A.; Zink, R. Electric Charging Demand Location Model—A User- and Destination-Based Locating Approach for Electric Vehicle Charging Stations. *Sustainability* **2019**, *11*, 2301. [CrossRef]
47. Ramachandran, A.; Balakrishna, A.; Kundzicz, P.; Neti, A. Predicting Electric Vehicle Charging Station Usage: Using Machine Learning to Estimate Individual Station Statistics from Physical Configurations of Charging Station Networks. *arXiv* **2018**, arXiv:1804.00714.
48. Sarker, M.R.; Pandžić, H.; Ortega-Vazquez, M.A. Electric vehicle battery swapping station: Business case and optimization model. In Proceedings of the 2013 International Conference on Connected Vehicles and Expo (ICCVE), Las Vegas, NV, USA, 2–6 December 2013; pp. 289–294. [CrossRef]
49. Mak, H.; Rong, Y.; Shen, Z. Infrastructure Planning for Electric Vehicles with Battery Swapping. *Manag. Sci.* **2013**, *59*, 1557–1575. [CrossRef]

50. Varga, B.O.; Sagoian, A.; Mariasiu, F. Prediction of Electric Vehicle Range: A Comprehensive Review of Current Issues and Challenges. *Energies* **2019**, *12*, 946. [CrossRef]
51. Rezvanizaniani, S.M.; Liu, Z.; Chen, Y.; Lee, J. Review and recent advances in battery health monitoring and prognostics technologies for electric vehicle (EV) safety and mobility. *J. Power Sources* **2014**, *256*, 110–124. [CrossRef]
52. Mruzek, M.; Gajdáč, I.; Kučera, L.; Barta, D. Analysis of Parameters Influencing Electric Vehicle Range. *Procedia Eng.* **2016**, *134*, 165–174. [CrossRef]
53. Berjoza, D.; Jurgena, I. Influence of batteries weight on electric automobile performance. *Eng. Rural Dev.* **2017**, *16*, 1388–1394. [CrossRef]
54. Weiss, M.; Cloos, C.K.; Helmers, E. Energy efficiency trade-offs in small to large electric vehicles. *Environ. Sci. Eur.* **2020**, *32*, 46. [CrossRef]
55. Iclodean, C.; Varga, B.; Burnete, N.; Cimerdean, D.; Jurchiş, B. Comparison of Different Battery Types for Electric Vehicles. *IOP Conf. Ser. Mater. Sci. Eng.* **2017**, *252*, 012058. [CrossRef]
56. Long, B.; Lim, S.T.; Ryu, J.H.; Chong, K.T. Energy-Regenerative Braking Control of Electric Vehicles Using Three-Phase Brushless Direct-Current Motors. *Energies* **2014**, *7*, 99–114. [CrossRef]
57. Xu, G.; Li, W.; Xu, K.; Song, Z. An Intelligent Regenerative Braking Strategy for Electric Vehicles. *Energies* **2011**, *4*, 1461–1477. [CrossRef]
58. Scheepmaker, G.M.; Goverde, R.M.P. Energy-efficient train control using nonlinear bounded regenerative braking. *Transp. Res. Part C Emerg. Technol.* **2020**, *121*, 102852. [CrossRef]
59. Rakov, V. Determination of optimal characteristics of braking energy recovery system in vehicles operating in urban conditions. *Transp. Res. Procedia* **2020**, *50*, 566–573. [CrossRef]
60. Oleksowicz, S.; Burnham, K.; Gajek, A. On the legal, safety and control aspects of regenerative braking in hybrid/electric vehicles. *Czas. Tech. Mechatronika Tech. Trans. Mech.* **2012**, *8*, 139–155.
61. Brady, J.; O'Mahony, M. Development of a driving cycle to evaluate the energy economy of electric vehicles in urban areas. *Appl. Energy* **2016**, *177*, 165–178. [CrossRef]
62. Hao, X.; Wang, H.; Lin, Z.; Ouyang, M. Seasonal effects on electric vehicle energy consumption and driving range: A case study on personal, taxi, and ridesharing vehicles. *J. Clean. Prod.* **2020**, *249*, 119403. [CrossRef]
63. Ma, R.; Heb, X.; Zheng, Y.; Zhoud, B.; Lu, S.; Wu, Y. Real-world driving cycles and energy consumption informed by large-sized vehicle trajectory data. *J. Clean. Prod.* **2019**, *223*, 564–574. [CrossRef]
64. Badin, F.; Le Berr, F.; Briki, H.; Dabadie, J.-C.; Petit, M.; Magand, S.; Condemine, E. Evaluation of EVs energy consumption influencing factors, driving conditions, auxiliaries use, driver's aggressiveness. *World Electr. Veh. J.* **2013**, *6*, 112–123. [CrossRef]
65. Çeven, S.; Albayrak, A.; Bayır, R. Real-time range estimation in electric vehicles using fuzzy logic classifier. *Comput. Electr. Eng.* **2020**, *83*, 106577. [CrossRef]
66. Zhao, X.; Ma, J.; Wang, S.; Ye, Y.; Wu, Y.; Yu, M. Developing an electric vehicle urban driving cycle to study differences in energy consumption. *Environ. Sci. Pollut. Res.* **2019**, *26*, 13839–13853. [CrossRef]
67. Jung, H.; Silva, R.; Han, M. Scaling Trends of Electric Vehicle Performance: Driving Range, Fuel Economy. *Peak Power Output Temp. Effect. World Electr. Veh. J.* **2018**, *9*, 46. [CrossRef]
68. De Gennaro, M.; Paffumi, E.; Martini, G.; Manfredi, U.; Scholz, H.; Lacher, H.; Kuehnelt, H.; Simic, D. Experimental investigation of the energy efficiency of an electric vehicle in different driving conditions. *SAE Tech. Pap.* **2014**, 1–11. [CrossRef]
69. Iora, P.; Tribioli, L. Effect of Ambient Temperature on Electric Vehicles' Energy Consumption and Range: Model Definition and Sensitivity Analysis Based on Nissan Leaf Data. *World Electr. Veh. J.* **2019**, *10*, 2. [CrossRef]
70. Meyer, N.; Whittal, W.; Christenson, M.; Loiselle-Lapointe, A. The impact of driving cycle and climate on electrical consumption & range of fully electric passenger vehicles. In Proceedings of the EVS26—International Battery, Hybrid and Fuel Cell Electric Vehicle Symposium, Los Angeles, CA, USA, 6–9 May 2012; pp. 1–11.
71. Donkers, A.; Yang, D.; Viktorović, M. Influence of driving style, infrastructure, weather and traffic on electric vehicle performance. *Transp. Res. Part D Transport. Environ.* **2020**, *88*, 102569. [CrossRef]
72. Jafari, M.; Gauchia, A.; Zhang, K.; Gauchia, L. Simulation and Analysis of the Effect of Real-World Driving Styles in an EV Battery Performance and Aging. *IEEE Trans. Transp. Electrif.* **2015**, *1*, 391–401. [CrossRef]
73. Rolim, C.C.; Gonçalves, G.N.; Farias, T.L.; Rodrigues, Ó. Impacts of Electric Vehicle Adoption on Driver Behavior and Environmental Performance. *Procedia Soc. Behav. Sci.* **2012**, *54*, 706–715. [CrossRef]
74. Jurecki, R.S.; Stanczyk, T.L. A Methodology for Evaluating Driving Styles in Various Road Conditions. *Energies* **2021**, *14*, 3570. [CrossRef]
75. Jurecki, R.S.; Stanczyk, T. Analysis of vehicle moving parameters in various road conditions. *Commun. Sci. Lett. Univ. Zilina* **2021**, *23*, F58–F70. [CrossRef]
76. Hu, K.; Wu, J.; Liu, M. Exploring the Energy Efficiency of Electric Vehicles with Driving Behavioral Data from a Field Test and Questionnaire. *J. Adv. Transp.* **2018**, *2018*, 1074817. [CrossRef]
77. Fetene, G.M.; Kaplan, S.; Mabit, S.L.; Jensen, A.F.; Prato, C.G. Harnessing big data for estimating the energy consumption and driving range of electric vehicles. *Transp. Res. Part D Transp. Environ.* **2017**, *54*, 1–11. [CrossRef]
78. Miri, I.; Fotouhi, A.; Ewin, N. Electric vehicle energy consumption modelling and estimation—A case study. *Smart Energy Technol.* **2021**, *45*, 501–520. [CrossRef]

79. Chew, K.W.; Yong, Y.R. Effectiveness Comparison of Range Estimator for Battery Electric Vehicles. *Adv. Automob. Eng.* **2016**, *5*, 1000128.
80. Kim, E.; Lee, J.; Shin, K.G. Real-time prediction of battery power requirements for electric vehicles. In Proceeding of the ACM/IEEE International Conference on Cyber-Physical Systems (ICCPS), Philadelphia, PA, USA, 8–11 April 2013; pp. 11–20.
81. Sarrafan, K.; Sutanto, D.; Muttaqi, K.M.; Town, G. Accurate range estimation for an electric vehicle including changing environmental conditions and traction system efficiency. *IET Electr. Syst. Transp.* **2017**, *7*, 117–124. [CrossRef]
82. Hong, J.; Park, S.; Chang, N. Accurate remaining range estimation for Electric vehicles. In Proceedings of the 21st Asia and South Pacific Design Automation Conference (ASP-DAC), Macao, China, 25–28 January 2016; pp. 781–786. [CrossRef]
83. All Answers Ltd. November 2018. Electric Vehicle Energy Consumption Modelling for Range Estimation. Available online: https://ukdiss.com/examples/electric-vehicle-energy-consumption-modelling-for-range-estimation.php?vref=1 (accessed on 22 March 2021).
84. Chew, K.W.; Gan, Y.H.; Leong, C.K. Contour Positioning System—New Traveling Distance Estimation Method for Electric Vehicle. *Appl. Mech. Mater.* **2013**, *284–287*, 451–455. [CrossRef]
85. Kuew, C.; Yong, W.; Rong, Y.; Morris, S. Simulation of a distance estimator for battery electric vehicle. *Alex. Eng. J.* **2015**, *54*, 359–371. [CrossRef]
86. Zhang, Y.; Wang, W.; Kobayashi, Y.; Shirai, K. Remaining driving range estimation of electric vehicle. In Proceedings of the 2012 IEEE International Electric Vehicle Conference, Greenville, SC, USA, 4–8 March 2012; pp. 1–7. [CrossRef]
87. Ferreira, J.; Monteiro, V.; Afonso, J. Dynamic range prediction for an electric vehicle. In Proceedings of the EVS27—International Electric Vehicle Symposium & Exhibition, Barcelona, Spain, 17–20 November 2013; pp. 1–11. [CrossRef]
88. Lee, C.; Wu, C. A Novel Big Data Modeling Method for Improving Driving Range Estimation of EVs. *IEEE Access* **2015**, *3*, 1980–1993. [CrossRef]
89. Bi, J.; Wang, Y.; Zhang, J. A data-based model for driving distance estimation of battery electric logistics vehicles. *J. Wirel. Commun. Netw.* **2018**, *2018*, 251. [CrossRef]
90. Ondruska, P.; Posner, I. Probabilistic attainability maps: Efficiently predicting driver-specific electric vehicle range. In Proceedings of the IEEE Intelligent Vehicles Symposium, Dearborn, MI, USA, 8–11 June 2014; pp. 1169–1174. [CrossRef]
91. Loyson, T.; Konworthy, J.; Austin, P.; Newman, P. The development of driving cycle for fuel consumption and emissions evaluation. *Transp. Res.* **1986**, *20*, 447–462.
92. Chindamo, D.; Gadola, M. What is the Most Representative Standard Driving Cycle to Estimate Diesel Emissions of a Light Commercial Vehicle? *IFAC Pap. OnLine* **2018**, *51*, 73–78. [CrossRef]
93. Gołębiewski, W.; Lisowski, L. Theoretical analysis of electric vehicle energy consumption according to different driving cycles. *IOP Conf. Ser. Mater. Sci. Eng.* **2018**, *421*, 022010. [CrossRef]
94. Andrade, G.M.S.D.; Araújo, F.W.C.D.; Santos, M.P.M.D.N.; Magnani, F.S. Standardized Comparison of 40 Local Driving Cycles: Energy and Kinematics. *Energies* **2020**, *13*, 5434. [CrossRef]
95. Regulation (EC) No 443/2009 of the European Parliament and of the Council of 23 April 2009 setting emission performance standards for new passenger cars as part of the Community's integrated approach to reduce CO 2 emissions from light-duty vehicles (Text with EEA relevance). *Off. J. Eur. Union* **2009**, *L 140*, 1–15.
96. Berzi, L.; Delogu, M.; Pierini, M. Development of driving cycles for electric vehicles in the context of the city of Florence. *Transp. Res. Part D Transp. Environ.* **2016**, *47*, 299–322. [CrossRef]
97. Zhao, X.; Yu, Q.; Ma, J.; Wu, Y.; Yu, H.; Ye, Y. Development of a Representative EV Urban Driving Cycle Based on a k-Means and SVM Hybrid Clustering Algorithm. *J. Adv. Transp.* **2018**, *2018*, 1890753. [CrossRef]

energies

MDPI

Article

Study on Image Correction and Optimization of Mounting Positions of Dual Cameras for Vehicle Test

Si-Ho Lee [1], Bong-Ju Kim [1] and Seon-Bong Lee [2,*]

[1] Department of Mechanical Engineering, Keimyung University, Daegu 42601, Korea; toto39690@gmail.com (S.-H.L.); 1952kbj@naver.com (B.-J.K.)
[2] Division of Mechanical and Automotive Engineering, Keimyung University, Daegu 42601, Korea
* Correspondence: seonbong@kmu.ac.kr; Tel.: +82-53-580-5476

Abstract: Among surrounding information-gathering devices, cameras are the most accessible and widely used in autonomous vehicles. In particular, stereo cameras are employed in academic as well as practical applications. In this study, commonly used webcams are mounted on a vehicle in a dual-camera configuration and used to perform lane detection based on image correction. The height, baseline, and angle were considered as variables for optimizing the mounting positions of the cameras. Then, a theoretical equation was proposed for the measurement of the distance to the object, and it was validated via vehicle tests. The optimal height, baseline, and angle of the mounting position of the dual camera configuration were identified to be 40 cm, 30 cm, and 12°, respectively. These values were utilized to compare the performances of vehicles in stationary and driving states on straight and curved roads, as obtained by vehicle tests and theoretical calculations. The comparison revealed the maximum error rates in the stationary and driving states on a straight road to be 3.54% and 5.35%, respectively, and those on a curved road to be 9.13% and 9.40%, respectively. It was determined that the proposed method is reliable because the error rates were less than 10%.

Keywords: autonomous vehicle; dual camera; real vehicle test; test scenario

Citation: Lee, S.-H.; Kim, B.-J.; Lee, S.-B. Study on Image Correction and Optimization of Mounting Positions of Dual Cameras for Vehicle Test. *Energies* **2021**, *14*, 4857. https://doi.org/10.3390/en14164857

Academic Editors: Guzek Marek, Rafał Jurecki and Wojciech Wach

Received: 5 July 2021
Accepted: 5 August 2021
Published: 9 August 2021

Publisher's Note: MDPI stays neutral with regard to jurisdictional claims in published maps and institutional affiliations.

1. Introduction

Driving automation, as defined by the Society of Automotive Engineers (SAE), is the internationally accepted standard. SAE provides taxonomy with detailed definitions for six levels of driving automation. These range from no driving automation (Level 0) to full driving automation (Level 5) [1]. Mass-produced vehicles have recently begun to be generally equipped with Level 2 autonomous driving technology. This technology provides drivers with partial driving automation and is called advanced driver assistance systems (ADAS). Among the examples of ADAS, adaptive cruise control (ACC) and lane-keeping assist system (LKAS) are Level 1 technologies, and highway driving assist (HDA) is a Level 2 technology.

The primary goal of autonomous driving technology is to respond proactively to unanticipated scenarios, such as traffic accidents and construction sites. This requires rapid and effective identification of the environment surrounding the vehicle. To achieve this, various sensors, such as light detection and ranging (LiDAR) and radar sensors, and cameras are used for detection [2]. Among these, cameras capture images containing a large quantity of information. This information enables object detection, traffic information collection, lane detection, among other tasks. Furthermore, cameras are more readily accessible compared to other sensors. Therefore, several studies have been conducted on the camera-based collection of environmental information and its processing.

With regard to the correction of camera images, Lee et al. [3] proposed a method to correct the radial distortion caused by camera lenses. In addition, Detchev et al. [4] proposed a method for simultaneously estimating the internal and relative direction parameters for calibrating measurement systems comprising multiple cameras.

With regard to camera-based lane detection, Kim et al. [5] proposed an algorithm to improve nocturnal lane detectability based on image brightness correction and the lane angle. In addition, Kim et al. [6] performed real-time lane detection using the lane path obtained based on the lane gradient and width information in conjunction with the previous frame. Choi et al. [7] proposed and validated a novel lane detection algorithm (namely random sample consensus (RANSAC) algorithm) by applying conventional lane detection algorithms. Kalms et al. [8] used the Viola-Jones object detection method to design and implement an algorithm for lane detection and autonomous driving. Wang et al. [9] achieved lane detection through pre-processing images and utilized the features extracted from the image pixel coordinates to detect lane departure using a stacked sparse autoencoder (SSAE). Andrade et al. [10] recommended a novel three-stage strategy for lane detection and tracking, comprising image correction and region of interest (ROI) set-up, edge detection.

With regard to monocular camera-based distance measurement, Bae et al. [11] proposed a method to measure the distance to the vehicle in front using the relationship between the lane and the geometry of the camera and the vehicle. A method suggested by Park et al. [12] involved measuring distances by training the distance classifier using the distance information obtained from LiDAR sensors in conjunction with the width and height of the bounding box corresponding to the detected object. Huang et al. [13] proposed a method to estimate inter-vehicle distances based on monocular camera images captured by cameras installed within a vehicle by combining the vehicular attitude angle information with the segmentation information. Moreover, Zhe et al. [14] constructed an area-distance geometric model based on the camera projection principle. They leveraged the advantages of 3D detection to combine the 3D detection of vehicles with the distance measurement model, proposing a robust inter-vehicle distance measurement method based on a monocular camera installed within a vehicle. Bongharriou et al. [15] combined vanishing point extraction, lane detection, and vehicle detection based on actual images and proposed a method to estimate distances between cameras and vehicles in front by correcting camera images.

Object detection and distance measurements using stereo cameras have also been researched extensively. Kim [16] proposed an algorithm to estimate vehicular driving lanes by generating 3D trajectories based on the coordinates of detected traffic signs, and Seo [17] proposed an improved distance measurement method using a disparity map. A method proposed by Kim et al. [18] comprises measuring distances by correcting image brightness and estimating central points of objects using two webcams. Furthermore, Song et al. [19] proposed a forward collision distance measurement method by combining a Hough space lane detection model with stereo matching. Additionally, Sie et al. [20] proposed an algorithm for real-time lane and vehicle detection and measurement of distances to vehicles in the driving lane by combining a portable embedded system with a dual-camera vision system. A method involving efficient pose estimation of on-board cameras using 3D point sets obtained from stereo cameras, and the ground surface was proposed by Sappa et al. [21]. Yang et al. [22] proposed a stereo vision-based system that detects vehicle license plates and calculates their 3D position in each frame for vehicular speed measurement. Zaarane et al. [23] proposed an inter-vehicle distance measurement system based on image processing utilizing a stereo camera by considering the position of vehicles in both the cameras and certain geometric angles. Cafiso et al. [24] proposed a system based on in-vehicle stereo vision and the global positioning system (GPS) to detect and assess collisions between vehicles and pedestrians. Wang et al. [25] proposed real-time object detection and depth estimation based on Deep Convolutional Neural Networks (DCNNs). Lin et al. [26] proposed a vision-based driver assistance system.

Several extensive studies have been conducted on the correction of camera images, road lane detection, and distance measurement. However, few studies have been conducted to optimize camera mounting positions and measure distances to objects in front of a vehicle. This gap is addressed in the present study by using testing and evaluation methods based on dual-camera-based image correction and lane detection. In this study, only theoretical concepts used in experiments are described in the computer vision phase for lane detection and the image distortion correction phase. In addition, focal length correction is performed after image distortion correction to reduce the effect of the change in detection distance caused by image distortion correction. Moreover, it investigates the three main variables related to the dual-camera installation. The experimental results according to installation height, camera baseline, and angle of inclination are presented. The parameters according to the camera's rotation axis are set as roll, pitch, and pan. The pitch is the installation angle of the camera, and the parallel stereo camera method applied in this study does not consider roll and pan. In addition, the parameters such as angle-of-view and focal length of the camera are excluded from the analysis because the two cameras constituting the dual-camera configuration had the same specifications. The actual test was conducted on the three presented variables to determine the optimal position of the dual camera. The three variables were tested for three values, respectively. Based on the obtained optimal value, the actual test was conducted on the theoretical equation. This study proceeds as follows.

1. Introduction of the lane detection algorithm.
2. Input image distortion correction, stereo rectification, and focal length correction.
3. Experimental evaluation of algorithm precision according to three variables for optimal dual-camera positioning.
4. Proposal of equations for calculating the distance between the vehicle and objects in front of it in straight and curved roads for test evaluation.
5. Applicability evaluation through real vehicle tests using the optimal camera position determined in step 3 and the distance calculation equation proposed in step 4.

2. Theoretical Background for Dual Camera-Based Image Correction and the Proposed Method of Distance Measurement

2.1. Road Lane Detection Method

2.1.1. Theoretical Background for Lane Detection

The ROI in a camera-captured image is the region containing the information relevant to the task at hand. As the range of the scenery captured by a camera affixed within a vehicle remains constant, the ROI in particular images must be obtained by removing the corresponding irrelevant regions.

Cameras usually capture images in the red, green, and blue (RGB) format, which comprises three channels. Grayscale conversion of such images produces monochromatic images, which comprise a single channel. As images converted via this method retain only brightness information, the amount of data to be processed is reduced by two-thirds, increasing the computational speed.

The canny edge detector is an edge detection algorithm that utilizes successive steps such as noise reduction, determination of the intensity gradient of the image, non-maximum suppression, and hysteresis thresholding [27]. Owing to its multi-step mechanism, it performs better than the methods that use differential operators (e.g., the Sobel mask).

The Hough transform is a method for transforming components in the Cartesian coordinate system to those in the parameter space [28]. Straight lines and points on the Cartesian coordinate system are represented by points and straight lines, respectively, in the parameter space. Thus, points of intersection between straight lines in the parameter space can be used to search for straight lines passing through a given set of points in the Cartesian coordinate system.

The hue, saturation, value (HSV) format is a color model that represents an image in terms of hue, saturation, and value. It is particularly effective for the facile expression of desired colors because its operational template agrees with the human mode of color recognition.

Perspective transform facilitates the modeling of homography using a 3 × 3 transformation matrix. The perspective of any image can be removed via a geometric processing method by relocating the pixels of the image.

The sliding window method uses a sub-level array of a certain size called a window and reduces the computational load for calculating the elements in each window in the entire array by reusing (rather than discarding) redundant elements.

The curve fitting method involves fitting a function to a given curve representing the input data. A polynomial function is most commonly used for this purpose. Furthermore, the input data can be approximated using a quadratic function by employing the least-squares approximation method.

2.1.2. Lane Detection Algorithm

Figure 1 depicts the algorithm used in this study for lane detection. It utilizes OpenCV, an open-source computer vision library, for image processing. After performing the lane detection algorithm, the detected lane information is used to calculate the distance to the object in front of the vehicle.

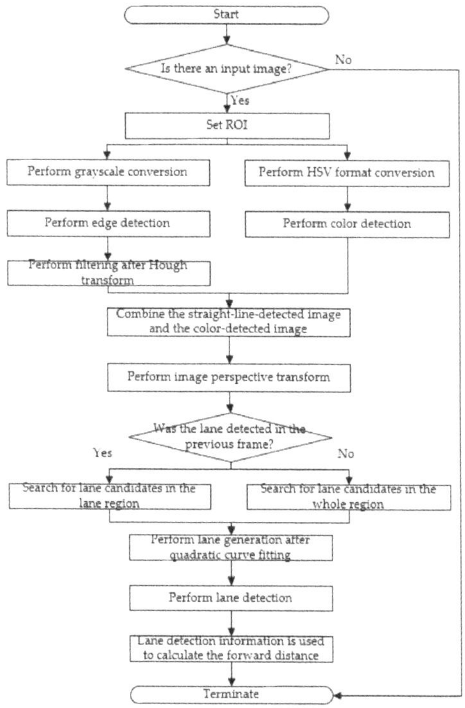

Figure 1. Flowchart of lane detection algorithm.

Figure 2a presents an input image captured by the fixed left camera. When the ROI corresponds to 20–50% of the image height and the resolution of the image is 1920 × 1080, it configures within a rectangular area having the following coordinate points as vertices: (0, 216), (1920, 216), (1920, 540), and (0, 540).

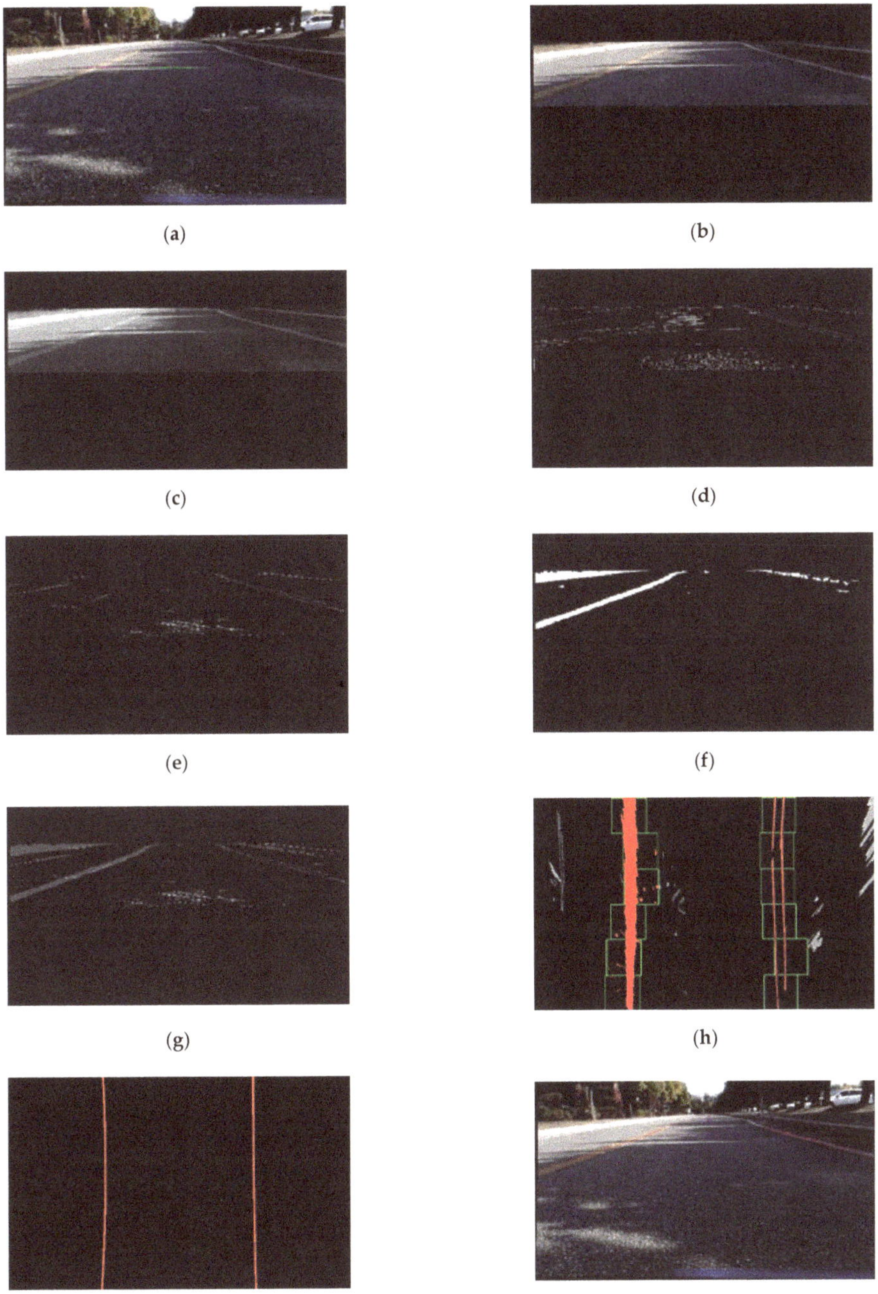

(a) (b) (c) (d) (e) (f) (g) (h) (i) (j)

Figure 2. Illustration of the steps involved in lane detection: (**a**) the input image, (**b**) selection of the appropriate ROI in the input image, (**c**) the image obtained via grayscale conversion, (**d**) the edge-detected image, (**e**) the image obtained via filtering following Hough transform, (**f**) the color-detected image after conversion into the HSV format, (**g**) the combination of the straight-line-detected image and the color-detected image, (**h**) the identified lane following the perspective transform, (**i**) the generated lane following quadratic curve fitting, (**j**) the final lane-detected image.

Figure 2c depicts the image obtained from the previous one via grayscale conversion. The single-channel image was generated by averaging the pixel values corresponding to the R, G, and B channels.

Figure 2d depicts the result of edge detection. A Gaussian filter was used to remove the noise, and the Canny edge detector was used to generate this edge-detected image. Then straight lines corresponding to the lane marks were obtained, as depicted in Figure 2e. The Hough transform was used to detect the edge components in the edge-detected image. Subsequently, straight lines corresponding to gradients with magnitudes of at most 5° were removed, resulting in the elimination of horizontal and vertical lines that were unlikely to correspond to the lane.

The yellow pixels were extracted from Figure 2b, and the result is depicted in Figure 2f. Following the conversion of the image from the RGB format to the HSV format, a yellow color range was selected. When the ranges of the hue channel, saturation channel, and value channel were normalized to the interval 0–1, the yellow pixels corresponded to values between 0–0.1, 0.9–1, and 0.12–1 for hue and saturation channels. One-third of the mean brightness of the image was used for the value channel. When the value of the pixel was within the range that was set, the value was set to 255; otherwise, it was set to zero.

Figure 2g depicts a combination of the image obtained by extracting straight lines to identify pixels corresponding to the lane candidates and that obtained by extracting the color. A combination was obtained by assigning weights of 0.8 and 0.2 to the images in Figure 2c,d, respectively.

Further, the lane candidates were obtained using the sliding window method after removing the perspective from the image presented in Figure 2g, and the output is depicted in Figure 2h. The image was captured in advance such that the optical axis of the camera was parallel to the road when the vehicle was located in the center of the straight road. The image can be warped so that the left and right lanes on a straight road are parallel. The coordinates (765, 246), (1240, 246), (1910, 426), and (5, 516) of the four points on the set of lanes visible within the ROI were relocated to the points (300, 648), (300, 0), (780, 0), and (780, 648) in the warped image to align these along straight lines. A square window comprising 54 pixels was selected, with a width and height that were one-twentieth and one-sixth, respectively, of those of the image. The window with the largest pixel sum was then identified via the sliding window method.

Subsequently, a lane curve was generated by fitting a quadratic curve to the pixels of the lane candidate, as depicted in Figure 2i. The quadratic curve fitting is performed using the least-squares method, and the positions of the pixels of the lane candidate are indicated by the six windows on the left and right lane marks in Figure 2h.

Finally, lane detection based on the input image was completed. The final result, obtained by applying the lane curve to the input image via perspective transform, is depicted in Figure 2j.

2.2. *Method for Calibrating Distance Measurement*

2.2.1. Image Distortion Correction

Images captured by cameras exhibit radial distortions due to the refractive indices of convex lenses and tangential distortions due to the horizontal leveling problem inherent to the manufacturing process of lenses and image sensors. Circular distortions induced by radial distortion at the edge of the image and elliptical distortions induced by the tangential distortion require correction. The values of pixels in the distorted image can be used as the values of the corresponding pixels in the corrected image by distorting the coordinates of each pixel in the image [29].

In this study, OpenCV's built-in functions for checkerboard pattern identification, corner point identification, and camera calibration were adopted for image processing. To correct the input image, a 6 × 4 checkerboard image was captured using the camera, its corner points were identified, and the camera matrix and distortion coefficients were calculated based on the points obtained. Figure 3a depicts the identification of the corner points in the original image and Figure 3b depicts the screen after the removal of distortion.

(a)

(b)

Figure 3. Checkerboard images utilized for distortion correction: (**a**) identification of the corner points in the checkerboard, (**b**) the output image.

2.2.2. Image Rectification

Parallel stereo camera configuration is the method that involves utilizing two cameras whose optical axes are parallel. It is particularly suitable for image processing because of the absence of vertical disparity [30]. In contrast, actual photographs require image rectification to correct the vertical disparity originating from the installation or internal parameters of cameras. This method corrects for an arbitrary object in the left and right images obtained with dual cameras to obtain equal coordinates for the height of images.

In this study, OpenCV's built-in stereo calibration and stereo rectification functions were adopted for image processing. In addition, the checkerboard image utilized during the removal of the image distortion was used to identify the checkerboard pattern and its corner points and calibrate the dual cameras (Figure 4a). As depicted in Figure 4b, the internal parameters, rotation matrix of the dual-camera configuration, and projection matrix on the rectified coordinate system can be obtained based on a pair of checkerboard images captured using the dual cameras.

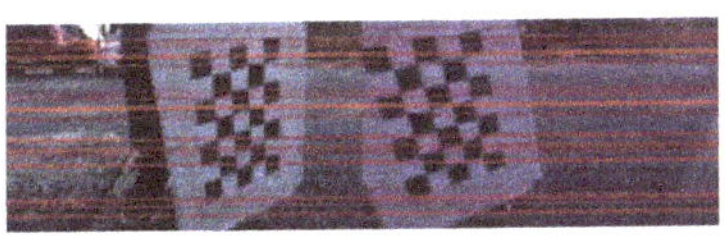

(a)

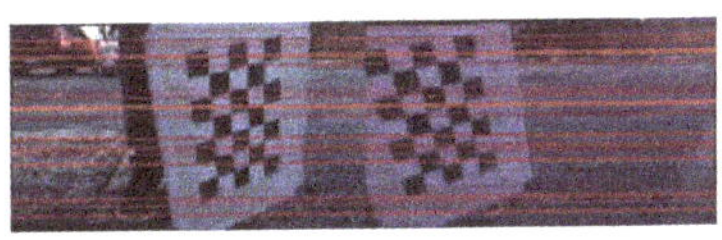

(b)

Figure 4. Image rectification: (**a**) the input image, (**b**) the output image.

2.2.3. Focal Length Correction

Dual cameras were installed collinearly such that the optical axes of the two cameras are parallel. Furthermore, the lenses were positioned at identical heights above the ground. The 3D coordinates of any object were calculated relative to the camera positions, based on the geometry and triangulation of the cameras depicted in Figure 5. It can be described as follows:

$$\begin{aligned} Z &= \frac{fb}{d} \\ X &= \frac{Z(xl+xr)}{2f} \\ Y &= \frac{Z(yl+yr)}{2f} \end{aligned} \tag{1}$$

where:

XYZ—the coordinates of the object; the local coordinate system with their origins at the center of dual cameras.

f—focal length.

b—baseline.

d—disparity.
xl, yl—the coordinates of the object in the left camera image plane.
xr, yr—the coordinates of the object in the right camera image plane.

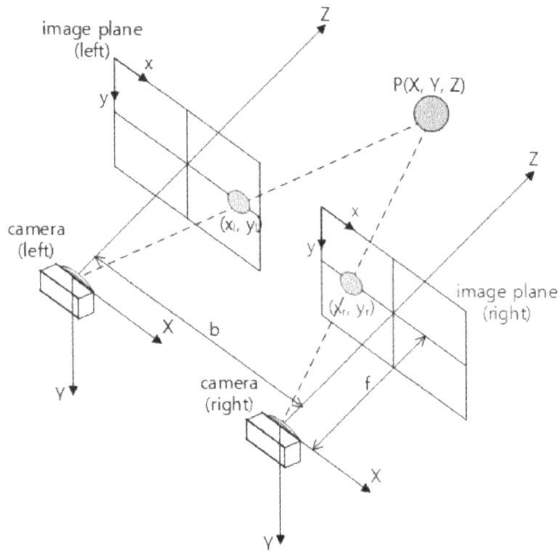

Figure 5. Parallel stereo camera model.

The focal length is an essential parameter in the calculation of the Z-coordinate. However, a problem with the use of inexpensive webcams is that some manufacturers do not provide details such as focal length. Further errors originating from image correction necessitate an accurate estimation of the focal length.

This can be achieved by employing curve-fitting based on actual data. Based on the relationship between distance and disparity, where Z_a is calculated from the equation:

$$Z_a = \alpha \frac{b}{d} + \beta \tag{2}$$

where:

XYZ—the coordinates of the object in the universal Cartesian coordinate system.
Z_a—distance to the object.
α, β—the coefficients obtained via the focal length correction.

During testing, images of objects installed at intervals of 0.5 m over the range of 1–5 m were captured. In addition, the differences between the X-coordinates of each object captured by the two cameras were recorded. Then, α and β were evaluated by fitting the curve described by the differences calculated via the least square method.

3. Optimization of the Mounting Positions of Dual Cameras

3.1. Configurations of Test Variables

3.1.1. Mounting Heights of Cameras

In the proposed equation, the distance is measured relative to the ground. It is evident that the mounting height of the cameras is inversely proportional to the fraction of the ground captured in the image. Therefore, the mounting height of the cameras wields a significant influence in the determination of the region captured in the image.

Heights of 30 cm, 40 cm, and 50 cm were considered. In the case of regular passenger cars, 30 cm was selected as the minimum value because their bumpers are at least 30 cm above the ground level. The maximum value was set to 50 cm because larger heights made it difficult to capture the ground within 1 m. Figure 6 depicts the input images corresponding to heights of 30 cm, 40 cm, and 50 cm where a baseline of cameras is 30 cm, and an angle of inclination of mounted cameras is 12°.

(a) (b) (c)

Figure 6. Input images corresponding to different heights: (**a**) corresponding to a height of 30 cm, (**b**) corresponding to a height of 40 cm, and (**c**) corresponding to a height of 50 cm.

3.1.2. Baseline of Cameras

Equation (1) is based on the geometry and triangulation of the cameras. Therefore, the baseline between the cameras significantly affects the measurement of distance.

Baselines of 10 cm, 20 cm, and 30 cm were considered in this study. First, 10 cm was selected as the minimum value because it was the smallest feasible baseline. Then, the baseline was increased three times at intervals of 10 cm to examine its influence. Figure 7 depicts the input images corresponding to baselines of 10 cm, 20 cm, and 30 cm, where the mounting heights of the cameras are 40 cm, and the angle of inclination of the mounted cameras is 12°.

(a) (b) (c)

Figure 7. Input images corresponding to various baselines: (**a**) corresponding to a baseline of 10 cm, (**b**) corresponding to a baseline of 20 cm, (**c**) and corresponding to a baseline of 30 cm.

3.1.3. Angle of Inclination of Mounted Cameras

The installation of cameras parallel to the ground reduces the vertical range and hinders close-range supervision of the ground. Therefore, it is essential to utilize an optimal angle of inclination during the installation of cameras.

Angles of 3°, 7°, and 12° were considered as feasible angles of inclination. First, 3° was selected as the minimum value owing to the difficulty of capturing the ground within a radius of 1 m at smaller angles of inclination from a height of 50 cm. The proportion of the road captured in the image increased as the angle was increased. However, vehicular turbulence or the presence of ramps was observed to affect the inclusion of the upper part of the road in the images. Further, 12° was selected as the maximum angle of inclination as it yielded images with the road accounted for 20–80% of the vertical range. Figure 8 depicts input images corresponding to angles of inclination of 3°, 7°, and 12°, where the mounting heights of the cameras are 40 cm and the baseline of the cameras is 30 cm.

(**a**) (**b**) (**c**)

Figure 8. Input images corresponding to different angles: (**a**) corresponding to an angle of 3°, (**b**) corresponding to an angle of 7°, and (**c**) corresponding to an angle of 12°.

3.2. Test Results for Optimization of Mounting Positions

During testing, dual cameras were mounted on the vehicle, and objects were installed at distances ranging between 1 m–5 m at intervals of 0.5 m on an actual road. Then, the differences between the X-coordinates captured by the left and right cameras were computed and substituted into Equation (2) to verify the precision. The calculated results are presented in Figures 9–11.

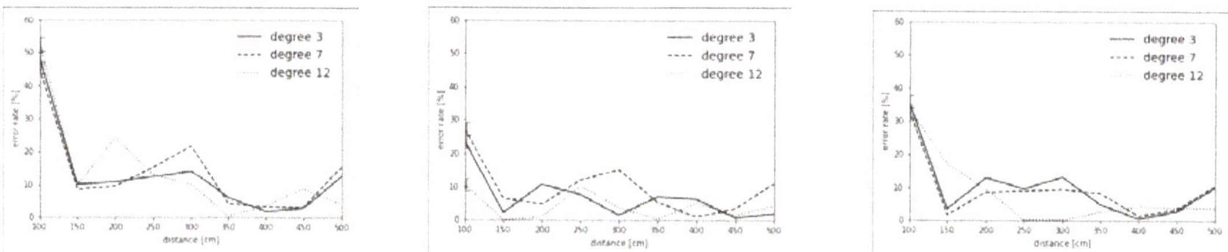

(**a**) (**b**) (**c**)

Figure 9. Test results with respect to varying baselines and angles at a height of 30 cm: (**a**) error rates corresponding to various angles and a baseline of 10 cm, (**b**) error rates corresponding to various angles and a baseline of 20 cm, (**c**) error rates corresponding to various angles at a baseline of 30 cm.

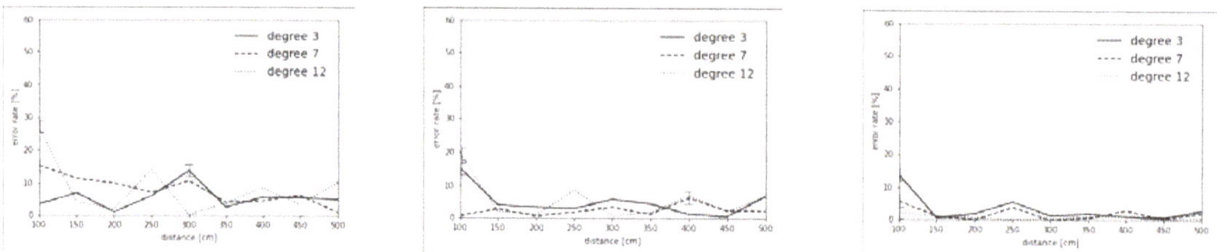

(**a**) (**b**) (**c**)

Figure 10. Test results with respect to varying baselines and angles at a height of 40 cm: (**a**) error rates corresponding to various angles and a baseline of 10 cm, (**b**) error rates corresponding to various angles and a baseline of 20 cm, (**c**) error rates corresponding to various angles and a baseline of 30 cm.

Figure 10 depicts the test results corresponding to a height of 40 cm. Figure 10a–c illustrates the variation in the degree of precision with respect to varying angles, corresponding to baselines of 10 cm, 20 cm, and 30 cm, respectively.

Figure 11 depicts the test results corresponding to a height of 50 cm. Figure 11a–c illustrates the variation in the degree of precision with respect to varying angles, corresponding to baselines of 10 cm, 20 cm, and 30 cm, respectively.

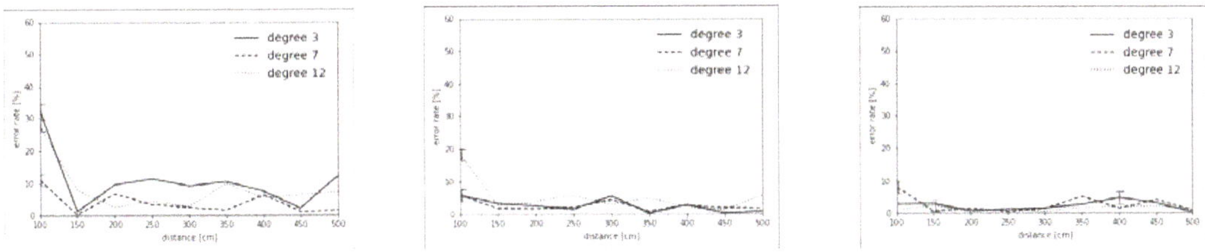

(a) (b) (c)

Figure 11. Test results with respect to various baselines and angles at a height of 50 cm: (**a**) error rates corresponding to various angles and a baseline of 10 cm, (**b**) error rates corresponding to various angles and a baseline of 20 cm, (**c**) error rates corresponding to various angles and a baseline of 30 cm.

Table 1 summarizes the result of Figures 9–11. It is evident from Figures 9–11 that the error rate exhibited a decreasing tendency as the angle was increased. Meanwhile, it tended to decrease as the baseline was increased. Finally, the error rate decreased when the height was increased from 30 cm to 40 cm, and it increased when the height was increased to 50 cm.

Table 1. Test results for optimization of mounting positions.

Height (cm)	Baseline (cm)	Angle (°)	Average Error Rate (%)	Maximum Error Rate (%)
30	10	3	13.28	48.62
		7	14.07	45.36
		12	14.15	53.04
	20	3	6.89	23.66
		7	9.64	27.3
		12	4.14	10.53
	30	3	10.34	35.83
		7	9.5	33.35
		12	8.34	33.99
40	10	3	5.55	13.77
		7	7.84	15.38
		12	8.18	26.9
	20	3	4.93	15.19
		7	2.38	6.17
		12	5.49	19.52
	30	3	3.34	13.9
		7	1.88	5.69
		12	0.86	2.15
50	10	3	10.62	32.49
		7	3.77	10.94
		12	8.19	27.67
	20	3	2.47	5.81
		7	2.45	5.84
		12	5.23	18.39
	30	3	2.15	4.65
		7	2.45	8.0
		12	1.32	2.34

Based on the aforementioned data, the best result was obtained corresponding to a height of 40 cm, a baseline of 30 cm, and an angle of inclination of 12°. In the next section, these values are used to validate the theoretical equation of distance measurement.

4. Proposed Theoretical Equation for Forward Distance Measurement

4.1. Measurement of the Distance to an Object in Front of the Vehicle on a Straight Road

The Z-coordinate of an object in front of the vehicle was obtained by using the coefficient α in Equation (2) (as evaluated via focal length correction) as the focal length and substituting it into Equation (1). However, during testing, the cameras were installed at an angle of inclination θ, to capture the close-range ground. That is, the optical axes of the cameras and the ground were not parallel during testing.

The Z-coordinate of the object relative to the position of the camera can be calculated considering the angle θ in Equation (3):

$$\begin{bmatrix} X_g \\ Y_g \\ Z_g \end{bmatrix} = \begin{bmatrix} 1 & 0 & 0 \\ 0 & -\cos\theta & -\sin\theta \\ 0 & -\sin\theta & \cos\theta \end{bmatrix} \begin{bmatrix} X \\ Y \\ Z \end{bmatrix} + h \begin{bmatrix} 0 \\ 1 \\ 0 \end{bmatrix} \tag{3}$$

where:

$X_g Y_g Z_g$—the coordinates of the object considering the angle of inclination of mounted cameras; the local coordinate system with their origins at the center of dual cameras.
θ—the angle of inclination of the mounted cameras.
h—mounting heights of the cameras.

On a straight road similar to that depicted in Figure 12, the calculation of the distance between the cameras and the object in front of the vehicle requires only an estimation of the longitudinal vertical distance. Therefore, Z_g can be considered to be the distance between the cameras and the object in front of the vehicle.

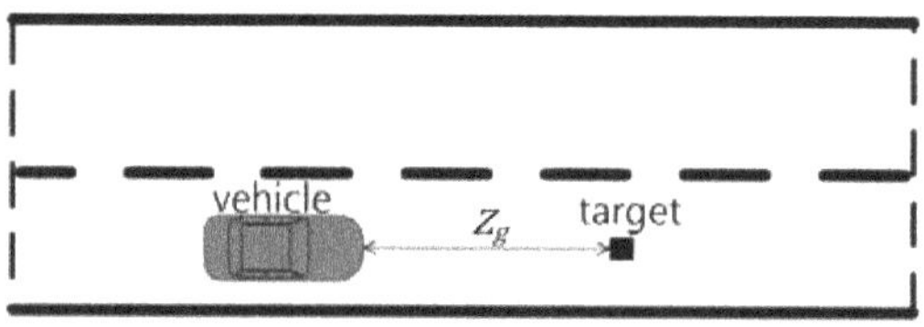

Figure 12. Distance to the object in front of the vehicle on a straight road.

4.2. Measurement of the Distance to an Object in Front of the Vehicle on a Curved Road

On a curved road similar to that depicted in Figure 13, the radius of curvature of the road should be incorporated into the measurement of the distance to the object in front of the vehicle. Therefore, after calculating the vertical distance using the object's X- and Z-coordinates, the distance to the object in front of the vehicle was calculated by considering the radius of curvature:

$$chord = \sqrt{X_g^2 + Z_g^2} \tag{4}$$

where:

chord—the vertical distance between the vehicle and object.

An angle φ is subtended at the center of the curvature and the vertical distance from the camera position to the object in front of the vehicle. An angle φ can be calculated as follows:

$$\varphi = \cos^{-1}\left(\frac{\sqrt{R^2 - \left(\frac{chord}{2}\right)^2}}{R}\right) \tag{5}$$

where:

φ—the angle subtended by the vehicle and the object at the center of curvature of the road
R—the radius of curvature of the road.

The length of the arc of the circle corresponding to the aforementioned *chord* was calculated using φ and *R*, by applying Equation (6):

$$\text{arc} = 2\pi R \cdot \frac{\varphi}{360} \quad (6)$$

where:

arc—the distance between the vehicle and the object along the curved road.

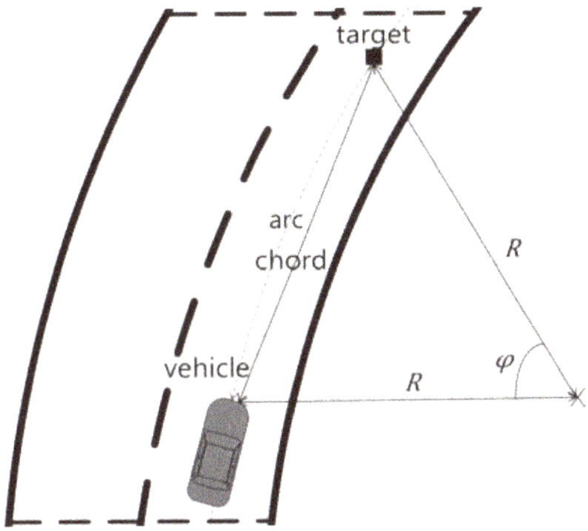

Figure 13. Distance to the object in front of the vehicle on a curved road.

4.3. Integrated Equation

In the preceding two subsections, equations have been proposed for distance measurement on straight and curved roads. The radius of curvature of the curved road is inversely proportional to the difference between the measured distances on the straight and curved roads. Therefore, if the radius of curvature is larger than a certain threshold, the curved road can be assumed to be approximately straight without a significant loss of accuracy. When the radius of curvature was 1293 m, an error rate of at most 0.1% was observed. Therefore, 1293 m was adopted as the aforementioned threshold in the proposed equation. This is written as follows:

$$Z_t = \begin{cases} Z_g & R > 1293\ \text{m} \\ arc & R \leq 1293\ \text{m} \end{cases} \quad (7)$$

where:

Z_t—the distance between the vehicle and the object in front of the vehicle.

5. Vehicle Test and Validation

5.1. Vehicle Used for Vehicle Test

The vehicle test was conducted to verify the accuracy of the forward distance measurement equation after mounting the dual camera setup at the optimized positions. H company's Veracruz (Figure 14) was used as the test vehicle, and its specifications are listed in Table 2.

Figure 14. Test vehicle.

Table 2. Specifications of the test vehicle.

Name	Specification
Veracruz (vehicle)	Overall length: 4840 mm Overall width: 1970 mm Overall height: 1795 mm Wheel base: 2805 mm Tread: 1670 mm

The dual cameras were mounted on the front bumper of the vehicle used for the test, as depicted in Figure 15. Each camera was a Logitech C920 HD Pro Webcam, and its specifications are listed in Table 3.

Figure 15. Test device.

Table 3. Specifications of cameras.

Name	Specification
C920 HD pro webcam (camera)	height: 43.3 mm width: 94 mm depth: 71 mm field of view: 78° field of view (horizontal): 70.42° field of view (vertical): 43.3° image resolution: 1920 × 1080 p focal length: 3.67 mm

5.2. Vehicle Test Location and Conditions

Because of safety considerations, the vehicle test was conducted within the Seongseo Campus of Keimyung University, located in Daegu Metropolitan City, Korea. Figure 16 depicts the straight and curved roads utilized for testing. The radius of curvature of the curved road was 69 m; it was calculated as the radius of a planar curve based on the design speed defined in article 19 of the rules for the structure and facility standards of roads, which are presented in Table 4. A curved road with a radius of curvature of at most 80 m was selected, considering the driving speed limit of 50 km/h in the city.

(a) (b)

Figure 16. Vehicle test roads: (**a**) straight road, (**b**) curved road.

Table 4. Minimum radius of curvature for curved roads according to the design standards.

Velocity (km/h)	Minimum Radius of Curve According to Maximum Slope (m)		
	6%	7%	8%
120	710	670	630
110	600	560	530
100	460	440	420
90	380	360	340
80	280	265	250
70	200	190	180
60	140	135	130
50	90	85	80

In the case of the straight road, stationary and driving states were classified using obstacles installed at intervals of 10 m between distances of 10 m and 40 m in front of the vehicle. In the case of the curved road, the two states were classified using obstacles installed at 5 m intervals between 6 m and 21 m in front of the vehicle along the center of the road and on the left and right lane markers. The entire test was conducted corresponding to a total of four cases.

The tests were repeated three times using the same equipment to acquire objective data. The environmental conditions during the experiments are summarized in Table 5. There was no variation in the weather.

Table 5. The environmental conditions.

Item	Condition
Road	flat, dry asphalt
Weather	sunny
Temperature (°C)	9–12

5.3. Test Results

The vehicle test was conducted corresponding to four cases of stationary and driving states on the straight and curved roads. Figure 17 depicts the post-correction images captured by the dual-camera setup used to measure the distance to the object in front of the vehicle. Table 6 summarizes the deviations of the theoretically calculated distances from the actual distances.

(a) (b)

Figure 17. Test result images: (**a**) on the straight road, (**b**) on the curved road.

Table 6. Test result analysis.

Scenario	Case	Real Distance (cm)	Calculated Distance (cm)	Error Rate (%)
Straight road, Stationary state	1	1000	1008	0.78
		2000	1975	1.24
		3000	2931	2.29
		4000	4032	0.81
	2	1000	1001	0.08
		2000	1960	1.98
		3000	3083	2.77
		4000	3925	1.87
	3	1000	1004	0.41
		2000	1973	1.36
		3000	3106	3.54
		4000	3956	1.09
Straight road, Driving state	1	1060	1053	0.65
		2060	2066	0.31
		3060	-	-
		4060	-	-
	2	1060	1004	5.32
		2060	2066	0.31
		3060	-	-
		4060	-	-
	3	1060	1080	1.86
		2060	2170	5.35
		3060	-	-
		4060	-	-
Curved road, Stationary state	1	620	614	0.96
		1120	1111	0.80
		1620	1602	1.11
		2120	2085	1.65
	2	620	632	3.63
		1120	1098	1.06
		1620	1630	1.26
		2120	2303	9.13
	3	620	562	7.89
		1120	1037	6.55
		1620	1618	0.50
		2120	2084	3.67

Table 6. *Cont.*

Scenario	Case	Real Distance (cm)	Calculated Distance (cm)	Error Rate (%)
Curved road, Driving state	1	620	562	9.40
		1120	1037	7.44
		1620	1746	7.79
		2120	-	-
	2	620	585	4.12
		1120	1063	4.28
		1620	1680	4.36
		2120	-	-
	3	620	562	9.39
		1120	1037	7.44
		1620	1746	7.79
		2120	-	-

In the case of the stationary state on the straight road, the objects placed at various distances in front of the vehicle were identified. The maximum error was observed to be 3.54%, corresponding to the 30 m point.

In the case of the driving state on the straight road, the objects at distances of 10 m and 20 m in front of the vehicle were identified, whereas those farther away were not. This can be attributed to factors such as vehicular turbulence, variations in illumination, and transmission of vibration to the cameras, caused by the driving state. The maximum error was observed to be 5.35% at the 20 m point.

In the case of the stationary state on the curved road, the objects at distances between 5 m and 20 m in front of the vehicle were identified. The maximum error was observed to be 9.13%, corresponding to the 20 m point.

In the case of the driving state on the curved road, the objects at distances between 5 m and 15 m in front of the vehicle were identified, whereas those at a distance of 20 m were not. Similar to the case of the straight road, this was attributed to factors such as vehicular turbulence, variations in illumination, and transmission of vibrations to cameras. The maximum error was observed to be 9.40%, corresponding to the 6 m point.

The test results demonstrate that the error in the measurement of the distance between the vehicle and objects in front of it increases when the object is detected inaccurately owing to factors such as vehicular turbulence, variations in illumination, and transmission of vibrations to the cameras. Furthermore, the error tends to be relatively large in the case of the driving state on a curved road compared with that on a straight road; this error is affected by the fixed radius of curvature used in the calculation process.

6. Conclusions

In this study, correction of camera images and lane detection on roads were performed for vehicle tests and evaluation. Furthermore, the mounting positions of cameras were optimized in terms of three variables: height, baseline, and angle of inclination. Equations to measure the distance to an object in front of the vehicle on straight and curved roads were proposed. These were validated via the vehicle tests by classifying stationary and driving states. The results are summarized below:

(1) Dual camera images were used for lane detection. The ROI was selected so as to reduce the duration required for image processing, and the yellow color was extracted to HSV channels. Then, the result was combined with a grayscale conversion of the input image. Following edge detection using the Canny edge detector, the Hough transform was used to obtain the initial and final points of each straight line. After calculating the gradients of the straight lines, the lane was filtered and determined.

(2) Height, inter-camera baseline, and angle of inclination were considered as variables for optimizing the mounting positions of the dual cameras on the vehicle. Vehicle tests were conducted on actual roads after mounting the dual cameras on a real vehicle.

The test results revealed that the error rate was the smallest (0.86%), corresponding to a height of 40 cm, a baseline of 30 cm, and an angle of 12°. Hence, this was considered to be the optimal position.

(3) Theoretical equations were proposed for the measurement of the distance between the vehicle and an object in front of it on straight and curved roads. The dual cameras were mounted on the identified optimal positions to validate the proposed equations. Vehicle tests were conducted corresponding to stationary and driving states on straight and curved roads. On the straight road, maximum error rates of 3.54% and 5.35% were observed corresponding to the stationary and driving states, respectively. Meanwhile, on the curved road, the corresponding values were 9.13% and 9.40%, respectively. Because the error rates were less than 10%, the proposed equation for the measurement of the distance to objects in front of a vehicle was considered to be reliable.

To summarize, the mounting positions of the cameras were optimized via vehicle tests using the dual cameras, and image correction and lane detection were performed. Furthermore, the proposed theoretical equation for measuring the distance between the vehicle and objects in front of it was verified via vehicle tests, with obstacles placed at the selected positions.

The aforementioned results are significant for the following reasons. These results establish that expensive equipment and professional personnel are not required for autonomous vehicle tests, enabling research and development focused on facilitating autonomous driving using only cameras as sensors. Furthermore, webcams with easy availability can also be applied without additional sensors to the testing and evaluation of autonomous driving. In the future, we expect tests to be conducted on ACC, LKAS, and HDA at the respective levels of vehicle automation.

Author Contributions: Conceptualization: S.-B.L.; methodology: S.-B.L.; actual test: S.-H.L. and B.-J.K.; data analysis: S.-H.L., B.-J.K. and S.-B.L.; investigation: S.-H.L.; writing—original draft preparation: S.-H.L., B.-J.K. and S.-B.L.; writing—review and editing: S.-H.L., B.-J.K. and S.-B.L. All authors have read and agreed to the published version of the manuscript.

Funding: This research was funded by the Ministry of Trade, Industry, and Energy and the Korea Institute of Industrial Technology Evaluation and Management (KEIT) in 2021, grant number 10079967.

Acknowledgments: This work was supported by the Technology Innovation Program (10079967, Technical development of demonstration for evaluation of autonomous vehicle system) funded by the Ministry of Trade, Industry, and Energy (MOTIE, Korea).

Conflicts of Interest: The authors declare no conflict of interest.

References

1. On-Road Automated Driving (ORAD) Committee. *J3016_202104: Taxonomy and Definitions for Terms Related to Driving Automation Systems for On-Road Motor Vehicles*; SAE International: Warrendale, PA, USA, 2021; pp. 4–29.
2. Kee, S.C. A study on the technology trend of autonomous vehicle sensor. *TTA J.* **2017**, *10*, 16–22.
3. Lee, S.; Lee, S.; Choi, J. Correction of radial distortion using a planar checkerboard pattern and its image. *IEEE Trans. Consum. Electron.* **2009**, *55*, 27–33. [CrossRef]
4. Habib, A.; Mazaheri, M.; Lichti, D. Practical in situ implementation of a multicamera multisystem calibration. *J. Sens.* **2018**, *2018*, 1–12. [CrossRef]
5. Kim, H.Y.; Lee, S.B. A study on image processing algorithms for improving lane detectability at night based on camera. *Trans. Korea Soc. Automot. Eng.* **2013**, *21*, 51–60. [CrossRef]
6. Kim, J.S.; Moon, H.M.; Pan, S.B. Lane detection based open-source hardware according to change lane conditions. *Smart Media J.* **2017**, *6*, 15–20.
7. Choi, Y.G.; Seo, E.Y.; Suk, S.Y.; Park, J.H. Lane detection using gaussian function based RANSAC. *J. Embed. Syst. Appl.* **2018**, *13*, 195–204. [CrossRef]
8. Kalms, L.; Rettkowski, J.; Hamme, M.; Göhringer, D. Robust lane recognition for autonomous driving. In Proceedings of the IEEE 2017 Conference on Design and Architectures for Signal and Image Processing, Dresden, Germany, 27–29 September 2017; pp. 1–6. [CrossRef]

9. Wang, Z.; Wang, X.; Zhao, L.; Zhang, G. Vision-based lane departure detection using a stacked sparse autoencoder. *Math. Probl. Eng.* **2018**, *2018*, 1–15. [CrossRef]
10. Andrade, D.C.; Bueno, F.; Franco, F.R.; Silva, R.A.; Neme, J.H.Z.; Margraf, E.; Omoto, W.T.; Farinelli, F.A.; Tusset, A.M.; Okida, S.; et al. A novel strategy for road lane detection and tracking based on a vehicle's forward monocular. *Trans. Intell. Transp. Syst.* **2019**, *20*, 1497–1507. [CrossRef]
11. Bae, B.G.; Lee, S.B. A study on calculation method of distance with forward vehicle using single-camera. In Proceedings of the Symposium of the Korean Institute of Communications and Information Sciences, Korea Institute of Communication Sciences, Jeju, Korea, 19–21 June 2019; pp. 256–257.
12. Park, M.; Kim, H.; Choi, H.; Park, S. A study on vehicle detection and distance classification using mono camera based on deep learning. *J. Korean Inst. Intell. Syst.* **2019**, *29*, 90–96. [CrossRef]
13. Huang, L.; Zhe, T.; Wu, J.; Wu, Q.; Pei, C.; Chen, D. Robust inter-vehicle distance estimation method based on monocular vision. *IEEE Access* **2019**, *7*, 46059–46070. [CrossRef]
14. Zhe, T.; Huang, L.; Wu, Q.; Zhang, J.; Pei, C.; Li, L. Inter-vehicle distance estimation method based on monocular vision using 3D detection. *IEEE Trans. Veh. Technol.* **2020**, *69*, 4907–4919. [CrossRef]
15. Bougharriou, S.; Hamdaoui, F.; Mtibaa, A. Vehicles distance estimation using detection of vanishing point. *Eng. Comput.* **2019**, *36*, 3070–3093. [CrossRef]
16. Kim, S.J. Lane-Level Positioning using Stereo-Based Traffic Sign Detection. Master's Thesis, Kyungpook National University, Daegu, Korea, February 2016.
17. Seo, B.G. Performance Improvement of Distance Estimation Based on the Stereo Camera. Master's Thesis, Seoul National University, Seoul, Korea, February 2014; pp. 1–36.
18. Kim, S.H.; Ham, W.C. 3D distance measurement of stereo images using web cams. *IEMEK J. Embed. Syst. Appl.* **2008**, *3*, 151–157.
19. Song, W.; Yang, Y.; Fu, M.; Li, Y.; Wang, M. Lane detection and classification for forward collision warning system based on stereo vision. *IEEE Sens. J.* **2018**, *18*, 5151–5163. [CrossRef]
20. Sie, Y.D.; Tsai, Y.C.; Lee, W.H.; Chou, C.M.; Chiu, C.Y. Real-time driver assistance systems via dual camera stereo vision. In Proceedings of the 2019 IEEE 89th Vehicular Technology Conference, IEEE, Kuala Lumpur, Malaysia, 28 April 2019; pp. 1–6. [CrossRef]
21. Sappa, A.D.; Dornaika, F.; Ponsa, D.; Geronimo, D.; Lopez, A. An efficient approach to onboard stereo vision system pose estimation. *IEEE Trans. Intell. Transp. Syst.* **2008**, *9*, 476–490. [CrossRef]
22. Yang, L.; Li, M.; Song, X.; Xiong, Z.; Hou, C.; Qu, B. Vehicle speed measurement based on binocular stereovision system. *IEEE Access* **2019**, *7*, 106628–106641. [CrossRef]
23. Zaarane, A.; Slimani, I.; Okaishi, W.A.; Atouf, I.; Hamdoun, A. Distance measurement system for autonomous vehicles using stereo camera. *Array* **2020**, *5*, 1–7. [CrossRef]
24. Cafiso, S.; Graziano, A.D.; Pappalardo, G. In-vehicle stereo vision system for identification of traffic conflicts between bus and pedestrian. *J. Traffic Transp. Eng.* **2017**, *4*, 3–13. [CrossRef]
25. Wang, H.M.; Ling, H.Y.; Chang, C.C. Object detection and depth estimation approach based on deep convolution neural networks. *Sensors* **2021**, *21*, 4755. [CrossRef]
26. Lin, H.Y.; Dai, J.M.; Wu, L.T.; Chen, L.Q. A Vision based driver assistance system with forward collision and overtaking detection. *Sensors* **2020**, *20*, 5139. [CrossRef]
27. Canny, J. A Computational approach to edge detection. *IEEE Trans. Pattern Anal. Mach. Intell.* **1986**, *PAMI-8*, 679–698. [CrossRef]
28. Ilingworth, J.; Kittler, J. A survey of the hough transform. *Comput. Vis. Graph. Image Process.* **1988**, *44*, 87–116. [CrossRef]
29. Lee, J.Y. Camera calibration and distortion correction. *Korea Robot. Soc. Rev.* **2013**, *10*, 23–29.
30. Kim, S.I.; Lee, J.S.; Shon, Y.W. Distance measurement of the multi moving objects using parallel stereo camera in the video monitoring system. *J. Korean Inst. Illum. Electr. Install. Eng.* **2014**, *18*, 137–145. [CrossRef]

energies

MDPI

Article

Analysis of the Effect of Mass Parameters on Motorcycle Vibration and Stability

Andrzej Dębowski

Faculty of Mechanical Engineering, Military University of Technology (WAT), 00-908 Warsaw, Poland; andrzej.debowski@wat.edu.pl

Abstract: This paper presents a vibration analysis method and an example of its application to evaluate the influence of mass parameters on torsional vibration frequencies in the steering system of a motorcycle. The purpose of this paper is to analyze to what extent vibration frequencies can change during their daily operation. These changes are largely due to the ratio of vehicle weight to driver and load. The complex dynamics make it very difficult to conduct research using simple models. It is difficult to observe the influence of individual parameters because they are strongly interrelated. This paper provides a description of the vibration analysis method, and the results are presented in the form of Bode diagrams and tables. On this basis, it was found that the driver, deciding on the way of using the vehicle and introducing modifications in it, influences the resonant frequencies of the steering system. Typical exploitation factors, on the other hand, do not cause significant changes, although they may contribute to increasing the sensitivity of the system to vibrations. The conducted analysis also showed some nonlinear changes in the dynamics of the system with linear changes of the parameter values.

Keywords: motorcycle dynamics; steering system; vibrations; wobble; shimmy; Bode plot; LabVIEW

Citation: Dębowski, A. Analysis of the Effect of Mass Parameters on Motorcycle Vibration and Stability. *Energies* **2021**, *14*, 5090. https://doi.org/10.3390/en14165090

Academic Editors: Guzek Marek, Rafał Jurecki and Wojciech Wach

Received: 18 June 2021
Accepted: 16 August 2021
Published: 18 August 2021

1. Introduction

Motorcyclists worldwide are 23% of all road traffic participants [1], and collisions with passenger cars are counted among common road accidents [2]. In the example of Poland, it can be said that the number of accidents of both cars and motorcycles is 1‰ of the number of registered cars and motorcycles. However, significant in the case of motorcycles is the fact that the average annual time of use of a motorcycle per year is much lower than that of a car. Based on [3,4], it can be concluded that motorcycles account for about 5% of the registered motor vehicles in Poland, and in 2019 alone, motorcyclists were involved in 8.6% of accidents, resulting in 8.6% of injuries and 14.2% of fatalities. Although the overall number of road accidents decreased by more than 30% between 2007 and 2019, the number of accidents involving motorcycles is still at the same level [3,4]. In contrast, the number of motorcyclist fatalities has increased.

The described state of affairs is largely due to the fact that motorcyclists belong to a group of vulnerable road users. Available elements of motorcyclist's clothing that affect the increase of passive safety—apart from the helmet—are not commonly used. On the one hand, it is affected by the comfort of riding (the weight of individual elements, the lack of freedom of movement, the time needed for the motorcyclist to get ready, poor ventilation in hot weather), on the other hand by the price of all these elements. Moreover, active systems have limited potential to improve safety, as one of the key parameters of a motorcycle affecting comfort is its mass and location of its center of mass. However, legal regulations have forced the use of the CBS/LBS system, which is the equivalent of the car ABS system. Currently, it is the only mandatory active system introduced in 2017 on newly registered motorcycles.

Steady-state motion disturbances causing vibration are not uncommon. Motorcycle drivers themselves often contribute to their occurrence. Occasionally the rider may lose control of the vehicle due to uneven pavement generating strong impulses to the wheel and causing severe vibration (wobble, kick-back). Long-lasting, although small, vibrations may also be produced that contribute to reduced safety, which may result in temporary perceptual disturbances and negative well-being as a result of their impact [5].

The vibrations that occur in motorcycles are characterized by different frequencies that result from natural vibrations, as well as from external excitations. The structure of a motorcycle has different natural frequencies, and it may happen that the frequencies resulting from external excitation cause resonance [6,7]. The effect of vibrations on humans in the frequency range from 0.1 to 100 Hz can cause particularly adverse effects due to the characteristic natural frequencies for organs and body parts (Figure 1). Numbness in the limbs and diminished sensation can be felt even during a ride of several hours and may persist for several days. The frequency values were taken from [6].

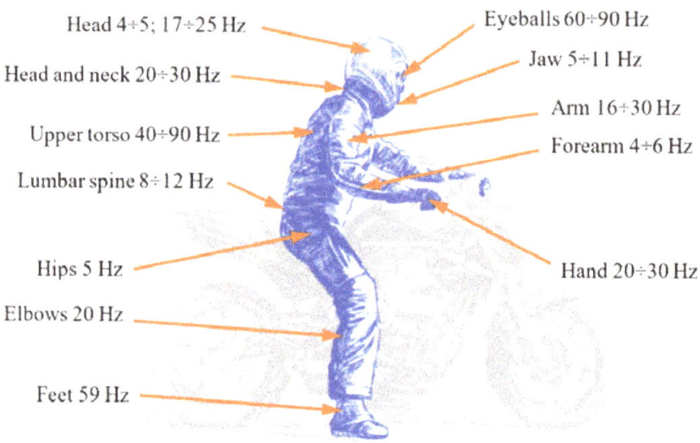

Figure 1. Natural frequencies of selected organs and parts of the human body.

In order to reduce the effects of vibration on humans, it is critical to study the source of vibration, look for methods to reduce it, and evaluate the effects on the body. The permissible values and the exposure time are defined in the relevant standard [8], but it is more important to limit the possibility of vibration generation. Nevertheless, the evaluation of the effects of vibrations on humans is the subject of numerous studies and they also concern vibrations produced in the structure of a motorcycle. As an example, [9] measured the effects of vibration on a motorcyclist on seven different surfaces, using four motorcycles that differed in mass and geometric parameters. The placement of acceleration sensors is crucial. Unfortunately, it is not clear from the article how it was concluded that the vibrations are transmitted from the road and not the drivetrain. This is important because the only sensors that were mounted are on the sprung masses. Additionally, filtration and possible methods to isolate significant frequencies from the road surface were not discussed. However, it has been shown that the vibrations transmitted to the driver are of such magnitude that even with short exposure times, they can cause adverse effects on the human body.

In order to validate the analyses based on mathematical modeling and computer simulation, motorcycle motion and vibration tests are conducted using special test stands. Unfortunately, this type of research is relatively rarely described in scientific publications. Therefore, attention was drawn to [10], which describes experiments with a separate steering system of a motorcycle whose tire-wheel rotates on a sliding belt. On the other hand, in [11,12], bench tests are conducted using a partially immobilized motorcycle, whose front wheel cooperates with a rotating drum reflecting the road surface. The analyses presented in the following section are conducted based on the mathematical model described in [13], where its verification is described. Data for verification were recorded on a test rig, the operation of which can be seen at [14].

With the development of computer methods, mathematical models have become widely used. The motorcycle model described in [15] allowed the determination of the forms of vibrations that occur for the case of driving a motorcycle without using hands, named by the author as capsize, weave, and wobble. This already shows that with relatively simple models, it is possible to analyze the dynamics of a motorcycle [16]. Subsequently, research was extended to include mainframe susceptibility by both Sharp, as described in [16,17], and Kane [18]. It was thus shown that the stability of the motorcycle, and thus the weave vibration, is significantly affected by the stiffness of the mainframe, while it has no effect on the other two.

Unfortunately, very few models have been developed for frequency analysis of motorcycle steering vibrations. In particular, [19], which is a comprehensive book on the dynamics of a motorcycle, should be mentioned here. The influence of various structural parameters of a motorcycle on vibration damping at different speeds was analyzed in detail. It was concluded that the moment of inertia of the steering system could be important for the stability of a motorcycle, which is included in the expression derived from the single mass model. However, this model neglects many other steering parameters whose specific values may also contribute to vibration. Other works include [10,20–22], in which there is a determination of the natural frequency of the steering system. On the other hand, in [23], a single-mass model of the steering system dynamics with sharp nonlinearities is presented and applied to test numerical procedures, which is particularly important in solving differential equations by iterative methods.

Among the methods used to study the vibrations and their graphical presentation, we can mention [19,24–26], in which the graphs of the real part describing the vibrations as a function of driving speed are presented. The curves created in this way illustrate the dynamic properties of the motorcycle in a wide range of speeds and its tendency to unstable behavior well. The root locus plot was also used in [19,24], which allows one to observe changes in the damping of particular types of vibrations as a function of speed. On the other hand, in [27], the possibility of analyzing steering system vibrations by means of phase plane was presented, which can also be useful in vibration analysis.

During daily use, the mass of a motorcycle can change significantly. While the weight of the rider will be taken into account at the design stage, the presence of a passenger will change the weight distribution and front -heel loading; also, the fuel tank itself, when fully filled, will be an additional factor. Some improvement in motorcycle stability can be achieved by fitting a torsional vibration damper to the steering system. However, whether or not it is included in the steering system, every motorcycle should have a natural tendency to handle steadily. For this reason, it is important to look for parameters that have a key influence on steering dynamics and how they affect vibration frequency. Therefore, a method of analysis will be presented that will allow a complex description of the system to extract relevant information to assist in the design of a motorcycle.

2. Method and Tools Used

When modeling vibrations that are occurring in mechanical systems, certain simplifying assumptions are introduced—most often to replace the nonlinear model with a linear model. This approach to modeling results from the fact that the methods of analysis of linear systems are well developed and effective. In addition, nonlinear systems are often stable in some limited neighborhood of the equilibrium point only, while linear systems, if stable, are globally and, moreover, asymptotically stable [28]. The basic forms of description of the dynamics of linear dynamical systems (apart from differential equations) include the operator transmittance $G(s)$ and the spectral transmittance $G(j\omega)$. These such forms of description will be used later in this paper.

The physical model (Figure 2) directly corresponds to the motorcycle steering system mounted in a special drum test rig presented in [13]. It was also validated on the basis of the results of measurements made on this test stand.

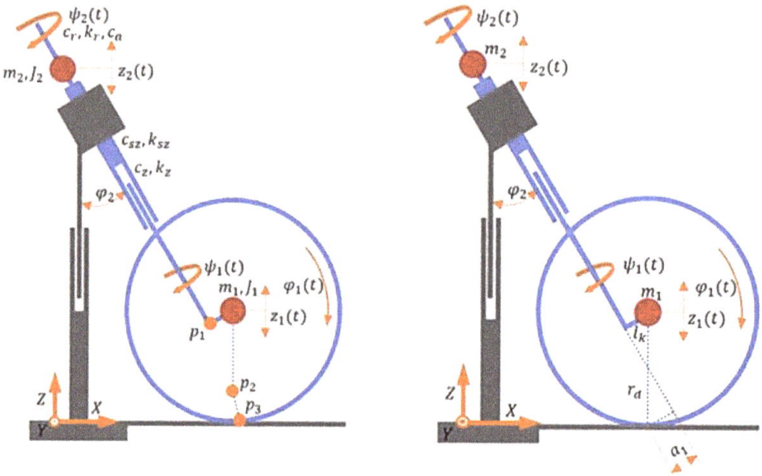

Figure 2. Physical model of a motorcycle steering system.

This model considers the key features of a motorcycle steering system. It has five degrees of freedom, which are the rotational motion of the wheel about the wheel axis, the angular motion of the handlebars and wheel about the steering axis of the head tube, and the vertical motion of the wheel and body. Thus, the longitudinal operation of the suspension and its torsional compliance is taken into account. A simplified model of wheel-road interaction was adopted, neglecting the phenomenon of tire relaxation length, with a simplified description of the stabilizing moment based on [29–31]. Despite the simplifying assumptions made, it represents the steering dynamics well and is sufficient to demonstrate the analysis method presented in this paper. However, if necessary, any other model can be used, as the procedure will not change. This method is particularly useful at the initial stage of a motorcycle or other vehicle because it allows the selection of parameters of the real system and to study the vibrations that arise. However, in the final stage of design and modeling of individual parts, the finite element method can be used, as presented in [32], on the design of an elastic component for a motorcycle.

List of designations appearing in Figure 2 and in the equations described, along with values that are taken as references:

A system of nonlinear differential equations (initial model) was obtained using Lagrange's equations of the second-order, and the most important formulas are presented below.

$$\frac{d}{dt}\left(\frac{\partial \mathcal{L}}{\partial \dot{q}_i}\right) - \frac{\partial \mathcal{L}}{\partial q_i} + \frac{\partial D}{\partial \dot{q}_i} = Q_{q_i}, \qquad (1)$$

$$\text{gdzie}: \ \mathcal{L} = T - U \text{ i } q_i = \varphi_1, \ \psi_1, \ z_1, \ z_2, \ \psi_2$$

The potential and kinetic energy are written by Equations (2) and (3), and the energy dissipation is written by Equation (4).

$$T = \frac{1}{2}\sum_{i=1}^{2} m_i\left(\dot{x}_i^2 + \dot{y}_i^2 + \dot{z}_i^2\right) + \frac{1}{2}\sum_{i=1}^{2}\left(I_{ix}\omega_{ix}^2 + I_{iy}\omega_{iy}^2 + I_{iz}\omega_{iz}^2 - 2\cdot\left(I_{ixy}\omega_{ix}\omega_{iy} + I_{iyz}\omega_{iy}\omega_{iz} + I_{izx}\omega_{iz}\omega_{iy}\right)\right), \qquad (2)$$

$$U = \frac{1}{2} \cdot k_z \left(\sqrt{(p_{1x} - x_2)^2 + (p_{1z} - z_2)^2} \right)^2 + U_{sz} = \frac{1}{2} \cdot k_{sz} (\psi_{1z} - \psi_{2z})^2 + \frac{1}{2} \cdot k_r (\psi_{2z})^2 + \frac{1}{2} \cdot \left(\sqrt{K_x (p_{2x} - p_{3x})^2 + K_y (p_{2y} - p_{3y})^2 + K_z (p_{2z} - p_{3z})^2} \right)^2, \tag{3}$$

$$D = \frac{1}{2} \cdot c_z \left(\sqrt{(\dot{p}_{1x} - \dot{x}_2)^2 + (\dot{p}_{1z} - \dot{z}_2)^2} \right)^2 + \frac{1}{2} \cdot c_{sz} (\omega_{1z} - \omega_{2z})^2 + \frac{1}{2} \cdot C_o \left(\sqrt{(\dot{p}_{2x} - \dot{p}_{3x})^2 + (\dot{p}_{2y} - \dot{p}_{3y})^2 + (\dot{p}_{2z} - \dot{p}_{3z})^2} \right)^2 + \frac{1}{2} \cdot c_r \left(\dot{\psi}_2 \right)^2 + \frac{1}{2} \cdot c_a \left(\dot{\psi}_2 \right)^2. \tag{4}$$

Relevant to the driving wheel dynamics is the stabilizing torque, which is written as follows:

$$M_{sk} = a_1 (F_y \cdot cos(\chi) + F_x \cdot cos(\chi) - F_z \cdot sin(\chi)), \text{ where}: \chi = \varphi_2 \, sin(\psi_1). \tag{5}$$

The parameter a_1 in Equation (5) is the actual overtaking distance and, taking into account the steering angle, its value can be calculated as follows:

$$a_1 = r_d \frac{cos(\psi_1) \cdot sin(\varphi_2)}{\sqrt{1 - (sin(\psi_1) \cdot sin(\varphi_2))^2}} - l_k. \tag{6}$$

The other components of generalized forces are:

$$M_{nk} = F_x \cdot z_1 + F_z \cdot e_x. \tag{7}$$

and M_{or}, which has a negative value, and the gravity forces $G_1 = g \cdot m_1$, $G_2 = g \cdot m_2$. The model also assumes the wheel slip is small, so the equation for the lateral force is expressed as follows:

$$F_y = K_\alpha \cdot \alpha_z. \tag{8}$$

and for the longitudinal force:

$$F_x = K_x^* \cdot S_x. \tag{9}$$

In the case under consideration, due to zero longitudinal slip, there will be no longitudinal force. For this reason, in the previously written equations also, some components related to the longitudinal force will be zero, but they are written for the sake of order since their absence could cause misunderstanding of the relations described.

Due to the adopted structure of the physical model, the obtained system of equations has a complex form. Therefore, symbolic transformation software (e.g., Maple, Octave) was used for their derivation as well as further transformations. This allowed the individual transmittances describing the ratio of the input signal to the output signal to be obtained and, most importantly, in the general form.

Ten different transmittances can be written for such a system, but four of them are important, namely: transmittance for vertical displacement of masses m_1 and m_2 at vertical excitation ($G_1(s) = \frac{Z_1(s)}{F_Z(s)}$, $G_2(s) = \frac{Z_2(s)}{F_z(s)}$) and the transmittance for the angular displacements of masses m_1 and m_2 under torsional moment excitation ($G_3(s) = \frac{\Psi_1(s)}{M_s(s)}$, $G_4(s) = \frac{\Psi_2(s)}{M_s(s)}$).

The process leading to the mentioned transmittances can be divided into different steps, which in order will be as follows:

1. Development of the mathematical model;
2. Linearization of the equations describing the system dynamics;
3. Writing the equations in the operator form and determining the transmittance;
4. Perform stability analysis of the system;
5. Breaking down of transmittance into summation form;

6. Frequency analysis of each of the members of transmittance.

On the other hand, analyses based on the transmittances obtained in this way were performed using the LabVIEW software and an application created for this purpose (Figures 3 and 4).

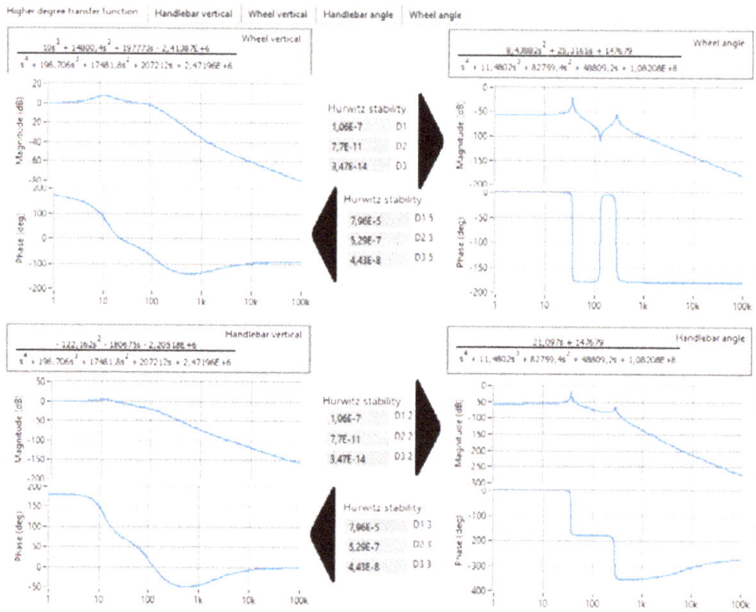

Figure 3. View of the steering frequency analysis program window.

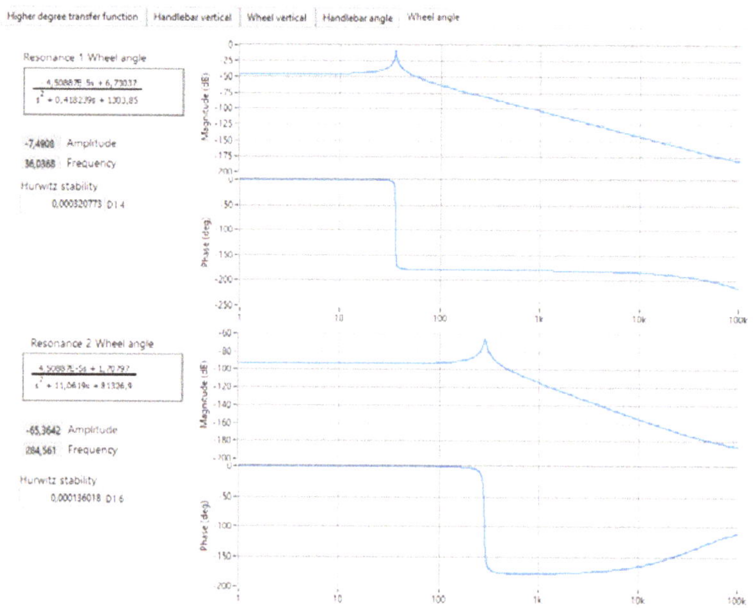

Figure 4. View of the program window for frequency analysis of individual transmittances.

It is worth noting that the obtained strictly proper transmittances are of a higher order than is the case for the analyzed dual-mass models by other authors. However, it should be noted that, unlike the aforementioned models, the one presented here has a feature that has a key influence on its dynamics—the angle of inclination of the head of the frame.

As already mentioned, LabVIEW software was used to develop a suitable tool for breaking the higher-order transmittance down into two lower-order transmittances and analyzing the resulting Bode diagrams.

While breaking the transmittances down into simple fractions is relatively straightforward in this case, doing it automatically is not. The first problem one may encounter is finding zero places of the characteristic polynomial. In LabVIEW, the appropriate Partial Fraction Expansion (PFE) function is available, but in special cases, it does not return the correct result; and this case is one of them. The incorrect result, in some cases, results directly from the way the function is implemented. The PFE function uses the Heaviside function to calculate the residues and poles, which is not applicable when the roots of the polynomial in the denominator are either double or complex, so only single real roots will give the correct results. For this reason, it was not possible to use the appropriate VI from the palette.

This difficulty can be overcome by using a MathScript Node and a suitably written script. In this case, the script for finding the roots of the denominator polynomial is limited to two lines, where the first is an array of the coefficients of the polynomial and the second is the roots command. In this way, four composite roots are obtained. The final form, which is interesting from the perspective of further analysis, is obtained as follows. Having the denominator of the transmittance of the form:

$$G(s) = \frac{a_1 s + a_0}{b_4 s^4 + b_3 s^3 + b_2 s^2 + b_1 s + b_0} \tag{10}$$

and taking the expected form:

$$G(s) = \frac{As + B}{(as^2 + bs + c)} + \frac{Cs + D}{(ds^2 + es + f)} \tag{11}$$

then, after writing the appropriate system of equations, one obtains the sought values of the coefficients a, b, c, d, e, f. Of course, a general form has been implemented in the program so that they are recalculated as the selected parameter changes (Figure 5).

The second problem is the determination of the numerator factors. To obtain them in an analytical way, it is enough to take any real number as s and solve a system of equations. In this case, it cannot be done because in LabVIEW, the operator s does not appear directly in the equation, but only the coefficients of the polynomial. Therefore, it is necessary to use an additional tool for symbolic transformations and determine specific values of parameters by solving a typical equation in search of numerator factors.

The third problem that can be encountered is the implementation of the calculated numerator factors—solving a system of equations depending on the transmittance yields extremely large equations that take up dozens of A4 pages (between 40 and 80 pages). The problem that arises, in this case, is that too many lines of text are needed to be placed in the MathScript Node. This structure only holds about 20 pages of equations. If there are too many characters, they will not be processed by this structure, and an additional file corruption and LabVIEW critical error may occur. In some cases, using the Formula Node structure is sufficient. The final form of the code that allows the breaking down of the fourth-degree transmittance into two second-degree transmittances is shown in Figure 5.

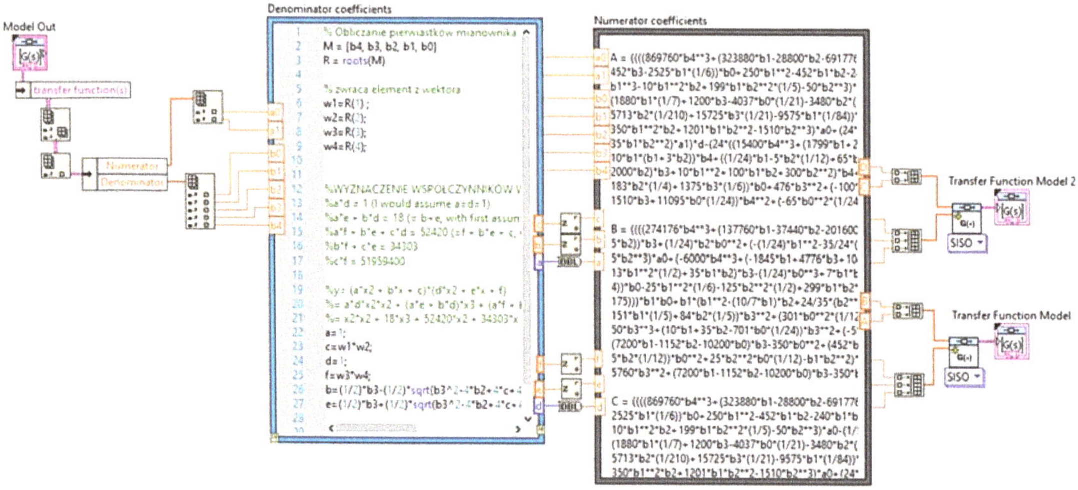

Figure 5. View of a window with a code fragment of the steering frequency analysis program.

The Formula Node structure takes three times as many characters, but even this may sometimes prove to be too few. However, this can be easily remedied by splitting the description of each coefficient into a single structure. This is also consistent with the idea of programming because the main code of the program should be divided into a subprogram.

Finally, the operation of the described program is presented in the diagram below (Figure 6). It is only one-quarter of the whole, but the remaining parts perform exactly the same calculations.

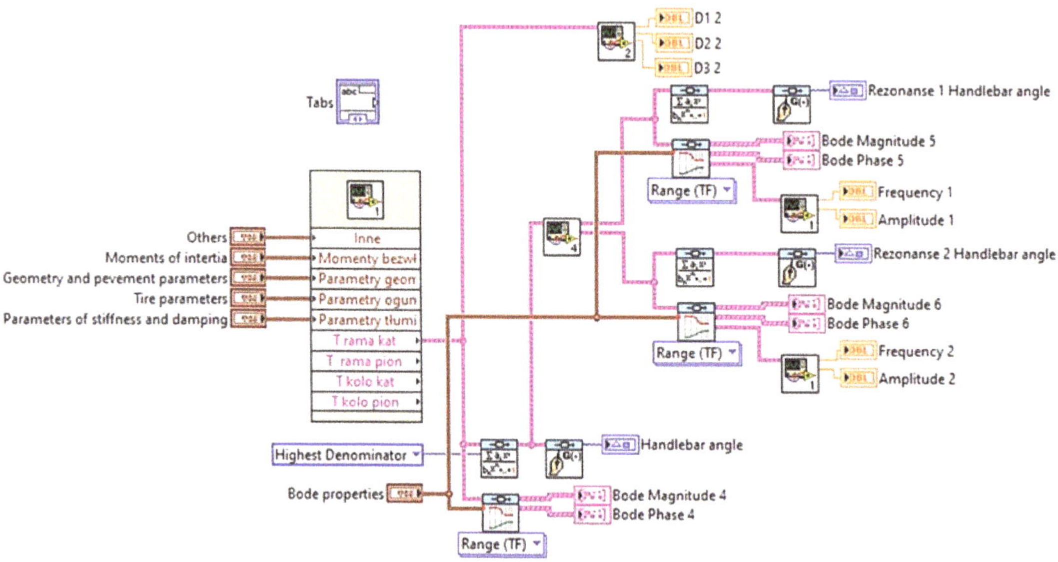

Figure 6. View of a window with a code fragment of the steering frequency analysis program.

With the program thus developed, the Bode plots summarized below were obtained along with the significant values read from the plots.

When the values of the parameter J_{z2} are varied (Table 1), the tendency of the system to decrease in vibration frequency with increasing values of the moment of inertia can be observed, but not in a completely linear manner. This is true for both resonance frequencies. On the other hand, an increase in the parameter J_{z2} causes a significant decrease in the amplitude of vibrations of the wheel while increasing the amplitude of vibrations of the steering wheel. Noteworthy is the effect of the value of this parameter on the stability of the system in different frequency ranges. While for low-frequency resonance, the stability is slightly improved, for high-frequency resonance, a deterioration of stability by exactly the same value is observed. The observed changes in the oscillation period are directly related to the frequency, and it becomes longer as the moment of inertia increases while the relative damping decreases. Changing the value of the analyzed parameter also has a significant effect on the overall damping properties of the system, which decrease as J_{z2} increases.

Table 1. The values of the parameters read from the graphs shown in Figure 7.

Changing Parameter	From Mas m_1					From Mas m_2				
	Frequency	Max. Amplitude	Stability	Oscillation Period	Relative Suppression	Frequency	Max. Amplitude	Stability	Oscillation Period	Relative Suppression
	(rad/s)	(dB)	(-)	(s)	(-)	(rad/s)	(dB)	(-)	(s)	(-)
0.2 kgm^2	46.88742	−15.5666	0.000310	0.021285	0.007291	312.4292	−63.5229	0.000141	0.003194	0.022028
0.3 kgm^2	40.65839	−10.9375	0.000313	0.024546	0.006386	294.1920	−64.3485	0.000138	0.003392	0.020282
0.4 kgm^2	36.38159	−7.58916	0.000315	0.027432	0.005746	284.7340	−65.3905	0.000136	0.003504	0.019382
0.5 kgm^2	33.21640	−4.96667	0.000316	0.030046	0.005264	278.9454	−66.4172	0.000135	0.003577	0.018834
0.6 kgm^2	30.75335	−2.81362	0.000317	0.032452	0.004885	275.0379	−67.3769	0.000134	0.003628	0.018465

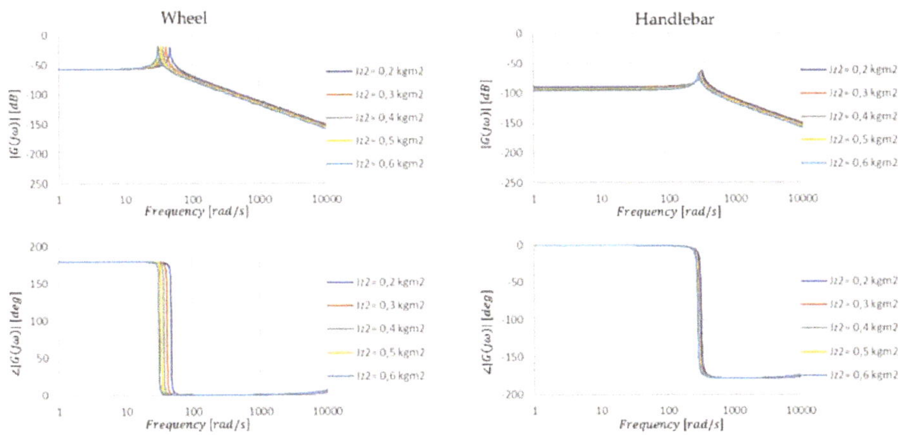

Figure 7. Amplitude and phase characteristics of the steering angle position transmittance when the parameter value is changed J_{z2}.

In the analyzed case, when the values of the parameter J_{x1} and J_{z1} are varied, the opposite situation to the one described earlier takes place, namely the change of the value of the wheel moment of inertia parameter affects the higher frequency vibrations to a greater extent. Changes in the value of the wheel moment of inertia parameter affect the resonance frequencies to a greater extent, and the nonlinearity of the changes is much more pronounced. The same relationship for stability is maintained for higher frequency resonance in this case; when increasing the value of moment of inertia increases, for lower frequency resonance, it decreases. The other values, such as vibration period and relative damping, have the same nature of changes, but their changes are more for the higher frequency resonance.

Previously, it could be observed (Figure 8, Table 2) that the moment of inertia of the wheel J_{y1} causes only slight changes in the low-frequency resonant vibration, while it affects the higher-frequency resonance. This case is similar (Figure 9, Table 3), while what is different is that the overall damping properties of the system no longer change so significantly.

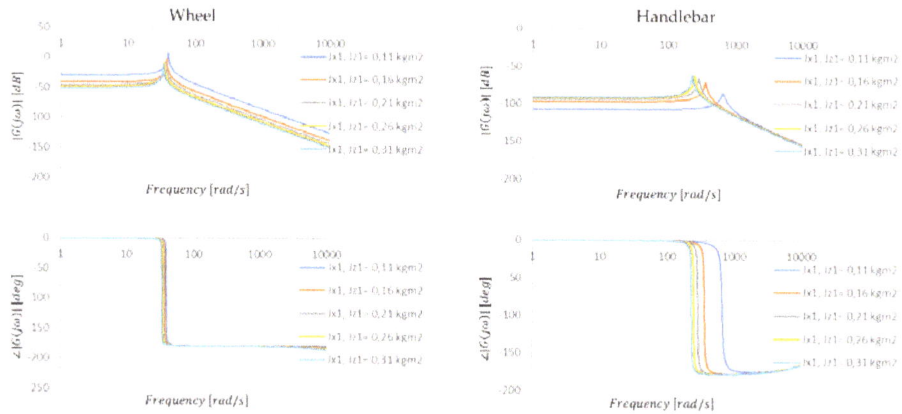

Figure 8. Amplitude and phase characteristics of the steering angle position transmittance when the parameter values are changed J_{x1} and J_{z1}.

Table 2. The values of the parameters read from the graphs shown in Figure 8.

Changing Parameter	From Mas m_1					From Mas m_2				
	Frequency	Max. Amplitude	Stability	Oscillation Period	Relative Suppression	Frequency	Max. Amplitude	Stability	Oscillation Period	Relative Suppression
	(rad/s)	(dB)	(-)	(s)	(-)	(rad/s)	(dB)	(-)	(s)	(-)
0.11 kgm^2	39.88838	7.73275	0.000320	0.025020	0.006398	656.7857	−85.4846	0.000131	0.001518	0.043121
0.16 kgm^2	38.02698	−3.21068	0.000318	0.026245	0.006051	358.2607	−70.5517	0.000133	0.002785	0.023963
0.21 kgm^2	36.38159	−7.58916	0.000315	0.027432	0.005746	284.7340	−65.3905	0.000136	0.003504	0.019382
0.26 kgm^2	34.91741	−10.3023	0.000313	0.028582	0.005476	248.8033	−62.6528	0.000138	0.004011	0.017207
0.31 kgm^2	33.60604	−12.2465	0.000311	0.029697	0.005237	227.0204	−60.9958	0.000140	0.004396	0.015926

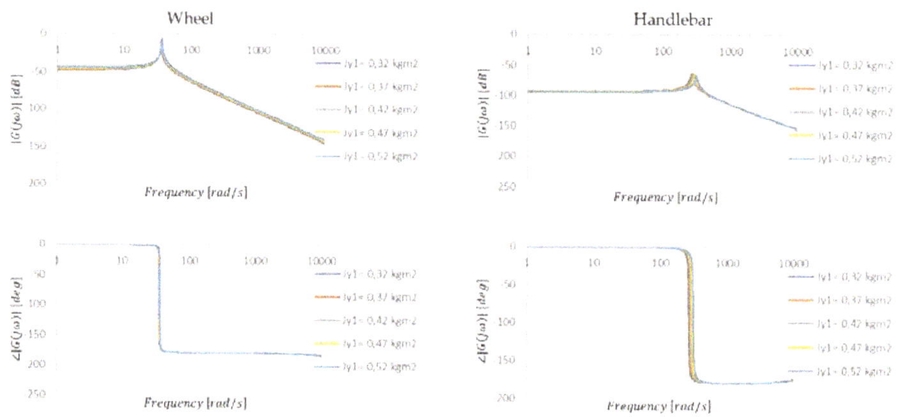

Figure 9. Amplitude and phase characteristics of the transmittance of the steering angle position when the parameter value is changed J_{y1}.

Table 3. The values of the parameters read from the graphs shown in Figure 9.

Changing Parameter	From Mas m_1 Frequency	Max. Amplitude	Stability	Oscillation Period	Relative Suppression	From Mas m_2 Frequency	Max. Amplitude	Stability	Oscillation Period	Relative Suppression
	(rad/s)	(dB)	(-)	(s)	(-)	(rad/s)	(dB)	(-)	(s)	(-)
0.32 kgm^2	35.62876	−9.07527	0.000314	0.028011	0.005607	264.219	−63.8366	0.000137	0.003777	0.018132
0.37 kgm^2	35.99977	−8.37082	0.000315	0.027723	0.005675	273.6848	−64.5580	0.000136	0.003646	0.018707
0.42 kgm^2	36.38159	−7.58916	0.000315	0.027432	0.005746	284.7340	−65.3905	0.000136	0.003504	0.019382
0.47 kgm^2	36.78380	−6.60721	0.000316	0.027138	0.005818	298.3647	−66.3273	0.000135	0.00335	0.020188
0.52 kgm^2	37.17963	−5.71695	0.000316	0.026843	0.005893	313.6558	−67.5049	0.000135	0.003181	0.021168

Looking at the graphs shown in Figure 10 and the data summarized in Table 4, which relate to the vertical displacement of the steering wheel, it can be concluded that low-frequency torsional vibration is only slightly transmitted to the vertical motion, while higher-frequency torsional vibration is transmitted to a greater extent. The system also shows much greater stability for vertical steering wheel vibration and damping than for torsional vibration. This is consistent with the actual behavior of the motorcycle, as no vertical self-excited vibration of the front of the motorcycle is observed, only torsional (wobble) vibration.

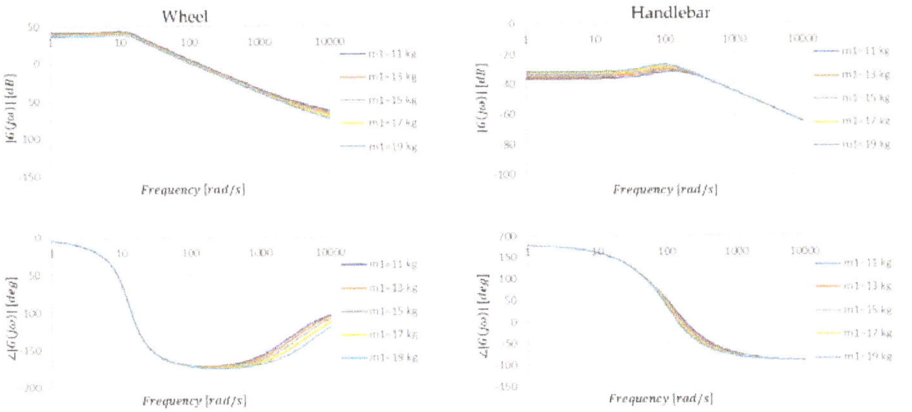

Figure 10. Amplitude and phase characteristics of the transfer function of the handlebar angle position when the value of parameter m_1 is changed.

Table 4. The values of the parameters read from the graphs shown in Figure 10.

Changing Parameter	From Mas m_1 Frequency	Max. Amplitude	Stability	Oscillation Period	Relative Suppression	From Mas m_2 Frequency	Max. Amplitude	Stability	Oscillation Period	Relative Suppression
	(rad/s)	(dB)	(-)	(s)	(-)	(rad/s)	(dB)	(-)	(s)	(-)
11 kg	9.752281	43.017107	0.071647	0.078168	0.458287	132.194115	−31.761882	0.012178	0.006956	0.875390
13 kg	9.746465	41.622516	0.071632	0.078237	0.457791	117.681195	−30.264714	0.012193	0.007562	0.806202
15 kg	9.740655	40.435734	0.071618	0.078306	0.457297	117.681195	−28.993801	0.012207	0.008122	0.751427
17 kg	9.734852	39.404514	0.071604	0.078375	0.456804	104.761575	−27.843054	0.012221	0.008647	0.706683
19 kg	9.729055	38.494112	0.071590	0.078444	0.456313	97.544519	−26.834981	0.012235	0.009141	0.669250

Compared to the previous graphs and values in the tables presented in Figure 11 and Table 5 show that the change of wheel mass m_1 causes only slight or even negligible changes in the analyzed values. On this basis, it can be concluded that the wheel mass does not significantly affect the torsional vibration despite some coupling of vertical and torsional motion through the advance section.

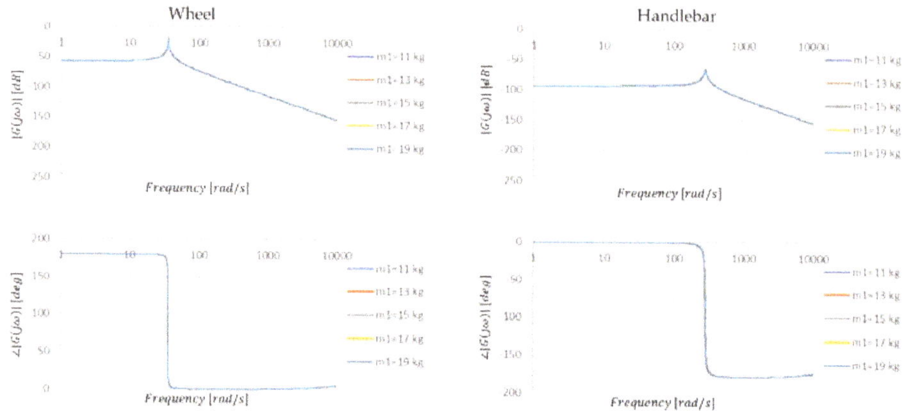

Figure 11. Amplitude and phase characteristics of the transmittance of the steering angle position when the parameter value is changed m_1.

Table 5. The values of the parameters read from the graphs shown in Figure 11.

Changing Parameter	From Mas m_1					From Mas m_2				
	Frequency	Max. Amplitude	Stability	Oscillation Period	Relative Suppression	Frequency	Max. Amplitude	Stability	Oscillation Period	Relative Suppression
	(rad/s)	(dB)	(-)	(s)	(-)	(rad/s)	(dB)	(-)	(s)	(-)
11 kg	36.458999	−18.859719	0.000316	0.027373	0.005771	288.253307	−64.962969	0.000136	0.003462	0.019600
13 kg	36.420204	−18.850010	0.000316	0.027403	0.005758	286.471864	−64.832420	0.000136	0.003483	0.019490
15 kg	36.381594	−18.840356	0.000315	0.027432	0.005746	284.734004	−64.704775	0.000136	0.003504	0.019382
17 kg	36.343169	−18.830756	0.000315	0.027461	0.005733	283.038060	−64.579941	0.000136	0.003525	0.019277
19 kg	36.304927	−18.821209	0.000315	0.027490	0.005721	281.382448	−64.457829	0.000136	0.003546	0.019174

The graphs presented in Figures 12 and 13 and the data in Tables 6 and 7 allow us to conclude that the change in mass m_2 has little effect on vertical vibration and negligible effect on torsional vibration of the handlebars. This thus contradicts claims among motorcyclists that mass distribution affects wobble vibration. It is interesting to note, however, that despite changes in mass m_2 and changes in stability amplitude, vibration period, and damping, the frequency does not change for higher frequency vibrations.

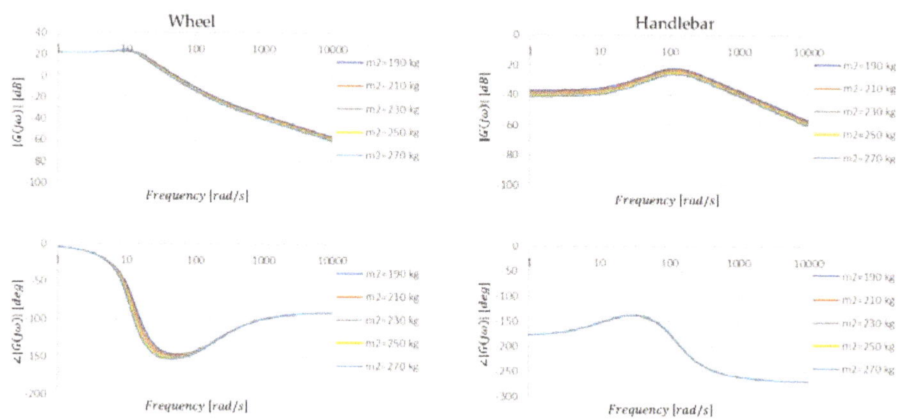

Figure 12. Amplitude and phase characteristics of the transmittance of the vertical position of the steering wheel with a change in the value of the parameter m_2.

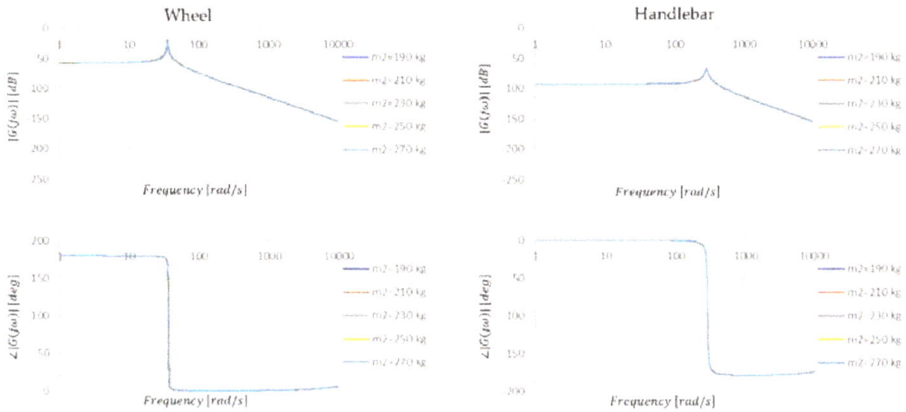

Figure 13. Amplitude and phase characteristics of the transmittance of the steering angle position when changing the value of the parameter m_2.

Table 6. The values of the parameters read from the graphs shown in Figure 12.

Changing Parameter	From Mas m_1					From Mas m_2				
	Frequency	Max. Amplitude	Stability	Oscillation Period	Relative Suppression	Frequency	Max. Amplitude	Stability	Oscillation Period	Relative Suppression
	(rad/s)	(dB)	(-)	(s)	(-)	(rad/s)	(dB)	(-)	(s)	(-)
190 kg	9.929478	23.353815	0.071281	0.070308	0.506918	117.681195	−21.872242	0.012544	0.008233	0.761811
210 kg	9.863696	23.577621	0.071469	0.074423	0.480156	117.681195	−23.015869	0.012356	0.008171	0.756067
230 kg	9.740655	23.806457	0.071618	0.078306	0.457297	117.681195	−24.020465	0.012207	0.008122	0.751427
250 kg	9.600872	24.033914	0.071739	0.081994	0.437463	117.681195	−24.916998	0.012086	0.008083	0.747599
270 kg	9.405021	24.256714	0.071839	0.085516	0.420033	117.681195	−25.726974	0.011986	0.008051	0.744385

Table 7. The values of the parameters read from the graphs shown in Figure 13.

Changing Parameter	From Mas m_1					From Mas m_2				
	Frequency	Max. Amplitude	Stability	Oscillation Period	Relative Suppression	Frequency	Max. Amplitude	Stability	Oscillation Period	Relative Suppression
	(rad/s)	(dB)	(-)	(s)	(-)	(rad/s)	(dB)	(-)	(s)	(-)
190 kg	9.929478	37.839789	0.071281	0.070308	0.506918	104.761575	−27.041897	0.012544	0.008233	0.761811
210 kg	9.863696	39.202653	0.071469	0.074423	0.480156	104.761575	−28.076808	0.012356	0.008171	0.756067
230 kg	9.740655	40.435734	0.071618	0.078306	0.457297	117.681195	−28.993801	0.012207	0.008122	0.751427
250 kg	9.600872	41.561546	0.071739	0.081994	0.437463	117.681195	−29.814093	0.012086	0.008083	0.747599
270 kg	9.405021	42.597395	0.071839	0.085516	0.420033	117.681195	−30.561990	0.011986	0.008051	0.744385

3. Discussion of Simulation Results

Complex models are not only difficult to interpret but also to identify a large number of parameters. The presented description of the program used shows not only the implementation of the method of analysis of system dynamics, but also the method of model reduction. This is a much simpler approach than the theory developed by Mandelstam to evaluate the coupling of partial sub-systems [33]. With Mandelstam's theory, it is possible to break the full model down into a group of partial models. Another approach is parametric simplification, which addresses Hadamard's postulate of continuity of solution changes with respect to model parameters [34]. However, both approaches are rarely used, because as a rule, partial models are extracted based on an a priori assumption about the weakness of the remaining interactions or the complete isolation of the system [33]. In the presented approach, the reduction can be obtained by breaking down the transmittance into a sum of factors. By omitting one of the lower order transmittances, a transmittance model of the system is obtained that is oriented towards the dynamics of the system in the range of selected vibration frequencies. However, this model is no longer in parametric form, which is indirectly related to the Abel–Ruffini theorem. In addition, depending on the values of the individual denominator factors, the breaking down into factors may or may not be performed, as the system will be unstable.

The system stability evaluation was performed using the Hurwitz method, but any other method can be used. This function has only been added to the program for your convenience and can be carried out separately using any other tool. The stability analysis indicates that the motorcycle's steering is at the limit of stability. The small values of the third Hurwitz sub-designator confirm that the torsional vibration can be induced by the forces from the road irregularities. It can also be observed that in order to improve the stability of the system to vibrations of one frequency, we cause a deterioration of stability for the other frequency.

The computational results, shown in Figures 7–13, although related to a selected portion of the simulation study, allow a number of interesting observations and conclusions to be drawn.

After breaking the higher-order transmittance down into two lower-order transmittances, pairs of the same values characterizing the lower- and higher-order resonance for specific transmittance pairs are obtained. For example, the transmittance under impulsive excitation of the steering wheel and handlebar. It naturally follows that the resonant frequencies for two connected masses will also be two. Therefore, in the future, there is no need to analyze four Bode diagrams but only two.

The presented method by separating the two resonance frequencies greatly facilitates the inference of the behavior of the system depending on the parameters describing it. In particular with respect to damping properties and stability. The analysis, apart from confirming the obvious truths, i.e., increasing resonant frequencies with decreasing element mass, allowed us to notice some nonlinear changes in the system dynamics with linear changes in parameter values (Figure 8). While the mass of the road wheel may change slightly, e.g., depending on a more massive tire design, the use of different materials for the wheel structure (magnesium, carbon fiber) will cause changes in both the resonance frequencies of the wheel and steering wheel and their phase shifts (Figure 10). Such modification can be done by the user themself, as wheel rim replacements are now available.

The change in road wheel mass manifests itself only slightly through its effect on torsional vibration (Figure 11), although from the values (Table 5) the changes are noticeable. Despite the great importance of the wheel advance section for the handling stability of the motorcycle, the impacts due to vertical loads are transferred to a small extent to the torsional vibrations. Although using the total mass of the motorcycle as an example, an interesting relationship can be observed. As the mass increases, the damping of the system for vertical vibration also increases, but the stability and relative damping for torsional vibration decreases, which may indicate that increasing the mass of the motorcycle will increase the likelihood of torsional vibration occurring. Interestingly they will be at a higher frequency than for an unloaded motorcycle (Table 7). Therefore, if vibrations occur with a more heavily loaded wheel, the motorcycle will be more difficult for the rider to stabilize.

4. Summary and Conclusions

This paper presents a method of the breaking down of a system model into simpler components and the analysis of vibrations of particular frequencies, which can be applied in dynamics studies of not only motorcycles but also other systems. A mathematical model written in the form of operator transmittance was used, and on this basis, Bode diagrams were determined. The difficulties encountered in breaking down higher-order transmittances into simple fractions using the LabVIEW environment have been described in detail.

The method can be easily applied to more complex objects, which only requires the use of another mathematical model describing the investigated object. The results obtained in the numerical form also make it easier to grasp small substitutions, which may not be directly visible on the frequency characteristics but may be useful at the initial stage of machine design. With this method, it is possible to perform the model reduction in a formal way so that all relevant parameters are included in the final form of the model. It is relatively easy to obtain a simplified formula describing the dynamics of the system for a selected frequency and then develop a control system.

The driver, by deciding how to use the vehicle and making modifications to it, has an influence on the resonant frequencies of the steering system. However, typical operating factors do not cause significant changes, although they may contribute to the system's sensitivity to vibration.

Funding: This research received no external funding.

Institutional Review Board Statement: Not applicable.

Informed Consent Statement: Not applicable.

Data Availability Statement: Data is contained within the article.

Conflicts of Interest: The authors declare no conflict of interest.

Abbreviations

X, Y, Z	main axes of the coordinate system;
$M_{nk}, M_b, M_{or}, M_{sk}, W_0$	a torque that adequately stimulates vibrations, driving from the drum and resistance to movement, steering, excitation from the road surface;
ψ_1, ψ_2	the steering angle respectively of the wheel and handlebar;
φ_1	wheel rotation angle;
φ_2	steering head angle (24 deg);
z_1, z_2	vertical displacement of the elements associated with the wheel and the frame, respectively;
J_1, J_2	the equivalent moment of inertia of the components associated with the wheel (0.21 kgm^2) and handlebar (0.4 kgm^2), respectively;
m_1, m_2, m	reduced weight of the components associated with the wheel (15 kg) and frame (230 kg), and sum of masses m_1 i m_2, respectively;
$c_r, c_a, c_{sz}, c_z, C_o$	damping coefficient of the driver's hands (0.2 Nms/rad), torsional vibration damper (0 Nms/rad), torsional damping of suspensions (3 Nms/rad), longitudinal damping of suspension (2.6 kNs/m), tire damping coefficient (150 Ns/m), respectively;
k_r, k_{sz}, k_z	stiffness coefficient of the driver's hands (1 kN/rad), torsional stiffness (7 kNm/rad), longitudinal stiffness (14 kN/m), respectively;
K_z, K_α, K_x	radial tire (190 kN/m), cornering stiffness coefficient (10 kN/rad), longitudinal stiffness coefficient (180 kN/m), respectively;
p_1, p_2, p_3	the point of suspension end, wheel rim, and wheel/road contact, respectively;
r_d	dynamic wheel radius (0.3 m);
l, l_1, l_k	wheelbase (1.35 m), distance between the center of mass and the front wheel's axis of rotation (0.6 m), offsetting the wheel axis from the control axis of the frame head (0.03 m);
w_{po}	tire profile height (0.08 m);
a_1	actual overtaking distance (0.1 m);
ω	angular velocity;
χ	wheel camber angle;
F_x, F_y, F_z	longitudinal, lateral, and vertical reaction forces, respectively;
e_x	displacement of normal force;
S_x	longitudinal slip.

References

1. Piantini, S.; Pierini, M.; Delogu, M.; Baldanzini, M.; Franci, A.; Mangini, M.; Peris, A. Injury Analysis of Powered Two-Wheeler versus Other Vehicle Urban Accidents. In *Proceedings of the 2016 IRCOBI Conference*; IRCOBI: Malaga, Spain, 2016. [CrossRef]
2. Prochowski, L.; Pusty, T. *Charakterystyka Obrotu i Unoszenia Motocykla po Uderzeniu w Bok Samochodu*; The Archives of Automotive Engineering (Archiwum Motoryzacji): Warszawa, Poland, 2012.

3. *Wypadki drogowe w Polsce w 2019 roku*; Technical Report; Biuro Ruchu Drogowego Zespół Profilaktyki i Analiz: Warszawa, Poland, 2020.
4. Stan bezpieczeństwa ruchu drogowego oraz działania realizowane w tym zakresie w 2019r. Available online: https://www.krbrd.gov.pl/baza-wiedzy/raporty-o-stanie-brd/ (accessed on 18 August 2021).
5. Dukalski, P.; Będkowski, B.; Parczewski, K.; Wnęk, H.; Urbaś, A.; Augustynek, K. Dynamics of the vehicle rear suspension system with electric motors mounted in wheels. *Eksploat. I Niezawodn. Maint. Reliab.* **2019**, *21*, 125–136. [CrossRef]
6. Nader, M.; Korzeb, J. *Przegląd Biomechanicznych Modeli do Oceny Oddziaływania Drgań na Organizm Ludzki, Materiały III Krajowego Sympozjum, Komputerowe Systemy Wspomagania Prac Inżynierskich w Przemyśle i Transporcie*; Politechnika Radomska: Zakopane, Poland, 1999.
7. Praca Zbiorowa, Tłumienie Drgań. 1997. Available online: https://tezeusz.pl/tlumienie-drgan-zbigniew-osinski?sortowanie=cena_malejaco (accessed on 18 August 2021).
8. International Organization for Standardization. ISO-2631—Mechanical vibration and shock. In *Evaluation of Human Exposure to Whole-Body Vibration*; ISO: New Dheli, India, 1997.
9. Shivakumara, B.S.; Sridhar, V. *Study of Vibration and Its Effect on Health of the Motorcycle Rider*; Online Journal of Health and Allied Sciences Peer Reviewed, Open Access; Free Online Journal Published Quarterly: Mangalore, South India, 2012; Volume 9. ISSN 0972-5997.
10. Di Massa, G.; Pagano, S.; Strano, S.; Terzo, M. A Mono-axial Wheel Force Transducer for the Study of the Shimmy Phenomenon. In Proceedings of the World Congress on Engineering, London, UK, 3–5 July 2013.
11. Grzegożek, W.; Weigel-Millert, K. Analiza możliwości wykorzystania badań stanowiskowych do oceny stabilności pojazdu jednośladowego. *Arch. Automot. Eng. Arch. Motoryz.* **2015**, *67*, 165–173.
12. Ślusarczyk, P. *Analiza Modelowa Stateczności Pojazdu Jednośladowego, Czasopismo Techniczne*; Wyd, P.K., Ed.; Zeszyt 7-M: Kraków, Poland, 2004; pp. 165–173.
13. Dębowski, A. *Analiza Możliwości Ograniczenia Drgań Skrętnych w Układzie Kierowniczym Motocykla, Rozprawa Doktorska*; Military University of Technology: Warszawa, Poland, 2019.
14. Available online: http://andrzejdebowski.wat.edu.pl/galeria.html (accessed on 14 April 2021).
15. Sharp, R.S. The stability and control of motorcycles. *J. Mech. Eng. Sci.* **1971**, *13*, 313–329. [CrossRef]
16. Sharp, R.S. The Influence of Frame Flexibility on the Lateral Stability of Motorcycles. *J. Mech. Eng. Sci.* **1974**, *16*, 117–120. [CrossRef]
17. Sharp, R.S.; Alstead, C.J. The influence of structural flexibilities on the straight-running stability of motorcycles. *Veh. Syst. Dyn.* **1980**, *9*, 327–357. [CrossRef]
18. Kane, T.R. *The Effect of Frame Flexibility on High Speed Weave of Motorcycles*; SAE Technical Paper 780306; SAE International: Warrendale, PA, USA, 1978.
19. Cossalter, V. *Motorcycle Dynamics*; LULU: Morrisville, NC, USA, 2006.
20. Cossalter, V.; Lot, R.; Massaro, M. *An Advanced Multibody Code for Handling and Stability Analysis of Motorcycles*; Springer Meccanica: Berlin/Heidelberg, Germany, 2011; Volume 46, pp. 943–958. [CrossRef]
21. De Falco, D.; Di Massa, G.; Pagano, S.; Strano, S. Wheel Force Transducer for Shimmy Investigation. In Proceedings of the World Congress on Engineering, London, UK, 1–3 July 2015.
22. Sharp, R.S.; Limbeer, D.J.N. On steering wobble oscillations of motorcycles. *J. Mech. Eng. Sci.* **2004**, *14*, 1449–1456. [CrossRef]
23. Żardecki, D.; Dębowski, A. Badania procedur obliczeniowych pod kątem zastosowań w symulacji drgań skrętnych w układzie kierowniczym motocykla z uwzględnieniem luzu i tarcia. *Arch. Automot. Eng.* **2014**, *64*, 79–95.
24. Pacejka, H. *Tyre and Vehicle Dynamics*; Elsevier: Waltham, MA, USA, 2012.
25. Sharp, R.S. *A Review of Motorcycle Steering Behaviour on Straight Line Stability Characteristics*; SAE Paper n 780303; SAE International: Warrendale, PA, USA, 1978. [CrossRef]
26. Sharp, R.S.; Watanabe, Y. Chatter vibrations of high-performance motorcycles, Vehicle System Dynamics. *Int. J. Veh. Mech. Mobil.* **2013**, *51*, 393–404.
27. Żardecki, D.; Dębowski, A. Metoda Analizy Drgań Skrętnych w Układzie Kierowniczym Motocykla na Płaszczyźnie Fazowej. *Arch. Automot. Eng.* **2017**, *72*, 137–154.
28. Praca Zbiorowa: Zarys Dynamiki i Automatyki Układów. 1991. Available online: https://bcpw.bg.pw.edu.pl/Content/767/PDF/01zdau_poczatek.pdf (accessed on 18 August 2021).
29. Adamiec-Wójcik, I.; Maczyński, A.; Wojciech, S. *Zastosowanie Metody Przekształceń Jednorodnych w Modelowaniu Dynamiki Urządzeń Offshore*; Wydawnictwo Komunikacji i Łączności: Warszawa, Poland, 2008.
30. Bhargava, A. Algorytmy. In *Ilustrowany przewodnik*; Wydawnictwo Helion: Gliwice, Poland, 2017.
31. Chang, Y.P.; El-Gindy, M.; Streit, A. Literature survey of transient dynamic response tyre models. *Int. J. Veh. Des.* **2004**, *34*, 354–386. [CrossRef]
32. Mann, V.; Dechwayukul, C.; Thongruang, W.; Srewaradachpisal, S.; Kaewpradit, P.; Kaew Apichai, W.; Bui, H.-T. Design and Fabrication a Lightweight Spring Made of Natural Rubber for a Motorcycle's Shock Absorber. *Int. J. Automot. Mech. Eng.* **2020**, *17*, 7758–7770. [CrossRef]

33. Stańczyk, T.L. *Metoda Modeli Częściowych Jako Podstawa Tworzenia Komputerowych Systemów Analizy Dynamiki Złożonych Układów Mechanicznych*; Rozprawa Habilitacyjna, Politechnika Świętokrzyska: Kielce, Poland, 1994.
34. Żardecki, D. Modelowanie i Badania Symulacyjne Dynamiki Ruchu Samochodu z Uwzględnieniem Luzu i Tarcia w Układzie Kierowniczym z Zastosowaniem Odwzorowań luz(...) i tar(...). In Proceedings of the Konferencja Między Teorią a Zastosowaniami: Matematyka w Działaniu, Poznań, Poland, 17–20 September 2014.

energies
MDPI

Article

Infrastructure Planning for Autonomous Electric Vehicles, Integrating Safety and Sustainability Aspects: A Multi-Criteria Analysis Approach

Konstantina Anastasiadou [1,*], Nikolaos Gavanas [2], Magda Pitsiava-Latinopoulou [1] and Evangelos Bekiaris [3]

1 School of Civil Engineering, Faculty of Engineering, Aristotle University of Thessaloniki, 54124 Thessaloniki, Greece; mpitsiav@civil.auth.gr
2 Department of Planning and Regional Development, School of Engineering, University of Thessaly, 38334 Volos, Greece; ngavanas@uth.gr
3 Hellenic Institute of Transport, Center for Research and Technology Hellas, 6th km Charilaou-Thermi Road, 57001 Thermi, Greece; abek@certh.gr
* Correspondence: kianasta@civil.auth.gr

Abstract: After the awareness-raising of recent years for coping with the global societal, economic and environmental challenges, the need for sustainable planning in the transport sector has become even more evident. Initiatives aiming at promoting sustainable and innovative mobility solutions, especially in urban areas where mobility needs are higher and transport problems are more intense, have been launched by different organizations around the world. In this context, autonomous electric vehicles are emerging as a promising solution; however, they are accompanied by new infrastructure requirements, along with safety concerns. Policymakers will be confronted with an array of choices, such as plug-in or wireless, dynamic or stationary charging and mixed flow with conventional vehicles or dedicated lanes, taking into account the uncertain impacts of innovation on safety and sustainability. Within this scope, these infrastructure alternatives are evaluated and prioritized, for the first time, in the present study, through the combined application of two hybrid multi-criteria analysis models, with the participation of experts. The analysis is based on a set of safety and sustainability criteria. Road safety and exposure to electromagnetic radiation emerge as the most important criteria, with the optimum solution—based on current data—consisting of plug-in charging and the circulation of autonomous electric vehicles in dedicated lanes.

Keywords: autonomous vehicles; electric vehicles; infrastructure planning; road safety; public health; sustainable mobility; transport policy; multi-criteria analysis

Citation: Anastasiadou, K.; Gavanas, N.; Pitsiava-Latinopoulou, M.; Bekiaris, E. Infrastructure Planning for Autonomous Electric Vehicles, Integrating Safety and Sustainability Aspects: A Multi-Criteria Analysis Approach. *Energies* **2021**, *14*, 5269. https://doi.org/10.3390/en14175269

Academic Editors: Guzek Marek, Rafał Jurecki and Wojciech Wach

Received: 18 July 2021
Accepted: 22 August 2021
Published: 25 August 2021

1. Introduction

The transport system plays a key role in socio-economic development by physically connecting the locations where various activities are conducted. Nonetheless, it is widely accepted that the transport system also produces external impacts on society and the economy, as well as on the environment. The intensification of climate change and air quality problems, and the depletion of natural resources during past decades, in combination with the ongoing increase in mobility demand and the challenges that persist regarding safety, accessibility and affordability, highlight even more the significance of decision-making concerning transport development within the framework of sustainability.

On the one hand, in Europe, mainly as a result of improvements in vehicle technology, roadway design and regulatory and policy interventions, the number of persons killed in road traffic accidents has continuously decreased since 2009 [1]. Nonetheless, the problem persists in densely populated European cities, where 38% of road fatalities take place, 70% of which involve vulnerable road users [2]. On the other hand, the transport sector is responsible for more than 25% of annual manmade greenhouse gas emissions (including international aviation but excluding maritime shipping) in the European Union (EU), with approximately 75% of this being due to road transport [3]. Apart from constituting a major contributor to climate change, transportation significantly contributes to air pollution (NO_x, particulate matter, etc.) and energy intensity, especially in urban areas, which are characterized by increased mobility needs, given the increasing concentrations of population and activity [3]. The adoption of innovative vehicle technologies with the potential to significantly reduce emissions and energy intensity is regarded as one of the most powerful means, being placed high in the EU sustainable mobility agenda [4,5].

In this framework, the evolution toward fully autonomous road vehicles is nowadays promoted as an innovative sustainable transport solution, mainly due to their ability to reduce road accidents that derive from human error on the part of the driver [6,7]. Nonetheless, the first evidence collected from the experimental application of autonomous driving in real-life conditions indicates that autonomous cars are not yet able to navigate safely in complex urban environments, while some scientists highlight the uncertain impact on road safety during the period when vehicles with different levels of automation will share the road network [8]. In terms of environmental sustainability, vehicles of automation level 5 [9], which are also connected and battery-electric (referred to as autonomous electric vehicles hereinafter), emerge as a promising sustainable mobility solution, as their efficiency in terms of air pollutants, greenhouse gas emissions and energy consumption is expected to be significantly higher than motorized vehicles of conventional technology under specific conditions [10].

In order to capitalize on the potential of autonomous electric vehicles to increase safety and decrease emissions and energy consumption in cities, decision-makers are charged with the task of developing not only the appropriate policy interventions for the vehicles but also the road network infrastructure. These interventions must aim at the maximization of positive and the minimization of negative impacts on aspects of socio-economic and environmental sustainability [11]. Given the novel character of the autonomous electric vehicles and the lack of previous wide-scale implementation of such technologies, decision-makers will have to step into uncharted territory in their effort to efficiently cope with dilemmas, such as choosing a plug-in or wireless charging infrastructure. Dynamic or stationary charging services? Road segments of mixed traffic flow, where autonomous vehicles share the roadway space with conventional vehicles or dedicated traffic lanes? Such issues need to be addressed in a holistic way, taking into account different sustainability criteria, primarily aiming at the enhancement of travel safety for all road users while ensuring the safety and public health of all residents, in terms of exposure to electromagnetic radiation.

With the purpose of contributing to the above task, the present research aims initially at an early-stage evaluation (due to the current lack of wide-scale implementation) of infrastructure alternatives for autonomous electric vehicles. These alternatives and the respective evaluation criteria are derived from the review of international literature, while a group of experts participated in the formulation of the final criteria list. The alternatives were finally evaluated through the combined application of two hybrid multi-criteria analyses (MCA) on the basis of a set of safety and sustainability criteria. The group of experts also provided the weighting of the criteria and the evaluation of the alternatives. The proposed combination of analyses was expected to lead to more reliable and valid estimations, compared to the application of each analysis separately.

The paper is structured as follows: Following the introductory part, the review of selected works for the evaluation of infrastructure requirements for autonomous electric vehicles is presented in Section 2. The methodological approach is described in Section 3. In Section 4, the combined analysis for the evaluation of infrastructure planning interventions for autonomous electric vehicles in urban areas is carried out, while Sections 5 and 6 refer to the interpretation of the results and to the study's relevant limitations, along with conclusions and future prospects, respectively.

2. Autonomous and Electric Vehicle Infrastructure Evaluation in Existing Literature

Several recent papers on the potential impacts of autonomous vehicles, especially in urban areas, can be found in the literature. The challenges for urban planners, derived from the potential impacts of autonomous vehicles on the location choices of people and businesses, the traffic and parking conditions, the requirements of pick-up and drop-off areas, as well as the need for integration to the energy and communication grids of smart cities, are highlighted in [12]. The expected changes in urban design and sustainability due to the advent of autonomous vehicles are discussed, among other aspects, in [13]. The expected impacts of autonomous and electric vehicles at social, economic, and environmental levels are included in [10], while the performance of these new technologies with regard to different sustainability criteria is also investigated. In [14], the expected impacts of autonomous vehicles on cities at various levels, such as road capacity and congestion, parking demand, land use, health, economy, the labor market, road infrastructure, and environment are studied. Scenario analysis to investigate potential implications for traffic, travel behavior and transport planning, on a time horizon up until 2030 and 2050 in the Netherlands, is conducted in [15]. The potential implications of automated vehicles at different levels, such as land use, transport infrastructure, energy consumption, air pollution and safety, are studied in [16]. The likely benefits and costs of autonomous vehicles, along with the potential impacts and implications for planning decisions, such as optimum road, parking and public transit supply, are investigated in [17]. However, no work has been published up to now concerning the systematic evaluation of road infrastructure alternatives for fully autonomous (automation level 5) and connected vehicles in the international research literature.

As regards the evaluation of infrastructure for electric vehicles, namely, the different charging systems for urban environments, a limited number of research papers can be allocated, mainly referring to electric buses, which are often based on assumptions due to the low maturity level of certain charging alternatives (such as dynamic inductive charging). In [18], for example, three different inductive charging systems (stationary, static and dynamic) for electric buses are compared, in terms of the initial investment cost (including those of the infrastructure and batteries), for two urban routes in Korea, stressing that such a comparison should be carried out again in the future when the cost data for these new technologies will be more reliable. The environmental and techno-economic feasibility of the dynamic charging of electric vehicles in urban and interurban road networks in the U.S.A. is assessed by [19], where a significant reduction in CO_2 emissions, partial reduction in VOC, CO, NO_x, PM10 and PM2.5, and an increase in SO_x, is estimated for dynamic charging. Moreover, the researchers of [20] present a new method for the performance evaluation of dynamic charging systems for two types of vehicles: a light-duty truck and a city car. A cost-benefit analysis (CBA) for three scenarios relating to plug-in charging, stationary inductive, and dynamic inductive charging of electric vehicles is conducted by [21]; however, they take into account only the acquisition cost of the vehicle, the charging cost, and the installation cost of the charging station. An estimate of the environmental and economic benefits stemming from dynamic charging systems use is included in [22]. Venugopal et al. conduct an economic evaluation of the sustainability of a future self-healing highway with integrated wireless electric vehicle charging (using renewable energy sources) in the Netherlands [23]. A CBA for the evaluation of the economic sustainability (in terms of initial investment and operation cost) of a dynamic charging system is conducted by [24], emphasizing the legal context for the use of this new technology. The study conducted by [25] models the benefits related to reducing the battery size, due to dynamic charging. The economic sustainability of a dynamic charging system is examined in [26], while a dynamic inductive charging system for electric buses is compared to a stationary inductive system, in terms of energy and pollutant emissions, by [27].

3. Materials and Methods

The use of multi-criteria analysis (MCA) is gaining more and more attention during recent years compared to conventional quantitative methods, such as the CBA, due to certain specific advantages [28,29]. There is no better or worse MCA method, but there is an appropriate MCA method for each problem [30].

In order to select the optimum solution in terms of infrastructure alternatives for autonomous electric vehicles for the promotion of sustainable urban mobility, a new decision-aiding methodology, as proposed in [10], is applied. The methodology mainly consists of the combined application of AHP-VIKOR and AHP-TOPSIS MCA models, with an additional condition that contributes to the "consolidation" of the optimum solution. Apart from the fact that AHP, VIKOR and TOPSIS methods can be easily understood and applied, the methodology in question is selected because it allows for the construction of a more solid background for optimum decision-making. The strengths of the engaged methods are capitalized, leading to more reliable and valid results, compared to a separate application of each method or model. At the same time, the in-depth comprehension of the problem and of the relevant parameters is ensured.

The main steps of the methodology are summarized below, while a more detailed description of the methodology can be found in [10].

Step 1: Scope—problem definition.

Step 2: Selection of the appropriate experts.

Step 3: Definition of the overall goal.

Step 4: Formulation of an initial list of alternatives, based on a literature review.

Step 5: Formulation of an initial list of evaluation criteria, based on a literature review.

Step 6: Formulation of the final list of alternatives by means of interviews with the chosen experts.

Step 7: Formulation of the final list of criteria applying, e.g., a modified Delphi.

Step 8: Hierarchy structure of the problem according to AHP [31].

Step 9: Design and distribution of questionnaires for pair-wise comparisons to the experts.

Step 10: Aggregation of the experts' judgments based on the "aggregation of individual judgments" method.

Step 11: Pair-wise comparison matrix derivation, as shown in Equation (1), based on AHP, both for the criteria and for the alternatives [31]:

$$A = \begin{bmatrix} a_{11} & a_{12} & \dots & a_{1n} \\ a_{21} & a_{22} & \dots & a_{2n} \\ \dots & \dots & \dots & \dots \\ a_{n1} & a_{n2} & \dots & a_{nn} \end{bmatrix}, \tag{1}$$

where $a_{ij} = w_i / w_j$ = element of matrix A (Equation (1)), representing the relative importance of the criterion or alternative (i) over the criterion or alternative (j) with regard to the overall goal achievement or with regard to each criterion respectively, with i = 1, 2, … , n and j = 1, 2, … , n ($a_{ij} = 1/a_{ji}$ and $a_{ii} = w_i/w_i = 1$).

w_i, w_j = weight coefficients of the criteria or of the alternatives (i) and (j) respectively.

Step 12: Normalization of the abovementioned pair-wise comparison matrices (each value is divided by the sum of values in the same column of the matrix).

Step 13: Extraction of criteria weights' (priority vector W) and alternatives' performance (priority vectors W) with regard to each criterion, based on AHP, applying Equation (2) [31]:

$$(A - \lambda_{max}) \cdot W = 0, \tag{2}$$

where A = pair-wise comparison matrix, $W = (w_1, w_2, \dots, w_n)^T$ = priority vector for each hierarchy level, λ_{max} = principal eigenvalue of matrix A.

Step 14: Calculation and control of the AHP consistency ratio (CR) using Equation (3) [31]:

$$CR = CI/RI < 0.10, \tag{3}$$

where CI = consistency index, calculated by Equation (4) [31]:

$$CI = \frac{\lambda_{max} - n}{n - 1}, \quad (4)$$

and RI = random consistency index (Table 1).

Table 1. Random consistency index values for n elements.

n	1	2	3	4	5	6	7	8	9	10
RI	0.00	0.00	0.58	0.90	1.12	1.24	1.32	1.41	1.45	1.49

Step 15: Decision matrix formulation, as shown in Equation (5) for n criteria and m alternatives, based on the extracted priority vectors of the alternatives with regard to each criterion, for the application of TOPSIS and VIKOR, aiming at the overall ranking of the alternatives:

$$D = \begin{matrix} \\ A_1 \\ A_2 \\ \ldots \\ A_m \end{matrix} \begin{matrix} \begin{matrix} C_1 & C_2 & \ldots & C_n \end{matrix} \\ \begin{bmatrix} x_{11} & x_{12} & \ldots & x_{1n} \\ x_{21} & x_{22} & \ldots & x_{2n} \\ \ldots & \ldots & \ldots & \ldots \\ x_{m1} & x_{m2} & \ldots & x_{mn} \end{bmatrix} \end{matrix}, \quad (5)$$

where x_{ij} = performance of the alternative A_i with regard to the criterion C_j, where i = 1, 2, …, m and j = 1, 2, …, n.

Step 16: Application of VIKOR [32] for the overall ranking of the alternatives:

Calculation of the best (f_j^*) and the worst (f_j^-) performance values for each criterion function, for the decision matrix formulated in step 15.

For benefit functions (where a higher value is better) [32]:

$$f^*_j = max_i(x_{ij}) \text{ and } f^-_j = min_i(x_{ij}), \quad (6)$$

and for cost functions (where a lower value is better) [32]:

$$f^*_j = min_i(x_{ij}) \text{ and } f^-_j = max_i(x_{ij}), \quad (7)$$

where i = 1, 2, …, m and j = 1, 2, …, n.

The calculation of group utility (S_i) and individual regret (R_i) values for each alternative A_i (i = 1, 2, …, m) are as follows [32]:

$$S_i = \sum_{j=1}^{n} (f_j^* - x_{ij}/(f_j^* - f_j^-), \quad (8)$$

$$Ri = max_j\left[w_j \cdot \left((f_j^* - x_{ij}/\left(f_j^* - f_j^-\right)\right)\right], \quad (9)$$

where w_j (j = 1, 2, …, n) gives the criteria weights.

Calculation of Q_i values for each alternative A_i (i = 1, 2, …, m) for v = 0.5 [32]:

$$Q_i = v \cdot \frac{S_i - S^*}{S^- - S^*} + (1 - v) \cdot \frac{R_i - R^*}{R^- - R^*}, \quad (10)$$

$$\text{where } S^* = min_jS_i, S^- = max_jS_i, R^* = min_jR_i \text{ and } R^- = max_jR_i. \quad (11)$$

Alternatives ranking based on S_i, R_i and Q_i values of each alternative (minimum value → best alternative and maximum value → worst alternative) and control of the conditions of acceptable advantage and of acceptable stability as defined in [32] for the reveal of the optimum solution.

Step 17: Application of TOPSIS [33] for the overall ranking of the alternatives:

If r_{ij} the elements of the decision matrix of step 15 (already normalized), the elements v_{ij} of the weighted normalized matrix are calculated as follows [33]:

$$v_{ij} = w_j \cdot r_{ij}, \quad (12)$$

where w_j is the weight of the criterion C_j (where j = 1,2, … , n) and $\Sigma w_j = 1$.

Calculation of the ideal (A^+) and negative-ideal (A^-) solution [33] is as follows:

$$A^+ = \{(\max_i v_{ij} \mid j \in J), \min_i v_{ij} \mid j \in J') \mid i = 1,2, \ldots ,m\} = \{v_1^+, v_2^+, \ldots , v_j^+, \ldots , v_n^+\}, \quad (13)$$

$$A^- = \{(\min_i v_{ij} \mid j \in J), \max_i v_{ij} \mid j \in J') \mid i = 1,2, \ldots ,m\} = \{v_1^-, v_2^-, \ldots , v_j^-, \ldots , v_n^-\}, \quad (14)$$

where J = {j = 1,2, … n and j refers to benefit criteria} and J′ = {j = 1,2, … n and j refers to cost criteria}.

Calculation of the "separation measure" (distance of each alternative from the ideal and the negative-ideal solution), applying the "Euclidian distance method" is as follows.

Euclidian distance of the alternative A_i from the ideal solution (S_i^+) [33]:

$$S_i^+ = \sqrt{\sum_{i=1}^{m} (v_{ij} - v_i^+)^2}, \text{ where } i = 1, 2, \ldots, m, \quad (15)$$

Euclidian distance of the alternative A_i from the negative-ideal solution (S_i^-) [33]:

$$S_i^- = \sqrt{\sum_{i=1}^{m} (v_{ij} - v_i^-)^2}, \text{ where } i = 1, 2, \ldots, m, \quad (16)$$

Calculation of the relative closeness c_i^+ to the ideal solution [33]:

$$\begin{gathered} c_i^+ = \frac{S_i^-}{(S_i^+ + S_i^-)}, \text{ where } 0 \le c_i^+ \le 1 \text{ for } i = 1, 2, \ldots, m \\ c_i^+ = 1 \text{ if } A_i = A^+ \text{ and } c_i^+ = 0 \text{ if} A_i = A^- \end{gathered} \quad (17)$$

Alternative ranking based on c_i^+ values (maximum c_i^+ value → best alternative).

Step 18: Control of the convergence of the derived results of the two models (AHP-VIKOR and AHP-TOPSIS):

- If both models yield the same optimum solution, the process is complete.
- In case of divergence between the two models, control of the additional 1/2 m condition, as described in [10], may be required.

4. Selection of Infrastructure Alternatives for Autonomous Electric Vehicles in Urban Areas

The methodology described in Section 3 was implemented to select the optimum solution regarding infrastructure alternatives for autonomous electric vehicles in urban areas, based on their evaluation according to safety and sustainability criteria. The implementation steps and the main output are presented below.

4.1. Definition of the Decision Problem

Autonomous vehicles are widely promoted as a new solution for the improvement of traffic and mobility conditions, with a significant reduction of accidents due to human error, while electromobility is considered as a prerequisite for the transition to low-carbon mobility, with the benefits expected to be much higher in the case of energy production by renewable and low-carbon resources [10,12,34–36]. Obviously, automation, connectivity, and electrification, apart from private cars, also concern means of public transport [37,38]. In the next few decades, it is expected that autonomous electric vehicles will gradually replace a significant percentage of conventional motorized road vehicles. This evolution will go along with the need to adjust current infrastructures and develop new ones in order to service autonomous and electric vehicles. In cities, policymakers will be called to choose the most appropriate infrastructure alternative, taking into account the overarching goal of promoting sustainable urban mobility. In the absence of relevant previous work, a preliminary evaluation of infrastructure alternatives for autonomous electric vehicles, in the context of sustainable urban mobility, is implemented to support decision-making by policymakers.

The following assumptions are made:

- The study area is a typical large-scale urban area.
- Infrastructure alternatives are mainly intended for passenger cars.
- Autonomous electric vehicles co-exist with conventional ones.
- The charging infrastructure is constructed and managed by the public sector.
- Plug-in (wired) charging is approximately 5 times faster than stationary wireless (contactless) charging and 10 times faster than charging via existing plug-in stations (based on the literature review mentioned in Sections 4.4 and 4.5 below, related to electric vehicles).
- There is no fare to enter the dynamic charging lane dedicated to autonomous electric vehicles, as well as to use stationary wireless charging infrastructure, within the framework of promoting these new technologies.
- The criterion of "traffic congestion", apart from the impact on user's time value, is also related to air pollutants and greenhouse gas emissions, as well as to natural resource consumption.

4.2. Selection of Experts

The selection of experts, both in terms of quality and in terms of quantity, is of extremely high importance for the application of such a methodology. The number of experts should be large enough to capture all the different aspects, but, at the same time, reasonable. The recommended number of experts participating in a group pair-comparison procedure is 8–15 [39,40]. In order to ensure that all the participants can express themselves independently and without fear of "exposure", as well as for equity reasons concerning their treatment, pair-wise comparisons are usually executed anonymously [39–41].

As regards their number, 15 experts participated in the present analysis, by means of interviews, for the definition of the evaluation criteria and the alternatives, on the basis of the initial list (based on a comprehensive literature review). As for the next stage of the criteria and alternative pair-wise comparisons, 12 experts participated in the process. Regarding their level of expertise, the criteria for their selection were the relevance of their studies to the evaluation subject, along with their years of experience and expertise in the field of transport, especially in relation to autonomous and electric vehicles.

4.3. Overall Goal

The overall goal is meeting the principles of sustainable urban mobility, which refers to the satisfaction of urban mobility needs, at the least possible economic, social and environmental cost [42]. The optimum compromise between social, environmental, and economic criteria is therefore sought by selecting the most appropriate road infrastructure alternatives for autonomous electric vehicles.

4.4. Alternatives

Concerning roadway infrastructure planning for autonomous (automation level 5) and connected vehicles, the examined alternatives comprise either mixed traffic roads, shared by vehicles of different automation levels or separate lanes for the circulation of autonomous vehicles. Given the absence of research related to the evaluation of such alternatives, due to the low maturity level of these technologies, the relevant alternatives, as well as the evaluation criteria, are mainly defined on the basis of literature related to the expected impacts of autonomous vehicles.

Concerning the types of electric vehicle charging systems, battery charging can be either wired (plug-in) or wireless. Wireless charging may be contactless or not, and can be realized either when the vehicle is parked (stationary) or when it is in motion (dynamic) [18,24,26,43,44].

The main advantages and disadvantages of wireless over wired (plug-in) electric vehicle charging are presented below.

Advantages of wireless over wired (plug-in) vehicle charging:

- A simple process, comfortable in bad weather conditions (e.g., rain) [45,46].
- Given that all the infrastructure parts are underground, the system is protected against stealing or vandalism [45].
- Given the underground installation, there are no visual intrusion and landscape "disruption" problems [45].
- Dynamic wireless charging is time-saving, as the vehicle can be charged when moving [24,43,47,48].
- The battery weight can be significantly reduced in the case of dynamic charging, also leading to vehicle weight reduction and, thus, a reduction in energy consumption and pollutant emissions (given the optimization of these systems in the future) [19,49–52].

Disadvantages of wireless over wired (plug-in) vehicle charging:

- Infrastructure requirements (both economic and technical) are much higher than in the case of wired charging [43,47,53,54].
- Safety questions are raised in the case of wireless charging, especially dynamic (when people will be in the vehicle during charging), as negative health impacts due to electromagnetic radiation may be caused, while there are safety concerns in the case of passengers with pacemakers, or animals that may be present between the charging infrastructure devices and vehicle equipment [49,55].
- Due to the high demand for wireless charging, the energy distribution network might not be sufficient, so a later need for upgrades is highly possible [47,56].
- Wired charging is characterized by less energy loss [45,46].

The initial list of infrastructure alternatives was derived from the literature review included in Section 2, as well as [57–62]. The final list of infrastructure alternatives was based on the review in combination with interviews with the experts. The following infrastructure alternatives were selected for evaluation:

- Mixed flow of conventional and autonomous electric vehicles (and charging at existing plug-in charging stations)—encoded as M.F.
- Lanes dedicated to autonomous electric vehicles, with plug-in charging stations beside the roadway, along the route—encoded as P.C.
- Lanes dedicated to autonomous electric vehicles, with stationary wireless charging stations beside the roadway—encoded as S.C.
- Lanes dedicated to autonomous electric vehicles, with dynamic wireless charging infrastructure along the route—encoded as D.C.

4.5. Evaluation Criteria

The initial list of evaluation criteria includes social, environmental, and economic criteria, based on the literature review included in Section 2 and in Section 4.4, as well as [63–67] (as already mentioned, in the case of autonomous vehicles, the literature is mainly related to the expected impacts, given the absence of previous work related to infrastructure evaluation). It is worth highlighting that the number of criteria should be sufficient to cover all aspects of the evaluation but also be reasonable, taking into account that the human mind may efficiently compare up to 7 $\pm$ 2 elements [68] in pairs. For this reason, a modified Delphi was then applied, as described in [10]. Briefly, the experts were asked to select (on the basis of the "7 $\pm$ 2 principle") the 7 most important criteria (in their opinion) from the list, as well as to add any other ones that might not have been included in the list, but would be among the 7 most important criteria for the evaluation of the 4 alternatives. The usual "consensus threshold" in traditional Delphi ranges from 50% to 97% [69], so a threshold of 75% was chosen for this modified Delphi application. Only 6 criteria were selected by at least 75% of the participants, while no other criterion was added. Thus, the final criteria list was the following:

- Construction, operation and maintenance infrastructure cost (much higher in case of dynamic wireless charging)—encoded as I.C.
- Impact on public health due to electromagnetic radiation (in case of dynamic wireless charging)—encoded as P.H.
- Road safety (e.g., high risk in case of mixed flow)—encoded as R.S.
- Traffic congestion (e.g., reduction in case of lanes dedicated to autonomous electric vehicles)—encoded as T.C.
- Charging time (e.g., dynamic wireless charging → less time-consuming)—encoded as C.T.
- Charging system energy efficiency (wireless—especially dynamic—charging → more energy-consuming)—encoded as E.E.

Other criteria that were included in the initial list, but were not selected by at least 75% of the experts, referred to equity concerns (e.g., lanes dedicated to autonomous electric vehicles, to the detriment of conventional ones), comfort in general and in adverse weather conditions, the adequacy of energy distribution in the case of increased demand and energy loss, visual intrusion, etc. It should be noted that the perception of these experts may change in a future iteration of the process when the maturity level of the engaged technologies will be higher.

4.6. Hierarchy Structure of the Problem, Based on AHP

The decision problem hierarchy is formed as shown in Figure 1. The alternatives are at the base, the criteria at the upper level, while the overall goal is at the top.

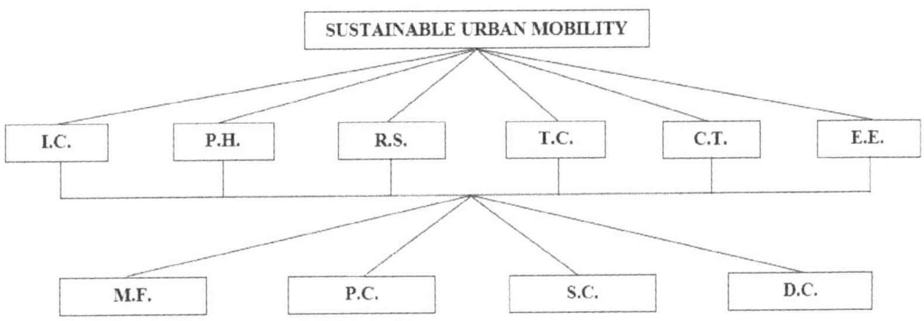

Figure 1. Hierarchy structure of the decision problem.

4.7. Criteria Weighting Based on AHP

After the formulation of the decision problem hierarchy, 15 criteria pair-wise comparisons (an indicative part is shown in Table 2), were executed by the group of experts, using the 9-level linear Saaty scale (Table 3).

Table 2. Indicative part of criteria pair-wise comparisons.

The Criterion on the Left is More Important Than the One on the Right (Select the Intensity of Relative Importance)								Equivalent Importance of the Two Criteria	The Criterion on the Right is More Important than the One on the Left (Select the Intensity of Relative Importance)								
Construction, operation and maintenance cost	9	8	7	6	5	4	3	2	1	2	3	4	5	6	7	8	9	Road safety

Table 3. Relative importance scale for the criteria.

Intensity of Importance	Definition
1	Equivalent importance of the two criteria
3	Moderate importance of the one over the other
5	Strong importance of the one over the other
7	Very strong importance of the one over the other
9	Extreme importance of the one over the other
2, 4, 6, 8	Intermediate values between the aforementioned ones

The experts' answers were aggregated by implementing the "aggregation of individual judgments" method, using the geometric mean-GEOM. MEAN (for a set of n numbers $x_1, x_2, \ldots, x_n$, geometric mean equals $\sqrt[n]{x_1 \times x_2 \ldots \times x_n}$) of the value attributed to each criterion. The AHP consistency ratio (CR: (Equation (3)) is used for the consistency control of the answers. The input data, based on the experts' answers and the geometric mean for each case, can be found in Table 4. When the criterion on the left is selected (Table 2), the value is used for the analysis exactly as it is. When the criterion on the right is selected, the reverse value of the selected one in Table 2 is used for the analysis.

Table 4. Expert judgments and geometric mean for the criteria pair-wise comparisons.

CRITERIA/EXPERTS	E1	E2	E3	E4	E5	E6	E7	E8	E9	E10	E11	E12	GEOM. MEAN
I.C. vs. P.H.	1/8	1/9	1/9	1/8	1/8	1/7	1/8	1/8	1/5	1/8	1/9	1/8	0.1276
I.C. vs. R.S.	1/8	1/8	1/9	1/6	1/7	1/5	1/8	1/7	1/5	1/6	1/7	1/6	0.1487
I.C. vs. T.C.	1/2	1/4	1	1/2	1/3	1	1	1/2	1	2	1	1/3	0.6609
I.C. vs. C.T.	1	2	2	1	1	3	2	1	2	3	1	2	1.6031
I.C. vs. E.E.	1/2	3	2	1	3	1/3	2	2	1	1/3	1	1/3	1.0243
P.H. vs. R.S.	1	1	1	2	1	2	1	1	1	2	2	3	1.3807
P.H. vs. T.C.	7	6	7	6	5	7	5	6	6	9	9	5	6.3744
P.H. vs. C.T.	9	7	8	8	7	9	9	9	9	9	9	9	8.4630
P.H. vs. E.E.	7	8	8	8	9	6	9	9	8	5	9	6	7.5418
R.S. vs. T.C.	7	6	7	4	7	9	5	7	6	7	7	4	6.1709
R.S. vs. C.T.	9	7	8	6	8	9	9	8	9	9	7	9	8.1019
R.S. vs. E.E.	7	8	7	6	9	6	9	8	9	5	7	5	7.0232
T.C. vs. C.T.	2	3	2	2	1	3	3	1	2	1/2	1	3	1.7151
T.C. vs. E.E.	1	6	2	2	5	1/3	3	1	1	1/4	1	1	1.3277
C.T. vs. E.E.	1/2	2	1	1	3	1/3	1	1	1/2	1/3	1	1/3	0.7859

The geometric mean values of Table 4 serve as input data for the AHP comparison matrix (Table 5), based on Equation (1), and for the normalized comparison matrix of Table 6, where the calculated criteria priority vector (criteria weights) and the respective consistency control are shown.

Table 5. Criteria pair-wise comparison matrix.

	I.C.	P.H.	R.S.	T.C.	C.T.	E.E.
I.C.	1	0.1276	0.1487	0.6609	1.6031	1.0243
P.H.	7.8349	1	1.3807	6.3744	8.4630	7.5418
R.S.	6.7236	0.7243	1	6.1709	8.1019	7.0232
T.C.	1.5131	0.1569	0.1620	1	1.7151	1.3277
C.T.	0.6238	0.1182	0.1234	0.5830	1	0.7859
E.E.	0.9763	0.1326	0.1424	0.7532	1.2723	1

Table 6. Normalized criteria pair-wise comparison matrix, priority vector (W), and consistency control.

	I.C.	P.H.	R.S.	T.C.	C.T.	E.E.	W
I.C.	0.0536	0.0565	0.0503	0.0425	0.0724	0.0548	0.0550
P.H.	0.4196	0.4426	0.4669	0.4101	0.3820	0.4032	0.4207
R.S.	0.3601	0.3205	0.3381	0.3970	0.3657	0.3755	0.3595
T.C.	0.0810	0.0694	0.0548	0.0643	0.0774	0.0710	0.0697
C.T.	0.0334	0.0523	0.0417	0.0375	0.0451	0.0420	0.0420
E.E.	0.0523	0.0587	0.0481	0.0485	0.0574	0.0535	0.0531
	λ_{max} = 6.0472 CI = 0.0094 CR = 0.0075 < 0.10						

4.8. Evaluation of Alternatives with Regard to Each Criterion, Based on AHP

The evaluation of the infrastructure alternatives concerning their performance in terms of each criterion is as follows. This is realized through pair-wise comparisons executed by the group of experts. In order to proceed with the pair-wise comparisons of infrastructure alternatives, the same methodological steps as were followed for the criteria pair-wise comparisons were implemented.

Following exactly the same process with the criteria, the alternatives are compared in pairs, using Saaty's 9-level scale, but this time, in terms of preference (Table 7). An indicative part of these pair-wise comparisons (6 pair-wise comparisons for each of the 6 criteria) is shown in Table 8. It should be noted that, in certain cases, in MCA, it might be meaningful to compare two alternatives in terms of a particular criterion, but not in terms of another. When a comparison between two alternatives in terms of a criterion is not meaningful, due to incompatibility or indifference, they are treated as if they were "equal", concerning their performance in terms of this criterion (attributed value: 1) [70].

Table 7. Relative preference scale for the alternatives.

Intensity of Preference	Definition
1	Indifference of preference
3	Moderate preference relation
5	Strong preference relation
7	Very strong preference relation
9	Absolute preference relation
2, 4, 6, 8	Intermediate values between the two adjacent judgments

Table 8. Indicative part of alternatives' pair-wise comparisons.

<table>
<tr><th colspan="19">With Regard to the Criterion "Construction, Operation and Maintenance Infrastructure Cost"</th></tr>
<tr><th></th><th colspan="8">The Alternative on the Left is Preferable to the One on the Right (Select the Degree of Relative Preference)</th><th>Indifference of Preference</th><th colspan="8">The Alternative on the Right is Preferable to the One on the Left (Select the Degree of Relative Preference)</th><th></th></tr>
<tr><td>Lanes dedicated to autonomous electric vehicles and plug-in charging</td><td>9</td><td>8</td><td>7</td><td>6</td><td>5</td><td>4</td><td>3</td><td>2</td><td>1</td><td>2</td><td>3</td><td>4</td><td>5</td><td>6</td><td>7</td><td>8</td><td>9</td><td>Lanes dedicated to autonomous electric vehicles and dynamic charging</td></tr>
</table>

The input data on the basis of the experts' answers for the infrastructure alternatives pair-wise comparison, as well as the geometric mean for each criterion, are shown in Table 9. The normalized infrastructure alternatives pair-wise comparison matrices, the priority vectors, and the consistency control for each criterion, can be found in Table 10.

Table 9. Expert judgments and geometric mean for the alternatives pair-wise comparison, with regard to each criterion.

Criterion I.C.: ALTERNATIVES/EXPERTS	E1	E2	E3	E4	E5	E6	E7	E8	E9	E10	E11	E12	GEOM. MEAN
M.F. vs. P.C.	6	7	5	4	7	6	5	7	7	7	4	6	5.8010
M.F. vs. S.C.	4	5	2	2	5	3	3	4	6	3	2	3	3.2808
M.F. vs. D.C.	9	8	9	4	7	9	9	9	9	9	9	8	8.0774
P.C. vs. S.C.	1/5	1/5	1/4	1/3	1/5	1/3	1/3	1/3	1/2	1/2	1/2	1/2	0.3279
P.C. vs. D.C.	5	5	6	3	5	3	5	3	3	3	4	3	3.8598
S.C. vs. D.C.	7	7	4	6	7	7	6	6	7	5	6	5	6.0003
Criterion P.H.: ALTERNATIVES/EXPERTS	**E1**	**E2**	**E3**	**E4**	**E5**	**E6**	**E7**	**E8**	**E9**	**E10**	**E11**	**E12**	**GEOM. MEAN**
M.F. vs. P.C.	1	1	1	1	1	1	1	1	1	1	1	1	1.0000
M.F. vs. S.C.	3	5	4	6	1	5	6	4	6	6	7	5	4.4122
M.F. vs. D.C.	9	9	9	9	1	7	9	8	8	9	8	7	6.9782
P.C. vs. S.C.	4	5	4	4	1	5	6	4	6	6	7	5	4.3691
P.C. vs. D.C.	9	9	9	8	1	7	9	7	8	9	4	7	6.4500
S.C. vs. D.C.	4	3	5	2	1	2	2	2	3	3	2	2	2.3890
Criterion R.S.: ALTERNATIVES/EXPERTS	**E1**	**E2**	**E3**	**E4**	**E5**	**E6**	**E7**	**E8**	**E9**	**E10**	**E11**	**E12**	**GEOM. MEAN**
M.F. vs. P.C.	1/7	1/9	1/4	1/3	1/3	1/4	1/4	1/5	1/3	1/3	1/5	1/7	0.2257
M.F. vs. S.C.	1/7	1/9	1/4	1/3	1/5	1/4	1/4	1/5	1/3	1/3	1/5	1/7	0.2163
M.F. vs. D.C.	1/7	1/9	1/4	1/3	1/7	1/4	1/4	1/5	1/3	1/3	1/5	1/7	0.2103
P.C. vs. S.C.	1	1	1	1	1	1	1	1	1	1	1	1	1.0000
P.C. vs. D.C.	1	1	1	1	1/3	1	1	1	1	1	1	1	0.9125
S.C. vs. D.C.	1	1	1	1	1/3	1	1	1	1	1	1	1	0.9125
Criterion T.C.: ALTERNATIVES/EXPERTS	**E1**	**E2**	**E3**	**E4**	**E5**	**E6**	**E7**	**E8**	**E9**	**E10**	**E11**	**E12**	**GEOM. MEAN**
M.F. vs. P.C.	1/5	1/7	1/4	1/3	1/7	1/4	1/5	1/5	1/3	1/4	1/5	1/4	0.2218
M.F. vs. S.C.	1/5	1/7	1/4	1/3	1/5	1/4	1/5	1/5	1/3	1/4	1/5	1/4	0.2281
M.F. vs. D.C.	1/5	1/8	1/4	1/3	1/3	1/4	1/5	1/5	1/4	1/4	1/5	1/4	0.2298
P.C. vs. S.C.	1	1	1	1	1	1	1	1	1	1	1	1	1.0000
P.C. vs. D.C.	1	1/3	1	1	1/3	1	1	1	1/2	1	1	1	0.7859
S.C. vs. D.C.	1	1/3	1	1	1/3	1	1	1	1/2	1	1	1	0.7859
Criterion C.T.: ALTERNATIVES/EXPERTS	**E1**	**E2**	**E3**	**E4**	**E5**	**E6**	**E7**	**E8**	**E9**	**E10**	**E11**	**E12**	**GEOM. MEAN**
M.F. vs. P.C.	1/4	1/5	1/5	1/6	1/5	1/6	1/6	1/4	1/5	1/5	1/5	1/4	0.2021
M.F. vs. S.C.	1/2	1/3	1/3	1/3	1/3	1/3	1/4	1/2	1/2	1/2	1/3	1/2	0.3853
M.F. vs. D.C.	1/9	1/9	1/8	1/9	1/9	1/9	1/8	1/9	1/9	1/9	1/9	1/8	0.1144
P.C. vs. S.C.	4	5	3	4	5	2	2	2	3	2	2	2	2.7982
P.C. vs. D.C.	1/6	1/5	1/4	1/4	1/7	1/2	1/2	1/3	1/4	1/3	1/3	1/4	0.2731
S.C. vs. D.C.	1/8	1/9	1/8	1/9	1/9	1/4	1/5	1/6	1/8	1/9	1/5	1/6	0.1445
Criterion E.E.: ALTERNATIVES/EXPERTS	**E1**	**E2**	**E3**	**E4**	**E5**	**E6**	**E7**	**E8**	**E9**	**E10**	**E11**	**E12**	**GEOM. MEAN**
M.F. vs. P.C.	1	1	1	1	1	1	1	1	1	1	1	1	1.0000
M.F. vs. S.C.	1	1	1	1	1	1	1	1	1	1	1	1	1.0000
M.F. vs. D.C.	1	1	1	1	1	1	1	1	1	1	1	1	1.0000
P.C. vs. S.C.	3	3	2	3	3	3	2	3	2	2	2	2	2.4495
P.C. vs. D.C.	5	5	4	5	5	5	4	4	4	5	3	4	4.3662
S.C. vs. D.C.	3	2	1	2	3	2	2	1	2	2	2	2	1.9064

Table 10. Normalized alternatives' pair-wise comparison matrices, priority vectors (W) and consistency control for each criterion.

Criterion: I.C.	M.F.	P.C.	S.C.	D.C.	W
M.F.	0.6246	0.5738	0.6870	0.4265	0.5780
P.C.	0.1077	0.0989	0.0687	0.2038	0.1198
S.C.	0.1904	0.3017	0.2094	0.3168	0.2546
D.C.	0.0773	0.0256	0.0349	0.0528	0.0477
		λ_{max} = 4.2546 CI = 0.0849 CR = 0.0962 < 0.10			
Criterion: P.H.	**M.F.**	**P.C.**	**S.C.**	**D.C.**	**W**
M.F.	0.4220	0.4195	0.4326	0.4149	0.4222
P.C.	0.4220	0.4195	0.4283	0.3835	0.4133
S.C.	0.0956	0.0960	0.0980	0.1421	0.1079
D.C.	0.0605	0.0650	0.0410	0.0595	0.0565
		λ_{max} = 4.0371 CI = 0.0124 CR = 0.0140 < 0.10			
Criterion: R.S.	**M.F.**	**P.C.**	**S.C.**	**D.C.**	**W**
M.F.	0.0675	0.0680	0.0653	0.0693	0.0675
P.C.	0.2992	0.3011	0.3019	0.3006	0.3007
S.C.	0.3122	0.3011	0.3019	0.3006	0.3040
D.C.	0.3211	0.3299	0.3309	0.3294	0.3278
		λ_{max} = 4.0005 CI = 0.0002 CR = 0.0002 < 0.10			
Criterion: T.C.	**M.F.**	**P.C.**	**S.C.**	**D.C.**	**W**
M.F.	0.0702	0.0635	0.0652	0.0820	0.0702
P.C.	0.3165	0.2862	0.2857	0.2805	0.2922
S.C.	0.3078	0.2862	0.2857	0.2805	0.2900
D.C.	0.3055	0.3641	0.3635	0.3569	0.3475
		λ_{max} = 4.0102 CI = 0.0034 CR = 0.0038 < 0.10			
Criterion: C.T.	**M.F.**	**P.C.**	**S.C.**	**D.C.**	**W**
M.F.	0.0579	0.0387	0.0347	0.0747	0.0515
P.C.	0.2864	0.1915	0.2520	0.1783	0.2270
S.C.	0.1502	0.0685	0.0900	0.0943	0.1007
D.C.	0.5056	0.7013	0.6233	0.6527	0.6207
		λ_{max} = 4.1449 CI = 0.0483 CR = 0.0548 < 0.10			
Criterion: E.E.	**M.F.**	**P.C.**	**S.C.**	**D.C.**	**W**
M.F.	0.2500	0.3792	0.2010	0.1209	0.2378
P.C.	0.2500	0.3792	0.4925	0.5278	0.4124
S.C.	0.2500	0.1548	0.2010	0.2304	0.2091
D.C.	0.2500	0.0868	0.1055	0.1209	0.1408
		λ_{max} = 4.2433 CI = 0.0811 CR = 0.0919 < 0.10			

4.9. Decision Matrix for the Application of VIKOR and TOPSIS for the Overall Ranking of the Alternatives

The priority vectors of the alternatives (Table 10), calculated with regard to each criterion with AHP, and showing the ranking of the alternatives for each criterion, serve as input data for the decision matrix, shown in Table 11. This is based on Equation (5), used for the application of VIKOR and TOPSIS, so that the final ranking of the alternatives can be derived. It should be noted that all the criteria are considered as benefit criteria (benefit functions), as the experts were asked which alternative is preferable to the other in the relevant questionnaires.

Table 11. Decision matrix for the application of VIKOR and TOPSIS.

	I.C.	P.H.	R.S.	T.C.	C.T.	E.E.
M.F.	0.5780	0.4222	0.0675	0.0702	0.0515	0.2378
P.C.	0.1198	0.4133	0.3007	0.2922	0.2270	0.4124
S.C.	0.2546	0.1079	0.3040	0.2900	0.1007	0.2091
D.C.	0.0477	0.0565	0.3278	0.3475	0.6207	0.1408

4.10. Application of VIKOR for the Overall Ranking of the Alternatives

Concerning VIKOR application, the f^*_j and f^-_j values (Equations (6) and (7)) are calculated on the basis of Table 11:

$f^*_j = \max_i(x_{ij}) = \{0.5780\ 0.4222\ 0.3278\ 0.3475\ 0.6207\ 0.4124\}$

$f^-_j = \min_i(x_{ij}) = \{0.0477\ 0.0565\ 0.0675\ 0.0702\ 0.0515\ 0.1408\}$

S_i, R_i and Q_i values for VIKOR (calculated according to Equations (8)–(11), for $v = 0.5$, adopting the criteria weights (priority vector W) of Table 6, can be found in Table 12. The corresponding ranking of the alternatives (minimum value → best alternative) can be also found in Table 12.

Table 12. S_i, R_i and Q_i values and relevant alternatives' ranking, based on the AHP-VIKOR model.

	M.F.	P.C.	S.C.	D.C.
S_i	0.5053	0.1382	0.5206	0.5288
R_i	0.3595	0.0475	0.3616	0.4207
Q_i	0.8879	0.0000	0.9103	1.0000
RankS_i	2	1	3	4
RankR_i	2	1	3	4
RankQ_i	2	1	3	4

According to Table 12, the alternative P.C. (lanes dedicated to autonomous electric vehicles with plug-in charging stations beside the road, along the route) has the minimum Q_i, so it is first in rank. After checking the satisfaction of the two conditions [32] of the acceptable advantage $(0.8879 - 0.000 = 0.8879 > 1/(m-1) = 1/(4-1) = 1/3 = 0.3333)$ and of the acceptable stability (the alternative P.C. is also first in rank by S_i and R_i), the alternative P.C. constitutes the optimum solution of new vehicle technologies in urban areas, according to the AHP-VIKOR model.

4.11. Application of TOPSIS for the Overall Ranking of the Alternatives

The weighted normalized decision matrix for the application of TOPSIS, shown in Table 13, is calculated on the basis of Table 11, according to Equation (12) and using the criteria weights (priority vector W) of Table 6.

Table 13. Weighted normalized decision matrix for TOPSIS application.

	I.C.	P.H.	R.S.	T.C.	C.T.	E.E.
M.F.	0.0318	0.1777	0.0243	0.0049	0.0022	0.0126
P.C.	0.0066	0.1739	0.1081	0.0204	0.0095	0.0219
S.C.	0.0140	0.0454	0.1093	0.0202	0.0042	0.0111
D.C.	0.0026	0.0238	0.1179	0.0242	0.0261	0.0075

The values of A^+ and A^- for TOPSIS are calculated according to Equations (13)–(14), as follows:

$A^+ = \{0.0318\ 0.1777\ 0.1179\ 0.0242\ 0.0261\ 0.0219\}$

$A^- = \{0.0026\ 0.0238\ 0.0243\ 0.0049\ 0.0022\ 0.0075\}$

S_i^+, S_i^- and c_i^+ values (calculated according to Equations (15)–(17)) for TOPSIS, as well as the alternatives ranking (maximum c_i^+ value → best alternative), are shown in Table 14.

Table 14. S_i^+, S_i^- and c_i^+ values and alternatives ranking based on the AHP-TOPSIS model.

	S_i^+	S_i^-	c_i^+	Ranking
M.F.	0.0989	0.1567	0.6130	2
P.C.	0.0321	0.1734	0.8437	1
S.C.	0.1360	0.0899	0.3979	3
D.C.	0.1573	0.0985	0.3851	4

As shown in Table 14, the alternative P.C. (lanes dedicated to autonomous electric vehicles with plug-in charging stations beside the road, along the route), having the maximum c_i^+, constitutes the optimum solution for new vehicle technologies in urban areas, according to the AHP-TOPSIS model.

4.12. *Reveal of the Optimum Solution*

According to the last step of the methodology described in Section 3, given the convergence (in terms of the optimum solution) of the results derived from the two models, the alternative solution of lanes dedicated to autonomous electric vehicles, with plug-in charging stations beside the road, along the route, constitutes the optimum choice in terms of road infrastructure for autonomous electric vehicles in urban areas, within the framework of safety and sustainability.

5. Results and Discussion

The application of the two MCA models resulted in the identification of an optimum solution for autonomous electric vehicles in urban areas, integrating safety and sustainability aspects, this being the alternative of lanes dedicated to autonomous electric vehicles with plug-in charging stations beside the road, along the route.

However, apart from the final result, derived as regards the optimum solution, it is also worth commenting on individual results relating to the criteria weighting, as well as to the performance of each alternative in terms of each criterion, derived through the decision-aiding methodology implementation. According to the criteria priority vector shown in Table 6, the criteria related to safety emerge as the most important ones. Specifically, public health, as it relates to electromagnetic radiation, and road safety are highlighted as the most important criteria, with the greatest weights among the other ones (42.07% and 35.95% respectively). Traffic congestion reduction (6.97%), infrastructure cost (5.50%), reduction in natural resources consumption due to low energy loss during charging (5.31%), and charging time (4.20%) are much lower in terms of importance.

According to the alternatives' priority vectors in terms of each criterion, as shown in Table 10, concerning the criteria of construction, operation and maintenance infrastructure cost, the first in ranking is the alternative of a mixed flow of conventional and autonomous electric vehicles, with the alternative of lanes dedicated to autonomous electric vehicles and stationary wireless charging following behind, the alternative of lanes dedicated to autonomous electric vehicles and plug-in charging stations beside the roadway coming next, and the alternative of dedicated lanes to autonomous electric vehicles and dynamic wireless charging along the route being the least preferred alternative for the criterion in question. As regards the criterion of public health related to electromagnetic radiation exposure, the alternatives of mixed flow and separate lanes with stationary plug-in charging are almost together in the first rank, while the alternative of separate lanes with stationary wireless charging follows behind, and the alternative of separate lanes with wireless charging has the lowest performance. Concerning the criterion of road safety, the alternative of a mixed flow has impressively low performance, while it does not make a significant difference for the other three alternatives (dynamic charging seems to have a slightly better performance, due to the fact that vehicle drivers will not have to stop beside the road to charge the vehicle). Concerning the criterion of traffic congestion, the alternative of circulation in separate lanes and dynamic charging comes first, with a significant difference from the last one, which is the mixed flow alternative, while the other two alternatives are in fact almost together in the second rank. As regards the criterion of charging time, the alternative of dynamic charging holds the first place, with impressively high performance; as expected, mixed flow with charging in existing stations is the least preferred, while separate lanes with plug-in charging and separate lanes with stationary wireless charging are ranked second and third, respectively. Finally, concerning the criterion of energy efficiency related to the charging system, the alternative of separate lanes and dynamic charging is the least preferred, while separate lanes with new plug-in charging stations hold the first place, and mixed flow with charging at existing stations and separate lanes are ranked second and third, respectively (this is to be expected, as energy loss is higher in the case of wireless, and especially dynamic, charging).

6. Conclusions

In the present study, infrastructure alternatives for autonomous electric vehicles are evaluated and prioritized, for the first time and at an early stage, due to the low maturity level of certain of these technologies (e.g., autonomous vehicles, dynamic charging, etc.), through the combined application of two hybrid multi-criteria analysis models, with the participation of experts. The experience drawn from conducting the research that is presented in this paper shows that the decision on the appropriate infrastructure for autonomous electric vehicles depends on the selection between alternatives with different maturity levels, in terms of technology readiness and real-life implementation. This fact increases the uncertainty in their evaluation. For example, there is relatively limited international experience in dynamic electric vehicle charging, which is until now at an experimental stage. The specific paper depicts the methodological framework that allows both for the comprehensive evaluation of different infrastructure alternatives within the current context and for the ability to update the evaluation in the future, taking into account the dynamic evolution of the relevant technologies, in order to support the decision-making process regarding the optimum solution relating to infrastructure for autonomous electric vehicles, in terms of safety and sustainability criteria.

Based on the current availability of data and information, plug-in (wired) charging facilities and the separate circulation of autonomous electric vehicles are revealed as the optimum solution. This selection is mostly due to the significantly high weight attributed to two aspects of safety for the users, i.e., public health, with the alternative of dynamic charging being linked to concerns due to exposure of the public to electromagnetic radiation, and road safety concerns, which relates to the avoidance of mixed traffic conditions for vehicles of different levels of automation.

In addition to prioritization and the optimum selection among the available alternatives, the methodology leads to the identification and prioritization of specific issues (e.g., electromagnetic radiation and road safety considerations) related to the overall concept of safety and sustainability that will emerge when the most technologically advanced of these alternatives will be ready for wide-scale implementation. Thus, the analysis offers useful insight to policymakers, highlighting important safety aspects that should be taken into account for the integrated and sustainable planning of transport infrastructure in the future.

It is therefore shown that these new technologies should be adopted with prudence and should focus on the appropriate design and operation of infrastructure in order to primarily ensure traffic safety and public health. Moreover, as technologies are advancing, the available infrastructure alternatives should be constantly evaluated concerning their safety and sustainability implications in a holistic, cross-disciplinary way that goes beyond the conventional traffic engineering and transport planning approaches.

Author Contributions: Conceptualization, K.A. and E.B.; methodology, K.A.; formal analysis, K.A.; investigation, K.A.; resources, K.A. and N.G.; data curation, K.A.; writing—original draft preparation, K.A., N.G., M.P.-L. and E.B.; writing—review and editing, K.A., N.G., M.P.-L. and E.B.; visualization, K.A., N.G., M.P.-L. and E.B.; supervision, M.P.-L. and E.B. All authors have read and agreed to the published version of the manuscript.

Funding: This research received no external funding.

Institutional Review Board Statement: Not applicable.

Informed Consent Statement: Not applicable.

Data Availability Statement: Data are contained within the article.

Conflicts of Interest: The authors declare no conflict of interest.

References

1. Eurostat. Road Accident Fatalities—Statistics by Type of Vehicle. Road Accident Fatalities, European Union. 2021. Available online: https://ec.europa.eu/eurostat/statistics-explained/index.php?title=Road_accident_fatalities_-_statistics_by_type_of_vehicle#Number_of_persons_killed_in_road_traffic_accidents_continuously_decreased_since_2009 (accessed on 10 July 2021).
2. Adminaité-Fodor, D.; Jost, G. *Safer Roads, Safer Cities: How to Improve Urban Road Safety in the EU*; PIN Flash Report 37; European Transport Safety Council: Brussels, Belgium, 2019.
3. EEA. *Transport: Increasing Oil Consumption and Greenhouse Gas Emissions Hamper EU Progress towards Environment and Climate Objectives*; European Environment Agency: Copenhagen, Denmark, 2020.
4. EC Website. Transport Emissions, A European Strategy for Low-Emission Mobility. 2016. Available online: https://ec.europa.eu/clima/policies/transport_en (accessed on 23 March 2021).
5. EC Website. Intelligent Transport Systems, Cooperative, Connected and Automated Mobility (CCAM). 2021. Available online: https://ec.europa.eu/transport/themes/its/c-its_en (accessed on 23 March 2021).
6. Ranft, F.; Adler, M.; Diamond, P.; Guerrero, E.; Laza, M. *Freeing the Road: Shaping the Future for Autonomous Vehicles*; A Policy Network Special Report; Policy Network: London, UK, 2016.
7. Fagnant, D.J.; Kockelman, K. Preparing a nation for autonomous vehicles: Opportunities, barriers and policy recommendations. *Transp. Res. Part A Policy Pract.* **2015**, *77*, 167–181. [CrossRef]
8. International Transport Forum. *Safer Roads with Automated Vehicles?* OECD: Paris, France, 2018.
9. SAE International. Taxonomy and Definitions for Terms Related to Driving Automation Systems for On-Road Motor Vehicles, June 2018. Available online: https://www.sae.org/standards/content/%20j3016_201806/preview/ (accessed on 8 March 2020).
10. Anastasiadou, K. Sustainable Mobility Driven Prioritization of New Vehicle Technologies, Based on a New Decision-Aiding Methodology. *Sustainability* **2021**, *13*, 4760. [CrossRef]
11. Backhaus, W.; Rupprecht, S.; Franco, D. *Road Vehicle Automation in Sustainable Urban Mobility Planning*; Rupprecht Consult—Forschung & Beratung GmbH: Koln, Germany, 2019.
12. Gavanas, N. Autonomous Road Vehicles: Challenges for Urban Planning in European Cities. *Urban Sci.* **2019**, *3*, 61. [CrossRef]
13. Cugurullo, F.; Acheampong, R.A.; Gueriau, M.; Dusparic, I. The transition to autonomous cars, the redesign of cities and the future of urban sustainability. *Urban Geogr.* **2020**, 1–27. [CrossRef]
14. Woudsma, C.; Braun, L. Tomorrow Has Arrived: Cities and Autonomous Vehicles. Pragma Council Discussion Paper. 2017. Available online: https://uwaterloo.ca/planning/sites/ca.planning/files/uploads/files/tomorrow_has_arrived_-_cities_and_autonomous_vehicles_pragma2017_cw_report1_opt.pdf (accessed on 1 July 2021).
15. Milakis, D.; Snelder, M.; van Arem, B.; van Wee, B.; Correia, G. Development and transport implications of automated vehicles in the Netherlands: Scenarios for 2030 and 2050. *Eur. J. Transp. Infrastruct. Res.* **2017**, *17*, 63–85.
16. Milakis, D.; van Arem, B.; Van Wee, B. Policy and society related implications of automated driving: A review of literature and directions for future research. *J. Intell. Transp. Syst.* **2017**, *21*, 324–348. [CrossRef]
17. Litman, T. Autonomous Vehicle Implementation Predictions: Implications for Transport Planning. Victoria Transport Policy Institute, 2018. Available online: https://www.vtpi.org/avip.pdf (accessed on 4 May 2020).
18. Jang, Y.J.; Jeong, S.; Lee, M.S. Initial Energy Logistics Cost Analysis for Stationary, Quasi-Dynamic, and Dynamic Wireless Charging Public Transportation Systems. *Energies* **2016**, *9*, 483. [CrossRef]
19. Quinn, J.C.; Limb, B.J.; Pantic, Z.; Barr, P.; Zane, R.; Bradley, T.H. Techno-economic feasibly and environmental impact of wireless power transfer roadway electrification. In Proceedings of the 2015 IEEE Wireless Power Transfer Conference (WPTC), Boulder, CO, USA, 13–15 May 2015; pp. 1–3. [CrossRef]
20. Deflorio, F.; Guglielmi, P.; Pinna, I.; Castello, L.; Marfull, S. Modeling and Analysis of Wireless "Charge While Driving" Operations for Fully Electric Vehicles. *Transp. Res. Procedia* **2015**, *5*, 161–174. [CrossRef]
21. Bansal, P. Charging of Electric Vehicles: Technology and Policy Implications. *J. Sci. Policy Gov.* **2015**, *6*. Available online: https://www.heliev.gr/wp-content/uploads/2020/02/Charging-of-Electric-Vehicles-bansal_new_ta3_1.2.2015_lb.pdf (accessed on 24 August 2021).
22. Limb, B.J.; Asher, Z.D.; Bradley, T.H.; Sproul, E.; Trinko, D.A.; Crabb, B.; Zane, R.; Quinn, J.C. Economic Viability and Environmental Impact of In-Motion Wireless Power Transfer. *IEEE Trans. Transp. Electrif.* **2019**, *5*, 135–146. [CrossRef]
23. Venugopal, P.; Shekhar, A.; Visser, E.; Scheele, N.; Mouli, G.R.C.; Bauer, P.; Silvester, S. Roadway to self-healing highways with integrated wireless electric vehicle charging and sustainable energy harvesting technologies. *Appl. Energy* **2018**, *212*, 1226–1239. [CrossRef]

24. Karakitsios, I.; Karfopoulos, E.; Madjarov, N.; Bustillo, A.; Ponsar, M.; Del Pozo, D.; Marengo, L. An Integrated Approach for Dynamic Charging of Electric Vehicles by Wireless Power Transfer—Lessons Learned from Real-Life Implementation. *SAE Int. J. Altern. Powertrains* **2017**, *6*, 15–24. [CrossRef]
25. Jeong, S.; Jang, Y.J.; Kum, D. Economic Analysis of the Dynamic Charging Electric Vehicle. *IEEE Trans. Power Electron.* **2015**, *30*, 6368–6377. [CrossRef]
26. Chen, Z.; Liu, W.; Yin, Y. Deployment of stationary and dynamic charging infrastructure for electric vehicles along traffic corridors. *Transp. Res. Part C Emerg. Technol.* **2017**, *77*, 185–206. [CrossRef]
27. Bi, Z.; Song, L.; De Kleine, R.; Mi, C.; Keoleian, G.A. Plug-in vs. wireless charging: Life cycle energy and greenhouse gas emissions for an electric bus system. *Appl. Energy* **2015**, *146*, 11–19. [CrossRef]
28. Macharis, C.; Bernardini, A. Reviewing the use of Multi-Criteria Decision Analysis for the evaluation of transport projects: Time for a multi-actor approach. *Transp. Policy* **2015**, *37*, 177–186. [CrossRef]
29. Browne, D.; Ryan, L. Comparative analysis of evaluation techniques for transport policies. *Environ. Impact Assess. Rev.* **2011**, *31*, 226–233. [CrossRef]
30. Bekiaris, E.; Nakanishi, Y. Economic Impacts of Intelligent Transportation Systems: Innovations and Case Studies. *Res. Transp. Econ.* **2004**, *8*, 639–640. [CrossRef]
31. Saaty, T.L.; Kearns, K.P. *The Analytic Hierarchy Process*; McGraw-Hill: New York, NY, USA, 1980.
32. Opricovic, S.; Tzeng, G.-H. Compromise solution by MCDM methods: A comparative analysis of VIKOR and TOPSIS. *Eur. J. Oper. Res.* **2004**, *156*, 445–455. [CrossRef]
33. Hwang, C.L.; Yoon, K. Multiple Attribute Decision Making. In *Lecture Notes in Economics and Mathematical Systems*; Springer: New York, NY, USA, 1981; Volume 186, ISBN 978-3-642-48318-9.
34. Olczak, P.; Kryzia, D.; Matuszewska, D.; Kuta, M. "My Electricity" Program Effectiveness Supporting the Development of PV Installation in Poland. *Energies* **2021**, *14*, 231. [CrossRef]
35. Wróblewski, P.; Drożdż, W.; Lewicki, W.; Dowejko, J. Total Cost of Ownership and Its Potential Consequences for the Development of the Hydrogen Fuel Cell Powered Vehicle Market in Poland. *Energies* **2021**, *14*, 2131. [CrossRef]
36. Wróblewski, P.; Kupiec, J.; Drożdż, W.; Lewicki, W.; Jaworski, J. The Economic Aspect of Using Different Plug-In Hybrid Driving Techniques in Urban Conditions. *Energies* **2021**, *14*, 3543. [CrossRef]
37. Bartłomiejczyk, M.; Połom, M. Possibilities for Developing Electromobility by Using Autonomously Powered Trolleybuses Based on the Example of Gdynia. *Energies* **2021**, *14*, 2971. [CrossRef]
38. Połom, M. Technology Development and Spatial Diffusion of Auxiliary Power Sources in Trolleybuses in European Countries. *Energies* **2021**, *14*, 3040. [CrossRef]
39. Curiel-Esparza, J.; Mazario-Diez, J.L.; Canto-Perello, J.; Martin-Utrillas, M. Prioritization by consensus of enhancements for sustainable mobility in urban areas. *Environ. Sci. Policy* **2016**, *55*, 248–257. [CrossRef]
40. Martin-Utrillas, M.; Reyes-Medina, M.; Curiel-Esparza, J.; Canto-Perello, J. Hybrid method for selection of the optimal process of leachate treatment in waste treatment and valorization plants or landfills. *Clean Technol. Environ. Policy* **2015**, *17*, 873–885. [CrossRef]
41. De Loë, R.C.; Melnychuk, N.; Murray, D.; Plummer, R. Advancing the State of Policy Delphi Practice: A Systematic Review Evaluating Methodological Evolution, Innovation, and Opportunities. *Technol. Forecast. Soc. Chang.* **2016**, *104*, 78–88. [CrossRef]
42. Anastasiadou, K.; Vougias, S. "Smart" or "sustainably smart" urban road networks? The most important commercial street in Thessaloniki as a case study. *Transp. Policy* **2019**, *82*, 18–25. [CrossRef]
43. Bi, Z.; De Kleine, R.; Keoleian, G.A. Integrated Life Cycle Assessment and Life Cycle Cost Model for Comparing Plug-in versus Wireless Charging for an Electric Bus System. *J. Ind. Ecol.* **2016**, *21*, 344–355. [CrossRef]
44. Jang, Y.J. Survey of the operation and system study on wireless charging electric vehicle systems. *Transp. Res. Part C Emerg. Technol.* **2018**, *95*, 844–866. [CrossRef]
45. Mazharov, N.D.; Hristov, S.M.; Dichev, D.A.; Zhelezarov, I.S. Some Problems of Dynamic Contactless Charging of Electric Vehicles. *Acta Polytech. Hung.* **2017**, *14*, 2017.
46. Moon, S.; Kim, B.-C.; Cho, S.-Y.; Ahn, C.-H.; Moon, G.-W. Analysis and Design of a Wireless Power Transfer System with an Intermediate Coil for High Efficiency. *IEEE Trans. Ind. Electron.* **2014**, *61*, 5861–5870. [CrossRef]
47. Bi, Z.; Keoleian, G.A.; Lin, Z.; Moore, M.R.; Chen, K.; Song, L.; Zhao, Z. Life cycle assessment and tempo-spatial optimization of deploying dynamic wireless charging technology for electric cars. *Transp. Res. Part C Emerg. Technol.* **2019**, *100*, 53–67. [CrossRef]
48. Chen, W.; Liu, C.; Lee, C.H.; Shan, Z. Cost-Effectiveness Comparison of Coupler Designs of Wireless Power Transfer for Electric Vehicle Dynamic Charging. *Energies* **2016**, *9*, 906. [CrossRef]
49. Bi, Z.; Kan, T.; Mi, C.; Zhang, Y.; Zhao, Z.; Keoleian, G.A. A review of wireless power transfer for electric vehicles: Prospects to enhance sustainable mobility. *Appl. Energy* **2016**, *179*, 413–425. [CrossRef]
50. Tan, L.; Guo, J.; Huang, X.; Liu, H.; Yan, C.; Wang, W. Power Control Strategies of On-Road Charging for Electric Vehicles. *Energies* **2016**, *9*, 531. [CrossRef]
51. Yin, A.; Wu, S.; Li, W.; Hu, J. Analysis of Battery Reduction for an Improved Opportunistic Wireless-Charged Electric Bus. *Energies* **2019**, *12*, 2866. [CrossRef]
52. De Marco, D.; Dolara, A.; Longo, M.; Yaïci, W. Design and Performance Analysis of Pads for Dynamic Wireless Charging of EVs using the Finite Element Method. *Energies* **2019**, *12*, 4139. [CrossRef]

53. Karakitsios, I.; Karfopoulos, E.; Hatziargyriou, N. Impact of dynamic and static fast inductive charging of electric vehicles on the distribution network. *Electr. Power Syst. Res.* **2016**, *140*, 107–115. [CrossRef]
54. Covic, G.A.; Boys, J.T. Inductive Power Transfer. *Proc. IEEE* **2013**, *101*, 1276–1289. [CrossRef]
55. Panchal, C.; Stegen, S.; Lu, J.-W. Review of static and dynamic wireless electric vehicle charging system. *Eng. Sci. Technol. Int. J.* **2018**, *21*, 922–937. [CrossRef]
56. Miller, J.M.; Jones, P.; Li, J.-M.; Onar, O.C. ORNL Experience and Challenges Facing Dynamic Wireless Power Charging of EV's. *IEEE Circuits Syst. Mag.* **2015**, *15*, 40–53. [CrossRef]
57. Funke, S.A.; Plötz, P.; Wietschel, M. Invest in fast-charging infrastructure or in longer battery ranges? A cost-efficiency comparison for Germany. *Appl. Energy* **2019**, *235*, 888–899. [CrossRef]
58. Anderson, J.M.; Kalra, N.; Stanley, K.D.; Sorensen, P.; Samaras, C.; Oluwatola, T.A. Autonomous Vehicle Technology: A Guide for Policymakers. Rand Corporation. 2016. Available online: https://www.rand.org/pubs/research_reports/RR443-2.html (accessed on 8 September 2020).
59. Zhu, W.-X.; Zhang, H. Analysis of mixed traffic flow with human-driving and autonomous cars based on car-following model. *Phys. A Stat. Mech. Appl.* **2018**, *496*, 274–285. [CrossRef]
60. Bagloee, S.A.; Tavana, M.; Asadi, M.; Oliver, T. Autonomous vehicles: Challenges, opportunities, and future implications for transportation policies. *J. Mod. Transp.* **2016**, *24*, 284–303. [CrossRef]
61. World Economic Forum in Collaboration with The Boston Consulting Group. Reshaping Urban Mobility with Autonomous Vehicles Lessons from the City of Boston. Available online: http://www3.weforum.org/docs/WEF_Reshaping_Urban_Mobility_with_Autonomous_Vehicles_2018.pdf (accessed on 8 July 2021).
62. Glancy, D.J. Autonomous and Automated and Connected Cars—Oh My! First Generation Autonomous Cars in the Legal Ecosystem. *Minn. J. Law Sci. Technol.* **2015**, *16*, 619. Available online: https://conservancy.umn.edu/bitstream/handle/11299/174406/619%20Glancy.pdf?sequence=1&isAllowed=y (accessed on 24 August 2021).
63. Covic, G.A.; Boys, J.T. Modern Trends in Inductive Power Transfer for Transportation Applications. *IEEE J. Emerg. Sel. Top. Power Electron.* **2013**, *1*, 28–41. [CrossRef]
64. Sdoukopoulos, A.; Pitsiava-Latinopoulou, M.; Basbas, S.; Papaioannou, P. Measuring progress towards transport sustainability through indicators: Analysis and metrics of the main indicator initiatives. *Transp. Res. Part D Transp. Environ.* **2019**, *67*, 316–333. [CrossRef]
65. Litman, T. Well Measured Developing Indicators for Sustainable and Livable Transport Planning, 19 March 2019, Victoria Transport Policy Institute. Available online: http://www.vtpi.org/wellmeas.pdf (accessed on 6 July 2019).
66. Marletto, G.; Mameli, F. A participative procedure to select indicators of policies for sustainable urban mobility. Outcomes of a national test. *Eur. Transp. Res. Rev.* **2012**, *4*, 79–89. [CrossRef]
67. Lima, J.P.; Lima, R.D.S.; da Silva, A.N.R. Evaluation and Selection of Alternatives for the Promotion of Sustainable Urban Mobility. *Procedia Soc. Behav. Sci.* **2014**, *162*, 408–418. [CrossRef]
68. Triantafyllou, E. *Multi—Criteria Decision Making Methods: A Comparative Study*; Kluwer Academic Publishers: Dordrecht, The Netherlands, 2000; ISBN 978-1-4757-3157-6.
69. Diamond, I.R.; Grant, R.C.; Feldman, B.M.; Pencharz, P.B.; Ling, S.C.; Moore, A.M.; Wales, P.W. Defining consensus: A systematic review recommends methodologic criteria for reporting of Delphi studies. *J. Clin. Epidemiol.* **2014**, *67*, 401–409. [CrossRef] [PubMed]
70. Deparis, S.; Mousseau, V.; Öztürk, M.; Pallier, C.; Huron, C. When conflict induces the expression of incomplete preferences. *Eur. J. Oper. Res.* **2012**, *221*, 593–602. [CrossRef]

energies

Article

Tactile Occupant Detection Sensor for Automotive Airbag

Naveen Shirur [1,2,*], Christian Birkner [1], Roman Henze [2] and Thomas M. Deserno [3]

[1] CARISSMA, Technische Hochschule Ingolstadt, Esplanade 10, 85049 Ingolstadt, Germany; christian.birkner@thi.de
[2] Institut für Fahrzeugtechnik, Technische Universität Braunschweig, Hans-Sommer-Straße 4, 38106 Braunschweig, Germany; r.henze@tu-braunschweig.de
[3] Peter L. Reichertz Institut für Medizinische Informatik, Technische Universität Braunschweig, Mühlenpfordtstraße 23, 38106 Braunschweig, Germany; thomas.deserno@plri.de
* Correspondence: naveen.shirur@carissma.eu; Tel.: +49-841-9348-3356

Abstract: Automotive airbags protect occupants from crash forces during severe vehicle collisions. They absorb energy and restrain the occupants by providing a soft cushion effect known as the restraint effect. Modern airbags offer partial restraint effect control by controlling the bag's vent holes and providing multi-stage deployment. Full restraint effect control is still a challenge because the closed-loop restraint control system needs airbag–occupant contact and interaction feedback. In this work, we have developed novel single and matrix capacitive tactile sensors to measure the occupant's contact data. They can be integrated with the airbag surface and folded to follow the dynamic airbag shape during the deployment. The sensors are tested under a low-velocity pendulum impact and benchmarked with high-speed test videos. The results reveal that the single sensor can successfully measure occupant–airbag contact time and estimate the area, while the contact position is additionally identified from the matrix sensor.

Keywords: automotive airbag; capacitive tactile sensor; occupant detection; passive safety; vehicle crash

Citation: Shirur, N.; Birkner, C.; Henze, R.; Deserno, T.M. Tactile Occupant Detection Sensor for Automotive Airbag. *Energies* **2021**, *14*, 5288. https://doi.org/10.3390/en14175288

Academic Editors: Guzek Marek, Rafał Jurecki and Wojciech Wach

Received: 27 July 2021
Accepted: 23 August 2021
Published: 26 August 2021

1. Introduction

In the event of severe vehicle collisions, the airbag deploys in 30–50 milliseconds and restrains the occupants providing a cushion effect [1]. An airbag offers optimum restraint effect when it deploys as designed for the situation; otherwise, there can be mortal injuries. A 20 ms late deployment can increase the risk of head injuries by 14% caused by the airbag and it also increases the risk of collision with the headrest [2]. Therefore, there is a need for control and optimization of the airbag deployment. Since the introductionof airbags, there have been many attempts to optimize the restraint effect and reduce the injuries for different crash situations by tuning various parameters such as airbag deployment time, early occupant coupling with the airbag, pressure dispersion direction and stage-wise deployment [2–5]. Mercedes-Benz developed PRE-SAFE® Impulse Side. In this technology, the occupant is pushed forward during the potential crashes and engaged with the restraint system to reduce the kinetic energy difference between the occupant and the restraint system. The technology achieved a 35% reduction in upper rib displacement in a standard pole test [3]. Kim et al. designed a low-risk deployment airbag with a protective wrap. It disperses the airbag pressure in lateral directions and reduces the force on the occupants [4]. The self-adaptive vent (SAV) is a useful optimization technique for keeping the airbag inflated for a longer protection time. When the airbag is fully deployed, the tether tightens and closes the vent holes. The airbag remains inflated for a longer time and protects the occupants [5].

The mentioned state-of-the-art technologies do not consider occupant detection and classification (size and position of impact). The occupant contact data with the airbag and the bag pressure feedback are essential to make the airbag self-adaptive. Ultrasonic sensors, capacitive sensors, seat sensors, infrared cameras and computer vision systems can detect the occupants [6–14]. Izumi et al. developed an occupant detection system using a far-infrared camera, which provides the occupant's thermographic images. A vital shortcoming of the IR camera is its difficulty in tracking the occupant after the airbag deployment. Due to the overlap, the occupant or the airbag's contour cannot be obtained from the thermographic images. Dust particles and temperature changes make occupant tracking further complicated [6].

Computer vision systems have evolved. The occupant's presence and position can be detected using a stereo camera. If an occupant is present, the geometry and position of the occupant's head can be calculated further [11]. Farmer et al. developed a vision-based system to classify (adult or child) and track the occupant's motion. The airbag can be suppressed if an infant seat or an adult who is critically out-of-position is detected [12,13]. Further, adaptive airbag deployment (trigger time) decisions can be made by tracking the occupant's head's position and deciding in-position and out-of-position situations [10]. Airbag deployment power can cause mortal injuries if the occupant is in an improper posture. Hence, posture estimation is also crucial for the airbag deployment strategy [12,14].

The vision systems discussed in the literature support only the airbag deployment strategy, for example, trigger time. The airbag deployment strategy is purely based on a set of pre-determined inputs from the machine vision. Airbag-induced injury mitigation requires continuous occupant motion and contact feedback with the airbag during the ride-down phase. Hence, an alternative solution is required.

There is significant technological advancement in capacitive sensing methods to detect the occupant. Blackburn et al. developed an occupant sensing apparatus using a variable capacitor that gives the occupant's position relative to the cockpit. Airbag trigger time can be decided based on the sensed position, but occupant–airbag interaction is not addressed [7]. Kithil et al. developed a capacitance sensing array mounted at different locations in the cockpit. The sensors provide the occupant position measuring the dielectric change between the plates. However, the array gives only the position. It is hard to address when the occupant contacts the airbag and differentiate between the occupant and the airbag [8]. White et al. designed a capacitive sensor using a rendered portion of the airbag. There is conductive paint on the bag surface connected to a capacitive sensor circuit outside the airbag. The sensor gives the occupant's contact time with the airbag, but to control the airbag power, we need first contact, out-of-position, and the occupant's area [9].

2. Problem Description and Research Contribution

The research discussed in the literature focuses mainly on optimizing the airbag for a long-standing time and less impact force. On the other hand, occupant detection systems aid only airbag deployment decisions based on size, motion and posture. The sensing methodologies are successful until the first contact with the airbag. There is a lack of occupant sensing methods for in-crash and post-crash phases. Further, to continuously adapt the airbag power and mitigate the injuries, we need airbag–occupant interaction data, which is still an open research opportunity. The interaction between the occupant and the airbag can be addressed by answering the following questions.

1. What are the first and total contact times?
2. What is the contact area?
3. Where is the contact position?

In this work, we have developed novel single and matrix tactile sensors to detect the occupant and measure the contact parameters, especially contact time and area. The research work answers the stated questions and bridges the gap between pre-crash and in-crash occupant monitoring.

3. Method Overview

This paper focuses on designing and testing capacitive tactile sensors, which answer the stated questions and provide more insight into the airbag–occupant interaction. The sensors consist of a conductive woven fabric connected to a conducting thread. Two sensor variants discussed in this paper are a single sensor, which gives occupant contact time and contact area, and a matrix sensor, which additionally provides the position. Figure 1 shows the sensors' configuration. The sensors are integrated with the airbag and follow the airbag's shape during the deployment. In this work, an airbag with sensors is tested under low-velocity pendulum impact. Understanding the sensor's behavior for different external parameter changes and the airbag deployment phases is crucial. We studied various deployment phases like the textile unfolding, time-to-first-gas, bag inflation and the cable capacitance effects [15]. Time-to-first-gas and inflation do not influence the sensor, whereas the unfolding event and cable capacitance significantly affect the sensor. Additionally, the sensor was benchmarked with contact times from the camera [15]. This paper's main objectives are sensor calibration, occupant's contact detection, position estimation and contact area calculation. The contact area and position enable adult-child classification with out-of-position cases, which is crucial to modulate the airbag's pressure.

The results of this work have a significant impact on the vent hole control. The vent hole opening can be controlled based on the occupant's data to optimize the restraint effect and minimize the injuries [16].

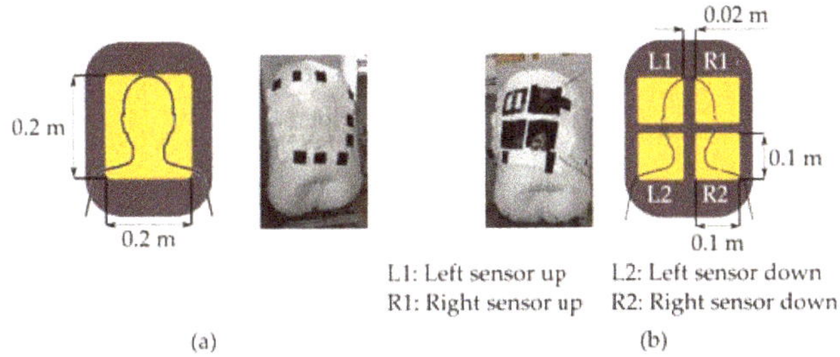

Figure 1. Sensors' configuration: (**a**) single sensor; (**b**) matrix sensor.

4. Capacitive Sensing

The human body has 100 to 300 pF capacitance, which is used to sense the occupant contact with the airbag [17]. In this work, projected self-capacitance theory in loaded mode is applied [18]. A single sensing electrode is installed behind the airbag surface, which acts both as transmitter and receiver [18–24]. It is called active capacitive sensing due to the single electrode [18]. The sensor is connected to a resistor. The sensor's voltage is measured using an RC circuit. Suppose the occupant comes near the sensor, the capacitance increases due to the active coupling resulting in voltage drop [21]. The familiar examples of self-capacitance applications are a touch screen of a cell phone, sliders and wheels, control buttons on home appliances and the on-board infotainment of an automotive system [20,21]. Figure 2 shows the analogy between a capacitor and a capacitive sensor. A capacitor comprises two electrodes separated by a dielectric material. If a conducting electrode is removed, it assumes a virtual ground through the human body. This phenomenon is known as projected capacitance [18,19]. In our case, the sensor forms a conductive layer and the airbag textile forms the dielectric layer. A conducting thread is used for the connection between the sensor and the external circuit.

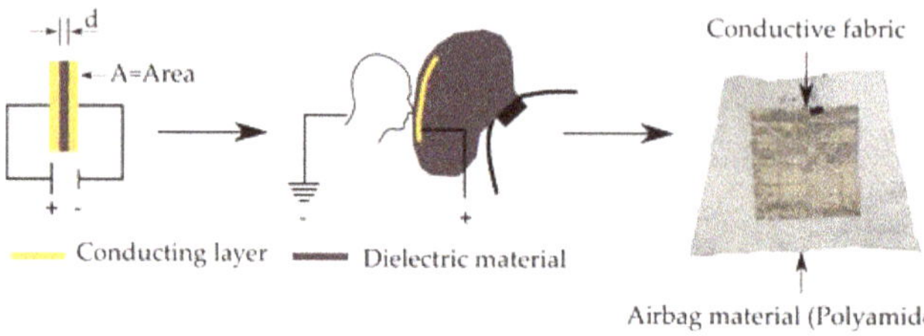

Figure 2. Sensor concept.

The sensor has a reference capacitance and voltage based on its geometry and dielectric values. Firstly, the reference value has to be measured to detect the touch. The voltage drop from the reference gives the touch. Figure 3 shows the capacitance variation as a function of occupant's distance from the sensor [22,25]. The capacitance increases when the occupant is in the sensor's proximity and observes a sharp gradient during the contact. There is a ΔC increase in capacitance, corresponding to ΔV voltage drop across the sensor.

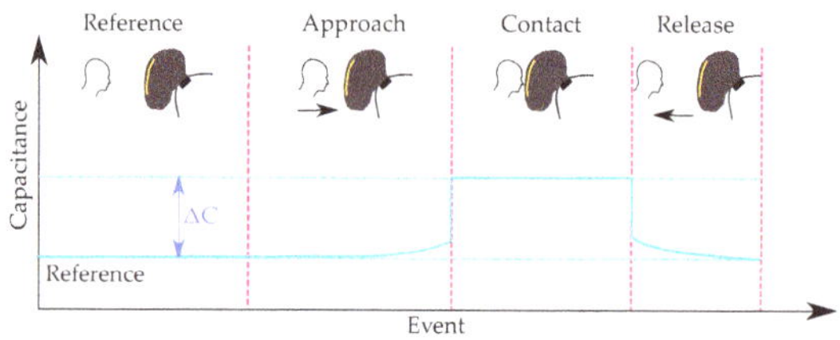

Figure 3. Capacitance variation as a function of occupant's distance from the airbag.

Equation (1) gives the voltage across the sensor [19,20,24,25]. The voltages across the capacitor and the supply are V_{cap} and V_s, respectively. t is the time elapsed after the supply and R is the resistor's resistance. The permittivities of vacuum and dielectric material are ε_0 and ε_r, respectively. A denotes the plate area (contact surface area between the occupant and the sensor), and d is the separation distance (Figure 2).

$$V_{cap} = V_s\left(1 - e^{\left(\frac{-td}{RA\varepsilon_0\varepsilon_r}\right)}\right) \tag{1}$$

The sensor voltage behaviour without and with touch is illustrated in Figure 4. The green and blue curves show the voltage (V_{cap}) across the sensor without and with touch, respectively, for a supply voltage (V_s). In a transient state, there is a voltage drop of ΔV_{cap}. The time constant T can be varied by varying the resistance. ΔV_{cap} is used as a contact detection parameter. Based on this value, the contact area can be calculated. When there are multiple sensors, an appropriate threshold can be applied to determine the first contact point.

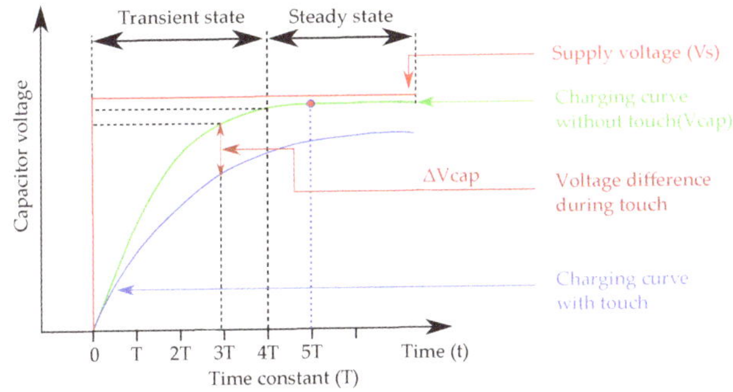

Figure 4. Capacitor charging curve with and without touch.

Equation (1) is implemented in Matlab® to simulate the sensor voltage behaviour as a contact area's function. In the theoretical simulation model, a 5 V source signal is given. The airbag fabric is 0.33 mm thick with $\varepsilon_r = 3.4$. A 220 kΩ resistor is used in series with the capacitor. Figure 5a shows the simulated sensor voltage as a function of the contact area. 5 V is taken as a reference value. If there is a contact, for example, 0.01 m^2, the voltage will be 2.5 V. The drop from 5 V to 2.5 V is the ΔV illustrated in Figure 5b. As the contact area increases, the voltage drop increases.

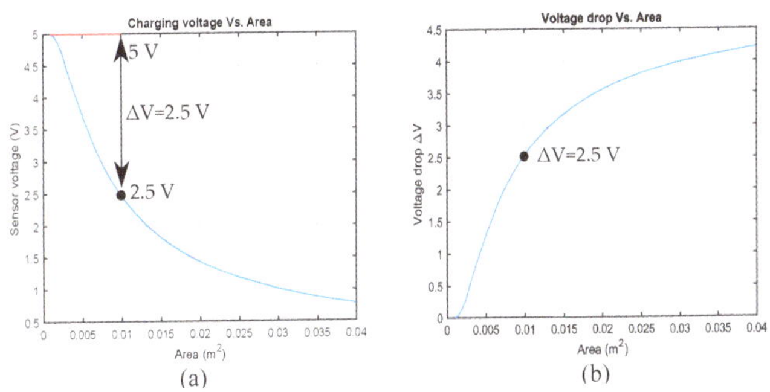

Figure 5. Sensor's voltage: (**a**) voltage–area relation; (**b**) voltage drop versus increase in contact area.

5. Method

5.1. Sensor Hardware and Circuit

The sensing surface consists of woven conductive fabric with a conductive thread (Figure 6). The fabric is a copper and nickel-plated nylon material that finds application in smart wearable technologies [26]. The thread is made of 30% stainless steel fibers [27].

The sensor dimensions are chosen to cover the full human face and partly the neck. The single sensor is a square with 200 mm sides and the matrix sensor has four individual sensors. Each sensor has a square shape with a 100 mm edge (Figure 1). A 20 mm gap is maintained between the sensors to avoid mutual capacitance and mutual touch.

Figure 7 shows the sensor's circuit diagram [21,23,28]. It is a simple RC circuit. The resistance has to be chosen based on the charging time. A smaller value requires an increased sampling rate; hence a 220 kΩ is selected. A 1 pF capacitor is connected to the ground to stabilize the sensor. The sensing surface is mounted on the airbag. The circuit is implemented using an Arduino development board as shown in Figure 6.

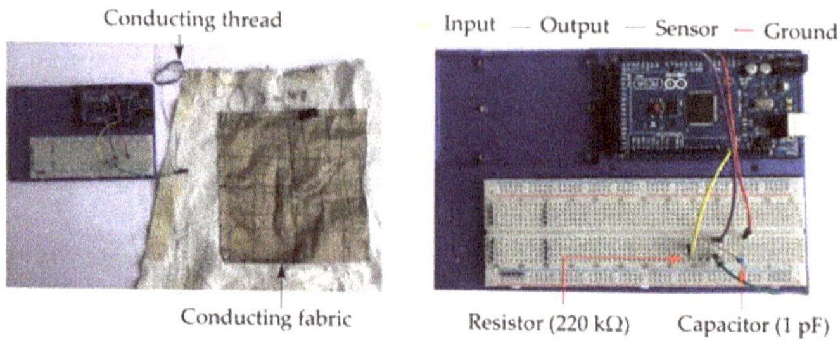

Figure 6. Hardware connections.

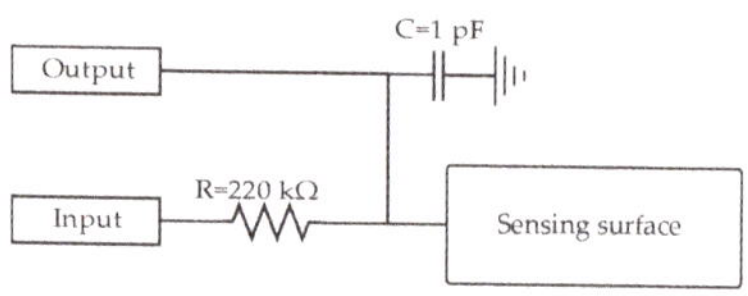

Figure 7. Sensing circuit.

5.2. Algorithm for Sensor Voltage Change Measurement

Digital input and analog output channels on the Arduino board are utilized to measure the voltage change. The channels are pulled down with internal resistors to avoid floating. A 5 V input signal is given for 136 microseconds. The voltage change across the sensor is measured from the analog channel. After the measurement, input and output channels are discharged and the next cycle is executed.

5.3. Test Bench

A low-impact velocity pendulum test bench is designed to test the sensors as it offers flexibility to change the parameters (Figure 8). It consists of a 2.03 m swinging arm and a human head-form of diameter 150 mm. The standard head mass of 4.5 kg is scaled in this work to 6.81 kg to increase the force. A 5 mm thick rubber sheet is glued to the head to damp the vibrations. Further, a thin aluminum sheet is attached to make the head conductive. The airbag is mounted rigidly and kept inflated. The measurement system (Table 1) includes a standard rotary encoder, an airbag pressure sensor and two high-speed cameras.

A conducting thread provides electrical contact between the pendulum head and the human body to simulate actual human body capacitance, which is kept constant throughout the calibration and testing.

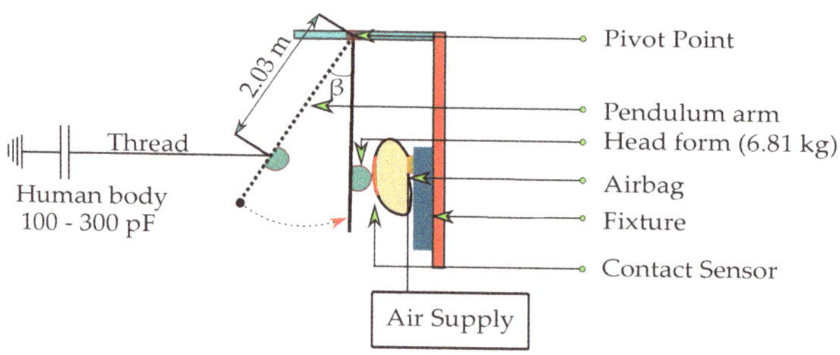

Figure 8. Low-impact velocity pendulum test bench.

Table 1. Instrumentation.

Parameter	Instrument	Sampling Frequency
Angle	Rotary encoder	250 kHz
Contact sensor voltage	Arduino Uno	800 Hz–8 kHz
Contact sensor voltage	Arduino Mega	800 Hz–8 kHz
Pressure	Pressure sensor (5 bar abs)	250 kHz
Trigger	Mechanical switch	250 kHz
Video	High-speed cameras	2500 fps [a]

[a] Frames per second.

5.4. Sensor Calibration

Squared conducting plates of length 40 mm to 200 mm are used to calibrate the single and the matrix sensors. The plates are pressed against the sensors and voltage drop is measured. Figure 9 shows an exemplary single sensor voltage drop for 200 mm square plate. The maximum voltage drop for the single sensor is 3.1 V. A mean reference voltage is calculated by averaging the first 300 samples. After many threshold iterations, a 4% threshold is applied to identify the touch. Suppose there is a touch, the voltage drops and reaches a minimum. A 20% threshold is used for the lower voltage because of the high noise. The mean value within a 20% voltage band (yellow curve in Figure 9) is calculated. The difference between mean reference and lower mean is ΔV, which gives contact time and the area.

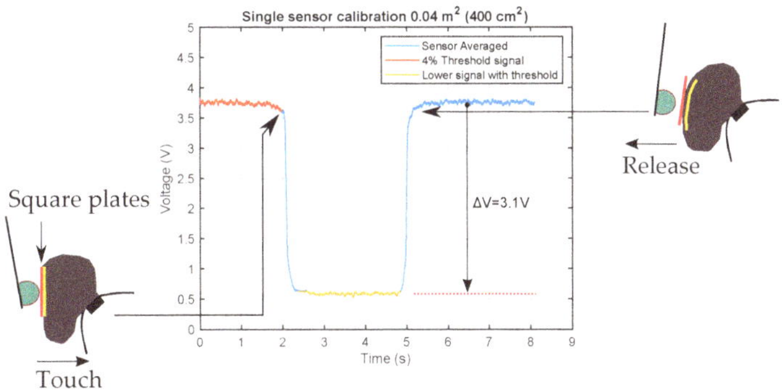

Figure 9. Voltage drop calculation method.

The voltage drops for different plates are obtained (Figure 10a). A non-linear 4th-order polynomial (in a least-square sense) equation is applied to the voltage drop data to fit the curve. The values of the contact areas are centered at zero and scaled to have a unit standard deviation, which improves the numerical properties of the polynomial. The fitted curve is interpolated using Piecewise Cubic Hermite Interpolating Polynomial (PCHIP) method to calculate the area from the voltage drop (Figure 10b). The curve's behavior matches the theoretical voltage drop calculation shown in Figure 5b. From the theoretical simulation model, the maximum voltage drop is close to 4.2 V. During the calibration, the maximum observed drop is 3.1 V. Cable effects, stray capacitances and environmental parameters contribute to the deviation.

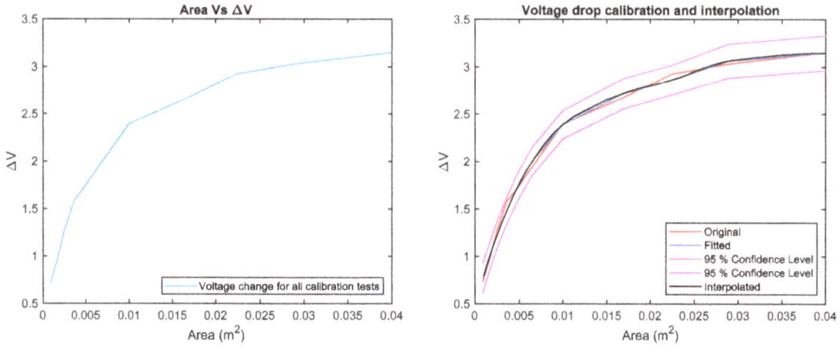

(a) (b)

Figure 10. Calibrated voltage drop curve for single sensor: (**a**) sensor voltage drop; (**b**) calibrated and interpolated voltage drop.

Similar to single sensor, individual sensors in the matrix are calibrated (Figures 11 and 12). The maximum voltage drop for all the sensors is between 2 V and 2.4 V, whereas the simulated voltage drop is close to 2.5 V.

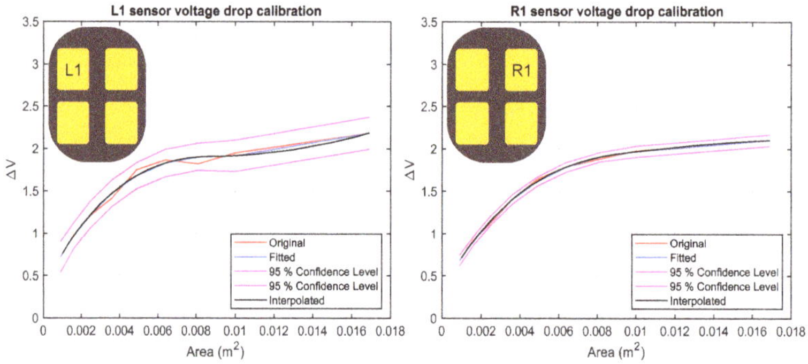

(a) (b)

Figure 11. Calibrated voltage drop curve for L1 and R1: (**a**) L1 sensor; (**b**) R1 sensor.

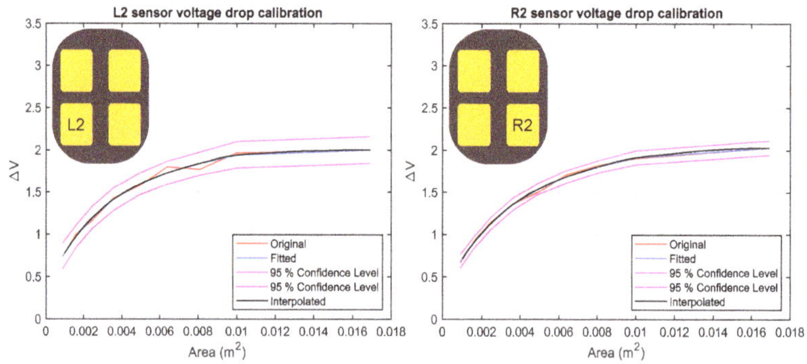

(a) (b)

Figure 12. Calibrated voltage drop curve for L2 and R2: (**a**) L2 sensor; (**b**) R2 sensor.

5.5. Hypothesis

It is hypothesized that the sensor voltage decreases with contact progression. At constant pressure and increasing velocity, the voltage drop increases with less contact time due to increased area. Further, the magnitude and peak time depend on the airbag pressure and the impact velocity. The hypothesis is tested by comparing the contact sensor results with the high-speed videos.

5.6. Experimental Design

Airbag pressure, occupant's impact velocity and the sensor position are the variables that decide the restraint effect in real crash situations. Therefore, we choose these parameters to design the experiments. Changing the pendulum position for out-of-position cases is a challenge to the test results' reproducibility; hence, the sensor's position is varied. The head's velocity is chosen based on the occupant's motion modeling during crashes [29]. Pressure values are chosen such that there is a perfect contact between the airbag and the sensor (Table 2).

Table 2. Single-sensor test matrix.

Test	Airbag Pressure (bar)	Swing Angle (°)	Impact Velocity (m/s)
1	1.34	30	2.30
2	1.41	45	3.41
3	1.46	60	4.46
4	1.39	70	5.11
5	1.18	30	2.30
6	1.15	45	3.41
7	1.11	60	4.46
8	1.13	70	5.11

The single sensor is tested by varying airbag pressure and impact velocity (Table 2) while keeping the sensor position and the impact point constant. The matrix sensor is tested with an approximately constant pressure (1.2 bar) and velocity (3.41 m/s) by varying the sensor's position to identify the position additionally. The matrix sensor has four individual sensors, which are geometrically symmetric on the airbag surface; hence, three experiments are carried out. Two experiments involve L1- and R1-centered impacts, which simulate out-of-position impact with respect to sensors' configuration and one in the middle of all sensors to simulate in-position impact.

5.7. Sensor Benchmark and Data Analysis Method

5.7.1. Head Depth Calculation from the Contact Sensor

The contact sensors are benchmarked with high-speed videos. Figure 13 shows an exemplary contact event for the single sensor. Since contact occurs before the trigger, the times before 0 s are negative. From the camera, the first touch is at −4.8 ms and peak displacement occurs at 56.4 ms. The total contact time is 61.2 ms. The contact time from the sensor is 64.3 ms. There is a 3.1 ms difference with a reasonable agreement between camera and contact sensor time.

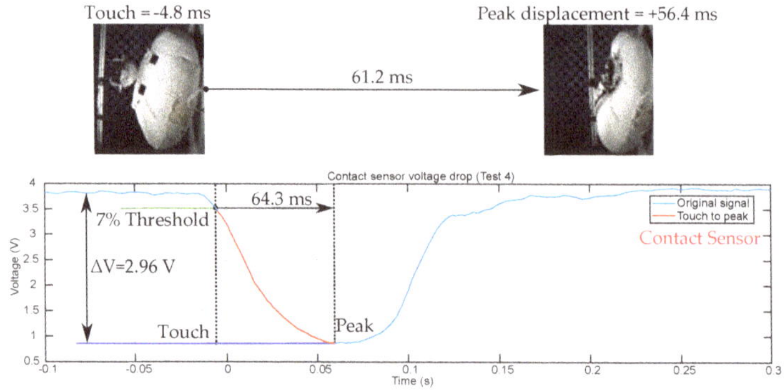

Figure 13. Sensor benchmarking.

Further, the contact sensor voltage drop at the peak displacement is 2.96 V. From the calibration curve (Figure 10) the area at 2.96 V is 0.0249 m^2. Since the head is a hemispherical form, the area obtained is the curved surface area of the hemisphere. The head depth (D_c in Figure 14) is calculated from the curved surface area, which is 0.063 m (63 mm).

5.7.2. Head Depth Calculation from the High-Speed Videos

The calculated depth from the contact sensor (Section 5.7.1) is compared with the depth obtained from the high-speed test video analysis. An open-source software (Tracker) is used for kinematic analysis. Initially, the pendulum's arm width (0.03 m) is calibrated in the video and the impact velocity is calculated. The impact velocities calculated from the swing angle and the video for an exemplary test are 5.11 m/s and 5.089 m/s, respectively. Then the peak head displacement is calculated. At the beginning of the contact, head depth (X) from the reference is 92.5 mm (Figure 14a). When the head is at peak displacement, the depth (X_1) from the reference is 36.7 mm (Figure 14b). The depth D_c is the head depth inside the airbag during the restraint phase, which is 55.8 mm. There is a 7.2 mm difference between the depth calculated from the contact sensor (63 mm from Section 5.7.1) and the video.

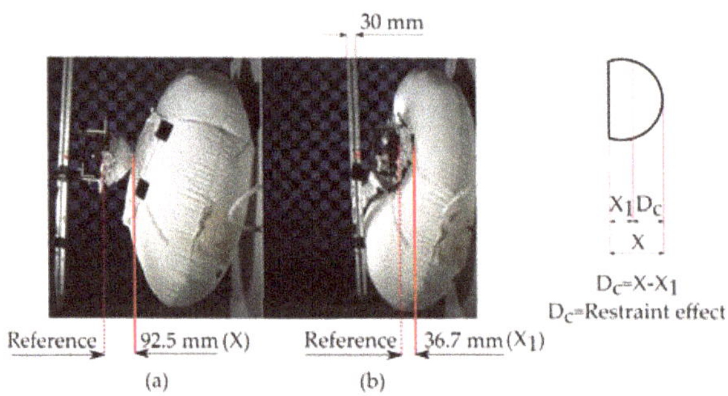

Figure 14. Depth calculation from high-speed video: (**a**) first contact; (**b**) peak displacement.

6. Results

6.1. Single Sensor

The sensor's voltage is a function of velocity and bag pressure. In the first set of experiments (tests 1 to 4), we observe that when the velocity is increased from a minimum of 2.30 m/s to 4.46 m/s, there is a major difference in voltage drop and contact time. From Figure 15a and Table 3 it can be seen that the voltage drop increases from 2.64 V (test 1) to 3.01 V (test 3) indicating an increase in the contact area. On the other hand, from Figure 15b, it can be noted that the contact time from the first touch to peak displacement decreases from 114 ms to 69.2 ms from the test 1 to 3. When the velocity is increased beyond 4.46 m/s, we observe no further drop in the voltage because the pendulum head covers the sensor completely at 4.46 m/s. Similar behaviour is observed for tests 5 to 8.

Further, Figure 15b shows the contact time comparison for the contact sensor and the camera for different tests. The contact time (first contact to peak depth) decreases with the increase in impact velocity. The maximum and minimum deviations from the camera are 17.76% (test 1) and 2.97% (test 3), respectively.

In the second set of experiments (tests 5 to 8), we reduced the bag's pressure from 1.4 bar to 1.2 bar. The bag pressure variation changes the impact positions dramatically due to its thickness in the inflated condition. The pendulum hits the bag even before achieving maximum velocity, which is a challenge to the reproducibility of the tests. Hence, we varied pressure such that impact always occurs at peak pendulum velocity. We observed similar behavior as in tests 1 to 4. There is no major deviation in the voltage drop values and contact times. From Figure 15a, we observe that the voltage drop for tests 1 and 5 (same velocity and different pressures) are approximately the same (highlighted in black box). Further, the drop behavior for 3.41 m/s (test 2 and 6) is also similar. After 4.46 m/s all the tests have the same behavior due to full contact between the pendulum and the sensor.

The contact area is calculated at the peak depth as the kinetic energy and sensor variations are low. Firstly, the depth (D_c in Figure 14) is calculated from the high-speed video. The voltage drop (ΔV_{cap}) from the sensor is compared with the calibration curve and the contact area, A, is calculated. Then the depth (D_s) is calculated. D_d is the difference between the depths obtained from the camera and the contact sensor. The deviation is calculated, keeping camera values as the reference (Table 3). The sensor has a minimum and maximum deviation of 13.32% and 16.41%, respectively.

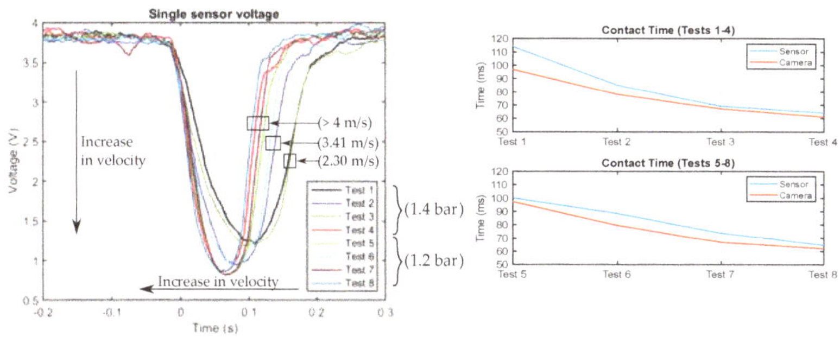

(a) (b)

Figure 15. Single sensor results: (**a**) sensor voltage drop; (**b**) contact time comparison for sensor and high-speed camera.

Table 3. Single sensor results.

Test	D_c (m)	ΔV_{cap} (V)	A (m^2)	D_s (m)	D_d (m)	% Deviation
1	0.0499	2.64	0.0147	0.0485	−0.0014	−2.80
2	0.0630	2.87	0.0227	0.0601	−0.0029	−4.60
3	0.0638	3.01	0.0266	0.0651	0.0013	2.03
4	0.0563	2.96	0.0252	0.0634	0.0071	12.61
5	0.0445	2.63	0.0144	0.0480	0.0035	7.86
6	0.0578	2.68	0.0157	0.0501	−0.0077	−13.32
7	0.0604	3.01	0.0266	0.0651	0.0047	7.78
8	0.0536	2.93	0.0244	0.0624	0.0088	16.41

6.2. Matrix Sensor

Figure 16 illustrates matrix sensor results for L1, R1 and middle impacts.

Impact at L1 sensor: The voltage drops for L1 and L2 are 2.56 V and 1.22 V, respectively. The pendulum does not touch R1 and R2.

Impact at R1 sensor: In this test, the sensor is moved to make R1-centered impact. R1 and R2 sensors record 2.40 V and 0.68 V, respectively. R1 has full contact while R2 has partial contact. L1 and L2 record no touch.

Impact in the middle of all the sensors: The sensors' voltage drop varies from 0.9 V to 1.4 V. The voltage drops are identical since the impact and the sensors are symmetrical.

First contact point estimation is crucial for the in-position and out-of-position decision. The first contact is detected when the voltage drops below 4% of the mean reference value. Table 4 shows the estimated first contact time from the tests. The first column is the impact position. During the L1 impact test, the head first touches the L1 sensor, detecting early touch at −3.9 ms. Once the airbag starts to deform, the head touches the L2 sensor at 21.8 ms, followed by R1 touch at 73.7 ms.

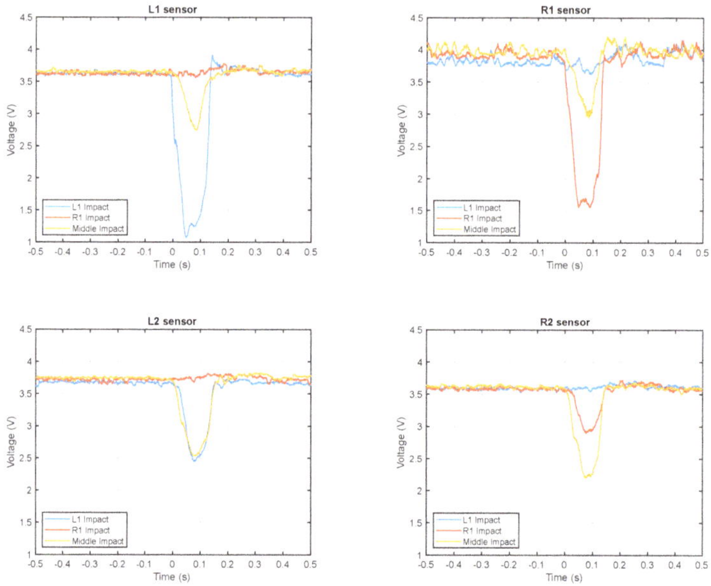

Figure 16. Matrix sensor tests with different impact points by changing sensor position.

Table 4. First touch identification for matrix sensor using threshold.

Impact Position	First touch			
	L1	**L2**	**R1**	**R2**
L1	−3.9 ms	21.8 ms	73.7 ms	NA
R1	NA	6.05 s	−3.8 ms	30.4 ms
Middle	21.9 ms	6.3 ms	8.7 ms	8.7 ms

NA = not available.

The contact area is calculated similarly to the single sensor (Table 5). The voltage drops of L1, L2, R1 and R2 for the L1 impact are 2.56 V, 1.22 V, 0.20 V and 0.07 V, respectively. These voltage drops are compared with the calibration curves and the corresponding contact area is calculated. A is the total contact area obtained by adding the gap area between sensors. The depth D_S is then determined and deviation from the camera is calculated. The deviation is close to 22%.

Table 5. Matrix sensor depth calculation for head-form impact tests.

Impact Position	D_c (m)	Gap (m^2)	A(m^2)	D_s (m)	% Deviation
L1	NA	0	0.01947	0.0556	NA
R1	0.0648	0	0.0169	0.0510	−21.29
Middle	0.0690	0.0084	0.01851	0.0542	−21.44

7. Discussion

Airbag performance is assessed through various test stages. Firstly, static deployment is performed to analyze the unfolding and filling behaviour, followed by linear impactor or pendulum impact tests. These tests are performed to analyze free-form body motion without vehicle deformation to assess airbag performance only. Further, sled tests are carried out on a rigid sled where vehicle motion and seatbelt restraint effects are considered. Finally, full-vehicle crash tests are performed to consider vehicle structural deformation, airbag displacement, and restraint effects. We have chosen pendulum tests in our work while it is practically feasible to change the parameters and provide scaled-down occupant free-form head kinematics and restraint effect. The experiments are cost and time-effective, hence better suited for first performance evaluation and hypothesis testing of the sensors. However, there are certain limitations of the test bench and experiments. The arm length limits the pendulum's impact velocity and, vibrations in the pendulum increase with arm length. Hence in our study, we have restricted the velocity to 5.11 m/s. Airbag pressure also has limitations. Pressure change increases the bag thickness, making pendulum impact before maximum kinetic energy, resulting in lower voltage drop and higher contact times. These limitations can be overcome by testing the airbag in a drop tower facility. The results of single and matrix sensors are further discussed in the following subsections.

7.1. Single Sensor

As hypothesized, the sensor voltage drops with contact progression and reaches a minimum when the head reaches peak depth. When the impact velocity is increased, the voltage drop increases due to the larger contact surface.

The deviation (Table 3) for low velocities is less as the sensor makes perfect contact with the head. When the velocity is high, the sensor flies and contacts different parts of the pendulum assembly, contributing to the deviation. This problem can be overcome by knitting the sensor on the airbag. On the other hand, the contact area is smaller when the head slides on the airbag beyond the sensor area. The drop increases when the velocity is increased. The deviations (Table 3) for the single sensor are reasonably acceptable due to dynamic irregular complex deployment. They can be further reduced by adequately integrating the sensor with the airbag. From the single sensor results (Table 3), it can be concluded that as the impact velocity increases, the area deviation also increases. In real-time moderate speed vehicle collisions, the deviation is acceptable.

Further, the contact times obtained from the sensor are in good agreement with the high-speed video times (Figure 15). With the increase in impact velocity, contact time decreases with a higher drop.

7.2. Matrix Sensor

The minimum deviation for the matrix sensor (from the camera depth) as a whole is 21.44% (Table 5), which means the measured area is smaller than the area obtained from the camera. There are several possible reasons. Firstly, the shape of the head is circular. When the head makes contact, the airbag wraps around the head, making contact with other parts at different time stamps. One solution to this problem is to provide a flat contact. This can be achieved by using a square plate. Although the square plate is not a real condition, it can ease the testing and analysis. The second reason is mutual capacitance and contact capacitance induced between the sensors when the object makes contact. A correction parameter can be incorporated in the occupant detection algorithm by testing individual sensors in the matrix. Mutual capacitance can be reduced by increasing the distance between the sensors.

Further, the experimental results answer the questions in Section 2.

- What are the first and total contact times?
  The contact time from the first contact to the peak can be estimated from both sensors (Figures 15 and 16). Irrespective of the contact position, the single sensor provides first contact time, total contact time, whereas the matrix sensor is position-specific. It gives contact parameters on different regions on the airbag. If there is a single-chambered airbag and out-of-position is not of interest, the single sensor can be preferred over the matrix. If region-specific times are required, the matrix is a choice of application.
- What is the contact area?
  Both sensors can estimate the area. The deviation for the matrix sensor (Table 5) is higher than the single sensor (Table 3). Hence, when the contact area is the only parameter of interest, the single sensor works better than the matrix sensor. A matrix sensor can be installed to estimate the overall area and the individual sensor area if a multi-chambered passenger airbag is used.
- Where is the contact position?
  When the impact position is the parameter of interest, then the matrix sensor plays a significant role. The position can be identified from the matrix sensor based on the threshold crossing time for different sensors in the matrix (Table 4). Early position estimation helps decide in-position and out-of-position, which is crucial information to control the individual chamber pressure.

Three main results—first contact point, time and position—significantly impact the vehicle's passive safety system (airbag or seat belt) during in-crash and post-crash phases. Each parameter can be used to tune the restraint system. Curtain airbags are usually multi-chambered with optional gas flow control between the chambers [30]. The integration of matrix sensors with such airbags enables inflation and exhaust pressure control of each chamber to define an optimal control strategy, which minimizes the occupant's rebound velocity.

Furthermore, sensor data also play a significant role in injury monitoring and rescue strategies. The vehicle can be used as a diagnostic space by installing accelerometers to monitor the respiration [31,32]. The head depth obtained (Table 3) can be combined with respiration data to correlate injuries and respiration. The detailed injury estimation analysis is beyond the scope of this paper. Finally, there is an eCall system in the vehicles which communicates the accident with vehicle data [33]. The diagnostic data and injury data can be integrated with the eCall system.

8. Conclusions

Airbag–occupant contact detection was an open research opportunity due to the short deployment time, sensor material and complex airbag shape during deployment. In this research, we have successfully developed tactile occupant detection sensors capable of estimating contact time with position and the area. The single sensor can be used when there is less probability of OOP occurrence, whereas the matrix sensor is suitable for bigger multi-chambered airbags with pressure control. The passive safety systems can be fully controlled and tuned from the sensor feedback. Injuries can also be estimated and communicated, which makes airbags smarter and adaptive to various crash scenarios.

Author Contributions: Conceptualization, N.S. and C.B.; methodology, N.S., C.B., T.M.D., R.H.; software, N.S.; validation, N.S. and C.B.; formal analysis, N.S., C.B., T.M.D., R.H.; investigation, N.S.; resources, N.S.; data curation, N.S.; writing—original draft preparation, N.S.; writing—review and editing, N.S., C.B., T.M.D., R.H.; visualization, N.S., C.B., T.M.D., R.H.; supervision, C.B., T.M.D., R.H.; project administration, C.B.; funding acquisition, C.B. All authors have read and agreed to the published version of the manuscript.

Funding: This work was financially supported by *Bayerisches Staatsministerium für Wirtschaft, Landesentwicklung und Energie* under the grant *IUK-1902-0007//DIK0102/01* and APC was funded by Technische Hochschule Ingolstadt.

Institutional Review Board Statement: Not applicable.

Informed Consent Statement: Not applicable.

Data Availability Statement: Data is contained within the article.

Conflicts of Interest: The authors declare no conflict of interest.

References

1. Hinch, J.; Hollowell, W.T.; Kanianthra, J.; Evans, W.D.; Klein, T.; Longthrone, A.; Ratchford, S.; Morris, J.; Subramanian, R. Airbag Technology in Light Passenger Vehicles. Available online: https://www.nhtsa.gov/sites/nhtsa.gov/files/rev_report.pdf (accessed on 27 August 2019).
2. Ionut Radu, A.; Cofaru, C.; Tolea, B.; Dima, D. Study regarding the influence of airbag deployment time on the occupant injury level during a frontal vehicle collision. *MATEC Web Conf.* **2018**, *184*, 01007. [CrossRef]
3. Schöneburg, R.; Paurevic, M.; Fehring, M.; Richert, J.; Bogenrieder, R. (Eds.) Evaluation of Occupant Protection Systems: From a Preventive to a Pre-Impacting Restraint System. In Proceedings of the 24th International Technical Conference on the Enhanced Safety of Vehicles (ESV), Gothenburg, Sweden, 8–11 June 2015.
4. Kim, D.E.; Park, W.I.; Shin, B.S.; Kang, M.C. Optimized low-risk deployment of a passenger airbag with a passenger protection wrap considering pressure dispersion. *Proc. Inst. Mech. Eng. Part D J. Automob. Eng.* **2017**, *231*, 27–34. [CrossRef]
5. Kim, Y.S.; Fischer, K.; Nayef, E.; Choi, H.H. (Eds.) Single Stage Driver Airbag Module Development for Out-of-Position. In Proceedings of the 23rd International Technical Conference on the Enhanced Safety of Vehicles (ESV), Seoul, Korea, 27–30 May 2013.
6. Izumi, T.; Saito, H.; Hagihara, T.; Hatanaka, K.; Sawai, T. Development of Occupant Detection System Using Far-Infrared Ray (FIR) Camera. Available online: https://sumitomoelectric.com/sites/default/files/2020-12/download_documents/69-10.pdf (accessed on 14 April 2020).
7. Blackbrun, B.K.; Gentry, S.B.; Mazur, J.F.; Steffens, C.E.; Bessonen, J.E.; Jarocha, W.M. Method and Apparatus for Sensing an Occupant Position Using Capacitance Sensing. U.S. Patent No. 5722686, 3 March 1998.
8. Kithil, P.W.; Barron, M.H.; McIntosh, W.C. Vehicle Occupant Sensing. U.S. Patent No. US6275146B1, 14 August 2001.
9. White, C.; Burdock, J.; Clancy, E. Airbag with Bag Mounted Sensor. U.S. Patent No. US2004/0090053 A1, 19 March 2004.
10. Krotosky, S.J.; Cheng, S.Y.; Trivedi, M.M., Eds. Face detection and head tracking using stereo and thermal infrared cameras for "smart" airbags: A comparative analysis. In Proceedings of the 7th International IEEE Conference on Intelligent Transportation Systems, Washington, WA, USA, 3–6 October 2004; IEEE: Piscataway, NJ, USA, 2004. [CrossRef]
11. Faber, P. Image-based Passenger Detection and Localization inside Vehicles. *Int. Arch. Photogramm. Remote. Sens.* **2000**, *33*, 230–237.
12. Mikic, I.; Trivedi, M. (Eds.) Vehicle Occupant Posture Analysis Using Voxel Data. In Proceedings of the Ninth World Congress on Intellgient Trasport Systems 2002, Chicago, IL, USA, 14–17 October 2002.
13. Farmer, M.E.; Jain, A.K. Smart Automotive Airbags: Occupant Classification and Tracking. *IEEE Trans. Veh. Technol.* **2007**, *56*, 60–80. [CrossRef]
14. yang, Y.; Zao, G.; Sheng, J. (Eds.) Occupant Pose and Location Detect for Intelligent Airbag System Based on Computer Vision. In Proceedings of the 2008 Fourth International Conference on Natural Computation, Jinan, China, 18–20 October 2008; IEEE: Piscataway, NJ, USA, 2008. [CrossRef]
15. Shirur, N.; Birkner, C.; Henze, R.; Deserno, T.M.; Dudhat, D. (Eds.) Effect of Airbag Deployment Phases on Tactile Occupant Sensor. In Proceedings of the 2020 XII International Science-Technical Conference AUTOMOTIVE SAFETY, Kielce, Poland, 21–23 October 2020; IEEE: Piscataway, NJ, USA, 2020. [CrossRef]
16. Zimmermann, R.E. Vent control as a means of enhancing airbag performance. *Shock Vib.* **2002**, *9*, 123–128. [CrossRef]
17. Jonassen, N. *Electrostatics*, 2nd ed.; The Springer International Series in Engineering and Computer Science; Springer Science+Business Media New York and Springer: Boston, MA, USA, 2002; Volume 700. [CrossRef]
18. Grosse-Puppendah, T.; Holz, C.; Cohn, G.; Wimmer, R.; Bechtold, O.; Hodges, S.; Reynolds, M.S.; Smith, J.R. (Eds.) *Finding Common Ground: A Survey of Capacitive Sensing in Human-Computer Interaction*; ACM: New York, NY, USA, 2017. [CrossRef]
19. Gray, T. *Projected Capacitive Touch*; Springer International Publishing: Cham, Switzerland, 2019. [CrossRef]

20. Touch Sensors Design Guide. Available online: http://www.farnell.com/datasheets/1504633.pdf (accessed on 2 February 2019).
21. Russo, T.V. Application Note, Atmel QTouch. Available online: https://www.microchip.com/content/dam/mchp/documents/OTH/ApplicationNotes/ApplicationNotes/Atmel-42094-QTouch-Schematic-and-Layout-Checklist_ApplicationNote_AT02259.pdf (accessed on 10 May 2019). [CrossRef]
22. Terzopoulos, N. Capacitive Touch Solutions for Wearable Applications. Available online: https://www.dialog-semiconductor.com/sites/default/files/capacitive_touch_solutions_for_wearable_applications_april_2017.pdf (accessed on 2 February 2019).
23. Fischer, D. Capacitive Touch Sensors. Available online: https://pdf4pro.com/view/capacitive-touch-sensors-fujitsu-global-e79f7.html (accessed on 5 March 2019).
24. Application Note: Capacitive Touch Sensing Layout Guidelines. Available online: https://www.mouser.com/pdfdocs/semtech-capacitive-touch-sensing-layout-guidelines.pdf (accessed on 6 May 2019).
25. Takamatsu, S.; Yamashita, T.; Imai, T.; Itoh, T. Fabric Touch Sensors Using Projected Self-Capacitive Touch Technique. *Sens. Mater.* **2013**, *25*, 627–634. [CrossRef]
26. Adafruit Woven Conductive Fabric. Available online: https://www.adafruit.com/product/1168 (accessed on 15 February 2019).
27. EXPTECH, Conductive Thread. Available online: https://www.exp-tech.de/zubehoer/sonstige/7419/conductive-thread-60g-stainless-steel?gclid=EAIaIQobChMI-L_1-42I5gIVUaqaCh2NOgvwEAQYBSABEgJ2KfD_BwE (accessed on 10 February 2019).
28. Arduino Capacitive Sensing Library. Available online: https://playground.arduino.cc/Main/CapacitiveSensor/ (accessed on 20 May 2019).
29. Lina, P.-V.; Jurkauskas, A. Research into Occupant'S Motion in Vehicles During Crashes. *Transport* **2004**, *19*, 184–190. [CrossRef]
30. Eung-Seo, K.; Dae-Young, K.; Hyeong-Ho, C.; Han-Il, B.; Seung-Hui, Y.; Seung-Man, K.; Dong-Jun, L.; Kwang-Soo, C. (Eds.) A Stufy of Curtain Airbag Design Factors for Enhanancement of Ejection Mitigation Performance. In Proceedings of the 22nd International Technical Conference on the Enhanced Safety of Vehicles (ESV), Washington, DC, USA, 13–16 June 2011.
31. Wang, J.; Warnecke, J.M.; Deserno, T.M. The Vehicle as a Diagnostic Space: Efficient Placement of Accelerometers for Respiration Monitoring uring Driving. *ICT Health Sci. Res.* **2019**, *258*, 206–210. [CrossRef]
32. Wang, J.; Warnecke, J.M.; Haghi, M.; Deserno, T.M. Unobtrusive Health Monitoring in Private Spaces: The Smart Vehicle. *Sensors* **2020**, *20*, 2442. [CrossRef] [PubMed]
33. Bonyar, A.; Geczy, A.; Krammer, O.; Santha, H.; Illes, B.; Kaman, J.; Szalay, Z.; Hanak, P.; Harsanyi, G. (Eds.) A review on current eCall systems for autonomous car accident detection. In Proceedings of the 2017 40th International Spring Seminar on Electronics Technology (ISSE), Sofia, Bulgaria, 10–14 May 2017; IEEE: Piscataway, NJ, USA, 2017. [CrossRef]

energies

MDPI

Article

Effects of All-Night Driving on Selective Attention in Professional Truck Drivers: A Preliminary Functional Magnetic Resonance Study

Stefan P. Gazdzinski [1,*,†], Marek Binder [2,*,†], Alicja Bortkiewicz [3,*], Paulina Baran [4] and Łukasz Dziuda [4]

[1] Department of Neuroimaging, Military Institute of Aviation Medicine, Krasinskiego 54/56, 01-755 Warszawa, Poland
[2] Institute of Psychology, Jagiellonian University, Ingardena 6, 30-060 Krakow, Poland
[3] Nofer Collegium, Nofer Institute of Occupational Medicine, św. Teresy od dzieciątka Jezus 8, 91-438 Lodz, Poland
[4] Department of Psychophysiological Measurements and Human Factor Research, Military Institute of Aviation Medicine, Krasińskiego 54/56, 01-755 Warszawa, Poland; pbaran@wiml.waw.pl (P.B.); ldziuda@wiml.waw.pl (Ł.D.)
* Correspondence: sgazdzin@wiml.waw.pl (S.P.G.); marek.binder@uj.edu.pl (M.B.); alicja.bortkiewicz@imp.lodz.pl (A.B.)
† The authors had equal contribution to the manuscript.

Abstract: Fatigue affects multiple aspects of cognitive performance among drivers. However, even after fatigue builds up, some are still able to maintain the level of behavioral performance. To evaluate these adaptations on the neural network level, we utilized functional magnetic resonance imaging (fMRI). Seventeen male professional drivers underwent two fMRI sessions, once while rested and once in a fatigued condition after 10-h of overnight driving. The cognitive task used in the study involved the detection of visual feature conjunctions, namely the shape and the color. There were no significant differences in the task performance between the conditions except for longer reaction times in the fatigued condition. However, we observed substantial differences in the activation patterns during the cognitive task involving selective attention between the conditions. On the global level, we observed a general decrease in activation strength in the fatigued condition, which appeared to be more pronounced in the left hemisphere. On the local level, we observed a (spatially) extended activation of the medial prefrontal regions in the fatigued condition, which reflected increased cognitive control mechanisms compensating for the diminished efficiency of mechanisms involved in meeting task demands.

Keywords: fatigue; selective attention; functional brain imaging

Citation: Gazdzinski, S.P.; Binder, M.; Bortkiewicz, A.; Baran, P.; Dziuda, Ł. Effects of All-Night Driving on Selective Attention in Professional Truck Drivers: A Preliminary Functional Magnetic Resonance Study. *Energies* **2021**, *14*, 5409. https://doi.org/10.3390/en14175409

Academic Editor: Hugo Morais

Received: 22 July 2021
Accepted: 24 August 2021
Published: 31 August 2021

1. Introduction

Driving a car is a complex behavior that requires diverse abilities, including perceptual, attentional, decision-making, and motor skills [1]. All-night driving leads the buildup of fatigue. Driver's fatigue is related to about 10% of the total number of accidents, and about 25% of single-vehicle accidents [1]. The impact of fatigued driving in the EU is significant: social costs (including healthcare) of all road accidents leading to injuries and fatalities are at least EUR 100 billion per year [2].

Fatigue is a cumulative process. It is related to a sustained activity and often results in impaired performance. The main symptoms are: difficulty in maintaining alertness and focus of selective attention, vigilance, and staying awake [3]. Understanding the mechanisms that modulate selective attention in fatigued subjects may provide implications for accident prevention, as this information may help with management of human errors and minimization of number of accidents and injuries [4]. For this reason, non-invasive measurements of brain activity can play a role in understanding the neural basis of driving ability [5].

The ability to attend to one's surroundings is of utmost importance to safety in an environment with competing stimuli; failure to perceive relevant stimuli due to decreased attentional control can result in accident and injury [4]. Alertness and selective attention are closely intertwined but separable dimensions of attention. Both play an important role in ensuring driving safety. According to the salience effort expectancy value (SEEV) attention allocation model, 90% of attention used to operate a vehicle has visual character [6]. Visual selective attention plays a special role in driving behavior/control, since each driver is confronted with a plethora of competing stimuli that must be recognized and processed quickly to ensure the coordinated responsiveness to all the environmental events occurring while driving a car. Inadequate allocation of attention was identified as one of major factors leading to road accidents [7].

According to feature integration theory [8], visual selective attention is based on two linked levels of representation. On the first level, visual features such as color or shape are represented in separate feature maps. The second level of representation, the master map of locations, encodes the current site of the attentional focus. The attentional selection works on the second level by binding visual features present in the site of attentional focus. Thus, the attentional selection of visual objects requires not only the correct registration of their features, but also their proper integration [9]. Earlier fMRI studies identified the nodes of the underlying neural network using feature and conjunction search tasks [10]. An object defined by multiple features is simultaneously processed by functionally specialized systems in the brain. Cognitive processes of visual short-term memory (color, shape, and their conjunctions) were shown to activate (in a load-dependent manner) the bilateral superior parietal lobule (Brodmann area 7), close to intraparietal sulcus [11]. The posterior parietal cortex was sensitive to visual short-term memory load for color and shape [11]; in particular, it was sensitive to feature and visual working memory load manipulations. Other studies found greater activity in the right parietal cortex at the junction of the intraparietal cortex and transverse occipital sulcus during a conjunction search, as compared to a single-feature search [9,12]. On the other hand, the prefrontal regions were sensitive to visual working memory load manipulation, but relatively insensitive to feature differences [13].

Acknowledging the complexity of the factors contributing to driver's fatigue and ensuing decrease in driving performance, in this study we focused on the modulatory influence of the fatigue on the attentional system of the brain. While the brain networks underlying alertness have been shown to be sensitive to fatigue, less is known about its influence on selective attention. We used a feature conjunction detection task in which the participants were required to respond to a specific combination of color and shape. Such a task specifically addressed the domain of attentional control, since feature integration in target detection tasks involves this type of top-down attentional processing. In our study, all participants performed the task in the fMRI scanner twice—in a fatigued condition, following substantial sleep deprivation during the preceding 24 h period, and in a rested condition, after several hours of ceaseless sleep. We predicted decreased neuronal activity in the nodes of the attentional control network associated with the fatigued condition, irrespective of the task performance.

2. Materials and Methods

This study was part of a larger project to detect early signs of fatigue to improve the safety of driving. Qualification for participation in the project was preceded by general medical, neurological, and ophthalmological examinations. Seventeen male, right-handed [4], professional drivers with a current medical examination to qualify for professional driving, aged 32.9 $\pm$ 4.4 years, with work experience in the profession of 12.6 $\pm$ 5.6 years, without self-reported chronic conditions that impaired sensory and cognitive functions and could lead to sleepiness during driving [2,3], took part in the randomized cross-over functional MRI examination. The sex of the participants reflected the fact that this profession is rarely chosen by females. Their body mass index (BMI = 28.7 $\pm$ 3.3) was above the recommended values [14]; only three participants had a proper body weight, eight were overweight, and six were obese. Nine participants had a higher education, five completed secondary school, two had vocational training, and one had junior high school). Their average number of working hours per month was 220 ± 36.

Each participant underwent fMRI scanning session twice; i.e., according to randomized, control trial methodology: (1) after a normal night of rest for the driver, and (2) after a 10 h, overnight period of driving under normal working conditions. The median interval between both sessions was seven days (range from one day to 98 days). In order to minimize the impact of the so-called learning effect (interfering variable) on the results, seven of the drivers were first tested at rest, while for the remaining 10, the first fMRI was after the 10 h driving period. All the tests were performed during morning hours to eliminate the influence of circadian rhythms on physiological functions.

All the procedures were approved by the Bioethics Committee of the Military Institute of Aviation Medicine, Warszawa, Poland (Decision No. 11/2015) and were performed in accordance with the ethical standards as laid down in the 1964 Declaration of Helsinki and its later amendments or comparable ethical standards. Prior to the study, all the participants gave written informed consent to all procedures and personal data processing for scientific purposes. All data was defaced and anonymized before analyses.

2.1. Assessment of Sleep Deprivation

Before the fMRI examinations, all participants completed a questionnaire inquiring about the length of sleep the previous night, the amount of time elapsed since last waking up, the number and length of naps in the last 24 h prior to the fMRI, and potential problems with maintaining awareness during the fMRI scanning session. The effects of sleep deprivation and drowsiness were assessed using the Sleep Deficiency Effects Scale (the CHICa scale) [15]. This self-report tool enables the measurement of four components of an individual's fatigue resulting from sleep deprivation, and its results are grouped into four subscales; i.e., impaired thermoregulation (cold subscale), disrupted appetite (hunger subscale), affective problems (irritation subscale), and decreased level of cognitive functioning (cognitive attenuation subscale). Finally, the symptoms of fatigue were also assessed using the Polish version of the Japanese Questionnaire (named as the Assessment of the Current Well-Being) [16,17]. This questionnaire evaluates the symptoms of fatigue based on three groups of symptoms; i.e., (1) decreased activity (e.g., sleepiness, weariness, heaviness of the body), (2) weakened motivation (e.g., irritability, inability to concentrate, apathy), and (3) physical fatigue (e.g., trembling of the limbs, back pain), as well as a general score indicating an overall level of fatigue.

2.2. fMRI-Visual Stimuli

The visual stimuli were pairs of filled line drawings of geometric figures (squares, triangles, circles) presented on an LCD screen (see Figure 1). Each figure was filled with one of four possible colors (red, green, yellow, or blue). Stimuli were presented over a grey background (RGB: 64,64,64). A fixation point was presented in the center of the screen (white-filled circle, diameter = 3 pixels). The stimulus consisted of a pair of geometric figures, which were shifted 150 pixels to the left or to the right in regard to the fixation point; this horizontal deflection subtended ±4.5 degrees of visual angle relative to the center of the screen. The LCD screen resolution was 1920 × 1080 pixels (InroomViewingDevice, NordicNeuroLab, Bergen, Norway). The stimuli were back-projected on the screen and viewed through a system of mirrors mounted inside the head coil. Each stimulus pair was presented for 200 ms. Following the presentation of the stimulus pair, the short irregular inter-trial interval of varying duration of 1300 ms to 2300 ms occurred. During this period, only the fixation point was visible.

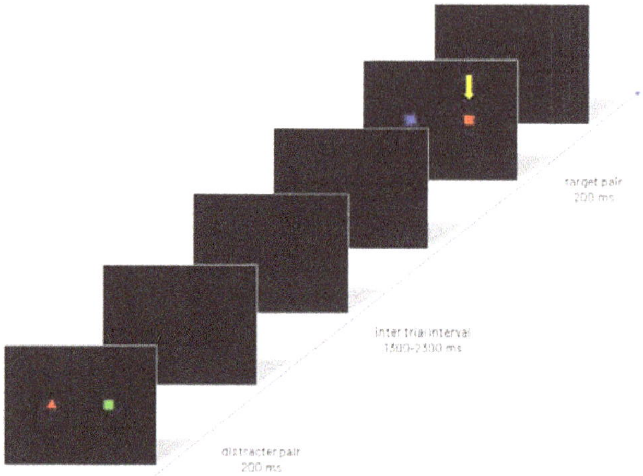

Figure 1. An outline of the cognitive task with two example trials separated by the inter-trial interval. In the actual experiment, 15 trials were presented during each 30 s block. The yellow arrow (not shown during the actual experiment) indicates the target stimulus. The relative size of the stimuli has been enlarged for illustrative purposes. Participants were instructed to respond with the left index finger to each instance of the distracter stimulus, and with the right index finger to each instance of the target stimulus.

2.3. Cognitive Task

The cognitive task used in the study involved the detection of visual feature conjunctions, namely the shape and the color. The study protocol was based on a blocked design, with each block lasting for 30 s. There was one type of an active block (repeated 6 times during the scanning session), which was interspersed with baseline blocks (7 repetitions). The functional scanning session always began and ended with the baseline block. All participants practiced the task before the scanning session (with three active blocks).

During the active block, 15 pairs of filled geometric figures appeared on the screen, presented on the left or the right side of the fixation point. Each active block was preceded with an instruction screen shown for 2000 ms containing the prompt "Detect the red square" in the center of the screen, displayed in yellow font. The volunteers had to detect the target stimulus in each pair, which was a specific combination of color and shape—a red square (see Figure 1 for the cognitive task outline). The participants had to react to this stimulus with right index finger using an NNL ResponseGrip (NordicNeuroLab, Bergen, Norway). They were instructed to respond with their left index finger to all other combinations of features and shapes, called distracter stimuli; for example, blue squares, red circles etc. Participants were required to respond in each trial.

To decrease participants' expectations about the probability of the target, the proportion of pairs containing the target stimulus varied in each block. In two blocks, the target probability was set to 0.8, in another two, it was decreased to 0.2, and in the remaining two blocks, it was 0.5. The order of the blocks was random. During the so-called baseline blocks, the participants' task was to keep their eyes on the fixation point.

The task difficulty was set to equalize the task performance during both the rested and fatigued conditions. In this way, detected differences between the rested and fatigued conditions could indicate modulation of task-related neural processing evoked by the alteration in tiredness, and not the confounding effects of varying subjective task difficulty.

2.4. Image Acquisition and Processing

The study was performed with a GE DISCOVERY MR750w with a 3.0 T field. For the presentation of the stimuli, an LCD screen with a horizontal angle of 24° was used, which was viewed by the participants through a system of mirrors in the head coil. The initiation of the procedure was synchronized with the first scanner TR by NNL SyncBox (NordicNeuroLab, Bergen, Norway https://nordicneurolab.com/product/responsegrips/ (accessed on 20 August 2021)). The participants responded with NNL grips (NordicNeuroLab, Bergen, Norway) by pressing the buttons under the index fingers of the left and right hand.

An 8-channel coil was used in all tests. Structural images were acquired with a 3D Fast Spoiled Gradient Echo (FSPGR BRAVO) sequence (TR = 8.496 ms, TE = 3.26 ms, TI = 450 ms, matrix 256 × 256 × 124, single voxel size: 0.9375 mm × 0.9375 mm × 1.2 mm). Functional images within the fMRI task used the echo planar (EPI) sequence with the following parameters: TR = 2000 ms, TE = 30 ms, 64 × 64 × 35 matrix, single voxel size: 3.125 × 3.125 × 3.5 mm, 196 repetitions, 4 dummy scans at the beginning of each scanning session.

2.5. Data Analyses

2.5.1. Behavioral Data

Behavioral data were compared between conditions (fatigued vs. rested) using paired *t*-tests. All tests were performed with IBM SPSS Statistics software.

2.5.2. Imaging Data

Imaging data analysis was performed using FSL 5.09, FEAT Version 6.0 [18]. Standard pre-processing steps were performed: correction of head movements with the MCFLIRT tool, and removal of non-brain voxels from images with the BET tool. Global intensity normalization of images with a single factor and a Gaussian-weighted least-squares straight line fitting, with sigma = 45.0 s, was performed. Registration and normalization results were visually inspected for each participant. Level 1 analysis was performed using the FILM tool with local autocorrelation correction. The main effect of active blocks was examined. The Z > 3.1 threshold was applied with the use of clusters and the corrected cluster significance threshold $p < 0.05$. Level 2 analysis was performed in the mixed-effects model, with a participant as a random factor; the significance of stimuli was determined by threshold Z > 3.1 with the use of clusters and adjusted significance threshold $p < 0.05$. The whole-brain main effects analyses were intended to reveal level 2 group means for each condition separately (i.e., the difference between active and baseline blocks). The correlations of the activations with reaction times were evaluated separately for each condition. The effect of condition was tested with a T-statistics contrast between conditions. We analyzed contrasts fatigued > rested as well as rested > fatigued.

3. Results

3.1. Behavioral—Fatigue Symptoms

The average sleep length during the night preceding the study was 1.7 ± 1.1 h (min = 0 h, max = 3 h) in the fatigued condition, as compared to 6.8 ± 1.1 h (min = 4 h, max = 9 h) in the rested condition. As some of the participants reported napping, we also measured time since last waking up. In the fatigued condition, the time was on average 8.2 ± 6.6 h (min = 2.5 h, max = 26 h), whereas in the rested condition, the times were the following: 2.9 ± 1.0 (min = 2 h, max = 5 h). The drivers had significantly more symptoms of sleepiness and fatigue in the fatigued condition than in the rested one, which confirmed the successful manipulation of their state (see Tables 1 and 2).

As shown in Table 1, the drivers participating in the study were more tired, and consequently reported more severe symptoms of fatigue in the assumed fatigued condition than in the rested one. Significant differences between these two experimental conditions were obtained in three subscales, as well as in the general score in the Japanese Questionnaire.

The results in Table 2 show that the reduced level of self-reported cognitive functioning of the drivers was the most symptomatic (the highest score on the cognitive-attenuation subscale in the fatigued condition compared to other subscales).

Table 1. Severity of fatigue symptoms reported in the Japanese Questionnaire (lower score = more severe reported symptoms). The general score was defined as the sum of the results from three subscales. All $p < 0.001$.

Symptoms	Fatigued Condition	Rested Condition
Decreased activity	3.1 ± 4.3	19.6 ± 7.3
Weakened motivation	1.3 ± 2.1	12.8 ± 6.9
Physical fatigue	1.4 ± 2.6	10.4 ± 7.9
General score	5.8 ± 7.3	42.8 ± 18.9

Table 2. Measures of sleep deprivation in rested and fatigued conditions on the CHICa subscales. All $p < 0.001$.

Fatigue Components	Fatigued Condition	Rested Condition
Impaired thermoregulation	3.1 ± 2.4	0.5 ± 1.0
Disrupted appetite	4.3 ± 2.5	0.8 ± 1.2
Irritation	5.4 ± 4.0	0.5 ± 1.5
Cognitive attenuation	10.3 ± 4.1	1.5 ± 2.4

3.2. *Behavioral—Task Performance*

There were no significant differences in the level of task performance between the fMRI sessions. The percentages of correct answers were 97 ± 3% for the rested condition and 95 ± 5% for the fatigued condition ($p = 0.06$). However, there were noted differences in the response speed. The mean response time was 13% longer for the fatigued condition; particularly for the right hand, the average response time increased from 644 ± 44 ms to 731 ± 130 ms ($p = 0.01$); whereas for the left hand, the mean response time for the rested condition was 657 ± 42 ms, while for the fatigued condition, the mean response time was 743 ± 113 ms ($p = 0.004$).

3.3. *Imaging Results (Analysis Level 2)*

3.3.1. Main Effects of the Task in Both Conditions

The results are displayed in Figures 2 and 3, as well as presented in Tables 3 and 4. In both conditions, we observed a mostly overlapping pattern of activations in the lateral frontal, lateral parietal, and medial frontal regions. In both conditions, we also observed overlapping clusters in the left lateral temporo-occipital regions. There were marked differences between both conditions. In the rested condition, the general extent of suprathreshold activation was notably wider and more bilateral than in the fatigued condition. The strongest effects of task in the rested condition were observed in the posterior parietal cortex, as well as in the premotor cortex, extending into the dorsolateral cortex, mainly in the right hemisphere. Bilateral activations were also observed in the anterior insulae and in the medial frontal regions (SMA). In contrast with the fatigued condition, we observed bilateral activations in the polar visual cortex, as well as in the fusiform gyrus.

Table 3. Main effects—rested condition. The table lists the approximate anatomical details of the suprathreshold clusters. Z-score represents the maximum Z value within the given cluster. X, Y, and Z values denote MNI coordinates of the maximum Z-value voxel. Cluster volume is given as a voxel count. Anatomical labels were identified using the Harvard-Oxford Cortical and Subcortical Atlases, Juelich Histological Atlas, and Talairach Daemon Labels as provided by FSL software (https://fsl.fmrib.ox.ac.uk/fsl/fslwiki/Atlases (accessed on 20 August 2021)). The characters within parentheses accompanying the anatomical labels indicate the hemisphere—left or right.

Anatomical Label	Cluster Volume	Z-Score (Maximum)	X	Y	Z
Supplementary motor area (L/R), medial frontal gyrus (L/R), inferior parietal lobule (R), postcentral gyrus (R)	9708	6.31	−2	4	52
Lateral occipital cortex (L), fusiform gyrus (L)	1425	4.74	−44	−70	−14
Inferior parietal lobule (L), postcentral gyrus (L)	1411	4.77	−32	−56	44
Putamen (R), thalamus (R), caudate (R)	1185	5.2	20	12	2
Anterior insula (L)	385	4.98	−36	14	14
Occipital pole (L)	327	4.24	−20	−100	−12
Thalamus (L), caudate (L)	300	4.66	−16	−8	16
Middle frontal gyrus (L), precentral gyrus (L)	292	4.82	−46	6	36
Occipital pole (R)	291	4.45	22	−92	−14
Precentral gyrus (L), superior frontal gyrus (L)	291	4.29	−32	−10	52

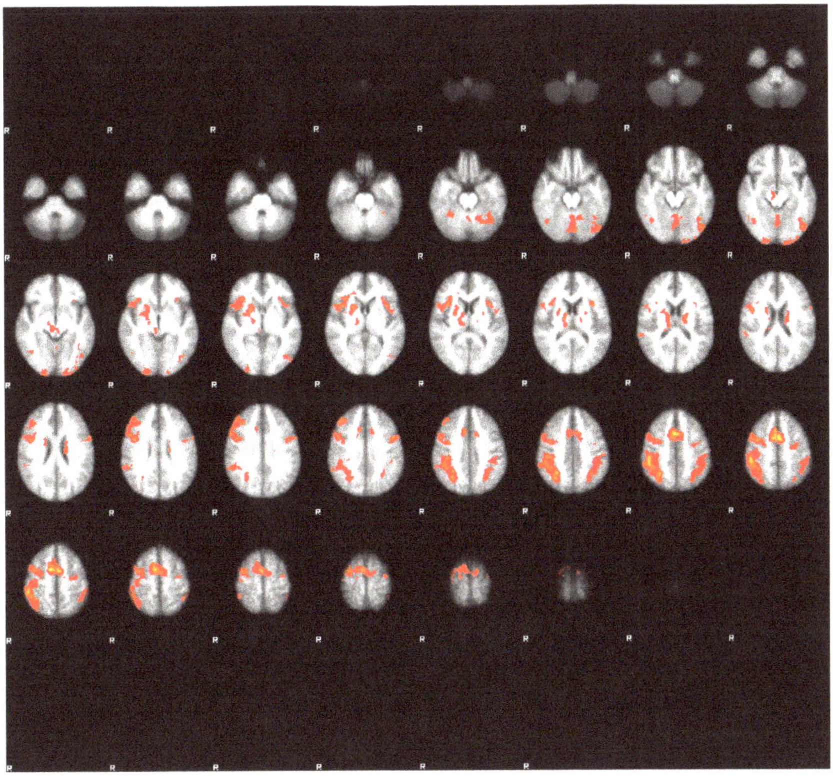

Figure 2. The main effects of the task in the rested condition. Images were thresholded at Z > 3.1. Note the extensive bilateral activation in the posterior parietal cortices, SMA, premotor cortex, and dorso-lateral prefrontal cortex. Activation in the middle temporal cortex was left-lateralized. The right side of the brain is denoted by letter R.

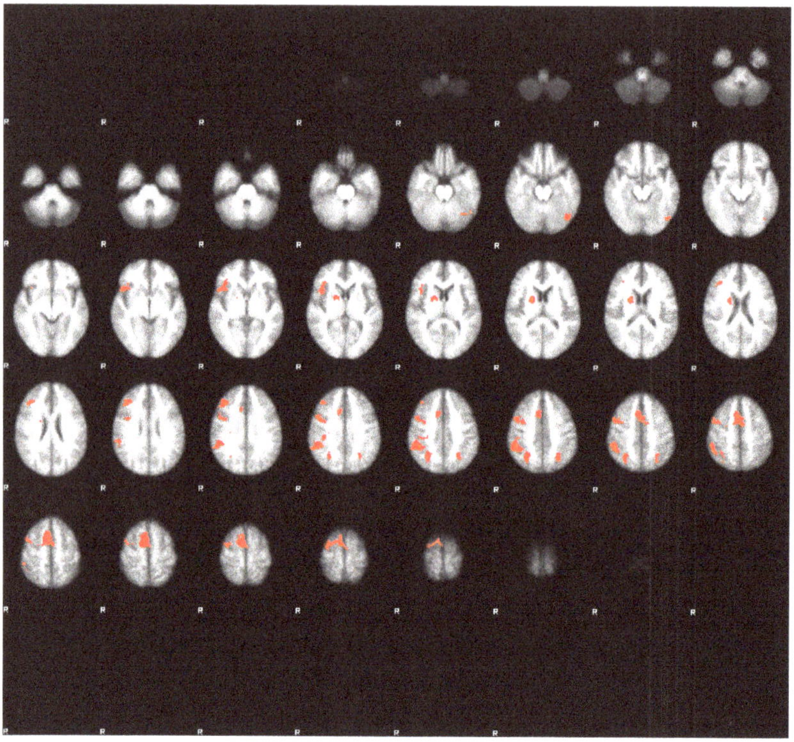

Figure 3. The main effects of the task in the fatigued condition. Images were thresholded at Z > 3.1. Note the more extensive activations in the right hemisphere of the brain (denoted by the letter R) in the posterior parietal cortices, supplementary motor area (SMA), premotor cortex, and dorso-lateral prefrontal cortex in comparison to the rested condition (see Figure 2). Activation in the middle temporal cortex maintained its lateralization in the left hemisphere.

Table 4. Main effects—fatigued condition. The table lists the approximate anatomical details of the suprathreshold clusters. Z-score represents the maximum Z value within the given cluster. X, Y, and Z values denote MNI coordinates of the maximum Z-value voxel. Cluster volume is given as a voxel count. Anatomical labels were identified using the Harvard-Oxford Cortical and Subcortical Atlases, Juelich Histological Atlas, and Talairach Daemon Labels as provided by FSL software. The characters within parentheses accompanying the anatomical labels indicate the hemisphere—left or right.

Anatomical Label	Cluster Volume	Z-Score (Maximum)	X	Y	Z
Supplementary motor area (R), precentral gyrus (R), middle frontal gyrus (R), medial frontal gyrus (R)	2348	4.57	46	6	34
Inferior parietal lobule (R)	967	4.85	48	−42	46
Inferior frontal gyrus (R)	372	4.45	40	30	2
Middle frontal gyrus (R)	358	4.41	40	34	34
Superior parietal lobule (R)	356	4.56	28	−66	44
Putamen (R)	286	4.72	22	2	16
Superior parietal lobule (L)	216	4.32	−28	−68	46
Fusiform cortex (L)	150	4.02	−48	−66	−20

In the fatigued condition, the extent of the suprathreshold clusters was notably smaller. Bilateral activations were seen only in the parietal regions, while in the frontal regions, the activations were mainly observed in the right hemisphere. In neither condition were the activations correlated with the reaction times.

3.3.2. Contrast Results

Direct contrast between both conditions revealed a suprathreshold cluster only in the contrast fatigued > rested. The cluster was located in the right superior frontal gyrus (see Figure 4c). Interestingly, this region did not exceed the Z = 3.1 threshold in either condition, and it appeared in the fatigued > rested contrast because of the negative Z values in the rested condition (see Figure 4a showing the unthresholded Z values) and positive Z values in the fatigued condition (Figure 4b shows the unthresholded Z values). Thus, the difference of activation of this region in both conditions was caused by low response of this region in the rested condition and the moderately elevated response in the fatigued condition.

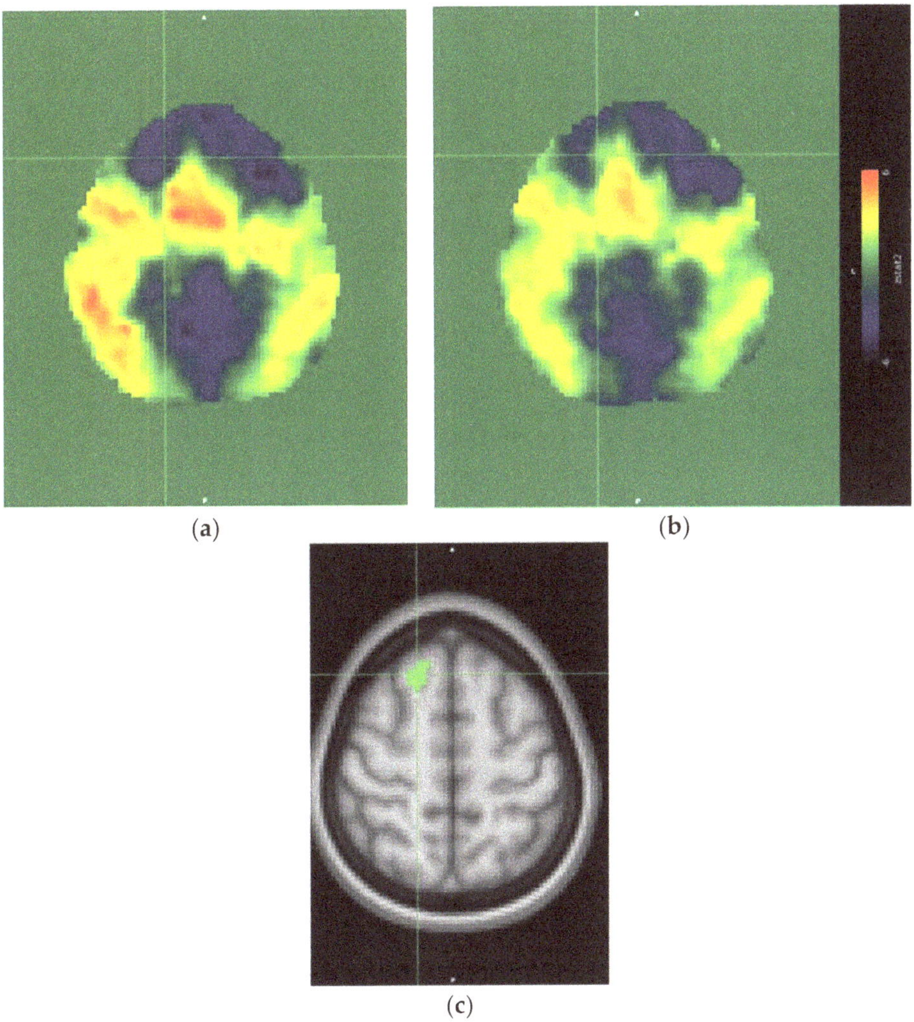

(a) (b) (c)

Figure 4. Unthresholded Z-statistics image of rested condition results (**a**), unthresholded Z-statistics image of the fatigued condition results (**b**). Note the larger extension of activations in the rested condition (**a**) as compared to the fatigued condition (**b**). The region of statistically significant differences; i.e., the contrast fatigued > rested conditions thresholded at Z > 3.1 (**c**). The crosshairs indicate the same voxel at the center of the suprathreshold cluster in the right superior frontal gyrus (MNI coordinates: x = 18, y = 26, y = 56, cluster volume = 136, Z-score (maximum—4.53)).

4. Discussion

In this study, we assessed the susceptibility of the networks involved in selective attention to the fatigue state induced by sleep restriction in professional truck drivers. In order to engage these networks, the visual feature detection cognitive task was used, in which the participants responded to the conjunctions of color and shape of visual stimuli. The difficulty level of the cognitive task was adjusted to ensure a commensurate performance in both arousal states.

Our manipulation of fatigue level via sleep deprivation proved successful. Both the CHICa scale and the Japanese Questionnaire revealed significant changes in self-reported arousal level due to the 10 h of overnight driving under normal working conditions. The results of the CHICa demonstrated that the reduced level of cognitive functioning in the fatigued condition was accompanied by the symptoms of impaired thermoregulation, disrupted appetite, and irritation. These results were consistent with the data obtained by the authors of this questionnaire themselves [15], according to which the symptoms of cognitive attenuation; i.e., problems with concentration, memory, logical thinking and understanding, lack of energy, and decreased accuracy at work, seemed to be the most characteristic and specific subjective manifestations of the sleep-deprivation state. It should also be noted, however, that while some physiological symptoms of sleep deprivation may be unpleasant for the driver (cold, hunger), they can be easily modified and fixed; whereas coping with emotional and cognitive problems is much more difficult, and consequently, they can endanger driving safety by influencing, for example, the driver's situational awareness in road traffic, as well as his sense of control behind the wheel [15].

The increase in response time in the fatigued condition was consistent with other studies [4,19,20]. The changes in response time can be interpreted as indicators of lowered alertness, accompanied by slowed processing of relevant stimuli without affecting the accuracy.

Regarding the whole-brain fMRI results in each condition, we observed a widespread activation spanning across the whole cortical mantle. The activated clusters in the rested condition were located in the regions consistently observed in the task involving selective attention (dorsal frontal areas, posterior parietal areas, medial frontal regions, and basal ganglia) [9–12]. Moreover, the additional suprathreshold clusters were observed in regions related to the visual ventral stream in the fusiform gyrus, as well as regions related to willed action in the anterior insula [21]. The fact that a similar pattern of activated regions involved in attentional selection was observed in both conditions suggests that in both situations, participants recruited basically the same brain networks. This observation was also supported by the lack of significant brain response differences in those networks as revealed by the direct contrast between the conditions.

However, at the same time we observed a notable decrease of the extent of activation in the fatigued condition, indicating a less-consistent brain response to the experimental task across participants. Interestingly, we observed a more lateralized pattern of activation in the fatigued condition, with the majority of clusters located within the right hemisphere. We surmised that this effect might be associated with lowered brain metabolism in the state of sleep deprivation [22]. Of course, this does not mean that truck driver's brain resembles a brain of a dolphin, yet it points to a lowered energy capacity of the fatigued brain, which can affect each hemisphere in a different manner [23]. The direct contrast between conditions revealed only one suprathreshold cluster, obtained in the fatigued > rested contrast. The reverse contrast did not reveal any significant differences. The cluster was located in the dorsal part of the prefrontal cortex. Interestingly, this region did not cross the significance threshold in any of the main effect contrasts for both the rested and fatigued conditions. However, a close inspection of the unthresholded images revealed that in the rested condition, this cluster was strongly deactivated, and in the fatigued condition, the response was positive, yet below the significance threshold. The inspection also revealed that the region discovered in the fatigued > rested contrast was a part of the larger cluster encompassing the medial frontal cortex—covering the SMA, pre-SMA, and parts of the superior frontal gyrus (see Figure 4). Previous research on the effects of cognitive working memory and attentional training research suggested that this region is a part of the cognitive control network, which is engaged in executive control over ongoing cognitive activity, and it is known that its responsiveness becomes significantly decreased over a training period [24,25]. Thus, in our case, the heightened activity of this region might suggest that a relatively simple task we used, involving simple visual feature conjunction detection, in the fatigued condition with lowered metabolic capacity due to sleep deprivation became a task that required more cognitive control resources in order to maintain the performance level. Thus, our results suggested that the influence in the arousal state of drivers on the cognitive level manifested as an increased demand for controlled attentional processing.

Although visually apparent, there were no statistically significant differences in activation of Brodmann area 7. An earlier study by Muto and colleagues [26] demonstrated that in young, healthy participants, there were no effects of one night of total sleep deprivation on attention; however, orienting and conflict resolutions were associated with significantly larger thalamic responses during sleep deprivation than during rested wakefulness. They concluded that sleep deprivation influenced different components of human attention non-selectively by affecting the structures maintaining vigilance or ubiquitously perturbing neuronal function. They further concluded that compensatory responses can counter these effects transiently by recruiting thalamic responses via that supporting thalamocortical function.

According to Johns [27], there is a continuous inhibitory interaction between a wake drive and a sleep drive, each of which involves integrated action of several different neuronal centers in the brain. At any time, the state of sleep or wakefulness depends on the comparative strengths of the total wake drive and the total sleep drive. Under the circumstances of fatigued driving, the driver can stay awake only by maintaining or increasing the secondary wake drive. Long-haul truck drivers are known to have developed methods to increase their wake drive, such as chewing gum, singing, or making frequent changes to their sitting position [27]. In our task, no behavioral influences on the wake drive were possible, as they would result in deterioration of data quality, which was not the case in our study. Therefore, the differences in brain activation between the rested and fatigued conditions reflected the adaptation mechanism involving increased cognitive control over task execution—in this case, voluntary focusing of attention on the incoming stimuli and the response selection.

Our research had certain limitations. The study was performed while the subjects were lying on a scanner table—a posture (position) that is known to increase subjective perception of sleepiness [27]. Therefore, some interactive effects of this position could not be excluded. Moreover, the small sample might have obscured smaller differences between experimental conditions. Furthermore, caffeine is known to have short-term effects on basal cerebral blood flow, and thus the activation strength of the brain [28]. Similarly, being sleepy was associated with lower cerebral perfusion than during normal functioning [23,29]. Therefore, future studies should evaluate cerebral perfusion to properly account for this phenomenon.

5. Conclusions

In conclusion, our research showed that the state of decreased arousal associated with sleep deprivation substantially changed the activation patterns during a cognitive task involving selective attention. When comparing the rested and fatigued conditions, these changes were twofold. On the global level, we observed a general decrease of activation strength in the fatigued condition that was more pronounced in the left hemisphere. On the local level, we observed an extended activation of the medial prefrontal regions in the fatigued condition that reflected an increased contribution of cognitive control mechanisms compensating for the diminished efficiency of mechanisms involved in selective attention and response selection.

Author Contributions: Conceptualization, M.B. and A.B.; methodology, M.B.; software, M.B.; validation, M.B.; formal analysis, M.B. and S.P.G.; investigation, P.B.; data curation, S.P.G.; writing—original draft preparation, S.P.G. and M.B.; writing—review and editing, A.B.; project administration, Ł.D.; funding acquisition, A.B. All authors have read and agreed to the published version of the manuscript.

Funding: This research was funded by the Polish National Centre For Research and Development, grant number PBS3/B9/29/2015 entitled: "Detector of early signs of fatigue as a part of improving the safety driving (Det)," project manager: Prof. Alicja Bortkiewicz.

Institutional Review Board Statement: The study was conducted according to the guidelines of the Declaration of Helsinki, and approved by the Bioethics Committee of the Military Institute of Aviation Medicine, Warszawa, Poland (Decision No. 11/2015, Date of approval: 17 June 2015).

Informed Consent Statement: Informed consent was obtained from all subjects involved in the study.

Data Availability Statement: Data are available upon request from the authors.

Acknowledgments: We thank Andrzej Gaździński for assistance in data processing and the radiology technicians for MRI data collection. We wish to extend our appreciation to all study participants who made this research possible.

Conflicts of Interest: The authors declare no conflict of interest.

References

1. Sung, E.J.; Min, B.C.; Kim, S.C.; Kim, C.J. Effects of oxygen concentrations on driver fatigue during simulated driving. *Appl. Ergon.* **2005**, *36*, 25–31. [CrossRef] [PubMed]
2. Kania, A.; Nastalek, P.; Celejewska-Wojcik, N.; Sladek, K.; Kosobudzki, M.; Bortkiewicz, A.; Siedlecka, J. Can alveolar hypoventilation due to kyphoscoliosis be a contraindication to driving? *Int. J. Occup. Med. Environ. Health* **2019**, *32*, 735–745. [CrossRef] [PubMed]
3. Soares, S.; Ferreira, S.; Couto, A. Driving simulator experiments to study drowsiness: A systematic review. *Traffic Inj. Prev.* **2020**, *21*, 29–37. [CrossRef]
4. Chandrakumar, D.; Keage, H.A.D.; Gutteridge, D.; Dorrian, J.; Banks, S.; Loetscher, T. Interactions between spatial attention and alertness in healthy adults: A meta-analysis. *Cortex* **2019**, *119*, 61–73. [CrossRef]
5. Schweizer, T.A.; Kan, K.R.; Hung, Y.W.; Tam, F.; Naglie, G.; Graham, S.J. Brain activity during driving with distraction: An immersive fMRI study. *Front. Hum. Neurosci.* **2013**, *7*, 53. [CrossRef]
6. Wickens, C.D.; McCarley, J.S. *Applied Attention Theory*; CRC Press: Boca Raton, FL, USA, 2019; p. 248. [CrossRef]
7. Werneke, J.; Vollrath, M. What does the driver look at? The influence of intersection characteristics on attention allocation and driving behavior. *Accid. Anal. Prev.* **2012**, *45*, 610–619. [CrossRef]
8. Treisman, A.M.; Gelade, G. A feature-integration theory of attention. *Cogn. Psychol.* **1980**, *12*, 97–136. [CrossRef]
9. Esterman, M.; Verstynen, T.; Robertson, L.C. Attenuating illusory binding with TMS of the right parietal cortex. *Neuroimage* **2007**, *35*, 1247–1255. [CrossRef]
10. Leonards, U.; Sunaert, S.; Van Hecke, P.; Orban, G.A. Attention mechanisms in visual search—An fMRI study. *J. Cogn. Neurosci.* **2000**, *12*, 61–75. [CrossRef]
11. Kawasaki, M.; Watanabe, M.; Okuda, J.; Sakagami, M.; Aihara, K. Human posterior parietal cortex maintains color, shape and motion in visual short-term memory. *Brain Res.* **2008**, *1213*, 91–97. [CrossRef] [PubMed]
12. Donner, T.H.; Kettermann, A.; Diesch, E.; Ostendorf, F.; Villringer, A.; Brandt, S.A. Visual feature and conjunction searches of equal difficulty engage only partially overlapping frontoparietal networks. *Neuroimage* **2002**, *15*, 16–25. [CrossRef]
13. Song, J.H.; Jiang, Y.H. Visual working memory for simple and complex features: An fMRI study. *Neuroimage* **2006**, *30*, 963–972. [CrossRef]
14. Janewicz, M.; Trejnowska, A.; Gaździński, S.; Wyleżoł, M. The potential influence of obesity on the ability to drive. *Pol. J. Aviat. Med. Bioeng. Psychol.* **2017**, *23*, 22–28. [CrossRef]
15. Oginska, H.; Mojsa-Kaja, J.; Fafrowicz, M.; Marek, T. Measuring individual vulnerability to sleep loss-the CHICa scale. *J. Sleep Res.* **2014**, *23*, 339–346. [CrossRef]
16. Paluch, R. Assessment of fatigue on the basis of subjective feelings–Japanese questionnaire. *Work Saf.* **1985**, *7–8*, 3–6.
17. Yoshitake, H. Three characteristic patterns of subjective fatigue symptoms. *Ergonomics* **1978**, *21*, 231–233. [CrossRef]
18. Jenkinson, M.; Beckmann, C.F.; Behrens, T.E.; Woolrich, M.W.; Smith, S.M. FSL. *Neuroimage* **2012**, *62*, 782–790. [CrossRef]
19. Pavelka, R.; Trebick, V.; Fialova, J.T.; Zdobinsky, A.; Coufalova, K.; Havlicek, J.; Tufano, J.J. Acute fatigue affects reaction times and reaction consistency in Mixed Martial Arts fighters. *PLoS ONE* **2020**, *15*, e0227675. [CrossRef]
20. Sant'Ana, J.; Franchini, E.; da Silva, V.; Diefenthaeler, F. Effect of fatigue on reaction time, response time, performance time, and kick impact in taekwondo roundhouse kick. *Sports Biomech.* **2017**, *16*, 201–209. [CrossRef] [PubMed]
21. Eckert, M.A.; Menon, V.; Walczak, A.; Ahlstrom, J.; Denslow, S.; Horwitz, A.; Dubno, J.R. At the Heart of the Ventral Attention System: The Right Anterior Insula. *Hum. Brain Mapp.* **2009**, *30*, 2530–2541. [CrossRef] [PubMed]

22. Thomas, M.; Sing, H.; Belenky, G.; Holcomb, H.; Mayberg, H.; Dannals, R.; Wagner, H.; Thorne, D.; Popp, K.; Rowland, L.; et al. Neural basis of alertness and cognitive performance impairments during sleepiness. I. Effects of 24 h of sleep deprivation on waking human regional brain activity. *J. Sleep Res.* **2000**, *9*, 335–352. [CrossRef]
23. Poudel, G.R.; Innes, C.R.H.; Jones, R.D. Cerebral Perfusion Differences between Drowsy and Nondrowsy Individuals after Acute Sleep Restriction. *Sleep* **2012**, *35*, 1085–1096. [CrossRef]
24. Schneiders, J.; Opitz, B.; Tang, H.; Deng, Y.; Xie, C.; Li, H.; Mecklinger, A. The impact of auditory working memory training on the fronto-parietal working memory network. *Front. Hum. Neurosci.* **2012**, *6*, 173. [CrossRef] [PubMed]
25. Schweizer, S.; Grahn, J.; Hampshire, A.; Mobbs, D.; Dalgleish, T. Training the Emotional Brain: Improving Affective Control through Emotional Working Memory Training. *J. Neurosci.* **2013**, *33*, 5301–5311. [CrossRef] [PubMed]
26. Muto, V.; Shaffii-Le Bourdiec, A.; Matarazzo, L.; Foret, A.; Mascetti, L.; Jaspar, M.; Vandewalle, G.; Phillips, C.; Degueldre, C.; Balteau, E.; et al. Influence of acute sleep loss on the neural correlates of alerting, orientating and executive attention components. *J. Sleep Res.* **2012**, *21*, 648–658. [CrossRef] [PubMed]
27. Johns, M.W. A sleep physiologist's view of the drowsy driver. *Transp. Res. Part F Traffic Psychol. Behav.* **2000**, *3*, 241–249. [CrossRef]
28. Gazdzinski, S.; Durazzo, T.C.; Jahng, G.H.; Ezekiel, F.; Banys, P.; Meyerhoff, D.J. Effects of chronic alcohol dependence and chronic cigarette smoking on cerebral perfusion: A preliminary magnetic resonance study. *Alcohol. Clin. Exp. Res.* **2006**, *30*, 947–958. [CrossRef] [PubMed]
29. Elvsashagen, T.; Mutsaerts, H.J.; Zak, N.; Norbom, L.B.; Quraishi, S.H.; Pedersen, P.O.; Malt, U.F.; Westlye, L.T.; van Someren, E.J.; Bjornerud, A.; et al. Cerebral blood flow changes after a day of wake, sleep, and sleep deprivation. *Neuroimage* **2019**, *186*, 497–509. [CrossRef]

energies

MDPI

Article

Assessment of Social Acceptance for Autonomous Vehicles in Southeastern Poland

Agnieszka Dudziak [1], Monika Stoma [1,*], Andrzej Kuranc [1] and Jacek Caban [2]

[1] Faculty of Production Engineering, University of Life Sciences in Lublin, 20-612 Lublin, Poland; agnieszka.dudziak@up.lublin.pl (A.D.); andrzej.kuranc@up.lublin.pl (A.K.)
[2] Faculty of Mechanical Engineering, Lublin University of Technology, 20-618 Lublin, Poland; j.caban@pollub.pl
* Correspondence: monika.stoma@up.lublin.pl

Abstract: New technologies reaching out for meeting the needs of an aging population in developed countries have given rise to the development and gradual implementation of the concept of an autonomous vehicle (AV) and have even made it a necessity and an important business paradigm. However, in parallel, there is a discussion about consumer preferences and the willingness to pay for new car technologies and intelligent vehicle options. The main aim of the study was to analyze the impact of selected factors on the perception of the future of autonomous cars by respondents from the area of Southeastern Poland in terms of a comparison with traditional cars, with particular emphasis on the advantages and disadvantages of this concept. The research presented in this study was conducted in 2019 among a group of 579 respondents. Data analysis made it possible to identify potential advantages and disadvantages of the concept of introducing autonomous cars. A positive result of the survey is that 68% of respondents stated that AV will be gradually introduced to our market, which confirms the high acceptance of this technology by Poles. The obtained research results may be valuable information for governmental and local authorities, but also for car manufacturers and their future users. It is an important issue in the area of shaping the strategy of actions concerning further directions of development on the automotive market.

Keywords: autonomous technology; road vehicles; road traffic; survey study

Citation: Dudziak, A.; Stoma, M.; Kuranc, A.; Caban, J. Assessment of Social Acceptance for Autonomous Vehicles in Southeastern Poland. *Energies* **2021**, *14*, 5778. https://doi.org/10.3390/en14185778

Academic Editors: Guzek Marek, Rafał Jurecki and Wojciech Wach

Received: 31 July 2021
Accepted: 11 September 2021
Published: 14 September 2021

1. Introduction

In the last decade, many modern technologies, solutions, products, etc., have been introduced in various areas of modern man's life, which is, among others things, an implication of implementing the assumptions of the Economy 4.0 project. All of these conceptions change lifestyle, preferences and expectations. This revolution also affected the automotive industry and widely understood transport. The automotive industry is currently undergoing a period of dynamic and permanent change. Over the past decade, the automotive and technology industries, supported by the interdisciplinary efforts of numerous research institutes around the world, have made a significant leap in introducing computerization into what has been purely human activity for over a century, i.e., driving. By introducing functions that seem simple or even basic today, such as anti-lock braking systems and traction control, and gradually introducing more complex functions, such as adaptive cruise control and autonomous parking assistance systems, we are moving closer to the era of unmanned vehicles.

Over the past few years, vehicle manufacturers and technology developers have worked to develop advanced automotive and communication technologies [1,2], modern construction materials, smart vehicle options and alternative types of fuels [3–5]. In addition to the development of engines powered by alternative fuels [5], including those powered by natural gas and hydrogen, we distinguish hybrid and electric vehicles [6,7]. Many car manufacturers work with technology vendors to improve the driving experience, especially in terms of safety and comfort aspects [8]. Hence, these technologies, along with the need to meet the needs of an aging population in developed countries, have given rise to the development and gradual implementation of the AV concept and have even made it a necessity and an important business paradigm [9]. It should be mentioned, however, that the first attempt to create driverless vehicles dates back to the early 1920s [10]; however, this concept gained momentum only in the 1980s, when scientists managed to develop automated highway systems [11]. This makes the way for the connection of semi-autonomous and AV to the road infrastructure, mainly in Germany and the United States in 1980–2000 [9].

Sensor-based intelligent/autonomous driving systems and advanced vehicle functions generate many potential benefits, the most important of which seems to be the reduction of the number and consequences of accidents (some authors even believe that these systems can virtually eliminate human error, which is a major contributor to road accidents [12]). In addition, multimedia platforms in combination with intelligent and autonomous driving systems can make the time spent in the car, on the one hand, more enjoyable, and on the other hand, more productive, as passengers can multitask while traveling [13]. Advanced communication systems embedded in cars can also lead to more efficient vehicle navigation and traffic flow, reducing congestion and eliminating critical bottlenecks [14]. It should also not be forgotten about reducing energy consumption and pollution.

However, in parallel, there is a discussion about consumer preferences and the willingness to pay for new car technologies and intelligent vehicle options. The speed of market penetration by these technologies, functions and solutions largely depends on whether consumers are interested in these technologies and options and, above all, whether they are willing to pay extra for them. It should be noted that although the idea of self-propelled vehicles without the participation of a driver has existed for decades, too high costs have hindered their mass production so far [15]. In addition, as shown by various studies, many potential users of this type of car have some concerns about their introduction and widespread use, pointing to the existence of several disadvantages, especially in the area of the sense of security, controllability and in the area of law, liability, ethics and cybersecurity.

Taking the above into account, the aim of the study is to analyze the impact of selected factors on the perception of the future of AVs by respondents from Southeastern Poland in terms of comparison with traditional cars. The main goal of the study was to identify the basic factors influencing the opinions of respondents about the social acceptance of AV, in particular the advantages and disadvantages of this solution.

The following research questions were formulated for a fuller implementation of the assumed objectives of the study:

1. Do the respondents, according to the gender grouping variable (men and women), express similar opinions about AVs?
2. Do the respondents have similar opinions on AVs due to the age grouping variable?
3. When it comes to autonomous cars, are there advantages over disadvantages in this concept?
4. How, according to the respondents, will the car market develop, and which cars will be the most popular in the near future?

AV technology can and, in principle, should be considered at the meeting point of many disciplines, such as transport sciences, electrical engineering, information technologies, software and hardware engineering, law, and even ethics and philosophy [8].

2. Literature Review

Currently the automotive industry is experiencing a revolution with the advent of advanced automotive technologies, intelligent vehicle options and alternative fuels. However, research into consumer preferences for this type of automotive technology is very limited. The implementation and penetration of advanced automotive technologies on the market and the planning of possible market adoption scenarios requires the collection and analysis of data on consumer preferences related to these new technologies. Hence, the research by Shin et al. [16] aimed to meet this need by offering a detailed analysis of consumer preferences for alternative fuel types and technological options, using data collected in the reported experiments conducted on a sample of consumers from six metropolitan cities in South Korea. The results indicated that there is significant heterogeneity in consumer preferences for different smart technology options, such as wireless internet, vehicle connectivity and voice command features, but relatively less diversity in preferences for intelligent in-car applications, such as real-time travel information on parking and road conditions [13].

The use of AVs, as suggested by many authors and experts not only from the automotive industry, generates many different potential benefits, both in an individual aspect-for users of this type of vehicle, as well as social and economic benefits. In relation to the first area, special attention is paid to the fact that AV have the ability to improve passenger comfort by eliminating the need to perform certain tasks related to driving. It is known that the process of driving a car is complex; in fact, several motor and cognitive tasks must be performed, sometimes in quick succession and sometimes simultaneously, with which drivers need to interact; they must also react to different vehicle parameters, driver and pedestrian behavior, all under different weather, lighting and road conditions [17].

In addition, they provide new mobility opportunities for groups of people who, until now, have been partially or completely excluded from participation in public life due to mobility restrictions (e.g., the elderly or disabled) [16,18,19]. The possibility of using AV does not require drivers to have a driving license. It should be added that the lack of the need to drive a vehicle and focus on the conditions and other road users means for some "drivers" more time to work, which in turn is beneficial for the economy.

Taking into account the area of social benefits, it should be emphasized that autonomous driving does not lead to a loss of safety or efficiency of road transport, but rather improves them [20]. It may also result in new business models, as driverless cars will enable car sharing at much higher levels than is currently the case; in the vast majority of cases, it will be cheaper to use these shared cars than your own, without the financial burden and hassle of owning a private car (frozen capital, taxes, insurance, repairs and parking difficulties) [21,22]. The appropriate and adequate information about the number of free places, their location, etc., can reduce even by 30% the traffic volume in some cases [23]. Sharing cars also means a significant reduction in the number of vehicles on the road, as well as less space and infrastructure needed for transport, which in turn means less money spent on it [24,25]. Shared autonomous vehicles (SAV) allow for a shorter travel time and better use of vehicles (reducing their number on roads) and the available urban infrastructure [26]. As shown in Reference [27], the travelers are exposed to a reduction in travel time once conventional transport modes are replaced by AVs. One SAV (shared autonomous vehicle) can replace eight conventional vehicles with acceptable average waiting time ranging from 7 to 10 min and usage of four seats (shared trip) [28].

Ball [29] emphasizes that the benefits of the widespread introduction of AV can be found in two schools [15,30–34]. In the first school, particular attention is paid to the potential of AV to stimulate additional travel. Without the need for a driver, vehicle passengers would have much more freedom to engage in other activities during the journey, which in turn would reduce travel costs and facilitate urban growth. AV would also increase the mobility of younger people (without a driving license), the elderly, the sick or the disabled. It is also emphasized that AV can generate more demand by increasing road capacity, as lanes could be narrowed and cars could come closer together [29].

In turn, the second school emphasizes the potential of AVs to reduce travel/mobility costs. Part of this may be due to the shift from private car ownership towards "mobility as a service" [33]. By replacing the fixed costs of owning a car with fees for access (time and mileage) to a fleet of shared AVs, the final cost of the vehicle's journey would consequently decrease. Moreover, since AVs would not require a parking space close to the destination, this would allow for the further development and filling of the former parking lots [35].

The introduction of AVs would also make it possible to increase road capacity by up to five times [36] and reduce road congestion (and, consequently, fuel consumption and environmental pollution [37,38]), as cars on the network would communicate with each other and local infrastructure to smoothly respond to changes in traffic and flexibly adapt to them [39–42]. Some estimates show that this would save the average commuter 42 h a year [43].

As shown by numerous studies and forecasts, AVs can cause the most intense transformation in urban transport systems and the greatest revolution in the practice of transport planning (proper design of parking lots, streets, networks of transit and paratransit services), since the appearance of the motor car over a hundred years ago [29]. However, attention should also be paid to the strategic implications for other road users-especially for pedestrians. In a risky situation, an AV slows down and gives way to pedestrians—this applies even to unmarked pedestrian crossings. Safer AVs also trigger a rational response from pedestrians and other road users. Pedestrians can pass with impunity with the certainty that the car will stop. The situation is slightly different from the point of view of a passenger in an automated car; as Ball [29] puts it, it would be "like driving a street full of five-year-old unaccompanied children".

No less important is the issue of road safety, which is reflected in the number of road accidents and collisions [15]. It is estimated that AVs can significantly reduce the number of accidents caused by human error, which currently account for over 90% of all road accidents (i.e., driving distracted, e.g., by a mobile phone, speeding, driving under the influence of alcohol/drugs, fatigue and spontaneous decision-making) [44], and consequently drastically reduce the number of fatalities, as well as accelerate emergency response. AVs communicate with each other better than drivers and can more effectively sense the presence of other road users and predict their behavior, as well as prevent them from falling asleep at the wheel [29]. Consequently, self-propelled vehicles can help save lives thanks to advanced avoidance and crash-response technologies. This is because AVs are able to detect the traffic environment, navigate through the software algorithm and control the movement of the vehicle without the driver's decisions and actions [45]. According to data from the World Health Organization (WHO), AVs can eliminate over 1.25 million fatalities from road accidents worldwide [46]. This is all the more important because, according to the WHO, over 1.2 million people die each year as a result of road accidents around the world, and this has a huge impact on health and development [47]. It should be added that road accidents and collisions generate various cost burdens with long-term consequences, affecting productivity, healthcare costs, legal and court costs, losses at the workplace, costs of emergency services, burdens related to congestion, costs of insurance and damaged property.

Moreover, other benefits in the social area include the following [48]:

- More effective real-time navigation and dynamic routing;
- More accessible, reliable and flexible carpooling in passenger transport;
- More efficient infrastructure through better vehicle control and coordinated actions;
- Greater savings of resources needed for infrastructure, including the construction of parking lots and roadways;
- More environmentally friendly vehicles and infrastructure (AVs have great potential to reduce energy consumption per mile [30]);
- Lower insurance costs.

Attention should also be paid to the potential impact on companies operating directly in the automotive industry, such as taxi services, new car dealerships and repair shops, which will evolve and adapt to the latest technologies. In the study of Berlin by Bischoff et al. [49], the authors examined the replacement of conventional taxis by autonomous taxis; they showed that one autonomous vehicle might replace 10 conventional taxis if the rides are shared, and six if the rides are not shared. From the point of view of the aftermarket, a greater number of AVs may also lead to an increased interest in accessories, such as, for example, advanced video technology providing entertainment to passengers, which in turn will affect the development of companies producing them.

Advanced systems, used in AVs, can also record and share driver data with insurance companies, thus helping to manage claims, estimate insurance premiums and reduce costs. These vehicles can also generate savings in the construction, transportation and logistics industries by providing information on vehicle downtime and cargo status, providing fleet managers with greater operational transparency [50].

Automation of vehicles, despite many potential advantages, may also generate some disadvantages. They can be, among others, issues related to the mentality of the society, such as user's resilience to giving up driving control, loss of situational awareness, loss of driving skills and privacy issues. Equally important are also technical and technological issues, i.e., increased vulnerability to software and hardware defects, as well as the possibility of cybercrime and greater terrorism potential.

Potential users of AVs also point to some understatements in the area of ethics and possible liability for damages; with the advent of the AV concept, a new ethical paradigm appears for man–machine, due to automated risk allocation during collisions.

Other negative aspects of the large-scale introduction of AV technology also include the need to build a completely new legal framework, the need to formulate different approaches to road control and enforcement, the loss of jobs focused on driving, the car navigation system's susceptibility to different types of weather, problems with communication with non-autonomous vehicles in mixed road traffic and the need to make large investments in the current road infrastructure to adapt to the new fully computerized requirements [51].

Other disadvantages of introducing AVs also include their potential negative impact on some enterprises and economies based on public transport, vehicle insurance and services. Insurance companies see the danger of AVs disrupting their business model. Moreover, it was indicated in Reference [38] that reducing the costs of driving a vehicle may increase traffic intensity.

3. Method

Consumer behavior on the market is a variable category and depends on many factors influencing the decision to choose products and their purchase. One of the most reliable methods of obtaining information on consumer behavior on the market are surveys. The research used in this study was conducted in 2019, among a group of 579 respondents, via the Internet. The original questionnaire was used, with the use of a deliberate sample, as the research concerned only respondents who are drivers or intend to become drivers in the future. The questionnaire used in the research consisted of two parts: the substantive part and the metric questions. The questions contained in the first part concerned many different aspects of the autonomous vehicle market, including the issues of knowledge and acceptance of autonomous vehicles, and opinions on the prospects of their development compared to other types of vehicles, as well as the benefits, disadvantages and barriers and challenges resulting from the introduction of this technology. In turn, the second part, i.e., metric, contained questions enabling the sociodemographic characteristics of the respondents according to various grouping variables (e.g., gender, age, place of residence and driving experience period). The questions in the survey were closed questions: single or multiple choice. In the case of some questions, it was also possible to provide your own answer (if none of the proposed variants of the answer reflected the attitudes of the respondent); however, it was always clearly indicated in the survey.

The results were prepared by taking into account the division of the respondents into four age groups: 19–25 (48% of respondents), 26–40 (24% of respondents), 41–60 (24% of respondents) and over 60 (4% of respondents). They came from the countryside (34% of the respondents) and from cities of various numbers: up to 100,000 (20% of respondents), 100,000–300,000 (10% of respondents) and over 300,000 inhabitants, 35% of respondents). Of the study participants, 38% were women and 62% were men.

In order to achieve the assumed goals, as well as to answer the research questions, the results obtained in the own questionnaire research were statistically analyzed. A correspondence analysis was used, thanks to which it was indicated how the vehicle market will develop in the near future in the opinion of the respondents, in terms of age and gender, i.e., how respondents from Southeastern Poland perceive the future of AVs in comparison with standard vehicles (SVs). Moreover, the results of research referring to the potential benefits and advantages, as well as disadvantages of introducing AVs onto Polish roads are presented; in order to be able to fully answer the research questions, two variables grouping respondents were used in the analyses, namely gender and age.

The research sample came from the area of Southeastern Poland; this region was selected due to two factors. Firstly, it is an area less prosperous than the western territories and is characterized by a less developed communication infrastructure that can be developed for the use of AVs. Secondly, due to the less developed road infrastructure, it was expected that the social distance to such solutions might be large.

Data analyses were carried out on the basis of the statistical processing software Statistica 13.3 and Excel 2007. In this work, the correspondence analysis method was used. It is a descriptive and exploratory technique for analyzing two-way and multi-way tables, containing certain measures that characterize the relationship between columns and rows. The obtained results provide information similar in nature to the results obtained in the case of factor analysis techniques and allow for the analysis of the structure of the qualitative variables that make up the table. The most common table of this type is the two-dimensional contingency table. In correspondence analysis, the frequencies in the contingency table are first standardized such that the relative frequencies are computed and, when summed across all fields (cells) of the table, they give 1.0. One way to show the goals of a typical analysis is to express relative frequencies in terms of the distance between individual rows or columns in a space with a small number of dimensions.

In addition, the study also used the Spearman rank-order correlation analysis. Spearman's rank correlation coefficient is used to describe the strength of the correlation of two features when features are qualitative, allowing for ordering according to the strength of this feature, or when features are quantitative, but their number is small. In the study, the ranks of the X and Y features were assigned in descending order. If some units of a feature have the same value, then these units are assigned identical ranks, calculating the arithmetic mean of the ranks for the same units.

4. Results

On the basis of the information obtained from the respondents, in the initial phase of the research procedure, an analysis of correspondence between three groups of characteristics was carried out, i.e., car type (five groups of answers), age of respondents (four groups of answers) and gender (two groups). Respondents were asked about the expected type of the most popular vehicles in Poland in the near future. The following vehicles are included: SV, hybrid vehicle (HV), electric vehicle (EV), AV and standard vehicle supported by autonomous solutions (SV+). In this questionnaire, the respondents did not distinguish between the driving method according to its propulsion (SV petrol or diesel, EV or HV), so vehicles marked as SV, EV, HV and SV+ are considered human-driven vehicles, and an AV is understood as a self-steering vehicle. In order to present the configuration of points representing the input data, a two-dimensional factor space was chosen. The first factor allows us to reproduce 63.01% of input data variation (i.e., total inertia), and the second factor 24.77% (see Table 1).

The largest contribution to the creation of the two-dimensional factor space by the type of vehicle was played by SV+ and SV-coordinate I, and EV and AV-coordinate II. On the other hand, women over 60 years of age (dimension I) and women aged 26–40 (dimension II) had the largest share in the creation of the two-dimensional factor space by age and gender (Figure 1).

Table 1. Information resources' factors.

Number of Dimensions	Eigenvalues and Inertia, Total Inertia = 0.15719 χ^2 = 90.067 df = 28 p = 0.00001				
	Singular Value	Eigenvalues	Percentage of Inertia	Cumulative Percentage	χ^2
1	0.314710	0.099043	63.01026	63.0103	56.75147
2	0.197308	0.038930	24.76721	87.7775	22.30709
3	0.111603	0.012455	7.92396	95.7014	7.13687
4	0.082199	0.006757	4.29857	100.0000	3.87159

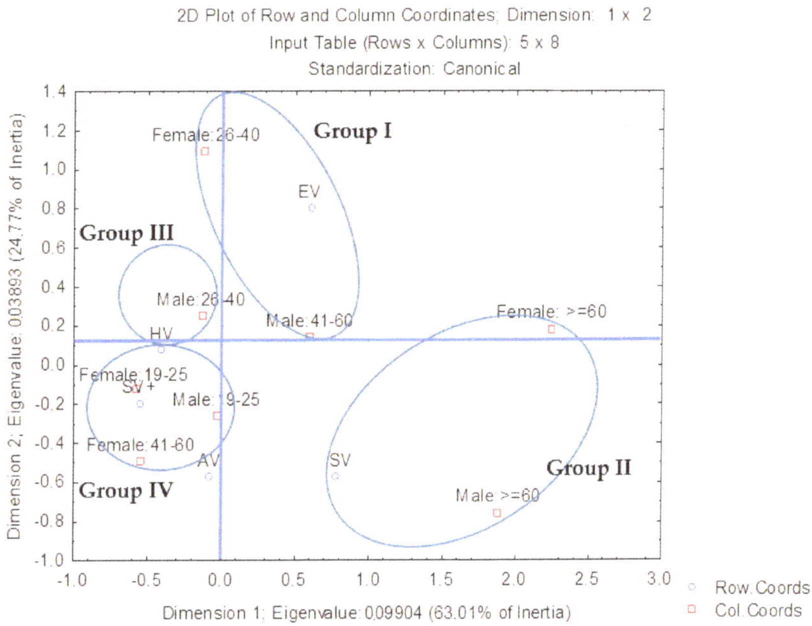

Figure 1. Correspondence analysis results between three groups of characteristics—vehicle type, age of respondents and gender.

There are four clear groups of vehicle types with the structure of indicators depending on the sex and age of the respondents (Figure 1), the first is made of EV, the second is SV, the third, with the most average structure, are HV, while the fourth is AV and SV+.

The greatest connection is between men aged 19–25 with SV+ and AV, it is a relatively strong link. Due to the value of the indicator-i.e., the age and sex of the respondents, the group in question differs from other preferred types of vehicle. An equally large link can be noticed between women aged 19–25 and SV+.

SV, on the other hand, correspond best with the group of women and men aged over 60. The age index of these respondents distinguishes this group of cars from the rest.

Another quite strong connection is between women aged 26–40 and men aged 41–60 with the group of EV.

At a later stage of the research procedure, we focused on the analysis of the benefits and advantages and-on the other hand-the disadvantages of using AV in the opinion of their potential users, using two grouping variables-gender and age, in order to identify possible differences in the perception of positive and negative aspects of the AV on Polish roads by women and men in different age groups.

Safer driving is one of the main driving forces in the development of AV and would be a prerequisite for their implementation on public roads in the future. In the research carried out using the proprietary questionnaire, 44% of respondents indicated that if they used AV, they would feel safer. However, most people surveyed are concerned about cybersecurity and the privacy of data about the internet technologies or services they currently use.

Besides safety, the introduction of AV can generate a number of other benefits. Hence, the respondents were asked to indicate other advantages related to this solution. The proposed answers included, among others: comfort (the ability to work and rest while driving, e.g., napping, receiving and writing e-mails, reading a book, etc.), more efficient use of time, greater safety (less accidents and collisions), less stress while driving a car, saving on operating costs, reducing road congestion, a solution for people who do not like driving a car, greater mobility (e.g., disabled, elderly), greater independence (e.g., disabled, elderly or unable to drive), enabling better access to services and more.

The conducted own research indicated gender differences and those resulting from the age of respondents. The results of this analysis are shown for women in Figure 2a and for men in Figure 2b.

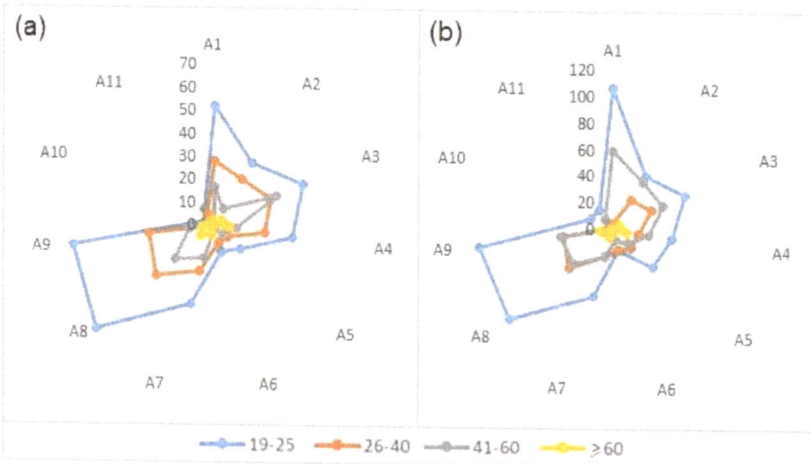

Figure 2. The advantages of the AV concept in the opinion of women (**a**) and men (**b**), A1—convenient solution, A2—efficient use of time, A3—safety, A4—less stress, A5—economy, A6—less congestion on the road, A7—solution for people who do not like driving a car, A8—mobility and A9—independence.

Young women aged 19–25 most often pointed to the benefits of mobility, independence and comfort; women aged 26–40, apart from mobility and independence, also pointed to the issue of comfort, but also safety (Figure 2a). On the other hand, women aged 41–60 paid attention primarily to safety, but also mobility. In turn, women over 60 years of age clearly indicated less stress while driving and, similarly to other groups, mobility, which can also be understood as greater independence. Interestingly, in all age groups, the least problem, according to the respondents, is access to services and the issue related to the reduction of parking spaces and less congestion on the road.

Compared to women, among the most frequently mentioned benefits related to AV, young men more often perceived comfort, independence and mobility, but also safety (group of men aged 19–25) (Figure 2b). On the other hand, men aged 26–40 most often indicated comfort, mobility and independence; men aged 41–60, apart from the issue of comfort, indicated the issue of effective use of time and safety. On the other hand, men over 60 indicated, similarly to women in this age group, less stress while driving and independence.

Table 2 presents the potential disadvantages of the AV concept perceived by women and men of different age groups. After analyzing the available literature, the proposed answers included boredom while driving; no pleasure in driving; unnecessary driving course; no driving habits; no control over electronics; elimination of many professions, e.g., taxi driver, professional driver, etc.; the threat of cybercrime in the area of vehicles; and others.

Table 2. Potential disadvantages of the AV concept in the opinion of women and men.

Gender	Women				Men			
Age	19–25	26–40	41–60	over 60	19–25	26–40	41–60	over 60
Boredom	22	10	10	2	62	12	32	5
No driving pleasure	46	28	25	6	112	34	55	6
Redundant driving course	26	10	8	0	22	14	16	2
Being used to driving a car	50	27	24	5	66	35	30	4
No control over electronics	70	44	28	5	110	56	48	5
Elimination of professions	62	32	12	4	106	30	26	4
Cybercrimes	50	28	16	2	106	44	48	0
Other	6	0	0	0	6	4	2	0

The conducted research shows (Table 2) that women in the group 19–25 indicated primarily no control over electronics, the potential elimination of professions (e.g., taxi drivers) and the possibility of cybercrimes. Women aged 26–40 spoke in a similar formula, pointing to no control over electronics and the elimination of professions, but at a similar level indicated the risk of cybercrime and the no pleasure in driving a car. On the other hand, women aged 41–60 declared no control over electronics, no driving pleasure, but also the risk of getting used to driving a vehicle. On the other hand, women over 60, among the biggest disadvantages, indicated the lack of driving pleasure, lack of control over electronics and the habit of driving a vehicle.

The research shows that men in the 19–25 age group indicated no pleasure in driving a car and no control over electronics, as well as the potential elimination of professions (e.g., taxi drivers) and the possibility of the so-called cybercrimes. Men aged 26–40 spoke in a similar formula, pointing to the no control over electronics and the risk of cybercrimes. In turn, men aged 41–60 declared no driving pleasure and no control over electronics, but also cybercrimes. On the other hand, men over 60, among the biggest disadvantages, indicated the no driving pleasure and no control over electronics, but also possible boredom while driving a car.

AVs seem to be an interesting solution for many respondents, but not entirely realistic. Table 3 below presents the analysis of the correlation of various variables with regard to associations relating to this type of solution. The performed Spearman correlation, as an analysis, allows us to correlate the variables on the ordinal scale. This is a kind of nonparametric correlation that is based on rank.

Table 3. Spearman rank-order correlations MD pairwise deleted marked correlations are significant at $p < 0.05$.

Variable	x_1	x_2	x_3	x_4	x_5	x_6	x_7
x_1	1.000000	0.248955	0.116724	0.167230	0.173011	−0.307904	**−0.372148**
x_2	0.248955	1.000000	−0.083286	**0.336166**	0.107645	**−0.345075**	−0.227059
x_3	0.116724	−0.083286	1.000000	−0.147148	0.175057	0.100478	0.056061
x_4	0.167230	**0.336166**	−0.147148	1.000000	−0.024476	−0.247382	−0.307910
x_5	0.173011	0.107645	0.175057	−0.024476	1.000000	0.011198	−0.076679
x_6	−0.307904	**−0.345075**	0.100478	−0.247382	0.011198	1.000000	**0.485749**
x_7	**−0.372148**	−0.227059	0.056061	−0.307910	−0.076679	**0.485749**	1.000000

Variables: x_1, innovative solution; x_2, more AV on the road in the future; x_3, many different barriers; x_4, more positive than negative sides; x_5, they still need to be refined; x_6, AV is a utopia; x_7, not an interesting concept.

The analysis shows that the correlation of the variables related to the fact that AVs are not interesting (x_7), and at the same time, utopia (x_6) is of great importance; the second quite strong correlation concerned the relationship between the statement that AVs, despite being an innovative solution (x_1), are not interesting for the respondents (x_7). Another correlation is the relationship between more AVs in the future on roads (x_2) and more positive sides than negative (x_4) and the fact that, although AVs are often treated as utopia (x_6), on roads, there will likely be more and more of them on the near future (x_2).

Figure 3 shows the approach to the correlation of individual variables presented in Table 3.

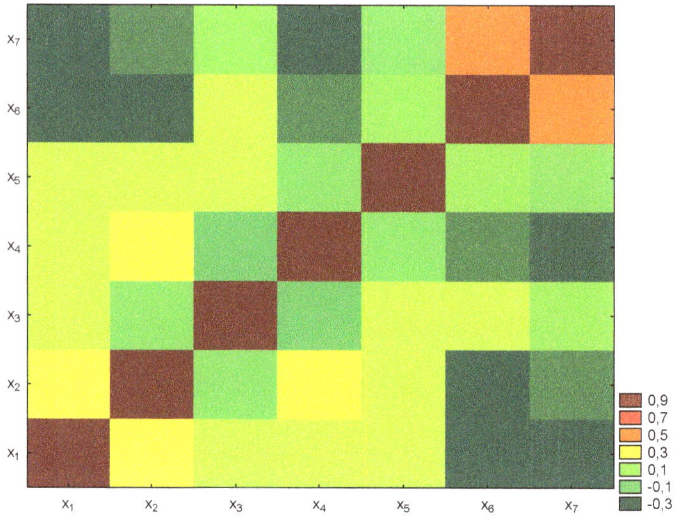

Figure 3. Three-dimensional sequential graph Spearman rank-order correlations in Workbook1 7v*7c.

The analysis shows that the respondents are quite cautious in making their judgments about this concept; they are quite skeptical about it, but they are aware that AV will enter the market, and, according to the analyses, most likely not so quickly. As it seems, the AV concept is realistic in the future, but, still, in relation to other types of vehicles traveling on Polish roads, it differs significantly from them in terms of its assumptions, infrastructure preparation and, above all, the mentality and culture of the society.

5. Discussion

AVs show great potential to radically change both the structure of cities and the dynamics of transport systems. Technologies such as advanced driver assistance systems (ADAS), e.g., lane keeping and adaptive cruise control, are already available on new vehicle models, and there is likely to be a gradual shift towards autonomy in the coming years.

The research results described in this paper show that changes to vehicle autonomy are generally well perceived by respondents and are only a matter of time. Similar statements can be found in the literature. Levinson and Krizek [32] suggest that full automation may be required in all new US vehicles by 2030 and that "human drivers will eventually be banned from public roads".

One of the most cited studies on AV adoption is the Litman study [25], which predicted that automated driving would become a standard feature of most new vehicles by 2050 and that AV would account for around 40–60% of vehicles, 80–100% of vehicle sales and 50–80% of vehicle travel. Litman also predicted that the beneficial effects of AV in increasing road safety and reducing traffic congestion would probably appear between 2040 and 2060. Bansal and Kockelman [39] predict that, by 2045, 24.8–87.2% of vehicle fleets will be accepted by self-propelled cars (SAE level 5). Their findings further revealed that AVs were viewed as a form of "somewhat low-risk" transport, and while there were concerns, there was little opposition to the prospect of their use on public roads.

The results obtained in the presented study regarding security as the main advantage of the possibility of using AV are similar to those achieved by Kyriakidis et al. [52], who found that gaining end-user trust in the issues of data security, protection and privacy will be critical to the widespread implementation of AV in the future.

Payre et al. [53] conducted a survey among 421 people. Automated driving in this group was unconditionally accepted by 68.1%, with higher acceptance depending on the type of driving, including motorway driving, driving in the presence of traffic congestion and automatic parking. Similar results were obtained in a study by Howard and Dai [54], in Berkeley, California. People in this study were most attracted to the potential benefits of safety, parking and multitasking [55], which is also reflected in the results of the research presented in this paper. In our survey, 68% of respondents (396 people) confirmed that AV will be gradually introduced to our market, which confirms the high acceptance of this technology by Poles.

Own research also showed that AVs are generally viewed in a positive light; in terms of qualitative risk perception, they have been assessed relatively well compared to the existing modes of transport, and there has been little opposition to them. It would therefore seem that the idea of AV on public roads is already accepted by many people. However, further findings indicate that considerable efforts are still needed to encourage public acceptance. As in the research by Hulse et al. [17], despite an obvious low negative, concerns were expressed not only relating to road safety issues. Moreover, significant associations have been found between perceived risk assessments or attitudes and various factors, including road user populations, gender and age, meaning that AV perceptions vary, something that was also reflected in our own findings.

Furthermore, Hohenberger et al. [56] found that emotions and affective responses to AV indicate gender differences in the willingness to use automated vehicles. In particular, men were found to be more likely to predict pleasure than anxiety about wanting to use AV. Similar conclusions were also given in Reference [17], which showed that, compared to cars operated by people, AVs were perceived differently depending on gender and age-men, and younger adults showed greater acceptance of them.

It should be emphasized that the respondents participating in the research conducted by the authors of this article also pointed to numerous negative views on the problem of functioning in the AV space. This corresponds to the results of the research by Schoettle and Sivak [57], wherein respondents similarly also revealed numerous concerns about AV travel. The greatest concerns were system or hardware failures with security implications. Moreover, participants were very concerned about the fact that the AVs did not give the option to take control of the vehicle, and they thought about the fact that other types of road vehicles are also autonomous. While there were some differences in the survey responses depending on the age of the participant (e.g., older participants were more likely than younger participants to say they would not drive AV), gender differences were detected in nearly all questions, with women being less convinced of AV than men. It should be added that References [10,57] present research of a more global nature, i.e., among the inhabitants of China, India, Japan, the United States, Great Britain and Australia. As can be seen, such fears and positions indicated in the own research among Polish respondents do not only occur locally, but show global trends. It can therefore be said that there are similar social concerns with regard to AV technology regardless of where you live.

In other studies that are available in the literature on the subject, respondents also stated that they could not afford the additional costs. It also took into account the risk context when unknown entities can gain access to data, i.e., through hacking; therefore, many participants in many studies conducted so far had clear concerns about privacy policy [58], which is also consistent with the results of own research.

At present, autonomous vehicles are still in the trial phase and have not yet been introduced into regular operation in any country. Scientific publications discussing organizational issues related to the functioning of transport systems, in which autonomous vehicles are present, are still mainly based on theoretical assumptions.

One of the problems in implementing of technology AV is the interaction between autonomous vehicles and other road users [59], including non-motorized ones, i.e., cyclists and pedestrians, who cannot easily be controlled by traffic control systems.

The sustainable development policy should primarily aim at eliminating the negative effects of motorization, such as congestion in cities, air pollution, the lack or unequal accessibility of transport, as well as the increasing costs of infrastructure maintenance. National governments should initiate close cooperation between vehicle manufacturers and local authorities in terms of strategies for implementing available technological solutions. However, the lack of precise visions of the technical and organizational model of automation means that local and regional authorities do not want to make specific planning decisions in this area yet. This is because of the high risk due to the many unknowns and sometimes because of more urgent investment needs [60]. The obtained results of our own and other social research may be helpful in developing assumptions for further development of urbanized areas and autonomous means of transport.

It should be assumed that the safety and reliability of AV will continue to increase with the progress of civilization and technology. Public infrastructure will begin to evolve to support AV. New business models will be created. Intelligent mobility will become a reality as cities become "smarter" and cars become autonomous. Companies and car manufacturers are ready for AV, while the biggest unknown is whether our society is ready for such a solution. Indeed, the trend of vehicle automation technology by major vehicle manufacturing industries is expected to move closer to highly automated or fully autonomous vehicles through technological advances in the robotics and artificial intelligence sectors. This is because, as also shown in our own research, AVs have great potential to increase mobility. In some cases, this situation may indicate the need to change from the existing infrastructure solutions to the habits of drivers, and it will certainly be a big revolution on the automotive market.

6. Conclusions

As a result of the research and analysis of their results, the following observations and conclusions were formulated.

Firstly, the analysis of the correspondence found that young people are most favorable to the concept of AV, and this applies to both women and men. EVs are in second place, dominating among women in the 26–40 age group and among men in the 41–60 age group. It was also noted that there are advantages over disadvantages to AV. Moreover, due to their age, the participants of the study had similar opinions on AV. Over 40% of respondents indicated that if AV was widely used, they would feel safer. On the other hand, most of the people interviewed are concerned about cybersecurity and privacy related to the technologies used. Among the benefits of using AV vehicles, the respondents indicated mainly comfort, more efficient use of time, greater safety and less stress related to driving. The issues of greater mobility and independence of people with disabilities, the elderly or those unable to drive a car turned out to be very important. The respondents paid less attention to access to services, savings in parking spaces or lower transport congestion.

Our own conducted research indicated some gender differences and those resulting from the age of the respondents. However, it is worth emphasizing that women aged 41–60 paid attention primarily to safety and mobility; in turn, women over 60 years of age clearly indicated less stress while driving. Among the disadvantages, women most often indicated the no control over electronics and control systems, while men, as well as the no control over the system, indicated the no pleasure in driving and the fear of becoming weaned from driving and the loss of this ability. Both women and men also noticed the social effect of eliminating the profession of taxi driver. Moreover, it was found that the local perception of the AV concept is positive and consistent with global trends in this area.

Author Contributions: Conceptualization, A.D., M.S. and J.C.; methodology, A.D. and M.S.; software, A.D.; validation, M.S., J.C. and A.K.; formal analysis, A.D. and J.C.; investigation, A.D., M.S. and J.C.; resources, A.D., M.S., A.K. and J.C.; data curation, A.D.; writing—original draft preparation, M.S., A.D., J.C. and A.K.; writing—review and editing, A.D., M.S. and J.C.; visualization, A.D. and A.K.; supervision, M.S. and A.D.; project administration, A.D. All authors have read and agreed to the published version of the manuscript.

Funding: Funded from the "Excellent science" program of the Ministry of Science and Higher Education as a part of the contract no. DNK/SP/465641/2020 "The role of the agricultural engineering and environmental engineering in the sustainable agriculture development".

Institutional Review Board Statement: Not applicable.

Informed Consent Statement: Not applicable.

Data Availability Statement: Not applicable.

Conflicts of Interest: The authors declare no conflict of interest.

References

1. Jia, D.; Ngoduy, D.; Vu, H.L. A multiclass microscopic model for heterogeneous platoon with vehicle-to-vehicle communication. *Transp. B Transp. Dyn.* **2019**, *7*, 311–335. [CrossRef]
2. Tucki, K.; Orynycz, O.; Świć, A.; Mitoraj-Wojtanek, M. The Development of Electromobility in Poland and EU States as a Tool for Management of CO_2 Emissions. *Energies* **2019**, *12*, 2942. [CrossRef]
3. Papis, M.; Jastrzębski, D.; Kopyt, A.; Matyjewski, M.; Mirosław, M. Driver reliability and behavior study based on a car simulator station tests in ACC system scenarios. *Ekspolatacja Niezawodn. Maint. Reliab.* **2019**, *21*, 511–521. [CrossRef]
4. Pauer, G.; Török, Á. Static System Optimum of Linear Traffic Distribution Problem Assuming an Intelligent and Autonomous Transportation System. *Period. Polytech. Transp. Eng.* **2018**, *47*, 64–67. [CrossRef]
5. Rimkus, A.; Žaglinskis, J.; Stravinskas, S.; Rapalis, P.; Matijošius, J.; Bereczky, Á. Research on the Combustion, Energy and Emission Parameters of Various Concentration Blends of Hydrotreated Vegetable Oil Biofuel and Diesel Fuel in a Compression-Ignition Engine. *Energies* **2019**, *12*, 2978. [CrossRef]
6. Šarkan, B.; Gnap, J.; Kiktova, M. The importance of hybrid vehicle in urban traffic in terms of environmental impact. *Arch. Automot. Eng. Arch. Motoryz.* **2019**, *85*, 115–122. [CrossRef]
7. Geca, M.; Wendeker, M.; Grabowski, L. A City Bus Electrification Supported by the Photovoltaic Power Modules. SAE Technical Paper, 2014-01-2898. Available online: https://saemobilus.sae.org/content/2014-01-2898/ (accessed on 8 August 2021).
8. Kirk, B. Connected Vehicles: An Executive Overview of the Status and Trends. *Globis Consulting*, 21 November 2011.
9. Bagloee, S.A.; Tavana, M.; Asadi, M.; Oliver, T. Autonomous vehicles: Challenges, opportunities, and future implications for transportation policies. *J. Mod. Transp.* **2016**, *24*, 284–303. [CrossRef]
10. Schoettle, B.; Sivak, M. *A Survey of Public Opinion About Autonomous and Self-Driving Vehicles in the US, the UK, and Australia*; University of Michigan Transportation Research Institute: Ann Arbor, MI, USA, 2014.
11. Ioannou, P. *Automated Highway Systems*; Springer Science & Business Media: New York, NY, USA, 2013.
12. Nelson, S.P. Challenges and Technologies: The Human Friendly Vehicle in 2030 and Beyond. Freescale Semiconductor Inc. 2014. Available online: http://www.freescale.com/files/training_pdf/VFTF09_AA106.pdf (accessed on 4 August 2021).
13. Shariff, A.; Bonnefon, J.-F.; Rahwan, I. Psychological roadblocks to the adoption of self-driving vehicles. *Nat. Hum. Behav.* **2017**, *1*, 694–696. [CrossRef]
14. Kraan, M.; Mahmassani, H.S.; Huynh, N. Traveler Responses to Advanced Traveler Information Systems for Shopping Trips: Interactive Survey Approach. *Transp. Res. Rec. J. Transp. Res. Board* **2000**, *1725*, 116–123. [CrossRef]
15. Fagnant, D.J.; Kockelman, K. Preparing a nation for autonomous vehicles: Opportunities, barriers and policy recommendations. *Transp. Res. Part A Policy Pract.* **2015**, *77*, 167–181. [CrossRef]
16. Shin, J.; Bhat, C.R.; You, D.; Garikapati, V.M.; Pendyala, R.M. Consumer preferences and willingness to pay for advanced vehicle technology options and fuel types. *Transp. Res. Part C Emerg. Technol.* **2015**, *60*, 511–524. [CrossRef]
17. Hulse, L.M.; Xie, H.; Galea, E.R. Perceptions of autonomous vehicles: Relationships with road users, risk, gender and age. *Saf. Sci.* **2018**, *102*, 1–13. [CrossRef]
18. Becker, F.; Axhausen, K.W. Literature review on surveys investigating the acceptance of automated vehicles. *Transportation* **2017**, *44*, 1293–1306. [CrossRef]
19. Meyer, J.; Becker, H.; Bösch, P.M.; Axhausen, K.W. Autonomous vehicles: The next jump in accessibilities? *Res. Transp. Econ.* **2017**, *62*, 80–91. [CrossRef]
20. Friedrich, B. The effect of autonomous vehicles on traffic. In *Autonomous Driving*; Springer: Berlin/Heidelberg, Germany, 2016; pp. 317–334. [CrossRef]

21. Johnson, B. *Disruptive Mobility—A Scenario for 2040*; Barclays Investment Bank: London, UK, 2015. Available online: https://www.investmentbank.barclays.com/content/dam/barclaysmicrosites/ibpublic/documents/investment-bank/global-insights/barclays-disruptive-mobility-pdf-120115-459kb.pdf (accessed on 20 May 2021).
22. Bartuska, L.; Labudzki, R. Research of basic issues of autonomous mobility. LOGI 2019-Horizons of Autonomous Mobility in Europe. *Transp. Res. Procedia* **2020**, *44*, 356–360. [CrossRef]
23. Škultéty, F.; Beňová, D.; Gnap, J. City Logistics as an Imperative Smart City Mechanism: Scrutiny of Clustered EU27 Capitals. *Sustainability* **2021**, *13*, 3641. [CrossRef]
24. Krueger, R.; Rashidi, T.H.; Rose, J.M. Preferences for shared autonomous vehicles. *Transp. Res. Part C Emerg. Technol.* **2016**, *69*, 343–355. [CrossRef]
25. Litman, T. Autonomous Vehicle Implementation Predictions: Implications for Transport Planning. Victoria Transport Policy Institute. 2021. Available online: https://www.vtpi.org/avip.pdf (accessed on 15 June 2021).
26. Duarte, F.; Ratti, C. The Impact of Autonomous Vehicles on Cities: A Review. *J. Urban Technol.* **2018**, *25*, 3–18. [CrossRef]
27. Hamadneh, J.; Esztergár-Kiss, D. The Influence of Introducing Autonomous Vehicles on Conventional Transport Modes and Travel Time. *Energies* **2021**, *14*, 4163. [CrossRef]
28. Hamadneh, J.; Esztergar-Kiss, D. Impacts of Shared Autonomous Vehicles on the Travelers' Mobility. In Proceedings of the 2019 6th International Conference on Models and Technologies for Intelligent Transportation Systems (MT-ITS), Cracow, Poland, 5–7 June 2019; pp. 1–9. [CrossRef]
29. Millard-Ball, A. Pedestrians, Autonomous Vehicles, and Cities. *J. Plan. Educ. Res.* **2016**, *38*, 6–12. [CrossRef]
30. Guerra, E. Planning for Cars That Drive Themselves: Metropolitan Planning Organizations, Regional Transportation Plans, and Autonomous Vehicles. *J. Plan. Educ. Res.* **2016**, *36*, 210–224. [CrossRef]
31. Heinrichs, D.; Cyganski, R. Automated Driving: How It Could Enter Our Cities and How This Might Affect Our Mobility Decisions. *disP Plan. Rev.* **2015**, *51*, 74–79. [CrossRef]
32. Levinson, D.; Krizek, K. The End of Traffic and the Future of Transport, 3rd ed.; Network Design Lab: 2015. Available online: http://www.princeton.edu/~{}alaink/Orf467F15/TheEndOfTraffic.pdf (accessed on 8 August 2021).
33. Milakis, D.; van Arem, B.; van Wee, B. *Policy and Society Related Implications of Automated Driving: A Review of Literature and Directions for Future Research*; Working Paper; Delft University of Technology: Delf, The Netherlands, 2015.
34. Xu, Z.; Wang, M.; Zhang, F.; Jin, S.; Zhang, J.; Zhao, X. PaTAVTT: A Hardware-in-the-Loop Scaled Platform for Testing Autonomous Vehicle Trajectory Tracking. *J. Adv. Transp.* **2017**, *2017*, 9203251. [CrossRef]
35. Nikitas, A. Automated cars: A critical review of the potential advantages and disadvantages of driverless technologies. In Proceedings of the First International Workshop on Smart Urban Mobility, Edinburgh, UK, 26–27 November 2015.
36. Fernandes, P.; Nunes, U.J.C. Platooning With IVC-Enabled Autonomous Vehicles: Strategies to Mitigate Communication Delays, Improve Safety and Traffic Flow. *IEEE Trans. Intell. Transp. Syst.* **2012**, *13*, 91–106. [CrossRef]
37. Greenough, J. 10 Million Self-Driving Cars Will Be on the Road by 2020. June 2016. Available online: http://uk.businessinsider.com/report-10-million-self-driving-cars-will-be-on-the-road-by-2020-2015-5-6?r=US&IR=T (accessed on 1 August 2021).
38. Krasniqi, X.; Hajrizi, E. Use of IoT Technology to Drive the Automotive Industry from Connected to Full Autonomous Vehicles. *IFAC-PapersOnLine* **2016**, *49*, 269–274. [CrossRef]
39. Bansal, P.; Kockelman, K.M. Forecasting Americans' long-term adoption of connected and autonomous vehicle technologies. *Transp. Res. Part A Policy Pract.* **2017**, *95*, 49–63. [CrossRef]
40. Dresner, K.M.; Stone, P. Sharing the Road: Autonomous vehicles meet human drivers. In Proceedings of the 20th International Joint Conference on Artificial Intelligence, Hyderabad, India, 6–12 January 2007; pp. 1263–1268.
41. Fajardo, D.; Au, T.-C.; Waller, S.; Stone, P.; Yang, D. Automated intersection control: Performance of future innovation versus current traffic signal control. *Transp. Res. Rec.* **2011**, *1*, 223–232. [CrossRef]
42. Hengstler, M.; Enkel, E.; Duelli, S. Applied artificial intelligence and trust—The case of autonomous vehicles and medical assistance devices. *Technol. Forecast. Soc. Chang.* **2016**, *105*, 105–120. [CrossRef]
43. Chan, C. Advancements, prospects, and impacts of automated driving systems. *Int. J. Transp. Sci. Technol.* **2017**, *6*, 208–216. [CrossRef]
44. Folsom, T.C. Social ramifications of autonomous urban land vehicles. In Proceedings of the 2011 IEEE International Symposium on Technology and Society (ISTAS), Chicago, IL, USA, 23–25 May 2011.
45. Xu, Z.; Zhang, K.; Min, H.; Wang, Z.; Zhao, X.; Liu, P. What drives people to accept automated vehicles? Findings from a field experiment. *Transp. Res. Part C Emerg. Technol.* **2018**, *95*, 320–334. [CrossRef]
46. World Health Organization. *Global Status Report on Road Safety 2015*; World Health Organization: Geneva, Switzerland, 2015.
47. Weber, M. Where to? A History of Autonomous Vehicles. 2014. Available online: http://www.computerhistory.org/atchm/where-to-ahistory-of-autonomous-vehicles (accessed on 15 July 2021).
48. Sparrow, R.; Howard, M. When human beings are like drunk robots: Driverless vehicles, ethics, and the future of transport. *Transp. Res. Part C Emerg. Technol.* **2017**, *80*, 206–215. [CrossRef]
49. Bischoff, J.; Maciejewski, M. Simulation of City-wide Replacement of Private Cars with Autonomous Taxis in Berlin. *Procedia Comput. Sci.* **2016**, *83*, 237–244. [CrossRef]
50. Nieoczym, A.; Tarkowski, S. The modeling of the assembly line with a technological Automated Guided Vehicle (AGV). *Logforum* **2011**, *7*, 35–42.

51. Nagy, S.; Csiszár, C. The quality of smart mobility: A systematic review. *Sci. J. Sil. Univ. Technol. Ser. Transp.* **2020**, *109*, 117–127. [CrossRef]
52. Kyriakidis, M.; Happee, R.; de Winter, J. Public opinion on automated driving: Results of an international questionnaire among 5000 respondents. *Transp. Res. Part F Traffic Psychol. Behav.* **2015**, *32*, 127–140. [CrossRef]
53. Payre, W.; Cestac, J.; Delhomme, P. Intention to use a fully automated car: Attitudes and a priori acceptability. *Transp. Res. Part F Traffic Psychol. Behav.* **2014**, *27*, 252–263. [CrossRef]
54. Howard, D.; Dai, D. Public perceptions of self-driving cars: The case of Berkeley, California. In Proceedings of the Transportation Research Board 93rd Annual Meeting; TRB 93rd Annual Meeting Compendium of Papers, Washington, DC, USA, 12–16 January 2014.
55. Zhang, W.; Guhathakurta, S.; Fang, J.; Zhang, G. Exploring the impact of shared autonomous vehicles on urban parking demand: An agent-based simulation approach. *Sustain. Cities Soc.* **2015**, *19*, 34–45. [CrossRef]
56. Hohenberger, C.; Spörrle, M.; Welpe, I.M. How and why do men and women differ in their willingness to use automated cars? The influence of emotions across different age groups. *Transp. Res. Part A Policy Pract.* **2016**, *94*, 374–385. [CrossRef]
57. Schoettle, B.; Sivak, M. *Public Opinion about Self-Driving Vehicles in China, India, Japan, the U.S., the U.K., and Australia*; University of Michigan Transportation Research Institute: Ann Arbor, MI, USA, 2014; pp. 1–31.
58. Maddox, J.; Sweatman, P.; Sayer, J. Intelligent vehicles? infrastructure to address transportation problems—A strategic approach. In Proceedings of the 24th International Technical Conference on the Enhanced Safety of Vehicles (ESV), Gothenburg, Sweden, 8–11 June 2015.
59. Moravčík, L. Typové Schválenie Autonómnych (Samojazdiacich) Vozidiel. *Perner's Contacts* **2020**, *15*, 2. [CrossRef]
60. Jaroszyński, M. Autonomous vehicles: Organizational scenarios and their threats and opportunities to transportation sustainability. *WUT J. Transp. Eng.* **2018**, *120*, 133–142. [CrossRef]

energies

Article

"Where Have I Heard It?" Assessing the Recall of Traffic Safety Campaigns in the Dominican Republic

Francisco Alonso [1], Mireia Faus [1], Cesáreo Fernández [2] and Sergio A. Useche [1,*]

[1] DATS (Development and Advising in Traffic Safety) Research Group, INTRAS (Research Institute on Traffic and Road Safety), University of Valencia, 46022 Valencia, Spain; francisco.alonso@uv.es (F.A.); mireia.faus@uv.es (M.F.)
[2] Department of Communication Sciences, University Jaume I., 12071 Castellón, Spain; cesar.fernandez@uji.es
* Correspondence: sergio.useche@uv.es

Abstract: Although traffic crashes are the eighth leading cause of death in the world, and are linked to vehicle and infrastructure-related factors, crash-related fatality rates are much higher in low-income countries. Particularly, the Dominican Republic is the country with the highest accident rate in the whole American continent. Therefore, in the past few years, public agencies have been developing different measures aimed at reducing traffic fatalities, including road safety campaigns. The aim of the present study was to assess the recalling of such campaigns among the Dominican population, which may serve as an additional indicator to evaluate their effectiveness in this and other countries of the region presenting similar traffic safety issues. For this cross-sectional study, a nationwide sample composed of 1260 people (50% males and 50% females) with a mean age of 39.3 years was used. The data were collected through personal interviews. Overall, the recall of traffic safety campaigns was found to be very low (9%); male drivers who were employed, possessed a driver's license and habitually drove were the ones who could commonly remember these campaigns. The results of this study suggest that further evaluation and follow-up could help to maximize the impact of future traffic campaigns and advertisements in the Dominican Republic, as well as in other emerging countries of the region with similar characteristics. Further, key segments of the population such as the female, young, less formally educated and non-driving populations should be also targeted for further actions in this regard.

Keywords: traffic safety; human factors; communication campaigns; mass media; transport planning; mobility; sustainability

Citation: Alonso, F.; Faus, M.; Fernández, C.; Useche, S.A. "Where Have I Heard It?" Assessing the Recall of Traffic Safety Campaigns in the Dominican Republic. *Energies* **2021**, *14*, 5792. https://doi.org/10.3390/en14185792

Academic Editor: Grzegorz Karoń

Received: 6 August 2021
Accepted: 10 September 2021
Published: 14 September 2021

1. Introduction

Traffic crashes represent a problem with a high economic, social and health-related impact on the population [1]. The latest data recorded by World Health Organization indicate that traffic accidents are the eighth leading cause of death worldwide, explaining almost 1.4 million fatalities yearly [2].

This situation is even more serious in emerging countries, if several existing shortcomings and gaps are considered. In brief, low- and middle-income countries (LMICs) are characterized by having an average low per capita income and a very limited advancement in their development and infrastructure and human development. Even so, economic growth, industrialization, and exportation have been rising with globalization dynamics, often implying unexpected (or *unplanned*) problems for key sectors such as transportation [3]. Colombia, Peru, Ecuador and the Dominican Republic are some noteworthy examples [4]. Emerging countries have a traffic accident death rate of 29.4 per 100,000 inhabitants compared to the world rate of 18.8 per 100,000 inhabitants [5].

For the past few years, the Dominican Republic has been the country with the highest number of fatal victims in traffic accidents. The death rate of traffic accidents fluctuates, but it has been stable at over 20 for every 100,000 inhabitants since 2008 [6]. These data make the Dominican Republic the country with the highest death rate from traffic accidents in the whole American continent [5].

Deaths related to traffic accidents are the second leading cause of external or violent death in the Dominican Republic. Thus, the data from the Permanent Observatory of Road Safety (Observatorio Permanente de Seguridad Vial—OPSEVI) have shown that the number of casualties caused by traffic accidents was 3204 in 2019 (the latest available data), with men between 15 and 29 years old as the population group with the highest rate [7].

In the past few years, from the National Institute of Traffic and Land Transportation (INTRANT, Domincan Republic), a set of measures have been developed, addressing the awareness of the population and the reduction in traffic crash rates in the country. Thus, new laws and norms have been approved within this scope, and actions have been carried out to improve the public infrastructure and the conditions of public transportation, emphasizing the road education of citizens [8].

In this sense, communications campaigns are very relevant since they are an appropriate way of transmitting changes happening in the traffic and road safety field to the Dominican Republic population [9]. Without appropriate use of the media and advertising campaigns, it will be difficult to achieve a real change in the attitudes and behaviors of road users [10,11]. For this reason, the communication plans of the traffic sector throughout the whole world tend to produce advertisements with a high emotional impact, using fear as their persuasive element [12] in addition to using messages or slogans that, with their repetition through different media, manage to be remembered by users.

In the Dominican Republic, different communication campaigns were developed in the years 2018 and 2019, aimed at both at-risk groups (pedestrians, cyclists, motorcyclists, etc.) and to the population as a whole. This is the case of "safe pedestrian action", "I use the bike", or "don't drive if you drank" [13]. The purpose of the traffic campaigns is to raise public awareness of road safety risk factors in order to reduce accidents and improve mobility, which will have a direct impact on reducing pollution and energy consumption [14].

Communication campaigns have been shown to have a certain level of effectiveness. It is difficult to isolate the impact of advertisements, as they are usually broadcast in conjunction with the implementation of other preventive measures [15]. However, previous research shows that campaign effectiveness significantly increases when they are used to complement regulations and sanctions [16,17]. Therefore, it is essential to make ads with appropriate features so that users remember the message and apply it in their journeys.

However, as has been previously mentioned, carrying out a campaign does not necessarily imply that the population will perceive and incorporate all the information that needs to be transmitted. Although recall is not the only important element in raising awareness among the population, it is a necessary first step in the process of acquiring appropriate road behavior. Therefore, the objective of this research is to find out to what extent the inhabitants of the Dominican Republic remember the communication and advertisement campaigns related to traffic, mobility, and road safety that have been carried out in recent years. Likewise, we will analyze whether there are population groups that are especially inclined to remember such advertisements, with the aim of improving the efficacy of future campaigns in the country.

After reviewing the literature and stating the objective of the study, the materials and methods used will be presented. Subsequently, the obtained results will be developed, and the data will be contrasted with other studies in this thematic area in the discussion. Finally, the most relevant conclusions will be presented.

2. Materials and Methods

2.1. Participants

The sample was composed of 1260 adult inhabitants of the Dominican Republic. The sample distribution was proportional to the population, according to the ONE census (National Statistical Office), by age (M = 39.3; SD = 15.37), gender, habitat and province [18]

(Table 1). To obtain the necessary representativeness, the minimum sample size should be about n = 680 if we assume a level of confidence of 99%, a maximum margin of error of 5% ($\alpha = 0.05$) and a beta β) of 0.20, which allows for an 80% power (Table 1). Participation was voluntary and anonymous. The personal information management was carried out in accordance with the current laws on data protection, complying with ethical requirements.

Table 1. Sociodemographic data of the study participants.

Variable	Category	Total		Male		Female	
		n	%	n	%	n	%
Age range	18–24	260	20.6%	136	21.6%	124	19.7%
	25–34	311	24.7%	146	23.2%	165	26.2%
	35–49	366	29.0%	193	30.6%	173	27.5%
	50–64	221	17.5%	108	17.1%	113	17.9%
	>65	102	8.1%	57	17.1%	55	8.7%
	Total	1260	100%	630	100%	630	100%
Level of studies	Cannot read or write	25	2.0%	14	2.2%	11	1.7%
	Can read and write but no studies	9	0.7%	3	0.5%	6	1%
	Primary studies, not completed	265	21%	138	21.9%	127	20.2%
	Primary studies, completed	80	6.3%	44	7%	36	5.7%
	High school, not completed	273	21.7%	145	23%	128	20.3%
	High school, completed	342	27.1%	177	28.1%	165	26.2%
	Technical training	6	0.5%	3	0.5%	3	0.5%
	University studies, not completed	153	12.1%	58	9.2%	95	15.1%
	University studies, completed	99	7.9%	43	6.8%	56	15.1%
	Graduate or Doctoral studies	8	0.6%	5	0.8%	3	0.5%
	Total	1260	100%	630	100%	630	100%
Do you drive a motor vehicle?	Yes	580	46.0%	458	72.7%	122	19.4%
	No	680	54.0%	172	27.3%	508	80.6%
	Total	1260	100%	630	100%	630	100%
Do you have a driver's license?	Yes	273	21.7%	244	38.7%	29	4.6%
	No	987	78.7%	386	61.3%	601	95.4%
	Total	1260	100%	630	100%	630	100%
Do you normally drive?	Yes	470	37.3%	392	62.2%	78	12.4%
	No	790	62.7%	238	37.8%	552	87.6%
	Total	1260	100%	630	100%	630	100%
Type of driver	Professional	202	35.1%	190	41.5%	12	9.8%
	Private	373	64.9%	264	57.6%	109	89.3%
	Total	685	100%	454	100%	121	100%

2.2. Design, Procedure and Instruments

The data reported in this study were collected through the National Survey on Mobility of the Dominican Republic from the year 2019 [19]. The questionnaire included issues related to the knowledge of institutions and traffic laws, public transportation, private transportation, on-foot movements, bike use, ITS systems and measures, among which the variables of the present research are included. It is a pioneer questionnaire in Latin America, where such exhaustive and complete instruments have not been administered. It began to be administered in 2018 to assess the situation in the country and detect the evolution and changes in the way Dominicans move as a result of the implementation of the various initiatives carried out by INTRANT.

The survey was administered through personal, in-person interviews with a duration of approximately 20 min. The sample was obtained during a time span from 24 November to 7 December 2019. The gathering of information was carried out through a CAPI system (Computer Assisted Personal Interviews) on tablets, recording and geo-referencing the interviews, with the aim of expediting the duration of the interview and minimize any recording mistakes.

In order to achieve the proposed objectives, the following variables were taken into account:

- Recall of traffic safety campaigns: obtained from the following questions. Do you remember any campaign or message on traffic, mobility and/or road safety? With one possible answer, yes/no; What was the topic? With an open answer (therefore, coded during the data processing phase); Who issued it? With an answer to be chosen among the options INTRANT/Presidency of the Republic/Private Company/Other; and through which media did you access this campaign/message? With a multiple-choice answer (Press/Television/Radio/Web/Social networks/Other).
- Sociodemographic variables and driving data: gender, age group, habitat, children, job situation, does he/she drive a motor vehicle? Does he/she habitually drive? Is he/she a professional driver? Does he/she have a driver's license? What type of license?

2.3. Data Processing

For this study, descriptive (frequency) analyses were carried out with the aim of describing and characterizing the general remembrance of traffic campaigns among the Dominican population. Moreover, Chi-square (χ^2) analyses were performed in order to obtain possible statistical associations with sociodemographic variables. Subsequently, and after basic statistical parameters were tested and met, a binary logistic regression (i.e., *Logit*) model was designed, using the fact of having recalled at least a traffic safety campaign as a dependent and binomial variable, where 1 = Yes and 0 = No, using basic demographic variables whose coverage is common to the whole population. Statistical analyses were carried out using the ©IBM SPSS (Statistical Package for Social Sciences) version 26.0 (Armonk, NY, USA).

2.4. Ethics

This type of study does not require ethical approval. No personal data were used, and the participation was anonymous, implying no potential risks for the integrity of our participants. However, the "Ethics Committee of Research in Social Sciences in Health" from the University Institute on Traffic and Road Safety of the University of Valencia (Spain) was consulted, certifying that the research responds to the general ethical principles relevant to research in Social Sciences. The Committee expressed its approval for the research. All participants gave their consent before taking part in the study and after receiving an explanation of its objectives, as well as all the previously mentioned considerations, from the research staff.

3. Results

The recall of these traffic safety-related communication campaigns or messages on traffic safety awareness in the Dominican Republic can be considered, at first glance, as very low, as only 9% of the sample answered positively to the corresponding question (Figure 1). More than half of the participants had seen the campaigns on TV (62.8%), and they had been produced by different organizations, among which the INTRANT stands out (35.4%).

Concerning the topic of the campaigns, there are not any that stand out. In Table 2, it can be observed that the use of seatbelts, safety and mobility, the don't drive if you drank campaign, and respect the laws are the most repeated elements, despite showing only a small difference from the rest. Moreover, it is remarkable that 21.2% of participants do not remember what the advertisement was about.

After analyzing the data depending on the different sociodemographic characteristics, we found that the only cases in which significant differences are produced are gender and job situation. Men and people who are currently employed are the ones who remember traffic campaigns the most (Table 3). Concerning the data related to driving, it is observable how people who drive any motor vehicle are those who remember this type of campaign the most. In relation to this, there is also a significant difference in the habitual driving and driver's license variables. On the other hand, no differences appear depending on the type of license, and neither on the type of driver (professional or private) (Table 3).

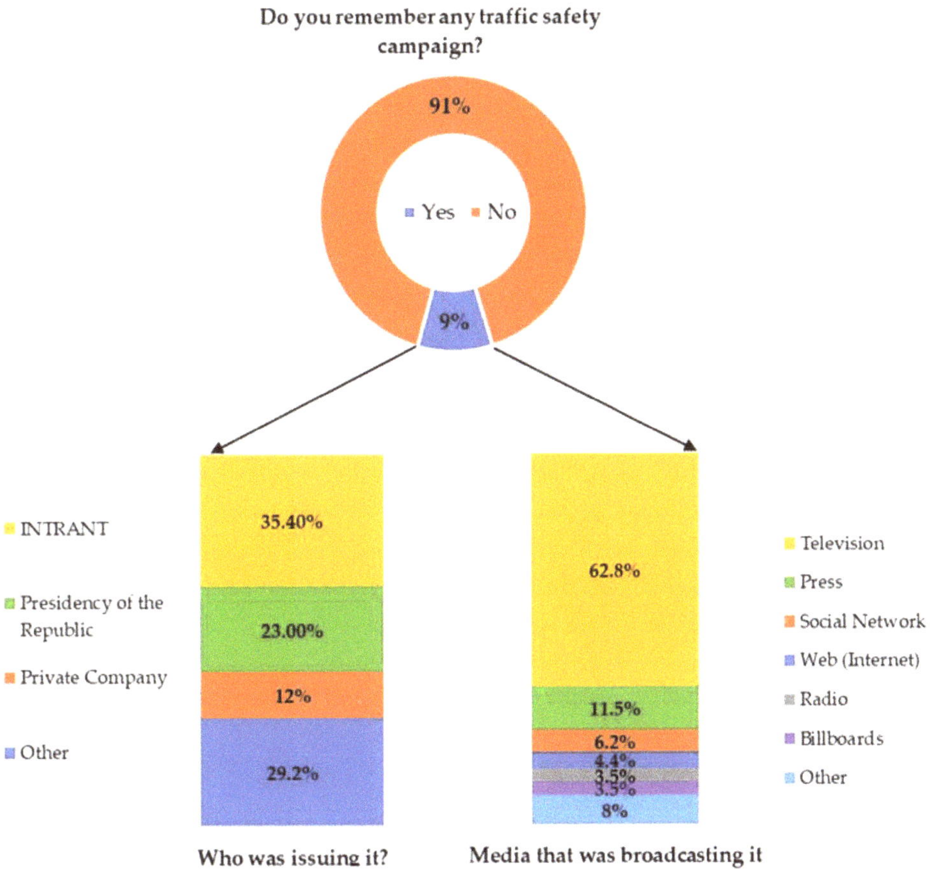

Figure 1. Overall data on the recall of traffic safety campaigns.

Table 2. Topics addressed by the traffic safety campaigns remembered by population.

Topic of the Campaign	Frequency	Valid Percentage
Use of the seatbelt	11	9.7
Safety and mobility	11	9.7
"Don't drive if you drank"	9	8.0
Respect the laws	9	8.0
Be careful	6	5.3
Drive slowly	5	4.4
Use of the cellphone	4	3.5
Accidents prevention	4	3.5
Transportation	3	2.7
Safety helmet	3	2.7
Land transportation	2	1.8
Yield sign	2	1.8
Easter holidays	2	1.8

Table 2. *Cont.*

Topic of the Campaign	Frequency	Valid Percentage
Allowed speed	2	1.8
Know how to drive	2	1.8
Pollution	2	1.8
Passengers	1	0.9
OMSA buses	1	0.9
Relieve traffic jams	1	0.9
Using the metro	1	0.9
Moto-taxi riders' training	1	0.9
Cyclists' corridor	1	0.9
Price of trip	1	0.9
Number of plate	1	0.9
Better traffic	1	0.9
Better roads	1	0.9
Color of cars	1	0.9
December advertisements	1	0.9
NS/NC	24	21.2
Total	113	100.0

Table 3. Memories of traffic campaigns depending on sociodemographic variables.

Variable	Category	Do You Remember Any Traffic Safety Campaign?				
		Yes		No		
		n	%	n	%	Test Value
Gender *	Man	72	11.4%	558	88.6%	$Chi^2(1) = 9.342$; $p < 0.010$ **
	Woman	41	6.5%	589	93.5%	
Age	18–24	20	7.7%	240	92.3%	$Chi^2(4) = 2.143$; $p = 0.709$
	25–34	29	9.3%	282	90.7%	
	35–49	35	9.6%	331	90.4%	
	50–64	17	7.7%	204	92.3%	
	>65	12	11.8%	90	88.2%	
Habitat	Urban	91	8.8%	938	91.2%	$Chi^2(1) = 0.107$; $p = 0.744$
	Rural	22	9.5%	209	90.5%	
Do you have children?	Yes	85	8.8%	886	91.2%	$Chi^2(1) = 0.238$; $p = 0.625$
	No	28	9.7%	261	90.3%	
Job situation *	Unemployed	26	5.7%	432	94.3%	$Chi^2(3) = 12.515$; $p < 0.010$ **
	Retired	4	9.8%	37	90.2%	
	Part-time employee	29	13.6%	184	86.4%	
	Full-time employee	40	10.3%	350	89.7%	
Do you drive a motor vehicle *	Yes	69	11.9%	511	88.1%	$Chi^2(1) = 11.288$; $p < 0.001$ ***
	No	44	6.5%	636	93.5%	
Do you have a driver's license *	Yes	42	15.4%	231	84.6%	$Chi^2(1) = 17.575$; $p < 0.001$ ***
	No	71	916	7.2%	92.8%	
Type of license	01	8	25%	24	75%	$Chi^2(5) = 3.658$; $p = 0.600$
	02	23	13.8%	144	86.2%	
	03	9	17.6%	42	82.4%	
	04	1	9.1%	10	90.9%	
	05	0	0%	1	0%	
Do you habitually drive *	Yes	56	11.9%	414	88.1%	$Chi^2(1) = 7.973$; $p < 0.010$ **
	No	57	7.2%	733	92.8%	
Type of driver	Professional	22	10.9%	180	89.1%	$Chi^2(1) = 0.106$; $p = 0.745$
	Private	44	11.8%	329	88.2%	

*** Significant at the level $p < 0.001$; ** Significant at the level $p < 0.010$; * Significant at the level $p < 0.050$.

If we focus on the media that were issuing the advertisement or campaign, it is interesting how the only sociodemographic variable reflecting significant results is age (Chi2(24) = 39.252; p = 0.026). In Table 4, we can observe that, despite television being the media from which all groups remember most campaigns, it comes out as especially remarkable for young people between 25 and 34 and people between 50 and 64. Likewise, people between 35 and 49 have had more access to campaigns through the press and the internet than the rest of groups.

Table 4. Media issuing the traffic campaigns remembered, depending on the age group.

Diffusion Media	**Age Range**									
	18–24		**25–34**		**35–49**		**50–64**		**>65**	
	n	%	n	%	n	%	n	%	n	%
Television	11	55.0%	22	75.9%	19	54.3%	12	70.6%	7	58.3%
Press	2	10.0%	1	3.4%	8	22.9%	1	5.9%	1	8.3%
Social network	0	0.0%	3	10.3%	1	2.9%	2	11.8%	1	8.3%
Web (Internet)	0	0.0%	2	6.9%	3	8.6%	0	0.0%	0	0.0%
Radio	0	0.0%	1	3.4%	1	2.9%	2	11.8%	0	0.0%
Billboards	3	15.0%	0	0.0%	0	0.0%	0	0.0%	1	8.3%
Other	4	20.0%	0	0.0%	3	8.6%	0	0.0%	2	16.7%
Total	20	100%	29	100%	35	100%	17	100%	12	100%

A binary logistic regression analysis was carried out taking into account the independent variables: sex, age, habitat, educational level and driving status (being or not a driver).

This significant model, conducted through a binomial logistic regression technique, was fitted using both demographic and road safety and demographic variables. The function defining the binary logistic regression model is presented below:

$$P(Y = 1; X_i) = e^{zi}/1 + e^{zi}$$

$$\text{where } z_i = \beta_0 + \beta_1 X_1 + \beta_2 X_2 + \cdots \ \beta_{10} X_{10} + \varepsilon$$

The variables included in the model were defined as:

$Y_i = 1$, which implies having recalled at least one traffic safety campaign performed in the country.

$Y_i = 0$, which implies not remembering any traffic safety campaign performed.

$X_1, X_2, \ldots X_n$: Independent (whether continuous or categorical–dummy) variables included in the model, where Sex (X_1; dummy variable), Age (X_2) and Educational level (X_3) represent the three independent variables of the model, as shown in Table 5.

Table 5. Logistic Regression Model (Logit). Dependent variable: Having remember at least one traffic safety campaign.

Variables in the Equation	**B**	**S.E.** [c]	**Wald**	**Df** [d]	**Sig.**	**Exp(B)** [e]
Sex (Male) [a]	0.777	0.212	13.436	1	<0.001 ***	2.175
Age	0.018	0.007	6.981	1	0.008 **	1.018
Educational level [b]	0.398	0.053	56.698	1	<0.001 ***	1.489
Constant	−5.842	0.516	128.385	1	<0.001 ***	0.003

Success categories = [a] Being a male, [b] Ranging from "No studies (Cannot read or write)" to "Graduate or Doctoral studies", as specified in Table 1; [c] Standard Error; [d] Degrees of Freedom; [e] Odds Ratio (OR) *** Significant at the level $p < 0.001$; ** Significant at the level $p < 0.010$.

The final solution presented an overall accuracy percentage of 91.0%, explained 19.1% of the variance among subjects (Nagelkerke's $R^2 = 0.191$) and showed a − 2Log-likelihood coefficient of 690.412. The basic parameters and variables included in the model are presented in Table 5, showing its Beta coefficients, significance level and odds ratio (OR). Three of the study variables had significant results (Table 5), indicating that sex (i.e., the fact of being a man), a greater age and educational level are significant predictors of the fact of remembering at least one of the road safety communication campaigns previously conducted in the Dominican Republic. The greatest Wald (χ^2), that is, the individual test statistic for each predictor variable, corresponded to the variable "Educational level" (Wald = 56.698), which has shown to be the most relevant demographic predictor among those included in the significant model.

4. Discussion

The core aim of this study was to assess to what extent inhabitants of the Dominican Republic remember traffic safety campaigns. In general terms, the results reflect a very low recall of advertisement campaigns in the scope of traffic, mobility and road safety, even below the data obtained in previous years (in 2018, the rate of remembrance was 11.2%) [20]. This implies a huge issue for the country, since it indicates that, at the moment, the population is not aware of traffic accidents being the second leading cause of violent death in the Dominican Republic, when citizens are actually essential to reverse the situation [7].

Thus, although in recent years, multiple communication campaigns have been carried out within the traffic and road safety sector in the Dominican Republic, the recall of these campaigns remains low. In Table 6, the main campaigns carried out during 2019 in the country, most of them broadcasted by different media, are shown. [21,22].

Table 6. Communication campaigns in the traffic sector in the Dominican Republic during 2019.

Campaign	Motto/Main Content
Deployment of police and health personnel at Christmas and Easter, with communication through the media and social networks.	Pact For Life
National campaign aimed at road safety for cyclists and sustainable transport.	I ride the bike
Campaign to promote the use of helmets and caution on the roads.	Ride Safe
Awareness-raising among citizens, with the aim of not driving under the influence of alcohol.	Take it Seriously: If you drink don't drive
Improvement of pedestrian mobility	Safe Pedestrian
Promoting the implementation of the Santo Domingo Tourist Corridor	Tourist Corridor
Preventing the risk of deaths and injuries related to the transportation of children to and from school.	Safe School Environments
Citizen information and education through the media and social networks.	Driver's Licenses
Promotion of safe traffic behaviors of citizens	Road Safety Education
Promotion of sustainable travel by foot, public transport and bicycle.	National Sustainable Mobility Week
Promotion of social responsibility and citizen engagement	Road Safety Week
Proposals for innovative solutions aimed at young people	HACKAMÉRICAS
Promotion of the Road Safety Education Park and its activities	Road Safety Education Park in Ciudad Juan Bosch
Promotion of safety and awareness activities	Holy union 2018 for your Values and your Safety
Citizen awareness-raising at points of high vehicular and pedestrian concentration	Discover Barahona
Implementation of various bicycles for the promotion of sustainable mobility	"United in Action for the Climate"
Road safety education and information for visitors	Seminar at the Book Fair

In this sense, it is convenient to remark that there is only one campaign that is remembered for its slogan. *This is Don't drive if you drank*, which refers to alcohol consumption when driving [22]. It is a message that was heavily emphasized during the year before the survey, and, in a way, it is positive that this slogan is one of the most remembered elements. In this sense, it would be useful to reflect on which variables are present in this campaign (and not in other ones) so that this situation may be reproduced.

The *Don't drive if you drank* campaign was broadcasted by different media, such as television, radio, newspapers, digital press and social networks. Likewise, there were acts and events related to this campaign in universities, where there was an attempt to raise young people's awareness of the detriments of alcohol consumption while driving (INTRANT, 2020b). This is important since it indicates that the message was strengthened enough for the majority of the population to access it, and therefore achieves an adequate spread, both of which are essential aspects of social marketing [23,24].

However, it must be distinguished between the recall of a campaign and its actual effectiveness. Biases can be generated and distort the recall of the advertisement, causing the message to be different from the one internalized by the users [25]. In this sense, it would be interesting to go deeper into further research on other elements that influence the effectiveness of a campaign, beyond recall, such as the change in attitudes and behaviors and the impact of the increase of knowledge on the viewer [26]. Therefore, the monitoring and evaluation of campaigns are also fundamental, being particularly important in the early years of the evolution of media campaigns in low- and middle-income countries [27].

The advertisements related to this campaign do not present especially violent images; they rather inform by providing a clear message. Maybe a higher impact could have been achieved had they used a more emotional component [28], as happens in other countries with similar campaigns, such as Spain [29] or Mexico [30]. In Spain, the highest peak of deaths caused by traffic accidents occurred at the end of the 1980s. At this time, the Dirección General de Tráfico (DGT) changed its approach to communication campaigns, starting to make advertisements with much more raw images, showing in a realistic way the terrible consequences that can result from irresponsible driving behavior. This contributed to the reduction in the accident rate that occurred during the 1990s in this country [12]. However, the communication strategy was changed in the following years because the prolonged high impact levels substantially reduced the effectiveness of the campaigns [31]. These results are repeated in advertisements that promote healthy behaviors beyond the transit sector (e.g., tobacco or safe sex), where it has been shown that high-impact campaigns could attract the viewer's attention but prevent learning and retention of public health problems [32]. For this reason, several studies indicate that the most appropriate formula is to intersperse campaigns with low and high visual impact so as not to habituate the user [33]. It is important to take as a reference the strategies and advertisements made in other countries because campaign messages that have proven to be effective in one environment can also be tested in other contexts (with prior adaptation), which could save resources and time [27].

On the other hand, the use of social networks is necessary because it makes the problem feel closer to the population, especially to young people, the group that uses them the most [34]. It is surprising that, according to the present research, young Dominicans do not have more memories of campaigns seen on social media. This can indicate that the way of conveying the information was not adequate or that the INTRANT social networks do not have enough engagement yet to influence the thoughts and behaviors of their followers [35]. Therefore, it is necessary to establish a good relationship with users through an accessible social network that will allow interaction with followers, with frequent publications, relevant and interesting information for Dominicans, encouraging users' participation, answering their questions and paying attention to their suggestions. These easy strategies can turn the INTRANT profile into a better communication source, with a more significant impact on the population and a very low economic cost [36].

However, this is a process that can take time. Therefore, future campaigns in the Dominican Republic must focus on television, since it is clearly the medium from which campaigns are remembered the most. This is interesting, since it is consistent with what happens in many countries around the world, where the television is still the preferred medium for advertising [37]. Thus, according to a study conducted by Nielsen IBOPE, in the Dominican Republic, open TV channels in 2019 reached more than 99.24% of all people in Santo Domingo and the province of Santiago. Viewers' consumption time increased to 115 min per day, consolidating television as the medium with the greatest impact and reach in the country. Largely because of this, 67% of the advertising investment in 2019 was made in this medium [38]. Regarding age groups, children (4–11 years) reached 15%, adolescents (12–17) 9.6%, young adults (18–34) 22%, adults (35–54) 30% and seniors over 55 years 23% [38]. Similar data are obtained in other emerging countries. In Latin America, television is widely consumed by the population as a whole, although more and more people opt for on-demand services, especially young people [39].

In this sense, despite the rise of social networks, this type of media is only useful for supporting the main advertisements issued on television, and this does not even apply to every case [40,41]. Thus, in future campaigns, the Dominican Republic could follow up with this strategy, issuing most information through television and using the rest of the media as a complement for establishing the new ideas and behaviors in the population.

Another expected element that should be remarked on is that the most influential variable in the recall of traffic campaigns is being a driver. All variables where a significant difference appeared between groups share this unifying element: driving a motor vehicle, being a habitual driver, and having a driver's license. Even in the case of the gender and job situation variables this link has been found, since those who drive the most in the Dominican Republic are men and people who have a job. This phenomenon can be explained because drivers may have more interest to pay attention to traffic advertisements in comparison with non-drivers. This is a key element in the attention to and retention of information [42,43].

Such circumstances may be useful for future campaigns, as it is drivers who cause most traffic accidents due to human error [44,45]. However, it should not be forgotten that pedestrians and motorists are the most vulnerable groups, being those who die the most from a traffic accident. Thus, the subject matter of the spots should be especially aimed at drivers, warning of the importance of the human factor, with special emphasis on distractions, alcohol and drug consumption and speeding due to their significant influence on accident rates [46,47].

In addition, emphasis should be placed on motorcyclists since, in the Dominican Republic, motorcycles represent more than 55% of the country's vehicle fleet [48]. This situation also occurs in many of the emerging countries of Latin America, especially because of motorcycle cabs (or moto-taxis), which are motorcycles that act as public transport two-wheeled vehicles for the movement of travelers [49]. These vehicles are involved in a high number of traffic accidents, and may be influencing variables such as drowsiness, fatigue and high workload [50]. Therefore, it would be relevant to act specifically on this risk group in future campaigns carried out in these countries.

5. Conclusions

The improvement of road safety and mobility in the Dominican Republic is necessary because of the high accident rates and deaths that have been happening in the country in recent decades [51]. The first step for users to remember the communication campaigns that are being carried out in the Dominican Republic is to increase their awareness of road safety and their appropriate behaviors on the road. This research establishes that the recall of the campaigns is very low. Hence, a change in the design and elaboration of the advertisements in this sector is necessary.

Thus, given the circumstances, the elaboration of future campaigns in the Dominican Republic might require extensive further evaluation and follow-up, a fact that could help to maximize the impact of future traffic campaigns and advertisements in the Dominican Republic, as well as in other emerging countries of the region with similar characteristics. Further, this study found that not all population segments in terms of sex, education and income have the same remembrance level, and gaps in terms of access to information channels and sources may possibly be enhancing it. In this regard, key segments of the population, such as the female, young, with lesser educational levels and non-driving population, that have been found as benefiting less from them should also be targeted for further campaigns and communicative strategies aimed at strengthening traffic safety.

Author Contributions: For this study, F.A. and M.F. conceived and designed the research and performed the data collection; S.A.U. supervised data collection and coding; S.A.U. and M.F. analyzed the data; C.F. and S.A.U. contributed with reagents/materials/analysis tools; M.F. and S.A.U. wrote and revised the paper. All authors have read and agreed to the published version of the manuscript.

Funding: This study was funded by the National Institute of Transit and Land Transportation (INTRANT) and its Permanent Observatory in Road Safety (OPSEVI; public agency of the Dominican Republic)-Grant number: 20170475. Additionally, this work was supported by the research grant ACIF/2020/035 (MF) from "Generalitat Valenciana". Funding entities did not contribute to the study design or data collection, analysis and interpretation or writing of the manuscript.

Institutional Review Board Statement: The study was conducted according to the guidelines of the Declaration of Helsinki, and approved by the Ethics Committee of Research in Social Sciences in Health" from the University Institute on Traffic and Road Safety of the University of Valencia (Spain) (protocol code HE000125101, approved on 25 October 2019).

Informed Consent Statement: Informed consent was obtained from all subjects involved in the study.

Data Availability Statement: The data will be available upon reasonable request to the corresponding author.

Acknowledgments: The authors wish to thank to Runa Falzolgher and Arash Javadinejad for the professional edition of the final version of the manuscript.

Conflicts of Interest: The authors declare no conflict of interest.

References

1. CNSS. Jóvenes los más Afectados por Accidentes de Tránsito en RD. Dominican Republic. 2020. Available online: https://www.cnss.gob.do/index.php/noticias/item/288-jovenes-los-mas-afectados-por-accidentes-de-transito-en-rd (accessed on 24 June 2021).
2. WHO. The Top 10 Causes of Death. 2018. Available online: https://www.who.int/news-room/fact-sheets/detail/the-top-10-causes-of-death (accessed on 19 June 2021).
3. Martínez, J.M.D. Desarrollo Económico, Países Emergentes y Globalización. *eXtoikos* **2014**, *14*, 5–14. Available online: https://dialnet.unirioja.es/servlet/articulo?codigo=5561995 (accessed on 15 June 2021).
4. PNUD. *Informe Sobre Desarrollo Humano 2019: Más allá del Ingreso, más allá de los Promedios, más allá del Presente*; Programa de las Naciones Unidas para el Desarrollo (PNUD): New York, NY, USA, 2019.
5. WHO. *Global Status Report on Road Safety*; 2018; ISBN 978-92-4-156568-4. Available online: https://www.who.int/violence_injury_prevention/road_safety_status/2018/en/ (accessed on 18 June 2021).
6. OISEVI. VII Informe Iberoamericano de Seguridad Vial (2015/2016). 2016. Available online: https://www.oisevi.org/a/images/files/informes/info-7.pdf (accessed on 20 June 2021).
7. OPSEVI & INTRANT. *Boletín Informativo OPSEVI*; Instituto Nacional de Tránsito y Transporte Terrestre: Santo Domingo, Dominican Republic, 2020. Available online: https://issuu.com/intrantrd/docs/bolet_n_informativo_-_opsevi (accessed on 17 June 2021).
8. INTRANT. Instituto Nacoinal de Tránsito y Transporte Terrestre: Noticias. Dominican Republic. 2020. Available online: https://www.intrant.gob.do/index.php/noticias/noticias?start=182 (accessed on 20 June 2021).
9. Perkins, H.W.; Linkenbach, J.W.; Lewis, M.A.; Neighbors, C. Effectiveness of social norms media marketing in reducing drinking and driving: A statewide campaign. *Addict. Behav.* **2010**, *35*, 866–874. [CrossRef] [PubMed]
10. Rubinson, J. Empirical evidence of TV advertising effectiveness. *J. Advert. Res.* **2009**, *49*, 220–226. [CrossRef]
11. Castillo, J.I.; Castro, M.; Pedregal, D.J. Efectividad de las Campañas de Publicidad Para Reducir la Siniestralidad vial en España: Un Análisis Econométrico. *XIV Encuentro Econ. Apl.* **2011**, 1–25. Available online: https://idus.us.es/handle/11441/61109 (accessed on 22 June 2021).
12. Segura-García, R. Evolución y Efectividad de los Spots de la GT. *Opción* **2015**, *31*, 1180–1200. Available online: https://www.redalyc.org/pdf/310/31045567062.pdf (accessed on 18 June 2021).
13. INTRANT. Instituto Nacional de Tránsito y Transporte Terrestre: Si Tomas no Manejes. 2020. Available online: https://www.intrant.gob.do/index.php/resultado-de-busqueda?searchword=SI%20TOMAS%20NO%20MANEJES&searchphrase=all (accessed on 15 June 2021).

14. Ma, X.; Li, X.; Kwan, M.P.; Chai, Y. Who could not avoid exposure to high levels of residence-based pollution by daily mobility? Evidence of air pollution exposure from the perspective of the neighborhood effect averaging problem (NEAP). *Int. J. Environ. Res. Public Health* **2020**, *17*, 1223. [CrossRef]
15. Wundersitz, L.; Hutchinson, T. What Can We Learn from Recent Evaluations of Road Safety Mass Media Campaigns? *J. Australas Coll. Road Saf.* **2011**, *22*, 40–50. Available online: https://search.informit.org/doi/abs/10.3316/INFORMIT.692489672234889 (accessed on 16 June 2021).
16. Staton, C.; Vissoci, J.; Gong, E.; Toomey, N.; Wafula, R.; Abdelgadir, J.; Zhou, Y.; Liu, C.; Pei, F.; Zick, B.; et al. Road traffic injury prevention initiatives: A systematic review and metasummary of effectiveness in low and middle income countries. *PLoS ONE* **2016**, *11*, e0144971. [CrossRef]
17. Diegelmann, S.; Ninaus, K.; Terlutter, R. Distracted driving prevention: An analysis of recent UK campaigns. *J. Soc. Mark.* **2020**, *10*, 243–264. [CrossRef]
18. Oficina Nacional de Estadística. *Censo Nacional de Población y Vivienda 2020 de la República Dominicana*, X ed.; Ministerio de Economía, Planificación y Desarrollo: Santo Domingo, Dominican Republic, 2018.
19. INTRANT. *National Mobility Survey of the Dominican Republic. Results REPORT 2019*; Instituto Nacional de Tránsito y Transporte Terrestre: Santo Domingo, Dominican Republic, 2020.
20. INTRANT. *National Mobility Survey of the Dominican Republic. Results REPORT 2018*; Instituto Nacional de Tránsito y Transporte Terrestre: Santo Domingo, Dominican Republic, 2019.
21. INTRANT. *Informe Acciones de Comunicación en la Implementación del Plan Estratégico Nacional Para la Seguridad Vial*; Instituto Nacional de Tránsito y Transporte Terrestre: Santo Domingo, Dominican Republic, 2020.
22. INTRANT. *Plan de Comunicación de Tránsito, Movilidad y Seguridad Vial en República Dominicana*; Instituto Nacional de Tránsito y Transporte Terrestre: Santo Domingo, Dominican Republic, 2020.
23. Janssen, M.M.; Mathijssen, J.J.; van Bon–Martens, M.J.; Van Oers, H.A.; Garretsen, H.F. Effectiveness of alcohol prevention interventions based on the principles of social marketing: A systematic review. *Subst. Abus. Treat. Prev. Policy* **2013**, *8*, 18. [CrossRef]
24. Helmig, B.; Thaler, J. On the effectiveness of social marketing—What do we really know? *J. Nonprofit Public Sect. Mark.* **2010**, *22*, 264–287. [CrossRef]
25. Seelig, D.; Wang, A.L.; Jaganathan, K.; Loughead, J.W.; Blady, S.J.; Childress, A.R.; Langleben, D.D. Low message sensation health promotion videos are better remembered and activate areas of the brain associated with memory encoding. *PLoS ONE* **2014**, *9*, e113256. [CrossRef]
26. Pol, E.; Vidal, T.; Romeo, M. Assumptions of attitudinal and behavioural change in advertising campaigns and environmental marketing. The four-spheres model. El modelo de las 4 esferas. *Estud. Psicol.* **2001**, *22*, 111–126. [CrossRef]
27. Mullin, S.; Prasad, V.; Kaur, J.; Turk, T. Increasing evidence for the efficacy of tobacco control mass media communication programming in low-and middle-income countries. *J. Health Commun.* **2011**, *16* (Suppl. 2), 49–58. [CrossRef] [PubMed]
28. Lennon, R.; Rentfro, R.; O'Leary, B. Social Marketing and Distracted Driving Behaviors among Young Adults: The Effectiveness of Fear Appeals. *Acad. Mark. Stud. J.* **2010**, *14*, 95–113. Available online: http://citeseerx.ist.psu.edu/viewdoc/download?doi=10.1.1.1071.3083&rep=rep1&type=pdf (accessed on 18 June 2021).
29. DGT. Dirección General de Tráfico: Campañas. España. 2020. Available online: http://www.dgt.es/es/la-dgt/campanas/ (accessed on 13 June 2021).
30. Barreño, R. Lanzan Campaña Para Concientizar la Combinación de Tomar o Mensajear y el Volante. El Occidental. 2019. Available online: https://n9.cl/mmmhn (accessed on 15 June 2021).
31. Castillo-Manzano, J.I.; Castro-Nuño, M.; Pedregal, D.J. How many lives can bloody and shocking road safety advertising save? The case of Spain. *Transp. Res. F Traffic Psychol. Behav.* **2012**, *15*, 174–187. [CrossRef]
32. Dahl, D.W.; Frankenberger, K.D.; Manchanda, R.V. Does it pay to shock? Reactions to shocking and nonshocking advertising content among university students. *J. Advert. Res.* **2003**, *43*, 268–280. [CrossRef]
33. Panda, T.K.; Panda, T.K.; Mishra, K. Does Emotional Appeal Work in Advertising? The Rationality behind Using Emotional Appeal to Create Favorable Brand Attitude. *IUP J. Brand Manag.* **2013**, *10*, 7. Available online: https://www.researchgate.net/publication/312470434_Does_Emotional_Appeal_Work_in_Advertising_The_Rationality_Behind_Using_Emotional_Appeal_to_Create_Favorable_Brand_Attitude (accessed on 21 June 2021).
34. Pikas, B.; Sorrentino, G. The Effectiveness of Online Advertising: Consumer's Perceptions of ads on Facebook, Twitter and YouTube. *J. Appl. Bus. Econ.* **2014**, *16*, 70–81. Available online: http://www.na-businesspress.com/JABE/pikas_abstract.html (accessed on 20 June 2021).
35. Lee, D.; Hosanagar, K.; Nair, H.S. *The Effect of Social Media Marketing Content on Consumer Engagement: Evidence from Facebook*; Stanford Graduate School of Business: Stanford, CA, USA, 2014; pp. 1–15.
36. Ashley, C.; Tuten, T. Creative strategies in social media marketing: An exploratory study of branded social content and consumer engagement. *Psychol. Mark.* **2015**, *32*, 15–27. [CrossRef]
37. Gómez-Domínguez, P. Era Digital y Televisión Autonómica: Un Estudio Comparativo de las Plataformas Web, Aplicaciones Móviles y Redes Sociales de TV3 y BBC One. 2016. Available online: https://hdl.handle.net/10171/41835 (accessed on 21 June 2021).
38. Nielsen IBOPE. *Análisis Comparativo Audiencia e Inversión en Medios Publicitarios en la República Dominicana 2015–2019*. 2020. Available online: https://www.resumenweb.com:8443/?year=2020 (accessed on 20 June 2021).

39. Ericsson. *TV and Media 2015: The Empowered TV and Media Consumer's Influence*. 2015. Available online: http://mb.cision.com/Public/15448/2245296/8c2cdb6ed1ea757f.pdf (accessed on 19 June 2021).
40. Saavedra-Llamas, M.; Papí-Gálvez, N.; Perlado-Lamo-de-Espinosa, M. Televisión y redes sociales: Las audiencias sociales en la estrategia publicitaria. *Prof. Inf.* **2020**, *29*, 2. [CrossRef]
41. Iniesta-Alemán, I.; Marta-Lazo, C.; Zaro-Becas, M.D.L.C. La inversión en publicidad, retos para la televisión del siglo XXI. *RETOS Rev. Cienc. Adm. Econ.* **2018**, *8*, 141–156. [CrossRef]
42. Bonetto, V.A.; Calderon, L.L. La Importancia de Atender a la Motivación en el Aula. *PsicoPediaHoy* **2014**. Available online: https://ri.conicet.gov.ar/handle/11336/33856 (accessed on 23 June 2021).
43. Nan, X. The influence of liking for a public service announcement on issue attitude. *Commun. Res.* **2008**, *35*, 503–528. [CrossRef]
44. Zhang, Z.; Turla, T.; Liu, X. Analysis of human-factor-caused freight train accidents in the United States. *J. Transp. Saf. Secur.* **2019**, 1–29. [CrossRef]
45. Gicquel, L.; Ordonneau, P.; Blot, E.; Toillon, C.; Ingrand, P.; Romo, L. Description of various factors contributing to traffic accidents in youth and measures proposed to alleviate recurrence. *Front. Psychiatry* **2017**, *8*, 94. [CrossRef] [PubMed]
46. Rolison, J.J.; Regev, S.; Moutari, S.; Feeney, A. What are the factors that contribute to road accidents? An assessment of law enforcement views, ordinary drivers' opinions, and road accident records. *Accid. Anal. Prev.* **2018**, *115*, 11–24. [CrossRef]
47. Martí-Belda, A.; Pastor, J.C.; Montoro, L.; Bosó, P.; Roca, J. Persistent Traffic Offenders: Alcohol Consumption and Personality as Predictors of Driving Disqualification. *Eur. J. Psychol. Appl. Leg. Context* **2019**, *11*, 81–92. [CrossRef]
48. DGII. Boletín Estadístico. In *Parque Vehicular 2019*; Dirección General de Impuestos Internos: Santo Domingo, Dominican Republic, 2019.
49. Rodríguez, D.A.; Santana, M.; Pardo, C.F. *La Motocicleta en América Latina: Caracterización de su uso e Impactos en la Movilidad en Cinco Ciudades de la Región*; CAF: Bogotá, Colombia, 2015. Available online: http://scioteca.caf.com/handle/123456789/754 (accessed on 16 June 2021).
50. Chumacero, P.; Peralta, J.; León, F. Carga de trabajo, somnolencia y accidentes de tránsito: Se potencian en conductores de mototaxi? *Rev. Med. Herediana* **2016**, *27*, 268–269. [CrossRef]
51. Díaz, N.; Del Carmen, R. Propuesta de Diseño de Proyecto de Educación vial Usando Plataforma Virtual en República Dominicana. Ph.D. Thesis, Universidad Nacional Pedro Henríquez Ureña, Santo Domingo, Dominican Republic, 2018. Available online: https://repositorio.unphu.edu.do/handle/123456789/1281 (accessed on 21 June 2021).

energies

MDPI

Article

A Hierarchical Autonomous Driver for a Racing Car: Real-Time Planning and Tracking of the Trajectory

Margherita Montani [1,*], Leandro Ronchi [1], Renzo Capitani [1] and Claudio Annicchiarico [2]

1 Department of Industrial Engineering of Florence, University of Florence, Via di Santa Marta 3, 50139 Firenze, Italy; leandro.ronchi@stud.unifi.it (L.R.); renzo.capitani@unifi.it (R.C.)
2 Meccanica 42 s.r.l., Via Ezio Tarantelli 15, 50019 Florence, Italy; claudio.annicchiarico@meccanica42.com
* Correspondence: margherita.montani@unifi.it

Abstract: The aim of this study was to develop trajectory planning that would allow an autonomous racing car to be driven as close as possible to what a driver would do, defining the most appropriate inputs for the current scenario. The search for the optimal trajectory in terms of lap time reduction involves the modeling of all the non-linearities of the vehicle dynamics with the disadvantage of being a time-consuming problem and not being able to be implemented in real-time. However, to improve the vehicle performances, the trajectory needs to be optimized online with the knowledge of the actual vehicle dynamics and path conditions. Therefore, this study involved the development of an architecture that allows an autonomous racing car to have an optimal online trajectory planning and path tracking ensuring professional driver performances. The real-time trajectory optimization can also ensure a possible future implementation in the urban area where obstacles and dynamic scenarios could be faced. It was chosen to implement a local trajectory planning based on the Model Predictive Control (MPC) logic and solved as Linear Programming (LP) by Sequential Convex Programming (SCP). The idea was to achieve a computational cost, 0.1 s, using a point mass vehicle model constrained by experimental definition and approximation of the car's GG-V, and developing an optimum model-based path tracking to define the driver model that allows A car to follow the trajectory defined by the planner ensuring a signal input every 0.001 s. To validate the algorithm, two types of tests were carried out: a Matlab-Simulink, Vi-Grade co-simulation test, comparing the proposed algorithm with the performance of an offline motion planning, and a real-time simulator test, comparing the proposed algorithm with the performance of a professional driver. The results obtained showed that the computational cost of the optimization algorithm developed is below the limit of 0.1 s, and the architecture showed a reduction of the lap time of about 1 s compared to the offline optimizer and reproducibility of the performance obtained by the driver.

Keywords: autonomous driving; trajectory planning; path tracking; sequential convex programming; linear programming; quadratic constraints; autonomous racing car

Citation: Montani, M.; Ronchi, L.; Capitani, R.; Annicchiarico, C. A Hierarchical Autonomous Driver for a Racing Car: Real-Time Planning and Tracking of the Trajectory. *Energies* **2021**, *14*, 6008. https://doi.org/10.3390/en14196008

Academic Editors: Guzek Marek, Rafał Jurecki and Wojciech Wach

Received: 8 July 2021
Accepted: 17 September 2021
Published: 22 September 2021

1. Introduction

In recent years, research on the automotive field has focused on making cars more and more able to make decisions autonomously due to the growth of electric and electronic technologies on modern road cars and to the possibility of providing increased safety and improved performance.

In the racing world, where the scenario is static and known, the perception system has a simple implementation, and good results have already been achieved, as shown, for example, in [1].

The main challenge is to improve performance through the use of models that represent the dynamics of the vehicle as faithfully as possible while ensuring an acceptable computational effort for real-time implementation.

The aim of this paper was to develop a trajectory planner able to update the trajectory online with the car progress on the track, trying to replicate the capabilities of a human driver. Then, the trajectory planning was embedded with a path tracking consisting of a Linear Quadratic Regulator (LQR) that provides the car's input signals: steering wheel angle, accelerator and brake pedals signals.

In this way, even if in a racing world scenario, real-time trajectory optimization can ensure a possible future implementation in the urban area where obstacles and dynamic scenarios could be faced.

To ensure a trajectory tracking able to reduce the lap time, the state-of-the-art offers different approaches both from the architecture and methodology point of view. From the architecture point of view, the research is divided between the ones that used a real-time single algorithm to plan and track the trajectory [2,3], others that preferred to separate the planner from the tracker by having the first offline and the second work in real-time [4], and others that divided trajectory planning and tracking but implemented them both in real-time [5–7]. The first architecture type, as said in [8], gives the advantage of incorporating tire force constraints in a straight-forward way. However, the integration of trajectory planning and tracking approaches can work for autonomous RC-cars but is difficult to scale to complex scenarios due to the non-convex nature of the constraints imposed by multiple vehicles. In addition, it implies a more complex tuning. Instead, there are many studies about the offline trajectory planner with good performances in terms of lap time reduction and car set-up, allowing to implement a vehicle model with more Degrees of Freedom (DoF) [9], adding also a friction coefficient mapping [4] and involving different methodologies. Some authors preferred to develop a geometrical solution by minimizing the curvature [8] or generating a racing line using professional driving techniques [10]. However, these techniques do not provide information about vehicle dynamics, despite the fact that the trajectory of least curvature ensures lap similar to the trajectory optimization techniques [8]. Therefore, other researchers preferred to face the problem as a kinodynamic problem by finding the optimal trajectory, minimizing the lap time and considering both the kinematic constraints of the track and the dynamic constraints of the vehicle [4,11]. Facing the problem from the kinodynamic point of view considers that the non-convexity of the trajectory does not allow a global solution [8,10] but a local one [12–14] using Sequential Convex Programming (SCP) methodology [7].

The study presented in this paper has involved the development of a trajectory planner that every 0.1 s provides the speed and trajectory that the car follows thanks to an implemented path tracking that controls the car motion every 0.001 s. The real-time implementation of both trajectory planning and path tracking is advantageous in terms of performance and safety. In fact, during the race, a professional driver is able to feel the car conditions and choose the trajectory that ensures the best performance. In the same way, a trajectory planner must provide the fastest trajectory for the dynamic and scenario conditions in which the vehicle is located. In [7], it was decided to use a point mass vehicle model and implement a Model Predictive Control (MPC) with the SCP approach. To simplify the algorithm without incurring in infeasibility and further reduce the computational time, it was chosen to solve the Linear Programming (LP) problem instead of the Quadratic Programming (QP) problem and to reduce the losses due to linearization, it was chosen to represent the car's GG-V with conical constraint equations. Furthermore, the weights and variables of optimization change with the track sequences, allowing the designer to change the constraints with changes in scenario and vehicle dynamics.

In Section 2, the architecture developed is shown. Then, in Section 3, the trajectory planner model and methodologies are explained with the hypothesis and the reference system used, followed by Section 4 where the path tracking developed is shown. In the last section, the tests performed will provide the results obtained by the algorithm in terms of the car's trajectory and lap time, comparing them with those obtained by the CarRealTime (CRT) MaxPerformance event, and by a professional driver in a real-time car simulator.

2. Trajectory Tracking Architecture

In Figure 1, the architecture developed is shown. The idea is to divide the trajectory planning from the path tracking instead of implemented a single MPC with a complex and detailed vehicle and tire model by ensuring a real-time implementation of the algorithm that the current technologies do not allow otherwise. Thus, an LP optimizer is implemented as trajectory planning, giving the reference position and speed to minimize the lap time every 0.1 s, and an optimum controller is implemented to allow the car to follow reference state values by defining the input steering wheel angle, throttle pedal and braking pedal every 0.001 s. These sample time values were necessary because an increase compared to the time of 0.1 s in providing the position and speeds that the vehicle must have is excessive compared to the speed of the vehicle resulting in a delay with respect to its position; from the tests undertaken, an increase in the time of 0.001 s of the controller was found to worsen the performance of the control.

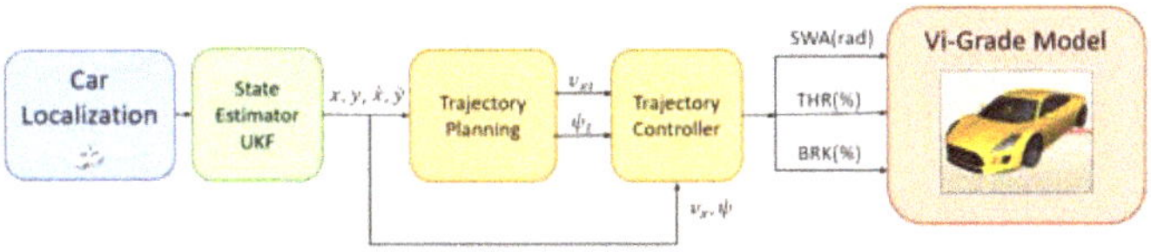

Figure 1. System architecture.

The trajectory planning and path tracking use the states of the car in the race track as feed-back. These states are estimated by a Localization Layer composed of an Unscented Kalman Filter (UKF) that uses the sensors measurements shown in Table 1.

Table 1. Variables measured by sensors.

Sensors	Variable	Units
IMU	Lateral and longitudinal acceleration, and yaw rate	m/s^2 – rad/s^2
GPS	Lateral and Longitudinal global position	m
Camera	handling angle	rad
Steering sensor	Steering angle	rad
Phonic wheel	wheels speed	m/s
Pressure sensors	Brake chamber pressures	Pa

All the layers shown in Figure 1 were implemented in Matlab-Simulink software and then compiled in C to be used in the real-time car simulator of Vi-Grade. The different control algorithms were integrated into the same Simulink model considering the rate transition between subsystem with different sample times. Since the difference in time step implies that the optimization layer provides the same value for 100 steps of the controller, an integration algorithm between the path planning and trajectory controller is developed. This algorithm ensures that the positions and velocities provided by the optimizer to the controller increase with the constant acceleration optimized, and after 100 steps, the vehicle achieves the position and velocity values expected by the optimizer. In the event that the vehicle fails to be in the expected condition, the change of values given by the change of optimizer steps needs to be compatible with the vehicle's possible acceleration limits based on the state the vehicle is in. The logic is shown in Equation (1):

$$V_x = \begin{matrix} V_{x0} - (V_{x0} - V_x)0.1 & \text{if } abs(V_{x0} - V_x) > 0.5, & \Delta t_o \neq 0 \\ V_{x0} + \Delta t_c a_x & \text{if } abs(V_{x0} - V_x) \leq 0.5, & \Delta t_o = 0 \end{matrix}$$

$$V_y = \begin{matrix} V_{y0} - (V_{y0} - V_y)0.1 & \text{if } abs(V_{y0} - V_y) > 0.5, & \Delta t_o \neq 0 \\ V_{y0} + \Delta t_c a_y & \text{if } abs(V_{y0} - V_y) \leq 0.5, & \Delta t_o = 0 \end{matrix}$$

$$X = \begin{matrix} X_0 - \Delta t_c V_x & \text{if } abs(V_{x0} - V_x) > 0.5, & \Delta t_o \neq 0 \\ X_0 - \Delta t_c V_x & \text{if } abs(V_{x0} - V_x) \leq 0.5, & \Delta t_o = 0 \end{matrix}$$

$$Y = \begin{matrix} Y_0 - \Delta t_c V_y & \text{if } abs(V_{y0} - V_y) > 0.5, & \Delta t_o \neq 0 \\ Y_0 - \Delta t_c V_y & \text{if } abs(V_{y0} - V_y) \leq 0.5, & \Delta t_o = 0 \end{matrix} \quad (1)$$

where V_x, V_y, X, Y are the speeds and positions that the car has to follow in the current time; V_{x0}, V_{y0}, X_0, Y_0 are the speeds and positions that the car had at the previous time step of the controller; $\Delta t_c, \Delta t_o$ are, respectively, the controller sample time of 0.001 s, and the difference between the actual and previous time of the optimizer in the sample time of the controller; 0.1 is a tuning value that reduced the speed step between the actual and the future value if it is higher than 0.5 m/s.

3. Trajectory Planning

To set-up the MPC optimization problems, the Gurobi solver is used, a commercial solver with parallel algorithms for large-scale linear programs, quadratic programs, and mixed integer programs. From the Gurobi optimization reference manual [15], the problem was implemented in the following form:

$$\underset{z}{minimize} \quad -t^T_{j(\tilde{p}^{(Hn)})}\boldsymbol{p}^{(Hn)} + \sum_{i=1}^{Hn}\Big(W_{(j(\tilde{p}^{(i)}))}\eta^{(i)} + R_{\chi F(j(\tilde{p}^{(i)}))}\delta^{(i)} + \cdots \quad (2)$$

$$\cdots + R_{\chi R(j(\tilde{p}^{(i)}))}\lambda^{(i)} + R_{\gamma(j(\tilde{p}^{(i)}))}\rho^{(i)}\Big)$$

$$\boldsymbol{x}^{(i)} = A\boldsymbol{x}^{(i-1)} + B\boldsymbol{u}^{(i-1)} \quad \forall i = 1, \cdots, Hn \quad (3)$$

$$N_{j(\tilde{p}^{(i)})}\boldsymbol{p}^{(i)} - \eta^{(i)} = diag(N^T_{j(\tilde{p}^{(i)})})T_{(j(\tilde{p}^{(i)})} \quad \forall i = 1, \cdots, Hn \quad (4a)$$

$$A_{acc}(\tilde{\boldsymbol{v}}^{(i)}, \boldsymbol{\theta}^{(k)}) \cdot \boldsymbol{u}^{(i)} \leq b_{acc}(\tilde{\boldsymbol{v}}^{(i)})c^2_{(j(\tilde{p}^{(i)})} \quad \forall i = 1, \cdots, Hn \quad \forall k = 1, \cdots, tn \quad (4b)$$

$$\left[R\tilde{r}ot_{(i-1,i)}\right]\begin{bmatrix}\boldsymbol{u}^{(i-1)} \\ \boldsymbol{u}^{(i)}\end{bmatrix} - \begin{bmatrix}\delta^{(i)} \\ \lambda^{(i)} \\ \rho^{(i)} \\ \rho^{(i)}\end{bmatrix} = 0 \quad \forall i = 1, \cdots, Hn \quad (4c)$$

$$\tilde{p}_x^{(i)} - v_{t\,max}\Delta t \leq p_x^{(i)} \leq \tilde{p}_x^{(i)} + v_{t\,max}\Delta t \quad \forall i = 1, \cdots, Hn \quad (5a)$$

$$\tilde{p}_y^{(i)} - v_{t\,max}\Delta t \leq p_y^{(i)} \leq \tilde{p}_y^{(i)} + v_{t\,max}\Delta t \quad \forall i = 1, \cdots, Hn \quad (5b)$$

$$-v_{t\,max} \leq v_x^{(i)} \leq v_{t\,max} \quad \forall i = 1, \cdots, Hn \quad (5c)$$

$$-v_{t\,max} \leq v_y^{(i)} \leq v_{t\,max} \quad \forall i = 1, \cdots, Hn \quad (5d)$$

$$-A_{Max}(\tilde{v}_t) \leq a_x^{(i)} \leq A_{Max}(\tilde{v}_t) \quad \forall i = 1, \cdots, Hn \quad (5e)$$

$$-A_{Max}(\tilde{v}_t) \leq a_y^{(i)} \leq A_{Max}(\tilde{v}_t) \quad \forall i = 1, \cdots, Hn \quad (5f)$$

$$\eta^{(i)}, \delta^{(i)}, \lambda^{(i)}, \rho^{(i)} \geq 0 \quad \forall i = 1, \cdots, Hn \quad (5g)$$

where the Objective Function (2), the Vehicle Model (3), the constraints used to represent the vehicle dynamic and track limitation (4), and the Bound Constraints (5) are shown and will be explained in the following subsections.

The problem is faced as an LP optimization, and its solutions are found by iterating the optimization every time step: the positions, $\tilde{p}^{(i)}$, velocities, $\tilde{v}^{(i)}$, and accelerations, $\tilde{u}^{(i)}$, are the state values optimized at the previous step; so the current iteration defines the optimum state, $x = [p_x, p_y, v_x, v_y]$, and input, $u = [a_x, a_y]$ for all the prediction horizon lengths. The variables weights of the Objective Function, $W_{(j(\tilde{p}^{(i)}))} - R_{\chi F(j(\tilde{p}^{(i)}))} - R_{\chi R(j(\tilde{p}^{(i)}))} - R_{\gamma(j(\tilde{p}^{(i)}))}$, are defined by a tuning process, and the prediction horizon was chosen to be composed of 120 time steps to ensure that the optimizer should be able to see the following curve when the longest straight begins, as tests have shown that the performance was satisfactory.

3.1. Models

To ensure a computational cost of 0.1 s, the vehicle was modeled as a point mass model, with the following assumption:

- The vehicle's side-slip and tires' slip angles are negligible. This means that the velocity vector is always tangent to the trajectory and it allows to ignore behavior such as drifting, spinning, or sliding;
- The tires' slippage is negligible;
- The longitudinal and lateral weight transfers are neglected in the vehicle model but will be considered in the constraint formulations;
- The racetrack is flat.

Therefore, referring to the coordinate system shown in Figure 2: x,y are the global coordinate system when stationary with respect to the racetrack; χ,γ are the vehicle coordinate system; and ψ is the vehicle yaw angle that indicates the tangent to the trajectory. The dynamic model was implemented in the discrete state space Equation (6) and detailed by Equation (7).

$$x^{(i)} = Ax^{(i-1)} + Bu^{(i-1)} \tag{6}$$

$$\begin{bmatrix} p_x^{(i)} \\ p_y^{(i)} \\ v_x^{(i)} \\ v_y^{(i)} \end{bmatrix} = \begin{bmatrix} 1 & 0 & \Delta t & 0 \\ 0 & 1 & 0 & \Delta t \\ 0 & 0 & 1 & 0 \\ 0 & 0 & 0 & 1 \end{bmatrix} \begin{bmatrix} p_x^{(i-1)} \\ p_y^{(i-1)} \\ v_x^{(i-1)} \\ v_y^{(i-1)} \end{bmatrix} + \begin{bmatrix} \frac{\Delta t^2}{2} & 0 \\ 0 & \frac{\Delta t^2}{2} \\ \Delta t & 0 \\ 0 & \Delta t \end{bmatrix} \begin{bmatrix} a_x^{(i-1)} \\ a_y^{(i-1)} \end{bmatrix} \tag{7}$$

where, in this case, i indicates the time-step index; $x^{(i)} \in \mathbb{R}^4$ is the state vector composed by the vehicle position and velocity vector in the global coordinate system, $\boldsymbol{p} = [p_x, p_y]$ and $v = [v_x, v_y]$; $u^{(i)} \in \mathbb{R}^2$ is the input vector composed by the vehicle acceleration vector in global coordinate system, $[a_x, a_y]$; $A \in \mathbb{R}^{4x4}$ is the system matrix and $B \in \mathbb{R}^{4x2}$ is the input matrix made explicit in Equation (7). From these equations, it is possible to deduce that to have a linear system, the time-step, Δt, must chose constant. Thus, the prediction horizon is discretized with constant time intervals, and the discrete model, shown in Equation (3), is then iterated forward in time from a time-step to the subsequent one.

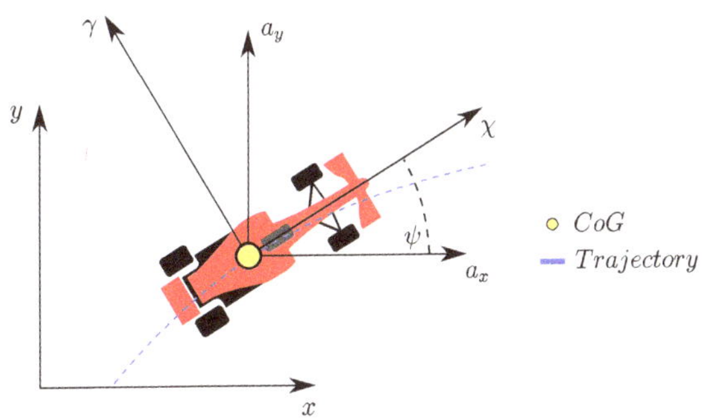

Figure 2. Global and vehicle coordinate systems.

3.2. Objective Function

The Objective Function proposed in this work is purely linear, as shown in Equation (2). This is a benefit because it reduces the optimization time and allows one to not worry about defining a positive objective matrix to ensure a feasible solution. It is composed of two strategies:

- Minimizing time strategy;
- Slack variables.

The first strategy aims to minimize the lap time, and the second one allows softening the hard limit of the constraints.

3.2.1. Minimizing Time Strategy

Since a constant discretization time Δt of the prediction horizon is chosen to make Equation (7) linear, maximizing the distance traveled in the direction of the tangent vector of the mid-line could be a strategy to cover the largest number of racetrack sectors. This strategy was expressed within the objective function as follows:

$$\underset{z}{minimize} \quad -t^T_{j(\tilde{p}^{(Hn)})}\begin{bmatrix} p_x^{(Hn)} \\ p_y^{(Hn)} \end{bmatrix} = -t^T_{j(\tilde{p}^{(Hn)})}\boldsymbol{p}^{(Hn)} \tag{8}$$

where it maximized the longitudinal and lateral distance between the current, $\boldsymbol{p}_{x,y}^{(Hn)}$, and previous, $\tilde{p}^{(Hn)}$, position points projected in the tangent direction, $t_{j(\tilde{p}^{(Hn)})}$, for all the points of the prediction horizon Hn.

3.2.2. Slack Variables

To avoid the infeasible problem, the slack variable strategy is used [16]. This methodology ensures avoiding that the constraints equations are not strictly equal to a constant limit but to an enlarged or reduced one by adding to the constraint a positive variable, weighted in the objective function. In this way, the designer can strengthen or relax the constraints by increasing or decreasing the associated weights.

$$\begin{aligned} \underset{z}{minimize} \quad & \sum_{i=1}^{Hn}\Big(W_{(j(\tilde{p}^{(i)}))}\eta^{(i)} + R_{\chi F(j(\tilde{p}^{(i)}))}\delta^{(i)} + \cdots \\ & \cdots + R_{\chi R(j(\tilde{p}^{(i)}))}\lambda^{(i)} + R_{\gamma(j(\tilde{p}^{(i)}))}\rho^{(i)}\Big) \end{aligned} \tag{9}$$

In Equation (9), $\eta^{(i)}$, $\delta^{(i)}$, $\lambda^{(i)}$ and $\rho^{(i)}$ are the slack variables and $W_{(j(\tilde{p}^{(i)}))}$, $R_{\chi F(j(\tilde{p}^{(i)}))}$, $R_{\chi R(j(\tilde{p}^{(i)}))}$, $R_{\gamma(j(\tilde{p}^{(i)}))}$ are the weights, respectively. In this work, by associating the weights to the nodes of the mid-line, $(j(\tilde{p}^{(i)}))$, it is possible to impose to the solver not only the ability to vary the constraints but also to decide whether one point of the racetrack is more important than another in terms of limit variations available, as a human driver can do.

3.3. Linear and Quadratic Constraints

To ensure optimized solutions compatible with vehicle dynamics and track limits, the vehicle accelerations and positions are limited by constraint Equations (4), which can be divided into three categories:

- Racetrack limitations, (4a);
- g-g limitations, (4b);
- jerk limitations, (4c).

3.3.1. Racetrack Limitations

In order to constrain the optimizer to provide output car positions within the path area, the optimization was solved by using the SCP approach as [17] suggested. Thus, the constraint is not the entire racetrack area, which involves a non-convex problem, but smaller racetrack sections where the problem is convex and so solvable. Therefore, it was decided to sample the middle line with points 1 m apart, $T_{Mid(j)}$, and to make these correspond to those intercepted on the inner and outer edge by the normal vector to the middle line and named, respectively, with $T_{IN(j)}$ and $T_{OUT(j)}$ in Figure 3. Then, it was possible to constrain the output states of the solver, $p^{(i)}$, within the path area, using the following equations:

$$\begin{aligned} \left(\boldsymbol{p}^{(i)} - \boldsymbol{T}_{Out(j(\tilde{p}^{(i)}))}\right)^T \left(-n_{j(\tilde{p}^{(i)})}\right) - \eta^{(i)} &= 0 \quad \forall i \in \{1, \cdots, Hn\} \\ \left(\boldsymbol{p}^{(i)} - \boldsymbol{T}_{In(j(\tilde{p}^{(i)}))}\right)^T \left(n_{j(\tilde{p}^{(i)})}\right) - \eta^{(i)} &= 0 \quad \forall i \in \{1, \cdots, Hn\} \\ \eta^{(i)} &\geq 0 \quad \forall i \in \{1, \cdots, Hn\} \end{aligned} \tag{10}$$

where $p^{(i)}$ is the position state vector found by the optimization in the current optimization step; $T_{Out(j(\tilde{p}^{(i)})}$ and $T_{In(j(\tilde{p}^{(i)})}$ are the limits given by the outer and inner edge referred to the solution found in the previous iteration step, $\tilde{p}^{(i)}$; $n_{j(\tilde{p}^{(i)})}$ is the normal direction to the mid-line, which allows the distance between the two points to be projected in the normal direction; and $\eta^{(i)}$ is the slack variable, which, being constrained in the Bound Constraints Equation (5g) as positive, forces the position of the vehicle to be less than zero making the racetrack constraints more or less rigid depending on the weight assigned to it in the Objective Function (2). Therefore, it is possible to vary the weight of the slack variable in a differentiated way according to the position of the car in the track, being able, for example, to soften it in the presence of a curb by allowing the car to go out of the limits of the track where possible to improve performance.

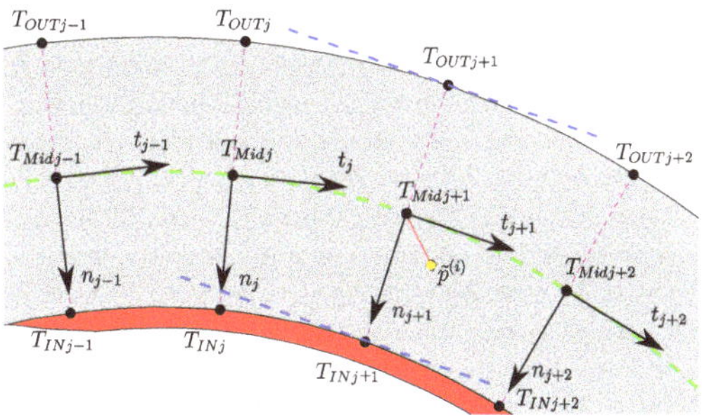

Figure 3. Representation of the racetrack linear approximation.

3.3.2. GG Limitations

In order to represent the exchange of forces occurring at the point of contact between wheel and road, it was decided to not implement a tire model but to bound the car's acceleration inside the GG diagram, measured at the center of gravity under steady-state conditions, such as [7,18,19].

The papers [20–22] show that the acceleration constraints area has the shape of two semi-ellipses: one stands for the forward acceleration, shown by the blue line in Figure 4a, and the other stands for backward acceleration, shown by the red line in Figure 4a.

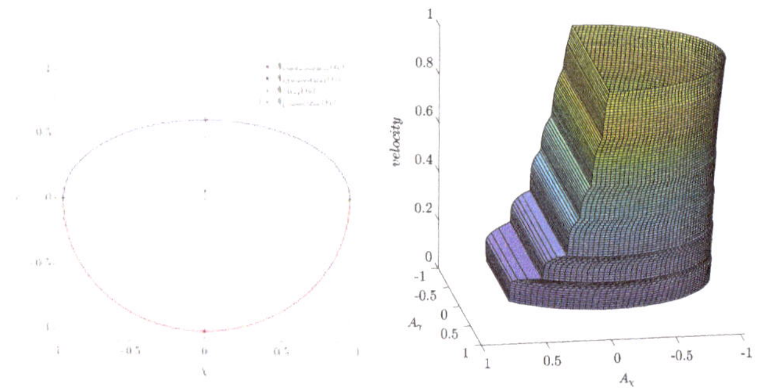

(a) (b)

Figure 4. Single GG diagram of the vehicle and its change with the vehicle speed. (**a**) Approximation method for the g-g diagram: two half ellipses for tire limits and a line for engine limit. (**b**) Autonomous driving car GG-V diagram, evaluated through simulation tests.

Empirically defining the maximum acceleration values, $A_{\gamma Max(v_t)}$, $A_{\chi backwardMax(v_t)}$ and $A_{\chi forwardMax(v_t)}$, the semi-axes of the ellipses are defined and so, the tire, aerodynamic and vehicle system limits are taken into account. Therefore, if the blue and red lines in Figure 4a show the limits just mentioned, the horizontal orange line shows the maximum performance that the car's engine can achieve in forward acceleration. In addition, these limits change with the vehicle speed because of the drag and engine influence [23,24]; thus, in Figure 4b, different GG diagrams were implemented in the function of the car speed obtaining the car GG-V diagram.

However, to better represent the vehicle dynamic and improve the performance, the idea was to implement the GG-V as quadratic constrains and not linearize its curves as is usually done, while maintaining the sample time under the specification of 0.1 s. However, due to the ability of the LP to handle only one convex quadratic constraint for each variable, a mixed approach was implemented: a conical constraint to reproduce the bigger, lower region associated with the vehicle braking capabilities exactly, and linear constraints to approximate the upper region, which is associated with the vehicle forward acceleration capabilities, being tested empirically, resulting in this region being well-described by the three orange dotted lines shown in Figure 4a. Therefore, the lower GG constraints are implemented as a negative semi-ellipse with the semi-axes values equal to the car's maximum deceleration. This inequality constraint is shown in the vehicle coordinate system by Equation (11) and in the global coordinate system by Equation (12).

$$\left(\frac{A_{\chi backward}}{A_{\chi backwardMax}(v_t)}\right)^2 + \left(\frac{A_{\gamma}}{A_{\gamma Max}(v_t)}\right)^2 - c^2 \leq 0 \tag{11}$$

$$\begin{aligned} a_x^2\left(\frac{v_x^2}{v_t^2}A_{\gamma Max}^2 + \frac{v_y^2}{v_t^2}A_{\chi backwardMax}^2\right) + a_y^2\left(\frac{v_y^2}{v_t^2}A_{\gamma Max}^2 + \frac{v_x^2}{v_t^2}A_{\chi backwardMax}^2\right) + \\ + a_x a_y\left(\frac{v_x v_y}{v_t^2}\left(A_{\gamma Max}^2 - A_{\chi backwardMax}^2\right)\right) - A_{\chi backwardMax}^2 A_{\gamma Max}^2 c^2 \leq 0 \end{aligned} \tag{12}$$

Instead, the upper GG constraints are represented as a set of lines tangent at the positive semi-ellipse. These lines are calculated by finding the tangents at the ellipse in points described by different angles, θ, thanks to Equation (13) in the coordinate vehicle system and Equation (14) in the global coordinates system.

$$\left(\frac{A_{\chi forward}}{A_{\chi forwardMax}(v_t)}\right)cos\theta + \left(\frac{A_{\gamma}}{A_{\gamma Max}(v_t)}\right)sin\theta - c \leq 0 \tag{13}$$

$$a_x(\frac{v_x}{v_t}A_{\gamma Max}cos\theta - \frac{v_y}{v_t}A_{\chi forwardMax}sin\theta) + a_y(\frac{v_y}{v_t}A_{\gamma Max}cos\theta + \\ + \frac{v_x}{v_t}A_{\chi forwardMax}sin\theta) - A_{\chi forwardMax}A_{\gamma Max}c \leq 0 \tag{14}$$

In all the equations shown, the ratio between the longitudinal speed component, v_x, and speed vector, $v_t = \sqrt{v_x^2 + v_y^2}$, is the cosine of the yaw angle ψ; and the ratio between the lateral speed component, v_y, and speed vector, $v_t = \sqrt{v_x^2 + v_y^2}$, is the sine of the yaw angle ψ. Both of these ratios are used to rotate the input vector $u = [a_x, a_y]$ in the global coordinate system and group the constraints in Equation (4b).

The variable c is a scalar coefficient that can expand or narrow the boundary of the GG diagram and change the tire performances representing degraded contact conditions due, for example, to rain or tire wear. c could be scaled in a global manner for all of the track or could be varied online in local sections being related to the position of the vehicle on the racetrack.

3.3.3. Jerk Limitations

Since the vehicle is modeled as a point mass, sudden variations in acceleration, $u = [a_x, a_y]$, can lead to the generation of an angular trajectory that does not represent a real trajectory of a racing car, which actually tends to be smooth. Usually, to avoid this problem, the quadratic variation of the accelerations is implemented as a term of the Objective Function to minimize it. In this study, to maintain the Objective Function as linear, the Jerk Limitations are implemented as constraint equations that require the jerks to be equal to positive slack variables. Therefore, the slack variables are inserted in the cost function as a linear term and a certain weight is assigned to them according to the point on the track where the car is located. In this way, the optimizer is able to determine the accelerations variation necessary to increase performance and maintain a smooth trajectory, and the designer can choose which direction of acceleration to favor in relation to where the car is located on the track, e.g., in a straight line the lateral acceleration can be forced to vary more slowly, favoring a faster variation of the longitudinal accelerations.

In Equation (4c), the strategy is shown, where $\left[\delta^{(i)}, \lambda^{(i)}, \rho^{(i)}, \rho^{(i)}\right]$, are the slack variables used to soften the zero hard limit given to the input vector variation, $\left[\boldsymbol{u}^{(i-1)}, \boldsymbol{u}^{(i)}\right]$, rotated in the global coordinate system by $\left[\tilde{Rrot}_{(i-1,i)}\right]$. In addition, it was chosen to implement different slack variables for different acceleration vectors diversifying the jerk of the various dynamic system: δ for forward longitudinal acceleration, λ for backward longitudinal acceleration, and ρ for lateral acceleration.

3.4. Bound Constraints

As usual, the optimization variables must be limited to effectively guide the solver in the optimization phases. This allows a faster search for the correct solution and reduces the possibility of facing the infeasible problem that results in a reduction in the optimization time. For this reason, the following trust regions were added to the solver: Equation (5a,b), which limit the positions; Equation (5c,d), which limit the speeds; Equation (5e,f), which limit the accelerations; and Equation (5g), which imposes a slack variable that is greater than zero.

4. Path Tracking

Regarding the path tracking, an optimum control was chosen to be implemented by decoupling the longitudinal and lateral vehicle dynamics and ensuring a low computational cost. The developed LQR has a sample time of 0.001 s to define the car input with a certain continuity as a driver will do. In this way, the path planning is embedded with a path tracking to compose the trajectory tracking with a dynamic vehicle model and a reduced time step.

4.1. Vehicle Model

To ensure a linear model to the LQR, the longitudinal and lateral vehicle dynamic are decoupled and modeled separately, obtaining a model where the longitudinal part defines the inputs pedals throttle and brake percentage, and the lateral part defines the steering wheel angle signal input.

4.1.1. Longitudinal Model

About the longitudinal model, it is supposed to have the single track vehicle model, shown in Figure 5, in which the longitudinal acceleration of the vehicle is set equal to the sum of the longitudinal tire forces:

$$a_x = \sum_{i=1}^{4} F_{xi} \tag{15}$$

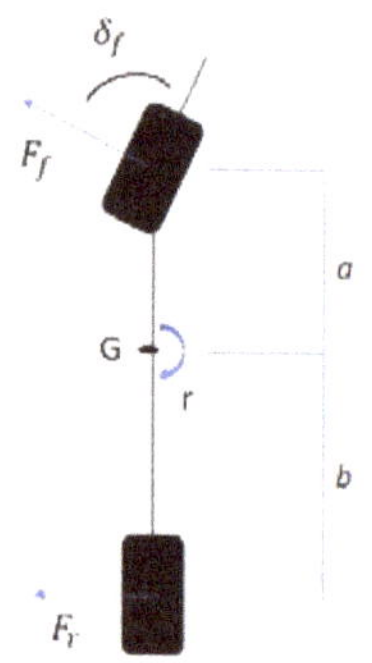

Figure 5. Single track vehicle model.

Considering the total longitudinal force that the car has to have as a percentage of the maximum tracking force that the engine and braking system can express, the longitudinal vehicle model could be implemented as:

$$\ddot{x} = \frac{CF_{max}}{m} \tag{16}$$

where $\ddot{x}$ is the vehicle longitudinal acceleration; C is the throttle or brake pedal percentage in the range [0–100](%); F_{max} is the maximum force that the engine and brake system can give; and m is the vehicle mass.

4.1.2. Lateral Model

Regarding the lateral model, the single track vehicle model shown in Figure 5 was used, considering the lateral and yaw dynamic balance given by [19], as follows:

$$\dot{v} = -\frac{(C_{yf} + C_{yr})}{mu} v + (-u + \frac{C_{yf}a - C_{yr}b}{mu}) r + \frac{C_{yf}}{m} \delta_f \tag{17}$$

$$\dot{r} = -\frac{(C_{yf}a - C_{yr}b)}{Ju} v + \frac{(C_{yf}a^2 + C_{yr}b^2)}{Ju} r + \frac{C_{yf}a}{J} \delta_f \tag{18}$$

where v and r are, respectively, the lateral speed component and yaw rate; C_{yf} and C_{yr} are the cornering stiffness tire parameters; u is the longitudinal speed component; a and b are, respectively, the front and rear wheelbase of the vehicle; m, J and R are the mass, vertical moment of inertia and wheel radius of the car.

4.2. LQR

Considering the longitudinal and lateral vehicle dynamics shown in the previous subsection, a LQR can be implemented by tracking the longitudinal and lateral vehicle speed and position found by the trajectory planning. In fact, the optimized longitudinal vehicle speed can be considered as the target speed that the vehicle has to have and subtracted from the actual longitudinal vehicle speed obtaining:

$$\dot{e_x} = \dot{x} - \dot{x_t} = \frac{de_x}{dt} \tag{19}$$

$$\ddot{e_x} = \ddot{x} - \ddot{x_t} = \frac{(C - C_t)F_{max}}{m} \tag{20}$$

where F_{max} and m are constant variable and $(C - C_t)$ is the throttle pedal percentage if the speed error, $\dot{e}_x$, is positive and the brake pedal percentage if it is negative, saturated from 0 to 100%.

Instead, to follow the lateral dynamics suggested by the path planning, the yaw and lateral vehicle position and speed are tracked, considering the errors between the actual vehicle values and the reference values exiting the trajectory planning. Thus, replacing the Equations (17) and (18) in (21) and considering the longitudinal vehicle speed, u, as a constant variable, the following was obtained:

$$\dot{e_y} = \dot{y} - \dot{y_t} = v + u\psi - v_t - u\psi_t \qquad \ddot{e_y} = (\dot{v} - \dot{v_t}) + u(r - r_t) \tag{21}$$

$$\ddot{e_y} = -\frac{(C_{yf} + C_{yr})}{mu}\dot{e_y} + \frac{(C_{yf} + C_{yr})}{m}e_\psi + \frac{(C_{yf}a - C_{yr}b)}{mu}\dot{e_\psi} + \frac{C_{yf}}{m}(\delta_f - \delta_{ft}) \tag{22}$$

$$\dot{e_\psi} = \dot{\psi} - \dot{\psi_t} = r - r_t \tag{23}$$

$$\ddot{e_\psi} = -\frac{(C_{yf}a - C_{yr}b)}{Ju}\dot{e_y} + \frac{(C_{yf}a - C_{yr}b)}{J}e_\psi + \frac{(C_{yf}a^2 + C_{yr}b^2)}{Ju}\dot{e_\psi} + \frac{C_{yf}a}{J}(\delta_f - \delta_{ft}) \tag{24}$$

Therefore, the dynamic state space has the following representation:

$$\begin{Bmatrix} \dot{e_x} \\ \ddot{e_x} \\ \dot{e_y} \\ \ddot{e_y} \\ \dot{e_\psi} \\ \ddot{e_\psi} \end{Bmatrix} = A \begin{Bmatrix} e_x \\ \dot{e_x} \\ e_y \\ \dot{e_y} \\ e_\psi \\ \dot{e_\psi} \end{Bmatrix} + B \begin{bmatrix} (C - C_t) \\ (\delta_f - \delta_f t) \end{bmatrix} \tag{25}$$

where the errors of longitudinal, lateral and yaw position and rate are the LQR states, $X = [e_x; \dot{e}_x; e_y; \dot{e}_y; e_\psi; \dot{e}_\psi]$, which must be minimized by defining the car input $U = [(C - C_t); (\delta_f - \delta_{ft})]$ as follows:

$$U = -KX \tag{26}$$

Here, K is the gains matrix calculated by solving the following offline optimization :

$$minimize \sum X^T QX + U^T RU \tag{27}$$

where the state and the input vectors are appropriately weighted by the matrices Q and R.

However, the state space formulation shown in Equation (25) does not ensure that the tracking errors converge to zero, even though the matrix $(Ax - BK)$ is asymptotically stable [25]. This is due to the omission of the term steady state:

$$B_2\dot{\psi}_{ss} \tag{28}$$

To ensure zero steady state errors, a feed-forward term is added to the state feed-back assuming that the steering controller is obtained by state feed-back plus a feed-forward term:

$$\delta = -Kx + \delta_{ff} \tag{29}$$

where δ_{ff} is:

$$\delta_{ff} = \frac{mu^2}{RL}\left[\frac{b}{C_{yf}} - \frac{a}{C_{yr}} + \frac{a}{C_{yr}}k_3\right] + \frac{L}{R} - \frac{b}{R}k_3 \tag{30}$$

where m, L, a, b are, respectively, the mass, the total wheelbase, the front and rear wheelbase of the car; R, C_{yf}, C_{yr} are the radius, the front and rear stiffness of the wheel; and k_3 is the third component of the matrix gain K calculated with Equation (27).

In addition, to compensate for the delay that the feed-back control produces in the actuation of the longitudinal input, a feed-forward term is added to the longitudinal feed-back state defining the acceleration and brake pedal control as:

$$C = -Kx + K_{ff}A_{xt} \tag{31}$$

where K_{ff} is the feed-forward gain found by a tuning process, and A_{xt} is the longitudinal acceleration target output from the path planning layer.

5. Experimental Results and Discussion

Two sessions of tests were carried out to validate the architecture proposed:

- Comparison with an offline motion planning;
- Comparison with a driver.

The first test session involved the comparison of the developed autonomous control architecture with an offline optimization algorithm developed by VI-Grade CRT. The intention of this comparison was to validate the optimization algorithm developed with a state-of-the- art algorithm and to show how having an online optimization of the trajectory could improve the performance. The test is composed of two phase: the first one involved co-simulation between the trajectory tracking developed in Matlab-Simulink and the car model of the dynamic simulation software CRT; and the second one involved the minimum curvature optimization of Calabogie racetrack and the execution of the max-performance event explained in more detail in the next subsection.

The simulations were carried out using a laptop PC with the characteristics, Hardware 1, given in Table 2 by setting the optimization as in Table 3.

The second test session had the intent of showing the real-time functionality of the entire structure by exploiting the static simulator shown in Figure 6a and located at Meccanica 42 s.r.l. In this way, the model developed in the Matlab-Simulink environment needed to be compiled in C in the simulator's hardware provided by VI-Grade, and a test model has to be built where the systems communication is simulated as would be in the car. The results obtained are then compared with the performance of a driver obtained at the same simulator to verify that the model-simplified but constraint-enhancing representation of dynamics achieves performance comparable to that of a professional driver, as per the intention of the paper.

(**a**) Car simulator.

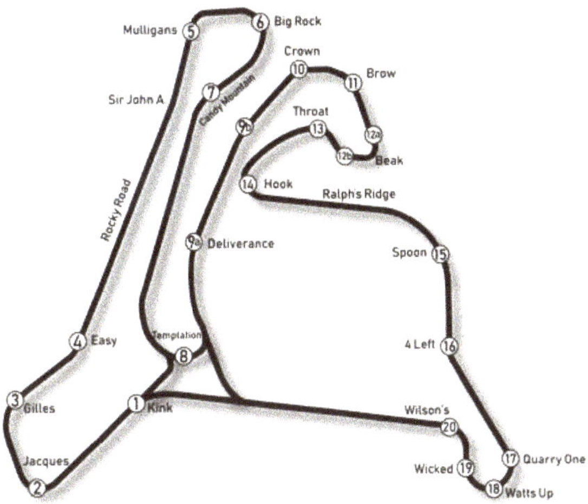

(**b**) Calabogie racetrack.

Figure 6. Tests set-up.

The hardware characteristics and computational performance of the simulator are shown in Table 2 under the name Hardware 2.

In Table 2, it can be seen that the path planning computational time that resulted for both the hardware implementations is below the required specification of 0.1 s.

Table 2. Hardware specifications.

	Hardware 1	Hardware 2
Operating System	Windows 10	Linux
CPU	i7-7700 HQ	Xeon E5-2667 v3
RAM	16 GB	32 GB
n° Cpu cores used	1	8
Single core frequency	2.6 to 3.5 GHz	3.2 GHz
Max computational time	0.07 s	0.06 s
Mean computational time	0.05 s	0.05 s

Table 3. Optimization algorithm settings.

Parameter	Value	Units
Prediction horizon length	120	step
n° optimization per step	1	
mid-line discretization	1	m

5.1. Comparison with Offline Motion Planning

These tests have involved the comparison of the architecture proposed with the max-performance event simulation of CRT suite. The max-performance simulation is used to define the dynamic speed limit profile on a given racetrack [26]. This offline simulation uses an iterative process where a specific static solver computes a velocity profile and then a dynamic solver verifies if the computed speed profile is feasible. Basically, the vehicle model used for static prediction has no suspensions and inherits all properties from the full CRT model. The effect of aero forces is considered and the effect of suspension jounce is taken into account by the presence of ride height maps, which link the dependency of ride heights to the vehicle velocity. The trajectory chosen to perform the max-performance event is that of minimum curvature of the Calabogie racetrack shown in Figure 6b. In the literature, this trajectory is often used as a reference, and it has been found that it is a trajectory that allows getting very close to the minimum lap time on a racetrack. For these tests, the simulation road is considered flat, neglecting the dynamics effects that the racetrack slopes and banks have on the performance.

In Figure 7, the car's longitudinal and lateral accelerations as a function of speed, GG-V, achieved by the car model controlled by the algorithm proposed and by the max-performance event are shown. In the same figures, the GG-V constraints implemented in the path planning are shown. By Figure 7b, it is possible to see that the constraints assumed in the trajectory tracking are representative of the car, and by Figure 7a, it is possible to see that the car controlled by the algorithm developed is able to keep the car inside the constraints imposed, even if with respect to the max-performance results, it goes through the transients more instead of staying within the limits of grip. This may be due to the inputs given to the vehicle being noisier than those of the offline control, as shown in Figure 8. Here, the engine torque, the chamber brake pressures and the steering wheel angle are almost similar in quantities with differences due to the different trajectory made, even if there is a visible difference in the amplitude and timing between the two configurations, i.e., the algorithm performed higher and delayed brake pressures with respect to the max-performance event. This is reflected in the speeds achieved by the car, which are almost similar, but the algorithm's one brakes later, maintaining the high speed for more time than the max-performance speed, resulting in a reduction of lap time, as shown in Table 4.

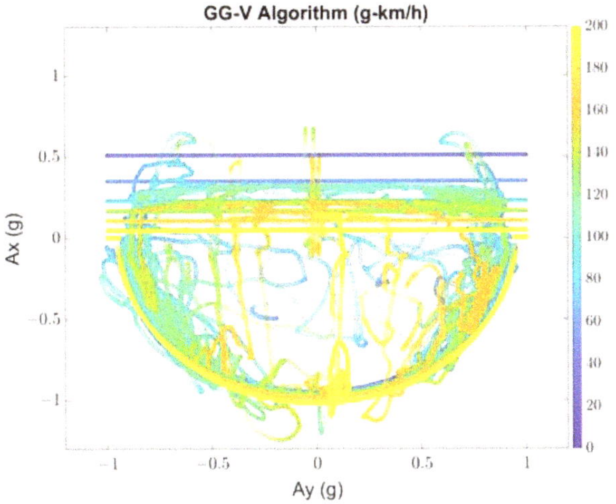

(**a**) Algorithm GG-V.

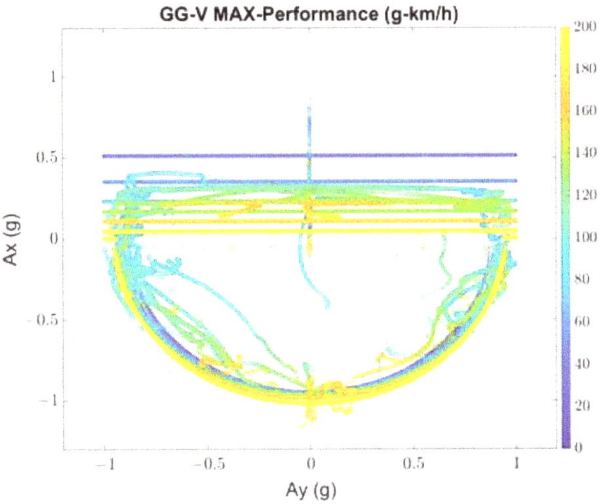

(**b**) Max-performance GG-V.

Figure 7. Acceleration performance comparison.

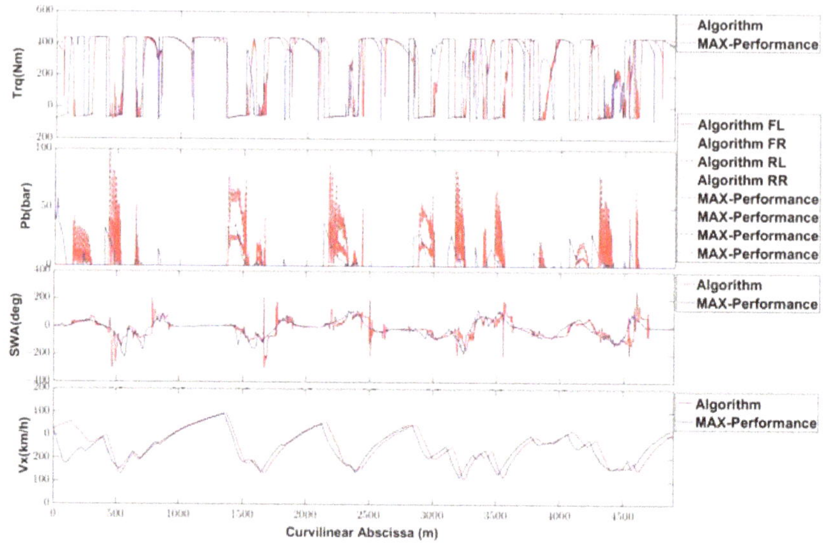

Figure 8. Comparison of the input given to the vehicle and the output longitudinal speed between the algorithm proposed and the max-performance event.

5.2. Comparison with Driver

This second session of tests involved the comparison of the performance achieved by a professional driver and the trajectory tracking proposed. To perform these tests, the algorithm was compiled for use in the real-time simulator environment, and the road used is no longer flat but includes the actual lateral banks and longitudinal slopes of the road. For this reason, the longitudinal and lateral acceleration components due to the pitch and roll body movements and road inclination were added in the acceleration constraints of the path planning knowing the inclination of the road while the car is moving on the track and the pitch and roll of the car due to the brake, acceleration and steering states calculated as proportional to the difference between the front/right and rear/left normal wheel forces.

Looking at Figure 9, where the GGs obtained by the vehicle driven by the driver and by the algorithm are shown, it can be seen that the algorithm is able to respect the acceleration constraints imposed, by guaranteeing the limits of the engine and the tires and replicating the results obtained by the driver.

In terms of performance, the results show that the algorithm developed is able to match the ones of the professional driver, although simplifications have been made to the vehicle model to keep the computational cost down, in fact, speed and input curves have the same trend and peak values, as shown in Figure 10. The lap times shown in Table 4 confirm that the trajectory tracking developed is able to replace the results of a human driver, and that the system integration performs well.

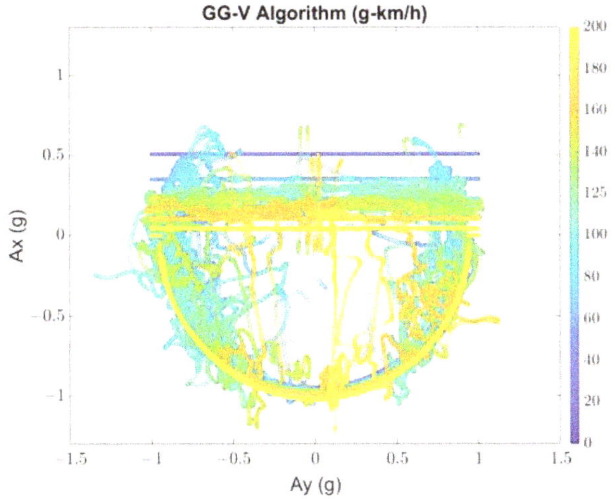

(**a**) Algorithm GG-V.

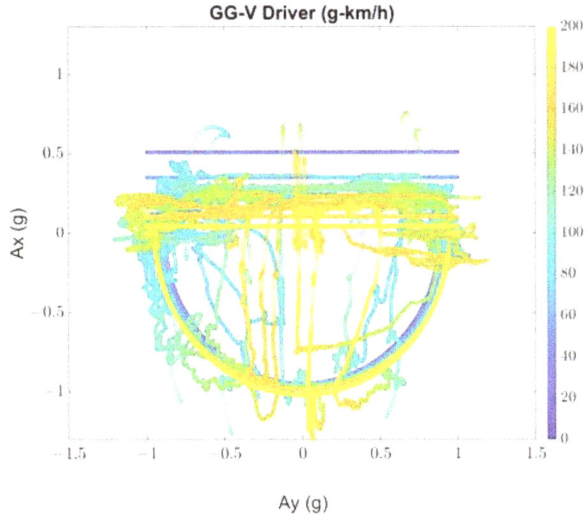

(**b**) Driver GG-V.

Figure 9. Accelerations performance comparison.

Table 4. Lap time performances.

	Lap Time (s)	
Test 1	Algorithm	151.0
	Max-performance	151.9
Test 2	Algorithm	150.7
	Driver	150.5

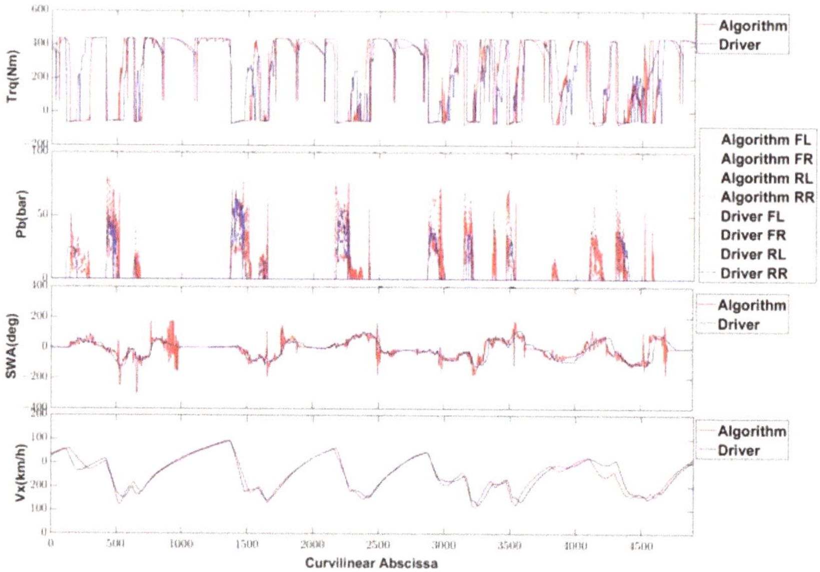

Figure 10. Comparison of the input given to the vehicle and the output longitudinal speed between the algorithm proposed and the driver.

5.3. Trajectory Comparison

In Figure 11, the trajectories made by the algorithm developed, max-performance event and driver are compared, highlighting the curves where the different strategies are more visible and impact the performance. In zoom 1, it is possible to see that the trajectory of least curvature achieved by the offline optimization exits the curve keeping the car on the outer side of the track, whereas the proposed path planning drives the car to the inner side of the track to anticipate and prepare the car for the following curve, as the driver does, too. In this track section, the max normal distance between the trajectory made by the algorithm with respect to the one made by the max-performance event is 5.86 m; whereas, the one made by the driver can be considered equal to the algorithm one. In zoom 2, the different approaches are more visible and impact more on the performance: the trajectory of least curvature maintains the inner side, whereas the algorithm developed goes to the outer side, achieving a max normal distance of 8.00 m; this behavior was proved by driver tests, which improves the performance of the car and reduces the lap time, even if the driver made a trajectory 3.20 m closer than the algorithm. However, in zoom 3, if at the beginning of the curves, the max-performance optimization maintains the car on the outer side, and the driver and online algorithm choose the narrowest trajectory, achieving a max normal distance of 5.60 m, and at the end of the curves, the algorithm follows the max-performance, enlarging the trajectory later of almost 6.00 m with respect to the driver losing longitudinal speed.

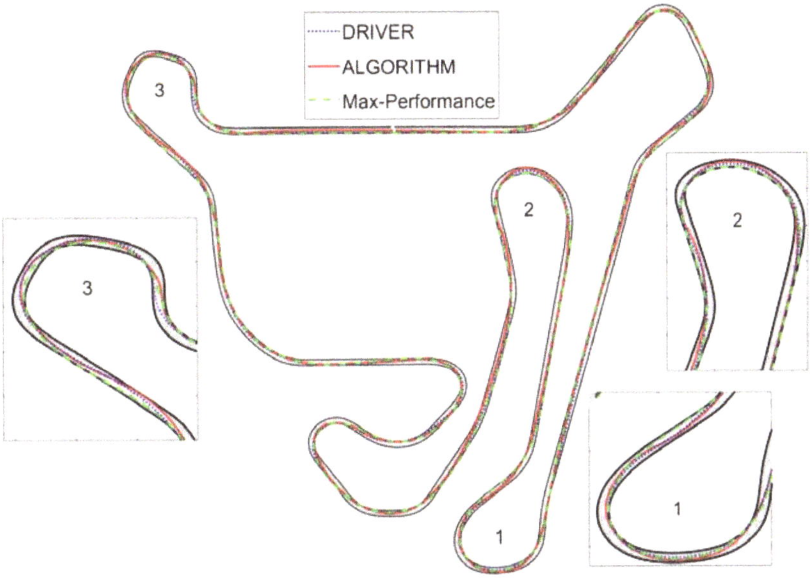

Figure 11. Comparison of the trajectory made by the algorithm proposed and the driver.

6. Conclusions

In this paper, we developed an autonomous racing car trajectory tracker that was able to update the optimized trajectory online to reduce the time lap as well as control the car, and we validated this by comparing it with the performance of offline path planning and a human driver.

The intention of the paper was to ensure the online functionality of an autonomous controller to match the decisions of a human driver instead of optimizing the trajectory offline considering a possible development for an urban area.

This aim was achieved by dividing the path planning algorithm from the trajectory control one and building a hierarchical control architecture, where the vehicle model complexity increased and the sample time decreased, i.e., the path planning uses a point mass vehicle model with a sample time of 0.1 s, and the path tracking uses a single-track vehicle model with a sample time of 0.001 s. To ensure a correct communication between the two systems and the car, the computational time of the trajectory optimization was chosen to be below 0.1 s: the division between the path planning and path tracking and the implementation of LP instead of QP ensured the respect of these constraints, as shown in Table 2. Furthermore, to increase the path planning representation of the vehicle dynamic, the GG-V constraints were implemented as a quadratic ellipse, and the weights and the constraints of the optimization could be changed as a function of the track sections, adapting the car to the characteristics and conditions of the track. The comparisons made with a state-of-the-art offline path-planner validated the optimizer developed and the better performance achieved in terms of lap time, reducing it by almost a second.

The comparison made with a professional driver can prove that even if the autonomous architecture developed uses a simplified vehicle model to reduce the computational cost, the car is able to replicate its performance, matching the lap time.

In the future the Anti-Lock Brake System developed and published in [27] will be added to the control architecture to ensure an improvement in stability performance.

Author Contributions: Conceptualization and methodology, M.M. and L.R.; validation and data curation, M.M. and L.R.; writing—original draft preparation, M.M.; writing—review and editing, R.C. and C.A.; project administration, R.C. and C.A. All authors have read and agreed to the published version of the manuscript.

Funding: This research received no external funding.

Institutional Review Board Statement: Not applicable.

Informed Consent Statement: Not applicable

Conflicts of Interest: The authors declare no conflict of interest.

Abbreviations

The following abbreviations are used in this manuscript:

LP	Linear Programming
LQR	Linear Quadratic Regulator
QP	Quadratic Programming
MPC	Model Predictive Control
SCP	Sequential Convex Programming
DoF	Degrees of Freedom
PID	Proportional-Integral-Derivative
CRT	Car-Real-Time
UKF	Unscented Kalman Filter

References

1. Dal Bianco, N.; Bertolazzi, E.; Biral, F.; Massaro, M. Comparison of Direct and Indirect Methods for Minimum Lap Time Optimal Control Problems. *Veh. Syst. Dyn.* **2019**, *57*, 665–696. [CrossRef]
2. Liniger, A. Path Planning and Control for Autonomous Racing. Ph.D. Thesis, ETH Zurich, Zurich, Switzerland, 2018.
3. Liniger, A.; Domahidi, A.; Morari, M. Optimization-Based Autonomous Racing of 1:43 Scale RC Cars: OPTIMIZATION-BASED AUTONOMOUS RACING. *Optim. Control Appl. Methods* **2015**, *36*, 628–647. [CrossRef]
4. Christ, F.; Wischnewski, A.; Heilmeier, A.; Lohmann, B. Time-Optimal Trajectory Planning for a Race Car Considering Variable Tyre-Road Friction Coefficients. *Veh. Syst. Dyn.* **2021**, *59*, 588–612. [CrossRef]
5. Verschueren, R.; De Bruyne, S.; Zanon, M.; Frasch, J.V.; Diehl, M. Towards Time-Optimal Race Car Driving Using Nonlinear MPC in Real-Time. In Proceedings of the 53rd IEEE Conference on Decision and Control, Los Angeles, CA, USA, 15–17 December 2014; pp. 2505–2510.
6. Verschueren, R.; Zanon, M.; Quirynen, R.; Diehl, M. Time-Optimal Race Car Driving Using an Online Exact Hessian Based Nonlinear MPC Algorithm. In Proceedings of the 2016 European Control Conference (ECC), Aalborg, Denmark, 29 June–1 July 2016; pp. 141–147.
7. Alrifaee, B.; Maczijewski, J. Real-Time Trajectory Optimization for Autonomous Vehicle Racing Using Sequential Linearization. In Proceedings of the 2018 IEEE Intelligent Vehicles Symposium (IV), Changshu, China, 26–30 June 2018; pp. 476–483.
8. Heilmeier, A.; Wischnewski, A.; Hermansdorfer, L.; Betz, J.; Lienkamp, M.; Lohmann, B. Minimum Curvature Trajectory Planning and Control for an Autonomous Race Car. *Veh. Syst. Dyn.* **2019**, 2, 1497–1527. [CrossRef]
9. Lot, R.; Dal Bianco, N. The significance of high-order dynamics in lap time simulations. In Proceedings of the 24th Symposium of the International Association for Vehicle System Dynamics (IAVSD 2015), Graz, Austria, 17–21 August 2015; Rosenberger, M., Plöchl, M., Six, K., Edelmann, J., Eds.; CRC Press: Boca Raton, FL, USA, 2016.
10. Theodosis, P.A.; Gerdes, J.C. Generating a Racing Line for an Autonomous Racecar Using Professional Driving Techniques. In Proceedings of the ASME 2011 Dynamic Systems and Control Conference and Bath/ASME Symposium on Fluid Power and Motion Control, ASMEDC, Arlington, VA, USA, 31 October–2 November 2011; Volume 2, pp. 853–860.
11. Perantoni, G.; Limebeer, D.J. Optimal Control for a Formula One Car with Variable Parameters. *Veh. Syst. Dyn.* **2014**, *52*, 653–678. [CrossRef]
12. Gerdts, M. A Moving Horizon Technique for the Simulation of Automobile Test-Drives. *ZAMM* **2003**, *83*, 147–162. [CrossRef]
13. Gerdts, M.; Karrenberg, S.; Müller-Beßler, B.; Stock, G. Generating Locally Optimal Trajectories for an Automatically Driven Car. *Optim. Eng.* **2009**, *10*, 439–463. [CrossRef]
14. Timings, J.P.; Cole, D.J. Minimum Maneuver Time Calculation Using Convex Optimization. *J. Dyn. Syst. Meas. Control* **2013**, *135*, 031015. [CrossRef]
15. Gurobi 8.1 Optimizer Reference Manual. 2018. Available online: https://www.gurobi.com/documentation/9.1/refman/index.html (accessed on 17 May 2020).
16. Boyd, S.P.; Vandenberghe, L. *Convex Optimization*; Cambridge University Press: Cambridge, UK, 2004.

17. Duchi, J. *Sequential Convex Programming;* Notes for EE364b; Stanford University: Stanford, CA, USA, 2018.
18. Subosits, J.K.; Gerdes, J.C. From the Racetrack to the Road: Real-Time Trajectory Replanning for Autonomous Driving. *IEEE Trans. Intell. Veh.* **2019**, *4*, 309–320. [CrossRef]
19. Guiggiani, M. *The Science of Vehicle Dynamics: Handling, Braking, and Ride of Road and Race Cars*, 2nd ed.; The Sciene of Vehicle Dynamics; Springer: Berlin/Heidelberg, Germany, 2018.
20. Segers, J. *Analysis Techniques for Racecar Data Acquisition*, 2nd ed.; SAE International: Warrendale, PA, USA, 2014.
21. Milliken, W.F.; Milliken, D.L. *Race Car Vehicle Dynamics;* SAE International: Warrendale, PA, USA, 1995.
22. Pacejka, H.B.; Besselink, I. *Tire and Vehicle Dynamics*, 3rd ed.; Elsevier/Butterworth-Heinemann: Amsterdam, The Netherlands, 2012.
23. Veneri, M.; Massaro, M. A Free-Trajectory Quasi-Steady-State Optimal-Control Method for Minimum Lap-Time of Race Vehicles. *Veh. Syst. Dyn.* **2019**, *58*, 1–22. [CrossRef]
24. Altche, F.; Polack, P.; de La Fortelle, A. A Simple Dynamic Model for Aggressive, near-Limits Trajectory Planning. *IEEE Intell. Veh. Symp.* **2017**, *IV*, 141–147.
25. Rajamani, R. *Vehicle Dynamics and Control;* Springer Science: Berlin/Heidelberg, Germany, 2006.
26. *VI-CarRealTime 18.2 Web Help Documentation;* VI-Grade GmbH: Darmstadt, Germany, 2018.
27. Montani, M.; Vitaliti, D.; Capitani, R.; Annicchiarico, C. Performance Review of Three Car Integrated ABS Types: Development of a Tire Independent Wheel Speed Control. *Energies* **2020**, *13*, 6183. [CrossRef]

energies

Article

Toward a Unified Model Approach for Evaluating Different Electric Vehicles

Wael Alosaimi [1], Md Tarique Jamal Ansari [2], Abdullah Alharbi [1], Hashem Alyami [3], Saquib Ali [4], Alka Agrawal [2] and Raees Ahmad Khan [2,*]

1 Department of Information Technology, College of Computers and Information Technology, Taif University, P.O. Box 11099, Taif 21944, Saudi Arabia; w.osaimi@tu.edu.sa (W.A.); amharbi@tu.edu.sa (A.A.)
2 Department of Information Technology, Babasaheb Bhimrao Ambedkar University, Lucknow 226025, Uttar Pradesh, India; tjtjansari@gmail.com (M.T.J.A.); alka_csjmu@yahoo.co.in (A.A.)
3 Department of Computer Science, College of Computers and Information Technology, Taif University, P.O. Box 11099, Taif 21944, Saudi Arabia; hyami@tu.edu.sa
4 Department of BCA, Azad Degree College, University of Lucknow, Lucknow 226002, Uttar Pradesh, India; saquibali99@gmail.com
* Correspondence: khanraees@yahoo.com

Abstract: Considering rising pollution as well as fuel expenses, it has now become critical to transition to a sustainable method of transportation. As a result, automakers have begun to spend on research and development in the electric vehicle (EV) industry. The amount of EVs has expanded rapidly in recent years. This is owing to new improved technology, particularly in electric motor engineering, as well as government initiatives to limit the level of environmental impact produced by combustion engines. Because EVs are powered by electricity, implementing their charging stations presents certain complications. In this paper, we have discussed the different types of EVs, such as BEVs, FCEVs, HEVs, PHEVs, and REHEVs. Even though the capacity of many of these electric car models has been substantially enhanced within the past few years, some challenges remain as a selection barrier for several customers. Considering these challenges, we have also implemented a fuzzy AHP-TOPSIS-based unified model to evaluate the different types of EVs. The study's technical importance is the identification of various evaluation factors, implementation of a unified model for measuring performance, and computation using the fuzzy MCDM technique. The outcomes of the unified model approach also were validated. We concluded that FCEVs are excellent for long journeys, and have the resources to cause minimal disruption.

Keywords: electric vehicles; renewable energy; fuzzy comprehensive evaluation; usage analysis; fuzzy logic

Citation: Alosaimi, W.; Ansari, M.T.J.; Alharbi, A.; Alyami, H.; Ali, S.; Agrawal, A.; Khan, R.A. Toward a Unified Model Approach for Evaluating Different Electric Vehicles. *Energies* **2021**, *14*, 6120. https://doi.org/10.3390/en14196120

Academic Editors: Guzek Marek, Rafał Jurecki and Wojciech Wach

Received: 9 September 2021
Accepted: 23 September 2021
Published: 26 September 2021

1. Introduction

Currently, the world is facing environmental degradation and an energy crisis as carbon emissions increase rapidly. A dramatic shift from internal combustion engine vehicles (ICEVs) to electric vehicles (EVs) can be observed in the automotive sector. Because petroleum is the principal fuel utilized in ICE vehicles, which are a significant contributor to the overall environmental catastrophe, EVs are the perfect alternatives [1,2]. An EV is one that runs on electricity rather than an internal combustion engine, which produces energy by consuming a mixture of oil and gases. As a result, EVs are seen as a potential alternative to current-generation cars to counter increasing pollution, environmental degradation, natural resource depletion, etc. [3]. Although there has been a very long period of the notion of electric vehicles, they have attracted significant interest in the last decade in the face of increasing carbon emissions and the other repercussions of fuel vehicles in the ecosystem.

As environmental issues continuously increase, governments across the globe have implemented numerous carbon dioxide and nitrogen oxide emission limits. From those perspectives, EVs, primarily based on electricity, would be able to soon replace traditional internal combustion engine vehicles, utilizing state-of-the-art electronic power systems, engine motors, electricity-generation systems, production of sustainable energy, as well as smart grids. EVs may be split into hybrid EVs (HEVs), plug-in HEVs (PHEVs), and EVs, based on the current design of the power generation and the system content. Industrialized nations have aggressively established numerous economic considerations in recent times in order to further support electrical engineering firms and research initiatives. Indeed, in the past 10 years, the power electronics industry and its infrastructure have grown rapidly [4,5]. Vehicles generate much carbon pollution that enters our natural surroundings, exposing us to pollution and global warming. An electric vehicle is a big step towards improving the quality of the environment effectively. EVs receive their energy from their rechargeable batteries. These batteries not only control the vehicles, but they are also utilized to power the lights and wipers. The batteries of electric automobiles have higher fuel economy and have lower fuel costs than a conventional petrol car. It is the same kind of battery that is commonly required when a gasoline engine is running. The advantages of electric vehicles are clear. With the development of new technology that promises to decrease the charging durations in minutes, increase the range, and achieve efficient security and technology, there has never been a greater moment to move to an EV. Figure 1 shows the several benefits of using EVs.

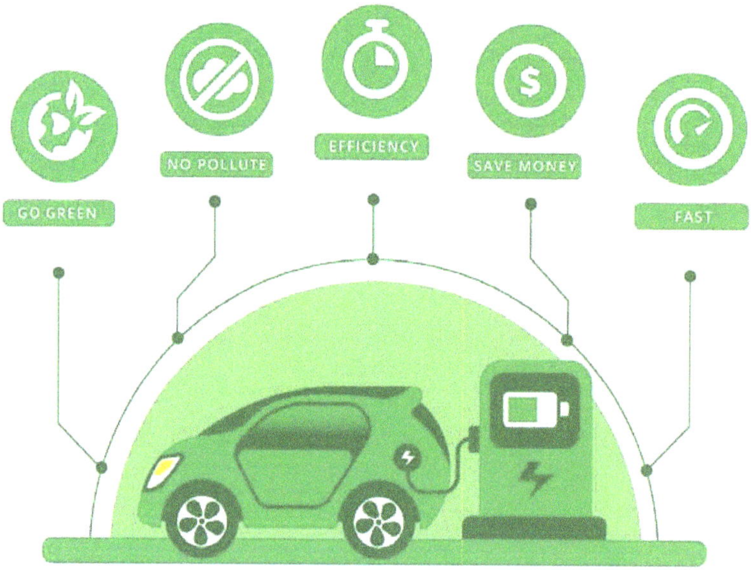

Figure 1. Several benefits of electric vehicles.

Consumers anticipate additional technological advancements and the introduction of new variants. Consumer behavior in the electric vehicle market is shifting from early buyers and technophile buyers to widespread adoption. Substantial advancements in technology, as well as a greater range of electric vehicle models on the market, have influenced consumer purchasing preferences. Automobiles will remain as a primary factor in energy requirements. China, India, and the Middle East are increasingly placing so many new automobiles on the road that the usage of oil for transportation fuel will continue to expand, and by 2035, it will require 12% more barrels compared to 2016. However, after 2025, there will still be a serious challenge for electric vehicles. Market shares are expanding tremendously above forecasts, with battery advancements boosting quicker than anticipated. In the present basic case, we perceive that the accumulation of EVs will be almost 100 million by 2035, with a daily fuel demand of approximately one to two million barrels per day. Figure 2 shows the intense growth in the electric vehicle market as per the report [6] published by Wood Meckenzie.

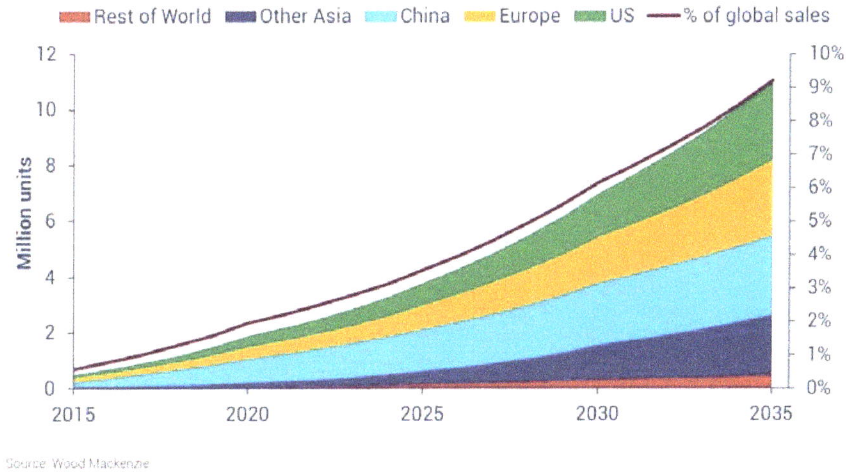

Figure 2. The dramatic growth in the electric vehicle market.

The emergence of the EV industry triggered a global economic transformation. Therefore, evaluation of different EVs' effectiveness is a significant and challenging task. There seems to be no ideal strategy for EV evaluation. Even a well-planned evaluation method may encounter difficulties. To meet this objective, multicriteria decision-making (MCDM) approaches are used in conjunction with the fuzzy set concept to establish a unified model for the effectiveness assessment of different EVs, given that each EV can have its own set of mechanisms and quality to evaluate. Furthermore, a lack of resource availability causes decision makers to make judgments under high ambiguity, resulting in unanticipated outcomes. As a result, dealing with ambiguous and contradictory information necessitates the use of a fuzzy-based unified model for collecting and organizing technical and analytical data. In this paper, we used a fuzzy-based unified model approach for evaluating the effectiveness of different EVs.

This research work is presented in different sections. Section 2 deliberates several similar existing pieces of literature. An overview of the different types of electric vehicles is discussed in Section 3, and Section 4 presents the method and results of this study. Section 5 recapitulates and concludes the research work.

2. Related Works

Wang et al. [7] presented an assessment of trust for the heterogeneous network of vehicles in sustainable development with electric vehicles. The benefits of low energy usage and high assessment precision were derived from the transport trust assessment compared to the standard trust evaluation process. Hashemnia and Asaei [8] analyzed various electric motors and compared the advantages of each motor with that which is more appropriate for EV deployments. The five basic types of electric engines were explored, including DC, induction, permanent synchronous magnet, switching reluctance, and brushless DC motors. In their study, they found that the induction motor technology had progressed more than the others, and that brushless DC and permanent magnet motors were much more appropriate than others for electric vehicles.

Prud'homme and Koning [9] presented a methodology in the form of a computerized model. It analyzed the expenses and efficiency of an electric vehicle in relation to a fuel-powered vehicle. This was a comparable assessment. It compared an electric automobile with a conventional car that provides approximately the same kind of service over a similar time. This was done from three main perspectives: customer costs, societal costs, and environmental impacts.

Iclodean et al. [10] demonstrated the flexibility of an electric car using four distinct battery types: lithium-ion (Li-ion), molten salt (Na-NiCl2), nickel metal hydride (Ni-MH), all with a similar reserve capacity of electrical power. The originality of this research was the application in a real-time computerized simulation of four different rechargeable batteries for EVs in a similar model, in order to assess the autonomy and effectiveness of these rechargeable batteries in the driving process.

Oh [11] discussed and determined which drivetrain arrangement was the best to use in a commercially obtainable test motor as a train for hybrid vehicles. The engine feature could be simulated, as well as the actual characteristics evaluated when used in the car for a distinct driving and operating condition. Qiu and Wang [12] carried out extensive research on the structure and operation of the electrically powered regenerative braking component of EVs. The contribution process and assessment methods provided by regenerative braking were addressed and assessed by the circulation of energy to enhance the energy effectiveness of EVs. They presented a methodology for the calculation of the renewable frequency contribution. Furthermore, a novel regenerative braking control approach was presented, termed the "Serial 2 control technique." In addition, as a contrast control approach, two control techniques were provided, namely the "parallel control strategy" and the "serial one control strategy".

Pfeiffer et al. [13] discussed the alternative time delay estimation (TDE) techniques. All options were evaluated by means of real data with EV energy trains. They focused not only on the correctness of the TDE, but also on computing performance to facilitate the operation of vehicle electronic control units (ECUs). Even modest noise, as well as offsets, were found in the measuring data in the recommended linear regression (LR) methodology, which were not suited for our purposes. The variance minimization (VM) technique is a good option. After the initial execution, it is not only noise-proof, but also very effective.

Song [14] presented an integrated framework to assess the consequences of various solutions for power management. Three energy management strategy (EMS) considerations were included in their suggested strategy. The first was the durability of the fuel cell. Fuel-saving was the second priority for assessing fuel efficiency, and was dependent on a dynamic algorithm created for optimal worldwide driving distances. The synthesis of weighted fuel-cell durability was the third priority for the EMS.

Wang et al. [15] presented an assessment indicator system for use patterns focused on data of the battery electric vehicle (BEV) to examine the use of car-sharing vehicles and private vehicles, in order to analyze their usability patterns. The assessment indicator system was built on the state transition strategy and defined the three-dimensional use pattern for BEVs. The time and space components of travels defined the time as and space properties of the pattern of use. The decisive dimension represented a decision-making pattern based on a perceptual psychological model as a reason for the state transformation at the microlevel.

Zhang et al. [16] investigated the requirements for charging stations while considering the plug-in electric vehicle (PEV) operational cost, as well as BEV feasibility. The area of research and PEV specifications were determined depending on the early cars used in the evolving trade market in California. An appropriate charging strategy based on 24 h travel trends was suggested to minimize operating costs. The findings demonstrated that the charging timelines were the main tool in minimizing PEV operating costs, while more charging locations offered to decrease advantages for plug-in hybrid electric vehicles (PHEVs).

This paper is unique in various ways [17–25]. First, in contrast to many other studies, our paper focuses on the expert-centric hierarchical structure for multicriteria decision making in the evaluation of different EVs. Second, this work presents a straightforward fuzzy-based unified methodology in the form of a computational model. This model helped to compare the efficiency of EVs to that of other types.

3. Different Types of Electric Vehicles

It is a very interesting opportunity to go shopping for cars, especially for people attempting to improve the ecosystem. The EV industry is changing fast, and one would then probably buy one of those five kinds of electrical vehicles (EVs) if they reached the conclusion that they wanted to buy or rent a car that is better for the atmosphere.

3.1. Battery Electric Vehicles (BEVs)

BEV denotes a battery electric vehicle that is operated by a battery-powered full electric engine. These are also called pure electric vehicles (PEVs) because they use only electricity as the primary source. The battery in these vehicles must be charged at regular intervals, often by connecting them to a charging station. One of the most significant barriers to BEV acceptance is "range anxiety" [26–33], which occurs when owners are concerned about being stuck in the middle of the highway with a completely depleted battery [17,18]. BEVs are capable of transforming about 80% of the power stored in the batteries into action. Teslas (all variants), the Nissan Leaf, and the Volkswagen e-Golf are a few examples of BEVs. Figure 3 shows the architectural diagram of battery electric vehicles.

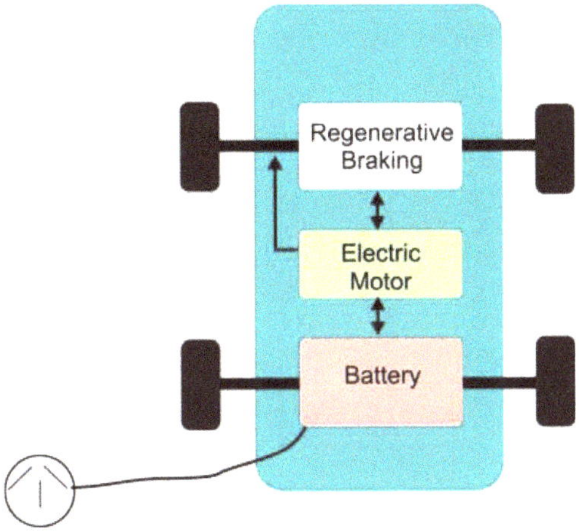

Figure 3. Architectural diagram of battery electric vehicles (BEV).

3.2. Fuel-Cell Electric Vehicles (FCEVs)

Fuel-cell electric vehicles (FCEVs) are different from other EVs. Fuel-cell EVs are powered by a fuel cell of hydrogen, and do not generate harmful emissions, only water vapor and warm air. In FCEVs, chemical power is transformed into electrical energy in the fuel cell; however, the hydrogen fuel is kept in a storage tank, therefore energy density and range are less likely to be an issue [19]. Like BEVs, FCEVs also primarily feature an electric motor, but employ a different storage and electricity supply technology. The propulsion battery in FCEVs is mostly substituted by the hydrogen tank, as well as by the chemical reaction, in which a number of fuel cells transform hydrogen into electricity as well as water vapor. The Toyota Mirai, Honda Clarity, and Hyundai Nexo are some examples of FCEVs. Figure 4 shows the architectural diagram of fuel-cell electric vehicles.

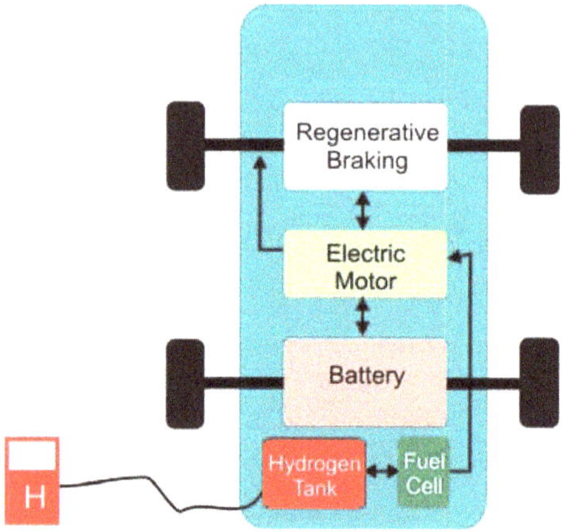

Figure 4. Architectural diagram of fuel-cell electric vehicles (FCEVs).

3.3. Hybrid Electric Vehicles (HEVs)

Hybrid electric vehicles are the most common type of EV. HEVs include a compact rechargeable battery that is not charged by plugging in, but instead by an inner combustion electric motor and/or the braking mechanism. The HEV is a multienergy system; unlike traditional vehicles that can only generate power, HEV batteries can both generate and absorb electricity. HEVs can already meet the needs of customers and therefore their numbers will increase at a quicker rate in the future. The key difficulty with HEVs is determining how to optimize the many sources of energy in order to achieve the optimum fuel economy or lowest pollution at the lowest cost [20]. There are various types of hybrids; however, on average, most are really battery-assisted automobiles instead of automobiles that are entirely powered by batteries. The Toyota Prius was first released in Japan in the late 1990s, and it made its way to the United States in 2001. Figure 5 shows the architectural diagram of hybrid electric vehicles.

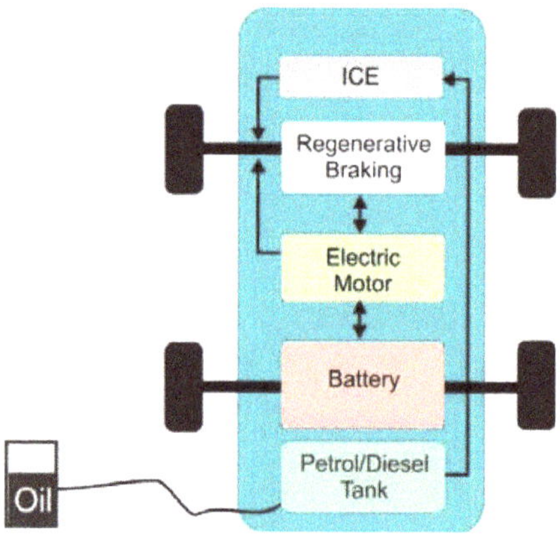

Figure 5. Architectural diagram of hybrid electric vehicles (HEVs).

3.4. Plug-In Hybrid Electric Vehicles (PHEVs)

Plug-in hybrid EVs (PHEVs) have relatively smaller rechargeable batteries than those of BEVs. The idea behind plug-in hybrids is to make short journeys powered by battery capacity. PHEVs are now becoming increasingly common. PHEVs are charged by either plugging into such an electric connection, or by using onboard energy generation. These vehicles have a limited selection in electric-only mode, and can operate at maximum throttle. PHEVs provide important fuel versatility. Although PHEVs contain a larger battery and a more robust motor compared to HEVs, the overall variety is still quite limited [21]. A plug-in hybrid, from a technical perspective, is basically a large hybrid with extra technology. The main distinction between a full hybrid and a plug-in hybrid (PHEV) is that the full hybrid EV battery is charged entirely by using internal combustion, whereas the plug-in hybrid's expanded traction battery is also charged by using a charging station. This means that these plug-in hybrids can only go about 100 km on battery without any internal combustion engine ignition. In particular for tiny towns and short round-trip commutes, this is a significant advantage. This version also permits the lowest potential CO_2 emissions between different hybrid systems. The Chevy Volt, Hyundai Ioniq PH EV, and VW Golf GTE are some examples of PHEVs. Figure 6 shows the architectural diagram of plug-in hybrid electric vehicles.

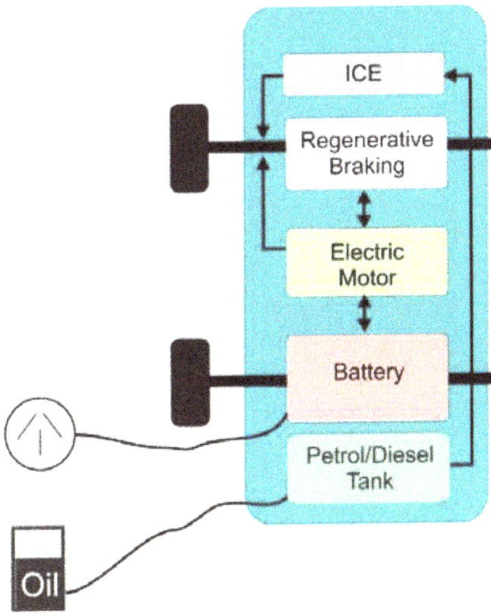

Figure 6. Architectural diagram of plug-in hybrid electric vehicles (PHEVs).

3.5. Range Extender Hybrid Electric Vehicles (REHEVs)

Range extender hybrid EVs are much like hybrid plug-in EVs; however, they are distinct in technology. REHEVs are most often regarded as PHEVs; however, REHEVs are often more powerful than PHEVs. The Chevy Volt is the most excellent example of a REHEV. This is arguably the appropriate choice for those who have experience in the EV market because it has a high all-battery portfolio and is driven by a combustion engine with which people are familiar. Figure 7 shows the architectural diagram of range extender hybrid electric vehicles.

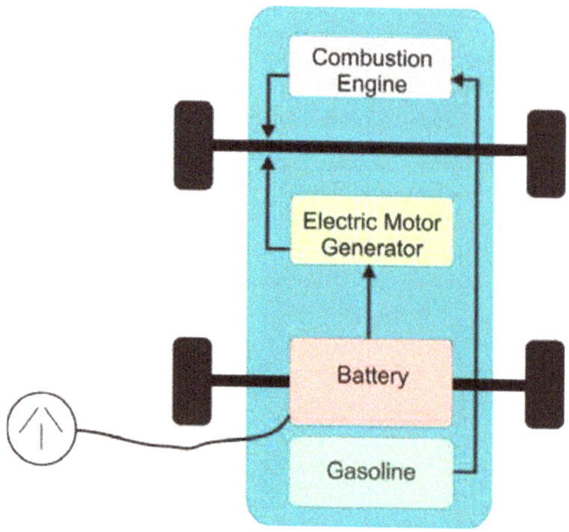

Figure 7. Architectural diagram of range extender hybrid electric vehicles (REHEVs).

4. Methods and Results

4.1. Hierarchical Design for the Evaluation of Different EVs

Electric vehicles are a revolutionary innovation that has yet to reach consumers outside of the "innovator" and "early adopter" groups in most regions [33]. Promoting a different and innovative technology creates hurdles, and the appropriate strategy may be quite beneficial in improving widespread approval. Under the challenges of energy efficiency and atmospheric pollution, several nations should reform their current energy utilization structure in order to minimize fuel energy demand and CO_2 emissions. Acceptance of EVs has the potential to decrease reliance on foreign oil energy while also addressing specific environmental pollution issues. When compared to a regular gas-powered automobile, EVs have a significantly higher purchase price, lower availability of charging facilities, and a longer charging time, making people reluctant to acquire an EV. In this paper, we used a fuzzy AHP-TOPSIS-based unified technique to evaluate the different types of EV alternatives such as BEVs, FCEVs, HEVs, PHEVs, and REHEVs, which are represented as T1, T2, T3, T4, and T5, respectively.

As shown in Figure 8, the four significant criteria at level one and corresponding sub-criteria at level two in the present method that contributed to the evaluation of different EVs were clearly recognized and constructed based on a survey of the literature, as well as input from several automobile specialists. The primary factors at level one that can have a substantial impact on EVs performance were divided into five categories; i.e., regulatory, technical, business, and design, denoted by S1, S2, S3, and S4, respectively. The regulatory level included three subfactors; i.e., government policies, traffic policies, and internal policies, denoted by S11, S12, and S13, respectively. Further, the technical level included four significant subfactors; i.e., efficiency, coverage, environmental, and safety, denoted by S21, S22, S23, and S24, respectively. Furthermore, the business level included three subfactors; i.e., consumer satisfaction, servicing, and investment, denoted by S31, S32, and S33, respectively. Lastly, the design level included three subfactors; i.e., battery, recyclable, and compatibility, denoted by S41, S42, and S43, respectively. Figure 8 illustrates the hierarchical structure used for the multicriteria decision making in this research. This hierarchical structure assisted in evaluating the performance of five alternatives.

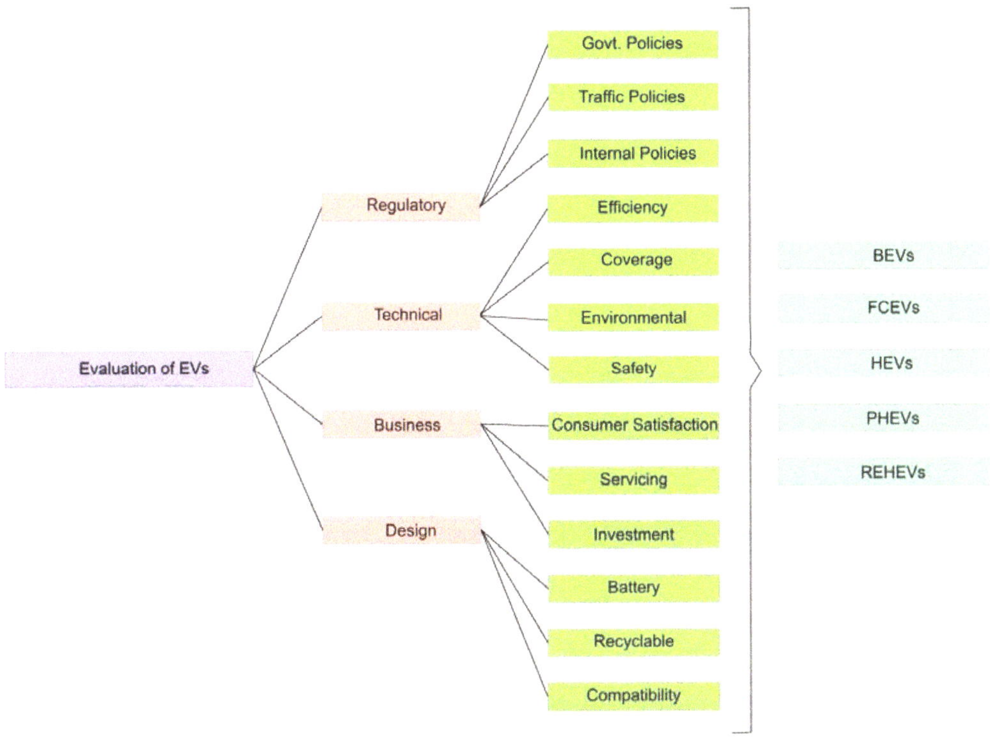

Figure 8. The hierarchy structure for the evaluation of different EVs.

4.2. Fuzzy AHP-TOPSIS Methodology

The analytic hierarchy process (AHP) was introduced by Saaty in 1990 [22]. Both numerical and subjective aspects were taken into account in the decision-making procedure. Given that AHP uses a discreet scale of 1 to 9, this technique is usually criticized because it does not include uncertainty during the decision-making procedure. The fuzzy-AHP approach has also been utilized in other disciplines to tackle multicriteria challenges. This method was used by Haq and Kannan [23] to choose the best supplier in a delivery chain. This was utilized by Huang et al. [24] for R&D shortlisting. For the selection of the most appropriate method of bridge building, Pan [25] employed this technique. For a staff-selection process, Güngör et al. [26] used this methodology. The fuzzy set theory, as developed by Zadeh, is a generalized variant of the classical set theory. It is an affiliate and assigns a grade of 1 to 10. In this paper, different types of EVs were classified using TOPSIS on the basis of characteristics [27].

In order to deal with uncertain numerical values in reality, Zadeh [34] invented fuzzy numbers. A fuzzy number is an amount for which a single-valued figure is not accurate, but imprecise. Classification of fuzzy numbers is a significant decision-making technique. Fuzzy decisions represent the effectiveness of different alternative models in the modeling of real-world situations by using fuzzy variables. Generally, a fuzzy-based approach is any system in which the variables vary over fuzzy values rather than real figures. These fuzzy values could reflect linguistic terms like "very small", "moderate", and so on, depending on how they are perceived in a specific scenario [35]. The defuzzification process is the technique of extracting a single value using the aggregated outcome of fuzzy numbers. This is used to convert the findings of fuzzy rules into a crisp output. In another word, defuzzification is accomplished through the use of a decision-making mechanism that determines the best crisp output from a fuzzy set.

A number of fuzzy criteria are used for a finite series of alternatives; i.e., the values of the alternatives are fuzzy figures. An additional process maps each m-tuple of fuzzy values into one fuzzy value, which is the alternative as per the entire set of criteria.

Let $A = \left\{\frac{A_i}{i} = 1, \ldots, n\right\}$ be a finite group of possibilities for decision making, and let $K = \left\{\frac{K_j}{j} = 1, \ldots, m\right\}$ be a finite group of fuzzy criteria, through which activity is regarded to be desirable. The estimates of the alternatives are fuzzy values. It must decide on this set of alternatives as a ranking challenge or decision problem. It is a two-stage technique:

(i) Compliance with all criteria aggregating judgments (fuzzy-numbers);

(ii) Ranking alternative decision making in relation to aggregating judgments.

In this research, two alternatives, the negative ideal solution (d−i) and positive ideal solution (d+i) were investigated. Further closeness coefficients (CCi) were calculated. We denoted CC−i as the degree of satisfaction in the i-th alternative and CC+i as the degree of gap in the i-th alternative. From a fuzzy collection of possible choices, we could evaluate which, as well as how, gaps must be closed in order to achieve ambitious goals and attain the ultimate findings. Closeness coefficients were used to rate all of the alternatives. Furthermore, CCi demonstrated the alternative closest to d+i and farthest from d−i. TOPSIS was determined by the choice of the ultimate solution or EVs that went beyond the perfect negative solution and were nearest to the ideal solution for the positive. The positive and the negative ideal solutions correspondingly had the highest advantages and lowest advantages. The final evaluations of the EVs were based on relative proximity to the optimal solution [28]. Figure 9 shows the integrated fuzzy AHP-TOPSIS methodology for the evaluation of different EVs.

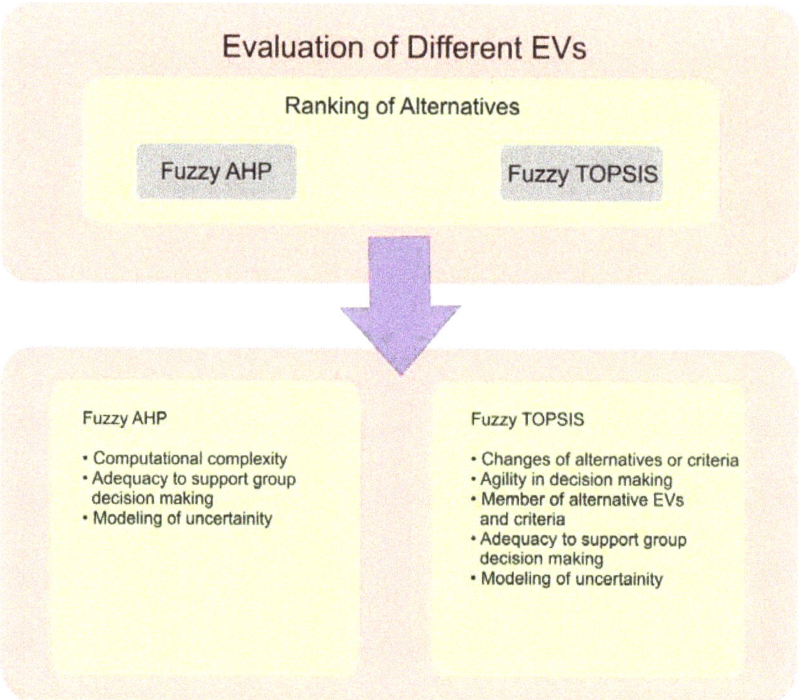

Figure 9. The fuzzy AHP TOPSIS methodology for the evaluation of different EVs.

4.3. Statistical Results

This integrated fuzzy AHP-TOPSIS methodology was used to evaluate the performance of different EVs. To acquire accurate information and insights, the investigators used comparative opinions of 75 automobile specialists from varied organization and scholarly backgrounds. It has been previously discussed that analyzing the performance of various EVs is extremely difficult in terms of competitive efficacy. The EVs were selected by using predetermined qualitative and quantitative assessment criteria during the process of different EVs' evaluation; however, the criteria demonstrated the requirement of the judgment, so herein unpredictability and fuzziness were included in the statistical and observational data evaluated by the decision makers with specific intelligence. A total of 75 automobile specialists, DM (k = 1, 2, 3,..., 75), were involved in analyzing the optimum available decision in linguistic variables. These 75 decision makers comprised 20 academics with 15 years of expertise, 20 researchers with 7 years of vehicle research experience, and 35 professionals from various automobile firms with 15 years of experience. The weights of the local criterion and subcriteria were derived using pairwise comparative matrices.

The aggregated fuzzify pairwise comparison matrix at Level 1 was formed, and can be seen in Table 1. The fuzzy-aggregated pairwise comparison matrix at Level 2 for regulatory, technical, business, and design is presented in Tables 2–5. For each second-layer aspect, the global weights were deliberate. These are tabulated in Tables 6–10. Further, Table 11 shows the overall weights and rankings of the methods. Table 12 presents the subjective cognition results for evaluators in linguistic terms. Table 13 shows the normalized fuzzy-decision matrix. Table 14 shows the weighted normalized fuzzy-decision matrix. In addition, with the support of the hierarchical structure, Table 15 and Figure 10 show the complete and final relative closeness of the alternatives.

Table 1. The aggregated fuzzify pairwise comparison matrix at Level 1.

	S1	S2	S3	S4
S1	1.000000, 1.000000, 1.000000	1.750254, 2.345258, 3.036563	1.485854, 1.956375, 2.526873	1.129628, 1.555351, 1.989625
S2	-	1.000000, 1.000000, 1.000000	0.576528, 0.786562, 1.168524	0.565263, 0.728568, 0.969954
S3	-	-	1.000000, 1.000000, 1.000000	0.628656, 0.816575, 1.075846
S4	-	-	-	1.000000, 1.000000, 1.000000

Table 2. The fuzzy-aggregated pairwise comparison matrix at Level 2 for regulatory.

	S11	S12	S13
S11	1.000000, 1.000000, 1.000000	0.237552, 0.287963, 0.367526	0.342154, 0.447785, 0.824763
S12	-	1.000000, 1.000000, 1.000000	0.661454, 1.172563, 1.693686
S13	-	-	1.000000, 1.000000, 1.000000

Table 3. The fuzzy-aggregated pairwise comparison matrix at Level 2 for technical.

	S21	S22	S23	S24
S21	1.000000, 1.000000, 1.000000	0.694154, 0.895356, 1.112485	0.234596, 0.287864, 0.364168	0.711256, 0.954163, 1.351257
S22	-	1.000000, 1.000000, 1.000000	0.493154, 0.642362, 1.241435	0.271354, 0.351565, 0.521635
S23	-	-	1.000000, 1.000000, 1.000000	1.085484, 1.329762, 1.558235
S24	-	-	-	1.000000, 1.000000, 1.000000

Table 4. The fuzzy-aggregated pairwise comparison matrix at Level 2 for business.

	S31	S32	S33
S31	1.000000, 1.000000, 1.000000	0.665365, 1.172384, 1.697465	1.157663, 1.447254, 1.704365
S32	-	1.000000, 1.000000, 1.000000	1.007762, 1.524765, 1.934368
S33	-	-	1.000000, 1.000000, 1.000000

Table 5. The fuzzy-aggregated pairwise comparison matrix at Level 2 for design.

	F41	F42	F43
S41	1.000000, 1.000000, 1.000000	1.197856, 1.588385, 2.156465	0.491541, 0.642285, 1.009958
S42	-	1.000000, 1.000000, 1.000000	0.224165, 0.295684, 0.427969
S43	-	-	1.000000, 1.000000, 1.000000

Table 6. The defuzzified pairwise comparison matrix.

	S1	S2	S3	S4	Weights
S1	1.000000	2.372530	1.981590	1.556640	0.392511
S2	0.421550	1.000000	0.824630	0.744770	0.152321
S3	0.504560	1.213520	1.000000	0.835090	0.202531
S4	0.642650	1.342880	1.203550	1.000000	0.252637

CR = 0.000602.

Table 7. The aggregated pairwise comparison matrix at Level 2 for regulatory.

	S11	S12	S13	Weights
S11	1.000000	1.173540	0.494564	0.275854
S12	0.852550	1.000000	1.172547	0.328627
S13	2.024340	0.853545	1.000000	0.395519

CR = 0.0488003.

Table 8. The aggregated pairwise comparison matrix at Level 2 for technical.

	S21	S22	S23	S24	Weights
S21	1.000000	0.892654	1.173554	0.994547	0.246313
S22	1.121242	1.000000	0.691526	0.372546	0.182575
S23	0.852562	1.447256	1.000000	1.298541	0.272112
S24	1.006624	2.688354	0.770435	1.000000	0.299000

CR = 0.034904.

Table 9. The aggregated pairwise comparison matrix at Level 2 for business.

	S31	S32	S33	Weights
S31	1.000000	1.172541	1.363652	0.382000
S32	0.853345	1.000000	1.491224	0.353026
S33	0.733754	0.670725	1.000000	0.255047

CR = 0.002506.

Table 10. The aggregated pairwise comparison matrix at Level 2 for design.

	S41	S42	S43	Weights
S41	1.000000	1.633244	0.691844	0.3259211
S42	0.612477	1.000000	0.303457	0.2731254
S43	1.447247	3.300347	1.000000	0.3112540

CR = 0.0052045.

Table 11. The overall weights and rankings of methods.

Level 1 Methods	Local Weights of Level 1	Level 2 Methods	Local Weights of Level 2	Overall Weights	Overall Ranks
S1	0.392511	S11	0.275854	0.108276	3
		S12	0.328627	0.128990	2
		S13	0.395519	0.155246	1
S2	0.152321	S21	0.246313	0.037519	12
		S22	0.182575	0.027810	13
		S23	0.272112	0.041448	11
		S24	0.299000	0.045544	10
S3	0.202531	S31	0.382000	0.077367	6
		S32	0.353026	0.071500	7
		S33	0.255047	0.051655	9
S4	0.252637	S41	0.325921	0.082340	4
		S42	0.273125	0.069000	8
		S43	0.311254	0.078634	5

Table 12. The subjective cognition results for evaluators in linguistic terms.

	T1	T2	T3	T4	T5
S11	5.3600, 7.3006, 8.7300	5.5500, 7.5500, 8.9100	0.6400, 2.2700, 4.2700	5.3600, 7.3600, 8.7300	4.1800, 6.0900, 7.6400
S12	3.7300, 5.5500, 7.2700	4.4500, 6.4500, 8.1800	1.6400, 3.5500, 5.5500	3.5500, 5.5500, 7.3600	5.0000, 7.0000, 8.4500
S13	2.3600, 4.2700, 6.2700	5.3600, 7.3006, 8.7300	5.5500, 7.5500, 8.9100	0.6400, 2.2700, 4.2700	5.3600, 7.3600, 8.7300
S21	4.8200, 6.8200, 8.5500	3.7300, 5.5500, 7.2700	4.4500, 6.4500, 8.1800	1.6400, 3.5500, 5.5500	3.5500, 5.5500, 7.3600
S22	5.5500, 7.5005, 9.2700	2.3600, 4.2700, 6.2700	2.4500, 4.2700, 6.2700	1.3600, 3.3600, 5.3600	4.4500, 6.4500, 8.1800
S23	4.2700, 6.2700, 8.1800	4.8200, 6.8200, 8.5500	4.6400, 6.6400, 8.5500	0.8200, 2.6400, 4.6400	4.4500, 6.4500, 8.2700
S24	5.3600, 7.3006, 8.7300	5.5500, 7.5500, 8.9100	0.6400, 2.2700, 4.2700	5.3600, 7.3600, 8.7300	5.7300, 7.7300, 9.2700
S31	3.7300, 5.5500, 7.2700	5.3600, 7.3006, 8.7300	5.5500, 7.5500, 8.9100	0.6400, 2.2700, 4.2700	5.3600, 7.3600, 8.7300
S32	2.3600, 4.2700, 6.2700	3.7300, 5.5500, 7.2700	4.4500, 6.4500, 8.1800	1.6400, 3.5500, 5.5500	3.5500, 5.5500, 7.3600
S33	5.3600, 7.3006, 8.7300	5.5500, 7.5500, 8.9100	0.6400, 2.2700, 4.2700	5.3600, 7.3600, 8.7300	4.4500, 6.4500, 8.1800
S41	3.7300, 5.5500, 7.2700	4.4500, 6.4500, 8.1800	1.6400, 3.5500, 5.5500	3.5500, 5.5500, 7.3600	4.4500, 6.4500, 8.2700
S42	2.3600, 4.2700, 6.2700	2.4500, 4.2700, 6.2700	1.3600, 3.3600, 5.3600	4.4500, 6.4500, 8.1800	5.7300, 7.7300, 9.2700
S43	4.8200, 6.8200, 8.5500	4.6400, 6.6400, 8.5500	0.8200, 2.6400, 4.6400	4.4500, 6.4500, 8.2700	5.1800, 7.1800, 8.8200

Table 13. The normalized fuzzy-decision matrix.

	T1	T2	T3	T4	T5
S11	0.3800, 0.6000, 0.8000	0.5400, 0.7500, 0.9200	0.5200, 0.7400, 0.9300	0.4200, 0.6900, 0.9900	0.5200, 0.7400, 0.9400
S12	0.5200, 0.7400, 0.9400	0.3800, 0.6000, 0.8000	0.5400, 0.7500, 0.9200	0.5200, 0.7400, 0.9300	0.4200, 0.6900, 0.9900
S13	0.3800, 0.6000, 0.8000	0.5200, 0.7400, 0.9400	0.5400, 0.7500, 0.9200	0.5200, 0.7400, 0.9200	0.2000, 0.4700, 0.7700
S21	0.3800, 0.6000, 0.8000	0.5400, 0.7500, 0.9200	0.5200, 0.7400, 0.9300	0.4200, 0.6900, 0.9900	0.5400, 0.7500, 0.9400
S22	0.5200, 0.7400, 0.9400	0.3800, 0.6000, 0.8000	0.5400, 0.7500, 0.9200	0.5200, 0.7400, 0.9300	0.4200, 0.6900, 0.9900
S23	0.3800, 0.6000, 0.8000	0.5200, 0.7400, 0.9400	0.5400, 0.7500, 0.9200	0.5200, 0.7400, 0.9200	0.2000, 0.4700, 0.7700
S24	0.3800, 0.6000, 0.8000	0.5400, 0.7500, 0.9200	0.5200, 0.7400, 0.9300	0.4200, 0.6900, 0.9900	0.5400, 0.7500, 0.9400
S31	0.5200, 0.7400, 0.9400	0.3800, 0.6000, 0.8000	0.5400, 0.7500, 0.9200	0.5200, 0.7400, 0.9300	0.4200, 0.6900, 0.9900
S32	0.3800, 0.6000, 0.8000	0.5200, 0.7400, 0.9400	0.5400, 0.7500, 0.9200	0.5200, 0.7400, 0.9200	0.2000, 0.4700, 0.7700
S33	0.3800, 0.6000, 0.8000	0.5400, 0.7500, 0.9200	0.5200, 0.7400, 0.9300	0.4200, 0.6900, 0.9900	0.5400, 0.7500, 0.9400
S41	0.5200, 0.7400, 0.9400	0.5400, 0.7500, 0.9200	0.3800, 0.6000, 0.8000	0.5400, 0.7500, 0.9200	0.5200, 0.7400, 0.9300
S42	0.3800, 0.6000, 0.8000	0.3500, 0.5800, 0.8100	0.5200, 0.7400, 0.9400	0.5400, 0.7500, 0.9200	0.5200, 0.7400, 0.9200
S43	0.5200, 0.7400, 0.9200	0.4600, 0.6700, 0.8600	0.3800, 0.6000, 0.8000	0.3500, 0.5800, 0.8100	0.4200, 0.6900, 0.9900

Table 14. The weighted normalized fuzzy-decision matrix.

	T1	T2	T3	T4	T5
S11	0.00000, 0.00200, 0.00900	0.00200, 0.00700, 0.02200	0.00200, 0.00700, 0.02400	0.00100, 0.00500, 0.01800	0.00300, 0.01100, 0.03600
S12	0.00300, 0.01200, 0.04100	0.00000, 0.00200, 0.00900	0.00200, 0.00700, 0.02200	0.00200, 0.00700, 0.02400	0.00100, 0.00500, 0.01800
S13	0.00300, 0.01200, 0.04200	0.00300, 0.01200, 0.04100	0.00300, 0.01200, 0.04100	0.00500, 0.01600, 0.04800	0.00500, 0.01600, 0.04900
S21	0.00000, 0.00200, 0.00900	0.00200, 0.00700, 0.02200	0.00200, 0.00700, 0.02400	0.00100, 0.00500, 0.01800	0.00200, 0.00900, 0.03800
S22	0.00300, 0.01200, 0.04100	0.00000, 0.00200, 0.00900	0.00200, 0.00700, 0.02200	0.00200, 0.00700, 0.02400	0.00100, 0.00500, 0.01800
S23	0.00300, 0.01200, 0.04200	0.00300, 0.01200, 0.04100	0.00300, 0.01200, 0.04100	0.00500, 0.01600, 0.04800	0.00500, 0.01600, 0.04900
S24	0.00000, 0.00200, 0.00900	0.00000, 0.00200, 0.00900	0.00200, 0.00700, 0.02200	0.00200, 0.00700, 0.02400	0.00100, 0.00500, 0.01800
S31	0.00300, 0.01200, 0.04100	0.00300, 0.01200, 0.04100	0.00300, 0.01200, 0.04100	0.00500, 0.01600, 0.04800	0.00500, 0.01600, 0.04900
S32	0.00000, 0.00200, 0.00900	0.00000, 0.00200, 0.00900	0.00200, 0.00700, 0.02200	0.00200, 0.00700, 0.02400	0.00100, 0.00500, 0.01800
S33	0.00300, 0.01200, 0.04100	0.00300, 0.01200, 0.04100	0.00300, 0.01200, 0.04100	0.00500, 0.01600, 0.04800	0.00500, 0.01600, 0.04900
S41	0.00000, 0.00200, 0.00900	0.00200, 0.00700, 0.02200	0.00200, 0.00700, 0.02400	0.00100, 0.00500, 0.01800	0.00200, 0.00900, 0.03800
S42	0.00300, 0.01200, 0.04100	0.00300, 0.01200, 0.04100	0.00500, 0.01600, 0.04800	0.00500, 0.01600, 0.04900	0.00100, 0.00500, 0.01800
S43	0.00300, 0.01200, 0.04200	0.00300, 0.01200, 0.04200	0.00200, 0.01000, 0.03700	0.00200, 0.00900, 0.03800	0.00100, 0.00500, 0.01800

Table 15. The closeness coefficients for the aspired level among the different alternatives.

Alternatives		d+i	d−i	Gap Degree of CC+i	Satisfaction Degree of CC−i
Alternative 1	T1	0.0451540	0.0545250	0.6125450	0.38712741
Alternative 2	T2	0.0564570	0.0365260	0.3562560	0.64714356
Alternative 3	T3	0.0464570	0.0548740	0.5698570	0.44421424
Alternative 4	T4	0.0451270	0.0452450	0.5754820	0.43471445
Alternative 5	T5	0.0346570	0.0215470	0.5535680	0.45485126

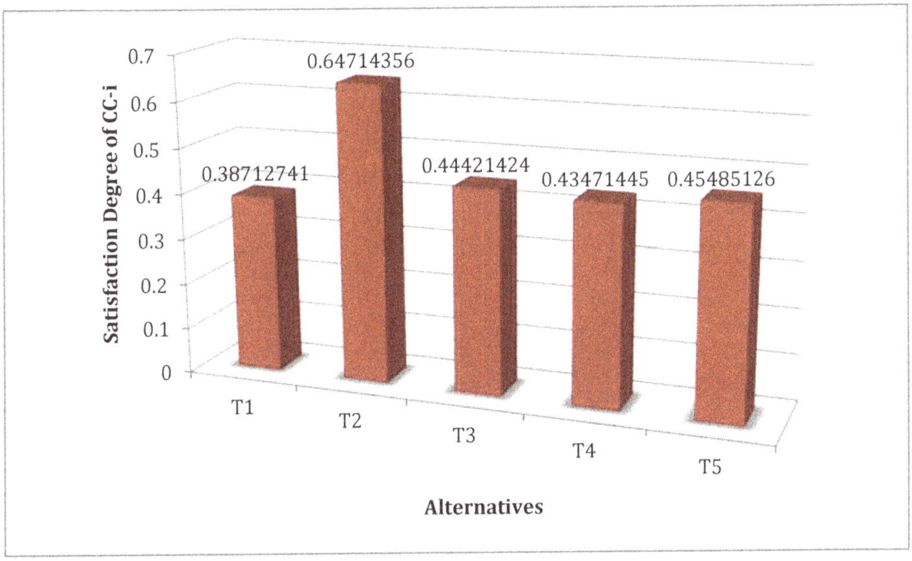

Figure 10. A graphical representation of the satisfaction degree of CC−i.

In this context, an evaluation of different electric vehicle alternatives was conducted for the inclusion of fuzzy AHP in the fuzzy TOPSIS; i.e., two major MCDM techniques. Although the proportional relevance of each aspect to the other can be expressed, the intricacies of subjective judgments in the description of the challenge were taken into consideration by fuzzy numbers. Ultimately, the suggested model was tested using a statistical method showing how the highest efficient electric vehicle type was chosen. The satisfaction degree (CC−i) of different alternatives was estimated as 0.38712741, 0.64714356, 0.44421424, 0.43471445, and 0.45485126 for T1, T2, T3, T4, and T5, respectively. As per the findings shown in Figure 10, the second alternative (T2) was highly effective and proficient among several other EV alternatives.

4.4. Comparison with the Classical AHP-TOPSIS Method

Whenever similar statistics are handled with different approaches, it produces contradictory interpretations [29]. Researchers have employed one or more techniques to check the correctness of anticipated methodology findings [30]. Therefore, in this study, we employed the classical AHP-TOPSIS approach [31] to estimate the findings using another approach and to check the effectiveness of consequences using fuzzy AHP-TOPSIS. The procedure of accumulating and projecting relevant information in classical AHP-TOPSIS is analogous to analyzing fuzzy AHP-TOPSIS without fuzzification. As a result, data were used in their actual numerical format to evaluate the different EV performances using traditional AHP-TOPSIS. Table 16 and Figure 11 show the differences in outcomes between fuzzy and classical AHP-TOPSIS. The findings produced using the traditional approach had a strong association with the ones produced using the fuzzy methodology. The outcomes of the comparative analysis were not as varied and distinct from one another; however, the precision of the findings varied. The correctness of the fuzzy-based methodology was higher and more accurate than that of the conventional methodology, as it was a more powerful approach over the classical AHP-TOPSIS. The fuzzy-based AHP-TOPSIS offered the capability of providing fuzzy set numbers for different parameters during the evaluation process.

Table 16. Comparison of the AHP-TOPSIS techniques.

Methods/Alternatives	T1	T2	T3	T4	T5
Fuzzy AHP-TOPSIS	0.38712741	0.64714356	0.44421424	0.43471445	0.45485126
Classical AHP-TOPSIS	0.38547400	0.64528700	0.44542700	0.43654800	0.46358700

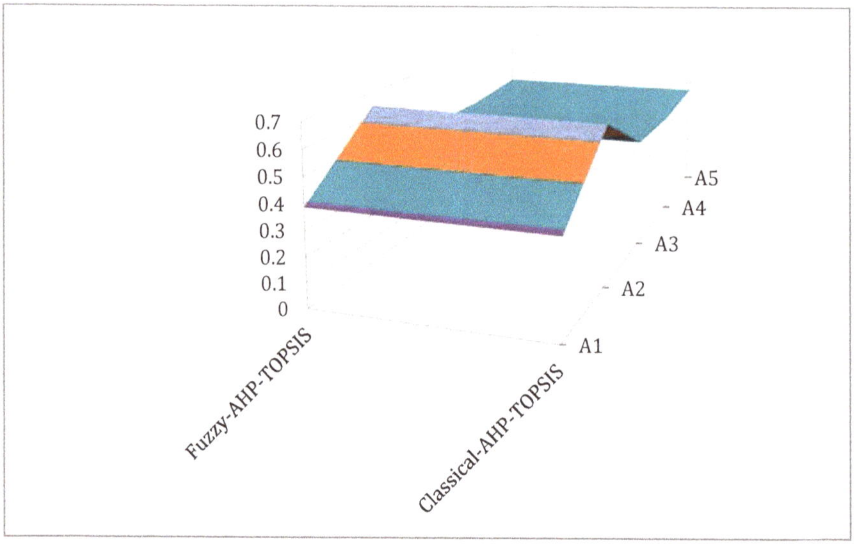

Figure 11. A graphical representation of the comparison of the AHP-TOPSIS techniques.

4.5. Sensitivity Analysis

The sensitivity evaluation was accomplished by altering the variables that influenced the report's correctness. During this statistical study, the sensitivity of the generated weights (variables) was assessed [32]. Following this, 13 variables were used throughout the analysis to evaluate sensitivity with the use of 13 experiments. The rate of satisfaction (CC−i) was calculated for each trial by taking into account weight alterations within each variable, whereas the weights of some other full variables were maintained constant by integrating the fuzzy AHP-TOPSIS methodology. The actual weights obtained in this research work are shown in the first row of Table 17. Alternative-2 (T2) had a significant satisfaction degree (CCi), as per the research findings. Thirteen experiments were performed, ranging from Experiment 1 to Experiment 13. In these 13 experiments, the obtained conclusions revealed that alternative-2 (T2) still had a higher satisfaction degree (CCi). Alternative-1 (T1) also was the lowest-weighted alternative in every trial. The variability in outcome suggested that alternative rankings were sensitive to weights. Table 17 and Figure 12 show the projected consequences.

Table 17. The sensitivity analysis.

Experiments	Weights/Alternatives		T1	T2	T3	T4	T5
Experiment-0	Original Weights	Satisfaction Degree (CC−i)	0.38712741	0.64714356	0.44421424	0.43471445	0.45485126
Experiment-1	S11		0.43576400	0.60004500	0.48962700	0.47718100	0.49393900
Experiment-2	S12		0.47776400	0.71004500	0.52912700	0.52018000	0.53493900
Experiment-3	S13		0.32806400	0.55804500	0.39112700	0.34048200	0.38563900
Experiment-4	S21		0.35976400	0.54044500	0.42412700	0.37798000	0.41803900
Experiment-5	S22		0.32916400	0.55554500	0.39612700	0.36368000	0.38383900
Experiment-6	S23		0.36076400	0.59104500	0.42712700	0.39817800	0.41783900
Experiment-7	S24		0.32916400	0.55554500	0.39612700	0.36368000	0.38383900
Experiment-8	S31		0.32916400	0.55554500	0.39612700	0.36368000	0.38383900
Experiment-9	S32		0.44446400	0.67604500	0.49912700	0.48167900	0.49743900
Experiment-10	S33		0.36076400	0.59104500	0.42712700	0.39817800	0.41783900
Experiment-11	S41		0.32916400	0.55554500	0.39612700	0.36368000	0.38383900
Experiment-12	S42		0.32176400	0.56004500	0.38362700	0.36018000	0.38293900
Experiment-13	S43		0.47816400	0.72654500	0.54062700	0.52768100	0.54343900

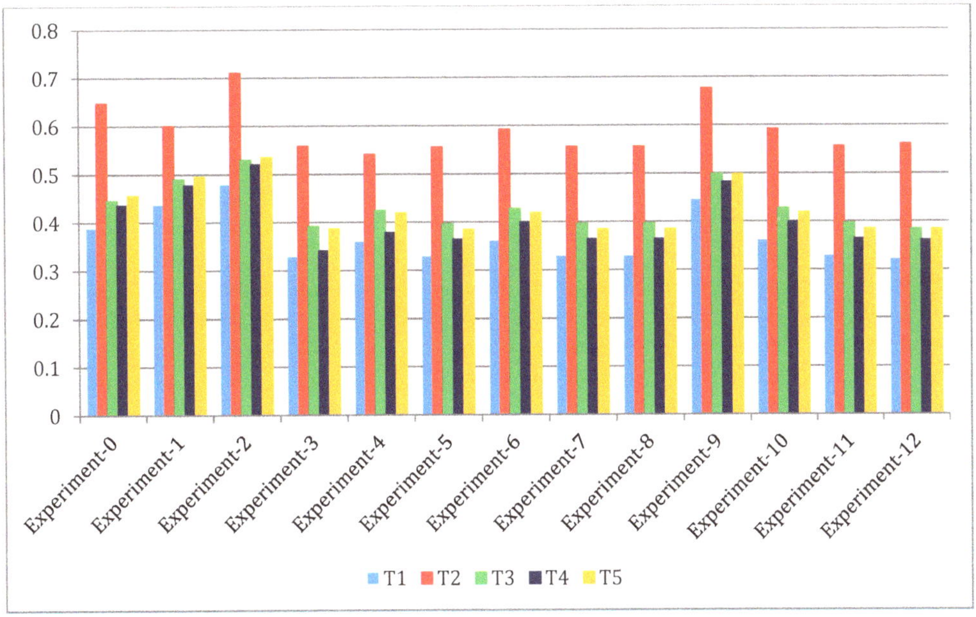

Figure 12. A graphical representation of the sensitivity analysis.

5. Conclusions

Although technological increases in worldwide transportation and society have enhanced life on this planet, they also have resulted in massive environmental devastation. As a result, people are paying close attention to the environment and its long-term sustainability. Renewable-power vehicles are one contributor to global challenges. BEVs have a reasonable consumption and good power generation, as long as the overall weight is not excessive. The vehicle weight depends on the number and capacity of the batteries installed. As a result, light BEVs that travel short distances have the highest efficiencies.

FCEV automobiles can store a greater amount of energy in comparison to their vehicle weight, and fuel-cell recharging can be done more speedily. FCEVs are thus ideal for long-distance travel and resources to create little interruption. The prospect of transportation will play a significant role in energy in systems centered on the interchange of modes of transportation, a future in which battery electric vehicles, as well as fuel-cell electric vehicles, will be supportive instead of combative.

Author Contributions: All authors contributed equally to the manuscript. All authors have read and agreed to the published version of the manuscript.

Funding: The project has been funded by Taif University, Kingdom of Saudi Arabia, under grant no: TURSP-2020/306.

Institutional Review Board Statement: Not applicable.

Informed Consent Statement: Not applicable.

Acknowledgments: This research was supported by Taif University Researchers Supporting Project Number TURSP-2020/306, Taif University, Taif, Saudi Arabia.

Conflicts of Interest: The authors declare no conflict of interest.

References

1. Pareek, S.; Sujil, A.; Ratra, S.; Kumar, R. Electric Vehicle Charging Station Challenges and Opportunities: A Future Perspective. In *2020 International Conference on Emerging Trends in Communication, Control and Computing (ICONC3)*; IEEE: Piscataway, NJ, USA, 2020; pp. 1–6.
2. Habib, S.; Khan, M.M.; Abbas, F.; Sang, L.; Shahid, M.U.; Tang, H. A Comprehensive Study of Implemented International Standards, Technical Challenges, Impacts and Prospects for Electric Vehicles. *IEEE Access* **2018**, *6*, 13866–13890. [CrossRef]
3. Shamshirband, M.; Salehi, J.; Gazijahani, F.S. Look-ahead risk-averse power scheduling of heterogeneous electric vehicles aggregations enabling V2G and G2V systems based on information gap decision theory. *Electr. Power Syst. Res.* **2019**, *173*, 56–70. [CrossRef]
4. Ansari, T.J.; Pandey, D.; Alenezi, M. STORE: Security Threat Oriented Requirements Engineering Methodology. *J. King Saud Univ. Comput. Inf. Sci.* **2018**, in press. [CrossRef]
5. Sahu, K.; Alzahrani, F.A.; Srivastava, R.K.; Kumar, R. Hesitant Fuzzy Sets Based Symmetrical Model of Decision-Making for Estimating the Durability of Web Application. *Symmetry* **2020**, *12*, 1770. [CrossRef]
6. Mackenzie, W. 2035: Can EVs Put the Brakes on Oil Demand? *Wood Mackenzie.* 2017. Available online: https://www.woodmac.com/news/editorial/2035-electric-vehicles-oil-demand/ (accessed on 10 September 2021).
7. Wang, T.; Luo, H.; Zeng, X.; Yu, Z.; Liu, A.; Sangaiah, A.K. Mobility Based Trust Evaluation for Heterogeneous Electric Vehicles Network in Smart Cities. *IEEE Trans. Intell. Transp. Syst.* **2021**, *22*, 1797–1806. [CrossRef]
8. Hashemnia, N.; Asaei, B. Comparative study of using different electric motors in the electric vehicles. In Proceedings of the 18th International Conference on Electrical Machines, Vilamoura, Portugal, 6–9 September 2008; pp. 1–5. [CrossRef]
9. Prud'Homme, R.; Koning, M. Electric vehicles: A tentative economic and environmental evaluation. *Transp. Policy* **2012**, *23*, 60–69. [CrossRef]
10. Iclodean, C.; Varga, B.; Burnete, N.; Cimerdean, D.; Jurchiş, B. *Comparison of Different Battery Types for Electric Vehicles*; IOP Publishing: Bristol, UK, 2017; Volume 252, p. 012058.
11. Oh, S.C. Evaluation of Motor Characteristics for Hybrid Electric Vehicles Using the Hardware-in-the-Loop Concept. *IEEE Trans. Veh. Technol.* **2005**, *54*, 817–824. [CrossRef]
12. Qiu, C.; Wang, G. New evaluation methodology of regenerative braking contribution to energy efficiency improvement of electric vehicles. *Energy Convers. Manag.* **2016**, *119*, 389–398. [CrossRef]
13. Pfeiffer, J.; Wu, X.; Ayadi, A. Evaluation of Three Different Approaches for Automated Time Delay Estimation for Distributed Sensor Systems of Electric Vehicles. *Sensors* **2020**, *20*, 351. [CrossRef]
14. Song, K.; Chen, H.; Wen, P.; Zhang, T.; Zhang, B.; Zhang, T. A comprehensive evaluation framework to evaluate energy management strategies of fuel cell electric vehicles. *Electrochim. Acta* **2018**, *292*, 960–973. [CrossRef]
15. Wang, W.; Zhang, Q.; Peng, Z.; Shao, Z.; Li, X. An empirical evaluation of different usage pattern between car-sharing battery electric vehicles and private ones. *Transp. Res. Part A Policy Pract.* **2020**, *135*, 115–129. [CrossRef]
16. Zhang, L.; Brown, T.; Samuelsen, S. Evaluation of charging infrastructure requirements and operating costs for plug-in electric vehicles. *J. Power Sources* **2013**, *240*, 515–524. [CrossRef]
17. Khan, M.; Kockelman, K.M. Predicting the market potential of plug-in electric vehicles using multiday GPS data. *Energy Policy* **2012**, *46*, 225–233. [CrossRef]
18. Tate, E.D.; Harpster, M.O.; Savagian, P.J. The Electrification of the Automobile: From Conventional Hybrid, to Plug-in Hybrids, to Extended-Range Electric Vehicles. *SAE Int. J. Passeng. Cars Electron. Electr. Syst.* **2008**, *1*, 156–166. [CrossRef]

19. Offer, G.; Howey, D.; Contestabile, M.; Clague, R.; Brandon, N. Comparative analysis of battery electric, hydrogen fuel cell and hybrid vehicles in a future sustainable road transport system. *Energy Policy* **2010**, *38*, 24–29. [CrossRef]
20. Shen, C.; Shan, P.; Gao, T. A Comprehensive Overview of Hybrid Electric Vehicles. *Int. J. Veh. Technol.* **2011**, *2011*, 1–7. [CrossRef]
21. Clement-Nyns, K.; Haesen, E.; Driesen, J. The Impact of Charging Plug-In Hybrid Electric Vehicles on a Residential Distribution Grid. *IEEE Trans. Power Syst.* **2009**, *25*, 371–380. [CrossRef]
22. Saaty, T.L. Decision making with the analytic hierarchy process. *Int. J. Serv. Sci.* **2008**, *1*, 83. [CrossRef]
23. Haq, A.N.; Kannan, G. Fuzzy analytical hierarchy process for evaluating and selecting a vendor in a supply chain model. *Int. J. Adv. Manuf. Technol.* **2005**, *29*, 826–835. [CrossRef]
24. Huang, C.-C.; Chu, P.-Y.; Chiang, Y.-H. A fuzzy AHP application in government-sponsored R&D project selection. *Omega* **2008**, *36*, 1038–1052. [CrossRef]
25. Pan, N.-F. Fuzzy AHP approach for selecting the suitable bridge construction method. *Autom. Constr.* **2008**, *17*, 958–965. [CrossRef]
26. Güngör, Z.; Serhadlıoğlu, G.; Kesen, S.E. A fuzzy AHP approach to personnel selection problem. *Appl. Soft Comput.* **2009**, *9*, 641–646. [CrossRef]
27. Alosaimi, W.; Ansari, T.J.; Alharbi, A.; Alyami, H.; Seh, A.; Pandey, A.; Agrawal, A.; Khan, R. Evaluating the Impact of Different Symmetrical Models of Ambient Assisted Living Systems. *Symmetry* **2021**, *13*, 450. [CrossRef]
28. Ansari, T.J.; Al-Zahrani, F.A.; Pandey, D.; Agrawal, A. A fuzzy TOPSIS based analysis toward selection of effective security requirements engineering approach for trustworthy healthcare software development. *BMC Med. Inform. Decis. Mak.* **2020**, *20*, 1–13. [CrossRef]
29. Kumar, R.; Alenezi, M.; Ansari, M.T.J.; Gupta, B.; Agrawal, A.; Khan, R. Evaluating the Impact of Malware Analysis Techniques for Securing Web Applications through a Decision-Making Framework under Fuzzy Environment. *Int. J. Intell. Eng. Syst.* **2020**, *13*, 94–109. [CrossRef]
30. Attaallah, A.; Ahmad, M.; Ansari, T.J.; Pandey, A.K.; Kumar, R.; Khan, R.A. Device Security Assessment of Internet of Healthcare Things. *Intell. Autom. Soft Comput.* **2021**, *27*, 593–603. [CrossRef]
31. Kumar, R.; Ansari, M.T.J.; Baz, A.; Alhakami, H.; Agrawal, A.; Khan, R.A. A Multi-Perspective Benchmarking Framework for Estimating Usable-Security of Hospital Management System Software Based on Fuzzy Logic, ANP and TOPSIS Methods. *KSII Trans. Internet Inf. Syst.* **2021**, *15*, 240–263.
32. Zarour, M.; Ansari, T.J.; Alenezi, M.; Sarkar, A.K.; Faizan, M.; Agrawal, A.; Kumar, R.; Khan, R.A. Evaluating the Impact of Blockchain Models for Secure and Trustworthy Electronic Healthcare Records. *IEEE Access* **2020**, *8*, 157959–157973. [CrossRef]
33. Viola, F. Electric Vehicles and Psychology. *Sustainability* **2021**, *13*, 719. [CrossRef]
34. Zadeh, L.A. Fuzzy sets as a basis for a theory of possibility. *Fuzzy Sets Syst.* **1978**, *1*, 3–28. [CrossRef]
35. Klir, G.J. From Classical Mathematics to Fuzzy Mathematics: Emergence of a New Paradigm for Theoretical Science. In *Fuzzy Logic in Chemistry*; Elsevier: Amsterdam, The Netherlands, 1997; pp. 31–63.

energies

MDPI

Article

Quasi-Static Research of ATV/UTV Non-Pneumatic Tires

Jerzy Jackowski [1], Marcin Żmuda [1,*], Marcin Wieczorek [1] and Andrzej Zuska [2]

[1] Department of Mechanical Engineering, Institute of Vehicles and Transportation, Military University of Technology (WAT), ul. gen. Sylwestra Kaliskiego 2, 00-908 Warsaw, Poland; jerzy.jackowski@wat.edu.pl (J.J.); marcin.wieczorek@wat.edu.pl (M.W.)

[2] Department of Automotive Vehicles and Transportation, Faculty of Mechatronics and Mechanical Engineering, Kielce University of Technology, al. Tysiąclecia Państwa Polskiego 7, 25-314 Kielce, Poland; a.zuska@tu.kielce.pl

* Correspondence: marcin.zmuda@wat.edu.pl

Abstract: The non-pneumatic tire (NPT) is a type of wheel whichdevelopment is related to the beginning of automotive development. The non-pneumatic tire (NPT) is a type of tire that does not contain compressed gases or fluid to provide directional control and traction. Nowadays, this type of wheel is more and more often used in special purpose vehicles, e.g., in military vehicles and working machines. The main feature of the non-pneumatic tire is a flexible support structure (including the part of the wheel between the tread and the rim). This paper presents the results of research aimed at determining the influence of the geometry of the NPT's (intended for All-Terrain Vehicle–ATV/Utility Task Vehicle–UTV) load-bearing structure on its quasi-static directional characteristics. The experimental tests included the determination of the radial stiffness of research objects on a non-deformable flat surface and on a single obstacle, as well as the determination of the degree of deformation for the elastic structure and belt. The significant influence of the elastic structure's shape and the elastomer, as the material forming the NPT, on its radial stiffness was revealed.

Keywords: non-pneumatic tire; airless tire; radial stiffness; spoke deformation

Citation: Jackowski, J.; Żmuda, M.; Wieczorek, M.; Zuska, A. Quasi-Static Research of ATV/UTV Non-Pneumatic Tires. *Energies* **2021**, *14*, 6557. https://doi.org/10.3390/en14206557

Academic Editors: Rafał Jurecki and Guzek Marek

Received: 17 August 2021
Accepted: 6 October 2021
Published: 12 October 2021

1. Introduction

The non-pneumatic tire's (NPT) characteristic feature is the transfer of directional forces of the vehicle to the ground without the need to keep gas or fluid inside under a certain pressure [1]. The obvious advantage of NPT is its resistance to going flat, which is typical for pneumatic tires and often leads to the accelerated necessity of their replacement and their troublesome utilization. [2].

Most current NPTs consist of the following main elements [3–6]: the rim, the supporting structure, and the tread. The NPT tread and rim perform the same functions as they do in a standard pneumatic tire. The supporting structure consists of a shear beam (band) and a flexible structure. The main task of the NPT shear beam is to copy the compressed air properties of a pneumatic tire [3]. In papers modelling the NPT, the shear beam is made of layers of inextensible material between which a core is located [7]. An elastomer (solid core) or a specific structure with a defined geometry can be used as the core. The type of core applied will have an impact on the forming of unit pressures. The solid core is characterized by a more even distribution of pressure in the contact area, while the defined geometry causes their accumulation at the points of core connection with the tread [8].

The task of flexible structure is to connect the wheel rim with the tread (being one part with the shear beam). The flexible structure can be made as single spokes [3,9] or with a cellular structure, e.g., hexagonal structure [10]. The transfer of normal and tangential forces (longitudinal and lateral) of the NPT in the contact area with the ground is carried out mainly by selecting the geometry of the wheel structure and the materials that form it (mainly elastomers). The type of applied structure determines the load transfer mechanism and affects the value of the displacements of the wheel center in relation to its outer shell under the action of the load with normal and tangential force. For example, loading with a normal force on an NPT with single spokes will cause them to buckle in the part located under the wheel axle and stress the spokes above the axle. This NPT design leads to the main part of the load being transferred by the spokes above the wheel axis [3], and the increase in the spoke curvature results in greater vertical displacements of the wheel center [11,12]. On the other hand, when the flexible NPT load-bearing structure has a cellular structure, the greater load is transferred to the lower part of the wheel [13].

Quasi-static research allows for the assessment of basic tire properties and precise observation of the NPT supporting structure deformation. Radial characteristics, i.e., the dependence of the radial load as a function of deflection, which provides information about the stiffness of the tire and, as a result, determines the smoothness of driving, derives from the cooperation of the wheel with the road surface, rolling resistance, and the state of load transferred to elements of the vehicle suspension [14,15].

The research presented in this paper is aimed at analyzing the deformation state of an NPT's supporting structure (with different designs and shapes) as a function of the radial load. The conducted research were aimed at comparing the elastic properties (in the radial direction) of NPTs with pneumatic tires. At this stage of the research, attention was focused on the comparison of the radial stiffness characteristics, which essentially determine the level of dynamic loads transferred in the ground–wheel–body system of the vehicle. The obtained research results will be used for simulation tests of a multi-body vehicle model.

The "objects, conditions and methodology of research" presents basic information about the tires selected for research. The tires (pneumatic and NPTs) are called research objects later in this article. The method of load selection and the type of ground, which are the test conditions, was presented. The methodology of determining the midline for the hysteresis loop is described, which was recorded during tests. The areas of the tire selected for static tests are presented.

The results of quasi-static tests aimed at determining the radial elasticity are presented in "results". The characteristics of the center line of the test objects were compared for a specific load range (three different normal loads) and different surfaces (flat rigid surface, single triangular obstacle). The analysis of the radial stiffness of the wheel as a function of the load is presented.

In the "discussion", the test results are analyzed and the deformation of the elastic structure under the wheel axis is compared. The areas of further research are presented.

2. Objects, Conditions and Methodology of Research

Two NPTs and one pneumatic tire for all-terrain vehicles (ATV) or utility terrain vehicles (UTV) were selected for the research:

- NPT_1 (non-pneumatic tire nr 1)—the flexible part of the support structure consists of 24 pairs of spokes;
- NPT_2 (non-pneumatic tire nr 2)—equipped with a cellular (hexagonal) flexible cell as a part of the supporting structure;
- PT (pneumatic tire)—diagonal tire with six ply carcass.

Basic information about the research objects is presented in Table 1. The choice of test conditions was guided by the destination use of the wheels and the range of normal load variability resulting there from F_Z. The range of the research was established on the basis of observation of changes in normal reactions under the wheels in the tests of an ATV vehicle weighing 460 kg. These tests were carried out in the variant vehicle without and with an additional load of approx. 103 kg, on horizontal ground and on inclined ground at an angle of 15°. On the basis of these measurements, the following test conditions were established: 1000 N, 2000 N, and 3000 N, with two operational inflation pressures of 300 kPa and 400 kPa selected for the PT.

Table 1. Basic information of research objects [16–19].

Research Object	Manufacturer's Designation	Overall Diameter (mm)	Width (mm)	Maximum Speed (km/h)
NPT_1	26x9N14 NHS	660.4	228.6	80.5
NPT_2	NPT 26X8.00-14 NHS	673.1	203.2	40.2 [1]
PT	AT 26x8-14	660.4	203.2	140.0

[1] manufacturer's guidelines for recommended maximum wheel speed.

The determination of the radial characteristics was carried out for two variants of non-deformable surfaces: flat and with a triangular obstacle. The dimensions of the selected obstacle (Figure 1) reflect the rough road with single roughness (small terrain obstacles).

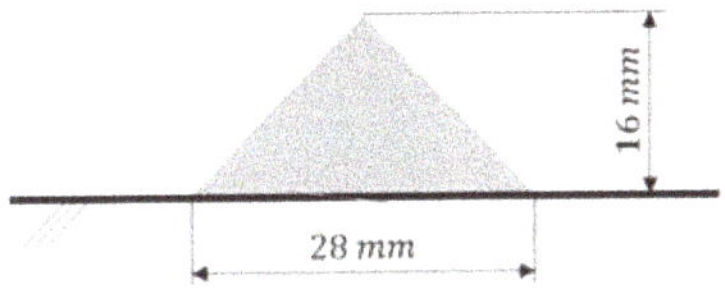

Figure 1. Triangular obstacle used in the wheel radial characterization research.

Determination of the radial stiffness characteristics included simultaneous registration of the normal load acting in the wheel axis and the deflection of the structure of the test object. In accordance with the adopted methodology, the load was increased to a value of 125% of the adopted normal load range and then unloaded. The own methodology [20] was used during the tests, which was developed on the basis of domestic and foreign standards for static tire research and on the basis of many years of authors' experience, also in cooperation with tire manufacturers. An example of the radial characteristic along with the direction of load changes is shown in Figure 2, where F_Z refers to normal load and z_W refers to vertical displacement of the wheel axis. On the basis of the obtained hysteresis loop, the course of the midline was determined, which was described by a fifth-order polynomial. Then, based on the course of the tangent to the center line at the point corresponding to 25%, 50%, 75%, and 100% of the assumed normal load, the value of the radial stiffness coefficient was calculated (Figure 2) [21]

$$k_Z = tg\alpha_X \tag{1}$$

where:

- k_Z—radial stiffness coefficient,
- α_X—the angle between the tangent line to the centerline and a horizontal line with the common point of intersection corresponding to the specified load.

Quasi-static wheels tests were carried out on the universal station for tire research shown in the Figure 3 (a detailed description of the station's capabilities is presented in [22]). Strain gauges placed under the measuring trolley platform are used to measure normal force. The accuracy of the normal force measurement is +/− 1N. An LVDT sensor with an accuracy of +/− 0.1 mm is used to measure the vertical displacement of the wheel axis.

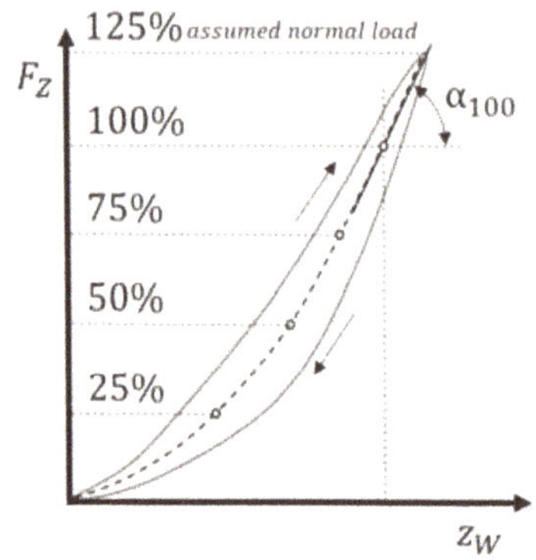

Figure 2. Graphical representation of the course of the center line (dashed line) and determination of the value of the radial stiffness coefficient (tan of the angle α) on the basis of the measured radial stiffness characteristic (solid line).

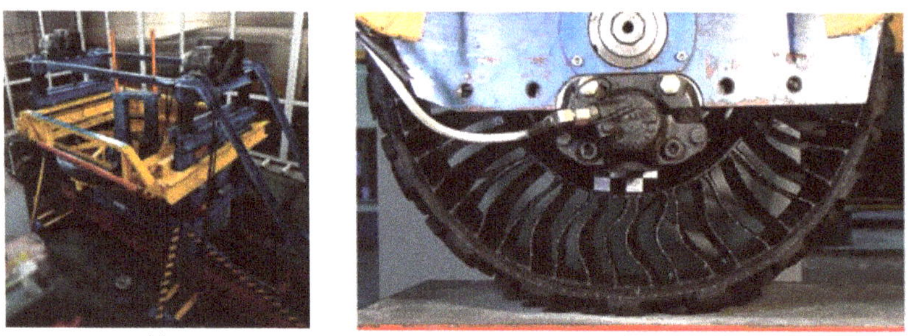

(**a**) (**b**)

Figure 3. Universal station for tire research. (**a**) general view, (**b**) NPT_1 during research.

Before starting the research, four evenly distributed points were selected around the perimeter of each object (which were the centers of contact of the test objects with the ground) positioned directly under the wheel axis and, additionally, at the top of the obstacle. Choosing the above-mentioned points, different arrangements of the spokes/cell arms of the elastic part of the NPT and projections of the tread pattern with respect to the center of contact with the ground were taken into account (Figure 4). The research under given conditions was repeated four times at each of these points.

Apart from measuring the normal force and vertical displacement of the wheel axis, the deformation of the elastic structure and shear beam were also recorded. Reference points were deposited on the mentioned elements (Figures 4 and 5). These points were used in the analysis of the video material using the TEMA Automotive software. After the video image was calibrated, the positions of the reference points before loading and for the maximum value of the applied normal load were determined. An example of the use of TEMA Automotive software is shown in Figure 5 (the right figure shows the points for displacement analysis).

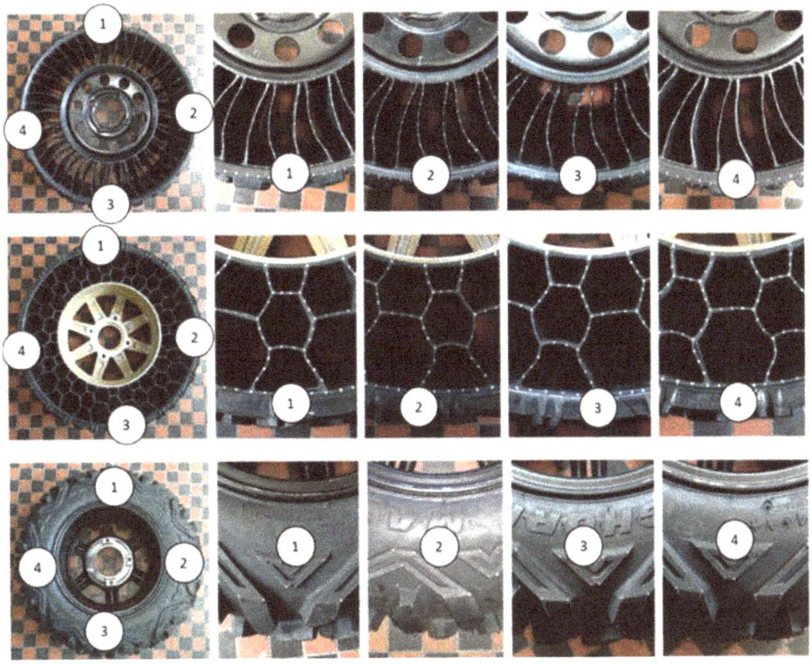

Figure 4. Cross-sections of the research objects in which the measurements of the radial characteristic were carried out (from the top in the lines NPT_1, NPT_2, and PT, respectively).

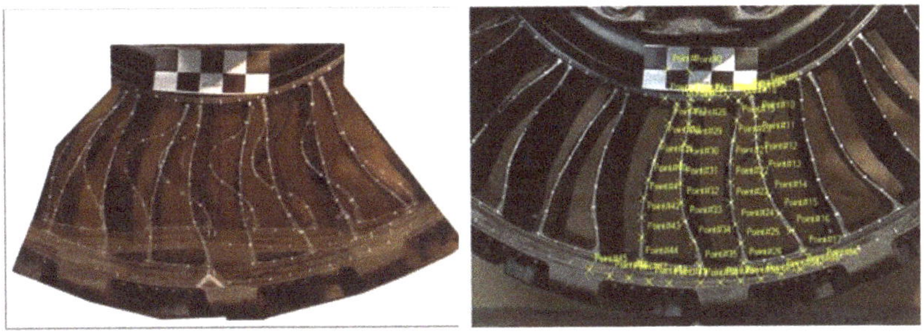

Figure 5. Summary of NPT_1 wheel images unloaded and loaded (image on the **left**), and preparation of reference points for the displacements analysis in the TEMA Automotive software (image on the **right**).

Unfortunately, due to the limitations of the universal station, the deformation analysis of the NPT's elastic structure covered the NPT's part between their axis and contact area.

3. Results

Figure 6 shows the hysteresis loop obtained in the research. The results for a normal load of 1000 N, 2000 N, and 3000 N were shown by continuous, dashed, and dotted lines, respectively. Different color lines were used on the same figure to distinguish the type of ground (flat, triangular obstacle). Moreover, in order to facilitate the analysis and improve readability, results were included in the same range of changes in the vertical displacement of the wheel axis and normal operating force.

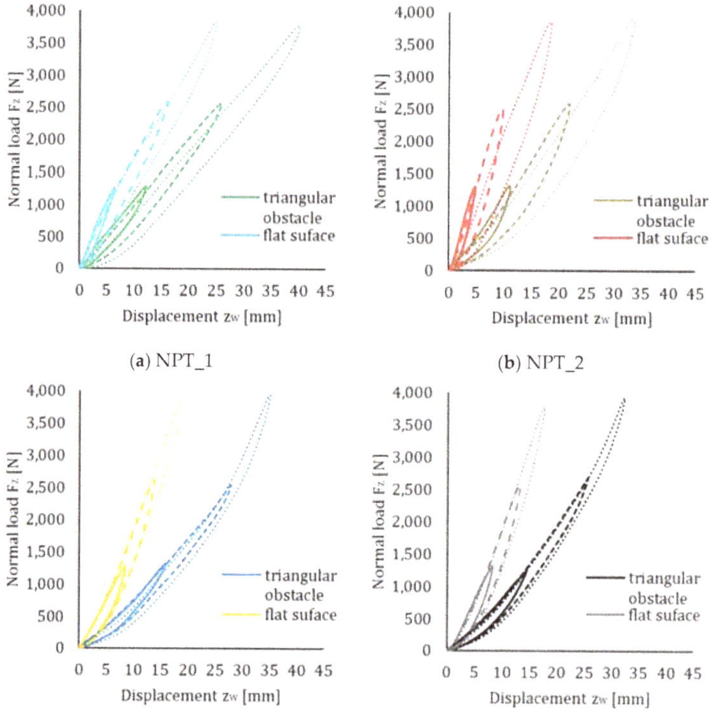

(a) NPT_1 (b) NPT_2

(c) PT (300 kPa) (d) PT (400 kPa)

Figure 6. Radial characteristic of research object on flat surface and triangular obstacle.

The first stage of analysis included determining the center lines and determining the stiffness coefficient of the research object depending on the adopted range of vertical load. The results are shown in Figure 7, and the estimated numerical values of the radial stiffness coefficients are summarized in Tables 2 and 3 (for a flat surface and a triangular obstacle, respectively) and in Figure 8 (maintaining the method of markings in Figure 6).

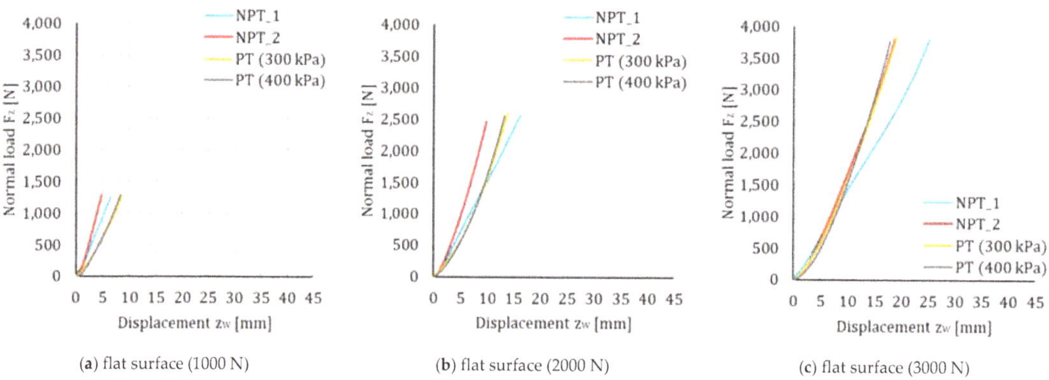

(a) flat surface (1000 N) (b) flat surface (2000 N) (c) flat surface (3000 N)

Figure 7. *Cont.*

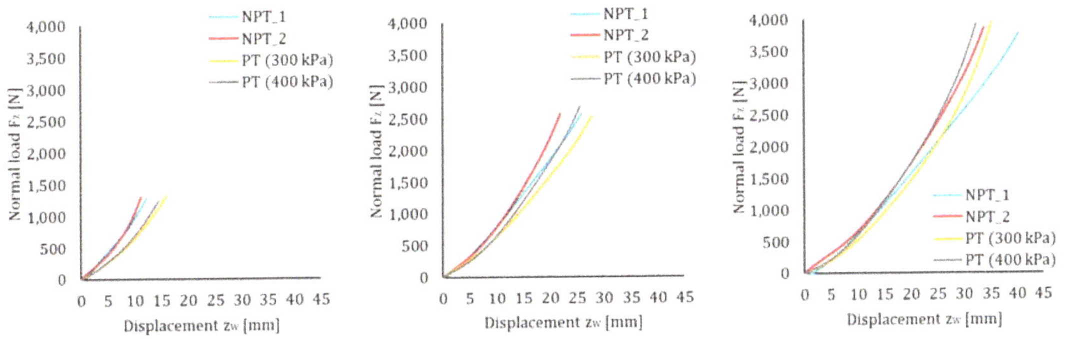

(**d**) triangular obstacle (1000 N) (**e**) triangular obstacle (2000 N) (**f**) triangular obstacle (3000 N)

Figure 7. Radial stiffness characteristics of research objects.

Table 2. Values of research objects radial stiffness on the flat surface.

Normal Load [N]		Radial Stiffness [N/m]			
		NPT_1	NPT_2	PT (300 kPa)	PT (400 kPa)
1000	25%	174,020 ± 7931	263,300 ± 10,695	130,349 ± 4688	142,043 ± 7133
	50%	201,423 ± 8331	316,383 ± 6578	168,680 ± 6823	160,095 ± 2574
	75%	198,265 ± 1615	345,044 ± 7132	196,030 ± 8950	188,368 ± 7584
	100%	191,715 ± 7190	359,262 ± 11,025	226,086 ± 7589	229,055 ± 12,808
2000	25%	163,057 ± 443	227,906 ± 5721	150,828 ± 1443	164,374 ± 4341
	50%	155,139 ± 204	270,951 ± 3921	196,501 ± 1392	216,262 ± 5222
	75%	148,790 ± 585	294,827 ± 5247	232,173 ± 2119	262,497 ± 3834
	100%	167,393 ±779	327,271 ± 5197	262,113 ± 4491	297,057 ± 6691
3000	25%	152,467 ± 2396	197,401 ± 4735	177,592 ± 2141	188,518 ± 8556
	50%	137,698 ± 921	219,405 ± 4601	215,389 ± 1339	239,282 ± 8387
	75%	139,993 ± 2570	225,558 ± 2562	240,071 ± 3890	275,726 ± 6260
	100%	179,758 ± 2990	246,176 ± 7376	266,583 ± 4918	312,840 ± 1349

Table 3. Values of research objects radial stiffness on the triangular obstacle.

Normal load [N]		Radial Stiffness [N/m]			
		NPT_1	NPT_2	PT (300 kPa)	PT (400 kPa)
1000	25%	93,326 ± 4474	73,212 ± 2644	62,836 ± 1409	65,545 ± 1593
	50%	91,481 ± 2039	98,165 ± 4029	81,851 ± 510	85,582 ± 902
	75%	111,877 ± 2801	148,508 ± 4109	98,898 ± 835	108,407 ± 1491
	100%	136,669 ± 3306	199,816 ± 9375	112,608 ± 2927	132,467 ± 4103
2000	25%	85,490 ± 3816	92,798 ± 3733	77,080 ± 253	85,426 ± 796
	50%	102,796 ± 1363	124,472 ± 4960	96,945 ± 903	113,298 ± 369
	75%	107,363 ± 987	145,738 ± 6690	103,057 ± 1313	125,631 ± 1781
	100%	111,697 ± 1863	169,508 ± 3907	116,728 ± 1072	140,195 ± 848
3000	25%	92,915 ± 2061	100,941 ± 3931	88,268 ± 1568	101,523 ± 623
	50%	98,760 ± 1196	105,916 ± 4786	110,533 ± 2477	126,068 ± 271
	75%	98,535 ± 494	121,795 ± 5610	145,009 ± 8305	147,586 ± 2164
	100%	113,349 ± 1250	153,718 ± 3901	199,243 ± 9681	194,120 ± 3578

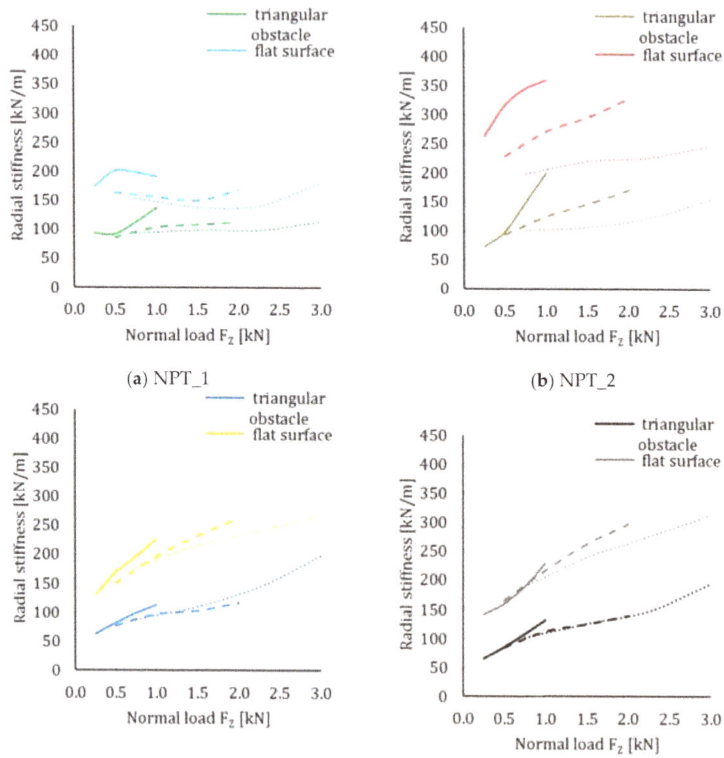

(a) NPT_1 (b) NPT_2

(c) PT (300 kPa) (d) PT (400 kPa)

Figure 8. Comparison of research objects radial stiffness on a flat surface and a triangular obstacle.

Figures 9 and 10 show the deformation of the elastic structure and shear beam of the NPTs. The dots correspond to the points marked on the observed parts of the research objects, and the colors distinguish different features: a rigid rim (red line), supporting structure (blue line—no load, yellow line—under load), and shear beam (gray line—no load, green line—under load). The displacements are given in the X–Y coordinates, and the non-deformable rim of the NPTs was taken as the reference line.

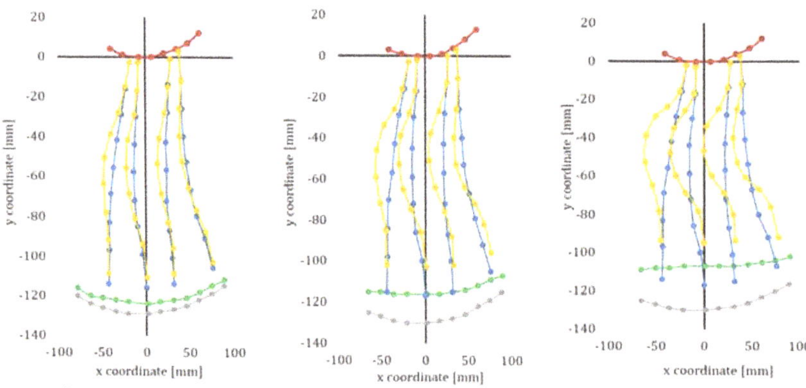

(a) flat ground—1000 N (b) flat ground—2000 N (c) flat ground—3000 N

Figure 9. *Cont.*

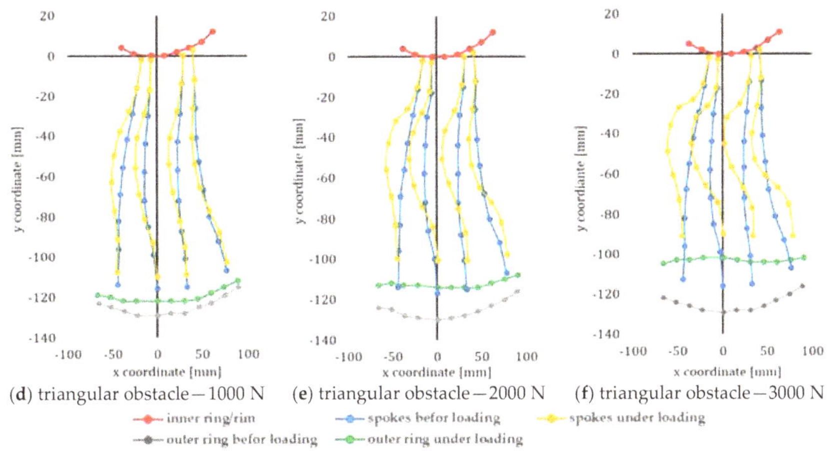

Figure 9. Deformation of the NPT_1 elastic structure between the wheel axis and different type of surface.

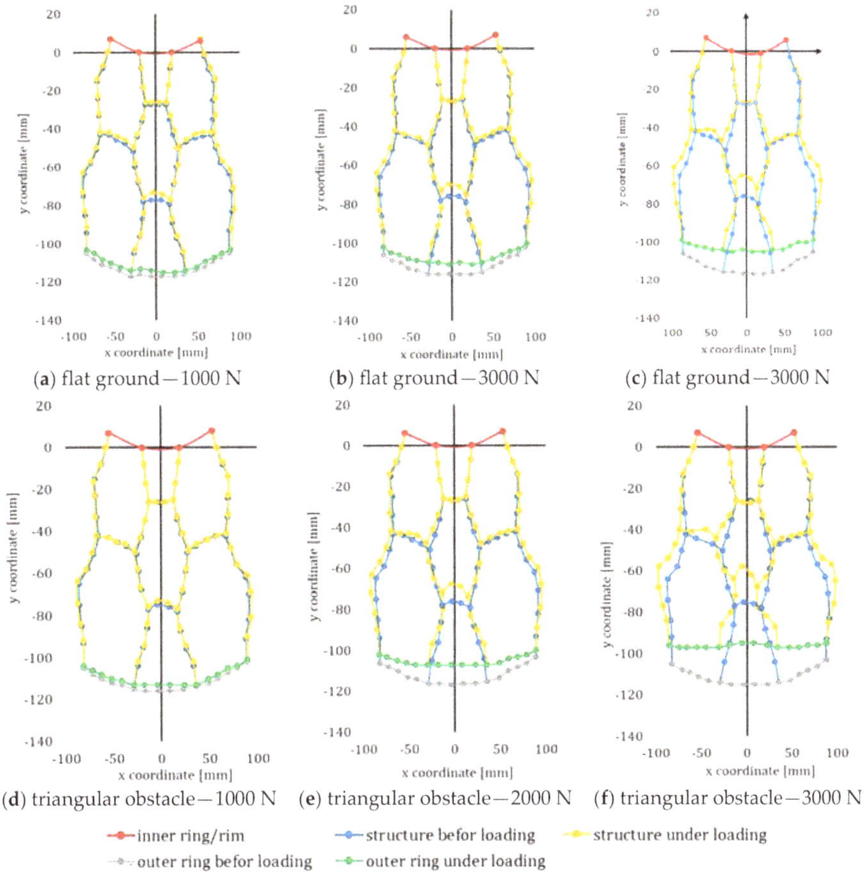

Figure 10. Deformation of the NPT_2 elastic structure between the wheel axis and different type of surface.

4. Discussion

Due to a different load transfer mechanism and differences in structure, the obtained results were analyzed in terms of similarities and differences.

The course of the radial elasticity characteristics of the tested wheels, shown in Figure 6, is similar. For each of the values of the assumed normal loads, the range of the corresponding vertical displacements of the wheel axis is similar. The influence of a triangular obstacle always increases the vertical deformation of wheels.

On a flat surface, the pneumatic tire, regardless of the inflation pressure value, showed similar hysteresis loop of the radial characteristics. For a load of 1000 N, this may be due to the fact that, for such a low load, the deformation element is the tread blocks. A similar phenomenon was observed for the NPT_1 and NPT_2, where similar hysteresis loops were obtained. For a load of 2000 N, the course of the NPT_1 hysteresis loop is similar to a pneumatic tire with an inflation pressure of 400 kPa (at load equal to 3000 N, the similarity was for the NPT_2 hysteresis loop).

It can also be seen that with increasing load, the hysteresis loop course recorded for the pneumatic tire coincide, while those determined for NPT are of a different character. As the maximum load of NPTs increased during the research, more sloping hysteresis loops were obtained. This phenomenon is intensified for both NPTs in the case of their interaction with the triangular obstacle and corresponds to the process of reducing rubber stresses under the influence of increasing deformation in subsequent measurement cycles.

The recorded hysteresis loops did not change their course until the load was increased to the next range. After the load change, the course was again fixed in the new position. This is due to the Mullins effect, which causes a change in mechanical properties under the influence of deformation and is characteristic of rubber and rubber-like materials [23–26]. This indicates that the differences in the determined hysteresis loop courses, apart from the design factor resulting from the structure of the elastic structure, are also influenced by the material (elastomer) from which it was made. This material is definitely significantly different in both considered NPTs.

Other differences are also visible. On flat ground and on the triangular obstacle, a pneumatic tire has a progressive characteristic (Figure 8c,d). Under the influence of a normal load of 1000 N, the highest values of the wheel axle displacement were recorded for the pneumatic tire compared to other test objects in this research set. In the case of NPTs, smaller displacements may be related to the resistance of the load-bearing structure to its deformation. Under the influence of the applied load of 2000 N and 3000 N for NPT_1 on a flat surface, the largest displacements among all research objects were recorded, which is confirmed by the stiffness comparison with Figure 8. Regardless of the ground type, NPT_1 is characterized by small variability of the radial stiffness value (except for the first set of loading Figure 8a). This may be related to the smooth "switching on" of consecutive spokes located above the axle to load carrying.

On flat ground, NPT_2 is characterized by significantly higher stiffness compared to other test objects (Figure 8b). This may be related to a different load transfer mechanism and the resistance of the supporting structure located between the axle and the ground surface, as well as the composite material (elastomer) applied. This is confirmed by the largest differences in the stiffness values for individual load sets, where a significant decrease in the NPT_2 stiffness was observed during increased load range (on both types of ground).

An important reason for the differences in the course of the radial stiffness as a function of the deflection of the NPTs is the design of their supporting (flexible) structure. A detailed analysis of the deformation of its spokes (Figure 9) under the NPT_1 axis indicates that they buckle. This type of deformation is not accidental and results mainly from the spoke curvature determined at the modeling stage. It is also worth noting that the first two points at each end of the spoke move mainly in the vertical plane, i.e., they do not change their position in relation to rigid rim and stiff shear beam.

Figure 10 shows the behavior of the NPT_2 supporting structure on flat ground (a, b, c) and against a triangular obstacle (d, e, f). The structure between the wheel axis and the ground carries loads that cause compression of its components. This is especially noticeable on the shorter section of hexagonal cells (along the Y = 0 coordinate). As a result, NPT_2 experiences less deformation on flat ground and a triangular obstacle compared to NPT_1. This property affects the size of the hysteresis loop and the radial stiffness of NPT. It was noticed that, as in the case of NPT_1, the arms of the NPT_2 hexagonal cells in the part of contact with the rim and the band are also only slightly deformed.

During analysis, attention was also paid to the energy values (hysteresis loop area), which is dissipated at the deformation of the research objects. However, no results were obtained that could be unequivocally associated with the structure of the NPTs support structure and compared with a pneumatic tire. An obstacle for this analysis is that there are a large number of factors that may co-determine the amount of hysteresis loop; these will be taken into account during further wheel tests. For example, a pneumatic tire has large, sparsely spaced tread blocks that can affect the determined amount of energy losses; high hysteresis losses with loads of up to 1000 N are primarily related to the internal friction of the compressed tread blocks. NPT_2 energy dissipation is probably mainly related to the shaping of the elastic structure and deformation of the part of the wheel located under the axis of its rotation, i.e., tread, shear beam, and spokes. The smallest energy losses were determined for NPT_1, whose spokes in the lower part were buckled (which practically did not affect the energy losses) and the stress in the upper part [3,12]. Lower values of hysteresis losses were also recorded in the research with an obstacle, which may be due to the fact that the research objects interacted only with the top of the obstacle, the remaining part of the tread area of the pneumatic tire, or the band of the NPTs that had no contact with the ground.

The conducted research and analysis and the revealed doubts have shown how complex NPTs are. The above will be used to formulate the scope of further work aimed at identifying NPTS properties, including:

1. determining the causes of such a different course of changes in the radial stiffness of these NPTs by extending the observation of deformations of the elastic structure by the area above the wheel axis and making an attempt to identify the details of the elastic and supporting structure and the materials from which they are made;
2. carrying out research while wheels are rolling, with simultaneous observation of temperature distribution and measurement of rolling resistance coefficient and forces occurring in the contact area;
3. checking the influence of the supporting structure on other parameters of the tire, such as longitudinal, lateral, and cornering stiffness;
4. assessing inclination tire angle dependency on the hysteresis behavior and the temperature influence on the overall NPTS dynamic behavior.

All the above-mentioned actions will provide greater credibility in reporting the properties of the NPTs and can be used, for example, in modeling the motion of the vehicle, including its energy consumption.

The authors of this work are currently conducting research in the scopes indicated above. Their results will be presented in subsequent publications.

Author Contributions: Conceptualization, J.J., M.W. and M.Ż.; methodology, J.J., M.W. and M.Ż.; software, J.J., M.W. and M.Ż.; validation, J.J., M.W. and M.Ż.; formal analysis, J.J., M.W., M.Ż. and A.Z.; investigation, J.J., M.W., M.Ż. and A.Z.; resources, J.J., M.W. and M.Ż.; data curation, J.J., M.W. and M.Ż.; writing—original draft preparation, J.J., M.W. and M.Ż. writing—review and editing, J.J., M.W. and M.Ż.; visualization, J.J., M.W. and M.Ż.; supervision, J.J.; project administration, J.J., M.W. and M.Ż.; funding acquisition, J.J., M.W. and M.Ż. All authors have read and agreed to the published version of the manuscript.

Funding: This work was co-financed by the Military University of Technology under research project UGB 884/2021. This research was also partially funded under the Project of the Ministry of National Defense of the Republic of Poland Program—Research Grant (GBMON/13-999/2018/WAT).

Institutional Review Board Statement: Not applicable.

Informed Consent Statement: Not applicable.

Data Availability Statement: The data presented in this study are available on request from the corresponding author.

Conflicts of Interest: The authors declare no conflict of interest.

References

1. 49 CFR 571.129. Code of Federal Standard No. 129; New Nonpneumatic Tires for Passenger Cars. Available online: https://www.govinfo.gov/app/details/CFR-2011-title49-vol6/CFR-2011-title49-vol6-sec571-129 (accessed on 21 April 2021).
2. Rowhani, A.; Rainey, T.J. Scrap tyre management pathways and their use as a fuel—A review. *Energies* **2016**, *9*, 888. [CrossRef]
3. Rhyne, T.B.; Cron, S.M. Development of a non-pneumatic wheel. *Tire Sci. Technol.* **2006**, *34*, 150–169. [CrossRef]
4. Ju, J.; Veeramurthy, M.; Summers, J.D.; Thompson, L. Rolling resistance of a nonpneumatic tire having a porous elastomer composite shear band. *Tire Sci. Technol.* **2013**, *41*, 154–173. [CrossRef]
5. Kwangwon, K.; Hyeoun, H.; Uddin, M.S.; Ju, J.; Kim, D.-M. *Optimization of Non-Pneumatic Tire with Hexagonal Lattice Spokes for Reducing Rolling Resistance*; SAE International: Pittsburgh, PA, USA, 2015. [CrossRef]
6. Jackowski, J.; Wieczorek, M.; Żmuda, M. Energy consumption estimation of non-pneumatic tire and pneumatic tire during rolling. *J. KONES Powertrain Transp.* **2018**, *1*, 159–168. [CrossRef]
7. Ju, J.; Summers, J.D.; Ziegert, J.; Fadel, G. Design of honeycomb meta-materials for high shear flexure. In Proceedings of the ASME 2009 International Design Engineering Technical Conferences & Computers and Information in Engineering Conference IDETC/CIE, San Diego, CA, USA, 30 August–2 September 2009; pp. 1–10. [CrossRef]
8. Ju, J.; Ananthasayanam, B.; Summers, J.D.; Joseph, P. Design of cellular shear bands of a non-pneumatic tire-investigation of contact pressure. *SAE Int. J. Passeng. Cars—Mech. Syst.* **2010**, 598–606. [CrossRef]
9. Rugsaj, R.; Suvanjumrat, C. Proper radial spokes of non-pneumatic tire for vertical load supporting by finite element analysis. *Int. J. Automot. Technol.* **2019**, *20*, 801–812. [CrossRef]
10. Vinay, T.; Marattukalam, K.J. Modeling and analysis of non-pneumatic tyres with hexagonal honeycomb spokes. *Int. J. Recent Technol. Mech. Electr. Eng. (IJRMEE)* **2015**, *2*, 19–24. [CrossRef]
11. Kucewicz, M.; Baranowski, P.; Małachowski, J. Airless tire conceptions modeling and simulations. In Proceedings of the 13th International Scientific Conference; Springer: Berlin/Heidelberg, Germany, 2017; pp. 293–301. [CrossRef]
12. Hryciów, Z.; Jackowski, J.; Żmuda, M. The influence of non-pneumatic tyre structure on its operational properties. *Int. J. Automot. Mech. Eng.* **2020**, *17*, 8168–8178. [CrossRef]
13. Kumar, A.S.; Kumar, R.K. Force and moment characteristics of a rhombi tessellated non-pneumatic tire. *Tire Sci. Technol.* **2016**, *44*, 130–148. [CrossRef]
14. Jackowski, J.; Luty, W.; Wieczorek, M. Analiza możliwości oszacowania wzdłużnej sztywności poślizgowej opon na podstawie statycznych badań laboratoryjnych. *J. KONES Powertrain Transp.* **2006**, *13*, 94–101.
15. Filipozzi, L.; Assadian, F.; Kuang, M.; Johri, R.; Alcantar, J.V. Estimation of tire normal forces including suspension dynamics. *Energies* **2021**, *14*, 2378. [CrossRef]
16. Opona Maxxis Bighorn 26x8-14 M917 6PR. Available online: http://www.sklepatv.pl/opona-maxxis-bighorn-26x8-r14-m917-6pr.html (accessed on 22 April 2021).
17. Opona Quad Atv Maxxis 26x8-14 TL 44N Big Horn M-917. Available online: https://intercars.pl/produkty/1831894-opona-quad-atv-maxxis-26x8-14-tl-44n-big-horn-m-917 (accessed on 22 April 2021).
18. Michelin®X®Tweel®Airless Radial Tire Family. Available online: https://tweel.michelinman.com/on/demandware.static/-/Sites-tweel-us-Library/default/dw7bce57cd/pdf/Michelin_Tweel_Full_Line_Brochure.pdf (accessed on 22 April 2021).
19. 14" Terrainarmor™ Npt Tires By Polaris®. Available online: https://polarisaccesorios.com/es/llantas-y-neumaticos/1461-14-terrainarmor-npt-tires-by-polaris.html (accessed on 22 April 2021).
20. Jackowski, J.; Sulej, S.; Wieczorek, M.; Prochowski, L. *Zestaw Metodyk Badań Ogumienia*; Wojskowa Akademia Techniczna: Warsaw, Poland, 1993.
21. Prochowski, L. *Mechanika Ruchu*; WKŁ: Warsaw, Poland, 2005; pp. 32–33.
22. Jackowski, J.; Wieczorek, M. Analysis of interaction between tire tread and road on the basis of laboratory test. In Proceedings of the 7th International Science Conference Transbaltica 2011, Vilnius, Lithuania, 5–6 May 2011; pp. 292–296.
23. Diani, J.; Fayolle, B.; Gilormini, P. A review on the Mullins effect. *Eur. Polym. J.* **2009**, *45*, 601–612. [CrossRef]
24. Mai, T.T.; Morishitaab, Y.; Urayama, K. Novel features of the Mullins effect in filled elastomers revealed by stretching measurements in various geometries. *Soft Matter* **2017**, *13*, 1966–1977. [CrossRef] [PubMed]
25. Krmela, J.; Krmelová, V. Dynamic experiment of parts of car tyres. *Procedia Eng.* **2017**, *187*, 763–768. [CrossRef]
26. Sandua, C.; Taherib, S.; Taheria, S.; Gorsich, D. Hybrid Soft Soil Tire Model (HSSTM). Part I: Tire material and structure modeling. *J. Terramechanics* **2019**, *86*, 1–13. [CrossRef]

energies

Article

Integrated Design of a Custom Steering System in Cars and Verification of Its Correct Functioning

Włodzimierz Choromański, Iwona Grabarek * and Maciej Kozłowski

Faculty of Transport, Warsaw University of Technology, 75 Koszykowa Street, 00-662 Warsaw, Poland; wlodzimierz.choromanski@pw.edu.pl (W.C.); maciej.kozlowski@pw.edu.pl (M.K.)
* Correspondence: iwona.grabarek@pw.edu.pl

Abstract: The subject of this article is the design of a nonstandard steering system in cars. The applied methodology takes into account universal design, ensuring the greatest possible adaptation of the steering system to potential users, and at the same time, thanks to the specific nature of the designed steering device, it also assumes a special approach allowing for individual adjustment of the steering system to the needs and limitations of drivers with lower-limb disabilities. It is implemented through the "custom design" methodology. This article presents the impact of the design features of the multifunction steering wheel on the correctness of driving, as well as the level of load on selected muscles of the upper limbs responsible for operating the steering wheel. The tests were carried out on a dynamic simulator of a motor vehicle using the electromyography (EMG) technique, which enables the measurement of muscle load. A systemic approach to training and verifying the skills of drivers using new HMI solutions is proposed.

Keywords: vehicle safety; driving simulator tests; steering system; EMG; universal design; custom design

Citation: Choromański, W.; Grabarek, I.; Kozłowski, M. Integrated Design of a Custom Steering System in Cars and Verification of Its Correct Functioning. *Energies* **2021**, *14*, 6740. https://doi.org/10.3390/en14206740

Academic Editors: Wojciech Wach, Guzek Marek and Rafał Jurecki

Received: 29 August 2021
Accepted: 8 October 2021
Published: 16 October 2021

1. Introduction

The driver-operated steering system is the basic equipment of the modern car and will remain so for a long time, despite the introduction of autonomous vehicle technology. According to the SAE classification (level of steering automation [1]), from L1 to L3 or even partially L4, the steering system will remain an element used by drivers. Persons with reduced mobility require a special adjustment of this system to the individual characteristics of their disability. This adjustment often requires additional, very expensive mechanical systems, thus cannot be considered for mass production. In addition, adjusting the parameters of the steering system to the individual characteristics of a fully able-bodied driver in modern cars is very limited and comes down to the tilt of the steering wheel or, by adjusting the seat, to changing the driver's distance from the steering wheel. The approach presented by the authors is based on universal design but complemented with additional individual adjustment of the steering system to specific users. The definition of "universal design" is included in the Convention on the Rights of Persons with Disabilities [2]: "Universal design" means the design of products, environments, programs, and services to be usable by all people, to the greatest extent possible, without the need for adaptation or specialized design. "Universal design" shall not exclude assistive devices for particular groups of persons with disabilities where this is needed. This additional help is implemented using the "custom design" methodology. If something is custom designed, it is made according to someone's special requirements. This approach also applies to people with special needs who, depending on their limitations, have individual needs, the fulfilment of which ensures their greater independence and the ability to function in society.

A detailed analysis of the division into a group of able-bodied and disabled people does not give a clear answer as to the border between these groups. This border is blurred.

Do we, for example, include the elderly with a number of age-related ailments in the group of disabled or nondisabled people? Failure to answer this question requires a special approach to the design of control devices used in vehicles. Speaking about universal design of the steering system, the authors require one basic condition to be met at the beginning, namely equipping the vehicle with a "steer by wire" system, i.e., the one that does not have a mechanical position between the steering wheel and the steering system [3]. The "steer by wire" technology was initially introduced in aviation (in particular when piloting large passenger jets), due to the high physical effort required from pilots using classic solutions of control devices. In the automotive industry, according to the authors, it is an indispensable solution in the universal design methodology. The elements of the steering system that should be subject to universal design and elements that should be additionally adapted to the individual characteristics of the driver with the use of "custom design" are defined below. They are divided into parametric and structural factors. In addition to classic adjustments (e.g., tilting the steering wheel, adjusting the driver's position in relation to the steering system by changing the position of the seat), the following can be mentioned:

(a) Parametric factors

- Gear ratio of the steering system:

$$i = \frac{\alpha\ angle\ of\ the\ steering\ wheel}{\beta\ steering\ angle\ of\ the\ car\ wheels} \tag{1}$$

- The resistance of the steering element (steering wheel or, e.g., joystick lever) during the maneuver (so-called "force feedback" [4]).
- The parameters of the damping filter, e.g., vibrations of the driver's hands.
- Geometric dimensions of the steering element.

(b) Structural factors

- Constructional solution for the steering system (steering wheel, multifunction steering wheel, joystick, sensor) for cooperation with bionic systems (e.g., bionic prosthesis after limb amputation).

The abovementioned systems must be consistent in the IT sense, e.g., integrated in the CAN bus. The steering system control calculations in the "steer by wire" technology should be performed by the host computer. The cost of adjusting the parametric and structural factors is relatively low. The question arises of how to verify the correct functioning of individual factors, and how to take into account the preferences of drivers. Correctness is understood as safety and comfort (as defined by drivers) in performing individual maneuvers. In the article, the authors do not analyze the problem of the reliability of operation of individual systems. It is a separate problem playing an important role in the "steer by wire" solutions, in particular in the automotive industry [5]. A separate problem, discussed in the article, is the training of drivers and verification of their skills. The article proposes a methodology for evaluating the functionality of new solutions consisting, inter alia, in:

- Use of dynamic vehicle simulators for conducting research.
- Analysis of the correctness of performing maneuvers.
- Performing an EMG measurement (surface electromyography) to analyze muscle activity and possible fatigue).
- Drivers' surveys.

The implementation of automated driving as well as of new interface solutions requires adequate training and, before that, an appropriate choice adequate to the driver's needs. A consequence of the changes in interface design will be the need for a selection and certification procedure for new equipment. The increasing level of automation in transport also foresees additional training for drivers to verify their, sometimes completely new, skills necessary when using nonstandard solutions. The authors propose their vision of a procedure for implementing new steering system designs, which defines the necessary conditions that must be met before a given solution is implemented in individual vehicles.

2. Materials and Methods

The main aim of the experiment was an attempt to introduce correctness measures for maneuvers performed with the use of a multifunction steering wheel and to propose an evaluation of the biomechanical factors of the driver through the use of the electromyography (EMG) [6] measurement. The tests were carried out in a dynamic physical simulator by Aerospace Industries (Figure 1). Partial results of these studies have been published in previous papers [7–9].

Figure 1. Car driving simulator of ETC PZL Aerospace Industries. Source: ETC PZL [10].

The description of the experiment conducted by the authors includes:

- Research sample.
- Research object.
- Research methodology.
- Assessment of the functionality of the steering wheel.

2.1. Research Sample

At this stage of the research, the group of participants in the experiment was limited to able-bodied drivers. The participation of disabled drivers required the adaptation of the dynamic simulator to enable the driver's cabin on a wheelchair to enter and exit the simulator. The study of this group of people is planned in the next stage. The research group consisted of 30 men aged 20–23. All drivers had a driving license. The second important limitation was the assumed number of test runs. Each driver performed two road tests, driving the vehicle using three steering wheels consecutively. In each test, the driver made three runs: with the classic steering system, and the ECO steering wheel with two gear ratios. In order to simplify the descriptions, in the tests in question, the following symbols were introduced: NOR (max rotation $\pm 720^{\circ}$) standard steering wheel, ECO 180 steering wheel, for which the max rotation was $\pm$ 180°, and ECO 120, for which the max rotation was $\pm 120^{\circ}$. The maximum turn of each steering wheel corresponded to a steering angle of the front wheels of $\pm 35^{\circ}$. The participants of the experiment did not constitute a representative sample for adult Polish citizens. It was a rather simple sample. With such a small sample, striving for a high "representativeness" of the sample would be even a methodological error. The dispersion of the results would probably render the obtained results worthless (they would be characterized by a very large dispersion).

2.2. Research Object

The tests were carried out in a simulator but were related to a specific vehicle solution. The mentioned solution is an "eco-car" developed under the [11] project (this car is shown in Figure 2a). The design of the vehicle was based on the principles of universal design, i.e., it was assumed that it will be an electric vehicle intended for rental by drivers of various abilities, including drivers using active (hand-powered) wheelchairs. The implementation of the universal design postulated required the design of a number of solutions ensuring the required functionality of the eco-car, such as:

- Lowered car floor/air suspension.
- Front seat folding and sliding system.
- Active wheelchair fixing system in the case of a driver with reduced mobility.
- Handles facilitating entry of the wheelchair from the side.
- Fold-out side steps to facilitate the entry of the wheelchair.
- Driver interface–remote control + touch screen + multifunction steering wheel.
- Adjustable control panel/with steering wheel and driver interface/–horizontal and vertical adjustment.

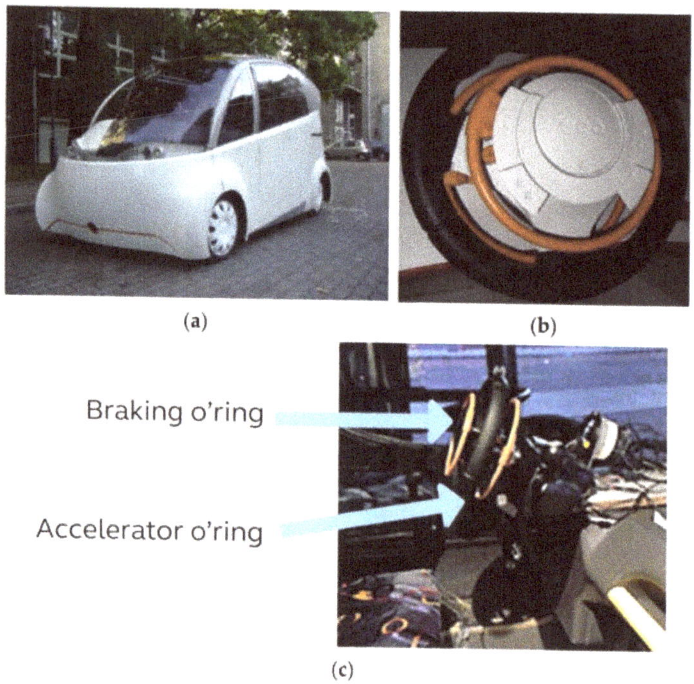

Figure 2. (**a**) Eco-car created within the [11] project, (**b**) multifunctional steering wheel the car is equipped with, (**c**) tested steering wheel installed in the simulator.

The car tested in the simulator was equipped with a designed multifunction steering wheel (see Figure 2b) to test its functionality. The versatility of the multifunctional steering wheel consists of the fact that, although it is aimed to be used by people with limited or complete lack of lower mobility, it does not exclude the possibility of steering the vehicle by able-bodied people. The tested eco-steering prototype is characterized by the fact that the main steering activities are performed only through the upper limbs, which is possible thanks to the steer-by-wire system of the designed eco-car. Two additional rings (hereinafter the "O-rings") that the steering wheel is equipped with are used for braking (smaller ring) and acceleration/maintaining a constant speed (larger ring). The orange color of both O-rings allows for their faster identification during use. In the case of this solution, the angle of rotation of the steering wheel was limited. People with paresis of the lower limbs often also have manual limitations of the upper limbs, in particular the hands, so in their case, limiting the range of steering wheel rotation may facilitate steering. Taking this into account, the steering gear ratio has been programmatically limited to a rotation angle of 180 or 120 degrees per full turn of the car's front wheels. In addition to O-rings, the steering wheel is equipped with function buttons located on its rim, which are operated without taking hands off the steering wheel. The vehicle with the said steering wheel was recreated in a dynamic simulator of motor vehicles.

2.3. Research Methodology

2.3.1. Simulator Research

The simulator tests concerned the analysis of two cases: the "route" type ride, shown in Figure 3a [7], and the "slalom" type ride, shown in Figure 3b [8].

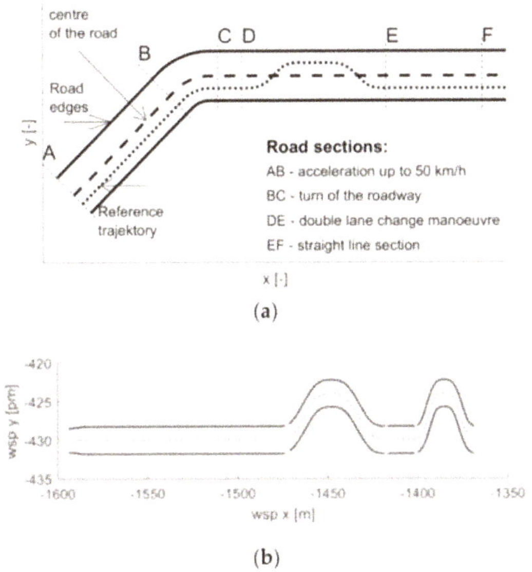

Figure 3. (**a**) "Route" type adapted from [7], (**b**) "slalom" type adapted from [8].

The "route" type ride consisted of straight sections connected by a circular curve (Figure 3a). The recording of the experiment began after reaching the speed of 50 km/h. In the second straight section, the driver had to maneuver around the obstacle (typical for the "moose" test). The "slalom" ride required a simultaneous change in speed and direction of driving, as shown in Figure 3b. In the first case, the driver's task was to follow the centerline of the lane; in the second, to follow the reference space-time line, i.e., the driver had to change speed in line with the reference space–time line (this required intensive use of acceleration and braking O-rings). Each of the tested drivers performed four runs: one training run and three runs with a steering wheel of the NOR type (the gear ratio in a typical car: two and a half turns of the steering wheel correspond to a turn of the wheels by approx. ±35 degrees), one with an ECO 180 steering wheel (the ratio of 180 degrees of the steering wheel rotation corresponds to approx. ±35 degrees of the wheels rotation), and the ECO 120 steering wheel (the ratio of 120 degrees of steering wheel rotation–approx. ±35 degrees of the wheels' rotation). The influence of the gear ratio on the "driving accuracy" was analyzed. The simulator screen is shown in Figure 4.

Figure 4. Simulator screen image: reference trajectory.

Defining the driving evaluation criteria was essential. Three criteria (measures) were used here: the first: the technical one, the second: the biomedical nature one based on the EMG measurement, and the third: based on the assessment of the experiment participants' survey.

The measurements for a maneuver's correctness were:

- Mean square deviations from the lane centerline (in relation to time or distance)

$$RMS1 = \left(\frac{1}{T}\int_0^T s^2(t)dt\right)^{1/2} or\ RMS2 = \left(\frac{1}{\Lambda}\int_0^\Lambda s^2(\lambda)d\lambda\right)^{1/2} \quad (2)$$

where T is the duration of the experiment, 5000 (s); Λ is route length; s is deviation from the lane centerline measured along the normal to the curve.

- Mean square deviations of the speed from the reference speed

$$RMS = \left(\frac{1}{T}\int_0^T deltaV^2(t)dt\right)^{1/2} \quad (3)$$

where T is the duration of the experiment, 5000 (s); $deltaV$ is deviation from the set speed.

The EMG analysis procedure was based on the signal obtained as a result of the applied surface electromyography.

2.3.2. Upper Limb Muscle Load Tests (EMG)

Noraxon [12] apparatus was used to measure the electromyography signal. The EMG signal was recorded with the use of surface electrodes. This signal, which gives a picture of muscle activity, is strongly dependent on internal (individual factors) and external [13] factors. So, it is not a signal with constant parameters (such as body temperature). Internal factors include conduction velocity of muscle fibers, thickness of the subcutaneous layer (fat content), and the proportions of a specific type of muscle fibers. External factors are related to the type of measuring electrodes used and the electrical resistance between the electrode and the skin. Moreover, the EMG signal depends on the geometric factors influencing the change in the length of the muscle (here: resulting from the change of body position and the technique of driving on the tested device). Among the factors influencing the EMG signal, the physical activity of the tested persons should also be specified. The analysis of the results of the experimental tests was carried out based on the most frequently used parameters of the EMG signal. These parameters are determined with respect to time (RMS amplitude) and frequency (MPF average frequency, determined following the Fourier transform).

The first type of analysis was used in the study because the EMG signal amplitude is an important indicator of muscle activity. The analysis of the EMG signal amplitude can be performed based on the root mean square (RMS) amplitude. The method of determining the RMS parameter is described in Relation (4). The RMS parameter is determined from signal fragments (so-called windows) of a specific length.

$$RMS = \sqrt{\frac{\sum_{i=1}^{n} X_i^2}{n}} \quad (4)$$

where n is the number of samples to be analyzed (window length); X_i is the value of the i-th sample.

The basic parameters of the experiment are defined below.

- Duration of the experiment: about 5000 (s).
- Sampling frequency: 1500 (Hz) (more than twice the Nyquist frequency).
- Transfer frequency of Noraxon apparatus equipped with an anti-aliasing filter: it is in the range of 10–500 (Hz).
- Duration of the experiment was divided into one-second time intervals, i.e., 1500 samples were obtained for one interval.
- In order to avoid errors, e.g., spectrum blur, the Hanning window was used for each interval.

The division into time intervals significantly influences the conditions of the analysis. In general, an EMG signal is not a stationary signal. It may show changes over the course of the experiment (e.g., a downward trend which is a measure of fatigue). It is assumed that the signal is locally stationary in the analyzed time intervals. The time intervals "overlap", which means that the obtained parameters (in this case, RMS) in adjacent intervals are strongly correlated with each other.

The subject of the analysis was six muscles from Table 1. The selection of muscles was made based on the literature analysis [14–21] as well as consultations with a physiotherapist [21].

Figure 5 shows the visualization of the muscles selected for the study.

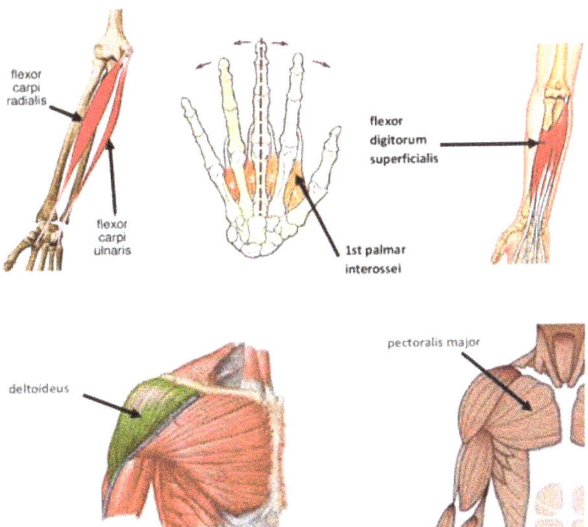

Figure 5. Visualization of the examined muscles/own work based on anatomical atlases.

Table 1. Analyzed muscles.

Muscles	Reference Level (μV) (the Highest RMS Value with a Time Window of 500 (ms))/X_{max}/
1. 1st palmar interossei ("interosseous"), MIE	1491
2. Anterior part of the deltoid muscle (deltoideus), NARA	867
3. Pectoralis major, PIER	271
4. Surface finger flexor/flexor digitorum superficialis ("finger extensor"), PROS	806
5. Wrist elbow flexor/flexor carpi ulnaris ("ulnaris flexor"), ULNA	308
6. Radial wrist flexor/flexor carpi radialis ("finger flexor")/flexor carpi radialis, ZGIN	756

EMG signals were measured using surface electrodes placed on the right limb of each participant, as shown in Figure 6. The signals were recorded while performing the assumed rides ("route", "slalom").

Figure 6. Location of electrodes for surface EMG measurement used in the experiment.

The obtained "raw" EMG results needed to be normalized. The most common method, known as MVC normalization, refers to the maximum voluntary contraction (Xmax; shown in Table 1) prior to starting the test measurements. Typically, MVC contractions are performed against static resistance. Ordinary (untrained) participants in the experiment may have difficulty reaching the MVC level as they are not used to such an effort. For patients who cannot, and should not, perform MVC due to damaged muscle structures, other methods of processing and analysis should be considered. Ultimately, the subject of the analysis in the experiment was the signal determined by the formula:

$$Xa = \frac{X_{measured} - X_{minimum}}{X_{maximum} - X_{minimum}} \quad (5)$$

where: $X_{minimum}$ represents EMG signal emitted at no load.

2.3.3. Evaluation of the Steering Wheel's Functionality

At this stage of research, it was necessary to consider how to verify the functionality of the universal steering wheel. In this case, the functionality of the steering wheel is understood as its influence on driving correctness and level of muscle load. This problem should be worked out using methods of statistical analysis based on the methods of experimental research. The properties of correctness of driving with a standard steering wheel with pedals were adopted as a reference level of the functionality. It was so because this type of steering wheel is a long-time, well-established solution. However, it needs to be remembered that the standard steering wheel does not make it possible to drive a vehicle for people with disabilities of the lower limbs. Thus, it will not find its application in the eco-car. Thus, the aim was to find an answer to the question of whether the newly designed ECO steering wheels (intended for use in the eco-car) are better or worse than the standard steering wheel, in terms of the analyzed test drive quality indicators. In order to achieve this goal, the following evaluation algorithm was adopted:

- Rides with the use of the defined types of steering wheels create the category of a priori event type (events determined before the measurement).
- The ride quality indicators characterizing the ride quality and muscle fatigue are dependent random events (events obtained from the experiment).
- The working hypothesis is the null hypothesis on the lack of dependence of the ride quality on the steering wheel type and the alternative hypothesis that such a relationship exists.
- Applying the confirmatory—confirming procedure to confirm the credibility of the null hypothesis (at the assumed level of significance) or the confirmatory—rejecting procedure to establish the credibility of the alternative hypothesis (at the assumed test power).
- Post hoc analysis (after the fact) to determine the value of the registered dependence (if any).

This algorithm, sketched very generally, required the use of a variety of statistical tests such as, for example, the Shapiro–Wilk test for normality, the Friedman's test of concordance (substitute for the ANOVA test for nonparametric hypotheses), and the Wilcoxon test for the relationship of paired observations. In the case of variables dependence confirmation, a simple post hoc analysis was performed, consisting of the classification of differences in the quality indicators of paired observations.

3. The Results of the Experiment

The following graphs show the results for one tested driver. As mentioned earlier, the sample included similar people, of similar age, skills, and one gender. This was reflected in relatively small differences in the obtained results, at the level of 10%. The presented results for a specific driver are therefore representative for the entire analyzed sample.

The h parameter is the RMS value calculated for time intervals (not for the entire duration of the experiment T). The physical interpretation of the h parameter is shown in Figure 7. Figures 8–10 show the course of the h parameter as a function of time for different types of steering wheels and the "route" type ride.

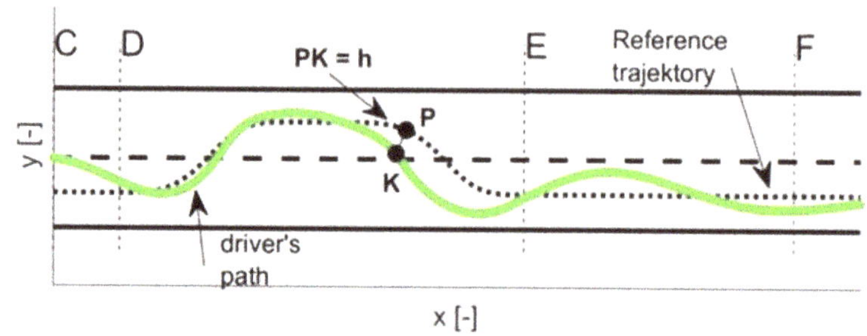

Figure 7. Interpretation of the h parameter. The parameter is measured in the radial direction to the reference trajectory and concerns the geometric center of the vehicle.

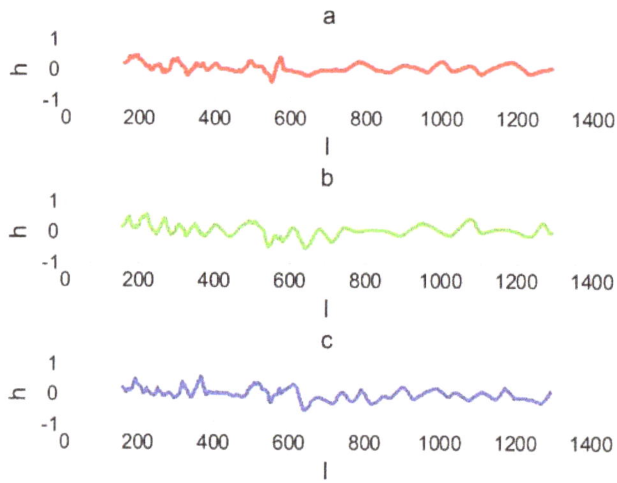

Figure 8. Values of the h parameter for the "route" type ride: (**a**) NOR steering wheel: red, (**b**) ECO 180 steering wheel: green, (**c**) ECO 120 steering wheel: blue.

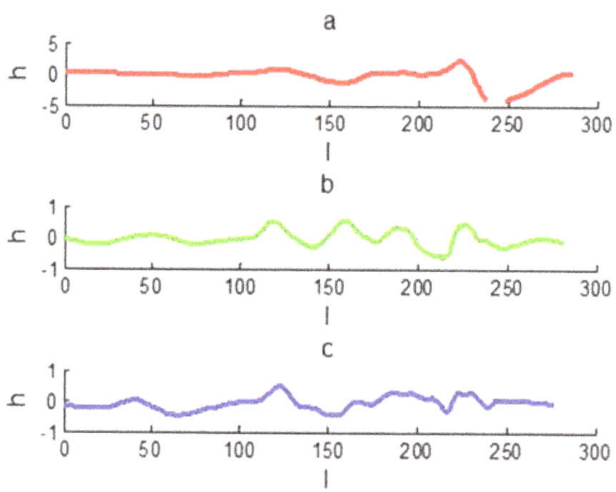

Figure 9. Values of the h parameter for the "slalom" ride: (**a**) NOR steering wheel: red, (**b**) ECO 180 steering wheel: green, (**c**) ECO 120 steering wheel: blue.

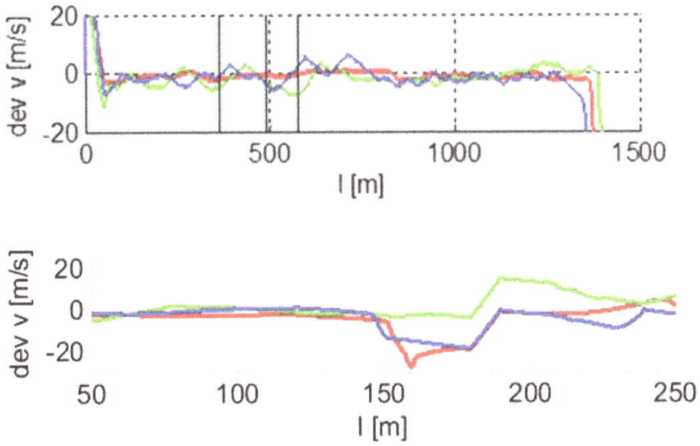

Figure 10. Graphs of deviations of the driving speed from the set value for three steering wheels and two types of rides: "route" and "slalom". The graphs are color-coded: red: NOR steering wheel, green: ECO 180 steering wheel, blue: ECO 120 steering wheel.

3.1. Analysis of the Correctness of Performing Maneuvers

The necessary condition for the ride to be considered correct (along with the maneuvers that everyone had to perform) is that the car does not leave the lane boundaries (one lane), and in the case of the moose test ("slalom" type ride), it does not leave the road (two lanes). This condition was met by all participants of the experiment. Then, the subject of detailed analysis was to determine the h deviation from the reference line.

The subject of comparative analyses in the group unit is the parameters of the characteristics of the runs made for three types of steering wheels: NOR, ECO 180, and ECO 120.

The charts relating to the "route" type ride show no particular differences. However, driving with a standard ratio steering wheel gives the smoothest ride. More sensitive steering wheels (ECO 180 and ECO 120) show a much greater "yawing" effect. The driver has more difficulty keeping the rectilinear motion. This effect is very clear and as expected. A similar effect also occurs during the "slalom" test ride. This situation is even more surprising since for more sensitive steering wheels it is reasonable to expect an increase in precision of performing a turning maneuver. However, also in this case, a deterioration of the values of the quality indicators describing the ability to follow the reference trajectory was observed. This effect may be related to the acquired habits of drivers resulting from the practice of driving cars with the standard steering wheel. It is difficult for the driver to adapt the correct angle of steering wheel rotation to the roadway's turn (curve) and speed of the car. This effect might not have been observed if drivers had received additional training with more responsive steering wheels. Figure 10 shows the course of the deviation of the speed from the set speed (changed while driving) for the "slalom" ride. For the "route" type ride, the entire ride was performed at a constant speed of V = 50 km/h. For the "slalom" ride, the driver had to follow the speed value displayed on the screen, varying in the range of 20–50 km/h.

While analyzing the obtained experimental waveforms of the lateral deviation from the track axis and deviations from the required driving speed of one of the drivers, a preliminary assessment was made. However, an unequivocal answer can only be obtained from the results of statistical analyses of experimental studies with the use of appropriate confirmatory—confirming or confirmatory—rejecting procedures described in our algorithm. So, having conducted the said analyses, it was found that eco steering wheels in the "route" test have poorer quality indicators than the standard steering wheel: they increase the yawing effect, the effect of uneven driving speed, and the effect of lateral distance from the designated drive line. However, no case of falling out of the permitted lane was registered for any of the steering wheels. Figure 11 shows the results of post hoc analysis for the value of the lateral deviation from the designated drive line. The differences in the number of paired observations are presented. The results are marked as follows: N: observations assigned to the standard steering wheel, 120: observations assigned to the ECO 120 steering wheel, 180: observations assigned to the ECO 180 steering wheel, 1: difference class N-120, 2: difference class N-180, 3: difference class 120–180. Dark bars indicate a positive result, and lighter ones a negative result. Since the higher value of the indicator constituting the observations determines the deterioration of the ride quality, the obtained result confirms a significantly lower functionality of the ECO steering wheels as compared to the standard steering wheel (for the indicator in question).

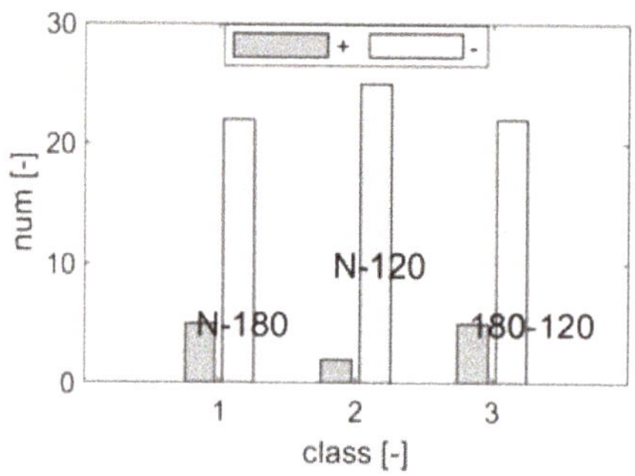

Figure 11. Graph of the number of observation classes for paired values of the effective transverse distance from the designated route line.

In the case of the "slalom" test, no statistically significant differences were found between the distributions of random variables of the paired observations. This result means that there is no reason to reject the hypothesis that the steering wheel functionality is identical.

3.2. Analysis of the Results Obtained from EMG Measurements

As before, the results for one tested driver are presented. Figure 12 shows the graphs of the EMG RMS signal for the NOR steering wheel, Figure 13 for the ECO 120 steering wheel, and Figure 14 for the ECO 180.

In the case of the standard NOR steering wheel, the most active muscles are PROS and ZGIN, i.e., the extensor and flexor of the fingers. Their activity is related to gripping the steering wheel.

The results presented in Figures 12–14 show, first of all, differences in the muscle load for particular types of steering wheels. For the standard steering wheel, for example, a smaller load on the first interosseous muscle can be observed, as compared to other types of steering wheels. In the case of ECO 120, we observe a statistically reliable increased activity of the flexor muscle of the fingers (in relation to the standard steering wheel). The greatest load on the wrist flexor is for the ECO 120 steering wheel, which is probably due to the so-called phenomenon of "yawing", i.e., the need to correct the vehicle movement due to excessive sensitivity of the steering system. A similar conclusion can be drawn for the pectoralis major muscle. The described conclusions were repeatable for approx. 80% of the analyzed sample of drivers. In the case of 20%, there were differences, which, however, did not have a constant trend. No physical fatigue of the muscles was observed in any of the cases, which would be reflected in the RMS signal trend. Perhaps it was also due to the relatively short duration of the experiment. The subjective assessment of the comfort of using a given type of steering wheel was assessed with the use of questionnaire surveys. The participants of the experiment had to indicate which steering wheel suited them best. The results were quite interesting: 57% of the participants indicated a steering wheel with a standard gear ratio, 23% a steering wheel with an ECO 120 gear ratio, and

20% a steering wheel with an ECO 180 gear ratio. In the case of people with disabilities, other results can be expected due to the frequent correlations between the impairment of the lower locomotor system and the partial impairment of the upper locomotor system, which prevent the use of a standard steering wheel (with pedals). Different preferences of the respondents in the choice of steering wheel prove that there is a need to adjust steering wheel parameters to the individual expectations of drivers.

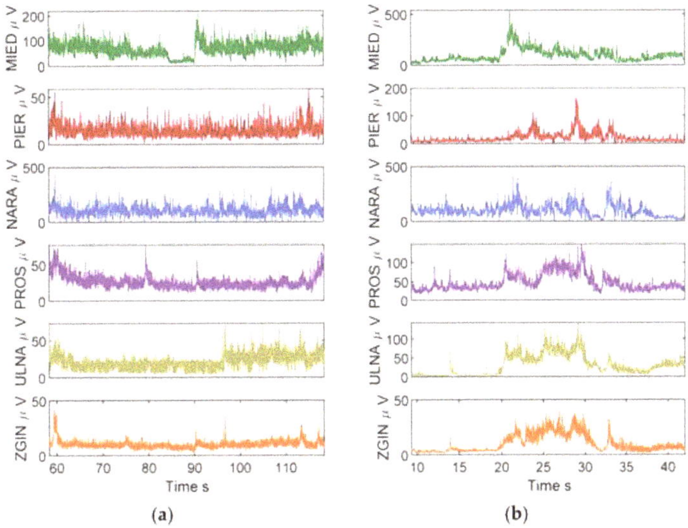

Figure 12. Graphs of the EMG RMS signal for the NOR steering wheel: (**a**) "route" type ride, (**b**) "slalom" type ride. The order of the graphs corresponds to the following muscles: MIE: "interosseous", PIER: "pectoralis", NARA: "deltoideus", PROS: "extensor", ULNA: "ulnaris", ZGIN: "flexor".

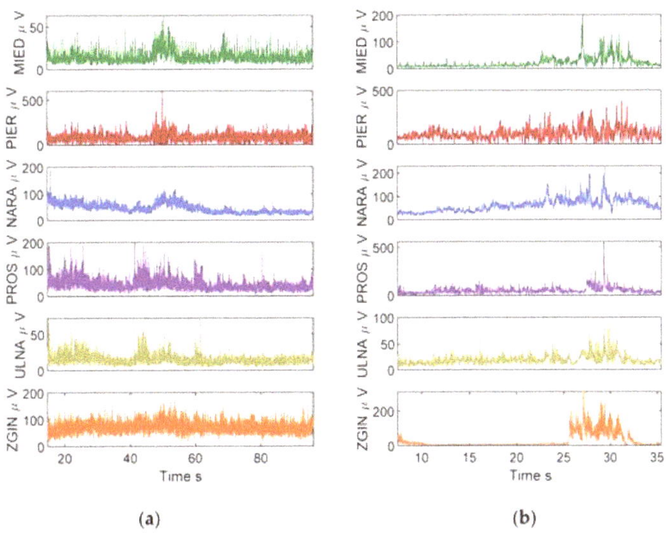

Figure 13. Graphs of the EMG RMS signal for the ECO180 steering wheel: (**a**) "route" type ride, (**b**) "slalom" type ride. The order of the graphs corresponds to the following muscles: MIE: "interosseous", PIER: "pectoralis", NARA: "deltoideus", PROS: "extensor", ULNA: "ulnaris", ZGIN: "flexor".

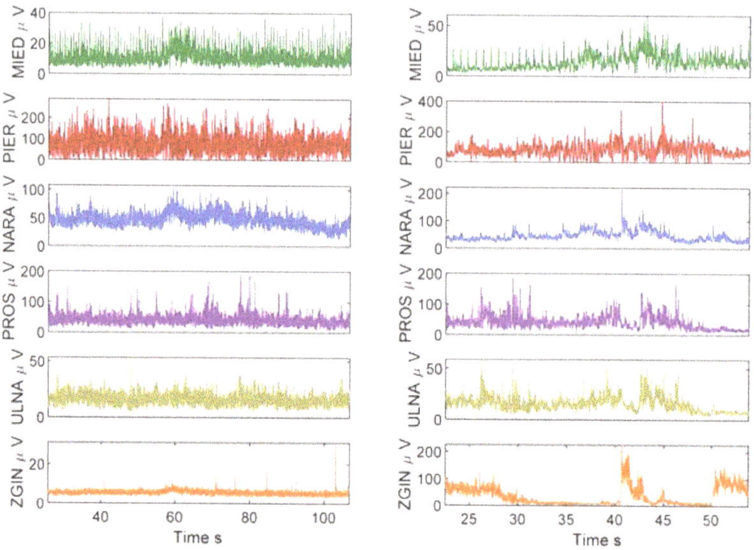

(a) (b)

Figure 14. Graphs of the EMG RMS signal for the ECO120 steering wheel: (**a**) "route" type ride, (**b**) "slalom" type ride. The order of the graphs corresponds to the following muscles: MIE: "interosseous", PIER: "pectoralis", NARA: "deltoideus", PROS: "extensor", ULNA: "ulnaris", ZGIN: "flexor".

However, a complete answer to the question about the possible change in muscle loads during the use of different types of steering wheels can only be obtained from a statistical analysis. In this case, its limitations should be described. First of all, the number of drivers for whom EMG tests were performed is much smaller than the number of drivers who performed simulator rides. This fact translates into a significant reduction in the test power in the case of the application of the confirmatory—rejecting procedure. In addition, the rides made were short (a few minutes), which turned out to be too short for the rides to generate significant muscle fatigue. For this reason, the collected data turned out to be insufficient to perform a full analysis. Only in the case of the study of the differences in the EMG RMS trend of the pectoral muscle (pectoralis) was there a statistically reliable result, confirming greater muscle effort when driving the car with ECO steering wheels. This situation is illustrated in Figure 15.

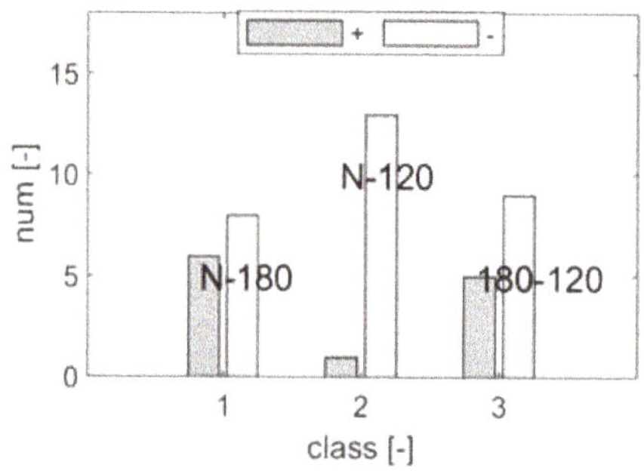

Figure 15. Number of observation classes of paired EMG RMS trend values of the pectoral muscle in the "route" test.

4. Critical Analysis of the Results

Based on the results of the experiment, it was possible to demonstrate that in terms of quality indicators, such as the criterion of deviation from the route line and "yawing", ECO steering wheels are less functional both in the "route" and "slalom" test. It was also possible to demonstrate (however, for only one muscle group) that the use of this steering wheel causes greater muscle fatigue. The authors of the experiment are aware of the need to conduct research that will enable them to broaden their knowledge on this subject, i.e.:

1. The ECO steering wheel is intended to be used by both able-bodied and disabled people. For this reason, its design differs from the standard solutions of the normal steering wheel. On the other hand, the research was carried out on a group of able-bodied drivers accustomed to using standard steering wheels. It should be repeated in the group of drivers with reduced mobility.
2. In addition, it should also be verified whether prior training could improve the experiment results for ECO steering wheels.
3. If the functionality of ECO steering wheels was to be compared with other solutions (e.g., joystick) in the future, it would be worth knowing which of the factors (removing the pedals or reducing the sensitivity of the steering gear system (gear ratio)) has a greater impact on reducing its functionality in relation to the normal steering wheel. However, the search for answers to the last question would require a significant extension of the scope of research by two new types of steering wheels: 1) normal gear without pedals, and 2) pedals and reduced gear ratio.

5. Nonstandard Steering System Designs and Related Certification and Driving License Examination Issues

Universal design and "custom design" necessarily require a re-evaluation of many issues related to the certification of new solutions and obtaining driving licenses. The problem becomes particularly complex when we consider that there is a breakthrough in road vehicle technology. It is primarily about introducing vehicles with various levels of automation (according to the SAE classification), including vehicles at various levels of autonomy. According to the authors, the issues analyzed in this paper can be of great help here. Below is an outline of the solutions based on three-layer diagrams. The basic assumptions for their creation can be formulated in the following points:

- New solutions, in a technical sense, should be subject to detailed reliability testing, e.g., steer by wire systems or technical solutions, such as a multifunction steering wheel.
- The proposal of new solutions should be subject to preliminary tests mainly through testing in dynamic simulators or biomechanical tests with the use of EMG, and in natural conditions on roads or closed tracks, to determine, inter alia, the scope of changes in the "feedback force" and gear ratios in the steering system.
- Trainings and examinations for persons with reduced mobility should, in some cases, be carried out in dynamic simulators.

Custom steering system designs increase the number of active users by adapting interface devices to the needs and limitations of drivers who are not fully able-bodied. This is possible if the vehicle is equipped with additional systems, such as drive by wire. The purpose of integrated design is, on the one hand, to provide a custom solution that is so intuitive and simple that it can be used by people with different constraints, while at the same time, there is some scope for individual customization. However, this requires close cooperation of designers with the vehicle end users in the design process. The next necessary stage is training with the use of the proposed interface solutions. Each driver would be given a diagram that defines the driver's rights. It would also be the same as the exam that was conducted. An example diagram is shown in Figure 16.

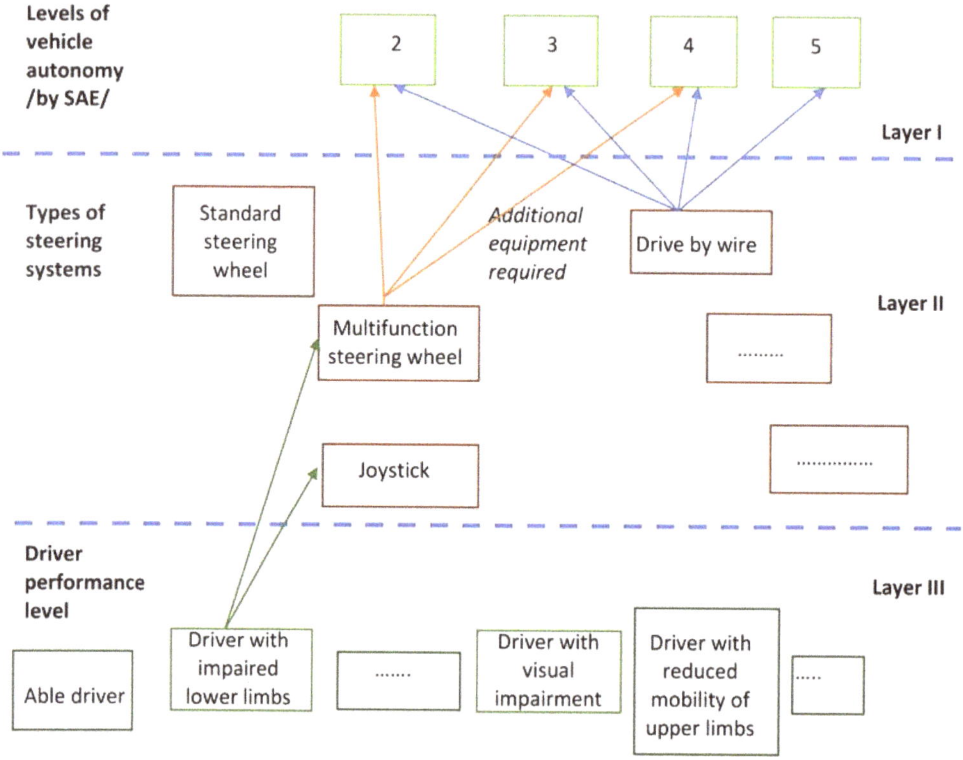

Figure 16. The concept of three-layer diagrams defining the driving license for a road vehicle.

Layer I in the case given in Figure 16 concerns the SAE classification. It can be seen how important it is to adopt the same (common on a global scale) classification. The work carried out, for example, under the UNECE World Forum for Harmonization of Vehicle Regulations (WP.29) on the regulation and harmonization of road regulations is of paramount importance here. Layer II, which will be developed along with the enhancement of new techniques, should include, among others, the type of steering system and additional equipment required. For example, in the case of steering the vehicle with brain waves, it is necessary to equip the cabin with screens that separate the driver's cabin from external electromagnetic interference. Layer III, on the other hand, contains a description of the disability. This list can be expanded and supervised by legal regulations. Figure 16 illustrates the following case:

The diagram concerns a person with a disability of the lower limbs (layer III). The driver can use a multifunction steering wheel or joystick. Additionally, the vehicle must be equipped with a drive by wire system. The driver is authorized to drive automation level 2, 3, or 4 according to the SAE classification. Level 5 means complete autonomy and will not have an interface (steering wheel) that would allow the driver to interfere with driving the car. The vehicle can only accept the trip's destination.

The above proposal should be considered as initially outlined. A number of additional problems should be taken into account when introducing it, e.g., how to eliminate the impact of the so-called simulator disease for research carried out on dynamic car simulators.

6. Conclusions

Providing people with various motor or sensory limitations with the possibility to drive vehicles will allow them to be more active in their social life and more independent.

The current practice consists of adapting the vehicle to the needs of a given person, using solutions available on the market that best suit the driver with a given disability. Nevertheless, the choice of the device does not always allow for efficient and easy steering. There is a need to search for solutions that are universal on the one hand and, on the other hand, still have a certain range of possibilities for individual adjustment. As part of the Eco-Mobility project, when designing a universal eco-car, it was equipped with a multifunctional steering wheel, which was anthropometrically verified, i.e., dimensioned to 95% of users. Only in experimental tests in a car simulator was it possible to evaluate the functionality of the new solution. The research was carried out on a sample of 30 men who did not have any motor limitations and who use standard steering wheels on a daily basis. The planned further tests will concern people using active wheelchairs, but as mentioned, it will require radical changes to the simulator's cabin in order to locate the wheelchair in it. Healthy drivers have specific driving habits and the short time spent on familiarizing themselves with the new steering wheel was not enough for them to feel its "sensitivity/responsiveness". In the case of people with disabilities of the lower limbs, there are usually also motor limitations of the upper limbs and, above all, the fingers. A smaller range of rotation of the ECO steering wheel and, at the same time, less movement of the limbs, in their case, may be an advantage and a positive feature of this solution.

The analysis of the test results showed the influence of the multifunction steering wheel parameters on the quality of maneuvers and the load on the driver's muscles. The obtained values of the adopted indicators confirmed a significantly lower functionality of ECO steering wheels in relation to the standard steering wheel in the "route" test. In the case of the "slalom" test, no statistically significant differences were found between the distributions of random variables of the paired observations. It can be assumed that the proposed solution of the multifunction steering wheel was not suited to the group of able-bodied drivers. Therefore, it seems that the thesis about the necessity to adapt the steering system to the individual characteristics of the driver is correct, in particular when it concerns people with reduced mobility. It can be obtained by integrating universal design and "custom design". This direction in the design process seems to be the only right one, as drivers with disabilities cannot use the standard steering system in most cases. On the other hand, the proposed solutions, apart from their reliability and safety, should minimize the stigmatization of users, i.e., they should be intended for all drivers with the possibility of minor improvements in special cases.

The development of the automotive industry, including the introduction of drive by wire technology, and the progressive automation of vehicles (partial autonomy) create completely new perspectives. In this case, the problem is also testing new solutions and granting permits to drive vehicles (in particular in relation to people with reduced mobility). Therefore, measures are taken to develop new rules for training drivers depending on the level of automation in the driven car. The three-layer diagrams proposed by the authors seem to be one of the interesting proposals defining the configuration of the steering system and the required skills obtained, for example, during training courses allowing driving at a specific level of automation and with an adapted steering system.

Author Contributions: Conceptualization, I.G., M.K. and W.C.; data curation, I.G.; formal analysis, M.K.; funding acquisition, W.C.; investigation, I.G., M.K. and W.C.; methodology, I.G., M.K. and W.C.; project administration, W.C.; resources, I.G. and W.C.; software, M.K. and W.C.; supervision, W.C.; validation, I.G., M.K. and W.C.; visualization, I.G. and M.K.; writing—original draft, I.G., M.K. and W.C.; writing—review and editing, I.G. and M.K. All authors have read and agreed to the published version of the manuscript.

Funding: European Regional Development Fund: Eco Mobility Project (UND-POIG.01.03.01-14-154).

Institutional Review Board Statement: The study was conducted according to the guidelines of the Declaration of Helsinki, and approved by the Ethics and Bioethics Committee Cardinal Stefan Wyszyński University in Warsaw (KEiB–6/2015).

Informed Consent Statement: Informed consent was obtained from all subjects involved in the study.

Data Availability Statement: The data presented in this study are available on request from the corresponding author.

Conflicts of Interest: The authors declare no conflict of interest.

References

1. SAE International Releases Updated Visual Chart for Its "Levels of Driving Automation" Standard for Self-Driving Vehicles. Available online: https://www.sae.org/news/press-room/2018/12/sae-international-releases-updated-visual-chart-for-its-%E2%80%9Clevels-of-driving-automation%E2%80%9D-standard-for-self-driving-vehicles (accessed on 16 August 2021).
2. Convention on the Rights of Persons with Disabilities. Available online: https://www.un.org/disabilities/documents/COP/crpd_csp_2017_4.pdf (accessed on 16 January 2021).
3. Polmans, K. Steer-By-Wire Systems—Safety, Comfort & Individuality, Auto Tech Review, Digital Edition. Available online: https://autotechreview.com/technology/steer-by-wire-systems-safety-comfort-individuality (accessed on 16 January 2021).
4. Mortazavizadeh, A.A.; Ghaderi, A.; Ebrahimi, M.; Hajian, M. Recent Developments in the Vehicle Steer-by-Wire System. *IEEE Trans. Transp. Electrif.* **2020**, *6*, 1226–1235. [CrossRef]
5. Fahami, H.; Zamzuri, H.; Mazlan, S. Development of Estimation Force Feedback Torque Control Algorithm for Driver Steering Feel in Vehicle Steer by Wire System: Hardware in the Loop. *Int. J. Veh. Technol.* **2015**, *2015*, 314597. [CrossRef]
6. Basmajian, J.V.; De Luca, C.J. *Muscles Alive Their Function Revealed by Electromyograph*; Williams Wilkins: Baltimore, MD, USA, 1985.
7. Choromański, W.; Grabarek, I.; Kozłowski, M. Research on an innovative multifunction steering wheel for individuals with reduced mobility. *Transp. Res. Part F Traffic Psychol. Behav.* **2019**, *2019*, 178–187. [CrossRef]
8. Kozłowski, M. Assessment of safety and ride quality based on comparative studies of a new type of universal steering wheel in 3D simulators. *Eksploat. I Niezawodn.* **2016**, *18*, 481–487. [CrossRef]
9. Grabarek, I.; Kozłowski, M.; Fiok, K.; Choromański, W. Zagadnienie oceny aktywności mięśni podczas kierowania nowym typem wielofunkcyjnej kierownicy w samochodzie elektrycznym. *Pr. Nauk. Politech. Warszawskiej. Transp.* **2017**, *119*, 137–148.
10. Aerospace Industries, sp. z o. o. Available online: https://www.ai.com.pl/ (accessed on 16 January 2021).
11. Eco-Mobilność. Available online: http://www.eco-mobilnosc.pw.edu.pl/ (accessed on 16 January 2021).
12. Konrad, P. The ABC of EMG. In *A Practical Introduction to Kinesiological Electromyography*; Noraxon Inc.: Scottsdale, AR, USA, 2005.
13. Balasubramanian, V.; Adalarasu, K. EMG-based analysis of change in muscle activity during simulated driving. *J. Bodyw. Mov. Ther.* **2007**, *11*, 151–158. [CrossRef]
14. Carrino, F.; Carrino, S.; Caon, M.; Angelini, L.; Khaled, O.A.; Mugellini, E. In-vehicle natural interaction based on electromyography. *Proc. Automot. UI* **2012**, *2012*, 12.
15. Nakamura, H.; David, A.; Mark, M. Is grip strength related to neuromuscular admittance during steering wheel control? In Proceedings of the 2011 IEEE International Conference on Systems, Man, and Cybernetics (SMC), Anchorage, AK, USA, 9–12 October 2011.
16. Park, J.; Shinsuk, P. Reduction of arm fatigue and discomfort using a novel steering wheel design. *Int. J. Precis. Eng. Manuf.* **2014**, *15*, 803–810. [CrossRef]
17. Pick, A.J.; Cole, D.J. Neuromuscular dynamics in the driver–vehicle system. *Veh. Syst. Dyn.* **2006**, *44*, 624–631. [CrossRef]
18. Abbink, D.A.; Mulder, M.; Van Paassen, M.M. Measurements of muscle use during steering wheel manipulation. In Proceedings of the 2011 IEEE International Conference on Systems, Man, and Cybernetics (SMC), Anchorage, AK, USA, 9–12 October 2011.
19. Hostens, I.; Herman, R. Assessment of muscle fatigue in low level monotonous task performance during car driving. *J. Electromyogr. Kinesiol.* **2005**, *15*, 266–274. [CrossRef] [PubMed]
20. Pick, A.J.; David, J.C. Measurement of driver steering torque using electromyography. *J. Dyn. Syst. Meas. Control* **2006**, *128*, 960–968. [CrossRef]
21. Wiszomirska, I. *Vivo Consultation of Muscle Selection and Electrode Placement*; Józef Piłsudski University of Physical Education in Warsaw: Warsaw, Poland, 2015.

integration of the accelerations recorded have been discussed as a method of obtaining information about the time histories of vehicle velocities and trajectory. Such a possibility has been shown, with potential difficulties having been emphasised. This possibility confirms to some extent the conclusions formulated by the authors of this paper in [27,28]. This paper presents the methods of determining the vehicle velocities and trajectory from EDR records on the one hand and, on the other hand, computational examples showing how some simplifications in EDRs' design can affect the accuracy of the time curves being reconstructed. In these terms, the research works reported here are a direct continuation of those described previously in [27–30], but updated mathematical models of the test specimens adopted have been used in this case.

The fact that EDRs of various types are commonly used in present-day motor vehicles and the possibility of using their records for the reconstruction of vehicle motion and, on the other hand, scanty literature addressing these issues fully justify the undertaking of research in this field.

2. Materials and Methods

In the research presented, simulation calculational methods were used, where an experimentally verified model of motor vehicle dynamics and models of EDR records and of record processing algorithms were employed. A block diagram of the method has been shown in Figure 1. In the method adopted, the motion simulation results are taken as accurate (reference) values, and EDR records are simulated on these grounds (EDR model). Based on these records, time histories of the parameters describing the vehicle motion (velocity, position) are reconstructed with data processing algorithms (denoted by DPM, i.e., Data Processing Methods) being used. The difference between the reconstructed and accurate values is a measure of the reconstruction error. Individual elements of the method (vehicle dynamics model, EDR records model, and EDR record processing method, i.e., DPM) will be presented in subsequent subsections).

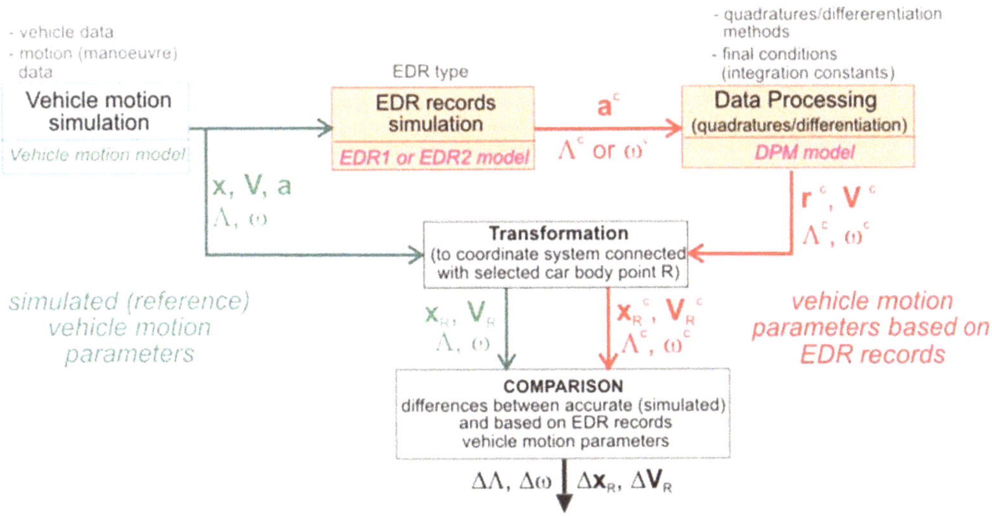

Figure 1. Method of estimating the accuracy of reconstruction of vehicle motion, based on EDR records (a, V, ω, r, Λ—component vectors of: acceleration, velocity, angular velocity, position, angles, respectively).

2.1. Vehicle Dynamics Model

The tests were carried out for a KIA Ceed SW motor car, provided with a McPherson strut front suspension system with an antiroll bar. The suspension system (together with steering system components) is fastened to a subframe, which in turn is fixed to the car body solid. The left and right rear wheel suspension systems are independent of each other (apart from being coupled together by an antiroll bar). Each of them consists of a spring element (helical spring), shock absorber, two transverse arms, trailing arm, and lateral control rod. On both sides, the wheel suspension systems are fixed in a part to a steel drawpiece, which plays the role of a subframe.

A simulation model of this vehicle has been presented in publications [31–33] (Figure 2). It consists of 9 mass elements: vehicle body solid (treated as a rigid body), 4 material particles O_1, O_2, O_3, and O_4, where the vehicle's "unsprung masses" have been concentrated (including road wheels in their translational motion), and 4 solids representing the rotating road wheels (exclusively in their rotational motion).

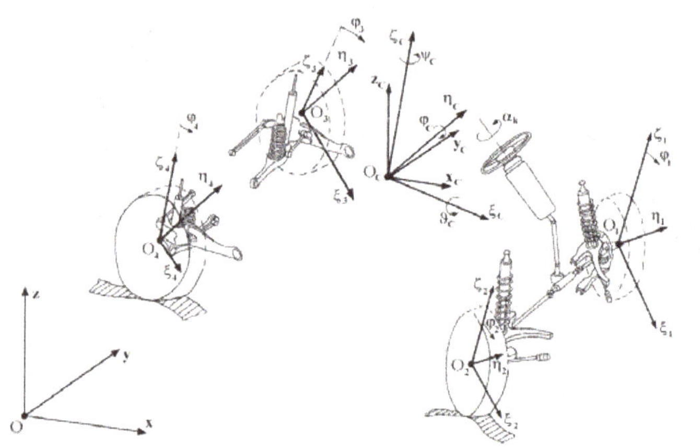

Figure 2. Physical model of a two-axle motor vehicle with independent front and rear wheel suspension systems, together with the coordinate systems adopted.

The following coordinate systems have been adopted:

- Oxyz—an inertial system, fixed to the road, with the Ox and Oy axes being horizontal and the vertical axis Oz pointing upwards;
- $O_Cx_Cy_Cz_C$—a moving non-inertial system, with its axes being respectively parallel to axes Ox, Oy, and Oz and with its origin situated at the centre of mass of the vehicle body solid Oc;
- moving coordinate systems fixed to the rigid bodies of the model, i.e., body solid ($O_C\xi_C\eta_C\zeta_C$) and four road wheels ($O_1\xi_1\eta_1\zeta_1$, $O_2\xi_2\eta_2\zeta_2$, $O_3\xi_3\eta_3\zeta_3$, $O_4\xi_4\eta_4\zeta_4$);
- Auxiliary systems, facilitating the defining of transformation matrices.

To describe the translational motion of the solids and material particles of the model, the positions of the centres of mass (O_C, O_1, O_2, O_3, O_4) of the said solids are used.

The axes $O_i\xi_i$, $O_i\eta_i$, $O_i\zeta_i$ (i = C, 1, 2, 3, 4) are treated as the principal central axes of inertia of the corresponding rigid bodies.

The rotation of the vehicle body solid about the fixed point O_C has been described with the use of "aircraft angles", also referred to as "quasi-Euler angles" [34–38]:

- Yaw angle ψ_C (rotation about the $O_C\zeta_C$ axis);
- Pitch angle φ_C (rotation about the $O_C\eta_C$ axis);
- Roll angle ϑ_C (rotation about the $O_C\xi_C$ axis).

The sequence of rotations has been adopted as identical to their listing order. The axes of individual rotations are treated as the principal central axes of inertia of the vehicle body solid.

The equations of motion have been derived with Lagrange equations of the second kind having been used. Prior to this, the following 14 generalised coordinates were adopted:

- $q_1 = x_{OC}$, $q_2 = y_{OC}$, $q_3 = z_{OC}$—coordinates defining the position of the centre of mass of the vehicle body solid (O_C) in the inertial reference system Oxyz;
- $q_4 = \psi_C$, $q_5 = \varphi_C$, $q_6 = \vartheta_C$—coordinates describing the rotation of the vehicle body solid about its centre of mass O_C; these are the quasi-Euler (aircraft) angles, i.e., yaw angle, pitch angle, and roll angle, respectively;
- $q_7 = \zeta_{CO1}$, $q_8 = \zeta_{CO2}$, $q_9 = \zeta_{CO3}$, $q_{10} = \zeta_{CO4}$—coordinates describing the motion of points O_1, O_2, O_3, O_4 relative to the vehicle body solid in the direction of axis $O_C\zeta_C$ of the $O_C\xi_C\eta_C\zeta_C$ coordinate system; to these points, the "unsprung masses" of the suspension system are reduced;
- $q_{11} = \phi_1$, $q_{12} = \phi_2$, $q_{13} = \phi_3$, $q_{14} = \phi_4$—angles of rotation of road wheels (front left and right and rear left and right wheel, respectively).

In the model, the steering system flexibility and directional stability of road wheels have been taken into account. The tyre-road contact forces and moments have been described with the HSRI-UMTRI model [39,40] having been used, extended by adding the IPG-Tire model of transient states of tyres [41] in the form as adopted in paper [37].

The model has been experimentally verified as satisfactorily applicable to typical vehicle motion tests recommended by ISO or ECE, including the calculational part of the steady-state circular driving tests (ISO 4138, [42]), tests with step input applied to the steering wheel (ISO 7401, [43]), straight-line braking tests (UN ECE Regulation No 13, [44]), with the ABS being inactive.

2.2. Model of EDR Records

The EDR design solutions that can be met in reality differ from each other in the number of the recorded components of the vectors that define the vehicle position. In these terms, two characteristic EDR types have been distinguished (Figure 3). The devices of the first type (denoted herein by EDR1) are similar (within the scope as considered here) to flight recorders: they record three acceleration components, i.e., longitudinal one a_w, lateral one a_p, and "vertical" one a_ζ, and three angles, i.e., yaw angle ψ_C, pitch angle φ_C, and roll angle ϑ_C, or the corresponding angular velocities. The devices of the second type (denoted herein by EDR2) can be frequently met in motor vehicles and are a simplified version in comparison with the aircraft solution. The vehicle motion is treated as a two-dimensional one, with the parameters recorded being two acceleration components, i.e., longitudinal one a_w and lateral one a_p, and one angle, i.e., yaw angle ψ_C, or its time derivative, i.e., yaw velocity.

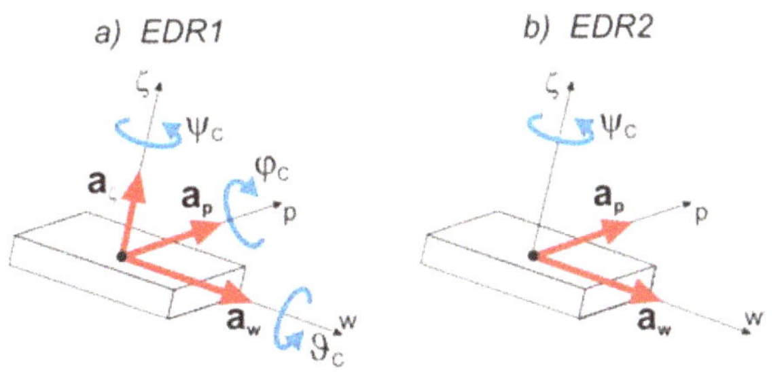

Figure 3. Two EDR types: (**a**) EDR1 (6 parameters); (**b**) EDR2 (3 parameters).

The assumptions adopted in the mathematical model of acceleration sensors' readings and the angular quantities have been described in dissertation [45] (and in [27] as well). An updated description can be found in [29,30]. Here, the main features of the description and fragments of its formal mathematical approach have been presented. The kinematic relations have been formulated by taking, as a basis, the model of kinematics shown in

Figure 4. The vehicle motion is treated as a composition of translational motion of the centre O_C of vehicle mass and of the rotation of vehicle body solid around point O_C. Thus, 6 degrees of freedom of the vehicle body solid (3 translations and 3 rotations) are taken into account. One more reference system has also been added to the coordinate systems mentioned above (see Section 2.1):

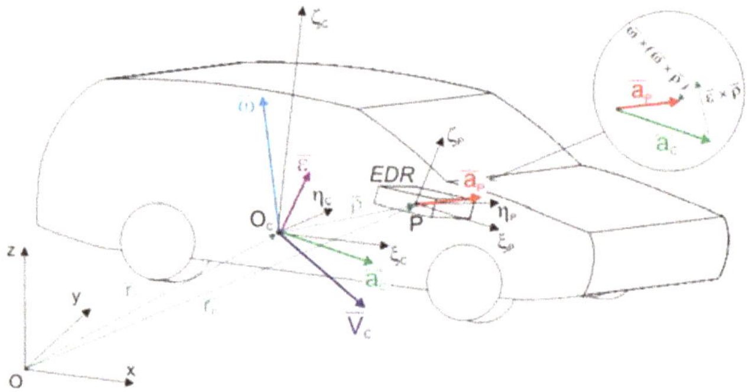

Figure 4. Model if the kinematics of motion of a vehicle provided with an EDR, where the set of sensors is placed at point P; notation: r—position; V—velocity; a—acceleration in translational motion; ω—angular velocity; ε—angular acceleration.

$P\xi_P\eta_P\zeta_P$—a moving one, fixed to the EDR; the $P\xi_P$, $P\eta_P$, $P\zeta_P$ axes define the directions of operation of the sensors of longitudinal and lateral acceleration and the yaw angle.

The vector symbols have the following meanings:

$\bar{r}_C \equiv \boldsymbol{x}_C = [x_C, y_C, y_C]^T$—Position of O_C in the inertial system Oxyz;

$\bar{r}_P \equiv \boldsymbol{x}_P = [x_P, y_P, y_P]^T$—Position of point P in the inertial system Oxyz;

$\bar{\rho} \equiv \boldsymbol{\rho} = [\xi_{CP}, \eta_{CP}, \zeta_{CP}]^T$—Position of point P in the $O_C\xi_C\eta_C\zeta_C$ system;

$\bar{\omega} \equiv \boldsymbol{\omega} = \left[\dot{\psi}_C, \dot{\phi}_C, \dot{\vartheta}_C\right]^T$—Angular velocity;

$\bar{\varepsilon} \equiv \boldsymbol{\varepsilon} = \left[\ddot{\psi}_C, \ddot{\phi}_C, \ddot{\vartheta}_C\right]^T$—Angular acceleration;

$\bar{V}_C \equiv \dot{\boldsymbol{x}}_C = [\dot{x}_C, \dot{y}_C, \dot{z}_C]^T$—Velocity of point O_C;

$\bar{a}_C \equiv \ddot{\boldsymbol{x}}_C = [\ddot{x}_C, \ddot{y}_C, \ddot{z}_C]^T$—Acceleration of point O_C;

$\bar{a}_P \equiv \ddot{\boldsymbol{x}}_P = [\ddot{x}_P, \ddot{y}_P, \ddot{z}_P]^T$—Acceleration of point P.

The kinematics of point P relative to the inertial system Oxyz is described as follows in matrix notation (point P is motionless in relation to the vehicle):

$$\text{position}: \ \boldsymbol{x}_P = \boldsymbol{x}_C + \boldsymbol{A} \cdot \boldsymbol{\rho} \tag{1}$$

$$\text{velocity}: \ \dot{\boldsymbol{x}}_P = \dot{\boldsymbol{x}}_C + \dot{\boldsymbol{A}} \cdot \boldsymbol{\rho} \tag{2}$$

$$\text{acceleration}: \ \ddot{\boldsymbol{x}}_P = \ddot{\boldsymbol{x}}_C + \ddot{\boldsymbol{A}} \cdot \boldsymbol{\rho} \tag{3}$$

where the matrix of rotation A (from the $O_C\xi_C\eta_C\zeta_C$ system to the Oxyz system) has the form (4):

$$A = \begin{bmatrix} \cos\psi_C \cdot \cos\phi_C & \cos\psi_C \cdot \sin\phi_C \cdot \sin\vartheta_C - \sin\psi_C \cdot \cos\vartheta_C & \cos\psi_C \cdot \sin\phi_C \cdot \cos\vartheta_C + \sin\psi_C \cdot \sin\vartheta_C \\ \sin\psi_C \cdot \cos\phi_C & \sin\psi_C \cdot \sin\phi_C \cdot \sin\vartheta_C + \cos\psi_C \cdot \cos\vartheta_C & \sin\psi_C \cdot \sin\phi_C \cdot \cos\vartheta_C - \cos\psi_C \cdot \sin\vartheta_C \\ -\sin\phi_C & \cos\phi_C \cdot \sin\vartheta_C & \cos\phi_C \cdot \cos\vartheta_C \end{bmatrix} \tag{4}$$

The positions of the sensors and their axes are defined by the positions of point P and coordinate axes of the $P\xi_P\eta_P\zeta_P$ system. The position of this system relative to the $O_C\xi_C\eta_C\zeta_C$ one is defined by its translation by vector $\bar{\rho}$ and by its rotation described by matrix $\boldsymbol{C}$ (see (5)). The rotations have been adopted in a way similar to that adopted previously, but in reverse order: ϑ_E (rotation about the longitudinal axis ξ_P), ϕ_E (rotation about the lateral axis η_P), and ψ_E (rotation about the "vertical" axis ζ_P)—see Figure 5b. Such a sequence of rotations is convenient because of the easy levelling of the sensor axes (oriented in relation to the vehicle).

$$C = \begin{bmatrix} \cos\psi_E \cdot \cos\phi_E & -\sin\psi_E \cdot \cos\phi_E & \sin\phi_E \\ \cos\psi_E \cdot \sin\phi_E \cdot \sin\vartheta_E + \sin\psi_E \cdot \cos\vartheta_E & -\sin\psi_E \cdot \sin\phi_E \cdot \sin\vartheta_E + \cos\psi_E \cdot \cos\vartheta_E & -\cos\phi_E \cdot \sin\vartheta_E \\ -\cos\psi_E \cdot \sin\phi_E \cdot \cos\vartheta_E + \sin\psi_E \cdot \sin\vartheta_E & \sin\psi_E \cdot \sin\phi_E \cdot \cos\vartheta_E + \cos\psi_E \cdot \sin\vartheta_E & \cos\phi_E \cdot \cos\vartheta_E \end{bmatrix} \quad (5)$$

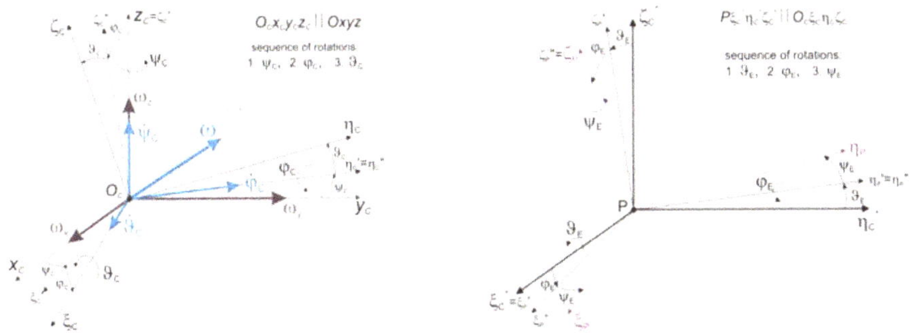

(a) (b)

Figure 5. Angular situation of the coordinate systems: (**a**) $O_C\xi_C\eta_C\zeta_C$ in relation to $Oxyz$; (**b**) $P\xi_P\eta_P\zeta_P$ in relation to $O_C\xi_C\eta_C\zeta_C$.

The basic component of the EDR model is the model of acceleration records. In this model, a system of transducer axes perpendicular to each other has been adopted, with the transducers being situated, as previously mentioned, freely in relation to the vehicle body. The fact has been taken into account that the acceleration transducers measure not only the real component of the acceleration of the EDR fixing point but also the pertinent components of the gravitational acceleration. Being inertia sensors, they measure a value proportional to the sum of the inertial and gravity force components acting along the sensor axis. The component resulting from the gravity force may be treated as the indication error.

Assuming a general formula for the sensor readings in the form (6):

$$\boldsymbol{a}^c = \left[a^c_w, a^c_p, a^c_z\right]^T \quad (6)$$

We may write that vector $\boldsymbol{a}^c$ is equal to:

$$\boldsymbol{a}^c = \boldsymbol{a}_{Pc} - \boldsymbol{g}_c \quad (7)$$

where:

$$\boldsymbol{a}_{Pc} = \left[a_{P\xi_P}, a_{P\eta_P}, a_{P\zeta_P}\right]^T = \boldsymbol{C}^{-1} \cdot \boldsymbol{a}_P = \boldsymbol{C}^{-1} \cdot \boldsymbol{A}^{-1} \cdot \ddot{\boldsymbol{x}}_P \quad (8)$$

$$\boldsymbol{g}_c = \left[g_{\xi_P}, g_{\eta_P}, g_{\zeta_P}\right]^T = \boldsymbol{C}^{-1} \cdot \boldsymbol{g}_\xi = \boldsymbol{C}^{-1} \cdot \boldsymbol{A}^{-1} \cdot \boldsymbol{g} \quad (9)$$

and $\boldsymbol{g} = [0, 0, -g]^T$—gravitational acceleration vector ($g = 9.81$ m/s^2).

Vectors $\boldsymbol{a}_{Pc}$ and $\boldsymbol{g}_c$ represent the acceleration of point P and gravitational acceleration, expressed in the reference system $P\xi_P\eta_P\zeta_P$ (in the $O_C\xi_C\eta_C\zeta_C$ system attached to the vehicle, the same acceleration vectors are denoted by $\boldsymbol{a}_P$ and $\boldsymbol{g}_\xi$, respectively). The acceleration sensors' readings have been graphically illustrated in Figure 6 (where the EDR2 unit has been used as an example for simplification).

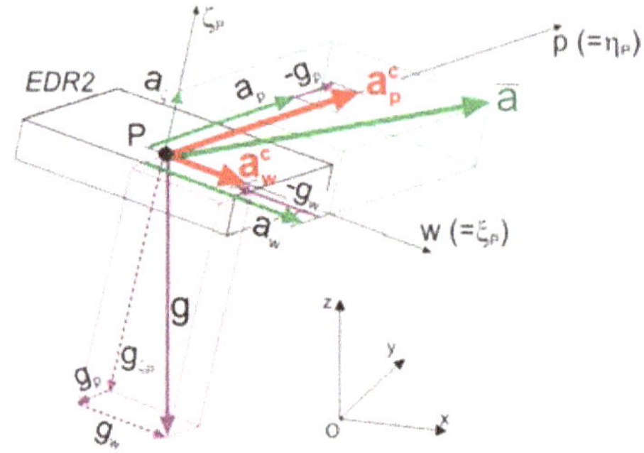

Figure 6. Graphic interpretation of the longitudinal and lateral acceleration readings ($a_w{}^c$ and $a_p{}^c$, respectively).

In the description presented, the sensors have been assumed to be free of intrinsic errors. A model applicable to the readings describing angular position (angles or angular velocities) has been developed as well. Its formal description can be found in [27,45].

2.3. Reconstruction of Motion (DPM Model)

The reconstruction of motion consists of appropriate (for the specific recorder configuration, e.g., EDR1 or EDR2) processing of the recorder's output signals. A schematic diagram of the data processing procedure has been shown in Figure 7. The procedure is based on the integration of acceleration records transformed to the inertial system attached to the road. The numerical integration ([46,47]) is carried out from a definite final instant corresponding to a known vehicle position and velocity (e.g., post-accident vehicle position specified by witnesses). If possible, transducer readings are adjusted by adding the values of gravitational acceleration components ($\Delta\mathbf{a}^c = -\mathbf{g}_c$, see Equation (7)). The results of successive quadrature operations define, as appropriate, the velocities and positions of point P of fixing the EDR, expressed in the inertial system Oxyz. Afterwards, they can be transformed in order to obtain time histories of the velocities and trajectories of any point in the vehicle body.

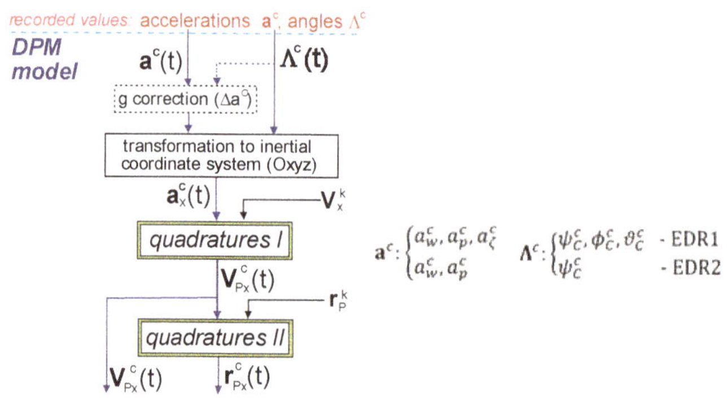

Figure 7. Schematic diagram of the processing of data recorded by the EDR: $\mathbf{a}_x$, $\mathbf{V}_x$, $\mathbf{r}_x$,—vectors of acceleration, velocity, and position, respectively, in the inertial reference system Oxyz; $\mathbf{V}^k$, $\mathbf{r}^k$—velocity and position at the final instant; "P" in the subscript indicates the values for point P of fixing the EDR; $\psi_C{}^c$, $\varphi_C{}^c$, $\vartheta_C{}^c$—values of angles ψ_C, φ_C, ϑ_C, determined from EDR records.

Figure 7 shows a schematic diagram of the processing of data produced by a recorder of angle values. In such a case, the angular velocities are obtained by numerical differentiation of the angles. If, on the contrary, angular velocities are recorded then the angle values are obtained by differentiation of time histories of the velocities.

3. Results

The calculations were carried out for data of a medium-class compact passenger car KIA Ceed SW, i.e., mass 1870 kg, wheelbase 2.65 m, and distance between the centre of vehicle mass and the front axle 1134 m, with simulating vehicle motion on dry, level, and ideally even road surface with good tyre-road adhesion (equivalent to asphalt concrete). A specific vehicle motion (manoeuvre) was forced by applying control signals representing time histories of the steering wheel angle and brake pedal effort (in the case of braking). For "constant sped" curvilinear motion tests, a driver model (driving torque controller) was activated to maintain constant vehicle speed.

Four typical vehicle motion cases were examined:

- Straight-line braking;
- Lane-change manoeuvre;
- Entering into a turn;
- Driving in a roundabout (circular motion).

In each of the cases, various intensities of the manoeuvres (measured by the level of longitudinal or lateral acceleration) were applied.

Both EDR types mentioned in Section 2.2 were taken into consideration: EDR1 (6 parameters of the motion) and EDR2 (3 parameters of the motion). For the analysis, an option was adopted where angles (instead of angular velocities) were used to define the angular position of the vehicle body solid. Moreover, to avoid the introduction of additional disturbing factors that might affect the calculation results, an assumption was made that the EDR sensors were placed at the centre of vehicle mass (see Figure 4, $P = O_C$) and the motion of the centre of the vehicle mass was analysed (see Figure 1, $R = P = O_C$). In addition to the above, an assumption was also made that the sensors had been calibrated for the vehicle standing on a horizontal road and loaded as in the tests. The frequency of recording the parameters under analysis was assumed as 20 Hz.

3.1. Straight-Line Braking

Figures 8 and 9 show the analysis results obtained for the case of braking the vehicle from an initial speed of 90 km/h with a deceleration of about 6 m/s^2 (this is approximately equal to the absolute value of the longitudinal acceleration of the centre of the vehicle mass). The vehicle braking was caused by applying a force to the brake pedal, where the driver reaction time and the reaction of braking system components were reflected in the time history of the brake pedal force. Figure 8a shows the actual (simulated, considered as "accurate") value of longitudinal acceleration $a_{P\xi}$ and the value recorded by the EDR ($a_W{}^c$). The difference between them (denoted by $\varphi a_W{}^c$) resulted from variations in the pitch angle ϕ_1 (see Figure 8b).

Figure 9 shows the reconstruction results obtained by using recorders of both types (EDR1 and EDR2). The time histories of vehicle velocity and position were reconstructed "backwards", i.e., from the known final vehicle position and velocity to the initial instant of the simulation of the vehicle motion. The data presented in Figure 9a,b concern vehicle velocity and position, respectively. The differences that can be seen at the beginning of the velocity and position curves illustrate the reconstruction errors, i.e., errors of determining the initial velocity (ΔV_0) and distance travelled (ΔS).

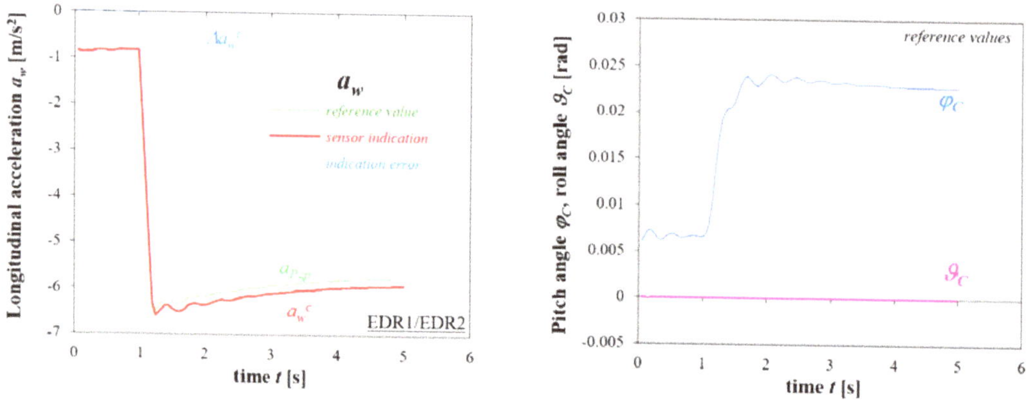

(a) (b)

Figure 8. Straight-line braking. Initial vehicle velocity 90 km/h, deceleration level 6 m/s^2. Time histories of: (**a**) longitudinal acceleration (accurate values and EDR sensor indications); (**b**) car body pitch and roll angles (accurate values).

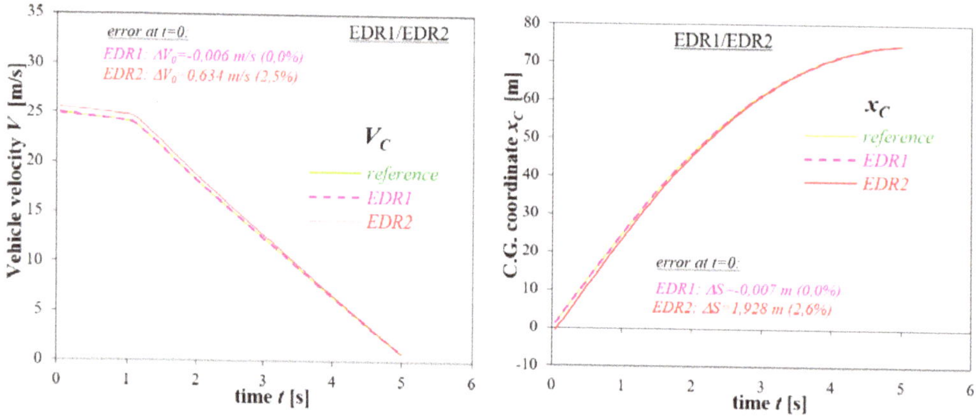

(a) (b)

Figure 9. Straight-line braking. Initial vehicle velocity 90 km/h, deceleration level 6 m/s^2. Reconstruction of time histories of: (**a**) vehicle velocity; (**b**) distance travelled by the vehicle. Reconstruction based on EDR1 and EDR2 records.

Very good conformity can be seen between the reconstructed and accurate values for the EDR1 device. This is thanks to the fact that in this case, the accelerometer indication error $\Delta a_w{}^c$ is automatically corrected in the EDR1 calculation algorithm, because the vehicle pitch angle, which causes this error, is known and, in consequence, the value of this error can be determined. In the case of the EDR2 device, the pitch angle is not known and the acceleration signals being integrated are burdened with the error arising from that. In the case under consideration, this translates into an overestimation of the values of initial velocity and distance travelled by the vehicle.

Analysis results were obtained for more cases, differing from each other in the pre-set initial vehicle velocity (50, 90, and 140 km/h) and braking intensity (deceleration level of about 2, 4, 6, and 8 m/s^2) have been summarised in Table 1. Additionally, the errors in estimating the initial velocity and stopping distance of the vehicle have been directly shown in the absolute and relative form in Figures 10 and 11.

Table 1. Straight-line braking. Summary of the accurate and reconstructed values of initial velocity and stopping distance of the vehicle for different preset levels of the initial velocity (50, 90, and 140 km/h) and braking intensity (deceleration level of about 8, 6, 4, and 2 m/s^2).

Initial Velocity V_0 [km/h]	Deceleration Level [m/s^2]	Initial Velocity V_0 [m/s]			Stopping Distance S_z [m]		
		Reference	EDR1	EDR2	Reference	EDR1	EDR2
50	8	13.871	13.869	14.235	25.691	25.686	26.384
	6		13.869	14.208	29.325	29.32	30.053
	4		13.871	14.163	36.512	36.512	37.289
	2		13.872	14.037	56.581	56.581	57.242
90	8	24.957	24.959	25.646	62.098	62.1	63.868
	6		24.951	25.591	73.367	73.36	75.295
	4		24.955	25.489	94.944	94.942	97.016
	2		24.955	25.248	171.381	171.384	173.411
140	8	38.797	38.795	39.855	129.788	129.787	133.499
	6		38.8	39.767	155.278	155.281	159.313
	4		38.799	39.587	200.958	200.965	205.152
	2		38.803	39.234	370.088	370.113	374.406

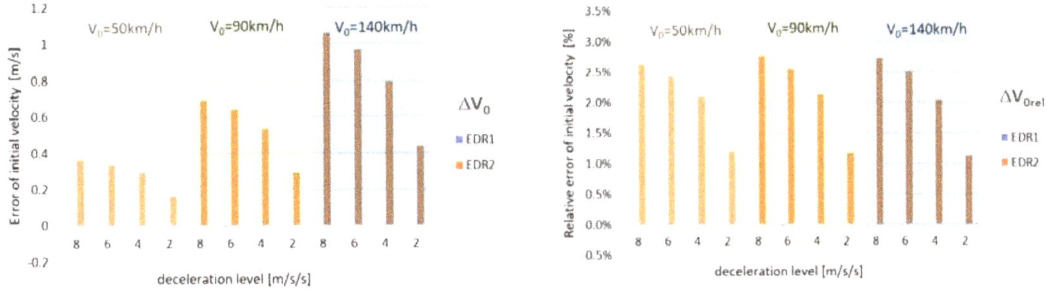

(a) (b)

Figure 10. Straight-line braking. Errors of the reconstruction of the initial velocity of the vehicle motion: (**a**) in the absolute form; (**b**) in the relative form.

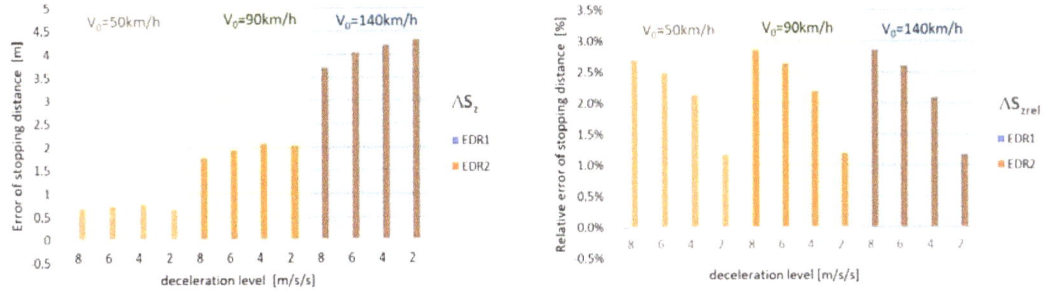

(a) (b)

Figure 11. Straight-line braking. Errors of the reconstruction of the vehicle stopping distance: (**a**) in the absolute form; (**b**) in the relative form.

In the case of EDR1, the errors are practically negligible (they arise from different integration frequencies in the vehicle motion simulation process and during the reconstruction of vehicle motion parameters). For the initial velocity and EDR2 records, growth in the relative and absolute errors with increasing braking intensity can be seen. Such an effect may be explained by the fact that an increase in braking deceleration values results in a growth in the pitch angle and, in consequence, in the longitudinal acceleration estimation error, which translates into bigger errors of the integration output signals. A somewhat different situation takes place for stopping distance errors. In this case, the absolute errors do not depend so visibly on deceleration values and a relatively more significant role is played by the initial velocity value. However, the situation becomes similar to that observed for velocity errors already when relative errors of estimating the distance travelled are concerned: higher braking intensity results in bigger, in percentage terms, relative errors of estimating the braking distance.

Nevertheless, it should also be emphasised that, from the practical point of view, the reconstruction errors in the case of EDR2, although being bigger, are still not very serious (they did not exceed 3 % in the tests under consideration). This is significantly influenced by the general inertial characteristics of the vehicle body and by the design and stiffness of the vehicle suspension system, which translate into relatively small changes in the pitch angle of the vehicle during the braking manoeuvre. In similar research works described in [27,28], the errors in estimating the initial vehicle velocity and stopping distance were bigger (even twice). Those results were obtained for another passenger car, which was also more susceptible to vehicle body pitch and roll.

3.2. Vehicle Lane-Change Maneuvre

In this test, the vehicle is to perform a manoeuvre similar to the lane-change one. The test manoeuvre was forced by applying a control signal representing the steering wheel angle in the form of one period of sinusoidal input (with reference to the Sinusoidal Input test according to ISO 7401) with a reaction time of 1 s. The sinusoid parameters (period and amplitude) were selected individually for a specific vehicle test velocity (50, 90, 140 km/h) so that the vehicle virtually changed a lane about 3.5 m wide (i.e., the centre of vehicle mass, often referred to as centre of gravity—C.G., changed its lateral position by about 3.5 m) and for a chosen intensity of the manoeuvre (low, medium, or high, with the manoeuvre intensity being defined by the level of the maximum lateral C.G. acceleration). The total manoeuvre observation time was 5 s. Figures 12 and 13 show the results obtained for the vehicle moving with an initial velocity of 90 km/h and performing the manoeuvre with an intensity defined as "medium" (with the maximum lateral acceleration level of about 5 m/s^2). Similarly, as in the case of braking, Figure 12a shows the actual (accurate) value of the lateral acceleration $a_{p\eta}$ and the value recorded by the EDR $a_p{}^c$. The difference between these two values (denoted by $\Delta a_p{}^c$) mainly results from changes in the roll angle ϑ_C and, to a lower degree, from changes in the pitch angle φ_C, shown in Figure 12b.

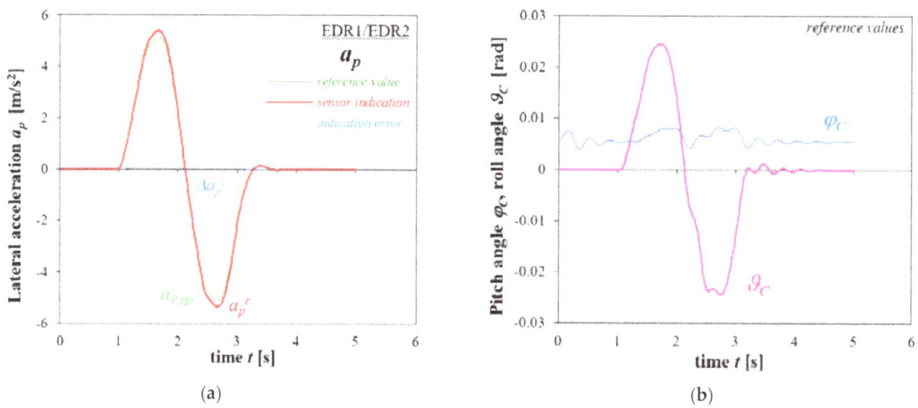

Figure 12. Lane-change manoeuvre, "medium" intensity level. Initial vehicle velocity 90 km/h. Time histories of: (**a**) lateral acceleration (accurate value and EDR sensor indication); (**b**) car body pitch and roll angles (accurate).

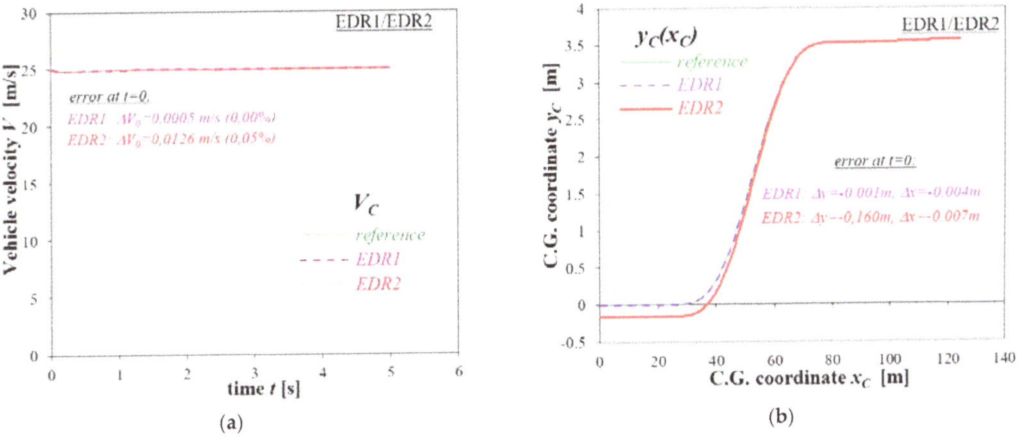

Figure 13. Lane-change manoeuvre, "medium" intensity level. Initial vehicle velocity 90 km/h. Reconstruction of: (a) vehicle velocity; (b) vehicle body C.G. trajectory. Reconstruction based on EDR1 and EDR2 records.

Figure 13 shows the results of reconstruction of a 5 s time period of the vehicle motion, obtained by using recorders of both types (EDR1 and EDR2). As previously, the time histories of vehicle velocity and position were reconstructed "backwards", i.e., from the known final vehicle position and velocity to the initial instant of the simulation of the vehicle motion. The data presented in Figure 13a,b concern vehicle velocity and C.G. trajectory, respectively. The differences that can be seen at the beginning of the velocity and position curves illustrate the reconstruction errors, i.e., errors of determining the initial velocity (ΔV_0) and distance travelled (ΔS).

Similarly, as in the case of braking, very good conformity can be seen between the reconstructed and accurate values for the EDR1 device. The reason for this fact is identical. The accelerometer indication errors are automatically corrected because the vehicle pitch and roll angles are known. In the case of the EDR2 device, the said angles are not known, and the acceleration signals being integrated are burdened with the errors arising from that. In the case under consideration, this translates into an overestimation of the values of the lateral vehicle displacement and a velocity estimation error.

Analysis results obtained for more cases, differing from each other in the pre-set initial vehicle velocity (50, 90, and 140 km/h) and manoeuvre intensity (three lateral acceleration levels: about 2.4–2.5, 5.0–5.4, and 7.1–8.2 m/s^2), only for the EDR2 device, in this case, have been summarised in Table 2. Additionally, the errors of estimating the initial velocity and distance travelled by the vehicle during the manoeuvre have been directly shown in the absolute and relative form in Figure 14. In quantitative terms, these errors may be considered quite small (tenths of percentage units), the reasons for which may be sought, as previously, in small changes in the roll angle (in this case) and in the resulting small errors of estimating the lateral acceleration. However, the trend of these errors rising with the intensity of this manoeuvre can be noticed here as well.

Table 2. Lane-change manoeuvre. Summary of the simulated (accurate) and reconstructed values of initial vehicle velocity, initial position estimation error, and distance travelled by the vehicle for different preset levels of the initial velocity (50, 90, and 140 km/h) and manoeuvre intensity (low, medium, high). Manoeuvre observation time: 5 s.

Preset Initial Velocity V_0 [km/h]	Intensity (a_{pmax} [m/s^2])	Initial Velocity V_0 [m/s]		Initial Position Error, EDR2 [m]		Distance Traveled S [m]	
		Reference	EDR2	Δx	Δy	Reference	EDR2
50	Low (2,4)	13.884	13.886	−0.0092	−0.1519	69.25	69.28
	Medium (5,4)		13.908	−0.0364	−0.1694	69.3	69.37
	High (8,2)		13.916	−0.0468	−0.2932	69.39	69.49
90	Low (2,4)	24.999	24.973	0.0883	−0.1458	124.56	124.49
	Medium (5,0)		25.004	−0.0072	−0.1597	124.49	124.51
	High (7,4)		25.017	−0.0371	−0.2323	124.24	124.31
140	Low (2,5)	38.971	38.942	−0.1508	−0.1554	174.96	175.12
	Medium (5,3)		38.963	−0.1884	−0.1692	174.71	174.91
	High (7,1)		38.991	−0.2427	−0.3774	173.82	174.08

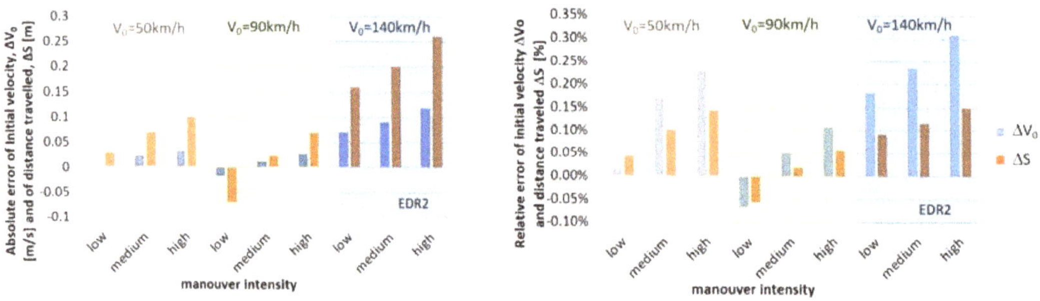

(a) (b)

Figure 14. Lane-change manoeuvre. Errors of the reconstruction of the initial vehicle velocity and of the distance travelled by the vehicle: (**a**) in the absolute form; (**b**) in the relative form. Reconstruction based on EDR2 records.

3.3. Entering a Turn

The test consists in dynamic making the vehicle under test follow a curve. Such a manoeuvre was forced by applying a control signal representing a constant steering wheel angle preceded by a linear angle raising period (with reference to the Step Input test according to ISO 7401), where the steering wheel angle values were selected individually for a specific vehicle test velocity (50, 90, 140 km/h) so that the vehicle performed the manoeuvre with various intensities defined by the maximum lateral C.G. acceleration level in the steady phase of the motion (2, 4, 6, 8 m/s^2). The angle-raising time was so adopted that the steering wheel rising-rate fell within the limits stipulated by the said ISO standard. The total manoeuvre observation time was 5 s. In this case, only the results obtained by using the EDR2 recorder have been presented (for the EDR1 device, the reconstruction results were practically identical with the accurate data). The reconstruction results have been presented in Figures 15–18, in the form as before, for two options of the manoeuvre, i.e., for initial velocities of 50 km/h and 140 km/h, and for the maximum lateral acceleration level of about 8 m/s^2.

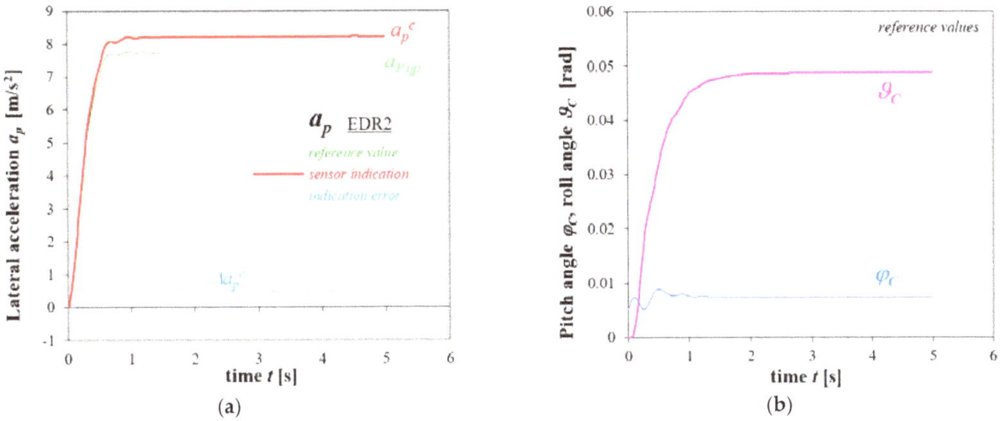

Figure 15. Entering a turn, intensity level 8 m/s^2. Initial vehicle velocity 50 km/h. Time histories of: (**a**) lateral acceleration (accurate values and EDR2 sensor indications); (**b**) car body pitch and roll angles (accurate values).

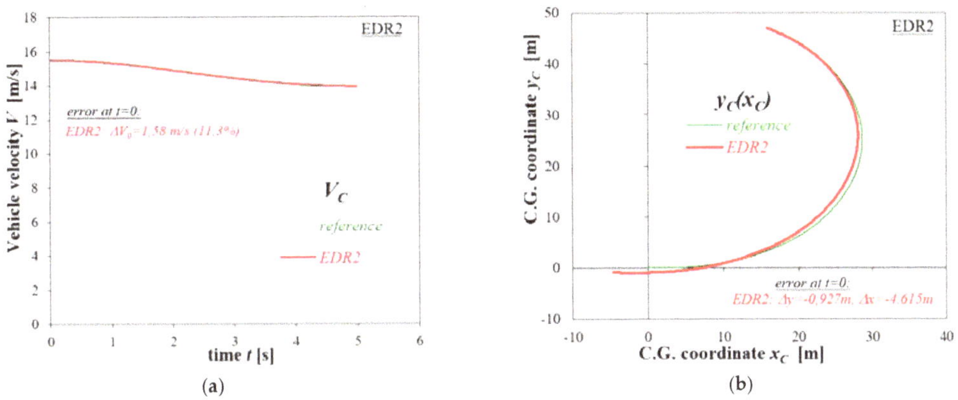

Figure 16. Entering a turn, intensity level 8 m/s^2. Initial vehicle velocity 50 km/h. Reconstruction of: (**a**) vehicle velocity; (**b**) vehicle C.G. trajectory; reconstruction based on EDR2 records.

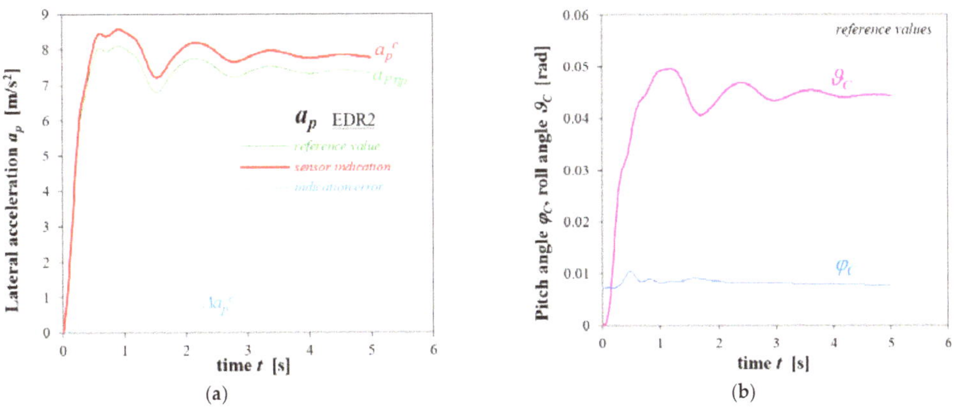

Figure 17. Entering a turn, intensity level 8 m/s^2. Initial vehicle velocity 140 km/h. Time histories of: (**a**) lateral acceleration (accurate values and EDR2 sensor indications); (**b**) Car body pitch and roll angle values (accurate).

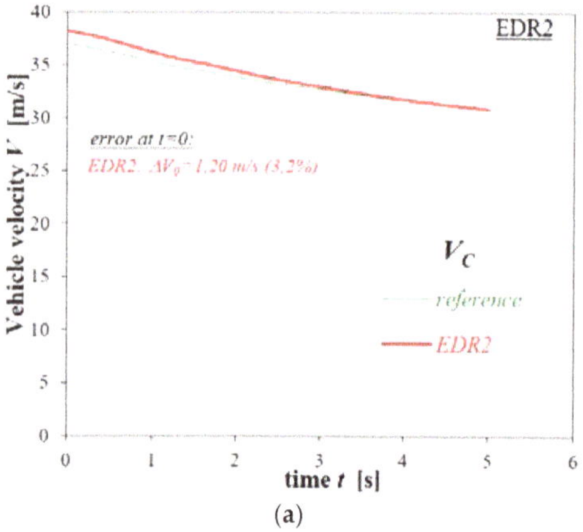

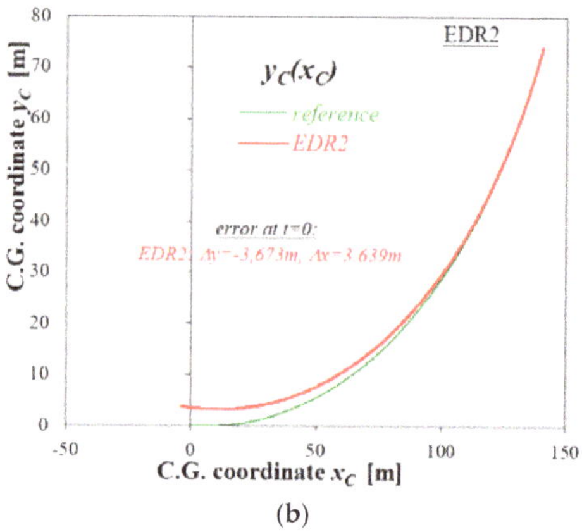

Figure 18. Entering a turn, intensity level 8 m/s^2. Initial vehicle velocity 140 km/h. Reconstruction of: (**a**) vehicle velocity; (**b**) vehicle C.G. trajectory; reconstruction based on EDR2 records.

Results of all the tests carried out with entering a turn have been presented in Table 3. For better illustration of the differences obtained, the errors of estimating the initial velocity and distance travelled by the vehicle during the manoeuvre have been directly shown in the absolute and relative form in Figure 19; Figure 20 presents the initial positions of the centre of vehicle mass (C.G.), reconstructed with EDR2 records being used.

Table 3. Entering a turn. Entering a turn. Summary of the simulated (accurate) and reconstructed values of initial vehicle velocity, errors of estimating the initial vehicle position, and distance travelled by the vehicle for different pre-set levels of the initial velocity (50, 90, and 140 km/h) and manoeuvre intensity (2, 4, 6, 8 m/s^2). Manoeuvre observation time: 5 s.

Preset Initial Velocity V_0 [km/h]	Intensity (a_p [m/s^2])	Initial Velocity V_0 [m/s]		Initial Position Error, EDR2 [m]		Distance Traveled S [m]	
		Reference	EDR2	Δx	Δy	Reference	EDR2
50	2	13.935	14.097	−0.5100	0.9233	69.37	69.66
	4		14.545	−1.8910	1.1858	69.5	70.73
	6		14.944	−3.0435	0.3791	69.61	71.82
	8		15.517	−4.6145	−0.9273	69.49	73.26
90	2	25.009	25.091	−0.2672	1.0112	124.75	124.89
	4		25.427	−1.3101	1.9083	124.62	125.45
	6		25.824	−2.5380	2.1633	124.34	125.98
	8		26.368	−4.2219	2.6445	123.88	126.56
140	2	37.068	37.210	−0.3776	1.1077	169.73	170.01
	4		37.475	−1.2059	2.1086	170.35	171.18
	6		37.804	−2.2332	2.6795	170,00	171.5
	8		38.265	−3.6725	3.6388	167.12	169.52

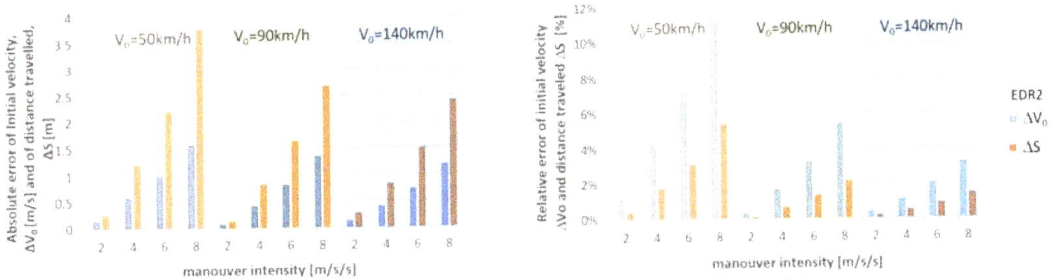

(a) (b)

Figure 19. Entering a turn. Errors of the reconstruction of the initial vehicle velocity and of the distance travelled by the vehicle: (**a**) in the absolute form; (**b**) in the relative form. Reconstruction based on EDR2 records.

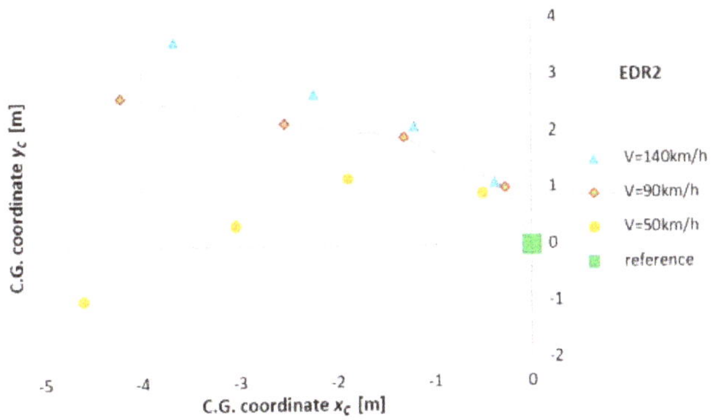

Figure 20. Entering a turn. Initial positions of the centre of vehicle mass (C.G.), reconstructed with EDR2 records being used.

When analysing the results obtained for this manoeuvre, it should be noted that the reconstruction errors are here markedly bigger than those occurring in the case of the lane-change manoeuvre. For the motion, parameters range under consideration, the errors of estimating the initial velocity or distance travelled by the vehicle reach from a few to even more than ten percent (Figure 19). The initial vehicle position obtained from the reconstruction may differ from the actual (accurate) one by a few meters (Figure 20). Noteworthy is the fact that the manoeuvre observation time was identical for both manoeuvre types (5 s). The reasons for such an effect may be sought in the fact that, similarly as it is in the case of braking and longitudinal deceleration, no reversal of the sign of vehicle acceleration and its error takes place (Figures 15 and 17), contrary to the situation that takes place during the lane-change manoeuvre, where the errors of the reconstruction (acceleration quadrature operations) generated in one phase of the manoeuvre are partly compensated by the reconstruction errors generated in the other manoeuvre phase.

Another factor worth consideration is the impact of manoeuvre intensity. Here, similarly as in other manoeuvres, the errors increase with rising manoeuvre intensity; this may be directly associated with an increase in the errors of the EDR acceleration records due to growing vehicle body pitch and roll angles.

The last regularity that can be noticed is a decrease in error values with increasing vehicle velocity, which distinguishes this manoeuvre from a similar case (in terms of absence of reversal of the acceleration sign), i.e., from the straight-line braking. To explain this finding, it is worth noticing that, apart from the initial phase of the manoeuvre, the vehicle motion under analysis is similar to the steady-state motion along a circle. Therefore, it is worth paying attention to another variable that characterises the vehicle motion, i.e., the yaw angle. Figure 21 presents the data collected that show the "angular distance" travelled by the vehicle in every case. When comparing the data plotted in this graph with the reconstruction errors presented previously in Figure 19, their qualitative convergence can be noticed: the higher the velocity, the smaller the change in the yaw angle in the manoeuvre under consideration. Based on the said qualitative convergence, a hypothesis may be formulated that the value of errors, especially the errors of the vehicle trajectory, are affected by the "angular distance" travelled by the vehicle. This aspect will be presented in broader terms in the next example, where the steady-state circular motion will be analysed.

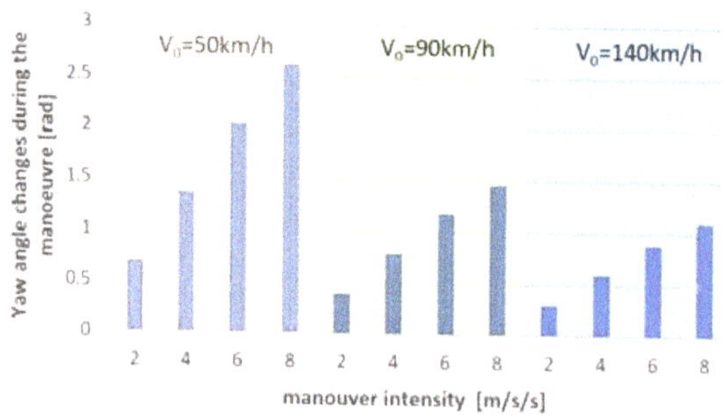

Figure 21. Entering a turn. Range of changes in the vehicle yaw angle during the manoeuvre.

3.4. *Driving in a Roundabout (Circular Motion)*

In such a test, the vehicle is driven with a constant speed along a curvilinear path close to a circle (to some extent, the test conditions are close to those of the ISO 4148 test [42]). The manoeuvre is forced by applying a constant non-zero steering wheel angle. The test parameters (vehicle speed, steering wheel angle) were so selected that, on the one hand, the vehicle moved with a prescribed lateral acceleration (2, 4, or 6 m/s^2) and, on the other hand, the vehicle path was in accordance with the roundabouts that can be met in real roads [48] (and simultaneously the actual radius of the vehicle path was in conformity with the recommendations of the said ISO 4138 standard). Actually, the vehicle path radius was set at about 30 m, and the values of the vehicle speed on the virtual traffic circle were about 30, 40, and 50 km/h. The manoeuvre was considered as completed when the vehicle travelled a "full circle", i.e., the vehicle yaw angle changed by 2π in the simulation.

This time, the trajectories reconstructed ("backwards" from the final vehicle position) have been presented as the first thing (see Figure 22). Figure 23 shows a reconstruction of the time history of the vehicle speed ("backwards" from the final instant). The direct causes of the visible differences between the accurate and reconstructed vehicle trajectories and speed vs. time curves mainly lie in errors of the EDR2 acceleration records (differences between the accurate and recorded acceleration values) shown in Figure 24.

At first, the differences between the accurate and reconstructed curves can be found to be similar to each other in qualitative terms: the differences increased in the first phase of the manoeuvre and decreased afterwards. This resulted in very good conformity between the accurate and reconstructed value of the vehicle speed and in quite good conformity between the accurate and reconstructed initial vehicle position (the vehicle C.G. coincided with the accurate trajectory). Here, reference may be made to the hypothesis proposed in the previous subsection about the dependence of errors on the "angular distance" travelled by the vehicle and, in consequence, about the "periodicity" of the errors. Therefore, the vehicle speed and position errors have been presented in Figure 25 not in the time domain but as functions of changes in the angular position (yaw angle) within the range from the final position (for which the change is equal to zero) to the initial position (where the change is equal to -2π). Additionally, the estimated values of the vehicle body C.G. path radius and the vehicle yaw velocity have been presented in Figure 26.

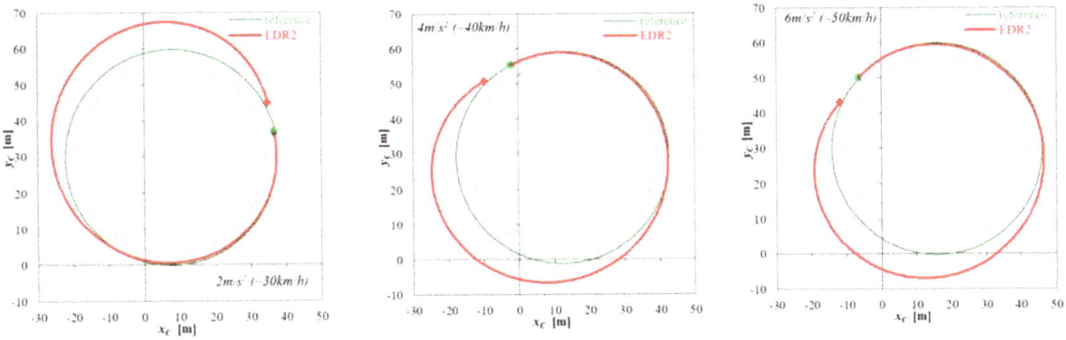

(a) (b) (c)

Figure 22. Driving in a roundabout, R = 30 m. Reconstruction of vehicle body C.G. trajectory based on EDR2 records. Lateral acceleration level: (a) 2 m/s^2; (b) 4 m/s^2; (c) 6 m/s^2. "•"—start of the reconstruction (end of the motion under analysis); "♦"—end of the reconstruction (reconstructed initial position); grey dotted lines represent edges of the roundabout carriageway 6 m wide.

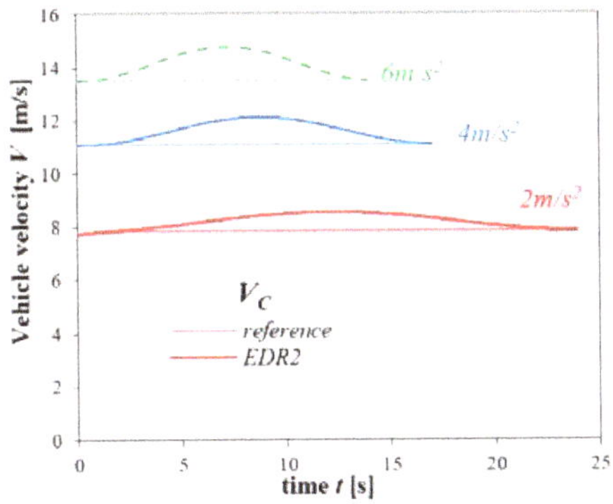

Figure 23. Driving in a roundabout, R = 30 m. Reconstruction of vehicle C.G. speed based on EDR2 records for three lateral acceleration levels: 2 m/s^2; 4 m/s^2; 6 m/s^2.

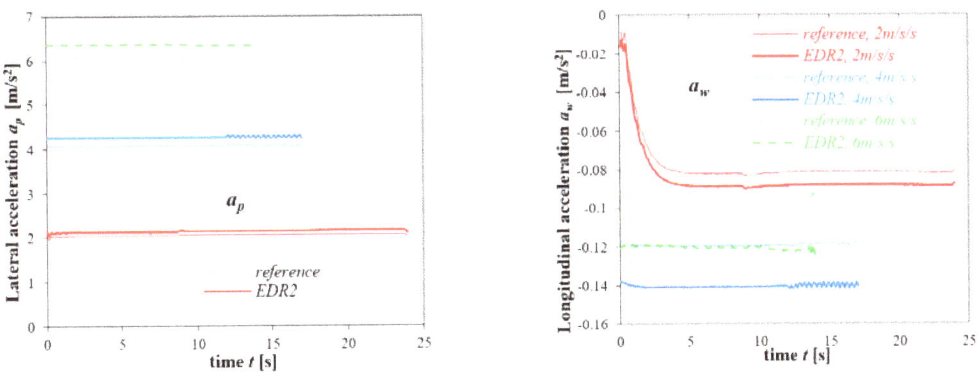

(a) (b)

Figure 24. Driving in a roundabout, R = 30 m. Comparison between time histories of accurate values and EDR2 sensor readings of: (a) lateral acceleration; (b) longitudinal acceleration.

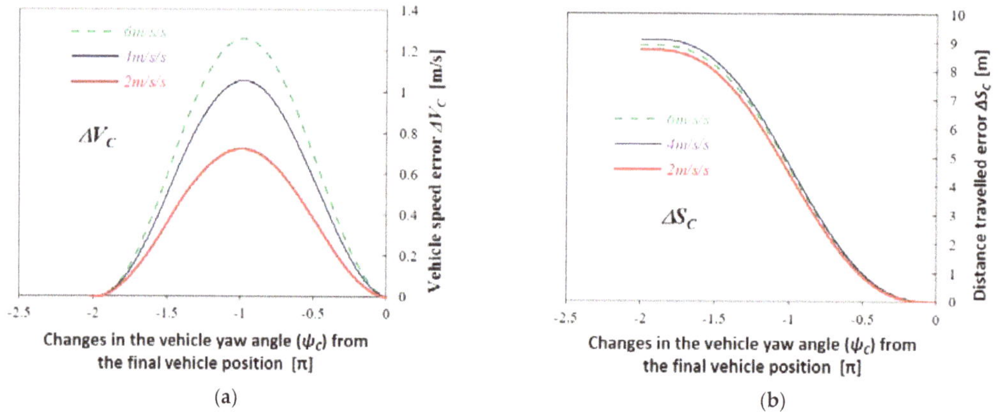

Figure 25. Driving in a roundabout, R = 30 m. Errors of: (**a**) vehicle body C.G. speed; (**b**) vehicle body C.G. distance travelled as a function of changes in the vehicle yaw angle (zero—vehicle final position, -2π—vehicle initial position).

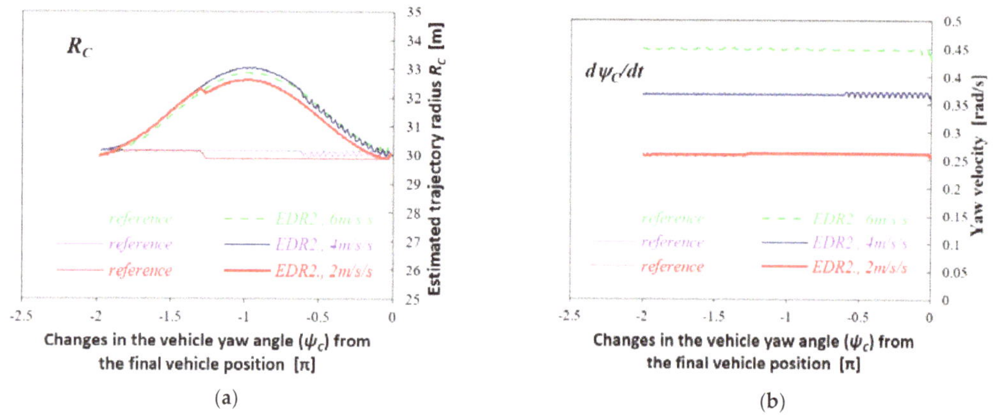

Figure 26. Driving in a roundabout, R = 30 m. (**a**) Estimated radius of the vehicle body C.G. trajectory; (**b**) Yaw velocity as a function of changes in the vehicle yaw angle (zero—vehicle final position, -2π—vehicle initial position).

It can be seen that on the interval from zero to $-\pi$, the speed reconstruction error is increasing; in consequence, the rate of growth in the error of estimating the distance travelled is increasing as well. On the interval where the vehicle has already "turned back", i.e., from $-\pi$ to -2π, the speed reconstruction error is decreasing to values close to zero; the distance error is still increasing, but the growth rate is decreasing. Since the vehicle yaw velocity is approximately constant (Figure 26b), the changes in the reconstructed vehicle speed value translate into changes in the reconstructed trajectory radius value (Figure 26a), i.e., growth in the radius on the yaw angle range from 0 to $-\pi$ and decline on the range from $-\pi$ to -2π.

Noteworthy is also the fact that the reconstruction errors are also affected by the error in estimating the longitudinal acceleration (Figure 24b). In the example presented, this can be seen, e.g., in Figures 25b and 26a, when the cases with 4 m/s^2 and 6 m/s^2 are compared with each other. For the manoeuvre with 6 m/s^2, the longitudinal acceleration level, in terms of its absolute value, is lower than that for the manoeuvre with 4 m/s^2 (see Figure 24b). This has contributed to the fact that in spite of a higher level of lateral acceleration, the errors in estimating the distance travelled are here somewhat lower (see Figure 25b). Similarly, the radius of the reconstructed trajectory is also somewhat smaller (see Figure 26a).

It may be assumed that during the next circling, this situation may repeat itself again and again. To illustrate this effect, Figure 27 presents the vehicle trajectory and speed curves for the simulation of driving in a runabout with a speed of about 50 km/h (lateral acceleration level of 6 m/s^2) for an "angular distance" exceeding the value of 2π as shown previously (the yaw angle was now 3.02 rad at the initial instant and 14.03 rad at the final instant, i.e., it changed by 11.01 rad $\cong 3.5\pi$). The vehicle speed error "oscillates" with a period of 2π, the distance error is increasing cycle after cycle, and the reconstructed trajectory "shifts" by a certain distance with every cycle.

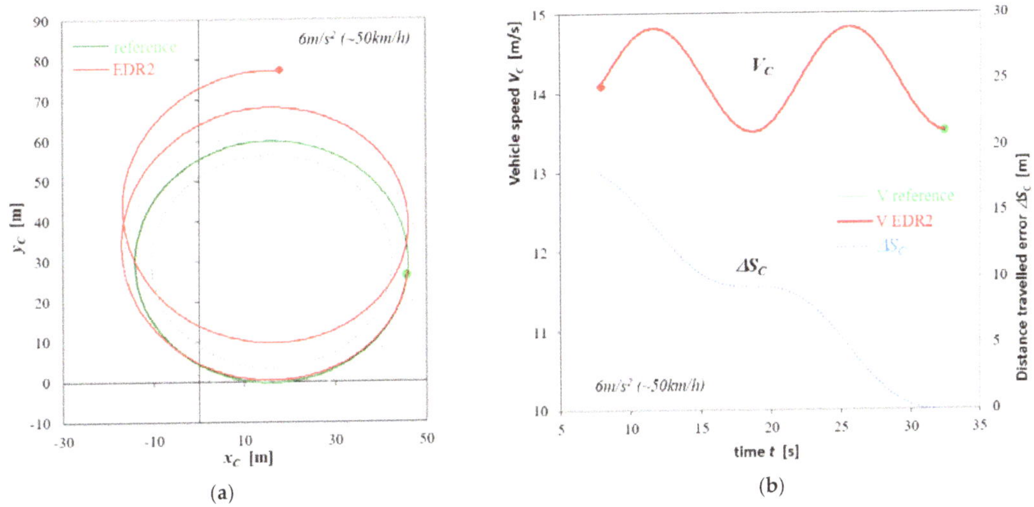

Figure 27. Driving in a roundabout, R = 30 m. Lateral acceleration level 6 m/s^2 (vehicle speed ca. 50 km/h). Reconstruction based on EDR2 records of: (**a**) vehicle body C.G. trajectory; (**b**) vehicle body C.G. speed and error of the distance travelled. "•"—start of the reconstruction (end of the motion under analysis); "♦"—end of the reconstruction (reconstructed initial position); grey dotted lines represent edges of the roundabout carriageway 6 m wide.

In the context of the above findings, therefore, attention should be paid to the fact that at the reconstruction of manoeuvres like the one discussed here (or the entering a turn, as discussed previously), a factor of considerable importance for the magnitude of errors in the event reconstruction based on records obtained by using EDR2-like devices is a change in the yaw angle during the manoeuvre. Depending on the scope of the change, the errors may be small or very big. For the vehicle trajectory, the positions determined by reconstruction may so differ from the actual ones that correct interpretation of the situation under analysis would become impossible (errors would be of the order of several meters, such that the vehicle position thus determined would be situated e.g., outside of the road area or of any realistic vehicle trajectory).

4. Discussion

The results obtained have been discussed in detail immediately after their presentation. To synthesise the findings in the context of the accuracy of the reconstruction of vehicle motion, the following may be stated.

- In the case of the EDR1-type devices (recording 6 vector components that describe the vehicle motion), the reconstruction results are practically identical with the reference (accurate) data.

- In the case of the EDR2-type devices, the velocities and trajectories may both slightly and significantly differ from the accurate data.
- The qualitative nature of the accident reconstruction errors depends, *inter alia*, on the vehicle motion type. In the case of straight-line braking, errors arise in determining the initial vehicle velocity and the distance travelled; when curvilinear motion is analysed, deviations from the accurate vehicle trajectory mainly occur.
- In quantitative terms, the errors may be influenced by the intensity of the manoeuvre. In general, the higher intensity is connected with larger reconstruction errors (arising from stronger angular movements of the vehicle body and, in consequence, from bigger errors in the accelerations recorded).
- Another factor that has an impact on the accuracy achieved is the period of time (or the length of the distance travelled) for which the vehicle motion is reconstructed. For short-duration manoeuvres (lasting less than 5 s), good accuracy can be achieved, even if the EDR2 devices are used (with reservations formulated in other conclusions). However, the errors increase with the increasing length of the period for which the motion is reconstructed. Attention should also be paid to the effects reported for the case of circular motion, i.e., a kind of cyclicity of the reconstruction errors.
- A quantitative influence on the accuracy may be exerted by the design and operational features of the vehicle. This concerns the properties that have a considerable impact on the angular movements of the vehicle body solid. The features of particular importance may be suspension system characteristics (e.g., suspension system design, spacing, and stiffness of springs, dry friction, etc.).

To some extent, especially as regards the quantitative findings, the results presented confirm the conclusions expressed in [27–29] as well as in [26]. The directions of further research on the accuracy of reconstruction of vehicle motion based on records obtained from EDR-type devices, concerning such issues as the impacts of current load and design features of the vehicle, accuracy of the measuring and recording apparatus, angular positioning of EDR sensors, or EDR configuration (e.g., recording of angular velocities instead of angles, recording frequency), can also be formulated.

5. Conclusions

The large-scale introduction of automotive "black boxes" into service will potentially bring many benefits. It will increase the resource of information about the course of an accident and will provide knowledge of the parameters of vehicle motion. Other advantages may consist in preventive influence on the driver, who will be less prone to risky behaviours in road traffic. The EDR solutions offered now in the automotive market are often a simplified version of the solutions having been used in aviation for many years. In the EDRs offered in the market, apart from those intended for research applications, the vehicle motion is generally treated as two-dimensional (as it is in the EDR2).

In the research carried out, the interest was focused on the concept of the EDR, without addressing the problems related to the measuring and recording apparatus (such as intrinsic sensor errors or errors arising in the recording process). The calculations carried out have shown that the said simplifications may result in significant errors in the reconstruction of vehicle motion, sometimes totally disqualifying the reconstruction results. This mainly applies to the reconstruction of vehicle trajectory; nevertheless, a result significantly differing from the actual value is also possible when the vehicle velocity is reconstructed. The basic reason for that lies in the fact that the devices of the EDR2 type do not provide information about some angular movements of the vehicle body solid, i.e., the pitch and roll angles. In the case of the EDR1 devices, no considerable reconstruction errors have been revealed.

Author Contributions: Conceptualization, M.G. and Z.L.; methodology, M.G. and Z.L.; software, M.G. and Z.L.; validation, M.G. and Z.L.; formal analysis, M.G. and Z.L.; investigation, M.G. and

Z.L.; writing—original draft preparation, M.G. and Z.L.; writing—review and editing, M.G. and Z.L. All authors have read and agreed to the published version of the manuscript.

Funding: This research received no external funding. APC was funded by Warsaw University of Technology.

Institutional Review Board Statement: Not applicable.

Informed Consent Statement: Not applicable.

Data Availability Statement: Not applicable.

Acknowledgments: The simulation model of motor vehicle motion and dynamics, used in this work, was built during the execution of a project carried out by the Faculty of Transport, Warsaw University of Technology (WUT) to commission of ETC-PZL Aerospace Industries Ltd. (being now owned by a US company Environmental Tectonics Corporation and AIRBUS Poland S.A.). The superior goal of the project was to build a simulator of driving emergency service vehicles within project No O ROB 0011 01/ID/11/1 "Simulator of driving emergency service vehicles during standard and extreme actions" (in the years 2012–2013). The experimental laboratory and road tests were carried out by teams of the Cracow University of Technology (managed by Wiesław Pieniążek), Automotive Industry Institute, now ŁUKASIEWICZ-PIMOT (managed by Dariusz Więckowski), and Military University of Technology (managed by Jerzy Jackowski), with active participation of the authors of this paper and Witold Luty (WUT, Faculty of Transport).

Conflicts of Interest: The authors declare no conflict of interest.

References

1. U. S. Department of Transportation, National Highway Traffic Safety Administration: Event Data Recorders. 49 CFR Part 563, Docket No. NHTSA-2004-18029. Available online: https://www.nhtsa.gov/fmvss/event-data-recorders-edrs (accessed on 19 August 2021).
2. A Proposed Rule by the National Highway Traffic Safety Administration on 12/13/2012, Federal Motor Vehicle Safety Standards; Event Data Recorders. Available online: https://www.federalregister.gov/documents/2012/12/13/2012-30082/federal-motor-vehicle-safety-standards-event-data-recorders (accessed on 19 August 2021).
3. A Proposed Rule by the National Highway Traffic Safety Administration on 02/08/2019, Federal Motor Vehicle Safety Standards; Event Data Recorders. Available online: https://www.federalregister.gov/documents/2019/02/08/2019-01651/federal-motor-vehicle-safety-standards-event-data-recorders (accessed on 19 August 2021).
4. Regulation (EU) 2019/2144 of the European Parliament and of the Council of 27 November 2019 on Type-Approval Requirements for Motor Vehicles and Their Trailers, and Systems, Components and Separate Technical Units Intended for Such Vehicles, as Regards Their General Safety and the Protection of Vehicle Occupants and Vulnerable Road Users. Available online: https://eur-lex.europa.eu/eli/reg/2019/2144/oj (accessed on 19 August 2021).
5. Malinová, K.; Kasanický, G.; Podhorský, J. Usage of Digital Evidence in the Technical Analysis of Traffic Collisions. *Transp. Res. Procedia* **2021**, *55*, 1737–1744. [CrossRef]
6. Iyoda, M.; Trisdale, T.; Sherony, R.; Mikat, D.; Rose, W. Event Data Recorder (EDR) Developed by Toyota Motor Corporation. *SAE Int. J. Trans. Saf.* **2016**, *4*, 187–201. [CrossRef]
7. Smith, G. Reconstructing Vehicle and Occupant Motion from EDR Data in High Yaw Velocity Crashes. In *SAE Technical Paper*; 2021-01-0892; SAE International: Warrendale, PA, USA, 2021. [CrossRef]
8. Vandiver, W.; Anderson, R.; Ikram, I.; Randles, B.; Furbish, C. Analysis of Crash Data from a 2012 Kia Soul Event Data Recorder. In *SAE Technical Paper*; 2015-01-1445; SAE International: Warrendale, PA, USA, 2015. [CrossRef]
9. Ruth, R.; Bartlett, W.; Daily, J. Accuracy of Event Data in the 2010 and 2011 Toyota Camry During Steady State and Braking Conditions. *SAE Int. J. Passeng. Cars—Electron. Electr. Syst.* **2012**, *5*, 358–372. [CrossRef]
10. Tsoi, A.; Hinch, J.; Ruth, R.; Gabler, H. Validation of Event Data Recorders in High Severity Full Frontal Crash Tests. *SAE Int. J. Trans. Saf.* **2013**, *1*, 76–99. [CrossRef]
11. Tsoi, A.; Johnson, N.; Gabler, H. Validation of Event Data Recorders in Side-Impact Crash Tests. *SAE Int. J. Trans. Saf.* **2014**, *2*, 130–164. [CrossRef]
12. Carr, L.; Rucoba, R.; Barnes, D.; Kent, S.; Osterhout, A. EDR Pulse Component Vector Analysis. In *SAE Technical Paper*; 2015-01-1448; SAE International: Warrendale, PA, USA, 2015. [CrossRef]
13. Xing, P.; Lee, F.; Flynn, T.; Wilkinson, C.; Siegmund, G. Comparison of the Accuracy and Sensitivity of Generation 1, 2 and 3 Toyota Event Data Recorders in Low-Speed Collisions. *SAE Int. J. Trans. Saf.* **2016**, *4*, 172–186. [CrossRef]
14. Atarod, M. Reconstruction of Passenger Vehicle Underride: An Analysis of Insurance Institute for Highway Safety Semitrailer Rear Underride Crash Data. In *SAE Technical Paper*; 2020-01-5091; SAE International: Warrendale, PA, USA, 2020. [CrossRef]
15. Fatzinger, E.; Landerville, J. Using Vehicle EDR Data to Calculate Motorcycle Delta-V in Motorcycle-Vehicle Lateral Front End Impacts. In *SAE Technical Paper*; 2020-01-0885; SAE International: Warrendale, PA, USA, 2020. [CrossRef]

16. Han, I. Verification and Complement of EDR Speed Data through the Analysis of Real World Vehicle Collision Accidents. *Int. J. Automot. Technol.* **2019**, *20*, 897–902. [CrossRef]
17. Bortles, W.; Biever, W.; Carter, N.; Smith, C. A Compendium of Passenger Vehicle Event Data Recorder Literature and Analysis of Validation Studies. In *SAE Technical Paper*; 2016-01-1497; SAE International: Warrendale, PA, USA, 2016. [CrossRef]
18. Martinesco, A.; Netto, M.; Neto, A.; Etgens, V. A Note on Accidents Involving Autonomous Vehicles: Interdependence of Event Data Recorder, Human-Vehicle Cooperation and Legal Aspects. *IFAC-PapersOnLine* **2019**, *51*, 407–410. [CrossRef]
19. Riexinger, L.; Sherony, R.; Gabler, H. Methodology for Estimating the Benefits of Lane Departure Warnings using Event Data Recorders. In *SAE Technical Paper*; 2018-01-0509; SAE International: Warrendale, PA, USA, 2018. [CrossRef]
20. Davis, G. Bayesian Estimation of Drivers' Gap Selections and Reaction Times in Left-Turning Crashes from Event Data Recorder Pre-Crash Data. In *SAE Technical Paper*; 2017-01-1411; SAE International: Warrendale, PA, USA, 2017. [CrossRef]
21. Scanlon, J.; Page, K.; Sherony, R.; Gabler, H. Using Event Data Recorders from Real-World Crashes to Investigate the Earliest Detection Opportunity for an Intersection Advanced Driver Assistance System. In *SAE Technical Paper*; 2016-01-1457; SAE International: Warrendale, PA, USA, 2016. [CrossRef]
22. Reust, T.; Morgan, J.; Smith, P. Method to Determine Vehicle Speed During ABS Brake Events Using Heavy Vehicle Event Data Recorder Speed. *SAE Int. J. Passeng. Cars—Mech. Syst.* **2010**, *3*, 644–652. [CrossRef]
23. Diacon, A.; Daily, J.; Ruth, R.; Mueller, C. Accuracy and Characteristics of 2012 Honda Event Data Recorders from Real-Time Replay of Controller Area Network (CAN) Traffic. *SAE Int. J. Trans. Saf.* **2013**, *1*, 399–419. [CrossRef]
24. Ruth, R.; Daily, J. Accuracy of Event Data Recorder in 2010 Ford Flex During Steady State and Braking Conditions. *SAE Int. J. Passeng. Cars—Mech. Syst.* **2011**, *4*, 677–699. [CrossRef]
25. Ruth, R.; Brown, T.; Lau, J. Accuracy of EDR During Rotation on Low Friction Surfaces. In *SAE Technical Paper*; 2010-01-1001; SAE International: Warrendale, PA, USA, 2010. [CrossRef]
26. Wach, W. Reconstruction of vehicle kinematics by transformations of raw measurement data. In Proceedings of the 2018 XI International Science-Technical Conference Automotive Safety, Casta, Slovakia, 18–20 April 2018; pp. 1–5. [CrossRef]
27. Guzek, M.; Lozia, Z. Possible Errors occurring during Accident Reconstruction based on Car "Black Box" Records Guzek Marek, Lozia Zbigniew. In *SAE Technical Papers*; No 2002-01-0549; SAE International: Warrendale, PA, USA, 2002. [CrossRef]
28. Guzek, M.; Lozia, Z.; Pieniążek, W. Accident Reconstruction Based on EDR Records—Simulation and Experimental Study. In *SAE Technical Papers*; No 2007-01-0729; SAE International: Warrendale, PA, USA, 2007. [CrossRef]
29. Guzek, M. *Uncertainty in the Analysis of Road Accidents*; Warsaw University of Technology: Warszawa, Poland, 2016. (In Polish)
30. Guzek, M. Vehicle motion reconstruction based on EDR/ADR records—simulation research. *IOP Conf. Ser. Mater. Sci. Eng.* **2018**, *421*, 032011. [CrossRef]
31. Lozia, Z. Vehicle dynamics simulation models of two emergency vehicles. In *Technical Transactions "Mechanics"*; Cracow University of Technology: Kraków, Poland, 2012; Volume 109, Issue 8/3-M; pp. 19–34. ISSN 0011-4561. (In Polish)
32. Lozia, Z. Examples of authorial models for the simulation of motor vehicle motion and dynamics. In *Proceedings of the Institute of Vehicles*; Warsaw University of Technology: Warszawa, Poland, 2015; Volume 104, pp. 9–27. ISSN 1642-347X.
33. Lozia, Z. Simulation testing of two ways of disturbing the motion of a motor vehicle entering a skid pad as used for tests at Driver Improvement Centres. *Arch. Automot. Eng.* **2016**, *72*, 111–125.
34. Kamiński, E.; Pokorski, J. The car theory. In *Dynamics of Suspensions and Drive Systems of Motor Vehicles*; WKŁ: Warszawa, Poland, 1983; ISBN 83-206-0348-X. (In Polish)
35. Dukkipati, R.V.; Pang, J.; Qatu, M.S.; Chen, G.S.; Shuguang, Z. *Road Vehicle Dynamics*; SAE International: Warrendale, PA, USA, 2008; ISBN 978-0-7680-1643-7.
36. Lozia, Z. *Driving Simulators*; WKŁ: Warszawa, Poland, 2008; ISBN 978-83-206-1663-7. (In Polish)
37. Lozia, Z. *The Analysis of the Movement of a Two-Axle Car on the Background of Modelling Its Dynamics; Scientific Papers of Warsaw University of Technology—Transport, Sheet 41*; Monograph, Warsaw University of Technology: Warszawa, Poland, 1998; ISSN 1230-9265. (In Polish)
38. Maryniak, J. The Dynamic Theory of Moving Objects; Scientific Papers of Warsaw University of Technology—Mechanics. Monograph, Warsaw University of Technology: Warszawa, Poland, 1975; ISSN 0137-2335. (In Polish)
39. Dugoff, H.; Fancher, P.S.; Segel, L. An analysis of tire traction properties and their influence on vehicle dynamic performance. In *SAE Technical Paper*; 700377; SAE International: Warrendale, PA, USA, 1970. [CrossRef]
40. Fancher, P.S., Jr.; Bareket, Z. Including roadway and tread factors in semi-empirical model of truck tyres. *Veh. Syst. Dyn.* **1993**, *21*, 92–107. [CrossRef]
41. Schieschke, R. The importance of tire dynamics in vehicle simulation. Presented at the Tire Society 9th Annual Meeting and Conference on Tire Science and Technology, Akron, OH, USA, 20–21 March 1990.
42. ISO 4138:2012 Passenger Cars—Steady-State Circular driving Behaviour—Open-Loop Test Methods. Standard of International Organization for Standardization. Publication date: 2012-06. Available online: https://www.iso.org/standard/54143.html (accessed on 19 August 2021).
43. ISO 7401:2011 Road Vehicles—Lateral Transient Response Test Methods—Open-Loop Test Methods. Standard of International Organization for Standardization. Publication date: 2011-04. Available online: https://www.iso.org/standard/54144.html (accessed on 19 August 2021).

44. Regulation No 13 of the Economic Commission for Europe of the United Nations (UN/ECE)—Uniform Provisions Concerning the Approval of Vehicles of Categories M, N and O with Regard to Braking, Annex 4. Available online: https://op.europa.eu/en/publication-detail/-/publication/0a43f880-d612-11e5-a4b5-01aa75ed71a1/ (accessed on 19 August 2021).
45. Guzek, M. Methods of Computational Error Determination in an Analysis of Selected Pre-Accident Situations in Road Traffic. Ph.D. Thesis, Warsaw University of Technology, Faculty of Transport, Warsaw, Poland, 2002. (In Polish)
46. Fortuna, Z.; Macukow, B.; Wąsowski, J. *Numerical Methods*, 7th ed.; WNT: Warszawa, Poland, 2015; ISBN 9788379262816. (In Polish)
47. Stoer, J. *Introduction to the Numerical Methods*; PWN: Warszawa, Poland, 1979; Volume 1. (In Polish)
48. Recommendation of the Minister of Infrastructure of February 2020. Patterns and standards. WRD-31-3. Guidelines for the Design of Road Intersections. Roundabouts. Available online: https://www.gov.pl/web/infrastruktura/wr-d (accessed on 19 August 2021). (In Polish)

energies

MDPI

Article

Analysis of the Impact of Countdown Signal Timers on Driving Behavior and Road Safety

Tomasz Krukowicz *, Krzysztof Firląg, Józef Suda and Mirosław Czerliński

Faculty of Transport, Warsaw University of Technology, Koszykowa 75, 00-662 Warszawa, Poland; krzysztof.firlag@pw.edu.pl (K.F.); jozef.suda@pw.edu.pl (J.S.); miroslaw.czerlinski@pw.edu.pl (M.C.)
* Correspondence: tomasz.krukowicz@pw.edu.pl

Abstract: This article describes the safety and effectiveness issues related to signal countdown timers (SCT). These devices are used in many countries around the world. The impact of these devices on road safety and the effectiveness of traffic lights is presented. During a literature review, more than 18 aspects of device use were recognized. The research involved measurements carried out at three intersections in Płock (Poland). The initial and final period of the green signal for vehicles was analyzed. Headways, incidences of vehicles passing through after the end of the green signal, and red-light violations were examined. Additionally, a fuel consumption analysis and a case study of a road crash in Szczecin (Poland) are presented. Problems related to signal countdown timers working during traffic light failure are described. The research shows different influences of signal countdown timers at various intersections. It was observed that SCTs increase the number of red-light violations and during the red-amber signal. On the other hand, the number of entries during the amber signal with SCTs is lower. A literature review also indicated that the use of SCT causes a reduction of start-up time (positive impact) and increases vehicle speed (negative impact). The article concludes that SCTs do not always fulfill their role in improving road safety and control efficiency. Conclusions can address various stakeholders, including drivers, road authorities, and traffic engineers.

Keywords: signal countdown timer; SCT; GSCT; RSCT; traffic signals; traffic lights; road safety; road traffic control; traffic engineering; driver behavior

Citation: Krukowicz, T.; Firląg, K.; Suda, J.; Czerliński, M. Analysis of the Impact of Countdown Signal Timers on Driving Behavior and Road Safety. *Energies* **2021**, *14*, 7081. https://doi.org/10.3390/en14217081

Academic Editor: Teresa Galvão Dias

Received: 9 October 2021
Accepted: 23 October 2021
Published: 29 October 2021

1. Introduction

1.1. Problems of Using Signal Countdown Timers

The first traffic light was installed in 1868 at a road junction in London [1]. It was based on a gas lamp. Electric traffic lights started to be used in the second decade of the 20th century. The very first, based solely on light signals, was used in 1914 in Cleveland. Signaling displayed the words "STOP" and "MOVE". The colors red and green were used for the first time as signals in 1917 in San Francisco. A third color, amber, was introduced in 1920 in Detroit, which improved traffic safety. In the first years of their operation, traffic light solutions were very different but, fortunately, they were unified. The signals used on roads were defined in international agreements such as the Geneva Convention on Road Traffic, signed on 19 September 1949 [2]. The convention allowed two-color signaling (red and green) and specified the meaning of individual signals. Similar provisions were agreed in the Convention on road signs and signals established at Vienna on 8 November 1968 [3], and allowed the use of directional signals. Poland signed both of these conventions. Drivers are obliged by the law to follow the indications of the traffic light. The rules set out in the above-mentioned international regulations indicate that drivers are not informed in advance about the signal change. Only a short period, lasting for one or a maximum of a few seconds during the amber, and red with the amber, signals notify drivers about the signal to be displayed next. Issues related to road safety at intersections are a current problem. Research conducted in Poland has investigated issues of road lighting, in particular pedestrian crossings [4], adaptation of road infrastructure to the needs of autonomous vehicles [5], road interchanges [6], safety at pedestrian crossings [7] and country highways [8]. There are fewer articles regarding traffic signals compared to other elements of traffic infrastructure, e.g., roadways, junctions, and interchanges. At the turn of the first and second decades of the 21st century, devices indicating how long a given signal would be displayed, called Signal Countdown Timers (SCT), began to be used. Their application took place on many continents, but in Poland, they were used was ahead of legislative changes (Figure 1). Regulations adapting their application were only introduced in 2017 [9]. However, there are still many problems with the use of these devices. In this article, the problem is reviewed from the perspective of research conducted around the world. Additional research was also carried out by the Road Traffic Control Team at the Faculty of Transport Warsaw University of Technology. Unexplored and unsolved problems require further research, which are indicated in the conclusions.

Figure 1. Signal Countdown Timer in Nowy Sącz, Poland (2016).

1.2. Literature Review

Issues related to drivers' reactions to changing signals have been the subject of many studies since the middle of the 20th century [10,11]. SCTs can be used both for the red signal (RSCT) and the green signal (GSCT) at crossroads [12]. The purpose of the SCT is to help the driver decide to stop or drive through at the amber signal and to start the vehicle column moving more efficiently at the beginning of the green signal. Research also shows that many drivers cross the traffic lights stopping line while displaying a red signal, which is called a red-light violation (RLV) [13].

The literature review was divided into main threads related to the behavior of drivers and pedestrians. Starting from the background of regulations in Poland (Section 1.2.1), the topic was investigated worldwide. Main problems related to vehicle traffic and SCT were categorized into subjects of capacity determination (Section 1.2.2), red light violation (Section 1.2.3), entering on the amber signal (Section 1.2.4), vehicle speed, braking characteristics, dilemma zone (Section 1.2.5), reaction time (Section 1.2.6), safety statistics (Section 1.2.7), GHG emissions (Section 1.2.8) and cycling (Section 1.2.9).

Some of the articles on SCT consider devices for pedestrians (in a shortcut later described as PSCT). The main problems covered by the research were pedestrians passing on the flashing green (Section 1.2.10), pedestrian red-light violations (Section 1.2.11), pedestrian speed (Section 1.2.12), and vehicle-pedestrian conflicts (Section 1.2.13).

1.2.1. Background of Regulations in Poland

The very first article related to the SCT situation in Poland was published in 2013 [14] before regulations were put in place [9]. Legally, at that time, the RSCT and GSCT were used, although their technical and functional requirements were not specified, and their use was not allowed. The article presented many legal and technical problems. It also presents issues related to road traffic engineering in designing traffic light control algorithms and signaling operation in the event of an SCT failure. Some of the problems described in this article have been resolved, while others are still relevant today.

As an extension to previous research, problems related to SCTs used in Poland have been further investigated [15]. Analyzed issues were divided into technical issues related to the assembly and supervision of signals, problems in road traffic engineering associated with the determination of displayed values and the effectiveness of traffic-actuated control, issues related to road safety, and legal issues. Although the article refers to the period in which the use of SCTs in Poland were prohibited, many problems had not been solved, nor been the subject of published research.

Due to enforcement of regulations [9], the use of SCTs in Poland was allowed from 1 July 2017, but the introduced provisions were criticized. One argument was that regulations omitted devices already installed [16]. However, it should be noted that the solutions used so far had not met safety requirements, so their locations and dimensions were not taken into account in the regulations. The limitation of SCTs to fix time control was criticized, and stricter rules on the location of these devices required.

Another literature review [17] was published after the implementation of regulations in Poland. Thirty-five bibliographic references were analyzed together with a statement of general opinion on selected factors. The methodology used for SCT research was described in another article [18]. The main goal of the research was to determine the impact of countdown timers on traffic conditions and the level of traffic safety.

1.2.2. Capacity Issues (Headway/Saturation Flow/Capacity/Start-Up Lost Time)

A 2005 study in Kuala Lumpur (Malaysia) [19] concerned the SCT for the green signal (GSCT). The research covered the period before and after installing the GSCT. The study indicated that the impact of installing the GSCT was small. Other research carried out in Kuala Lumpur [20] concerned the influence of SCTs on vehicle capacity at six intersections. The time intervals between vehicles moving at the start of the queue were examined (headways). Only passenger vehicles were investigated. Measurements with heavy vehicles were rejected. The study showed that saturation flow for a junction with an SCT is greater than for a junction without an SCT.

Issues related to intersection capacity were also examined in the research at intersections in Bangkok [21]. Start-up delay and headway in the initial period of the green signal (six vehicles) and the later period were examined. In the case of headway, no differences were found for the subsequent period.

Similar studies were conducted in India [22] concerning mixed traffic in conditions of queuing. Two-wheeled vehicles, rickshaws, and buses were distinguished. It was emphasized that the drivers in India showed a lack of discipline. Video recordings were analyzed with the use of Matlab. The study showed reduced time intervals at the beginning and end of the green signal using SCT.

Headway was also analyzed during start-up of a vehicle column at a junction with an RSCT [23]. The research was conducted in Guangzhou (China). The results showed a reduction in headway and increased capacity due to the use of RSCT by approx. 5% at night and approx. 10% during the day, and an even greater reduction in headway standard deviation.

Similarly, different technical solutions were explored in [24]. The comparison applies to standard traffic lights, signals with different types of light rules, and numerical SCTs. The research was conducted in Brazil. The studies showed no statistically significant effect on the headway of different types of SCT in relation to standard signaling.

Study [25] investigated the effect of RSCTs on the headway/saturation flow. The research was conducted with the SCT turned on and off for 24 h and showed that the use of the RSCT reduced start-up time by 22%. The statistically significant impact concerned only the first vehicle in the queue, regardless of the period of the day. In some cases, reduction of start-up delay was even more significant. Use of an RSCT in Bangkok reduced it by 33% [21]. Research was also carried out in New Delhi at three intersections [26] where was no influence of GSCT on saturation flow, whereas RSCT reduced start-up lost time.

Other research [12] was carried out in Kayseri (Turkey) at the approaches of two intersections where an SCT was used for the red signal (RSCT) and green signal (GSCT) at the crossroads. The study determined the start-up time distributions of the first vehicle in the queue after the green signal was displayed. A statistically significant positive effect on the start of the queue of vehicles was demonstrated but, simultaneously, an increase in the number of crossing the intersection before the green signal was noted. In study [27], the psychological aspect was not considered when analyzing drivers' behavior during vehicle start-up, pointing to the possibility of obtaining a start-up delay of 0 s, which contradicts the conclusions from [12], indicating the random nature of this phenomenon.

Study [28] also concerns starting vehicles at traffic signals. Tests were carried out in Changchun (China) at six intersections with quite long cycles (100–150 s). The research concerned an RSCT and covered the first three vehicles in the queue. In the regression models used for the analysis, nine factors related to the intersection and traffic influencing start-up delay and headway were considered and showed a reduction in both values after using the RSCT.

1.2.3. Red Light Violation (RLV)

One of few studies analyzing the long-term impact of GSCT on traffic was conducted in Singapore [29]. The number of red signal violations was examined before installing the GSCT and after installation for 1.5, 3, and 7.5 months. The research was conducted at different times of the day and on different days of the week. In the first two post-installation periods, RLVs (number of red-light violations) were lower than before the GSCT was installed. However, after 7.5 months, the number of RLVs returned to the number before the GSCT installation.

In Kuala Lumpur (Malaysia) study [19], the number of entries on the red signal was two times smaller when the GSCT was installed. However, the need for further research before using these devices was indicated. In other research conducted in Kuala Lumpur [20], the rate of RLVs was higher in the case of using an SCT.

In Bangkok, a survey was conducted with over 300 drivers constantly moving around the city [30]. An essential factor was that the cycle times used in this city were 120–240 s, twice as high as the typical values used in Poland. The research showed only a slight change in saturation flow, while the start-up lost time decreased significantly. For other vehicles, as in [25], no differences were found. The absence of SCTs increased the number of vehicles entering during the red signal. The questions in the survey related to reductions of driver frustration, turning off the engine when parked, and using the waiting time for the green signal.

A study in Chennai, India [31] examines the effects of the RSCT and GSCT across four vehicle types. An increase in red signal violations was noticed in the final period, especially among rickshaw drivers and two-track vehicle drivers. On the other hand, the number of entries at the beginning of the green signal decreased among all road users. Concerning headway, the entry time of the first vehicle was significantly reduced, and the differences were small for subsequent vehicles.

Article [32] presents research carried out at three junctions in New Delhi (India). Signal programs with cycle lengths of 165–180 s were in place at the intersections. The number of red-light violations in the final period of a red signal increased. In article [33], 44 bibliographic items regarding SCT research and other solutions influencing red signal violations are presented. Conclusions concerning the use of SCT are both positive and negative. The main difference is in the use of GSCT and RSCT. Research carried out in New Delhi at three intersections indicated that both types of SCT (GSCT and RSCT) affected red-light violations [26].

Article [34] concerns a study carried out in Hyderabad, India, at two intersections. The research included start-up loss time and capacity. Turning off the SCT increased the number of red-light violations by about 5%. The research revealed higher speeds at intersection approaches with a working SCT.

The research in [35] was carried out following the methodology described in [18] and [36] concerning a GSCT. Crossing at the beginning of the red signal at three intersections was analyzed. A significant effect of GSCT in reducing the number of red-light violations by buses and trucks was also observed.

Research in Europe conditions is presented in article [37] including SCTs installed at two intersections in Cotonou, Greece and involved vehicle column start-up and drivers' behavior at the end of the green signal. The results showed that there were practically no early starts at intersections without an SCT. When an RSCT was used, the number increases by approx. 24%. The use of a GSCT also increased the number of vehicles stopping at the end of the green signal. At the same time, the proportion of vehicles passing on the red signal decreased, but the proportion accelerating during the amber signal increased.

A study from Ljubljana (Slovenia) [38] covered one intersection. The article presents the verification of three hypotheses concerning start-up lost time, amber violations, and red-light violations. A survey among drivers was also conducted. Concerning signal violations, a greater number of entries during the amber signal occurred with the GSCT turned off, as did red-light violations. However, with respect to violation of the red-amber signal (before the green signal), a greater number of cases occurred with the RSCT turned on. The results showed positive feedback from drivers about the SCT. However, about 30% of people indicated that SCTs can be distracting.

In Poland, the first research on driving behavior at an SCT was carried out in 2014, although SCTs were used before 2014 at intersections. Research carried out by the Municipal Roads Authority in Grudziądz [39] used a speed camera and a red-light violation camera and found that the launch of the SCT increased the number of entries on the red signal, although these offenses were registered and the drivers were punished for them. Similarly, with the inclusion of an SCT, the proportion of vehicles driving over the speed limit increased.

1.2.4. Entering on Amber Signal

In a study carried out in Zabrze (Poland) it was observed that disabling the GSCT increases the number of vehicles entering the intersection during the amber signal and at the beginning of the red signal [40]. Research in [41] concerned the influence of a pedestrian SCT (PSCT) on driver's behavior. It was observed that SCTs increased the number of vehicles passing through on an amber signal and reduced the number of passages on a red signal shortly following the amber.

A study in Kayseri (Turkey) [12] observed a reduction in the number of vehicles entering during the amber signal with a working SCT, except during congestion and with significant delays at the approach. The authors concluded it was important to consider the behavior of drivers at the end of the green signal when assessing appropriate times for each signal phase. The conclusion that the RSCT did not affect traffic safety is contradictory to the findings of other authors.

A different study dealt with the termination of the green signal [42]. The research was conducted in Changsha (Hunan Province, China) at four intersections. The signaling programs had cycles of 100–128 s. Three drivers' reactions were studied: stopping before the stop line, passing through on the amber signal, and passing on the red signal. Binary logistical regression analysis was used. The study showed a greater number of vehicles passing on the amber and red signals after applying the GSCT.

The research methodology in [43] is similar to that in [42] and the same research grounds were used. Logistic regression was used to analyze the data. The probability of stopping as a function of speed and distance from the stop line was examined. The study was extended to the analysis of the method of crossing the intersection on the red signal. It was confirmed that the use of a GSCT increased the number of entries on the amber signal.

As well as GSCTs and RSCTs, there are also displays showing the duration of the amber signal. A study of these in Harbin, China, is presented in [44]. Such displays reduced the number of vehicles entering during the second, third, and fourth seconds of the amber signal. A display with the function of showing the duration of the amber signal did not change the speed of vehicles, unlike the GSCT, which caused an increase.

1.2.5. Vehicle Speed, Braking Characteristics, and the Dilemma Zone

The issue of traffic safety from the point of view of the mechanics of vehicle movement when an SCT is turned off is described in [27]. Analyzing mathematical relations, it was found that the GSCT did not increase the vehicle speed above the minimum. A psychological factor was not included in the article. Research in [45] showed a reduction in the speed of vehicles reaching a crossing during the yellow signal. A negative impact on the speed of vehicles at the approach was found in Kayseri (Turkey) [12]. Vehicle speed at intersections with a GSCT was also analyzed in New Delhi [46]. The proportion of vehicles exceeding the speed limit at the intersection entrance was higher (approx. 4% compared to approx. 13%). On the other hand, the speed of vehicles crossing the intersection with a GSCT during the amber signal was almost twice as high.

An analysis of an SCT functioning in the Philippines is presented in [47]. As part of the study, drivers were surveyed in Manila and at intersections in Quezon and Pasig. It was observed that the use of countdown timers caused no low-speed vehicle passages in the final period of the amber and green signals, while there were soft brakings and stops of cars on the amber signal and starts from the stopping line before the green signal. Research also shows that drivers' declared and real behavior during the amber and green signals differed significantly. For example, drivers declared that they decelerated during an amber signal when they were accelerating or moving at a constant speed.

An analysis covering decisions made by drivers while driving through an intersection with and without an SCT was conducted in China [48]. A logistic model was used for the research, taking into account the probabilities of such behaviors as acceleration, braking, and speed maintenance. Determined from the time displayed on the GSCT, the distance from the stop line and the vehicle's speed was assessed. The study showed that at intersections with an SCT, the proportion of stopping or slowing vehicles was greater than at intersections without an SCT, while the percentage of accelerating vehicles decreased.

Differences in drivers' behavior at traffic lights with an SCT and a CCTV system in summer and winter were examined in [49]. The research was conducted in Changchun, China, at eight intersections where traffic violation cameras were installed. The tests covered 90 m long sections of intersection approaches. Based on video observations, braking speeds and decelerations were determined. It was observed that in winter, drivers braked at a greater distance from an intersection equipped with a GSCT, while at intersections not equipped with a GSCT, drivers braked when an amber signal was displayed.

GSCT tests were also carried out with the use of a vehicle simulator [50]. The study included an analysis of drivers' behavior in the dilemma zone and braking deceleration. With a GSCT in place, drivers who were in the dilemma zone decided to stop more often, and at the same time, braking was performed with less deceleration. Research in [25] also showed a shortening of the dilemma zone. The influence of GSCT on vehicle traffic parameters used in car-following models (the method used to determine how vehicles follow one another on a roadway) was also investigated [51]. Numerical simulation showed that GSCT significantly affected drivers' behavior before traffic lights and should be considered during simulations.

The technique used for measurements has also developed over the years. Article [36] describes SCT tests performed with the use of UAVs. Video recording from the air has proven successful and, as a result of the study, a solution was obtained to reduce the start-up lost time (by 16–39%) and increase the speed of crossing the intersection (by 10–28%).

1.2.6. Reaction Time

Another influence of the SCT is a shortening of the brake perception-reaction time. This is indicated by research in [52] carried out in Harbin, China, at six intersections with an amber signal duration of 3 or 4 s and a cycle length of 107–170 s. The presence of a GSCT significantly reduced the driver's reaction time to changing signals. The article analyzed reaction time distribution, and the median reaction time was reduced by 30%.

A broader issue was described in the article [53] concerning various solutions about the end of the green signal such as a green blinking signal and an RSCT. The article presents five models concerning the driver's decision to drive or stop. The analysis concluded that the driver's reaction with the application of GSCT was faster than that resulting from the use of a flashing green signal, and the slowest response occurred in the case of the standard sequence of signals (green/amber/red).

1.2.7. Safety (Number of Crashes, Vehicle-Vehicle Conflicts)

Safety characteristics of SCTs were investigated in [54] for not only a GSCT and an RSCT but also CCT a (continuous countdown timer—displaying all signals) The literature was researched using the PRISMA method and 79 references were analyzed, of which 14 were thoroughly investigated. It was found that the effect of countdown timers is dependent on the type of device, and different parameters in different studies indicated improvement or lack thereof. In particular, the impact of SCTs on road safety was not unequivocal.

Article [55] describes the advantages and disadvantages of SCTs and presents survey results. The opinions of the Ministry of Infrastructure are also presented, and the cities using these devices are listed (albeit, in a manner inconsistent with applicable law). Article [56] covers an analysis of SCTs installed in Toruń (Poland) at four intersections. The number of crashes on intersections was used as a measure of safety. The test results did not show any improvement or deterioration in traffic safety.

In other research, standard traffic lights, signals with different types of light rules, and numerical SCTs were compared. Traffic safety studies were carried out using the pre and post analysis method. The results show a reduction in the number of crashes by approximately 35% [24].

Research in Słupsk (Poland) [57] included only questionnaire studies on the opinions of road users in terms of safety, driving comfort, and driving behavior. However, as noted in [47], there were significant discrepancies between the survey results, and the article did not contain a verification.

The values displayed on the SCT impact the driver and can be analyzed in conjunction with road crashes. This issue is described in article [58], which includes a case study for a selected intersection in Krakow. However, the article's conclusions indicate that further research is needed related to traffic safety with the use of an SCT. Study [59] considers the time display when modeling traffic with a cellular automaton. The model was implemented for a two-lane road section.

A detailed analysis of the causes of GSCT-related road incidents is presented in [60]. The research was carried out for five intersections in Delhi (India). The intersections were characterized by a cycle length of 220–240 s, and characteristics related to red light violations and conflicts between road users were determined as part of the research. The use of a GSCT reduced the number of red-light violations and the number of conflicts by almost half.

The influence of RCST and GCST on road safety in China was also analyzed [61]. In 2000–2007 such devices were installed at 1036 intersections in China. The tests were conducted in Taipei in four periods: preinstallation, 1.5 months after installation, and 3 and 4.5 months after installation. The research showed a negative impact of the GCST on the dilemma zone and traffic safety.

The psychological aspect of using SCT is critical, and can be studied based on statistics gathered by surveys. One such survey took into account the gender, age, and driving experience of drivers. It included 32 questions regarding opinions on various SCT solutions, driver's behavior, such as engine shutdowns when stopped at traffic lights, and the scope of the displayed information [62].

Innovative technical solutions in the field of RSCT are presented in [63]. Various graphic SCT solutions were presented in which the passage of time was displayed as an LED ruler, a circle segment, and a change in the diameter of the signal. The proposed solutions were tested on a group of 12 road users and assessed in terms of the perception of the information provided and assessment of the impression of the device. Doubts as to the practical application of the solutions were raised, because of difficulties regarding meeting the requirements of the norm [64].

1.2.8. GHG Emissions

The impact of SCT on greenhouse gas (GHG) emissions was investigated in [65]. The study was conducted in Haeundae-gu, Korea. As part of the study, a driver behavior model was developed and 340 trips were made to test the effect of SCT on such behavior. Then, a model of the Haeundae-gu road network was made in the Vissim program, and based on the simulation results, the GHG emission from idling vehicles was determined. It was shown that with the use of an SCT it is possible to significantly reduce greenhouse gas emissions resulting from a reduction in the number of stops and the possibility of turning off the engine.

1.2.9. Cycling

A study of the impact of an RSCT on cycle traffic was analyzed in [66]. The study was conducted in Groningen, the Netherlands. Research showed that cyclists observe the SCT and adjust their riding style to the displayed value. After using an SCT, a small number of cyclists approached the intersection during the amber signal, and eye-tracking studies indicate that they observed the SCT from a distance of about 60–70 m. However, when standing in front of a signal head, eyesight is more focused on the SCT than on the signal head.

1.2.10. Pedestrian Passing on Flashing Green

Research [67] conducted in China was also related to pedestrian decision-making before crossing an intersection. The study was carried out at crossings equipped with PSCTs. Pedestrians were divided into groups according to their behavior. While the study did not directly address the impact of PSCTs on road traffic, the methodology could be used for this purpose.

A study done in Korea concerning numerical and graphical SCTs monitored the number of pedestrians entering the road during a flashing green signal. A reduction in the number of pedestrians passing during a flashing green signal was indicated, which is not allowed in Korea [68].

Extensive PSCT studies have been conducted in Sydney, which concluded that the use of PSCT did not reduce the number of entries during the "Flashing Don't Walk" (blinking) signal and reduced the proportion of late finishers (pedestrians entering the crossing in the last seconds of the green signal) [69]. Other research based in Sydney [70], indicated an increase in the proportion of pedestrians entering the crossing during the "Flashing Don't Walk" signal after the launch of a PSCT. Similar results were obtained in a study [71], also carried out in Australia. An increase in the number of entries to the pedestrian crossing was observed during the "Flashing Don't Walk" signal.

A study conducted in the USA [72], indicated that PSCTs increased the number of pedestrians entering during a "Flashing Don't Walk" signal. Similar results were described in research carried out in India [73], indicating an increase in the number of pedestrians entering the crossing in the final period of the green signal and during the yellow signal. Research [45] showed a reduction in the proportion of pedestrians reaching a pedestrian crossing during the "Flashing Don't Walk" signal and shortening of the dilemma zone.

Research conducted in the United Arab Emirates [74] showed that after using a PSCT, the number of people crossing according to regulations increased, i.e., crossing at the end of the "Flashing Don't Walk" signal. The research also included a survey in which the majority of respondents supported the use of the PSCT.

In other research, PSCTs were analyzed concerning the time counted down during green and red signals [75]. The study was conducted at five intersections in Shanghai. Signal cycle lengths ranged from 128 to 200 s. The use of a PSCT reduced the number of pedestrians over 50 years old crossing during a flashing green signal. In all age groups, the number of pedestrians finishing the crossing during the flashing green signal increased. The reaction time of pedestrians at the beginning of the green signal also decreased, especially for people aged 50+.

A study in China researching children's' behavior showed no influence on the ratio of entries during the green blinking signal. Still, slightly more children finished crossing during the blinking green signal. The use of a PSCT did not affect crossings started during the blinking green signal [76].

A team from the Silesian University of Technology in Poland researched PSCTs [40]. The research was carried out in Zabrze. With the PSCT turned off, the proportion of pedestrians entering the crossing was greater during the flashing green signal (allowed in Poland) and at the beginning of the red signal.

1.2.11. Pedestrian Red-Light Violations

Research held in China [77], investigated the behavior of pedestrians, dividing them into four groups. The results showed a statistically significant increase in the proportion of pedestrians obeying signaling when a PSCT was used. Children's behavior related to PSCT is described in article [76]. The PSCT counts down the duration of the green flashing signal and the duration of the red signal. The research was conducted in Jinan (China) at two intersections. During the study, it was found that at an intersection equipped with a PSCT, the proportion of children entering during the red signal was greater.

Other studies conducted in Europe (Greece, Thessaloniki) [78] indicate numerous incorrect crossings of pedestrians regardless of various factors. The proportion of pedestrians crossing during the red signal in this study are greater than those presented in other articles.

PSCT research was also conducted on two intersections in New Delhi [79]. The use of a PSCT significantly increased the number of pedestrians crossing during the red signal. At crossings without a PSCT, the proportion of people crossing during the green signal was higher, while after installing a PSCT, the percentage of people crossing during the amber signal also increased. The use of a PSCT did not affect the waiting time of pedestrians before the crossing or times of arrival at the pedestrian crossing.

In Poland, the inclusion of a PSCT reduced the number of entrances at the beginning of the red signal but resulted in pedestrians to intrude on the crossing at the end of the red signal [40].

1.2.12. Pedestrian Speed

A study prepared in Sydney indicates that the speed of pedestrians passing through the crossing increases with a PSCT in place [69]. During the "Flashing Don't Walk" signal, pedestrians move at a higher speed than when the PSCT is off. Nevertheless, pedestrians positively assessed the use of PSCT in the survey [70]. Similar results were obtained in a study [71], also carried out in Australia. An increase in the speed of pedestrians at the crossing in the final period of the green signal was observed [71]. In Shanghai, the speed of pedestrians who started crossing on the green signal after 5 s increased [75]. An increase was also observed in the research described in [45]. In research conducted in New Delhi [79], no influence of PSCT on the speed of pedestrians crossing the road was found.

1.2.13. Pedestrian Safety (Vehicle-Pedestrian Conflicts)

A study in Sydney indicated that it is not possible to unequivocally assess the impact of a PSCT on the number of pedestrian-vehicle conflicts [69]. Studies conducted in Canada do not show a statistically significant influence of PSCT on the number of pedestrian-vehicle collisions [80].

The safety aspects of PSCTs studied at 106 intersections with traffic lights in North Carolina, USA [81]. There was a statistically insignificant reduction in the number of road incidents with pedestrians after installing PSCTs. In contrast, the total number of road incidents decreased in a statistically significant manner.

PSCTs were also researched in a study related to road traffic accidents of older people [82]. The research was conducted at 190 intersections, and included the analysis of road incidents before and after installing a PSCT. A radius of 250 ft from the intersection was assumed as the intersection area. The study showed a reduction in the total number of incidents injuries after installing the PSCT.

However, when PSCTs were used in India [73] and China [76], the number of pedestrian-to-vehicle conflicts was more significant. Canadian studies [83] showed that PSCTs increased the number of all crashes with pedestrians by 7.5%. There was no significant influence on the number of crashes with injuries or fatalities.

Research was also conducted in the USA on the use of pedestrian buttons at intersections with a PSCT [84], which showed the use of pedestrian buttons reduced the number of traffic incidents with pedestrians. The use of a PSCT alone, without buttons for pedestrians, did not bring about any effect on improved safety.

The behavior of pedestrians using crossings with pedestrian islands in Belgrade (Serbia) has been investigated [85]. A PSCT had a positive effect on reducing the number of pedestrians entering the first pedestrian crossing, but did not affect pedestrians leaving the second crossing.

Study [86] shows that the use of a PSCT had a statistically significant effect on the number of rear collisions and the number of incidents involving pedestrians. The analysis of PSCT assembly costs and road incident costs shows that the use of PSCT is economically justified.

1.2.14. Literature Review Summary

The above literature analysis shows that there are significantly different results concerning the effects of STCs. Table 1 presents the essential information about each of the studies.

A '+' sign indicates a positive impact on road traffic (understood as improvement of safety, improvement of indicators related to safety, or improvement of measures of effectiveness). For example, this symbol denotes a decrease in headway or an increase in saturation flow. The symbol '−' indicates a negative impact on road traffic, while symbol 0 shows no significant effect. In the "others" column, L stands for literature analysis and research methodology development, S survey research, T theoretical analysis, M modeling or data collection for modeling, C comparison SCT with similar devices.

Table 1. Summary of research results on SCT (set chronologically and divided into SCT for drivers and pedestrians).

Bibliography	Headway/Saturation Flow/Capacity	Start-Up Lost Time	Red light Violation before the Onset of Green Signal	Red Light Violation at the Beginning of the Red Signal	RLV at the Start of a Red-Long Time after Installing the GSCT	Entering on Amber Signal	Vehicle Speed	Braking Characteristics	Dilemma Zone	Number of Crashes	Pedestrian Reaction Time	Pedestrians Passing on Flashing Green (or Amber or Flashing Don't Walk) Signal	Pedestrian's Red Light Violation	Pedestrian Speed	Number of Conflicts Vehicle-Pedestrian or Vehicle-Vehicle	Driver Reaction Time	GHG Emissions	Adjusting the Technique of Cycling to the SCT Indications	Others
[68]												+							
[41]						–													
[69]												0		+	0				
[81]										+									
[70]												–		+					
[71]												–		+					
[75]											+	+		+					
[72]												–			0				
[80]										0					0				
[77]												–	–						
[73]												–			–				
[76]												0	–	0	+				
[45]									+					+					
[40]			–									+	0						
[79]												–	–	0					
[74]												–							S
[83]										–									
[84]										0									
[86]										+									
[82]										+									
[85]																			
[67]																			L
[19]	0			+															
[29]				+	0														
[12]		+	–			+	–												

Table 1. *Cont.*

Bibliography	Headway/Saturation Flow/Capacity	Start-Up Lost Time	Red light Violation before the Onset of Green Signal	Red Light Violation at the Beginning of the Red Signal	RLV at the Start of a Red-Long Time after Installing the GSCT	Entering on Amber Signal	Vehicle Speed	Braking Characteristics	Dilemma Zone	Number of Crashes	Pedestrian Reaction Time	Pedestrians Passing on Flashing Green (or Amber or Flashing Don't Walk) Signal	Pedestrian's Red Light Violation	Pedestrian Speed	Number of Conflicts Vehicle-Pedestrian or Vehicle-Vehicle	Driver Reaction Time	GHG Emissions	Adjusting the Technique of Cycling to the SCT Indications	Others
[20]	+			−															
[27]																			T
[48]							+												
[25]	0	+																	
[22]	+																		
[61]	0	+	0						−										
[30]	0	+	−																S
[28]	+	+																	
[42]				+		−													
[21]	0	+																	
[31]	0	+	−	+															
[14]																			L
[38]	0		−	+		+													S
[43]				−		−													
[23]	+																		
[55]																			S
[56]										0									
[37]			−	+			−												
[53]																			C
[46]				+			−												
[47]			−			+													S
[44]						+	0												
[63]																			C
[24]	0									+									C
[51]																			M

Table 1. *Cont.*

Bibliography	Headway/Saturation Flow/Capacity	Start-Up Lost Time	Red light Violation before the Onset of Green Signal	Red Light Violation at the Beginning of the Red Signal	RLV at the Start of a Red-Long Time after Installing the GSCT	Entering on Amber Signal	Vehicle Speed	Braking Characteristics	Dilemma Zone	Number of Crashes	Pedestrian Reaction Time	Pedestrians Passing on Flashing Green (or Amber or Flashing Don't Walk) Signal	Pedestrian's Red Light Violation	Pedestrian Speed	Number of Conflicts Vehicle-Pedestrian or Vehicle-Vehicle	Driver Reaction Time	GHG Emissions	Adjusting the Technique of Cycling to the SCT Indications	Others
[15]																			L
[54]																			L
[32]	0	+							+										
[52]																+			
[39]			−	−			−												
[57]																			S
[49]								+											
[62]																			S
[26]	0	+	−	+															
[50]								+	+										
[16]																			L
[17]																			L
[18]																			L
[58]										0									L
[59]																			M
[36]		+					−												
[33]																			L
[65]																	+		
[35]				+															
[60]															+				
[34]	0	+		+			−												
[66]																		−	
No. of '+'	4	11	0	10	0	4	1	2	3	4	1	3	0	5	2	1	1	0	0
No. of '0'	11	0	1	0	1	0	1	0	0	4	0	2	1	2	3	0	0	0	0
No. of '−'	0	0	9	3	0	3	6	0	1	1	0	7	3	0	1	0	0	1	0
Rating	4	11	−9	7	0	1	−5	2	2	3	1	−4	−3	5	1	1	1	−1	0

Based on the number of articles, a rating was calculated considering publications with a positive impact of an SCT on a given factor with a weight of '+1' and a negative impact with a weight of '−1'. On this basis, it can be concluded that in terms of positive effects (rating 4–11) of the use of SCT, the most frequently mentioned are:

- reduction of start-up lost time,
- reduced red light violations at the onset of the red signal,
- decreased headway/increase saturation flow or capacity.

The factors assessed negatively (score from −3 to −9) in order of the most frequently mentioned:

- increased red light violations before the onset of the green signal,
- increased vehicle speed,
- increased incidences of pedestrians crossing on a flashing green (or amber or Flashing Don't Walk) signal,
- increased pedestrian red-light violations.

The remaining factors were examined in individual studies. The results were inconsistent, or no significant influence of SCT was demonstrated.

Due to the often-divergent results, research gaps remain. Most of the research was conducted in large urban centers. It is therefore important to conduct research in smaller cities and outside the cities. There are visible differences in drivers' behavior between individual countries, related to, for example, cultural norms. Research carried out in Poland is sparse, so it is advisable to conduct research to a greater extent. It is interesting to consider whether the results obtained mainly in Asia, America, and Australia will be similar in Poland. There are no studies related to traffic safety during failures in the use of SCTs or the traffic light system.

1.3. Article Content

The article is structured as follows. Section 1 presents an in-depth literature review and analysis of the research carried out so far. Section 2 presents the research methodology of the conducted research. Section 3 presents the results of the field research. The latter part of the article (Section 4) discusses the results in relation to the literature analysis and previous research. Unexplored issues related to the functioning of SCTs are indicated. Selected case studies associated with SCTs are also presented. The aim of the paper is to conduct a comprehensive analysis of the literature, to conduct field research carried out covering many criteria, and to identify previously unexplored issues related to SCTs that have not been described in previous publications. The article is summarized in the conclusions presented in Section 5.

2. Methods

2.1. Place and Method of Conducting Research

Field measurements were carried out to experimentally evaluate the impact of countdown timers on the movement of vehicles. The research aimed to assess the validity of the use of countdown timers and their impact on the driver's decision to brake and start the vehicle. The research included making direct observations of traffic at the approaches of selected intersections. The study was conducted in Płock, Poland (Masovian Voivodeship). Płock is a county city with a population of approximately 120,000. Traffic lights are installed at 57 intersections; 32 are equipped with SCTs and 11 with PSCTs. An SCT for vehicles performs the functions of an RSCT and GSCT.

The following factors were taken into account when selecting the study site:

- the presence of queues of vehicles at the approaches,
- fixed-time control,
- presence of an SCT for vehicles,
- the possibility of installing the camera in a way that allows observation of the approach and indications of signaling devices and SCT.

Because of the above requirements, three intersections were selected:

1. The intersection of F. Kobylińskiego Avenue with I. Łukasiewicza Street;
2. The intersection of I. Łukasiewicza Street with Miodowa and Tysiąclecia Streets;
3. The intersection of F. Kobylińskiego Avenue with Bielska and Jachowicza Streets.

Later in the study, intersections were marked with the numbers mentioned above. The location of intersections on the Płock map is shown in Figure 2.

Figure 2. Location of measurement objects in the city of Płock. (Background map © OpenStreetMap contributors, [87]).

All intersections have four approaches, with traffic lights operating in a fixed-time mode. There are pedestrian crossings at each of the intersections at all legs. Moreover, during the morning and afternoon rush hours, queues of vehicles of an appropriate length form at intersections.

At each of the intersections equipped with an SCT, measurement points were selected to observe the approach. The measurements were made using an image recording method. Vehicles crossing the stop line at the selected approach were recorded [88,89]. A Sony camera was used, which was placed on a tripod at a convenient observation point. The camera was set in a place that did not attract the attention of drivers, and as little visible to the drivers as possible. The field of view for the camera used for the measurements was 155 degrees and included the stop line, a distance of about 10 m in front of the stop line, and signals. The focal length of the camera was f = 29.8–298.0 mm (16:9). Films were recorded in HD quality. Time was superimposed on the image with an accuracy of 0.1 s. Vehicles on all lanes were visible.

Based on previous measurements of vehicle traffic, the duration of the morning and afternoon peaks was determined: morning peak hours 6:45 a.m. to 8:45 a.m. and afternoon peak hours 2:45 p.m. to 4:45 p.m. The research was carried out in both rush-hour peaks and the off-peak period (11:30–13:30). Traffic at junction No. 1 was monitored in the morning rush hour, in the off-peak period at junction No. 2, and in the afternoon rush-hour at junction No. 3. A total of 12 h of recordings were made and were used in full to compile the results.

During the measurements, the weather conditions were very good (sunny, rainless). Due to administrative procedures related to obtaining appropriate approvals, measurements were taken on 19 July and 20 July 2017. This is the period of the summer holidays. The test results were analyzed manually using a method of observation and analysis of the obtained video material.

Unfortunately, the traffic volumes and lengths of queues during the holiday season (the Road Administration agreed to such date of the study) were not high. Therefore, only six vehicles from the queue were registered for intersection No. 2, data for eight vehicles were recorded for intersection No. 1, and for intersection No. 3 observations only nine vehicles were included.

Courtesy of the Płock City Hall, the research was conducted in the same conditions with the SCT turned on and off. This made it possible to compare the behavior of drivers in the following days. Both drivers' behavior at the green timer (GSCT) and the red timer (RSCT) were tested. The research focused on the following three issues:

- respect for the red signal by drivers,
- headway during vehicle column start-up,
- time of entering the intersection after the end of the green signal.

Apart from speed, which could not be measured by the adopted method, these are the most frequently studied factors related to the functioning of a GSCT.

2.2. Characteristics of Intersections

At intersection No. 1, measurements were made at the eastern approach. There is one 3.25 m wide lane for straight ahead, a left-turn lane, and a right-turn lane. The traffic organization at intersection No. 1 is shown in Figure 3.

Figure 3. Traffic organization at intersection No. 1. (Background map Head Office of Geodesy and Cartography).

During measurements at the intersection, signaling programs with a cycle duration of 100 s (06:30–08:00) and 95 s (08:00–08:30) operated. The green signals had a time of 36 s and 32 s, respectively.

At intersection No. 2, measurements were also carried out at the eastern approach. There are no marked lanes at the approach, and its width is 5.8 m. The traffic organization at intersection No. 2 during the measurements is shown in Figure 4. Currently, the traffic organization has been changed.

Figure 4. Traffic organization at intersection No. 2. (Background map Head Office of Geodesy and Cartography).

A signaling program with a cycle duration of 90 s was in operation during measurements at the intersection. The green signal duration was 30 s.

At intersection No. 3, measurements were made at the western approach. There are two 3 m wide lanes for straight-ahead driving. Besides, there is a left-turn and right-turn lane. The traffic organization at intersection No. 3 is shown in Figure 5.

Figure 5. Traffic organization at intersection No. 3. (Background map Head Office of Geodesy and Cartography).

During the measurements at the intersection, signaling programs with a cycle duration of 100 s (06:30–08:00) and 95 s (08:00–08:30) operated. The green signals had a duration of 35 s and 26 s, respectively.

Table 2 shows the results of traffic measurements at the intersections. The number of vehicles other than passenger cars was minimum. During the research, the influence of the type of vehicles on the results was not considered. Vehicles other than passenger cars were not analyzed.

Table 2. The volume of vehicles on intersections during the measurement period (2 h).

Intersection	Vehicle	Volume (per 2 h)	
		GSCT Off	GSCT On
1	Passenger cars	707	507
	Trucks	12	6
	Buses	10	5
	Two-wheelers	5	4
	PCE	753	531
2	Passenger cars	272	366
	Trucks	2	4
	Buses	9	11
	Two-wheelers	3	2
	PCE	297	399
3	Passenger cars	564	507
	Trucks	9	8
	Buses	5	3
	Two-wheelers	4	8
	PCE	593	532

2.3. Data Analysis

After the recordings were made, the material was divided according to the SCT function (on or off). Then, the recordings were prepared for analysis. The analysis was performed separately for each of the recorded parameters:

- respect for the red signal by drivers,
- headway during vehicle column start-up,
- times of entering the intersection after the end of the green signal.

Statistical data analysis was performed using the R system [90,91]. The charts were prepared using the ggplot2 package [92,93]. The methodological steps of data analysis are shown in Figure 6.

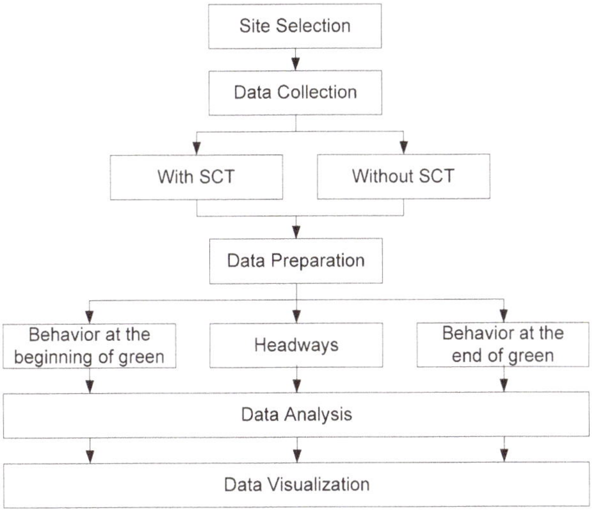

Figure 6. Flowchart of data analysis.

2.4. Other Methods Used for SCT Research

The following simplified formula is used in traffic modeling computer programs such as Synchro [94] and Transyt to analyze fuel consumption:

$$F = TotalTravel \cdot k1 + TotalDelay \cdot k2 + Stops \cdot k3 \quad (1)$$

where: $k1 = 0.075283 - 0.0015892 \cdot Speed + 0.000015066 \cdot Speed^2$, $k2 = 0.7329$, $k3 = 0.0000061411 \cdot Speed^2$, F = fuel consumed in gallons, *Speed* = cruise speed in mph, *TotalTravel* = vehicle miles traveled, *TotalDelay* = total signal delay in hours, *Stops* = total stops in vehicles per hour. Measurements in metric units were converted to imperial units. Based on fuel consumption, it was also possible to estimate the emission of pollutants CO, NOx, and VOC.

Parameters such as Delay and Stops are strictly related to the way traffic lights function [95]. An analysis of the SCT literature showed that they may have a potential impact on factors influencing fuel consumption (Table 1):

- headway/saturation flow,
- vehicle speed,
- start-up lost time.

An analysis of the research subject matter showed that many of the problems related to SCTs have not yet been investigated. Many road authorities are not aware of the risks associated with the use of SCTs.

The use of SCT in traffic-actuated control reduces the effectiveness of traffic control. The use of long countdown times makes it impossible to shorten this time at any time and speed up the passage a road user. However, it could be possible for traffic reasons, for example, when the vehicle stream has ended at one of the approaches. A display of the phase duration is only possible with fixed-time control. This solution was adopted in [9]. However, there are known cases of disabling traffic-actuated control and working in the fixed-time control mode because of an SCT installation. Such a situation should be assessed unfavorably due to control efficiency in off-peak hours. Although some studies indicate a slight improvement in control efficiency locally by reduction of headway [20,22,23,28], and globally [65], there are no comparative studies between fixed-time control with SCT vs. actuated control. It is particularly interesting how the change of control would affect efficiency in a period of lower traffic and using algorithms with priority for public transport.

The use of such algorithms allows shorter travel times, but also benefits in terms of energy consumption [96]. Problems with displayed time values may also occur with fixed-time signaling due to offsets changing during the exchange of signaling plans on coordinated arterial roads. The use of SCT requires developing special control programs and extension of the time of offsets. It is also impossible to display the values on the SCT in the case of control algorithms allowing for the extension of a certain phase without time limits (until a notification occurs). An example of such signaling is the all-red algorithm and the "return of green signal to the main road".

Efficiency of control, depending on the SCT solution used, can be tested using simulation methods. It is possible to analyze the countdown of all signals, only selected ones, or to limit the countdown period. Such methods make it possible to determine the number of implementations of a given phase and its duration depending on road conditions. This allows assessing the impact of additional control limits on measures of effectiveness.

Additionally, ref. [3] specified in Art. 23, that "Subject to the provisions of paragraph 12 of this Article, the only lights which may be used as light signals for regulating vehicle traffic, other than those intended solely for public transport vehicles, are the following (...)". Use of an SCT is, therefore, contrary to the provisions laid down by the signatories of this Convention.

The behavior of road users in an SCT failure mode situation is completely unexplored. Such situations include the shortening of the signal (red or green) in relation to the displayed time, as well as lengthening of these signals with respect to the time displayed on the SCT. A failure mode may also occur in the event of bulb burnout or other failures of the traffic control system. In this case, the signaling is switched to a flashing amber signal. This seems to be the most dangerous situation when using a GSCT. The driver assumes they will pass the intersection on a green signal, while due to a signal failure, they must give way to pedestrians or vehicles. The behavior of other road users must also be considered, such as braking, or an attempt to drive through the intersection during a signal failure. Traffic lights are derived from the signals used on railways, where the priority is traffic safety in the event of a failure [97]. Similarly, in the case of road traffic, the introduction of a new solution in traffic control should be preceded by detailed tests of traffic lights in failure mode situations.

The behavior of drivers causing crashes, or forcing other road users to avoid collisions, is very rare, since this is possible only through monitoring systems. It is necessary to record SCT indications, signals of traffic lights, and vehicle traffic. Such recordings can be analyzed as a case study. The number of crashes on a single intersection is usually low, which does not allow for statistical analysis. A case study of a road incident is presented in this article.

Scientists must consider that the benefits of an SCT can be assessed in different ways, e.g., by extending the red and amber signal to 2 s. Such a signal was used in Poland in the years 1990–1994 [98]. Similarly, the amber signal could also be extended. In the years 1990–2003, the amber signal in Poland had a duration depending on road speed. In the case of the green signal for trams, this can be extended if there is a tram in the dilemma zone. Such a solution is used in Warsaw, e.g., at the intersection of Marymoncka and Zabłocińska streets.

3. Results

3.1. Respect for the Red Signal by Drivers

During the measurements, the number of vehicles crossing the intersection was determined. Then, the number of vehicles passing during individual signals was monitored and broken down into:

- green signal—correct behavior,
- amber signal—the passage is allowed, if the vehicle is close to the stop line, it is beneficial to reduce the number of crossings on this signal,
- red signal (entries after the amber signal)—incorrect behavior,

- red signal with amber (entries before the green signal)—wrong behavior.

The proportion of particular behaviors were determined in relation to the number of all crossings through the intersection. The results are presented in Figure 7. The results are divided into individual intersections and the SCT status (On/Off). It should be noted that for the turned-off SCTs, not a single red-amber signal entry was recorded. For intersections No. 1 and No. 2, the proportion of entries on the green signal decreased after switching on the SCT, while for intersection No. 3, it increased. However, the proportion of entries both on red and red with amber signals increased for this intersection. At all intersections, the percentage of entries during the amber signal decreased.

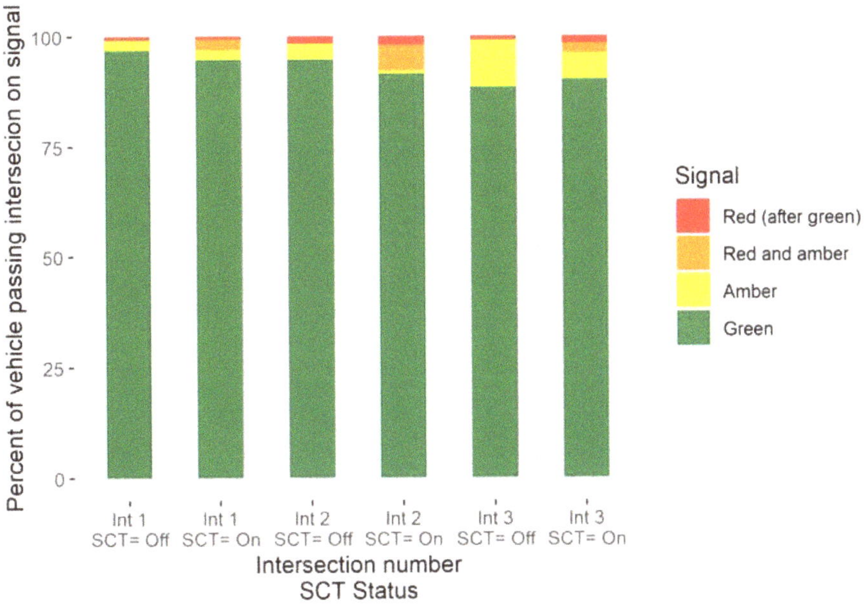

Figure 7. Percentage of vehicle entries on each signal.

3.2. Headway

Based on video recordings, the times of entry of subsequent vehicles from the queue were marked. The time of entry of the first vehicle was determined from the time of displaying the green signal to crossing the stop line. The mean values of the headway are presented in Figure 8.

In the chart, the number of the intersection is highlighted by color, and the RSCT status is marked with a symbol. The vehicle number in the queue is shown on the horizontal axis. Additionally, the following ranges are presented using background color:

- red background—minimum and maximum values from the measurements presented in the literature with the RSCT disabled;
- blue background—minimum and maximum values from the measurements presented in the literature with the RSCT enabled;
- yellow background—80% confidence interval (0.1 quantile–0.9 quantile) from the studies for intersections in Poland presented in [99].

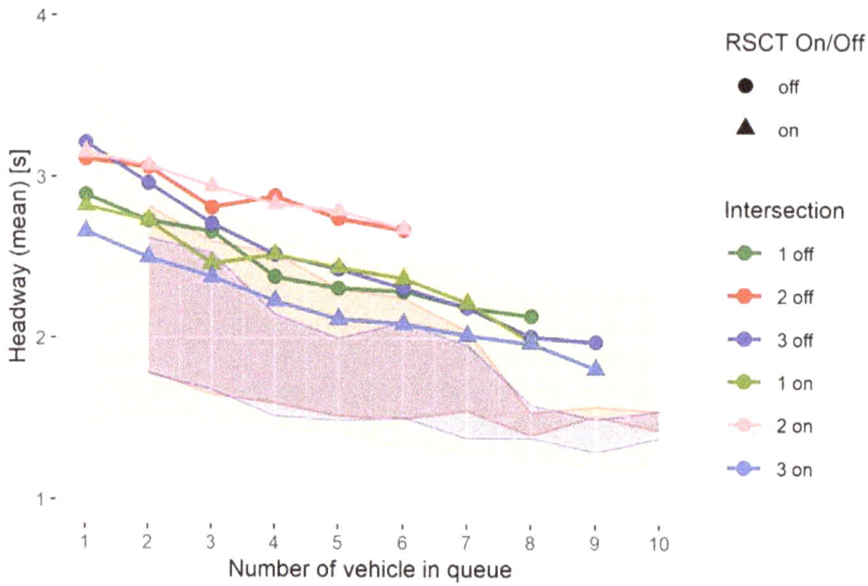

Figure 8. Headways between vehicles during queue start (description of filled areas given in text).

3.3. Amber Light Running

For vehicles entering the intersection after the end of the green signal, the time from the end of the green signal was recorded. The results were broken down into intersections and GSCT status. The distribution of entry times is presented in Figure 9.

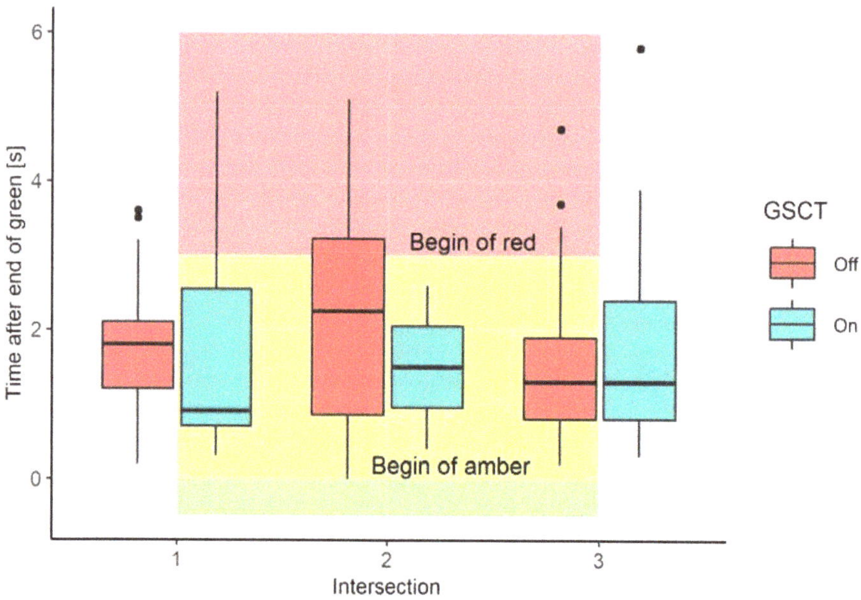

Figure 9. Boxplots of the time of entry after the end of the green signal.

Additionally, for the entire population of registered entries after the end of the red signal, a histogram of the entry time measured from the end of the green signal was developed, as shown in Figure 10. Both graphs show the moment of the signal change to red. The red color refers to the situation with the GSCT turned off, and blue color with the GSCT turned on.

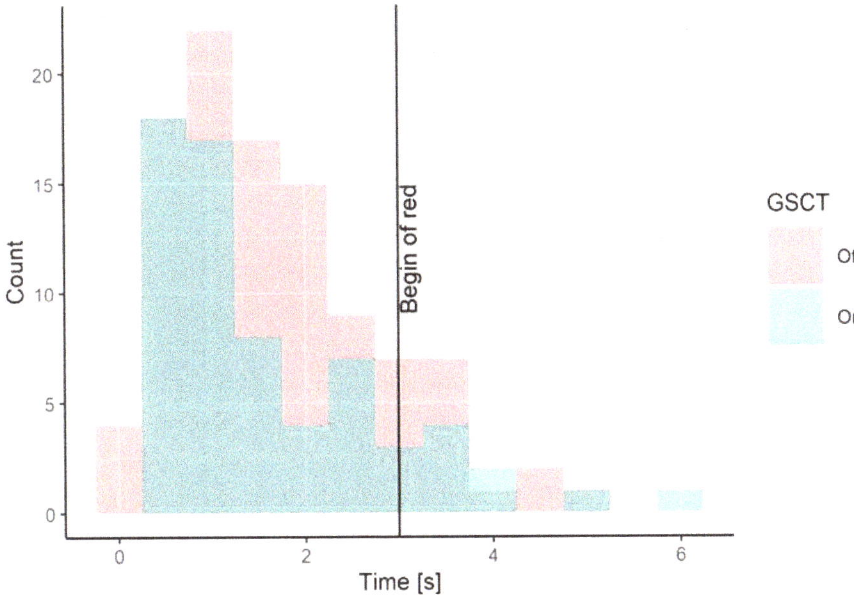

Figure 10. Histogram of entry times after the end of the green signal (data from all intersections).

3.4. Fuel Consumption Analysis

For fuel consumption analysis, a simplified model of intersection No. 3 in the Synchro 7 program was prepared [94] (Figure 11).

With the use of this model, calculations of the efficiency indicators related to fuel consumption were made:

- headway/saturation flow;
- total delay;
- stops;
- average speed;
- consumed fuel;

Results of the analysis are presented in Section 4.5 (Discussion section).

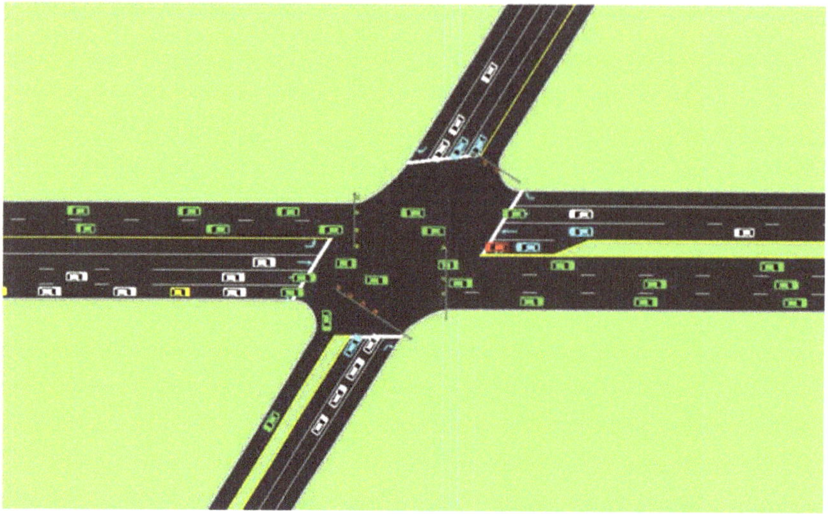

Figure 11. Model of intersection No. 3 in Synchro 7 and SimTraffic.

3.5. Case Study of a Road Crash on an Intersection with SCT

Some of the SCTs installed at intersections in Poland work independently of other devices of this type and are not connected to the signal controller. The SCT analyzes the displayed signals from the signal cable, calculates the signals' duration, and then displays the values in the next cycle on this basis. This device is only intended for fixed-time control. The principle of operation of such a device results in displaying erroneous indications in the event of a change in the duration of signals. Such changes take place, for example, when the signaling program is changed. A case study of a road crash with such a mechanism in Szczecin (Poland) is presented below. The event took place in 2016 at the intersection of Niepodległości Avenue and Żołnierza Polskiego Square. SCTs with an RSCT function and a GSCT were installed on both intersection approaches, located on mast arms to the right of the signals.

Before the event, approach A SCT had counted green, and approach B SCT had counted red (Figure 12a). After reaching the duration of the green signal, as in the previous cycle, the GSCT at approach A was off, while the green signal continued, most likely due to a change in the signaling program. At approach B, there was a countdown for the RSCT. There was little traffic at approach A (Figure 12b).

After counting down the RSCT time at approach B, the vehicles started to move even though the signal heads were showing a red signal (Figure 12c). This could be caused by too early a start, when the RSCT was displaying, for example, 1 s to display a green signal, or the situation described in [66], indicating that more drivers, before starting, were watching the RSCT than the actual signal head. As a consequence of such driver behavior and the way the SCT functioned at the intersection, a road accident occurred between the vehicle marked with a red arrow and the vehicle marked with a green arrow (Figure 12d).

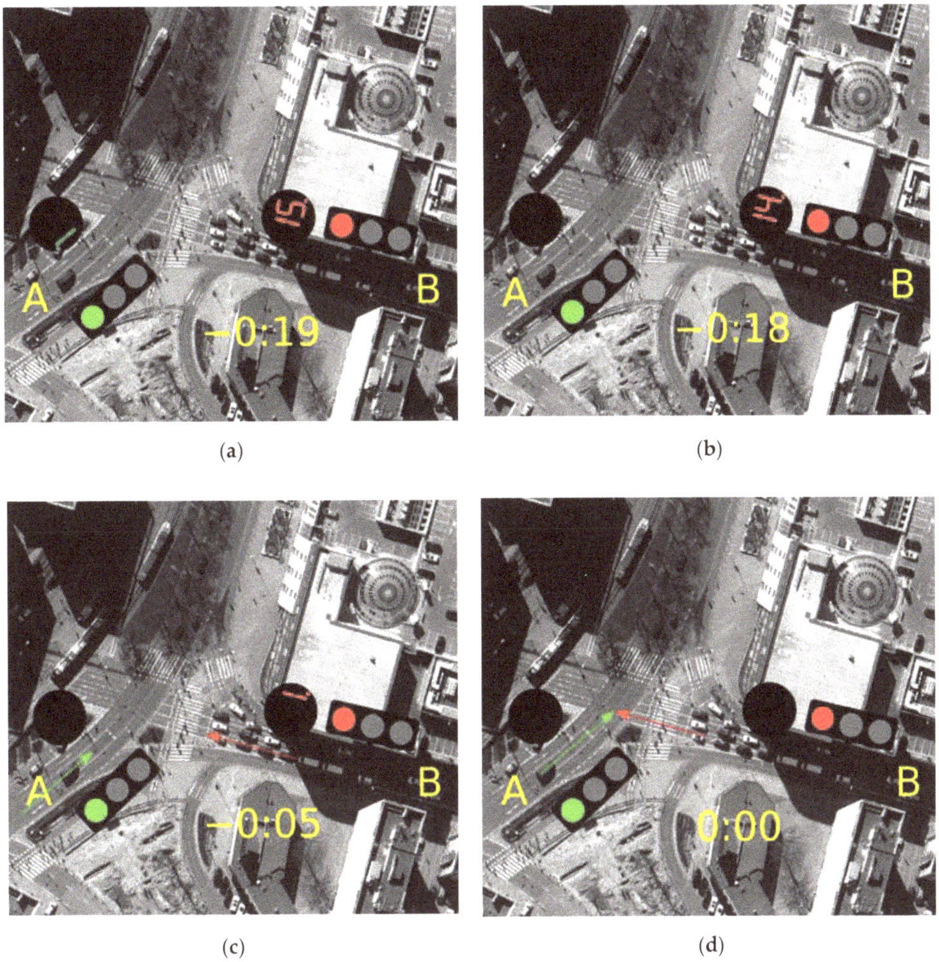

Figure 12. Sequence of events related to the crash: (**a**) GSCT and RSCT at intersection approaches; (**b**) extinction of GSCT at approach A; (**c**) extinction of RSCT at approach B, early start of vehicles at approach B; (**d**) road accident. (Background map Main Office of Geodesy and Cartography).

4. Discussion

4.1. Influence of SCT on Driver's Behavior

If the SCT is installed and working at the intersection, the driver has more information about the working of traffic lights. The hypothesis is that SCTs influence the behavior of drivers. Devices have been installed in Płock since 2011 (six years before conducting the study). This means that the drivers moving around the city already had learned the daily rules by which these devices operate. The SCT in Płock turned off 5 s before the beginning of the red signal. The day-to-day tests increased the value of the obtained data under similar road conditions.

4.2. Red Light Violation

The first issue examined was the proportion of vehicles crossing during a signal other than green (Figure 6). After switching on the SCT at intersections 1 and 2, the number of vehicles crossing on other signals than green increased. On the other hand, at junction 3, the opposite situation was observed. When analyzing this situation in more detail, it was noticed that in all cases, after switching on the GSCT, the number of vehicles entering an intersection on the amber signal decreased. This is consistent with other research [12,38,44,47]. The results of the research on the red and amber signal are similarly confirmed by previous studies [26,30,31,37–40,47]. An increase in the number of entries during the red and amber signals was also observed in Płock. No such cases were observed with the RSCT turned off, but they occurred at each intersection with the RSCT turned on. The increased proportion of crossing on the green signal at intersection No. 3 can be explained by the large proportion of entries on the amber signal with the GSCT turned off. The improvement of the results at intersection No. 3 is mainly due to the reduction of a large number of entries during the amber signal.

4.3. Headways

In the case of headways measurements (Figure 8), the following phenomena were observed:

1. The recorded headways were quite high in comparison to other studies. For all intersections, they were around the maximum values recorded in other studies for RSCT off, and above the values recorded in other studies for RSCT turned on [12,14,68].
2. The headway values were exceptionally high for intersection No. 2. In other studies, conducted at Polish intersections, they did not fall within the 80% confidence interval (0.1 quantile–0.9 quantile) [99].
3. At intersections No. 1 and No. 2, no significant changes in headway were observed whether the RSCT was turned on or off.
4. A reduction in headway and, consequently, an increase in saturation flow and capacity was observed only at intersection No. 2.

Exceptionally high headway values for junction No. 2 were most likely because there were no marked lanes at the 5.8 m wide approach, which increases the headway and significantly impacts capacity; this was taken into account in the capacity calculation methods [95]. Headways for subsequent vehicles decreased according to a similar pattern at other intersections, but these were much higher values than those recorded at different intersections. Intersection No. 2 is also located outside of the network of national and provincial roads.

The headway values at intersections No. 1 and No. 3 were also high compared to foreign surveys [12,14,68], and higher than the values recorded during other surveys carried out in Warsaw (the capital of Poland). They are above the median for studies conducted in Polish cities, similar to the results obtained at rural intersections [99]. A possible reason is the size of the city of Płock and the length of the trip related to it. Studies conducted outside Poland were often conducted in metropolises with a population of millions, such as Bangkok [30], Shanghai [21], Kuala Lumpur [19,20], Kayseri [12], Sydney [69], and New Dehli [46], but Płock has only 120,000 inhabitants and is a county city.

Improvement of traffic conditions after the use of RSCT was observed only at intersection No. 3. This is consistent with the results of studies presented in the literature (11 cases of no effect [19,21,24–26,30–32,34,38,61] and four cases of improvement [20,22,23,28]). Improvement was observed only at the largest of the examined intersections.

4.4. Entering after the Green Signal

The boxplot analysis (Figure 9) with the green signal entry times indicates that the median time decreased after the application of a GSCT (intersection No. 1 and No. 2) or remained the same (intersection No. 3). For intersections No. 1 and No. 3, the maximum values of these times increased, which is an unfavorable situation. The speed of the vehicles was not recorded in our study, but this situation may be related to an increase in speed at the inlets equipped with GSCTs, which was observed in previous research: [12,37,44,46,48]. For intersection No. 2, the number of registered entries after the green signal was only two, which does not allow statistical analysis. At that intersection, there was the lowest traffic volume among the examined intersections.

The chi-square (Pearson) test for compliance with a normal distribution was performed for the data. The test results for individual populations are presented in Table 3.

Table 3. Pearson test of compliance of travel times after the green signal.

Intersection	*p*-Value	
	GSCT On	GSCT Off
1	0.0006666	0.0316
2	$<2.2 \times 10^{-16}$	0.9098
3	0.002506	0.00177

p-Value < 0.05 indicates that the population is likely not normally distributed. For this reason, the comparison of the distribution of entry times was carried out using the nonparametric Wilcoxon-Mann-Whitney test. The results are presented in Table 4.

Table 4. Pearson test of compliance of the travel times after the green signal with the GSCT on (t_{on}) and off (t_{off}).

Intersection	*p*-Value/Alternative Hypothesis		
	$t_{on} \neq t_{off}$	$t_{on} < t_{off}$	$t_{on} > t_{off}$
1	0.3567	0.8289	0.1784
2	0.5297	0.7712	0.2648
3	0.8024	0.4012	0.6013
All data	0.4013	0.8003	0.2006

The test results show no indication of a shift between the distributions of entry with GSCT on and GSCT off. However, there is a noticeable shift in the modal value of the distribution to the right for the situation with the SCT off (Figure 10).

4.5. Change in Fuel Consumption during Use of SCT

The effect of saturation flow on the road capacity, and consequent fuel consumption, was direct. Studies in 11 cases indicated that the use of SCT does not affect saturation flow. Research made by the Faculty of Transport at Warsaw University of Technology indicated that such an impact only occurs in Płock at one intersection out of three. Only four studies [20,22,23,28] showed an improvement in capacity. Typical initial saturation flow values, depending on the capacity calculation method, were 1700–1900 veh./h. In the research, the speed of vehicles was measured at intersections or directly before them. However, the average speed over a longer distance and maintaining a constant speed significantly impacts fuel consumption. Therefore, it is not possible to conclude the effect of the SCT on fuel consumption based on these measurements.

On the other hand, start-up lost time applies only to the first vehicle in the queue. The effect on increasing the capacity and, consequently, on fuel consumption is smaller the longer the time of the green signal. For intersection No. 3, an analysis of the impact of the headway change on indicators related to fuel consumption was carried out. The analysis included vehicles in the 4th–9th position in the queue due to the large variation in the headway in the initial period of the green signal. Stabilization of saturation flow during start-up of the vehicle column is a typical phenomenon. For the situation with SCT turned off, the mean headway was 2.24 s, and with the SCT turned on was 2.04 s. This corresponds to a saturation flow 1607 veh./h and 1764 veh./h (for one lane), respectively. However, these are the actual values of saturation flow, considering inter alia, lane width, heavy vehicle occurrence, approach inclination and lane position. For analytical needs, the change of the starting saturation flow in the range of 1600–2000 veh./h was modeled. For such data, selected measures of effectiveness were calculated using the Synchro program. The results are presented in Table 5.

Table 5. Fuel consumption at the intersection depending on the output saturation flow.

Saturation Flow veh./h	Headway s	Total Delay/Vehicle s/veh.	Stops	Average Speed km/h	Fuel Consumed L
1600	2.25	40	2536	22	299
1650	2.18	38	2523	23	294
1700	2.12	37	2505	23	290
1750	2.06	36	2489	24	286
1800	2.00	35	2470	24	283
1850	1.95	33	2457	24	280
1900	1.89	33	2442	25	278
1950	1.85	32	2430	25	275
2000	1.80	31	2416	25	273

The results presented in the table refer to the fuel consumption of all vehicles in the model within the intersection with inflow sections. The results are the sum of all four approaches, not only for the SCT test approach. They show that reducing the headway (and consequently increasing the saturation flow) has a beneficial effect on fuel consumption. However, this effect was observed only at one of the three examined intersections. Similarly, headway variability occurs between different surveys, as shown in Figure 8.

4.6. Threats Resulting from Improper Operation of the SCT: A Case Study

A published regulation [9] introduced safety requirements for SCT use including device operation supervision and switching off signaling in a failure mode. Switching off the SCT should take place within 0.3 s from the occurrence of a failure mode situation. However, the lack of a displayed value on the SCT, also in a failure mode, may contribute to the occurrence of a road accident. It should be considered whether it is not justified to switch the signaling to the warning operation mode (amber flashing signal) in the case of SCT failure and then start the signaling with the SCT turned off.

Since the introduction of the regulations stated in [9], the number of SCTs installed in Poland has decreased, as the offered products did not meet the new requirements. SCT supervision within standards [100,101] compliant with national requirements is quite complex. This requires designing the device and communication with the controller from scratch.

5. Conclusions

Based on the conducted research, it is impossible to state unequivocally that the use of SCTs is beneficial for road traffic safety and efficiency. It has been observed that SCTs increase the number of red-light violations (crossing after the end of the amber signal) and on the red-amber signal. This is a negative situation and is confirmed by other studies. On the other hand, the number of crossings on the amber signal with SCTs is lower, which is also consistent with the conclusions of other studies. The modal value of the distribution of vehicle entry times after the end of the green signal with the GSCT turned on is shifted to the left, but the tests performed do not indicate that the shift of the entire distribution is statistically significant. The study results show that the impact of SCTs on the effectiveness of traffic control depends on the geometry of the intersection. The literature review also indicates that the use of SCTs results in reduction of start-up lost time (positive impact) and increases vehicle speed (negative impact). Just as the effect of using an SCT on efficiency varies, the effect on fuel consumption can vary. If the MOEs improve in a given situation, it is also possible to observe a reduction in fuel consumption. It should be noted that assistive steering devices are positively assessed, but confidence in them varies from country to country [102]. In the event of their use, particular attention should be paid to the intelligibility of the messages conveyed.

The limitations of the study are the short measurement period and the date of research (during holidays), due to the consent of the road administration to turn off the SCT. Other limitations include conducting the study in a medium-sized city on roads with an urban traffic and a small number of heavy goods vehicles. Further tests on approaches with long queues are advisable to ensure more significant measurements. Research should also be carried out in larger cities and on rural roads, especially national roads.

The conclusions of the article can be addressed to various stakeholders. It is important for drivers to be aware that they should always follow the indications of the signal heads, not the SCT. There are known situations where the indications of these two devices are divergent, which can cause road crashes. Road authorities and local authorities need to point out that the use of SCTs does not always bring the expected effect. For road traffic engineers, it is important to know the impact of SCT on traffic safety and control effectiveness. This allows assessment of the SCT impact on traffic and making a rational decision whether or not to use an SCT. Research shows that an SCT does not always fulfill its role in improving road safety and control efficiency.

Other unexplored problems are associated with SCT use related to drivers' behavior on intersections with traffic lights functioning in a failure mode. The case study shows that such a situation may be an indirect cause or a factor contributing to a road accident. For ethical reasons, these issues must be investigated safely. This is not possible in regular road traffic. Such situations can only be tested with a vehicle simulator.

Author Contributions: Conceptualization, T.K. and J.S.; data curation, T.K. and K.F.; methodology, T.K. and J.S.; visualization, T.K.; writing—original draft, T.K. and K.F.; writing—review & editing, T.K., K.F., M.C. and J.S. All authors have read and agreed to the published version of the manuscript.

Funding: This research was funded by Warsaw University of Technology.

Institutional Review Board Statement: Not applicable.

Informed Consent Statement: Not applicable.

Data Availability Statement: The data presented in this study are available on request from the corresponding author. The data are not publicly available due to privacy issues.

Acknowledgments: The authors would like to thank Agata Krezyman and Kamil Gąska for making video recordings and for work put into their analysis. The authors would like to thank the initiator of this research Marek Buda, who passed away in 2017. Thanks also to the Office of the City of Plock for agreeing to conduct the research.

Conflicts of Interest: The authors declare no conflict of interest.

References

1. The History and Meaning of Colored Traffic Lights. Available online: https://www.idrivesafely.com/defensive-driving/trending/history-and-meaning-colored-traffic-lights (accessed on 24 June 2021).
2. Convention on Road Traffic Geneva, 19 September 1949. In Proceedings of the United Nations Conference on Road and Motor Transport, Geneva, Switzerland, 28 August–19 September 1949. Available online: https://treaties.un.org/doc/Treaties/1952/03/19520326%2003-36%20PM/Ch_XI_B_1_2_3.pdf (accessed on 15 June 2021).
3. Convention on Road Signs and Signals Done at Vienna on 8 November 1968. In Proceedings of the United Nations Economic and Social Council Conference on Road Traffic, Austria, Vienna, 7 October–8 November 1968. Available online: https://treaties.un.org/doc/Treaties/1978/06/19780606%2000-35%20AM/CTC-xi-b-20-searchable.pdf (accessed on 15 June 2021).
4. Tomczuk, P.; Chrzanowicz, M.; Mackun, T.; Budzyński, M. Analysis of the Results of the Audit of Lighting Parameters at Pedestrian Crossings in Warsaw. *Arch. Transp.* **2021**, *59*, 21–39. [CrossRef]
5. Magyari, Z.; Koren, C.; Kieć, M.; Borsos, A. Sight Discances at Unsignalized Intersections: A Comparison of Guidelines and Requirements for Human Drivers and Autonomous Vehicle. *Arch. Transp.* **2021**, *59*, 7–19. [CrossRef]
6. Budzynski, M.; Tubis, A.; Rydlewski, M. Preliminary Safety Assesment of Polish Interchanges. *Arch. Transp.* **2021**, *58*, 99–113. [CrossRef]
7. Budzynski, M.; Gobis, A.; Guminska, L.; Jelinski, L.; Kiec, M.; Tomczuk, P. Assessment of the Influence of Road Infrastructure Parameters on the Behaviour of Drivers and Pedestrians in Pedestrian Crossing Areas. *Energies* **2021**, *14*, 3559. [CrossRef]

8. Ostrowski, K.; Budzynski, M. Measures of Functional Reliability of Two-Lane Highways. *Energies* **2021**, *14*, 4577. [CrossRef]
9. *Ordinance of the Minister of Infrastructure and Construction of May 24, 2017 Amending the Ordinance on Detailed Technical Conditions for Road Signs and Signals and Road Safety Devices and the Conditions for Their Placement on Roads*; Journal Of Laws 2017; Item 1062; Ministry of Infrastructure and Construction: Warsaw, Poland, 2017. Available online: http://isap.sejm.gov.pl/isap.nsf/download.xsp/WDU20170001062/O/D20171062.pdf (accessed on 15 June 2021). (In Polish)
10. Gazis, D.; Herman, R.; Maradudin, A. The Problem of Amber Signal Light in Traffic Flow. *Oper. Res.* **1960**, *8*, 112–132. [CrossRef]
11. Bąk, R. *Ocena Wpływu Strefy Dylematu Na Bezpieczeństwo Ruchu Na Skrzyżowaniach z Sygnalizacją Świetlną Za Pomocą Miar Pośrednich*; Politechnika Krakowska, Wydział Inżynierii Lądowej: Kraków, Poland, 2017.
12. Fujita, M.; Suzuki, K.; Yilmaz, C. Behavior and Consciousness Analyses on Effect of Traffic Signals Including Countdown Device for Vehicles. *J. East. Asia Soc. Transp. Stud.* **2007**, *7*, 2289–2304. [CrossRef]
13. Retting, R.; Williams, A.; Greene, M. Red-Light Running and Sensible Countermeasures: Summary of Research Findings. *Transp. Res. Rec.* **1998**, *1640*, 23–26. [CrossRef]
14. Krukowicz, T. Liczniki czasu w sygnalizacji drogowej—za i przeciw ich stosowaniu. *TTS Tech. Transp. Szyn.* **2013**, *20*, 1337–1343.
15. Buda, M.; Folwarski, T.; Krukowicz, T. Problemy stosowania liczników czasu w drogowej sygnalizacji świetlnej. *Transp. Miej. I Reg.* **2016**, *9*, 3–10.
16. Wengel, M. Countdown Timers at Intersections with Traffic Lights—Evaluation of the Regulation Journal of Laws of 31.5.2017 Item 1062. *Arch. Motoryz.* **2018**, *80*, 115–126. [CrossRef]
17. Miłaszewicz, B. *The Issue of Using Countdown Timers at Intersections with Traffic Lights: A Literature Review*; IEEE: Casta, Slovakia, 2018.
18. Sobota, A.; Karoń, G.; Żochowska, R.; Kłos, M. Methodology for Research on Traffic Safety at Signalized Intersections with Countdown Timers. *Sci. J. Sil. Univ. Technol. Ser. Transp.* **2018**, *100*, 191–201. [CrossRef]
19. Kidwai, F.A.; Ibrahim, M.R.; Karim, M. Traffic Flow Analysis of Digital Count down Signalized Urban Intersection. *Proc. East. Asia Soc. Transp. Stud.* **2005**, *5*, 1301–1308.
20. Ibrahim, M.R.; Karim, M.; Kidwai, F. The Effect of Digital Count-Down Display on Signalized Junction Performance. *Am. J. Appl. Sci.* **2008**, *5*, 479–482. [CrossRef]
21. Raksuntorn, W. The Effects of Countdown Signals on Intersection Capacity. *Int. Trans. J. Eng. Manag. Appl. Sci. Technol.* **2012**, *3*, 159–165.
22. Sharma, A.; Vanajakshi, L.; Rao, N. Effect of Phase Countdown Timers on Queue Discharge Characteristics under Heterogeneous Traffic Conditions. *Transp. Res. Rec.* **2009**, *2130*, 93–100. [CrossRef]
23. Wenbo, S.; Zhaocheng, H.; Xi, X.; Feifei, X. Exploring Impacts of Countdown Timers on Queue Discharge Characteristics of Through Movement at Signalized Intersections. *Procedia-Soc. Behav. Sci.* **2013**, *96*, 255–264. [CrossRef]
24. Spigolon, L.; Bezerra, B.; Bastos, J.T.; Ferraz, A.; Battistelle, R. Comparison between Common Traffic Lights and Three Types of Traffic Lights with Visual Cycle—A Safety and Capacity Analysis. *J. Transp. Lit.* **2015**, *9*, 5–9. [CrossRef]
25. Limanond, T.; Chookerd, S.; Roubtonglang, N. Effects of Countdown Timers on Queue Discharge Characteristics of through Movement at a Signalized Intersection. *Transp. Res. Part C Emerg. Technol.* **2009**, *17*, 662–671. [CrossRef]
26. Biswas, S.; Ghosh, I.; Chandra, S. Influence of Signal Countdown Timer on Efficiency and Safety at Signalized Intersections. *Can. J. Civ. Eng.* **2017**, *44*, 308–318. [CrossRef]
27. Li, K.; Dong, S.; Sun, J.; Yu, X. Study on the Influence of Signal Countdown Device on Traffic Safety of Intersections. In Proceedings of the 2009 International Conference on Measuring Technology and Mechatronics Automation, Zhangjiajie, China, 11–12 April 2009. [CrossRef]
28. Yu, K.; Fujita, M.; Suzuki, K. Behavior Analysis of Vehicle Starts under Countdown-Type Traffic Signal through Survey in China. *World Acad. Sci. Eng. Technol.* **2011**, *58*, 250–255.
29. Lum, K.M.; Halim, H. A Before-and-after Study on Green Signal Countdown Device Installation. *Transp. Res. Part F Traffic Psychol. Behav.* **2006**, *9*, 29–41. [CrossRef]
30. Limanond, T.; Prabjabok, P.; Tippayawong, K. Exploring Impacts of Countdown Timers on Traffic Operations and Driver Behavior at a Signalized Intersection in Bangkok. *Transp. Policy* **2010**, *17*, 420–427. [CrossRef]
31. Sharma, A.; Vanajakshi, L.; Girish, V.; Harshitha, M.S. Impact of Signal Timing Information on Safety and Efficiency of Signalized Intersections. *J. Transp. Eng.* **2012**, *138*, 467–478. [CrossRef]
32. Biswas, S.; Ghosh, I.; Chandra, S. Effect of Countdown Timers on Traffic Characteristics and Driver Behaviors at Signalized Intersections. In Proceedings of the Conference Transportation Research Board 95th Annual, Washington, DC, USA, 10–14 January 2016.
33. Elias, S.; Ghafurian, M.; Samuel, S. Effectiveness of Red-Light Running Countermeasures: A Systematic Review. In Proceedings of the 11th International Conference on Automotive User Interfaces and Interactive Vehicular Applications, New York, NY, USA, 21–25 September 2019; pp. 91–100.
34. Jatoth, J.; Singh, N.; Mehar, A. Evaluating the Performance of Signalized Intersection with Signal Countdown Timer. *Int. J. Intell. Transp. Syst. Res.* **2020**, *19*, 180–190. [CrossRef]
35. Kłos, M.J.; Sobota, A.; Żochowska, R.; Karoń, G. Effects of Countdown Timers on Traffic Safety at Signalized Intersections. *Trans. Transp. Sci.* **2020**, *11*, 19–27. [CrossRef]

36. Pamuła, W.; Kłos, M.J. Evaluation of Changes in Drivers Behaviour Due to Introduction of Countdown Timers at Signalized Intersections Using UAV Data. In *Nodes in Transport Networks—Research, Data Analysis and Modelling, Proceedings of 16th Scientific and Technical Conference "Transport Systems. Theory and Practice 2019", Katowice, Poland, 16–18 September 2019*; Springer: Cham, Switzerland, 2019; pp. 115–124.
37. Papaioannou, P.; Politis, I. Preliminary Impact Analysis of Countdown Signal Timer Installations at Two Intersections in Greece. *Procedia Eng.* **2014**, *84*, 634–647. [CrossRef]
38. Rijavec, R.; Zakovšek, J.; Maher, T. Acceptability of Countdown Signals at an Urban Signalized Intersection and Their Influence on Drivers Behaviour. *Promet-Traffic Transp.* **2013**, *25*, 63–71. [CrossRef]
39. Rządkowski, Ł.; Lazarczyk, S. *Report Signal Heads Countdown Timers Research into the Impact of Time Counters Remaining Until the Change of the Green Signal Displayed on the Traffic Light on Road Safety*; Zarząd Dróg Miejskich: Grudziądz, Poland, 2014. (In Polish)
40. Sobota, A.; Klos, M.J.; Karoń, G. The Influence of Countdown Timers on the Traffic Safety of Pedestrians and Vehicles at the Signalized Intersection. In *Intelligent Transport Systems and Travel Behaviour, Proceedings of 13th Scientific and Technical Conference "Transport Systems. Theory and Practice 2016", Katowice, Poland, 19–21 September 2016*; Sierpiński, G., Ed.; Springer International Publishing: Cham, Switzerland, 2017; pp. 13–21.
41. Huey, S.; Ragland, D. *Changes in Driver Behavior Resulting from Pedestrian Countdown Signals*; University of California: Berkeley, CA, USA, 2007.
42. Long, K.; Han, L.D.; Yang, Q. Effects of Countdown Timers on Driver Behavior after the Yellow Onset at Chinese Intersections. *Traffic Inj. Prev.* **2011**, *12*, 538–544. [CrossRef] [PubMed]
43. Long, K.; Liu, Y.; Han, L.D. Impact of Countdown Timer on Driving Maneuvers after the Yellow Onset at Signalized Intersections: An Empirical Study in Changsha, China. *Saf. Sci.* **2013**, *54*, 8–16. [CrossRef]
44. Fu, C.; Zhang, Y.; Peng, T. Effects of Countdown Timer on Yellow Light Running: An Empirical Study. *J. Traffic Logist. Eng.* **2015**, 3. [CrossRef]
45. Abioye, O. *Evaluation of the Usage of Pedestrian Countdown Signals at Intersections*; FAMU-FSU College of Engineering, Department of Civil Engineering: Tallahassee, FL, USA, 2016.
46. Devalla, J.; Biswas, S.; Ghosh, I. The Effect of Countdown Timer on the Approach Speed at Signalised Intersections. *Procedia Comput. Sci.* **2015**, *52*, 920–925. [CrossRef]
47. Felicio, G.P.; Grepo, L.C.; Reyes, V.F.; Yupingkun, L.C. Traffic Light Displays and Driver Behaviors: A Case Study. *Procedia Manuf.* **2015**, *3*, 3266–3273. [CrossRef]
48. Wu, W.; Juan, Z.; Jia, H. Drivers' Behavioral Decision-Making at Signalized Intersection with Countdown Display Unit. *Syst. Eng. Theory Pract.* **2009**, *29*, 160–165. [CrossRef]
49. Huang, M.; Fujita, M.; Wisetjindawat, W. Countdown Timers, Video Surveillance and Drivers' Stop/Go Behavior: Winter versus Summer. *Accid. Anal. Prev.* **2017**, *98*, 185–197. [CrossRef]
50. Islam, M.R.; Wyman, A.A.; Hurwitz, D.S. Safer Driver Responses at Intersections with Green Signal Countdown Timers. *Transp. Res. Part F Traffic Psychol. Behav.* **2017**, *51*, 1–13. [CrossRef]
51. Yu, S.; Shi, Z. Analysis of Car-Following Behaviors Considering the Green Signal Countdown Device. *Nonlinear Dyn.* **2015**, *82*, 731–740. [CrossRef]
52. Fu, C.; Zhang, Y.; Bie, Y.; Hu, L. Comparative Analysis of Driver's Brake Perception-Reaction Time at Signalized Intersections with and without Countdown Timer Using Parametric Duration Models. *Accid. Anal. Prev.* **2016**, *95*, 448–460. [CrossRef] [PubMed]
53. Huang, H.; Wang, D.; Zheng, L.; Li, X. Evaluating Time-Reminder Strategies before Amber: Common Signal, Green Flashing and Green Countdown. *Accid. Anal. Prev.* **2014**, *71C*, 248–260. [CrossRef]
54. Fu, C.; Zhang, Y.; Qi, W.; Cheng, S. Effects of Digital Countdown Timer on Intersection Safety and Efficiency: A Systematic Review. *Traffic Inj. Prev.* **2016**, *17*, 98–103. [CrossRef] [PubMed]
55. Kempa, J.; Bebyn, G. Wyświetlacze czasu na sygnalizatorach sygnalizacji świetlnej. *Logistyka* **2014**, *6*, 5364–5370.
56. Kempa, J.; Bebyn, G. Doświadczenia z funkcjonowania wyświetlaczy czasu na sygnalizatorach w Toruniu. *Logistyka* **2014**, *6*, 5351–5355.
57. Zielińska, L. Analysis of the Validity of Uing Time Counters at road Intersections with Traffic Lights on the Example of Słupsk. *Transp. Miej. I Reg.* **2017**, *2*, 23–26. (In Polish)
58. Metelski, A. Analysis of Selected Methodological Problems Regarding the Examination of Traffic Events at Road Intersections. *Arch. Automot. Eng.–Arch. Motoryz.* **2018**, *82*, 75–85. [CrossRef]
59. Małecki, K.; Iwan, S. Modeling Traffic Flow on Two-Lane Roads with Traffic Lights and Countdown Timer. *Transp. Res. Procedia* **2019**, *39*, 300–308. [CrossRef]
60. Paul, M.; Ghosh, I. Influence of Green Signal Countdown Timer on Severe Crash Types at Signalized Intersections Due to Red Light Violations. *Transp. Lett.* **2020**, *12*, 528–539. [CrossRef]
61. Chiou, Y.-C.; Chang, C.-H. Driver Responses to Green and Red Vehicular Signal Countdown Displays: Safety and Efficiency Aspects. *Accid. Anal. Prev.* **2010**, *42*, 1057–1065. [CrossRef] [PubMed]
62. Pan, F.; Zhang, L.; Ma, C.; Li, H.; Yang, J.; Liu, T.; Wang, F.; Chai, S. Impact of Vehicular Countdown Signals on Driving Psychologies and Behaviors: Taking China as an Example. *J. Adv. Transp.* **2017**, *2017*, 5838520. [CrossRef]

63. Frank, A.; Schneider, F.; Meschtscherjakov, A.; Stadon, J. Advanced Traffic Light Interface: Countdown Timers to Increase User Experience. In Proceedings of the Adjunct Proceedings of the 7th International Conference on Automotive User Interfaces and Interactive Vehicular Applications, Nottingham, UK, 1–3 September 2015; pp. 56–61.
64. *PN-EN 12368:2015-07 Traffic Control Equipment—Signal Heads*; Polish Committee for Standardization: Warsaw, Poland, 2015.
65. Kim, M.; Kim, H.K. Investigation of Environmental Benefits of Traffic Signal Countdown Timers. *Transp. Res. Part D Transp. Environ.* **2020**, *85*, 102464. [CrossRef]
66. Nygårdhs, S. Cyclists' Adaptation to a Countdown Timer to Green Traffic Light: A before-after Field Study. *Appl. Ergon.* **2021**, *90*, 103278. [CrossRef] [PubMed]
67. Long, X.; Zhou, M.; Zhao, H.; Song, Y. Pedestrian Crossing Decision during Flashing Green-Countdown Signal for Urban Signalized Intersection. *J. Transp. Saf. Secur.* **2021**, 1–21. [CrossRef]
68. Kim, K.W.; Kim, Y.; Seo, H.Y. An Evaluation of Pedestrian Countdown Signals. *KSCE J. Civ. Eng.* **2002**, *6*, 533–537. [CrossRef]
69. Levasseur, M.; Brisbane, G. *Trial of Pedestrian Countdown Timers—RTA—NSW Government*; ARRB Group: Sydney, Australia, 2011.
70. Cleaver, M.; Hislop, J.; Roos, M.; Fernandes, R.; Prendergast, M.; Brisbane, G.; McTiernan, D.; Levasseur, M. An Evaluation of Pedestrian Countdown Timers in the Sydney CBD. In Proceedings of the Australasian Road Safety Research, Policing and Education Conference 2011, Perth, WA, Australia, 6–9 November 2011.
71. McTiernan, D.; Levasseur, M.; Brisbane, D. Pedestrian Countdown Timers—Two Australian Trial Methods and Results. In Proceedings of the 25th ARRB Conference, 2012, Perth, WA, Australia, 23 September 2012.
72. Supernak, J.; Verma, V.; Supernak, I. Pedestrian Countdown Signals: What Impact on Safe Crossing? *Open J. Civ. Eng.* **2013**, *3*, 39–45. [CrossRef]
73. Biswas, S.; Ghosh, I.; Chandra, S. Effect of Traffic Signal Countdown Timers on Pedestrian Crossings at Signalized Intersection. In Proceedings of the 3rd Conference of Transportation Research Group of India (3rd CTRG), Kolkata, West Bengal, India, 17–20 December 2015. Available online: https://www.researchgate.net/publication/311355751_Effect_of_traffic_signal_countdown_timers_on_pedestrian_crossings_at_signalized_intersection (accessed on 15 June 2021).
74. Balwan, M.; Jaison, J.; Fellendorf, M. Impacts of Countdown Timers on Pedestrian Behavior—Case Study in UAE. *J. Traffic Logist. Eng.* **2017**, *5*, 2. [CrossRef]
75. Ma, W.; Liao, D.; Bai, Y. Empirical Analysis of Countdown Signals on Pedestrian Behaviour. *Proc. ICE-Transp.* **2015**, *168*, 15–22. [CrossRef]
76. Fu, L.; Zou, N. The Influence of Pedestrian Countdown Signals on Children's Crossing Behavior at School Intersections. *Accid. Anal. Prev.* **2016**, *94*, 73–79. [CrossRef]
77. Xiong, H.; Xiong, L.; Deng, X.; Wang, W. Evaluation of the Impact of Pedestrian Countdown Signals on Crossing Behavior. *Adv. Mech. Eng.* **2014**, *6*, 518295. [CrossRef]
78. Anapali, I.; Basbas, S.; Nikiforiadis, A. Pedestrians' Crossing Dilemma during the First Seconds of the Red-Light Phase. *Soc. Sci.* **2021**, *10*, 213. [CrossRef]
79. Biswas, S.; Ghosh, I.; Chandra, S. Effect of Traffic Signal Countdown Timers on Pedestrian Crossings at Signalized Intersection. *Transp. Dev. Econ.* **2017**, *3*, 2. [CrossRef]
80. Richmond, S.; Willan, A.; Rothman, L.; Camden, A.; Buliung, R.; Macarthur, C.; Howard, A. The Impact of Pedestrian Countdown Signals on Pedestrianemotor Vehicle Collisions: A Quasi-Experimental Study. *Inj. Prev.* **2014**, *18*, 155–158. [CrossRef] [PubMed]
81. Pulugurtha, S.S.; Desai, A.; Pulugurtha, N.M. Are Pedestrian Countdown Signals Effective in Reducing Crashes? *Traffic Inj. Prev.* **2010**, *11*, 632–641. [CrossRef] [PubMed]
82. Boateng, R.A.; Kwigizile, V.; Miller, J.S.; Oh, J.-S. A Justification for Pedestrian Countdown Signals at Signalized Intersections: The Safety Impact on Senior Motorists. *J. Transp. Health* **2019**, *14*, 100617. [CrossRef]
83. Escott, B.G.; Richmond, S.A.; Willan, A.R.; Ravi, B.; Howard, A.W. The Impact of Pedestrian Countdown Signals on Single and Two Vehicle Motor Vehicle Collisions: A Quasi-Experimental Study. *Int. J. Inj. Control Saf. Promot.* **2017**, *24*, 429–434. [CrossRef] [PubMed]
84. Boateng, R.A.; Kwigizile, V.; Oh, J.-S. A Comparison of Safety Benefits of Pedestrian Countdown Signals with and without Pushbuttons in Michigan. *Traffic Inj. Prev.* **2018**, *19*, 588–593. [CrossRef] [PubMed]
85. Maric, B.; Lipovac, K.; Nešić, M.; Đerić, M. The Influence of a Countdown Display On Pedestrian Behavior at a Signalized Pedestrian Crossing Equipped with a Pedestrian Refuge Island. *Transp. Res. Procedia* **2021**, *55*, 1720–1728. [CrossRef]
86. Amjadi, R. *Safety Evaluation of Pedestrian Countdown Signals*; FHWA: McLean, MA, USA, 2019.
87. Königbauer, M. OSM Geofabrik Universal QML Style. Available online: https://github.com/mkoenigb/OSM-Geofabrik_Universal-QML-Style (accessed on 15 June 2021).
88. Gąska, K. *Wpływ Wyświetlaczy Czasu Na Podjęcie Decyzji o Hamowaniu Pojazdu*; Politechnika Warszawska, Wydział Transportu: Warszawa, Poland, 2017.
89. Krezyman, A. *Wpływ Wyświetlaczy Czasu Na Rozruch Kolumny Pojazdów*; Politechnika Warszawska, Wydział Transportu: Warszawa, Poland, 2017.
90. R Core Team. *R: A Language and Environment for Statistical Computing*; R Foundation for Statistical Computing: Vienna, Austria, 2021. Available online: https://www.R-project.org/ (accessed on 15 June 2021).
91. Long, J.; Teetor, P. *R Cookbook*, 2nd ed.; O'Reilly: Sebastopol, CA, USA, 2019; ISBN 978-0-596-80915-7.
92. Wickham, H. *Ggplot2: Elegant Graphics for Data Analysis*; Springer: New York, NY, USA, 2016; ISBN 978-3-319-24277-4.

93. Chang, W. *R Graphics Cookbook*, 2nd ed.; O'Reilly: Sebastopol, CA, USA, 2021; ISBN 978-1-4493-1695-2.
94. Husch, D.; Albeck, J. *Synchro Studio 7 Traffic Signal Software-User Guide*; Trafficware: Sugar Land, TX, USA, 2006; ISBN 0-9742903-3-5.
95. *Highway Capacity Manual 2010*; Transportation Research Board: Washington, DC, USA, 2010; Volumes 1–4, ISBN 978-0-309-16077-3.
96. Czerepicki, A.; Krukowicz, T.; Górka, A.; Szustek, J. Traffic Light Priority for Trams in Warsaw as a Tool for Transport Policy and Reduction of Energy Consumption. *Sustainability* **2021**, *13*, 4180. [CrossRef]
97. Szmel, D.; Zabłocki, W.; Ilczuk, P.; Kochan, A. Method for Selecting the Safety Integrity Level for the Control-Command and Signaling Functions. *Sustainability* **2019**, *11*, 7062. [CrossRef]
98. *Traffic Light Instruction, Annex 2 to the Ordinance of the Ministers of Transport and Maritime Economy and Internal Affairs of June 6, 1990*; (M. P. 24 Poz. 184 z 1990 r.); Official Gazette of the Government of the Republic of Poland: Warsaw, Poland, 1990. (In Polish)
99. Ostrowski, K.; Chodur, J. Charakterystyka Strumienia Ruchu w Początkowym Okresie Sygnału Zielonego Na Skrzyżowaniu z Sygnalizacją. *Transp. Miej. I Reg.* **2012**, *5*, 3–9.
100. *PN-EN 12675:2002 Traffic Signal Controllers—Functional Safety Requirements*; Polish Committee for Standardization: Warsaw, Poland, 2002.
101. *PN-EN 50556:2011 Road Traffic Signal Systems*; Polish Committee for Standardization: Warsaw, Poland, 2011.
102. Alonso, F.; Faus, M.; Esteban, C.; Useche, S.A. Is There a Predisposition towards the Use of New Technologies within the Traffic Field of Emerging Countries? The Case of the Dominican Republic. *Electronics* **2021**, *10*, 1208. [CrossRef]

energies

MDPI

Article

Solar-Powered Active Road Studs and Highway Infrastructure: Effect on Vehicle Speeds

Richard Llewellyn *, Jonathan Cowie and Grigorios Fountas

Transport Research Institute, Edinburgh Napier University, Edinburgh EH10 5DT, UK; j.cowie@napier.ac.uk (J.C.); g.fountas@napier.ac.uk (G.F.)
* Correspondence: r.llewellyn@napier.ac.uk; Tel.: +44-(0)131-455-3755

Abstract: Vehicle speeds have a direct relationship with the severity of road crashes and may influence their probability of occurrence. Solar-powered active road studs have been shown to have a positive effect on driver confidence, but their impact on vehicle speed in conjunction with other road features is little understood. This study aims to address this gap in knowledge through a case study of a 20 km section of a strategic major road featuring a variety of highway infrastructure features. Before-and-after surveys were undertaken at 21 locations along the route using manual radar speed measurement. Analysis of nearly 10,000 speed measurements showed no statistically significant change in mean speeds following the implementation of the road studs. Linear regression models are proposed for two different posted speed limits, associating road features with expected vehicle speed. The models suggest that vehicle speeds are chiefly influenced by merges, curves, gradients, and ambient light conditions. The findings of this study should provide confidence that active road studs may be implemented without a negative impact on speed-related safety. The work also provides further expansion of the evidence base describing the effect of highway infrastructure features on vehicle speeds.

Keywords: active road studs; highway infrastructure; vehicle speeds; road safety; road features

Citation: Llewellyn, R.; Cowie, J.; Fountas, G. Solar-Powered Active Road Studs and Highway Infrastructure: Effect on Vehicle Speeds. *Energies* **2021**, *14*, 7209. https://doi.org/10.3390/en14217209

Academic Editors: Marek Guzek, Rafał Jurecki and Wojciech Wach

Received: 17 October 2021
Accepted: 30 October 2021
Published: 2 November 2021

Publisher's Note: MDPI stays neutral with regard to jurisdictional claims in published maps and institutional affiliations.

1. Introduction

Road traffic crashes are the eighth leading cause of death globally with an estimated 1.35 million killed and 50 million injured each year [1]. Speed management has a long association with road safety and is enshrined within contemporary international road safety strategies such as the safe systems approach [2,3]. Such strategies are founded upon the basis that the consequences of crashes at higher speeds are a simple matter of physics: the greater the change in velocity, the greater the energy dissipation and, subsequently, the higher the severity of injury [4]. On this basis, understanding the relationships between drivers, vehicles, road infrastructure, and speed continues to be an important goal of road safety research.

Early case-control studies in speed research suggested that the greater the differential in the speed of a vehicle from the average of all the traffic, the higher the risk of a crash [5,6]. Self-reporting studies have also found that drivers found to be travelling faster are more likely to have a history of crashes [7]. More recent research has tended to look at the road level (including segments, sections, intersections, and corridors), rather than individual driver level, not least because of the difficulties of categorically associating speed as a causation factor in individual crashes [8]. The question then becomes one of how the frequency and severity of the crashes vary with mean speed, to which the answer has been extensively explored; before-and-after studies of measures, such as changes in the posted speed limit, traffic calming interventions, or increased enforcement have resulted in linear, power, and asymptotic relationships associating higher mean speeds with increased crash rates [9–14]. Therefore, from the point of view of the road designer, road features that result in unintended increases in mean speed are unlikely to be desirable.

In several countries across the world, reflective road studs are used as a measure to assist drivers in low light by highlighting road features, such as lanes, carriageway edges, curvature, and junctions. They rely on the reflection of the vehicle's headlight beam back to the driver to provide a preview of oncoming features. However, such a system is limited in range by the power of the vehicle headlights and can be further affected by weather and the physical condition of the studs themselves. Active Road Studs (also known as Internally Illuminated Raised Pavement Markers) are a recent development which seeks to address such issues. The studs feature high-powered LEDs which significantly increase the preview time of the road ahead and have been suggested as being of assistance to drivers in several road scenarios [15,16]. They have also been shown to have a positive impact on driver confidence [17]. However, if increased confidence were to result in a corresponding increase in speed, this could potentially compromise any road safety gains made. On this basis, the effect of active road studs on vehicle speeds in the context of other road features is of particular interest.

The relationship between active road studs and speed has been the subject of several simulator studies exploring the behaviour of drivers with and without their implementation. One such study investigated thirty-six participants split evenly across three age groups over a 37 km generic, rural, single carriageway route, with matching sections of no studs, passive (i.e., traditional reflective) studs, and active studs [18]. The simulation showed statistically significant increases in mean speed along the route between the no-stud and the passive- or active-stud sections. However, no statistically significant difference was found between passive and active road studs. Looking at corners alone, the same results were found for right-hand curves. For left-hand curves, an increase was found in mean speed for the youngest and oldest driver groups. Drivers were also found to brake more strongly for both studded conditions on corners, but there was no significant difference in braking between active and passive studs.

A later simulator study compared the speed of drivers on a route with no studs, active studs on curves, and full road lighting on curves [19]. In this case, the studs were activated by an approaching vehicle and switched off once the vehicle had passed. The distance between studs was also varied; larger distances between studs were applied on the approach, with smaller distances on the curve itself. Twenty participants drove the route, comprising sixteen straight and sixteen curved sections. On the straight sections, higher mean speeds were recorded with both the stud and the lighting treatment when compared with the unlit route; however, no significant difference was found between the mean speeds of the studded and lit conditions. Overall, faster speeds were found on the approach to curves than within them. Noting that this study was based on right-hand driving, in right-hand curves no differences between the three speeds were found. For left-hand curves, the speed on the approaches was higher with the lighting than with the studs or no treatment.

Regarding the speed effects of active road studs in real-world applications, a before-and-after video study of an undefined length of road was undertaken between two bends in Victoria, Australia [20,21]. The reductions in mean speed before and after installation in each direction were found to be 1.2 km/h (0.7 mph) and 3.1 km/h (1.9 mph), respectively. However, the change was only statistically significant in the latter case, and the methods of control for other confounding factors were not detailed. A review of the literature, practitioner feedback, and manufacturers undertaken in the United States included speed effects on highway junctions and links in its scope [22]. Whilst the reporting of reduced speed was suggested as a possible outcome in the review, it notes that the data were limited and could not be considered conclusive.

The research to date on the relationship between active road studs and speed is limited to simulator-based studies or very small field installations. Furthermore, the focus of the existing work has been on plain tangent and curve sections and has not covered potential mean-speed changes at other road features, such as junction approaches, merges, and dual carriageway sections. The present work attempts to address this gap in knowledge. On this basis, the aim of this research was to measure the choice of speed by drivers, using real-world rural junctions and links and to determine whether changes in the speed of vehicles may be associated with the installation of active road studs when compared to other road features. Building on the findings of previous studies, the stated hypothesis is that the implementation of active road studs results in no change in speed when compared to their traditional reflective equivalent and that other road features are associated with differences in mean speed. The objectives were to:

- quantify the speed choice of drivers on typical rural-road infrastructure;
- survey and record factors in the road environment that may affect speed choice, including active road studs;
- identify and assess the effect of such factors against actual vehicle speeds; and
- identify the response to such factors over a longer time frame.

The work described here forms part of a larger study investigating the effects of active road studs on driver behaviour, such as lane discipline, gap acceptance, and driver confidence. The focus in this element is the effect of active road studs on vehicle speeds. Section 2 describes the materials and methods adopted, including the background to the case study chosen, the procedure used for obtaining the speed and infrastructure data, and the statistical methods used for the analysis. Section 3 states the results of the route and individual site surveys and the development of the linear regression model. The findings are discussed in the context of the existing literature in Section 4, with conclusions and implications for practitioners identified in Section 5.

2. Materials and Methods

2.1. Case Study Background

The A1 trunk road (a route of national strategic importance) runs between Edinburgh and London in the UK. The road is constructed to dual carriageway or motorway standard for most of its length, but the section straddling the border between Scotland and England is more mixed in its composition. It also carries a lower volume of traffic than other parts of the route. Although complying with modern alignment standards due to extensive upgrading in the latter half of the 20th century, it mainly consists of single carriageway, with a posted speed limit of 60 mph (97 km/h). Short sections of 2 + 1 carriageway and higher speed 70 mph (113 km/h) dual carriageway are provided at intervals to provide overtaking opportunities. The speed limits for heavy goods vehicles (HGVs) are 20 mph lower than the posted limit at 40 mph (64 km/h) and 50 mph (80 km/h) for the single and dual carriageway sections, respectively. The route is particularly rural in nature, carrying national strategic traffic along with linking local towns and villages. During the hours of darkness, there is very little ambient light due to its location away from major population centres. Street lighting is provided at certain points, but the route is predominantly unlit.

The case study for this research is a 20 km section of the A1 in Scotland immediately north of its border with England. The route here has a modest crash rate relative to other Scottish trunk roads; this has been associated with fixed-enforcement-camera treatment [23]. Nevertheless, local communities have continued to raise safety concerns, prompting a review by the national roads authority. The review found that the legibility of the route during hours of darkness, particularly around junctions, could be problematic. An improvement scheme was proposed [24] comprising the installation of 4200 solar-powered active road studs to highlight the approaches to nine junctions and two intermediate carriageway sections. The active road studs were effectively a like-for-like replacement of the reflective road studs which were already in situ. In accordance with the UK traffic sign regulations, white, red, amber, and green studs were used to indicate the centreline, nearside and offside edge lines, and the merge/diverge sections, respectively. Examples of typical active road stud installations implemented as part of the scheme are shown in Figure 1.

(a) (b) (c) (d)

Figure 1. Example active stud installations: (**a**) dual to single carriageway merge; (**b**) single carriageway curve; (**c**) single carriageway junction; (**d**) dual carriageway junction.

2.2. Methodological Considerations

Two methods were considered for the measurement of speeds on the treated section of the route. The first option was the use of automatic traffic counters. Such counters have the benefit of being able to record a large amount of data over a continuous period, resulting in a large sample size and coverage of a range of conditions. With careful installation, they can be installed relatively inconspicuously to avoid any potential influence on the results. However, as an observer is not usually continuously present, confounding factors, such as incidents and inclement weather conditions, can mean collected speed data may not be fully representative. The accuracy of the automatic traffic counters can also be affected by irregular vehicle trajectories, detector spacing, scanning time, and multiple vehicles in the detection zone [25,26]. The traffic counters also record the speeds of all vehicles passing, irrespective of flow density. As a result, individual recorded speeds may be affected due to the presence of other vehicles and may not reflect a true speed choice by the drivers.

The alternative method considered for measuring speeds was a manual radar survey. Such surveys are labour-intensive and are more limited in the choice of location due to the need for the observer to remain inconspicuous whilst not compromising the safety of the survey staff or other road users. They can also be prone to human error or bias, particularly where the observer is inexperienced. However, manual speed surveys allow for a truer representation of free mean speeds to be recorded as the observer can exercise judgement regarding whether speeds are likely to be by choice; for example, this can be achieved by recording speeds only where a certain headway between vehicles is exceeded and by the monitoring of other factors, such as the illumination of brake lights. Other benefits of the continuous presence of an observer include the monitoring of weather conditions and the accounting for any unusual incidents during the survey period. On this basis and in the interests of potentially greater precision, the manual radar survey was selected as the preferred method for this work.

2.3. Manual Radar Survey Procedure

Manual survey locations were established on each approach to the treated junctions and mid-link on the curve and merge sections. A pre-survey reconnaissance visit was undertaken to determine the exact location of each site in order to consider the health and safety of the enumerator, the covertness of the survey, and the potential accuracy of the recording. Specific concerns at each location included the angle of the radar beam, the depth of field, the field of view to the traffic flow, and the presence of objects which may have resulted in unwanted radar reflection. In total, 21 sites were established. Prior to each survey being undertaken, weather forecasts and traffic-incident information were checked and rescheduled if conditions were considered sub-optimal. This ensured that weather and unexpected traffic conditions were controlled for all surveys being undertaken during 'neutral' conditions, i.e., good visibility, no precipitation, light or no winds, generally dry road surfaces, and normal traffic flows. A typical radar survey arrangement is shown in Figure 2.

Figure 2. Typical radar speed survey arrangement.

To provide an adequate sample size, the surveys were carried out for a minimum of one hour at each location, or longer if a minimum threshold of sixty vehicles had not been reached. Only vehicles subject to free flow conditions were recorded, defined as where a headway of greater than three seconds existed, as used in other similar studies (e.g., [7]. Wherever practicable, every vehicle travelling in such conditions was recorded. The speeds were recorded directly into a spreadsheet during the survey itself, with cosine corrections applied if local factors required a larger angle of measurement. Each site was surveyed once during the hours of daylight and once during the hours of darkness. The exercise was repeated three times over a 30-month period: before installation of the active road studs, six months after installation, and two years after installation. The surveys were undertaken in November or February to ensure that sunset times and ambient lighting conditions were controlled to be as similar as possible.

2.4. Additional Data Sources

To investigate other road factors that may influence speed, a combination of Ordnance Survey (UK national mapping agency) vector data, and visual observation of the infrastructure was used. The mapping data was used to derive the vertical and horizontal alignment of the route along with the carriageway widths at each speed-survey location. The gradients on the approach to the survey sites were derived from the vertical alignment profile and presented as a percentage; uphill gradients were recorded as positive and downhill gradients as negative, respectively. Curvature values were taken initially as an absolute centreline radius in metres; these were subsequently recoded into three binary variables corresponding with the desirable minimum, one-step, and two-step relaxation design values from the UK highway design manual [27]. Any radii exceeding 2880 m were considered tangent sections. The carriageway widths were measured at either end of the zone covered by the radar gun; these values were then averaged and divided by two to provide an effective carriageway half-width for the survey location.

Visual inspection at each survey site provided the posted speed limit (PSL), the presence of street lighting, the presence of 2 + 1 carriageway configuration, the locations of junctions or merges, and other features which may affect speed choice. Binary variables were created to represent each feature accordingly. Two further variables were created with respect to the presence of enforcement cameras; one represented the distance since the last camera with the direction of flow, the other the distance since the last camera passed facing either direction. A variable was also coded representing the length of the active road stud installation on the approach to the survey site. Each speed case recorded during the surveys was appended with the previously described variables corresponding to the survey site in question; binary variables were also included for whether the case had been recorded during daylight or darkness and whether it was before or after the implementation of the active road studs.

2.5. Statistical Analysis

Quantitative statistical analysis of the speed data was undertaken using the Statistical Package for the Social Sciences (SPSS) v26 (IBM). For the each of the datasets, from before, after, and two years after, descriptive statistics were produced. Z-scores were calculated and used to check for outlying cases of more than three standard deviations from the mean. Levene's test was used to test for equality of variance, and differences in mean were tested for using Welch's t-test. The null hypothesis was examined using a two-tailed test with a level of confidence (α) of 0.05.

At an infrastructure level, descriptive statistics were produced for the 15 variables of interest. Pearson's correlation coefficient was used to assess the association between road-environment factors and speed. Checks were made on the speed-data variables prior to the calculation of the correlation coefficients to determine the suitability of each variable for testing. All variables under test were at the interval or nominal level. Values of between 0.5 and 1 were taken as a high correlation, 0.3 to 0.49 as medium, and 0.1 to 0.29 as small.

Linear regression models were created for the dependent variable of speed and the independent variables of road and environmental features. For model estimation, the Ordinary Least Squares (OLS) approach was used [28]. Two models were created, one for sections of road with a 60 mph posted speed limit, the other for the remaining sections with a 70 mph limit. The stepwise regression function in SPSS was used to create the models. Prior to the process commencing, checks for normality and tests of collinearity were used from the output of the correlation testing previously described. No variable was excluded from the initial list of independent variables. Exclusion of variables was undertaken on a stepwise basis, with p-values of 0.05 used for entry and 0.10 used for exclusion.

3. Results

3.1. Descriptive Statistics

A sample size of 9135 speed cases was achieved across the survey periods of before, after, and two years after, with similar proportions recorded under each scenario. The recordings were found to approximate a normal distribution with no significant skewness or kurtosis by visual check. There was some evidence to suggest that the distributions could be slightly bimodal, which is likely to be due to the different speed limits between cars and heavy goods vehicles. However, as individual vehicle-type data had not been collected during the speed surveys, it was not possible to refine the data any further. On this basis, the datasets were retained for further analysis unchanged. The descriptive statistics for the total number of cases under the three scenarios are shown in Table 1.

Table 1. Descriptive statistics by scenario and light conditions for route.

Scenario	n	Min Speed (mph)	Max Speed (mph)	Mean Speed (mph)		Standard Deviation (mph)
				Value	Standard Error	
Before Treatment	3297	27	97	56.89	0.155	8.902
Light	1779	29	90	56.51	0.207	8.748
Dark	1518	27	97	57.35	0.233	9.062
After Treatment	3032	28	104	56.53	0.153	8.405
Light	1597	28	95	56.01	0.201	8.035
Dark	1435	34	104	57.11	0.231	8.764
Two Years After	2856	35	92	56.01	0.150	8.020
Light	1577	35	86	55.76	0.194	7.687
Dark	1279	35	92	56.33	0.235	8.404

3.2. Change in Route Mean and Mean Excess Speed

The significance testing for the changes in the global mean speed along the route is shown in Table 2 for the 60 mph and 70 mph posted speed limits, respectively. The F and t values for Levene's test for equality of variance and Welch's test for equality of means are shown, respectively, along with the associated probability (p) values and the degrees of freedom (df). For the 60 mph posted speed limit sections, the changes in the calculated overall mean speed ($\Delta\bar{v}$) after the implementation of the active road studs were negligible. During the hours of darkness in both scenarios, the mean speeds were reduced compared with the situation before installation. In the hours of daylight, a negligible reduction was recorded six months after installation, followed by a small increase after two years. No scenario was recorded with a significance level less than 0.05; therefore, in terms of the global mean speeds of the route, the null hypothesis cannot be rejected for the 60 mph posted speed limit sites.

For the speed distributions at the 70 mph posted speed limit sites, a reduction in mean speed of 1 mph overall was apparent six months after the treatment, or 1.5 mph during the hours of darkness. This reduced in magnitude slightly after two years, but a decrease was still apparent. In the former case, the results were recorded at a significance level of less than 0.05, and the null hypothesis may be rejected. Furthermore, a statistically significant change in variance could be observed during the six months after the treatment in all lighting conditions, although this was no longer the case two years after implementation.

The significance testing for changes in mean excess speeds (i.e., speeds greater than the posted limit) is shown in Table 3 for the 60 mph and 70 mph posted speed limit sections, respectively. The changes in the mean excess speed after the implementation of the active road studs were generally negligible. The differences in any change in mean excess speed between light and dark conditions were found to be broadly similar in both magnitude and direction in each scenario. Only the change in mean excess speed in the 60 mph scenario after two years approached significance; for the other scenarios, the null hypothesis cannot be rejected.

Table 2. Change in mean speed by posted speed limit.

PSL	Scenario	Equality of Variance		Equality of Means			
		F	*p*	t	df	*p*	$\Delta\bar{v}$ (mph)
60 mph	After Treatment	0.252	0.616	0.543	4139	0.587	−0.117
	Light	4.669	0.031 [a]	0.247	1894	0.349	−0.071
	Dark	2.183	0.140	0.454	1880	0.650	−0.149
	Two Years After	9.281	0.002 [a]	0.353	4301	0.724	−0.072
	Light	13.394	0.000 [a]	−0.615	2302	0.539	0.167
	Dark	0.307	0.579	1.033	1965	0.302	−0.318
70 mph	After Treatment	10.289	0.001 [a]	2.808	2137	0.005 [a]	−1.094
	Light	6.656	0.010 [a]	1.628	1130	0.104	−0.839
	Dark	4.639	0.031 [a]	2.576	971	0.010 [a]	−1.527
	Two Years After	1.333	0.248	1.581	1634	0.114	−0.695
	Light	1.845	0.175	0.622	892	0.534	−0.351
	Dark	0.114	0.736	1.684	742	0.093	−1.163

[a] Denotes significant result ($p < 0.05$).

Table 3. Change in mean excess speed by posted speed limit.

PSL	Scenario	Equality of Means			
		t	df	*p*	$\Delta\bar{v}$ (mph)
60 mph	After Treatment	0.618	481	0.536	−0.228
	Light	1.396	219	0.164	−0.583
	Dark	0.059	261	0.953	−0.033
	Two Years After	1.983	474	0.048 [a]	−0.650
	Light	1.523	234	0.132	−0.619
	Dark	1.184	239	0.237	−0.062
70 mph	After Treatment	1.700	314	0.090	−0.938
	Light	1.048	124	0.297	−0.708
	Dark	1.604	181	0.111	−1.357
	Two Years After	0.713	267	0.476	−0.396
	Light	1.038	140	0.301	−0.624
	Dark	0.220	132	0.826	−0.200

[a] Denotes significant result ($p < 0.05$).

3.3. *Change in Mean Speed by Site*

Changes in the local mean speed on an individual site basis along the route were assessed; the results of the analysis are shown in Table 4. The results suggest that the null hypothesis cannot be rejected at most sites. Five sites showed statistically significant changes in mean speeds immediately after treatment, increasing to nine sites after two years. The changes were mixed in direction and magnitude. Immediately after the stud installation, a decrease in mean speeds was seen at four sites and an increase at one. Two years after the installations, six sites had decreased in mean speed and three had increased. The largest statistically significant decrease in mean speed was 3.5 mph and the largest increase was 4.6 mph.

example, application of the power model [11] would suggest the reduction in mean speed during darkness measured here could reduce the number of fatal crashes by up to 10%.

Whilst significant global mean speed changes were restricted only to the 70 mph PSL sections of the route, mean speed changes at selected individual sites across both speed limits were found. No site could realistically be considered the same; each had its own geometric and topographic characteristics. In many cases, the implications of such characteristics are intuitive. For example, it would be reasonable to expect drivers to traverse curves of one- and two-step design relaxations at a slower speed in general than those of a desirable minimum radius, if only in the interests of occupant comfort. On this basis, it would appear that any relationship between mean speed and road features is a complex one, where the active road stud may be a contributory variable in certain circumstances. The subsequent regression analysis undertaken provides some insight into how the active road stud, along with other variables, may be associated with mean speed.

In the 60 mph PSL regression model, the greatest effect on speed was the presence of a merge. This could be a result of drivers travelling faster than usual to complete passing manoeuvres before the end of overtaking lanes. Whilst this result was statistically significant, it is noted that only two merge sites existed in the dataset, one of which was on a single carriageway divided by a central reserve, which is a relatively unusual arrangement. On this basis, it is suggested that other factors may contribute to this increase. Variables contributing to reductions in speed, particularly those related to gradients and curves, would seem intuitive; it seems quite possible that the speed of the vehicles may be affected by increased engine strain on the gradients and the discomfort and lower visibility experienced on the sharper curves. Where street lighting is not present, the increase in speed in the hours of darkness found in this study is consistent with other research findings [29,33].

In terms of the active road studs, the presence of the studs was found to be insignificant in the 60 mph PSL model. However, the length of a treated approach variable was retained. The decrease in mean speed associated with an increased length of stud treatment is a potentially interesting finding. However, it is noteworthy that the studs were installed only on the approaches to junctions, with the length of installation being proportionate to the importance of the junction itself. On this basis, it could be that drivers may, at least partially, have already adjusted their speed in anticipation of the conditions ahead, and the association with stud treatment length was merely coincident. Given that all the 60 mph PSL sites were at junctions, unfortunately it was not possible to control for this possible effect with the data gathered.

In the 70 mph PSL model, again the presence of darkness was associated with higher speeds, more so in fact than the former. Significantly, the presence of the active road stud appeared to influence speed, with a reduction of 1 to 1.5 mph being implied. In the case of the length of the stud installation, the opposite to the 60 mph PSL model occurs; in this scenario, speeds increase as the length of treated approach increases. It is known that the dual carriageway sections of this road are mostly used to provide overtaking opportunities. On this basis, it may be that longer sections of dual carriageway are more attractive for overtaking and therefore result in higher speeds, which could explain this apparent pattern.

The other variable of significance in the 70 mph PSL model was road half-width. In this case, increases in half-width appeared to result in increased speeds. It has been suggested that the additional steering workload, along with a better perception of speed, is associated with slower speeds in narrower lane widths [34,35]. Such association could be expected during hours of daylight but less so during hours of darkness, where a driver's field of view and depth of field are restricted by the capability of the vehicle headlights. It could be that the presence of active road studs might enhance a driver's awareness and perception of speed, leading to this association.

Some limitations in the present work are acknowledged. Most of the survey sites were located on the approach to junctions as they were the focus of the original installation. It could be that drivers already moderate their speed when in such circumstances, in which case any change in mean speed may be suppressed. The work here was also entirely random in its sampling, and specific vehicle types were not recorded. Whilst this means the results are likely to be valid for typical traffic composition, the effects may be different between vehicle categories. This study also controlled for weather, with all the results obtained during benign conditions. One aspect in which active road studs are thought to be of particular benefit is during times of poor visibility. On this basis, their effect on mean speeds under such conditions could be worthy of further investigation.

5. Conclusions

The aim of this research was to measure the choice of speed by drivers, using real-world rural junctions and links and to determine whether changes in the speed of vehicles may be associated with the installation of active road studs. The null hypothesis stated that drivers do not change their speed in the presence of active road studs, and there is no difference in mean speeds before and after installation. Based on these findings, looking at the entire route, a universal rejection of the null hypothesis across all tests is not possible; there is no consistent correlation between mean speed and the presence of the active road studs at all sites. However, there appears to be some specific situations in which the null hypothesis could be rejected. Most notably, on the 70 mph PSL dual carriageway sections, the presence of studs may contribute to mean speed reduction as part of a wider group of variables.

The results of this work are likely to be of interest to road safety engineers, specifically as the improved visibility resulting from the installation of active road studs appears unlikely to result in any corresponding increase in the mean speed of vehicles. This is a positive finding, as it means that any other safety benefits found, such as improved driver confidence or better lane discipline, would not be offset by the risk of greater crash severity through increased speed. On higher-speed dual carriageway sections, the presence of the stud may even result in a reduction in mean vehicle speeds. In road safety terms, this could mean that the severity of crashes may be decreased in the presence of the stud. This would appear to be a particularly positive outcome, given that these are the highest speed sections—and therefore potentially the highest crash severity—on the road network.

Author Contributions: Conceptualization, R.L. and J.C.; methodology, R.L.; software, R.L.; validation, J.C. and G.F.; formal analysis, R.L.; investigation, R.L.; resources, R.L.; data curation, R.L.; writing—original draft preparation, R.L.; writing—review and editing, R.L., J.C. and G.F.; visualization, R.L.; supervision, J.C. and G.F.; project administration, R.L. All authors have read and agreed to the published version of the manuscript.

Funding: This research received no external funding.

Data Availability Statement: Not applicable.

Acknowledgments: The authors would like to thank Amey for facilitating the inspection of the installation of the works during construction and Transport Scotland for the provision of the background information regarding the scheme.

Conflicts of Interest: The authors declare no conflict of interest.

References

1. *World Health Organisation Global Status Report on Road Safety 2018*; World Health Organisation: Geneva, Switzerland, 2018.
2. *International Transport Forum towards Zero: Ambitious Road Safety Targets and the Safe System Approach*; OECD: Paris, France, 2008; ISBN 9789282101957.
3. *PIARC-World Road Association Road Safety Manual*; PIARC (World Road Association): Paris, France, 2019.
4. Joksch, H.C. Velocity Change and Fatality Risk in a Crash—A Rule of Thumb. *Accid. Anal. Prev.* **1993**, *25*, 103–104. [CrossRef]
5. Cirillo, J.A. Interstate System Accident Research Study II, Interim Report II. *Public Roads* **1968**, *35*, 71–75.
6. Solomon, D.H. *Accidents on Main Rural Highways: Related to Speed, Driver, and Vehicle*; US Department of Transportation, Federal Highway Administration: Washington, DC, USA, 1964.
7. Taylor, M.C.; Lynam, D.A.; Baruya, A. *The Effects of Drivers' Speed on the Frequency of Road Accidents*; Transport Research Laboratory: Crowthorne, UK, 2000.

8. Hauer, E. Speed and Safety. *Transp. Res. Rec.* **2009**, *2103*, 10–17. [CrossRef]
9. Elvik, R.; Christensen, P.; Helene Amundsen, A. *Speed and Road Accidents an Evaluation of the Power Model*; Transportøkonomisk Institutt: Oslo, Norway, 2004.
10. Finch, D.J.; Kompfner, P.; Lockwood, C.R.; Maycock, G. *Speed, Speed Limits and Accidents*; Transport Research Laboratory: Wokingham, UK, 1994.
11. Nilsson, G. Traffic Safety Dimensions and the Power Model to Describe the Effect of Speed on Safety. Ph.D. Thesis, Lund University, Lund, Sweden, 2004.
12. Nilsson, G. The Effects of Speed Limits on Traffic Crashes in Sweden. In Proceedings of the International Symposium on the Effects of Speed Limits on Traffic Crashes and Fuel Consumption; Organisation for Economy, Co-Operation, and Development (OECD), Dublin, Ireland, 6–8 October 1981.
13. Pantangi, S.S.; Fountas, G.; Anastasopoulos, P.C.; Pierowicz, J.; Majka, K.; Blatt, A. Do High Visibility Enforcement Programs Affect Aggressive Driving Behavior? An Empirical Analysis Using Naturalistic Driving Study Data. *Accid. Anal. Prev.* **2020**, *138*, 105361. [CrossRef] [PubMed]
14. Pantangi, S.S.; Ahmed, S.S.; Fountas, G.; Majka, K.; Anastasopoulos, P.C. Do High Visibility Crosswalks Improve Pedestrian Safety? A Correlated Grouped Random Parameters Approach Using Naturalistic Driving Study Data. *Anal. Methods Accid. Res.* **2021**, *30*, 100155. [CrossRef]
15. Boys, J.T.; Green, A.W. Intelligent Road-Studs-Lighting the Paths of the Future. *Trans. Inst. Prof. Eng. N. Z. Gen. Sect.* **1997**, *24*, 33–40.
16. Shahar, A.; Brémond, R. Toward Smart Active Road Studs for Lane Delineation. In Proceedings of the Transport Research Arena (TRA) 5th Conference: Transport Solutions from Research to Deployment, Paris, France, 14–17 April 2014.
17. Llewellyn, R.; Cowie, J.; Maher, M. Active Road Studs as an Alternative to Lighting on Rural Roads: Driver Safety Perception. *Sustainability* **2020**, *12*, 9648. [CrossRef]
18. Reed, N. *PPR143 Driver Behaviour in Response to Actively Illuminated Road Studs: A Simulator Study*; Transport Research Laboratory: Crowthorne, UK, 2006.
19. Shahar, A.; Bremond, R.; Villa, C. Can Light Emitting Diode-Based Road Studs Improve Vehicle Control in Curves at Night? A Driving Simulator Study. *Lighting Res. Technol.* **2018**, *50*, 266–281. [CrossRef]
20. Styles, T.; Cairney, P.; Studwick, G.; Purtill, S. Effects of Self-Activated Pavement Markers on Driver Behaviour. *Road Transp. Res.* **2004**, *13*, 98–100.
21. Styles, T.; Cairney, P.; Studwick, G.; Purtill, S. Trial and Evaluation of Internally Illuminated Pavement Markers. In Proceedings of the Road Safety Research, Policing and Education Conference, Sydney, Australia, 24–26 September 2003; Volume 2, pp. 550–555.
22. Voigt, A.P.; Carson, J.L.; Tydlacka, J.; Gray, L.S. *Applications of Illuminated, Active, In-Pavement Marker Systems*; Transportation Research Board: Washington, DC, USA, 2008; ISBN 978-0-309-27955-0.
23. Rutherford, N. More than 6000 Caught Speeding on A1 through Borders. Available online: https://www.bbc.co.uk/news/uk-scotland-south-scotland-33568848 (accessed on 8 February 2021).
24. Scottish Borders Council Minutes of the Berwickshire Area Forum-3 September 2015. Available online: https://scottishborders.moderngov.co.uk/ieListDocuments.aspx?CId=183&MID=252#AI2933 (accessed on 17 April 2020).
25. Ki, Y.K.; Baik, D.K. Model for Accurate Speed Measurement Using Double-Loop Detectors. *IEEE Trans. Veh. Technol.* **2006**, *55*, 1094–1101. [CrossRef]
26. Taghvaeeyan, S.; Rajamani, R. Portable Roadside Sensors for Vehicle Counting, Classification, and Speed Measurement. *IEEE Trans. Intell. Transp. Syst.* **2014**, *15*, 73–83. [CrossRef]
27. Design Manual for Roads and Bridges, CD 109 Revision 1: Highway Link Design. Available online: https://www.standardsforhighways.co.uk/prod/attachments/c27c55b7-2dfc-4597-923a-4d1b4bd6c9fa (accessed on 10 February 2021).
28. Washington, S.; Karlaftis, M.; Mannering, F.; Anastasopoulos, P. *Statistical and Econometric Methods for Transportation Data Analysis*; Chapman and Hall/CRC: New York, NY, USA, 2020; ISBN 0429244010.
29. Assum, T.; Bjørnskau, T.; Fosser, S.; Sagberg, F. Risk Compensation-the Case of Road Lighting. *Accid. Anal. Prev.* **1999**, *31*, 545–553. [CrossRef]
30. Fountas, G.; Fonzone, A.; Gharavi, N.; Rye, T. The Joint Effect of Weather and Lighting Conditions on Injury Severities of Single-Vehicle Accidents. *Anal. Methods Accid. Res.* **2020**, *27*, 100124. [CrossRef]
31. Winston, C.; Maheshri, V.; Mannering, F. An Exploration of the Offset Hypothesis Using Disaggregate Data: The Case of Airbags and Antilock Brakes. *J. Risk Uncertain.* **2006**, *32*, 83–99. [CrossRef]
32. Triggs, T.J.; Berenyi, J.S. Estimation of Automobile Speed under Day and Night Conditions. *Hum. Factors J. Hum. Factors Ergon. Soc.* **1982**, *24*, 111–114. [CrossRef] [PubMed]
33. Jägerbrand, A.K.; Sjöbergh, J.; Martins, M.; Pereira, N.; Reis, R.; Tomaz, E.; Jägerbrand, A.K.; Sjöbergh, J. Effects of Weather Conditions, Light Conditions, and Road Lighting on Vehicle Speed. *SpringerPlus* **2016**, *5*, 505. [CrossRef] [PubMed]
34. WAARD, D.D.; Jessurun, M.; Steyvers, F.J.; Reggatt, P.T.; Brookhuis, K.A. Effect of Road Layout and Road Environment on Driving Performance, Drivers' Physiology and Road Appreciation. *Ergonomics* **1995**, *38*, 1395–1407. [CrossRef] [PubMed]
35. Godley, S.T.; Triggs, T.J.; Fildes, B.N. Perceptual Lane Width, Wide Perceptual Road Centre Markings and Driving Speeds. *Ergonomics* **2004**, *47*, 237–256. [CrossRef] [PubMed]

energies

Article

Development of Comfort and Safety Performance of Passenger Seats in Large City Buses

Ivan Kernytskyy [1], Yevheniia Yakovenko [2], Orest Horbay [2], Maryana Ryviuk [3], Ruslan Humenyuk [3], Yaroslav Sholudko [3], Yurii Voichyshyn [3], Łukasz Mazur [1], Piotr Osiński [1], Konstantin Rusakov [1] and Eugeniusz Koda [1,*]

[1] Institute of Civil Engineering, Warsaw University of Life Sciences, 159 Nowoursynowska St., 02-76 Warsaw, Poland; ivan_kernytskyy@sggw.edu.pl (I.K.); lukasz_mazur@sggw.edu.pl (Ł.M.); piotr_osinski@sggw.edu.pl (P.O.); konstantin_rusakov@sggw.edu.pl (K.R.)

[2] Department of Automotive Engineering, Lviv Polytechnic National University, 12 Stepana Bandery St., 79000 Lviv, Lviv Oblast, Ukraine; yevhenija.yakovenko@lpnu.ua (Y.Y.); ogorbay@polynet.lviv.ua (O.H.)

[3] Lviv National Agrarian University, 1 Volodymyra Velykoho St., 30831 Dubliany, Lviv Oblast, Ukraine; maryanaryv.lnau@ukr.net (M.R.); hrv.lnau@ukr.net (R.H.); Shyv.lnau@ukr.net (Y.S.); yurij.voichyshyn@lpnu.ua (Y.V.)

* Correspondence: eugeniusz_koda@sggw.edu.pl

Abstract: A bus seat needs to be designed ergonomically for better seating comfort. The present study is intended to develop a cost-effective ergonomic bus seat design based on seat comfort and safety demands. As part of the proposed seat design procedure, seating comfort analysis, identifying preferred design features, and developing a seat design are included. An analysis of the bus seat back and seat pan profiles was conducted. Based on the results of the comfort analysis, the authors identified the preferred design features of bus seats during the design identification process. An improved bus seat prototype was developed based on selected design features in the design development stage. Seating comfort analyses were used to compare the achieved seat with the reference seat. The seat design developed in the present study may be applicable for various types of bus public transport.

Keywords: public transport; bus seat; seating comfort analysis; FEM model; structural safety assessment; ergonomics

Citation: Kernytskyy, I.; Yakovenko, Y.; Horbay, O.; Ryviuk, M.; Humenyuk, R.; Sholudko, Y.; Voichyshyn, Y.; Mazur, Ł.; Osiński, P.; Rusakov, K.; et al. Development of Comfort and Safety Performance of Passenger Seats in Large City Buses. *Energies* **2021**, *14*, 7471. https://doi.org/10.3390/en14227471

Academic Editors: Guzek Marek, Rafał Jurecki and Wojciech Wach

Received: 16 October 2021
Accepted: 6 November 2021
Published: 9 November 2021

1. Introduction

Seating comfort is one of the most important indicators of automotive seat performance [1–3]. Around the world, there have been many studies on seating comfort, including car seats, truck seats [4], and bus and train seats [5]. A seat that is comfortable in static conditions may have poor dynamic characteristics that make it uncomfortable on the road. The profile of a bus seat needs to be designed ergonomically for various body sizes of passengers. Most automotive seats are not designed according to anthropometry, neither are the automakers willing to invest resources in designing components appeasing human ergonomics [6].

Nowadays, urban space, especially in large cities, faces several challenges resulting from the permanently increasing number of inhabitants. One of the urgent issues to be solved is a safe, comfortable, and quick way of commuting for the residents of urban agglomerations. However, along with overloaded and rapidly expanding urban space, along with modernization of the existing road infrastructure, the growth of traffic level is also significant [7–11].

A natural way of solving such a challenge is to develop collective transport and encourage the still undecided citizens to use it. Residents choosing the mode of transport in urban agglomerations analyze various factors, among which the largest impact on their decisions are safety and comfort (which affect visual, thermal, acoustic, and vibration aspects) [12]. Research studies have already shown that passenger comfort is a key factor in the choice of the means of transport, and its improvement may attract more users of urban transport [13–17]. Following investments in the safety and comfort of collective transport, the quality of life in contemporary cities may be improved. On the other hand, increased road safety could be achieved in many other ways, too (Figure 1). Passenger protection while travelling is one of them, i.e., the design of safe and comfortable seats along with safety belts. A seat's dimensions must be designed to suit the anthropometry of passengers when designing ergonomic seats. Optimum passenger seat design according to passenger anthropometry can decrease fatigue and discomfort during long periods of sitting [18]. The passengers of the vehicle must be comfortable, as the discomfort can result in fatigue, which can lead to a condition of body imbalance, since the passenger seat used does not usually correspond with the wearer's anthropometry (non-ergonomic). When it comes to the passenger seat design sector, an ergonomic factor and aspect will ensure greater comfort and less fatigue for passengers [19]. According to The Harvard School of Public Health, ergonomics is the science, art, and application of technology that aims to harmonize or balance between all facilities used at work and rest with human abilities and limitations so that the overall quality of life can be increased [20].

Figure 1. Examples of ways to improve road safety.

A literature review on the comfort of bus driver seats indicates several medical problems resulting from their long-term use [21–23]. Bus drivers are struggling with such problems as sleepiness, cramps, muscle fatigue, difficult circulation, spine pain, depression, stiffness, pain, numbness in the spine, and other musculoskeletal problems that may become chronic [24–30]. For that reason, to reduce these medical problems resulting from a malformed bus seat, further research should be carried out on the existing bus seat profile. Bearing that that in mind the authors tried to fill the knowledge gap by fully understanding the mechanisms of stress and strength distribution for a specific seat construction design, the present research focuses on specific solid plastic frame construction, which has not been fully studied in the available scientific literature. The study concerns bus seats occupied by an adult (50 years old) and a young (10 years old) passenger, concerning different seat solid plastic frame mounting scenarios.

A comparative analysis of existing bus seat profiles and their evaluation may be presented in the following aspects: (i) comfort evaluation and (ii) ergonomic evaluation. In the comfort evaluation part, 48 participants evaluated seven parts (headrest, upper-back support, lumbar support, seatback bolster, hip support, thigh support, and seat pan bolster) [25]. There are 12 components of the bus seat that directly affect passenger comfort, including seatbelts, armrests, recliners, headrests, dorsal support, lumbar support, side back support, seatback overall support, hip support, thigh support, and seat pan overall support. In addition, 17 ergonomic evaluation measures (such as reachability, controllability, tactile sensation, grip sensation, adjustability, size appropriateness, shape appropriateness, cushioning, and overall comfort) were selected to be assessed based on literature reviews [31–33].

In addition to comfort, safety is an essential requirement. The subject of traffic safety has been investigated by many researchers in the context of cars, buses, and coaches [34,35]. Rupp et al. [36] found that the fracture tolerance of the femur is 7.59 kN. In the US, the National Highway Traffic Safety Administration (NHTSA) specified a maximum femur load of 10 kN for a male dummy of the 50th percentile [37]. Leg femurs on both sides experienced 5.2 kN loads. Although within the safe limit, this model predicted a high pelvis load. Due to the dummy posture, the knees are the main point of contact between the seat back and the femurs. The pelvis injury tolerance was determined based on peak pelvis acceleration according to Haffner [38]. Using 130 g of acceleration on the pelvis, he proposed that the occupant's pelvis can suffer serious injury. When the setback impact occurs, the model predicted a peak acceleration of 33 g for the pelvis [39]. There was a similar interval between the peak acceleration of the pelvis and the peak load of the femur. However, it is below the safe threshold of injury for the pelvis acceleration level. Mertz and Patrick [40] conducted a study showing showed that the human chest is capable of bearing a distributed load of 49 g. This value should not exceed 60 g, according to the FMVSS 208 test.

The development of the Bus Safety Standards (BSS) in different countries included seat testing, both in computer simulations and sledge testing (which replicates the collision forces in a repeatable way, but for testing just the seat in isolation and not the entire vehicle). This testing compared traditional low-back seats against medium (taller) back seats and high (for example coach style) back seats. In rear-facing seats, the BSS encourage high back seats. The additional weight of the different seats makes them difficult to implement throughout the entire bus. European Union Transport Politics aims to emphasize the importance of using public transport rather than private transport while focusing on the safety of the passengers and reducing pollution [41,42]. The World Health Organization (WHO) acknowledges this fact in its annual report on global road safety, where they advocate for better public transport that is safe, accessible, and affordable because it is vital to increase safety in urban areas where traffic has become more crowded [43–45].

For the purpose of the present research, the comfort and safety of solid plastic frame seat performance were evaluated based on stress deformation and kinematic behavior. The seats' frame structure and the material used in the modeling and analyses are considered not to be fully investigated in the available literature, making the research performance the biggest motivation.

2. Materials and Methods

Seating comfort analysis can be performed using vibration evaluation, electromyography, electroencephalography, oxygen saturation, posture-image analysis, spinal loading, computer-aided engineering (CAE), pressure, temperature and humidity monitoring, etc. [46–48].

The experimental, purpose-built physical model of the bus that had a front seat and two rows of seats is presented in Figure 2. The model concerned a dummy hybrid representing a 50-year-old man (M50) and a dummy representing a 10-year-old child (P10). The position of the dummies, the seats, and the safety belts were adopted based on the assumed real event scenario. All the characteristic dimensions could be followed in detail in Table 1.

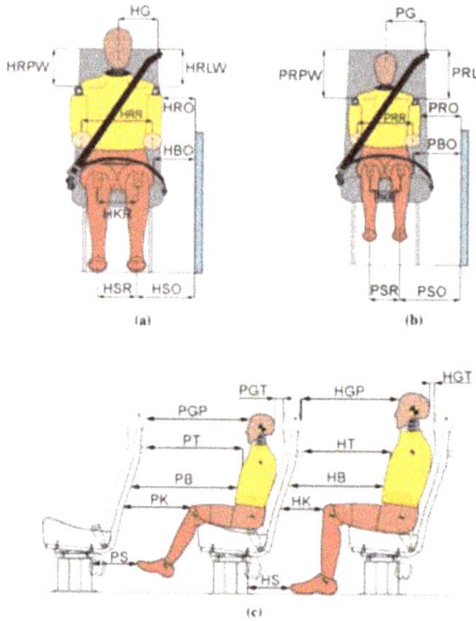

Figure 2. Dimensions describing the dummy position concerning the seat and safety belt location: (**a**) front view M50, (**b**) front view P10, (**c**) side view (modified after [49]).

Table 1. Distances describing the position of the dummies.

Specimen	Front View		Side View	
	Symbol	Distance (mm)	Symbol	Distance (mm)
M50 dummy	HG	100	HGT	220
	HBPW	225	HGP	350
	HBR	50	HT	430
	HRR	205	HB	490
	HRO	40	HK	100
	HBO	85	HS	295–300
	HKR	175		
	HSR	245		
	HSO	155		
P10 dummy	PG	90–100	PGT	0
	PBPW	330	PGP	515
	PBR	330	PT	540–545
	PRR	70	PB	575–580
	PRO	282	PK	255
	PBO	105	PS	40
	PKR	150		
	PSR	85		
	PSO	250		

The Mathematical Dynamics Models for Applications (MADYMO) software program uses the method of multibody system dynamics for the formulation of a numerical model. Here, a chain of rigid bodies is linked together using kinematic pairs to copy or model an object. Several physical variables determine the bodies as well as the kinematic joints, which make it possible to solve the equations governing movement. There are three parameters used to describe rigid bodies: mass, moment of inertia, and center of gravity. The kinematic pairs defined specify the bodies that are joined with each kinematic pair. In addition, the coordinate location of those bodies is determined as well.

A general equation, Equation (1), is used to describe the relationship between the position of a local coordinate system and the global coordinate system, where X_i is the matrix of the coordinates of the position vector, r_i is the matrix of the coordinates of the vector joining the beginnings of both coordinate systems, A_i is the matrix of direction cosines, and x_i is the matrix of the coordinates of the vector of local displacement of the coordinate system.

$$X_i = r_i + A_i x_i \tag{1}$$

The numerical model for the solution to the general equation of motion Equation (2) uses a modified single-phase Euler's method with a constant time step t_s:

$$M\ddot{x} + C\dot{x} + Kx_t = P_t \tag{2}$$

where M—mass matrix; C—matrix defining the damping of the system; K—rigidity matrix; x_t—displacement; P_t—matrix defining external loads applied to the system. This matrix is usually adopted in the form of the so-called proportional damping (depending on K and M matrices).

There are 3 types of mechanical models, in which fuzzy models and neural networks are not used because of their non-linearity. Unfortunately, due to the nature of the experiment, the ramification cannot be used to achieve the desired damping force in an open control system. In this case, the Bouc–Wen model was used. As described by Spencer et al. [50], the model is commonly used to elaborate the MR damper hysteretic characteristics. In the present experimental study, the controlled MR damper has high vibration restitution when compared with the passive MR damper. A controlled MR damper presents the better performance by 24% of road-holding vehicles when compared with fully active and 22% on an ideal semi-active suspension system [51].

The hysteretic behavior of the MR damper was depicted with the help of the Bouc–Wen model [50]. A scheme of the mechanical diagram of the MR damper is shown in Figure 3. The dampers can be used as comfort and noise protection elements, too [41–54]. Figure 3 presents the general view of the soundproof partition of the bus motor. The cross-section shows the bus septum, where 1—engine compartment, 2—passenger compartment, 3—DVA, 4—internal part of the partition. The partition has an elastic fastening, which is the external part of the sound absorber of reinforcing elements. A damper on the driver's seat is a required feature. However, transport comfort can be increased by fitting dampers to the passenger seats as well.

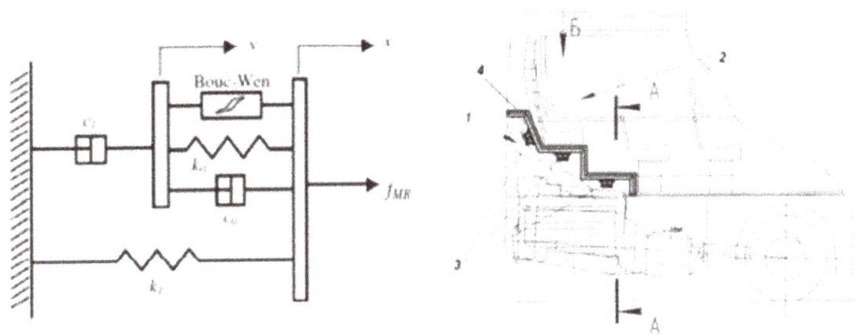

(a) (b)

Figure 3. Mechanical model of MR damper (**a**), and general appearance of a bus soundproofing system (**b**).

From the empirical point of view, the mechanical model is governed by the following equations

$$\dot{y} = \frac{1}{c_0 + c_1}\left[\alpha z + k_0(x - y) + c_0\dot{x}\right] \tag{3}$$

$$\dot{z} = -\gamma|\dot{x} - \dot{y}||z|^{n-1}z - \mu(\dot{x} - \dot{y})|z|^n + A(\dot{x} - \dot{y}) \tag{4}$$

$$f_{MR} = c_1\dot{y} + k_1(x - x_0). \tag{5}$$

Based on dimensional differences, 12 bus seats were classified based on their features for the purpose of the study. Accordingly, lumbar supports were classified in reference to dimensional differences between extracted centerline and seat pan lines measured for the measured bus seat profiles. This means that the lumbar was shaped for ≥25 mm (mean) or flattened to achieve the lumbar shape of <25 mm on the center line, and the designed side lumbar support was shaped for ≥45 mm or flattened to achieve the side lumbar support shape <45 mm than the side lumbar support on the seat pan lines (Figure 4) [48].

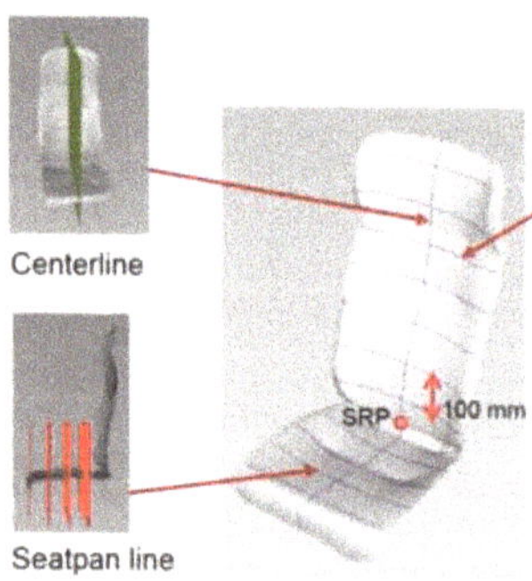

Figure 4. Critical cross-sections and outlines of a bus seat profile.

3. Results and Discussion

Unlike nowadays, in the past, the seat designs were made mostly of the tubular profile. However, stamped designs of passenger bus seats still need further experimental analyses. In this case, a stamped construction with a thickness of 2 mm, made of Steel 20, was used in the 3D modeling (Figure 5). The software allowing stress distribution evaluation was the one presenting the traditional approach for solving and presenting the distribution of internal stresses in various structures. Dynamic analysis was conducted to determine the stress deformation and kinematic behavior of the mechanism with consideration of the elastic–plastic behavior of individual parts of the mechanism.

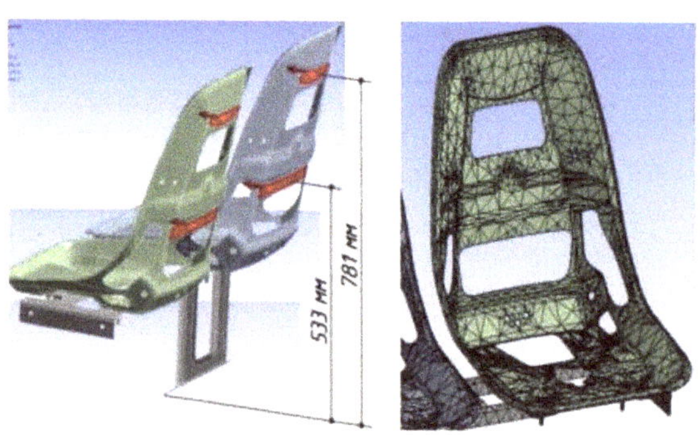

(a) (b)

Figure 5. Seat model made of pressed sheet: (**a**) solid model; (**b**) Finite Element Method (FEM) model.

While building the model, the first step was to reduce weight, which weakens the supporting frame, and to increase its compliance, the model seat was cut out. The cut-out is very important for obtaining the necessary movements at checkpoints following the requirements of static tests proposed by the Economic Commission for Europe of the United Nations Regulation No. 80 [55].

The loading height H1 of load P1 was 781 mm concerning the base surface, and the loading height H2 of load P2 was 533 mm, respectively. The values of loads P1 and P2 applied to the back seat were parallel to the horizontal plane and the longitudinal axis of the bus and could be calculated as follows in Equations (6) and (7):

$$P_1 = \frac{1000}{H_1} \pm 50 = \frac{1000}{0.781} + 50 = 1330N \tag{6}$$

$$P_2 = \frac{2000}{H_2} \pm 100 = \frac{2000}{0.533} + 100 = 3852N \tag{7}$$

A further performance of the research on the seat frame conforming to the requirements of UNECE Regulation No. 80 [55] is the analysis of the structure strength presented in the form of a solid-state model (solid model). This would ensure higher accuracy of the calculation and would allow taking into account the plastic deformations (allowing physical non-linearity). The main advantage of the solid model is the possibility of transition from the nodal connections of the rods to the actual volumetric connections of pipes and other elements of the design corresponding to the actual scenario. In contrast to the beam performance, the solid-state model is much more complex in terms of its construction and analysis, because the model needs to be split into sets of sizes and configurations.

The study of three variants of the seats loaded with the following scenarios was carried out: with a stand (Figure 6a), with a side fixing console (Figure 6b), and with an additional handrail, which was attached to the roof frame (Figure 6c).

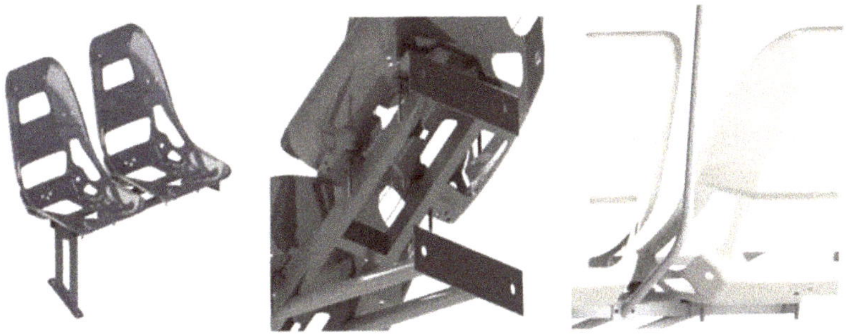

(a) (b) (c)

Figure 6. Three-dimensional model of a seat with a stand (**a**), a side control fixing (**b**), and a handrail attached to the roof frame (**c**).

The first example (Figure 6a) consists of a vertical rack, which was made of a 2 mm thick profile and with a hole in the center to save material; however, it was not reducing the structure's stiffness. The bottom of the rack is welded to a 3 mm thick plate with two holes of 10 mm in diameter, under a bolted connection. In the model, the welded connection of the plate is simulated, which acts as the flange fastening the base to the wheel arches of the bus and the rack's frame (Figure 6a).

The second scenario (Figure 6b) is relevant for low-entry and low-floor vehicles, allowing efficient bus floor maintenance. Its base is not fixed to the racks on the floor but is firmly tied to the side of the bus through two supporting circular pipes of 20 × 2 mm in size (Figure 6b). These pipes are welded to a steel plate (265 × 50 × 3 mm in size) with two holes for a bolted connection of 10 mm. The flanges are attached to the side of the seat at a height of 200 mm.

Starting from the fact that the second scenario (Figure 6b) is characterized by higher optimization potential, in comparison with the classical variant (Figure 6a) with a stable floor mount, which has a much smaller distance between the mount supports, and therefore a significantly greater moment of action from the forces, it is advisable to strengthen it by the introducing an additional rail attached to the roof frame (Figure 6c). It is worth noting that it is often possible to find a construction variant with a handrail attached to the frame of one of the seats. Thus, as a result of loads following UNECE Regulation No. 80, different values of displacement and absorbed energy are obtained for the back and the right-sided seats.

In general, each node has 6 degrees of free space (three linear movements and three angles of rotation), but for a particular task, certain degrees of free space may not be possible. For instance, in the solving process of deforming a flat five-point fastening of the base of the seat loaded by plane forces, the basic functions describing the displacement of the peak plane can be omitted, since their value is zero.

The description type includes here the topology of an element (flat triangle, flat quadrilateral, a polygon with curved sides, bulk element-tetrahedron, hexagon, etc.). For such shapes, the basic functions (a quadrilateral with 6 degrees of free space in the node, quadrilateral with 2 degrees of free space in the node, etc.) and formulas for calculating elastic energy are used. The solution of a set of equations of FEM (finite elements method) by the prediction of displacement will be the components of nodal displacements of the discrete design of a passenger seat.

In general, the base model with a stand consists of 19 constituent elements. Its weight is 15.1 kg. The number of finite elements of the FEM model approaches 40,000 and the number of nodes is 85,376. The results of the calculations indicate that despite being identical to one another, the left and right seats are divided into different amounts of finite elements (FE): the left consists of 16,160 elements joined by 33,265 nodes and the right consists of 16,189 items with 33,309 nodes. Similar conditions can be observed for absolutely identical M8 bolts fastening the base. There may be several explanations: firstly, the Ansys Workbench algorithm generates a grid every time individually (adapting to the set conditions of the calculation), and may start, for example, not from the seat mounting area, but its upper part, showing signs of an artificial neural network. Secondly, the base to which the seat is fixed is not symmetrical (different degrees of detail in its various areas).

The model of the seat with a console mount consists of 19 constituent elements and is heavier by 0.6 kg. The number of end-points of the FEM model increases to 42,699, and the number of the corresponding nodes is almost 5000 units.

All elements that are part of the seat model have nonlinear characteristics of the material, including its physical nonlinearity (Nonlinear Effects = Yes). For example, one of the M8 bolts consists of 198 elements and 408 knots, and the corresponding nut consists of 66 elements and 473 knots. The contact sensitivity parameter that is automatically set in the Ansys Workbench environment is 3.4712 mm (Tolerance Value). In general, the model is more complex, concerning the rack sample, and it requires a more detailed breakdown of key elements: the degree of detail in the partition is determined by the Relevance parameter, which is assumed to be 0.35 for the model under study (as part of the Ansys Workbench software functionality, the maximum value for Relevance is -100). Compared with the first scenario (Figure 5a) and despite the absence of a rack mount to the floor, the overall height of the model (Length Y) increased from 791.09 to 1184 mm. The difference in overall height is due to the presence of a high rail, which is fixed to the system of railings on the roof of the cabin. Compared with the second scenario (Figure 5b), the number of FEs increased by 7%, and the number of nodes increased by 7.5%, respectively.

The weight of the model with a handrail is 16.6 kg, and the number of constituent elements of assembly increased not as much. Data on the FEM-wire rail were as follows: the number of FE railings is 3792, and the number of nodes is 7426. Its dimensional height was 971.5 mm, and the weight of the whole model was 0.9 kg.

During the computer simulation of seat frame tests, the behavior of the component elements with the corresponding steel–steel friction pairs has also been determined. For steel–steel pairs, the coefficient of friction is set to 0.2, which corresponds to the typical value in rest. In addition to the bolted joints, all contact pairs have a friction coefficient of 0.2. The hard type bonded (motionless) joint provides a bolted connection (for example, Bonded—M8 bolt to bolt connection).

It should be noted that the statement about the direct correlation of the growth of relevance to the accuracy of the calculation is false. In the split configuration on the FE, the Relevance Center parameter is set to Fine, which means an improved detail of the element at the curves of the model and complex transitions (bolted connections, etc.). The minimum value of the edge length of the finite element was 5,8838e−004 mm.

Firstly, the stress values of the base model with the rack were analyzed. The maximum value reached was 431.74 MPa. The rack was fixed using the bolted connection at the left bracket, which was mounted along with the movement of the seat to the base in its front part. At the same time, the maximum stress value of mounting bolts was 199.43 MPa. In the place where the seat frame was fixed with a bolted connection, the stress value was 240 MPa. The constituent elements reached maximum stresses at different times, which is presented in Figure 7.

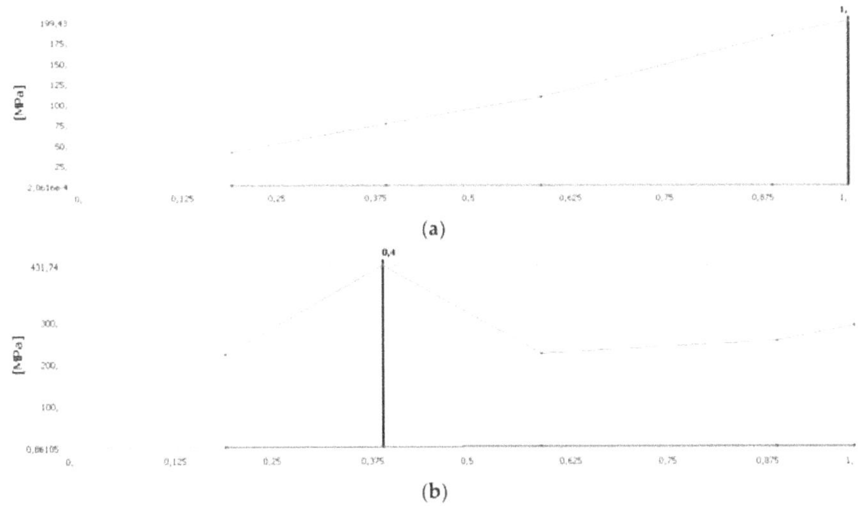

Figure 7. Tension diagrams depending on load steps: (**a**) nut; (**b**) bolt.

When considering the stress–strain state of the design at the stage of full absorption of the impact energy presented in Figure 7, the maximum stress value, equal to 374.74 MPa, is recorded in the frame of the right-hand drive of the bus in the area of the seat bend. This is as expected due to the greatest bending moments passing through this area. The stress values in the rack vary from 5 MPa in the flange with an increase of 45 MPa in the bolted joints and up to 270 MPa in the rack (Figure 8). Structurally, the seat leg is a welded structure consisting of a pressed sheet with a thickness of 2 mm and a flat flange fastened to the floor with a thickness of 3 mm.

Together with the base of the seat (welded square tube frame), the rack and flange mounted to the bracket of the seat arch forms a single inseparable structure. The structure demonstrated a maximum stress value of 339.81 MPa, which is fixed in the right seat mount bracket.

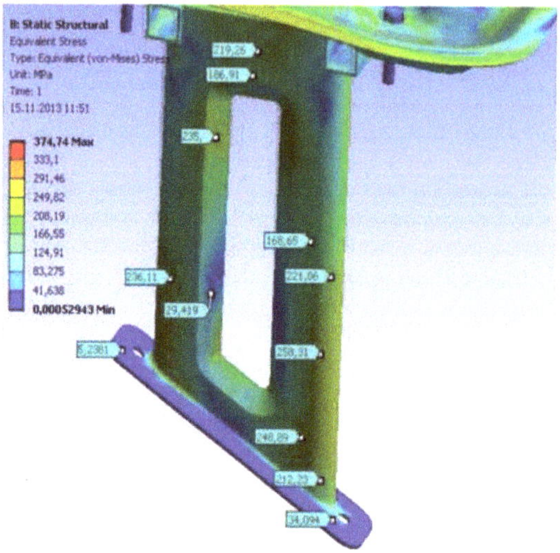

Figure 8. Strength map of the rack.

The variant of the seat base, designed for the installation in low-entry buses, has a one-way mount attached to the bus sidewall, and therefore, in theory, it has a lower equilibrium than the previous design. The maximum value of stresses recorded during the loading process was 434.33 MPa. It corresponds to the contact area of the anterior nut with the left seat frame. Such conditions and detection of the maximum stress design fully correspond with the previous case. In both cases, the nut will withstand the force reactions transmitted from the bolt while maintaining the design integrity. The maximum stress value is 390.02 MPa, which is 3–5% more compared to the previous version. This value corresponds to the area of the left-hand drive of the seated bus, which is the opposite of the seat with a stand (the right seat was more loaded).

In a reverse case, the value of the maximum stress occurs at the base of the seat, with 305.81 MPa versus 339.81 MPa for the rack variant. Although the area where the extremum of stress is fixed coincides with the same seat mount bracket. Thus, the bracket is in lower load conditions as part of the console construction. Even though the absolute value of stresses is lower in the console variant, the average value of stresses is 30–35% higher when referring to 270–300 MPa for pipe profiles (25 × 2 mm) against 220–250 MPa for a variant providing a stable mounting.

In general, the console structure in terms of stresses has a sufficient margin of safety, but a significant part of the constituent elements is outside the yield strength of the material (Steel 20–250 MPa). This is because the seat performs the bearing function, which may result in increased displacement under regular loads following UNECE Regulation No. 80.

For the third variant of the base with a console fastening and additional armor with a stress map, a decrease by 5% in the value of stresses was noticed. The maximum value corresponds to the area of the left seat bend. In general, the relative equilibrium of the construction of the seat together with an additional handrail may be estimated: the oscillation of the voltage value is at 30–40% between the model components, meaning there are no overloaded elements or those with an excess safety margin.

Although the seat base showed a maximum stress value of 5 MPa more than in two other cases, the average value of stresses in round profiles, 20 × 2 mm in size, at 15–20% is lower due to stress redistribution. A significant contribution to stress redistribution has a vertical additional handrail, absorbing a significant part of the energy when struck. The voltage map of this handhold shows the effectiveness of its work: the maximum stress value was 266.77 MPa. As expected, this value is fixed in the area of the seat mounting to its base.

4. Conclusions

Bearing in mind that there is a number of factors influencing the general performance of bus seats, the present paper aimed to focus on an aspect considered crucial in urban public transportation, which is safety and comfort. The comfort analyses were performed using purpose-built models on seat profiles occupied by an adult (M50) and a young (M10) passenger. The safety analyses were based on analyzing the frame strength and stress distribution for three variants of seat mounts: a seat with a stand, a side control seat fixing, and a seat-supporting handrail attached to the roof frame.

All considered models were solid-based and created using computer-aided design and a computer-aided engineering software environment; then, they were imported into the software package for Finite Element Method analysis. In general, the base model with a stand consists of 19 constituent elements. The number of finite elements of the Finite Element Method model approached 40,000, and the number of nodes exceeded 80,000.

The simulation of different scenarios for seat loading complies with Economic Commission for Europe of the United Nations Regulation No. 80 [54]. The common seat frame types with various mounting types used in standard or low floor buses were tested, giving a promising correlation between the simulation and test results according to the Economic Commission for Europe of the United Nations Regulation No. 80. For different seat mounting scenarios, the achieved maximum stresses were 431.74 MPa for a base model, 374.74 MPa at the stage of full absorption of the impact energy, and 339.8 MPa for the rack and flange mounted to the bracket of the seat arch. For the variant of the seat base, which was designed for installation in low-floor buses, the maximum value of stresses recorded during the loading process was 434.33 MPa.

The present study is a primary investigation focusing on particular events concerning two passengers with assumed seating locations and coordinates. The study should be further developed to evaluate seating comfort and safety for all passengers located at different seating spots. Bearing in mind that the results of the present study could be applied in urban public transportation only, the outcomes would allow understanding how to improve the comfort zone preventing potential physical discomfort or injuries.

Author Contributions: Conceptualization, I.K., O.H. and E.K.; methodology, O.H. and M.R.; software, Y.Y., M.R., R.H. and Y.S.; validation, Y.V., P.O. and K.R.; formal analysis, I.K. and O.H.; investigation, I.K., Y.Y., M.R., R.H., Y.S. and Y.V.; resources, I.K., O.H. and M.R.; data curation, P.O. and K.R.; writing—original draft preparation, O.H. and P.O.; writing—review and editing, P.O. and E.K.; visualization, R.H., Y.S., Y.V. and Ł.M.; supervision, E.K. All authors have read and agreed to the published version of the manuscript.

Funding: This research received no external funding.

Institutional Review Board Statement: Not applicable.

Informed Consent Statement: Not applicable.

Data Availability Statement: Not applicable.

Conflicts of Interest: The authors declare no conflict of interest.

References

1. Da Silva, L.; Bortolotti, S.L.V.; Campos, I.C.M.; Merino, E.A.D. Comfort model for automobile seat. *Work* **2012**, *41*, 295–302. [CrossRef] [PubMed]
2. Molenbroek, J.; Albin, T.; Vink, P. Thirty years of anthropometric changes relevant to the width and depth of transportation seating spaces, present and future. *Appl. Ergon.* **2017**, *65*, 130–138. [CrossRef]
3. Pijls, R.; Galetzka, M.; Groen, B.H.; Pruyn, A.T. Comfortable seating: The influence of seating comfort and acoustic comfort on customers' experience of hospitality in a self-service restaurant. *Appl. Ergon.* **2019**, *81*, 102902. [CrossRef] [PubMed]

4. Iftekhar, H.; Khan, U.; Asghar, R.M.W.; Qadeer, M.A.; Umair, A.; Nawab, M.; Hamdani, S.T.A. Study of comfort performance of novel car seat design for long drive. *Proc. Inst. Mech. Eng. Part D J. Automob. Eng.* **2020**, *234*, 645–651. [CrossRef]
5. Brunoro, C.; Sznelwar, L.I.; Bolis, I.; Abrahão, J. Contributions of ergonomics to the construction of bus drivers health and excellence in public transport and at work. *Work* **2012**, *41*, 30–35. [CrossRef] [PubMed]
6. Gowtham, S.; Ramnaath, M.; Sudharsan, S.; Kumar, B.V.L.; Praneeth, V.; Dinesh, S.; Subramaniyam, M. Seating comfort analysis: A virtual ergonomics study of bus drivers in private transportation. *IOP Conf. Ser. Mater. Sci. Eng.* **2020**, *912*. [CrossRef]
7. Castañeda, K.; Sánchez, O.; Herrera, R.F.; Pellicer, E.; Porras, H. BIM-based traffic analysis and simulation at road intersection design. *Autom. Constr.* **2021**, *131*, 103911. [CrossRef]
8. Gu, W.; Mei, Y.; Chen, H.; Xuan, Y.; Luo, X. An integrated intersection design for promoting bus and car traffic. *Transp. Res. Part C Emerg. Technol.* **2021**, *128*, 103211. [CrossRef]
9. Muro, F.-J.M.; Skorin-Kapov, N.; Pavon-Marino, P. Revisiting core traffic growth in the presence of expanding CDNs. *Comput. Netw.* **2019**, *154*, 1–11. [CrossRef]
10. Oskarbski, J.; Birr, K.; Żarski, K. Bicycle Traffic Model for Sustainable Urban Mobility Planning. *Energies* **2021**, *14*, 5970. [CrossRef]
11. Sołowczuk, A. Effect of Traffic Calming in a Downtown District of Szczecin, Poland. *Energies* **2021**, *14*, 5838. [CrossRef]
12. Litman, T.; Fitzroy, S. *Distance-Based Vehicle Insurance Feasibility. Costs and Benefits*; Victoria Transport Policy Institute: Victoria, BC, Canada, 2018.
13. Dell'Olio, L.; Ibeas, A.; Cecin, P. The quality of service desired by public transport users. *Transp. Policy* **2011**, *18*, 217–227. [CrossRef]
14. Atombo, C.; Dzigbordi, W.T. Indicators for commuter's satisfaction and usage of high occupancy public bus transport service in Ghana. *Transp. Res. Interdiscip. Perspect.* **2021**, *11*, 100458.
15. Barabino, B.; Eboli, L.; Mazzulla, G.; Mozzoni, S.; Murru, R.; Pungillo, G. An innovative methodology to define the bus comfort level. *Transp. Res. Procedia* **2019**, *41*, 461–470. [CrossRef]
16. Barone, V.; Festa, D.C.; Mongelli, D.W.E.; Tassitani, A. Comfort Index CI(bus): A methodology to measure the comfort on board. *Procedia Comput. Sci.* **2018**, *134*, 439–444. [CrossRef]
17. Karekla, X.; Fang, C. Upper body balancing mechanisms and their contribution to increasing bus passenger safety. *Saf. Sci.* **2021**, *133*, 105014. [CrossRef]
18. Rao, M.J.; Sivapirakasam, S.P.; Phanindra, K.H.; Vishnu, B.R. Prediction and analysis of semi low floor bus driver seat vibrations by using the response surface methodology. *Int. J. Heavy Veh. Syst.* **2021**, *28*, 409–422. [CrossRef]
19. Fittipaldi, F.; Ranzo, P.; &Veneziano, R. Ergonomic Design Approach for an Urban Bus. In Proceedings of the International Conference on Applied Human Factors and Ergonomics, New York, NY, USA, 25–29 June 2021; Springer: Cham, Switzerland, 2021.
20. Nguyen, T.; Nguyen-Phuoc, D.Q.; Wong, Y.D. Developing artificial neural networks to estimate real-time onboard bus ride comfort. *Neural Comput. Appl.* **2021**, *33*, 5287–5299. [CrossRef]
21. Jianghong, Z.; Long, T. An evaluation of comfort of a bus seat. *Appl. Ergon.* **1994**, *25*, 386–392. [CrossRef]
22. Jonsson, P.M.; Rynell, P.W.; Hagberg, M.; Johnson, P.W. Comparison of whole-body vibration exposures in buses: Effects and interactions of bus and seat design. *Ergonomics* **2014**, *58*, 1133–1142. [CrossRef]
23. Lim, S.; Chung, M.K.; Jung, J.; Na, S.H. The Effect of Lumbar Support Prominence on Driver's Comfort and Body Pressure Distribution. In Proceedings of the Human Factors and Ergonomics Society Annual Meeting. *Proc. Hum. Factors Ergon.* **2000**, *44*, 308–311.
24. Chung, Y.-S.; Wong, J.-T. Developing effective professional bus driver health programs: An investigation of self-rated health. *Accid. Anal. Prev.* **2011**, *43*, 2093–2103. [CrossRef] [PubMed]
25. Park, J.; Lee, H.; Choi, Y.; Park, K.; Kim, M.; You, H. Development of an Ergonomic Bus Seat Profile Design Protocol. *Proc. Hum. Factors Ergon. Soc. Annu. Meet.* **2014**, *58*, 1825–1828. [CrossRef]
26. Anund, A.; Fors, C.; Ihlström, J.; Kecklund, G. An on-road study of sleepiness in split shifts among city bus drivers. *Accid. Anal. Prev.* **2018**, *114*, 71–76. [CrossRef] [PubMed]
27. Heberle, S.M.; Lorini, C.; Rosa, M.S.; Barros, N. Evaluation of bus driver exposure to nitrogen dioxide levels during working hours. *Atmos. Environ.* **2019**, *216*, 116906. [CrossRef]
28. Anto, E.O.; Owiredu, W.; Adu, E.A.; Obirikorang, C.; Fondjo, L.A.; Annani-Akollor, M.E.; Acheampong, E.; Asamoah, E.A.; Roberts, P.; Wang, W.; et al. Prevalence and lifestyle-related risk factors of obesity and unrecognized hypertension among bus drivers in Ghana. *Heliyon* **2020**, *6*, e03147. [CrossRef]
29. Golinko, V.; Cheberyachko, S.; Deryugin, O.; Tretyak, O.; Dusmatova, O. Assessment of the Risks of Occupational Diseases of the Passenger Bus Drivers. *Saf. Health Work* **2020**, *11*, 543–549. [CrossRef]
30. Wang, X.; Wang, K.; Huang, K.; Wu, X.; Huang, W.; Yang, L. The association between demographic characteristics, personality, and mental health of bus drivers in China: A structural equation model. *Physiol. Behav.* **2021**, *229*, 113247. [CrossRef]
31. Kim, J.; Na, H.; Cho, D.; Shin, Y.; Park, S.; Kim, J. Development of Questionnaire for Automobile Seat Comfort Evaluation. *J. Kor. Soc. E Sens.* **2010**, *13*, 381–390.
32. Kolich, M. Automobile seat comfort: Occupant preferences vs. anthropometric accommodation. *Appl. Ergon.* **2003**, *34*, 177–184. [CrossRef]
33. Smith, D.R.; Andrews, D.M.; Wawrow, P.T. Development and evaluation of the Automotive Seating Discomfort Questionnaire (ASDQ). *Int. J. Ind. Ergon.* **2006**, *36*, 141–149. [CrossRef]

34. Hamid, A.I.; Li, Q.M. New definitions of deformation index for the measurement of bus survival space in crash. *Proc. Inst. Mech. Eng. Part D J. Automob. Eng.* **2018**, *233*, 2108–2119. [CrossRef]
35. Law, T.H.; Daud, M.S.; Hamid, H.; Haron, N.A. Development of safety performance index for intercity buses: An exploratory factor analysis approach. *Transp. Policy* **2017**, *58*, 46–52. [CrossRef]
36. Rupp, J.D.; Reed, M.P.; Van Ee, C.A.; Kuppa, S.; Wang, S.C.; Goulet, J.A.; Schneider, L.W. The tolerance of the human hip to dynamic knee loading. *Stapp Car Crash J.* **2002**, *46*, 211–228. [CrossRef] [PubMed]
37. Kleinberger, M.; Sun, E.; Eppinger, R. *Development of Improved Injury Criteria for the Assessment of Advanced Automotive Restraint Systems. II*; National Highway Traffic Safety Administration, U.S., Department of Transportation: Washington, DC, USA, 1998; p. 180.
38. Haffner, M. *Synthesis of Pelvic Fracture Criteria for Lateral Impact Loading*; Technical Paper; SAE International: Warrendale, PA, USA, 1985.
39. Sharma, S.; Sharma, S.; Gupta, U.; Joshi, R.; Pawar, S. Finite Element Analysis and Validation of Bus Seat Structure as per AIS023: Safety Features Evaluation of Bus Seat using Hybrid III Dummy. *SAE Tech. Pap. Ser.* **2015**, *1*. [CrossRef]
40. Mertz, H.J.; Patrick, L.M. *Strength and Response of the Human Neck*; SAE Technical Paper; SAE International: Warrendale, PA, USA, 1971.
41. Martinez, L.; Espantaleón, M.; de Loma-Ossorio, M.; Alcalá, E.; Torres, C.R. Adult and Child Dummies Tests for Safety Assessment of Seated Occupants in Urban Bus Collisions. In Proceedings of the 25th International Technical Conference on the Enhanced Safety of Vehicles (ESV) National Highway Traffic Safety Administration, Detroit, MI, USA, 5–8 June 2017.
42. Peters, S.E.; Grogan, H.; Henderson, G.M.; Gómez, M.A.L.; Maldonado, M.M.; Sanhueza, I.S.; Dennerlein, J.T. Working Conditions Influencing Drivers' Safety and Well-Being in the Transportation Industry: "On Board" Program. *Int. J. Environ. Res. Public Health* **2021**, *18*, 10173. [CrossRef] [PubMed]
43. Mousavi, S.M.; Osman, O.A.; Lord, D.; Dixon, K.K.; Dadashova, B. Investigating the safety and operational benefits of mixed traffic environments with different automated vehicle market penetration rates in the proximity of a driveway on an urban arterial. *Accid. Anal. Prev.* **2021**, *152*, 105982. [CrossRef]
44. Peters, S.E.; Trieu, H.D.; Manjourides, J.; Katz, J.N.; Dennerlein, J.T. Designing a Participatory Total Worker Health® Organizational Intervention for Commercial Construction Subcontractors to Improve Worker Safety, Health, and Well-Being: The "ARM for Subs" Trial. *Int. J. Environ. Res. Public Health* **2020**, *17*, 5093. [CrossRef]
45. Sorensen, G.; Dennerlein, J.T.; Peters, S.E.; Sabbath, E.L.; Kelly, E.L.; Wagner, G.R. The future of research on work, safety, health and wellbeing: A guiding conceptual framework. *Soc. Sci. Med.* **2021**, *269*, 113593. [CrossRef]
46. Tan, C.F.; Chen, W.; Delbressine, F.; Rauterberg, M. Objectifying discomfort seat measurement for next generation truck driver's seat. *Int. Fed. Aut. Eng. Soc.* **2008**, *28*, F2008-SC-028.
47. Halder, P.; Mahmud, T.; Sarker, E.; Karmaker, C.; Kundu, S.; Patel, S.; Setiawan, A.; Shah, K. Ergonomic considerations for designing truck drivers' seats: The case of Bangladesh. *J. Occup. Health* **2018**, *60*, 64–73. [CrossRef] [PubMed]
48. Dama, K.; Babu, V.S.; Rao, R.N.; Madhava, Y. A Review on Automotive Seat Comfort Design. *Int. J. Eng. Res.* **2015**, *4*, 678–683. [CrossRef]
49. Jamroziak, K.; Joszko, K.; Wolanski, W.; Gzik, M.; Burkacki, M.; Suchon, S.; Zielonka, K. Experimental and modelling research on coach passengers' safety in frontal impacts. *Arch. Civ. Mech. Eng.* **2020**, *20*, 1–13. [CrossRef]
50. Spencer, B.F., Jr.; Dyke, S.J.; Sain, M.K.; Carlson, J.D. Phenomenological Model for Magnetorheological Dampers. *J. Eng. Mech.* **1997**, *123*, 230–238. [CrossRef]
51. Hatwalane, S.; Kothavale, B.S.; Ohol, A.; Girish, P. Review of Driver Seat Suspension using MR Fluid Damper. *Int. J. Curr. Eng. Technol.* **2011**, *4*, 333–336.
52. Kernytskyy, I.; Diveyev, B.; Hlobchak, M.; Horbaj, O.; Kopytko-Kernytska, M.; Zachek, O. Optimization of the impact multi-mass vibration absorbers. *Sci. Rev. Eng. Environ. Sci.* **2017**, *26*, 394–400. [CrossRef]
53. Horbay, O.; Diveyev, B.; Kernytskyy, I.; Humenyuk, R. Some components of safety and comfort of a car: Current Problems of Transport. In Proceedings of the ICCPT 2019, Ternopil, Ukraine, 28–29 May 2019.
54. Snitynskyy, V.; Diveyev, B.; Horbay, O.; Humenyuk, R.; Kernytskyy, I.; Kokhana, T.; Koruniak, P. Impact and particle vibration absorbers optimal design. *Acta Sci. Pol.—Arch. Bud.* **2020**, *19*, 93–102. [CrossRef]
55. UNECE Regulation No. 80. Uniform provisions concerning the approval of seats of large passenger vehicles and of these vehicles with regard to the strength of the seats and their anchorages. E/ECE/324/Rev.1/Add.79/Rev.2. United Nations. GE.12-24958. 2012. Available online: https://eur-lex.europa.eu/legal-content/EN/ALL/?uri=CELEX:42013X0824(01) (accessed on 3 April 2021).

energies

Article

Influence of Clearance on the Rocker Arm Pin on the Steerability and Stability of the Vehicle Motion

Krzysztof Parczewski * and **Henryk Wnęk**

Department of Internal Combustion Engines and Vehicles, Faculty of Mechanical Engineering and Computer Science, University of Bielsko-Biala, Willowa Str. 2, 43-309 Bielsko-Biala, Poland; hwnek@ath.bielsko.pl
* Correspondence: kparczewski@ath.bielsko.pl

Abstract: The article presents an analysis of the impact of a malfunction resulting from excessive clearance on the rocker arm pin of the front suspension on the vehicle's steerability. The first part of the article presents an analysis of the influence of the clearance on the rocker arm pin on the geometry of the suspension and steering system. The occurrence of forces acting on the rocker arm pin in various phases of the vehicle motion was analyzed. To assess the vehicle's steering, the vehicle's response time to sudden steering wheel movement was used. The vehicle's response time to sudden movement of the steering wheel was used to assess the vehicle's steerability. The second part presents the results of bench tests and traction tests of a vehicle equipped with a specially made measuring rocker arm with the possibility of simulating a clearance. The tests were carried out on a class B passenger car in selected road tests. The results of measurements obtained for the roadworthy vehicle and the vehicle with the rocker arm with clearance were compared. The influence of the clearance on the rocker arm pin on the change of vehicle steerability in steady and dynamically changing conditions was analyzed. The test results show the effect of clearance on vehicle steering and on the vehicle steerability. The study tried to determine to what extent the clearance on the rocker arm affects the vehicle's steerability and thus the safety in road traffic.

Keywords: clearance on the rocker arm pin; vehicle steerability; steerability tests; road safety

Citation: Parczewski, K.; Wnęk, H. Influence of Clearance on the Rocker Arm Pin on the Steerability and Stability of the Vehicle Motion. *Energies* **2021**, *14*, 7827. https://doi.org/10.3390/en14227827

Academic Editor: Guzek Marek

Received: 22 October 2021
Accepted: 17 November 2021
Published: 22 November 2021

1. Introduction

The steerability of the vehicle determines the ease and certainty with which the driver can put the vehicle on the intended trajectory and keep it on this track [1,2]. Increased clearance in the steering system makes it difficult to maintain the intended direction of travel. Based on the motion parameters of the car and the characteristics of the steering system, the driver anticipates the vehicle's behavior and adjusts the angle and speed of the steering wheel rotation during turning maneuvers and while maintaining a straight-line direction of travel. The effects of the driver's actions are the longitudinal, lateral, and inclination movement of the vehicle. The flexibility and clearance in the steering system affect the "confidence" of driving the car, and thus the steering.

Wear of the steering elements and steering linkage components causes an increase in clearance in the steering system, which makes it difficult to maintain the desired direction of travel and increases the "inaccuracy" of steering the vehicle's direction of movement.

The element susceptible to wear and damage is the ball pin of the rocker arm. Wear is caused by friction in the joint during its rotation resulting from the vertical movement of the suspension and the turning of the wheels. Its accelerated wear may occur when driving on uneven surfaces, in an environment with high dustiness or saltiness, or as a result of damage to the seal, which protects against the ingress of dirt and water into the ball joint. Often, damage to the seal can be caused mechanically, for example by a stone impact while driving. If the rubber cover of a joint is damaged, water and debris will rapidly degrade the joint, causing loss of grease and corrosion of the metal components. Sudden damage to the rocker arm pin may also occur while driving over large unevenness. In this situation, the ball pin may even slip out of the socket and the vehicle may lose its steerability. In addition, the operating parameters such as incorrect suspension geometry and unbalanced wheels also affect accelerated wear or damage to the pin [3,4].

The clearance in the ball joint can cause misalignment, uneven tire wear, sometimes the pulling of the steering wheel to one side, and/or suspension noise. The permissible wear of the connector depends on the application of the vehicle. In the old literature, there is information stating that the clearance in the ball joint cannot exceed 1 mm, but today car manufacturers say that ball joints used in car suspensions should not have a noticeable clearance. The only way to check if a joint is worn is to refer to the service specifications of the vehicles and measure the clearance (both axial and sideways). Load-carrying ball joints wear out the fastest. In double-wishbone suspensions, the lower ball joint carries the load and is more prone to wear.

In suspensions where the spring is mounted on the upper control arm, the upper ball joint carries the load and is more susceptible to accelerated wear. The lower ball joint in this type of suspension is unloaded, so any clearance indicates that it should be replaced.

In the case of a MacPherson strut suspension, the strut carries the load and the lower ball joint is unloaded.

The clearance in the joint is created as a result of wear of the mating surfaces, and the spring washer causes the clearance to be canceled in the axial direction. Side clearances are not canceled. The systems used to reduce side slip of the tire [5] in the event of clearance in the ball joint will cause instability of this system's operation, causing the occurrence of oscillations.

During the tests, efforts were made to determine the conditions necessary to reset the clearance on the rocker arm pin, describe the impact of the clearance on the direction of vehicle movement, and assess the deterioration of the steerability.

2. Factors Influencing the Dynamic Characteristics of Car Control

The vehicle control is influenced by factors depending on the vehicle structure, including its steering system, the tires and their load, the side inclination of the vehicle, and traffic conditions: the quality of the surface and the value of the tire adhesion coefficient. They are generally defined by the concept of dynamic vehicle control characteristics. This characteristic is influenced, among other things, by:

- steering angles,
- tire sideslip angles,
- clearance in the steering system.

The driver decides the value of the steering angles by turning the steering wheel. However, a vehicle's path is also influenced by factors depending on the design and motion parameters of the vehicle and the forces transmitted between the tires and the road. As outlined above, the factors that affect the trajectory are steering wheel rotation angle, tire and suspension characteristics, vehicle stiffness and damping, steering component compliance and damping, wheel alignment geometry, mass distribution and vehicle mass moments of inertia, rolling resistance, and the conditions of grip as well as the dynamics of the turning maneuver [6–9].

The steering angles result from the angle of rotation of the steering wheels, the gear ratio and characteristics of the steering system, its flexibility and damping. They have a fundamental effect on the vehicle's path.

The factors influencing the tire sideslip angles are related to the parameters resulting from the vehicle structure and the variables determining its motion. These factors can be grouped as follows:

- The tire: its stiffness, profile height, and tread condition. They mainly affect the tire deflection and its cooperation with the road, and they depend on the pressure force acting on the wheel and its inclination angle [10,11].
- Suspension: vertical and tilting stiffness of the suspension, flexibility of suspension components, and damping. Suspension stiffness affects the vehicle tilt angles in

relation to the longitudinal (X) and transverse (Y) axes as well as the forces acting on the wheel axles. The compliance of the metal–rubber suspension elements is selected during the construction of the vehicle in order to reduce understeer and oversteer of the vehicle. The torsional flexibility of the car body was neglected due to its small, compared to suspensions, impact on the vehicle path [5,7,8,12].

- Vehicle: location of the center of mass, mass moments of inertia with respect to the vehicle axis [13–15].
- Conditions of cooperation between the tire and the road: conditions of adhesion between the tire and the road resulting from the type of surface, its condition (dry or wet), longitudinal and transverse slip of the wheel, or rolling resistance. They directly affect the tire sideslip angles [5,13].
- Car motion parameters: steering wheel rotation angle (steered wheels turning angles), travel speed, and longitudinal and lateral acceleration. These are the quantities that characterize the motion and steering of the vehicle [13,16].
- Car traffic conditions: surface quality and longitudinal and side inclination of the road.

All of the above-mentioned factors affect the vehicle's trajectory. The steering angles are also affected by clearance in the steering system. The occurrence of clearance affects the direction of movement of the vehicle.

3. Assessment of the Vehicle's Steerability

So far, no methodology for assessing vehicle steerability with analytical methods has been developed. Most often, subjective evaluations based on the opinion of the driver are used to determine the vehicle's steerability. A number of studies and attempts to find a method for determining controllability have been carried out. In most cases, they consisted in conducting experimental tests of the vehicle movement and then the following parameters were analyzed: the vehicle path, deviation from the optimal track, or the biological response of the driver in the form of heart rate, sweating, or other parameters related to the size of the driver's work was measured. In the described research, attempts were made to change the characteristics of the vehicle's movement and examined the biological response, and on this basis, the conditions in which the vehicle is easier to control were distinguished. Based on the research, the influence of selected parameters on the vehicle's steerability was determined [16].

Hofmann et al. [17] investigated the vehicle response time t_r to the yaw rate and the deviation from the optimal vehicle trajectory, and on this basis assessed the relationship between the track error level and the time t_r. The response time t_r is the vehicle's yaw rate response to the control pulse.

Based on the research, the most favorable vehicle response time to t_{r_opt} = 0.2 s was determined (Figure 1). This means that the vehicle is more difficult to steer with longer vehicle response times.

The response time t_r of the vehicle was considered to be an important factor in determining the steerability of the vehicle [16–18]. It depends on cornering stiffness of tires, the weight of the vehicle, and the speed of its movement:

$$t_r = \frac{mV}{2\left(K_f + K_r\right)} \cdot \left(\frac{k^2}{l_f l_r}\right) \approx \frac{mV}{2\left(K_f + K_r\right)} \tag{1}$$

where: m—vehicle weight, V—driving velocity, k—vehicles radius of inertia, l_f—distance of the center of gravity from the front axle, l_r—distance of the center of gravity from the rear axle, K_f—cornering stiffness of the front axle tires, and K_r—cornering stiffness of the rear axle tires.

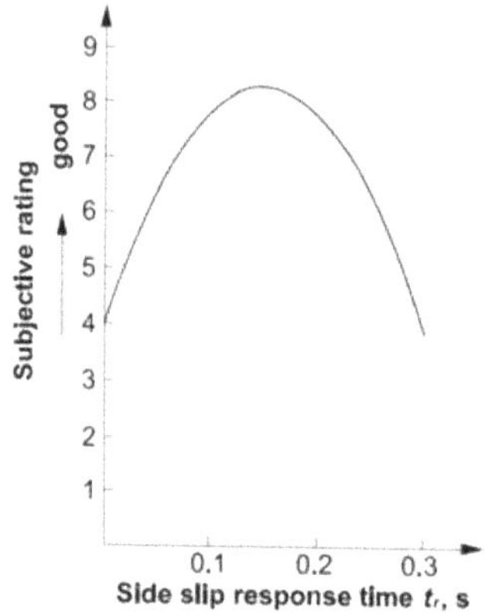

Figure 1. Assessment of vehicle steerability depending on the vehicle response time [16,17].

Similarly, the influence of the response time on the lateral acceleration a_y and the vehicle sideslip angle β was analyzed.

Weir et al. [19], based on their own and other authors' research, determined the ranges in which it is considered that the vehicle is characterized by good steerability, and the drivers were divided into two groups: medium-skilled and high-skilled drivers. These ranges partially overlap, and this area was considered optimal. At the end of the article was shown the vehicle response times to the steering wheel turn.

The relations presented above were used by the authors to evaluate the behavior of the vehicle with increased clearance on the rocker arm pin.

4. Assessment of the Influence of Clearance in the Suspension Arm Pin on the Vehicle Control

Driving the wheels in the correct position during the turning maneuver requires the use of a rotary connection of the rocker arm with the steering knuckle. In McPherson suspensions, the steering knuckle is connected to the control arm by means of a ball joint mounted to the control arm. The articulation wears out with use of the vehicle. The resulting clearance is manifested by the knocking of the suspension and deterioration of the driving quality. The occurrence of clearance on the rocker arm pin causes that during vehicle movement, depending on the maneuvers performed: acceleration, braking, or turning, and when driving on a road with a side slope or on a bumpy road, the clearance will be reset depending on the direction of the side force acting. A clearance of ~1 mm causes the steering angle to change by about 0.7° (which corresponds to the steering wheel turning angle by ~11°). In addition, the action of a force with changing direction generates the formation of dynamic forces acting on the pin and accelerates its degradation. In extreme cases, the ball joint may burst.

It was assumed that the clearance on the rocker arm of a size greater than 1 mm causes noise and deterioration of the steering so noticeable that it is usually removed during repair. For this reason, a 1 mm clearance was used during the tests.

5. Analysis of the Distribution of Forces Acting on the Pin of the Suspension Rocker Type McPherson

In the case of a McPherson type suspension, the virtual axle of the steering knuckle is tilted from the vertical. The position of this axis is determined by the inclination angle σ and the caster angle τ of the kingpin. The loading force acting on the suspension causes the spring to deflect and the pressure force at the contact between the wheel and the road. The force component acts on the rocker arm pin and has a direction defined by a straight line passing through the steering knuckle pin and the axis of mounting the rocker arm in the body. The direction of the force is towards the outside of the vehicle. The described force causes the clearance on the pin to be reset towards the outside of the vehicle. The amount of lateral force is also influenced by the inclination angle of the rocker arm. In the case of vehicle movement on a rectilinear path, this force is counterbalanced by the tensile force of the rocker arm. During turning maneuvers or when driving on an inclined road, there is a lateral force that will cause, depending on the direction of the vehicle movement, either the clearance on the rocker arm pin to be cancelled or the rocker arm tensile force to be increased. Figure 2 shows the distribution of forces loading the wheel.

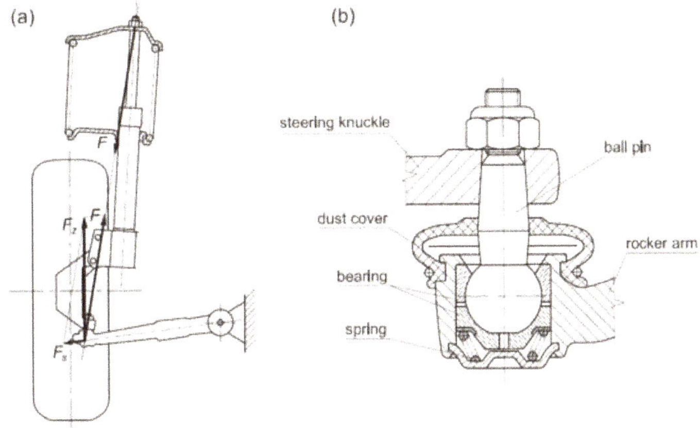

Figure 2. Forces acting on the rocker arm pin (**a**), rocker arm pin (**b**): F—force acting on the bodywork, F_s—side force, F_z—vertical force.

The transverse component F_s of the force F acting on the bodywork while the vehicle is in motion causes stretching of the rocker arm (the clearance is reset to the outside of the vehicle).

While driving, the forces acting on the contact between the wheel and the road will change the direction of the resultant force acting on the pin. Deleting the clearance will change the wheel turning angle. Figure 3 shows the lateral forces acting on the contact area between the wheel and the road, occurring during a turning maneuver.

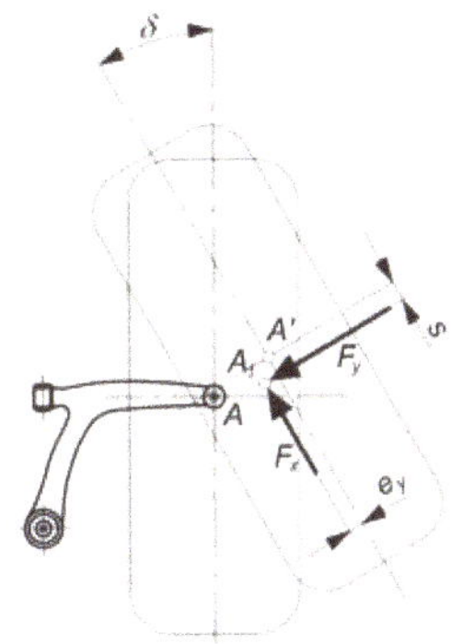

Figure 3. Forces acting on the rocker arm pin during the turning maneuver: F_x—longitudinal force, F_y—lateral force, e_y—lateral displacement of the center of the tire footprint, s—displacement of the center of the tire footprint in the longitudinal direction, δ—the steering angle of the wheel.

Figure 4 shows the change in the steering angle caused by the clearance on the rocker arm pin.

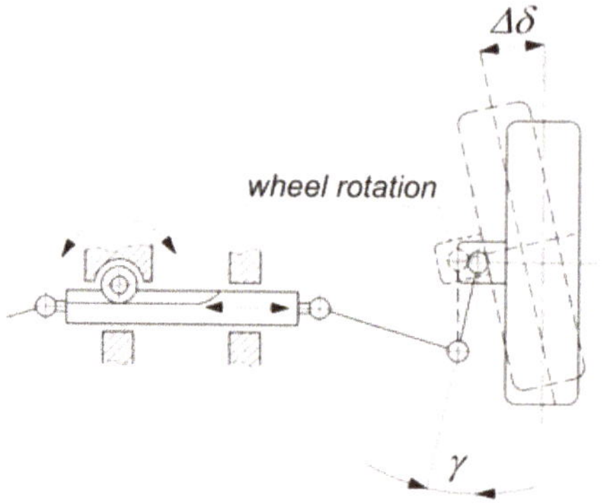

Figure 4. Change of the steering angle caused by clearance on the rocker arm pin: $\Delta\delta$—changing the steering angle of the wheel, γ—steering arm inclination angle.

Figure 5 shows the distribution of forces acting on the rocker arm in straight and curved motion, during vehicle acceleration and braking (for a positive wheel swing radius). The reset of the clearance of the rocker arm pin will depend on the direction of the driving or braking force as well as on the steering angle of the wheel and its direction of rotation.

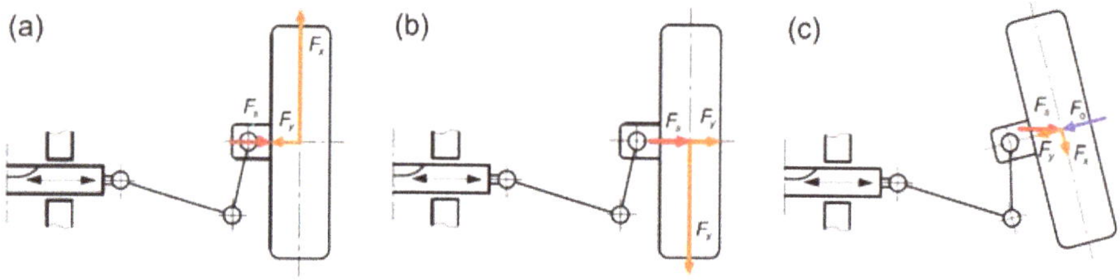

Figure 5. Forces acting on the rocker arm pin during various cases of vehicle motion (for a positive swing radius): (**a**) acceleration, (**b**) braking, (**c**) driving along a road curve: F_S—side force, F_y—lateral force, F_y—longitudinal force, F_S—centrifugal force.

6. Research on the Influence of Play on the Rocker Arm Pin on the Vehicle's Steerability Characteristics

The research on the effect of clearance on the rocker arm pin was divided into two stages. The first was carried out under stationary conditions and the second during road tests under real load conditions.

A specially prepared measuring rocker was installed in the vehicle, generating clearance, simulating the appearance of a clearance on the pin. In order to achieve a similar effect as in the case of clearance on the ball bolt, the rocker arm was cut and assembled in such a way that the part of the rocker arm with the ball joint could slide over the part of the control arm that is mounted to the car body. The parts of the rocker arm allowed for the mutual displacement of these elements by 1 mm in the lateral direction. A sensor Keyence GT is attached to the rocker arm that allows determining the moment of resetting the clearance—shown in the photo (Figure 6).

The tests were carried out using a vehicle from the B segment (unloaded weight ~1000 kg, wheelbase ~2.3 m). The measuring arm was mounted in the right front suspension of the car.

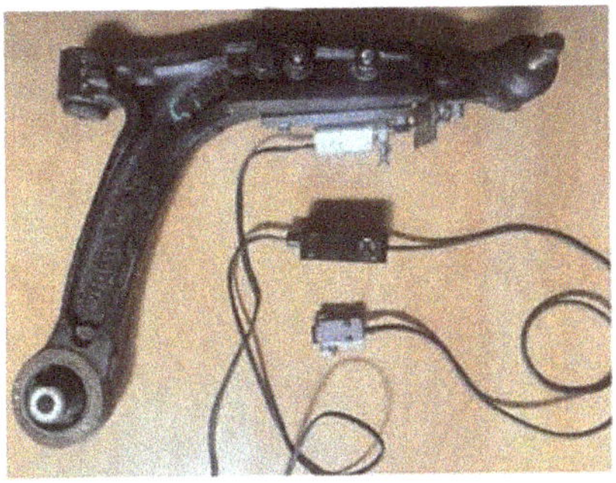

Figure 6. View of the measuring arm.

6.1. Bench Research

The bench tests were carried out in stationary conditions. The vehicle was stationary on the platform. The load acting on the front suspension was generated by a linkage system and measured directly at the point where the forces were introduced into the suspension (Figure 7).

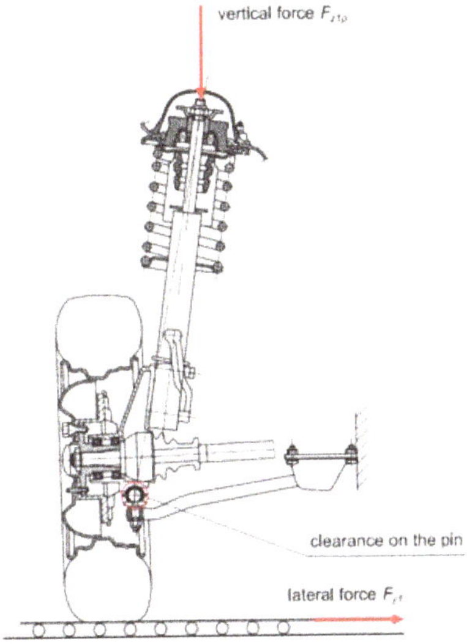

Figure 7. Measurement of the required lateral force to eliminate the rocker arm clearance.

The wheel was placed on a sliding base that allowed it to move freely in the transverse direction. A lateral force was exerted to the wheel. After overcoming the horizontal component of the wheel loading force, the rocker arm clearance was reset. The values of vertical and lateral forces as well as the moment at which the clearance was canceled (changes in the distance between the rocker arm mounting point and the steering knuckle pin) were measured and recorded.

The lateral force causing the clearance to be canceled was measured for different wheel load conditions. The result of the measurements was the characteristic of erasing the clearance as a function of the wheel load. The diagram of the measuring stand is shown in Figure 8.

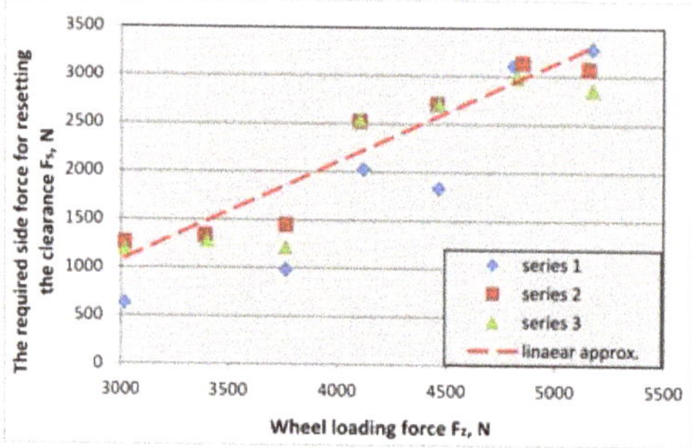

Figure 8. Characteristics of resetting the rocker arm clearance.

Figure 8 shows the course of the lateral force acting on the contact area between the tire and the road, required to eliminate the clearance on the rocker arm pin. The obtained mileage was approximated by a straight line, and the spread of the force values obtained decreases with increasing load on the suspension.

6.2. Road Tests

The field tests were carried out with the use of the same vehicle (unloaded vehicle weight ~1000 kg, wheelbase ~2.3 m) [20,21]. To assess the vehicle's steerability, three road tests modeled on ISO standards were used:

- the first, during which the vehicle was moving along a circular track at a set speed [9,22]. During this test, the vehicle was driven under steady-state driving speed and steady steering wheel angle.
- the second, during which the vehicle performed a single lane change maneuver at a set speed.
- the third, during which the vehicle was suddenly braked on a flat, straight road section. The vehicle speed at the beginning of braking was ~70 km/h.

The first two tests are characterized by high lateral accelerations, both when driving on a circular track under steady traffic conditions and when changing lanes. During the third attempt, the lateral force was practically non-existent.

During the tests, speed and trajectory, longitudinal and lateral accelerations, steering wheel rotation angle, vehicle body roll and yaw angle, roll and yaw speeds, wheel speeds, brake pedal force, and pressure in wheel brake circuits were measured.

For this purpose, the vehicle was equipped with measuring devices: Correvit S-CE (Corrsys, Schwalbach, Germany) measuring head, MSW (Corrsys-Datron) measuring steering wheel, VBOX3i SL GPS measuring system with inertial module IMU04 (Racelogic, Buckingham, Great Britain), ADXL203 acceleration sensors (Analog Devices, Norwood, MA, USA) with a range of 1.7 g, MPX200 pressure sensors (Peltron, Warsaw, Poland), CL23 brake pedal force sensor (ZEPWN, Warsaw, Poland), incremental encoders with a resolution of 1024 pulses/rev (HEEDS, China), and CRS03 gyro sensors (Silicon Sensing Systems Japan, Hyogo, Japan). Measurement data was recorded with a frequency of 100 Hz in the AD-32 (Grapol Electronic, Warsaw, Poland) [15,20,23] and VBOX Racelogic [24] measurement systems.

6.3. Measurements and Evaluation of Vehicle Steerability

On the basis of road tests, the described parameters characterizing the vehicle movement were determined. The results for the ride on the circular track, single lane change, and braking tests on a straight road section are presented below. The tests were carried out for the following vehicle: roadworthy (the possibility of a clearance in the rocker arm pin is blocked) and with a testing rocker that simulates a clearance on the pin. The measuring rocker arm is mounted in the front right wheel suspension.

6.4. The Ride on the Circular Track Test

During the test, the vehicle was moving along a track with a certain radius at a set speed.

Due to the fact that the vehicle was driven by the driver, efforts were made to approximate the traffic parameters assumed for the test. Attempts were made to keep the driving speed the same for both tests (roadworthy vehicle and with clearance in the rocker arm pin). The diagram below (Figure 9) shows the vehicle trajectory with a noticeable change in the trajectory resulting from the erasure of the rocker arm clearance, obtained from the GPS satellites of the VBOX system.

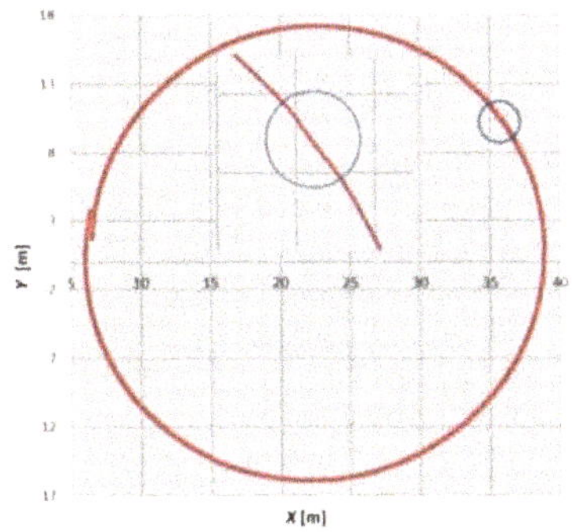

Figure 9. Change of the trajectory caused by resetting the rocker arm clearance while driving on a circular track (driving at a speed of ~30 km/h within a radius of ~22 m).

When driving on a circular track, the angle of rotation of the steering wheel is practically unchanged. Despite the very small difference in the angle of rotation of the steering wheel, there are slight differences in the values of the yaw angle and lateral acceleration.

6.5. Single Lane Change

During the test, the vehicle performed a lane change maneuver at a set driving speed. Due to the fact that the vehicle was driven by the driver, efforts were made to approximate the traffic parameters assumed for the test. Attempts were made to keep the driving speed the same for both tests (roadworthy vehicle and simulated slack). Figure 10 shows the vehicle trajectory with a noticeable change in the trajectory resulting from the erasure of the rocker arm clearance. This trajectory was obtained from the VBOX system.

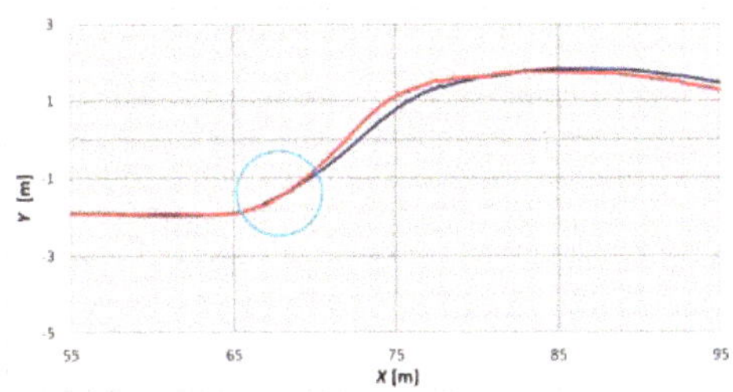

Figure 10. Change of the trajectory caused by resetting the rocker arm pin clearance (driving at ~40 km/h).

Figure 11 shows the angle of rotation of the steering wheel during a lane change maneuver at a travel speed of 40 km/h.

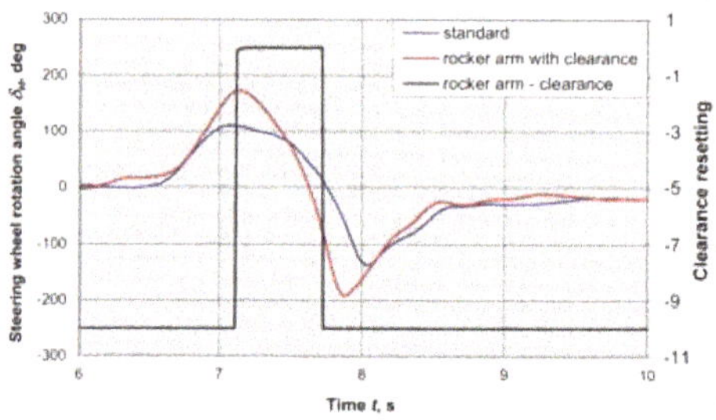

Figure 11. The angle of rotation of the steering wheel during the lane change maneuver (driving at ~40 km/h).

Figure 11 shows the moment of resetting the rocker arm pin clearance. The blue line shows the steering wheel rotation angle of a roadworthy vehicle, and the red line shows the vehicle with the increased clearance of the rocker arm pin. Although the angles of the steering wheel rotation clearly differ, the course of the vehicle motion path is very similar, and the yaw velocity and lateral acceleration values are similar. At lower speeds, the clearance effect was less noticeable.

6.6. Braking on a Straight Road Section

During the test, the vehicle was braking from the initial speed of ~70 km/h.

Due to the slight curvature of the track (Figure 12), the vehicle slightly turns to the right in both cases of braking. The beginning of braking is marked with the blue line. The clearance was cancelled at the moment of the change from vehicle acceleration to its braking.

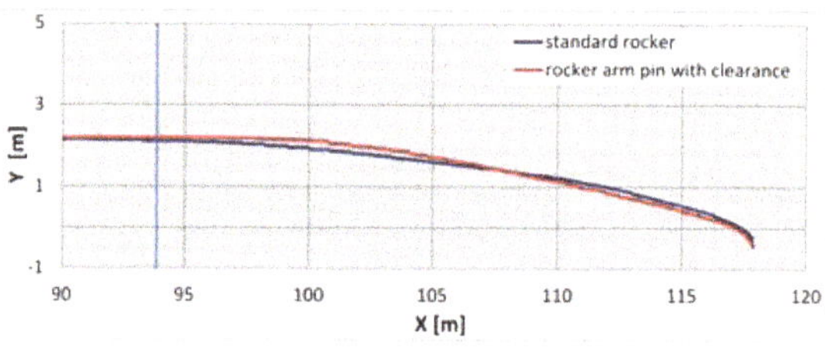

Figure 12. Change of the trajectory caused by resetting the rocker arm clearance during braking (blue line—braking start).

Figure 13 shows the steering wheel rotation angle during the braking maneuver. In the initial phase of braking, with a slight movement of the steering wheel, the clearance of the rocker arm is reset. Increasing the braking intensity changes the direction of the lateral forces on the wheel, and the clearance is canceled in the opposite direction.

Lateral acceleration in a roadworthy vehicle with increased clearance in the rocker arm pin is slight and may result from the operation of the ABS system.

The conducted tests were used to determine the influence of the rocker arm pin clearance on the vehicle's steerability.

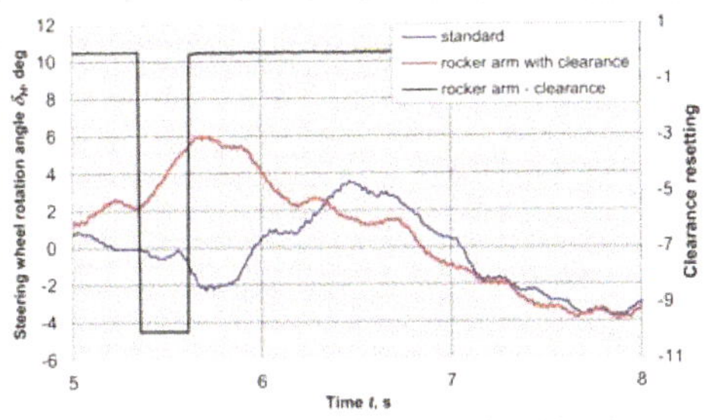

Figure 13. The angles of rotation of the steering wheel during the braking maneuver (initial speed ~70 km/h).

7. Influence of Rocker Arm Pin Clearance on Vehicle Steerability

Due to the oblique position of the steering knuckle, the clearance cancellation takes place after the lateral force component required for the clearance cancelled is achieved. The king-pin inclination angle and the instantaneous wheel load affect the amount of force necessary to eliminate the clearance. Additionally, the fact of accelerating or braking the wheel in the suspension with clearance in rocker arm pin will also influence its size.

Based on the analysis of measurement data, it can be concluded that the easiest way to reset the clearance is as a result of changing the direction of the steering wheel rotation.

On the basis of the research, attempts were made to determine the influence of the rocker arm pin clearance on the vehicle's steerability. For this purpose, the above-described relationships and charts were used. The results of a single lane change maneuver test were used to present changes in vehicle steerability. The comparison shows that the vehicle response time to the yaw rate in the case of a roadworthy vehicle is 0.32 s and in the case of vehicles with a clearance in the rocker arm pin is 0.37 s. Figure 14 shows the response time of a roadworthy vehicle, during a lane change, marked with a blue point, and a vehicle with a clearance on the rocker arm pin marked with a red point.

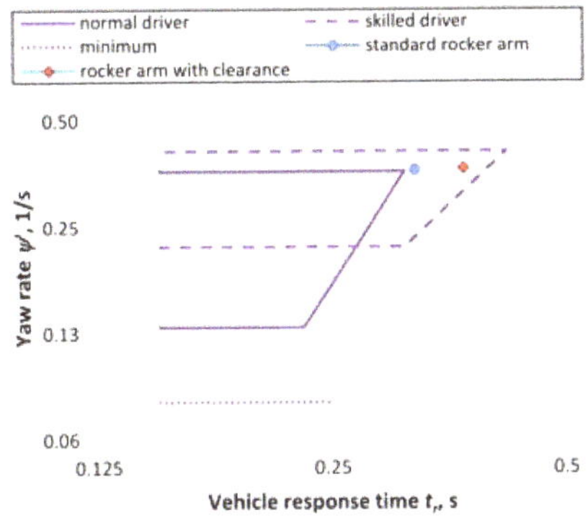

Figure 14. Vehicle steerability depending on yaw speed and vehicle response time [16,17] and own research.

When assessing the vehicle's handling, it can be concluded that both measured values of the vehicle response time are within the acceptable range, while for a vehicle that is in good working order, the response time is shorter, close to the value recommended for a driver with average skills. The response time of a vehicle equipped with a rocker arm simulating the clearance on the rocker arm pin is greater and lies in the field where a skilled driver will find the vehicle steerable, while for a medium-skilled driver, driving such a vehicle may be difficult.

The clearance in the control arm joint causes the vehicle's response time to the movement of the steering wheel to be extended, which clearly indicates a deterioration of the vehicle's maneuverability.

Relating the values of the vehicle response time to Equation (1), it can be assumed that the clearance in the rocker arm joint causes an effect similar to the reduction of the tire's cornering stiffness. In such a case, we can propose a modification of the equation to the form:

$$t_r \approx \frac{mV}{2\left(\alpha K_f + K_r\right)} \tag{2}$$

The coefficient was introduced in the equation, changing the cornering stiffness of the front tires $\alpha < 1$. The size of the coefficient will depend on the amount of clearance. Based on the research, the value of the coefficient $\alpha \cong 0.7$ was established.

The presented analyses show that the increased clearance of the rocker arm pin is particularly visible when analyzing the direction of vehicle movement. If the steering wheel rotation angle is kept constant, it causes a change of the direction of travel at the moments of overcoming road unevenness, and in the case of turning maneuvers, when changing the direction of the steering wheel rotation, it will cause the necessity to increase this angle by the amount resulting from the clearance.

8. Conclusions

The effect of the increased rocker arm clearance on the vehicle steerability is presented above. The control arm with the introduced clearance was mounted in the suspension of the front right wheel. In the tests that were carried out, the wheel was an outer wheel, i.e., a weighted wheel.

Based on the dimensional analysis, it was found that the 1 mm clearance causes the wheel steering angle δ_w to change by about 0.7°, which corresponded to a change of the steering wheel rotation angle δ_H by ~11° in the tested car. Increasing the clearance value of the rocker arm pin in will increase the value of this angle.

The tests were carried out in stationary conditions on a test stand and during road tests of the car. The bench tests made it possible to determine the required lateral force causing the clearance to be canceled depending on the wheel load (characteristics of erasing the clearance). This characteristic is approximately linear, the amount of force required to eliminate the clearance increases with the load on the wheel. The minimum value of the force required to remove the clearance results from the kingpin inclination angle and the wheel suspension load.

The road tests were carried out in the conditions of vehicle motion characterized by high values of lateral acceleration and during the braking process on a rectilinear section of the road. On the basis of road tests, it was established that obtaining the required lateral force to eliminate the clearance occurs relatively quickly when the direction of travel is changed as a result of changing the steering angle of the steered wheels and when the direction of the drive transmission is changed (acceleration-braking).

The increased clearance of the rocker arm pin, in the case of a lateral force sufficient to eliminate the clearance, causes the change of the steering angle of the wheel, regardless of the angle of rotation of the steering wheel. In most cases, resetting the clearance occurs at the initial moment of turning the steering wheel, but it can also occur when the road is sloping, negotiating bumps or driving on surfaces with different coefficients of traction under the wheels of different sides of the vehicle.

The analyses show that the increased clearance of the rocker arm pin, as a rule, causes a change in the direction of the vehicle movement during various maneuvers. This causes the feeling of "uncertainty" in steering the vehicle and increasing the vehicle's response time to maneuvers. The analyses show that the steering uncertainty is not strongly felt by the driver, yet it requires the driver to correct the steering wheel angle. There was no significant influence of the rocker arm play on the vehicle stability. However, increasing the clearance may make the steering uncertainty more perceptible.

The measure of vehicle steerability, which appears in most of the works, is the vehicle's response time to the movement of the steering wheel. The lateral acceleration a_y and the yaw velocity $\dot{\psi}$ are taken into account for the analysis. Using this method of assessing the vehicle steerability, the influence of clearance in the control arm joint on the vehicle steerability was presented. After introducing a clearance of ~1 mm, the vehicle response time increased by ~20%.

Based on the research, it was found that, for a driver with medium skills, the increased clearance of the rocker arm (1 mm) will cause difficulties with steering the vehicle, especially during sudden steering maneuvers.

The presented analysis shows that the increased clearance of the rocker arm pin affects the vehicle's steerability and extends the vehicle's response time to steering maneuvers, and this affects road safety.

Author Contributions: Conceptualization, K.P. and H.W.; methodology, K.P. and H.W.; software, K.P. and H.W.; validation, K.P. and H.W.; formal analysis, K.P. and H.W. investigation, K.P. and H.W.; resources, K.P. and H.W.; data curation, K.P. and H.W.; writing—original draft preparation, K.P. and H.W.; writing—review and editing, K.P. and H.W.; visualization, K.P. and H.W.; supervision, K.P. and H.W.; project administration, K.P. and H.W.; funding acquisition, K.P. and H.W. All authors have read and agreed to the published version of the manuscript.

Funding: This research received no external funding.

Institutional Review Board Statement: Not applicable.

Informed Consent Statement: Not applicable.

Data Availability Statement: Not applicable.

Acknowledgments: The authors thank the management of FCA Poland, Tychy Assembling Plant for enabling the vehicle to be tested and for any assistance in their implementation.

Conflicts of Interest: The authors confirm that this article content has no conflict of interest.

References

1. Pieniążek, W.; Więckowski, D. *Badania Kierowalności i Stateczności Pojazdów Samochodowych*; Państwowe Wydawnictwo Naukowe: Warszawa, Poland, 2020; pp. 19–98.
2. Świder, P. *Teoria Ruchu Samochodu: Part I*; Wydawnictwo Politechniki Krakowskiej: Kraków, Poland, 2012; p. 143.
3. Muscă, I.; Românu, I.C.; Gagea, A. Preliminary study of friction in automotive ball joints. *IOP Conf. Ser. Mater. Sci. Eng.* **2020**, *724*, 012020. [CrossRef]
4. Dziubak, T.; Dziubak, S.D. Experimental study of filtration materials used in the car air intake. *Materials* **2020**, *13*, 3498. [CrossRef] [PubMed]
5. Deng, B.; Shao, K.; Zhao, H. Adaptive Second Order Recursive Terminal Sliding Mode Control for a Four-Wheel Independent Steer-by-Wire System. *IEEE Access* **2016**, *21*, 75936–75945. [CrossRef]
6. Andrzejewski, R. *Stabilność Ruchu Pojazdów Kołowych*; WNT: Warszawa, Poland, 1997; pp. 121–150.
7. Litwinow, A. *Kierowalność i Stateczność Samochodu*; Wydawnictwa Komunikacji i Łączności: Warszawa, Poland, 1975; pp. 11–24.
8. Lozia, Z. Rollover of the biaxial vehicle (untripped case). *Logistyka* **2014**, *4*, 1–8, (On CD).
9. Pieniążek, W. Wybrane zagadnienia badania stateczności i kierowalności samochodów. *Zeszyty Naukowe Instytytu Pojazdów* **2010**, *79*, 29–43.
10. Pacejka, H. *Tire and Vehicle Dynamics*, 2nd ed.; SAE International: Warrendale, PA, USA, 2006; pp. 1–60.
11. Parczewski, K.; Wnęk, H. The tire characteristics of physical models used to investigate vehicles lateral stability. *Proc. Inst. Mech. Eng. Part D J. Automob. Eng.* **2015**, *229*, 1419–1426. [CrossRef]
12. Reński, A. *Bezpieczeństwo Czynne Samochodu: Zawieszenia Oraz Układy Hamulcowe i Kierownicze*; Oficyna wydawnicza Politechniki Warszawskiej: Warszawa, Poland, 1992; pp. 298–301.
13. Adamiec-Wójcik, I.; Awrejcewicz, J.; Grzegożek, W.; Wojciech, S. Dynamics of articulated vehicles by means of multibody methods. In *Dynamical Systems—Mathematical and Numerical Approaches*; TU of Łódź: Łódź, Poland, 2015; pp. 11–20.
14. Gillespie, T.D. *Fundamentals of Vehicle Dynamics*; SAE International: Warrendale, PA, USA, 1992; pp. 284–291.
15. Parczewski, K.; Wnęk, H. Analiza wpływu parametrów masowych na stateczność ruchu samochodu ciężarowego w oparciu o badania mobilnego modelu pojazdu. *Postępy Nauki i Techniki* **2012**, *14*, 208–223.
16. Abe, M. *Vehicle Handling Dynamics, Theory and Application*, 2nd ed.; Elsevier BH: Amsterdam, The Netherlands, 2015; pp. 267–279.
17. Hoffmann, E.R.; Joubert, P.N. The effect of changes in some vehicles handling variables on driver steering performance. *Hum. Factors* **1966**, *8*, 245–263. [CrossRef] [PubMed]

18. Hu, L.; Fang, S.; Yang, J. Study of the Vehicle Controllability and Stability Based on Multi-body System Dynamics. *Open Mech. Eng. J.* **2014**, *8*, 865–871. [CrossRef]
19. Weir, D.H.; Di Marco, R.J. Correlation and evaluation of driver/vehicle directional handling data. *SAE Tech. Pap.* **1978**, *780010*. [CrossRef]
20. Parczewski, K.; Wnęk, H. Using mobile scaled vehicle to investigate the truck lateral stability. *Eksploatacja i Niezawodnosc* **2013**, *4*, 414–420.
21. Parczewski, K. Exploration of the shock-absorber damage influence on the steerability and stability of the car motion. *J. KONES Powertrain Transp.* **2011**, *18*, 331–338.
22. Parczewski, K. *Analiza Możliwości Wykorzystania Modelu Fizycznego Pojazdu do Oceny Stateczności Ruchu Pojazdów Wielkogabarytowych*; Treatise 52; University of Bielsko-Biala: Bielsko, Poland, 2014.
23. Grąziewicz, K.; Pokorski, J. *Układ pomiarowy AD-32*; Grapol Electronic: Warszawa, Poland, 2015.
24. Racelogic. VBOX Manual. 2015. Available online: https://en.racelogic.support/VBOX_Automotive/Product_Info/VBOX_Data_Loggers/VBOX_3i_Range (accessed on 10 May 2020).

energies

MDPI

Article

Validation of Vehicle Driving Simulator from Perspective of Velocity and Trajectory Based Driving Behavior under Curve Conditions

Liang Chen [1], Jiming Xie [1], Simin Wu [1], Fengxiang Guo [1,*], Zheng Chen [1] and Wenqi Tan [2]

1 Faculty of Transportation Engineering, Kunming University of Science and Technology, Kunming 650504, China; cl@kust.edu.cn (L.C.); xiejiming@stu.kust.edu.cn (J.X.); wusm@stu.kust.edu.cn (S.W.); chen@kust.edu.cn (Z.C.)

2 College of Information and Smart Electromechanical Engineering, Xiamen Huaxia University, Xiamen 361024, China; tanwq@hxxy.edu.cn

* Correspondence: guofengxiang@kust.edu.cn; Tel.: +86-871-65920131

Abstract: With their advantages of high experimental safety, convenient setting of scenes, and easy extraction of control parameters, driving simulators play an increasingly important role in scientific research, such as in road traffic environment safety evaluation and driving behavior characteristics research. Meanwhile, the demand for the validation of driving simulators is increasing as its applications are promoted. In order to validate a driving simulator in a complex environment, curve road conditions with different radii are considered as experimental evaluation scenarios. To attain this, this paper analyzes the reliability and accuracy of the experimental vehicle speed of a driving simulator. Then, qualitative and quantitative analysis of the lateral deviation of the vehicle trajectory is carried out, applying the cosine similarity method. Furthermore, a data-driven method was adopted which takes the longitudinal displacement, lateral displacement, vehicle speed and steering wheel angle of the vehicle as inputs and the lateral offset as the output. Thus, a curve trajectory planning model, a more comprehensive and human-like operation, is established. Based on directional long short-term memory (Bi–LSTM) and a recurrent neural network (RNN), a multiple Bi–LSTM (Mul–Bi–LSTM) is proposed. The prediction performance of LSTM, MLP model and Mul–Bi–LSTM are compared in detail on the validation set and testing set. The results show that the Mul–Bi–LSTM model can generate a trajectory which is very similar to the driver's curve driving and have a preferable generalization performance. Therefore, this method can solve problems which cannot be realized in real complex scenes in the simulator validation. Selecting the trajectory as the validation parameter can more comprehensively and intuitively reflect the simulator's curve driving state. Using a speed model and trajectory model instead of a real car experiment can improve the efficiency of simulator validation and lay a foundation for the standardization of simulator validation.

Keywords: vehicle driving simulator; curve driving behavior; validation; multiple bi-directional long short-term memory (Mul–Bi–LSTM)

Citation: Chen, L.; Xie, J.; Wu, S.; Guo, F.; Chen, Z.; Tan, W. Validation of Vehicle Driving Simulator from Perspective of Velocity and Trajectory Based Driving Behavior under Curve Conditions. *Energies* **2021**, *14*, 8429. https://doi.org/10.3390/en14248429

Academic Editors: Guzek Marek, Rafał Jurecki and Wojciech Wach

Received: 16 October 2021
Accepted: 8 December 2021
Published: 14 December 2021

1. Introduction

In recent years, with the development of virtual reality technology towards practical applications, vehicle driving simulators have gradually become an important development direction in simulating the "people-vehicle-road-environment" system of road traffic research [1,2]. A vehicle driving simulator is a typical representative of a human-vehicle-road-environment simulation system. It exploits electronic computer images, with the support of electronic control and other technical support, to conduct manned simulation and research, including vehicle driving behavior, dynamic performance and traffic systems. The driving simulator has the distinguishing features of high safety, anticipated reproducibility, strong exploitability and low cost [3–6]. It can provide a safe environment for driving research and conveniently formulate a research approach related to driving behavior strategies. In particular, the experimental conditions feature a wide range and can be adjusted, and are also easy to change. Furthermore, the experimental data can be processed online, and classified and stored by a computer, which provides great convenience for the consequent statistical analysis [7,8]. Until now, driving simulators have been widely applied in research on intelligent control of vehicles, road traffic facilities and intelligent transportation systems [9,10]. They have become an effective auxiliary means and research tool for studying human efficiency, civil engineering, traffic engineering, psychology and related fields [11–15]. However, a vehicle driving simulator can only partially simulate an extremely complex transportation system. Meanwhile, each vehicle driving simulator experiment [16–18] involves effectiveness of experimental design, the rationality of content selection and representativeness of the selection of the sample (subjects), as well as the accuracy of the scene design, all of which may affect the experimental results, and even make the simulation results inconsistent with real situations. Therefore, how to validate the performance of a vehicle driving simulator deserves to be further investigated.

Until now, the simulator validation method has changed with the development of techniques and the extension of its applications. In the initial stage, the validation of simulators focuses on the physical characteristics of the vehicle. In the intermediate stage, the validation of simulators focuses on virtual images and vehicle kinematics. At present, the validation of simulators focuses on the characteristics of human-computer interaction, which is mainly due to the leading of cutting-edge technologies, such as human-computer driving intelligent vehicle and advanced driver assistance systems (ADAS). Bham et al. [7] combined the objective evaluation and subjective evaluation to validate the driving simulator by comparing the simulation results with the real operation data collected by the global positioning system (GPS) along the highway and the video records of specific locations in the specific working area. Meuleners et al. [8] recruited 47 testers to participate in experiments and suggested they drive along a selected route on the real road and a virtual road in the driving simulator. Then, the driving simulator was calibrated by evaluating the behaviors of the driver in observing the left, right, front-viewing behavior, the speed of the intersection, the speed of following the vehicle and the observance of traffic lights and stop signs under two conditions. John A. Groeger et al. [19] compared the drivers' behavior characteristics of simulator driving and the real car driving under different curvature curves based on real vehicle experiment and two simulator simulation experiments. Through the analysis of vehicle speeds, position steering behavior and other parameters on curved road sections with different radii, the correlation of these parameters between the real vehicle and the simulator is attained to complete the effectiveness calibration of the driving simulator. Yanning Zhang et al. [20] calibrated the simulator from the perspective of absolute and relative validity. The simulator calibration was attained by setting up two experimental scenarios of free driving and following driving, and using mathematical statistics, regression methods, Wilcoxon test and non-parametric methods to compare the test point speed, driving distance, vehicle position, reaction time, risk perception and other parameters.

In terms of the current driving simulator validation methods, statistical testing is the main method, especially the Tukey and Anova tests. Speed, vehicle position, acceleration and deceleration are often considered as the validation parameters. By designing experimental scenes, conducting comparative experiments between vehicle and simulators is the main way to validate simulators. However, in the simulator validation experiment of road perception and driving interaction, the representativeness of complex road selection is quite critical. In addition, it is difficult to achieve the selection of complex roads and guarantee the safety of the experiment, as well as the efficiency and spread ability of the experiment being consequently low. According to the analysis of the above methods, considering the role of the road in the overall traffic, this paper selects the curve road as the criterion to evaluate the performance of the vehicle driving simulator, as the curve road condition can preferably evaluate the driving characteristics. In the comparison study of different curve radii, vehicle speed and trajectory are two main parameters that can comprehensively reflect driving characteristics [21,22]. On this account, this article validates the effectiveness of driving simulators based on speed from reliability and validity aspects, where reliability is an indicator of the consistency of test results, and validity accounts for the accuracy of test results [23]. In the curve test, the driver judges and manipulates the vehicle to pass the curve with different radii at certain speeds, and the speed value of each sampling point on the curve is equivalent to the test on the driver [24]. After that, the experiment's quality, i.e., the reliability of the experimental results, is evaluated. From this point of view, the reliability and validity theory can be used as a reference [25]. The driving trajectory is the specific performance of the driver's strategy and behavior of manipulating the vehicle. The similarity of the trajectory is an important reference index for investigating the steering behavior of the driver [26]. When analyzing the position parameters of the vehicle on the curve, the trajectory similarity model can be used to compare the simulator experiment. After the validation parameters and validation methods are determined, a method of using theoretical models instead of real vehicle experiments is proposed for easy acquisition of calibration data. In summary, this method can solve the problem that real complex scenes are difficult to realize in the simulator validation. Selecting the trajectory as the validation parameter can more comprehensively and intuitively reflect the simulator's curve driving state. Using speed model and trajectory model instead of real car experiment can improve the efficiency of simulator validation and lay a foundation for the standardization of simulator validation [27,28].

Driving simulator technology has the advantage of being able to collaborate with multiple external vehicles at different speeds. It can also provide complex traffic environments and extreme test conditions. It further offers the advantage of safe and fast verification of motion strategies under complex operating conditions. Based on the above advantages, in recent years some car manufacturers and research institutes have started to experiment with driving simulators to promote research into automotive intelligence technologies. In particular, they have played an important role in the development and advancement of driverless car technology. For example, in January 2017, the US Department of Transportation designated 10 pilot units for testing unmanned vehicle technology, four of which were research institutes using driving simulators as a means of testing [29]. Therefore, as the advantages of driving simulators for applications in complex conditions increase, the need for driving simulator effectiveness under equivalent conditions is also increasing. The method proposed in this paper addresses the prediction of vehicle trajectories under extreme curves, small sample size and multi-dimensional metrics conditions. The proposed validation method addresses the reliability, robustness and generalisation of the simulator as a vehicle for the above research. The main contributions of this study adding to the literature are summarized as follows.

(1) A simulator validation method is designed under curved road conditions. By designing an experimental scenario for the simulator and conducting systematic analysis of the parameters and model measurement of the parameters, a comprehensive conclusion of simulator validation is attained. The method can effectively remedy the problem of insufficient research on the validation of simulators under complex road conditions.

(2) A simulator vehicle curve trajectory planning model is constructed, which leverages the Bi–LSTM neural network and employs the characteristic parameters of the driving behavior on the curves as input and the lateral offset of the vehicle trajectory as output. The model can replace the desired trajectory of a vehicle driving under complex road conditions, thereby reducing the cost of real vehicle parameter acquisition and improving the efficiency of simulator validation.

The remainder of this paper is structured as follows. Section 2 details the methods, and introduces the whole design framework. Section 3 explains the experimental data, and Section 4 discusses improvement of simulator curve track planning model, followed by the conclusion drawn in Section 5.

2. Methods

2.1. The Whole Design Framework

By designing curves with different radii as experimental scenes, selecting the speed and trajectory as validation indicators, and exploiting the Kronbach's α reliability, split-half reliability, r reliability, criterion-related validity and Cohen's d exponent as the evaluation criteria [18], this paper completes the validation of the speed indicator of the driving simulator, and leverages the cosine similarity method to complete the validation of the driving simulator's trajectory. After converting the trajectory data into state variables, a body of sequence data of the driver's trajectory characteristics can be obtained. Based on the analysis of experimental data, the establishment of a curve trajectory planning model, i.e., a multi-bidirectional long-short term memory (Mul–Bi–LSTM) recurrent neural network (RNN), driven by human-like operation data is proposed [30,31]. Bi–LSTM is currently widely used in the prediction of sequence data, such as pedestrian trajectory prediction, intrusion detection, traffic speed prediction and traffic flow prediction [32–34]. Since the vehicle trajectory is sequence data and features a strong regularity, the Bi–LSTM neural network data-driven method enables to establish a vehicle trajectory planning model with high performance. The model results manifest that the proposed validation model can better replace the expected trajectory of a vehicle driving under complex road conditions and reduce the cost of collection of actual vehicle parameters. The whole design framework is shown in Figure 1. As can be found, the whole validation is conducted from the perspectives of speed and trajectory, which are respectively evaluated from reliability, validity, similarity and trajectory fit. In addition, a data-driven trajectory planning model is constructed to facility trajectory design.

2.2. Experimental Design

In this paper, the vehicle data are obtained through experiment under different horizontal curves, including longitudinal vehicle speed, longitudinal displacement, lateral displacement, longitudinal acceleration, vehicle position, accelerator pedal and brake pedal. The analysis of speed under curves, driving trajectory, theoretical vehicle speed and expected trajectory approximation are intensively analyzed to study the effectiveness of driving simulator based on vehicle-road interaction. As Table A1, a total of 27 drivers with different genders, ages, driving years and driving mileages are selected as the test subjects in the experiment, and they are numbered 1–27. Among them, all drivers are with corrected visual acuity of 1.0 and can skillfully complete the driving tasks. The experiment recruited 16 skilled drivers, which were test numbers 1, 2, 3, 4, 5, 7, 10, 11, 13, 15, 16, 17, 18, 19, 21, and 26, including 11 male drivers and 5 female drivers. There are 11 new drivers, and the test numbers are 6, 8, 9, 12, 14, 20, 22, 23, 24, 25, and 27, including 7 male drivers and 4 female drivers. Drivers' age, driving experience, annual vehicle kilometers traveled total mileage, number and gender all meet the basic conditions of driving simulation experiments [35–37]. The experimental section is set as a two-way highway with no central separation zone and a length of nineteen kilometers. The design speed is 40 km/h and the lane width is 3.5 m. The experimental road types are composed of straight road sections and different horizontal curved road sections. The straight-line section is set between each characteristic road section (different horizontal curved sections), with a length of 700 m, so that the driver can adjust their speed to enter the next curve with the expectation speed as the entry speed. The curve consists of a symmetrical basic type and simple type, mainly involving straight lines, circular curves and transition curves.

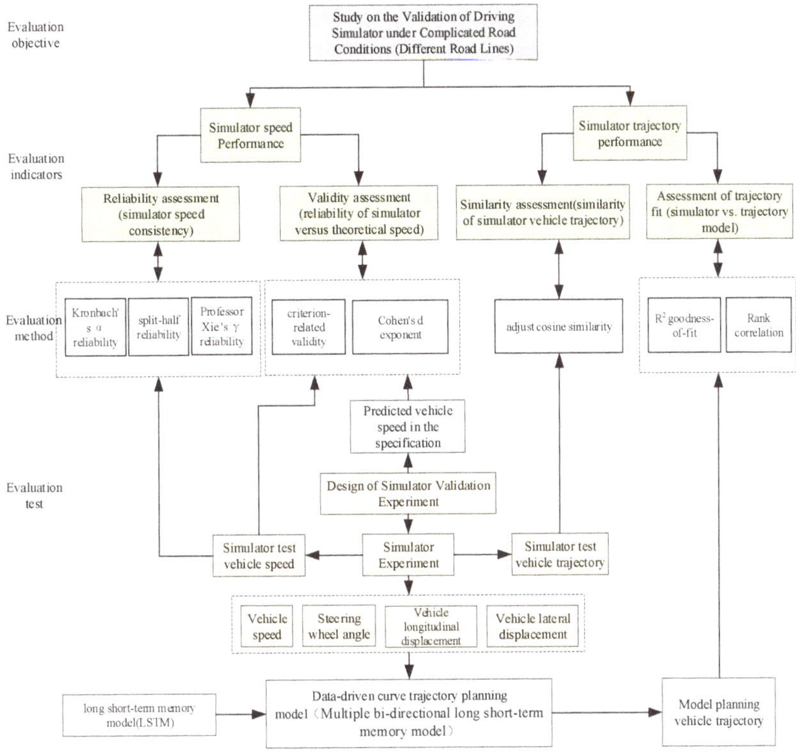

Figure 1. The whole design framework.

The circular curve setting considers the driving characteristics and the design speed to specify the minimum curve radius. According to *Design Specification for Highway Alignment* [38], the design radii include 55 m, 150 m, 250 m, 350 m, 450 m and 550 m and the transition curve is set to 35 m, 85 m, 135 m, 185 m, 235 m and 285 m, considering centrifugal acceleration, driver operation response and reaction time, as well as visual conditions. Finally, a road design software is employed to devise the curve shape of the experimental characteristic curve. The virtual road environment scene is built based on our self-developed software [39]. The driving simulator system consists of two parts: the cockpit and the console. The cockpit portion was composed of an actual Jetta automobile equipped with all the necessary sensors and a monitoring system. Three cameras and one dual voice intercom system were installed in the cockpit, and the facial expressions, movements of the hands and feet, and sounds were transmitted to the recording system and the monitor to be stored for further study. Figure 2 shows the KMRTDS driving simulator and the test while driving. The visual scenario was presented on a large screen, providing an approximately 150° horizontal view [40].

Figure 2. The KMRTDS driving simulator and the test while driving.

2.3. Experimental Data Preprocessing

The driving simulator test can collect corresponding experimental data including simulation time, vehicle longitudinal acceleration, longitudinal velocity, lateral velocity, lateral acceleration, displacement, yaw rate, vehicle mileage, steering wheel angle, accelerator, pedal, brake pedal and gear steering. In order to facilitate the analysis of the driver's driving behavior in a curved road, it is necessary to preprocess the collected raw driving data. Since the start and end points of each trajectory data are slightly different, and the difference in vehicle speed leads to misalignment between the different original trajectory data. In the data processing, the form pile number is used as the curve landmark, and the curve landmark is set in each curve. A series of points is marked on the lane dividing line of the experimental road, and thus the distance between adjacent virtual landmarks is equal. The virtual landmarks are only used for calculations. Considering the length of the road and the number of sampling points at the same time, the interval of virtual landmarks in this paper is 1 m. The sampling interval of the tested driving simulator is 50 Hz (20 ms), and the sampling points will be different according to the speed of the vehicle. When analyzing the similarity of the vehicle trajectory, the experimental data should be distinguished according to the speed when entering the corner. As shown in Figure 3, for ease of calculation, a Cartesian coordinate system needs to be established by two mutually perpendicular coordinate axes, usually called the x-axis and y-axis, and each axis points to a specific direction. The middle y-axis is tangent to the lane dividing line, and the x-axis is perpendicular to the y-axis and points to the inner lane. The intersection of the two coordinate axes is the origin. The coordinate axes of these two different lines determine a plane, called xy-plane, i.e., the Cartesian plane. For easier understanding, this paper quotes the concept of lateral offset, which is the distance between the projection point of the vehicle center on the xy plane and the form pile number. The specific calculation for the lateral offset can be formulated as:

$$D_P = \sqrt{(X_{CAR} - X_Z)^2 + (Y_{CAR} - Y_Z)^2}, \quad (1)$$

where D_P is offset, X_{CAR} is Longitudinal displacement of the vehicle, X_Z is longitudinal displacement of pile number, Y_{CAR} is vehicle lateral displacement, and Y_Z is longitudinal displacement of pile number.

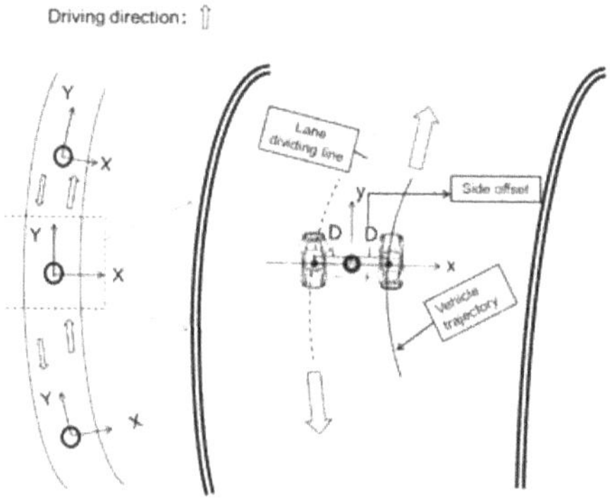

Figure 3. Driving simulation track diagram.

3. Experimental Data Analysis

3.1. Speed Reliability Evaluation of Drivers in Curved Road Tests

This paper randomly selects experimental curves 6 and 9 for specific analysis [37–39]. For curve 6 (R = 150 m), according to the speed value of the vehicle entering the curve, the drivers are classified at 40 km/h, 50 km/h and 60 km/h, respectively, and there are seven testers who turn the curve at a speed of 60 km/h, and the driver numbers are 5, 7,

11, 13, 19, 20 and 23. The testers who pass the curve road with 50 km/h are No. 4, 8, 9, 10, 15, 16, 17, 18, 21, 26 and 27. There are 8 testers, i.e., No. 1, 2, 6, 12, 14, 22, 24 and 25, who cross the corner with 40 km/h. According to the speed analysis, it can be found that the drivers who turn at 60 km/h are generally skilled drivers. Only tester 23 is classified as a new driver but with 4 years of driving experience of 50,000 kms. In this research, this can be classified as skilled driver. In the 50 km/h cornering vehicle, only testers 8, 9 and 27 are novice drivers, and in the 40 km/h cornering vehicle, drivers 6, 12, 14, 22, 24 and 25 are identified as novices. This further manifests as better performance of the skilled drivers when cornering. The related results are shown in Figure 4. As can be observed, the vehicle entering a curve with the speed of 50 km/h and 60 km/h exhibits better speed change law, i.e., first deceleration and the consequent acceleration, and the vehicle entering the curve with the speed of 40 km/h basically presents the characteristic of gradual acceleration after entering the curve. This can be explained by the reason that the initial speed of 50 km/h and 60 km/h is mainly adopted by skilled drivers, and it can also be proved that skilled drivers possess better turning maneuverability. The novice drivers mainly take the speed of 40 km/h to cross the curve. Through analyzing the forward movement data, it can be found that the novice drivers tend to slow down before the curve due to tight cornering, therefore, the vehicle speed is reduced to the designed speed in the turning phase. Overall, the skilled drivers show a higher speed when entering curve 6. By contrast, when the novice drivers enter the curve, the speed trend basically maintains the characteristics of deceleration first and consequent acceleration.

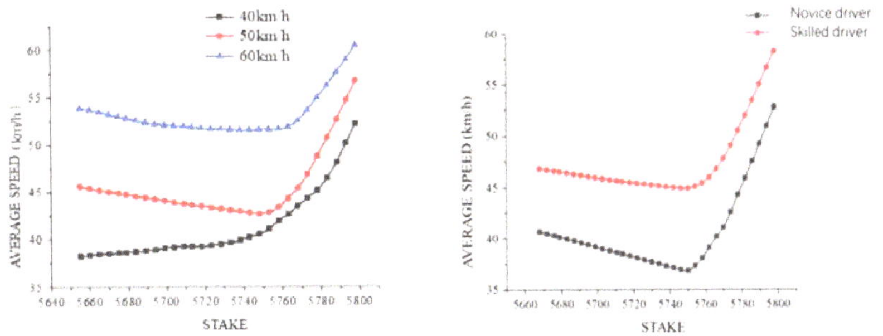

(a) (b)

Figure 4. Speed analysis under curve 6. (**a**) Average velocity distribution; (**b**) Novice and skilled velocity distribution.

For curve 9 (R = 55 m), the drivers are classified according to the critical turning speeds of 30 km/h, 40 km/h and 50 km/h. Tester 11, who turns with the speed of 60 km/h, can be excluded due to higher risk driving; and there are 8 testers, including testers 7, 13, 16, 20, 22, 24, 25 and 26, turning with the speed of 50 km/h. Among them, the testers 20, 22, 24 and 25 are novice drivers. There exist 13 testers, including testers 1, 4, 5, 6, 8, 12, 14, 15, 17, 18, 19, 21 and 23, turning with the speed of 40 km/h, and among them the testers 6, 8, 12, 14, and 23 are novice drivers. Testers 3, 9, 10 and 27 turn with the speed of 30 km/h, and tester 2 turn with the speed of 20 km/h. Since there is only one case for the tester with high speed of 60 km/h and low speed of 20 km/h, it is not discussed and instead considered as abnormal data. As shown in Figure 5, the vehicles entering a curve at the speed of 40 km/h and 50 km/h exhibit a better speed change law, i.e., first deceleration and following acceleration. The vehicles entering a turn at the speed of 30 km/h basically present gradual acceleration after entering the curve. According to the analysis of testers who use different speeds to enter a curve, there is no specific distribution law between skilled drivers and novice drivers. Through the analysis of the forward movement data, the drivers who enter the curve with the speed of 30 km/h tend to slow down before the curve, and the speed drops to the minimum value at the turning phase. On the whole, the turning speed of skilled drivers looks basically the same that of the novice drivers in curve 9, and the skilled drivers generally maintain the operations of first deceleration and the following acceleration, while the novice drivers raise more significant deceleration trend.

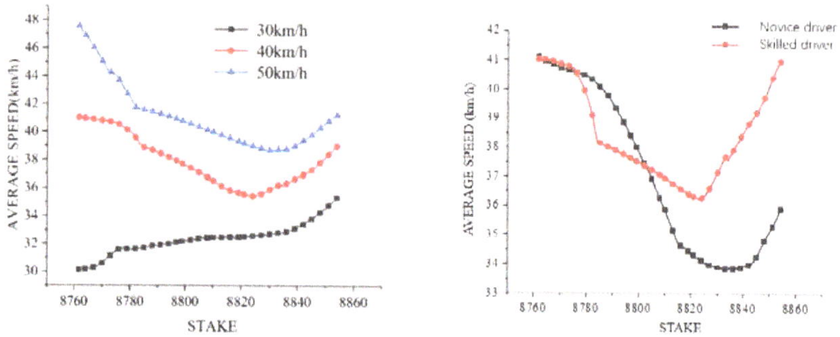

Figure 5. Trajectory analysis of curve 9: (**a**) average velocity distribution; (**b**) novice and skilled velociy distribution.

Based on the analysis of the vehicle speed, the reliability coefficients of curves 6 and 9 are further obtained. According to the validation method of the curve reliability, the Kronbach's α reliability, split-half reliability and r reliability were selected to evaluate the simulator speed [25], as shown in the following equations. The Kronbach's α reliability can be presented as:

$$\alpha = \frac{k}{k-1}\left[\frac{S_X^2 - \sum S_{X_i}^2}{S_X^2}\right] = \frac{k}{k-1}\left[1 - \frac{\sum S_{X_i}^2}{S_X^2}\right] \quad (2)$$

where k is the number of form stakes, S_X^2 is the total variance of the sampled vehicle speed, and $S_{X_i}^2$ is the speed variance of the driver at the X_i form stake. The split-half reliability can be expressed by Spearman-Brown coefficient, as:

$$r_{xx} = \frac{2 \times \frac{\sum X_1 X_2 / n - \overline{X_1 X_2}}{S_{x_1} S_{x_2}}}{1 + \frac{\sum X_1 X_2 / n - \overline{X_1 X_2}}{S_{x_1} S_{x_2}}} = \frac{2(\sum X_1 X_2 / n - \overline{X_1 X_2})}{S_{x_1} S_{x_2} + (\sum X_1 X_2 / n - \overline{X_1 X_2})} \quad (3)$$

where r_{xx} is the reliability coefficient of the simulator curve experiment, X_1 is the odd-numbered pile speed sum of the curve, X_2 is the even-numbered pile speed sum of the curve, $\overline{X_1}$ is the mean value of the odd-numbered pile speed sum of the curve, $\overline{X_2}$ is the mean value of the sum of even-numbered station speeds in the curve, n is the number of test drivers, S_{X_1} is the standard deviation of the odd-numbered station speed sum and S_{X_2} is the standard deviation of the even-numbered station speed sum. The r reliability [41] can be formulated, as:

$$r = 1 - \frac{\sqrt{k \sum_{i=1}^{k} S_i{}^2 - S^2 \times 3.92}}{\sqrt{k-1}(X_{\max} - X_{\min})} \quad (4)$$

where S^2 is the total variance of the sum of speeds in the curve, $S_i{}^2$ is the variance of the speed under the i form of stake, k is the number of form of stakes, $X_{\max}$ and $X_{\min}$ is the highest and lowest values of the sum of speeds. In terms of these coefficients, the result is shown in Table 1. By selecting the experimental data of curves with different radii and obtaining the reliability coefficient values according to the reliability model, it is found that 3 types of reliability coefficients for curve 6 meet the experimental reliability requirements; whereas for the curve 9, the α reliability coefficient and Spearman-Brown coefficient meet the requirements, but the r reliability coefficient is low. Note that when evaluating the reliability of the test, the r coefficient should be generally higher than 0.70, the other reliability coefficients should be generally higher than 0.80 [41–43]. Therefore, the data consistency of curve 9 is biased simply from the perspective of r reliability coefficient. Meanwhile, the reliability coefficients of the data are improved after data cleaning. Therefore, the difference between testers is a key factor that should be taken into consideration. Furthermore, through the comparison of reliability coefficients, the reliability of the tested driving simulator in small-radius turns is low, due to the insufficient follow-up of the steering scene when testers turn in the simulator. In this context, the testers lack a full understanding of the lateral scene and cannot accurately complete the driving operation.

Table 1. Reliability coefficient results of curve 6 and cure 9.

Reliability	Curve 6 (Outliers Included) (R = 150 m)	Curve 6 (Outliers Excluded) (R = 150 m)	Curve 9 (Outliers Included) (R = 55 m)	Curve 9 (Outliers Excluded) (R = 55 m)
α reliability	0.908	0.932	0.807	0.807
Spearman-Brown coefficient	0.912	0.982	0.885	0.894
r reliability	0.714	0.844	0.59	0.61
Mean reliability	0.845	0.919	0.761	0.770

3.2. Verification of Driving Speed in Curve Test

Based on the basic information of the experimental road, the operating speed is theoretically calculated according to the speed prediction model detailed in the specification [44], as:

$$v_{middle} = -244.123 + 0.6v_{in} + 40\ln(R_{now} + 500), v_{in} \in [30, 120], R_{now} \in [55, 600] \quad (5)$$

$$v_{out} = -183.092 + 0.7v_{middle} + 30\ln(R_{front} + 500), v_{middle} \in [30, 120], R_{front} \in [55, 600] \quad (6)$$

where v_{in} is the running speed at the entrance of the curve, v_{middle} is the running speed at the midpoint of the curve, v_{out} is the running speed at the exit of the curve, R_{now} is the radius of current curve and R_{front} is the radius of the curve to be driven into when the front is a straight line, $R_{front} = 600$ m . If $R_{front} > 5R_{now}$, R_{front} is set to $5R_{now}$.

Through the calculation and comparison of the experimental data, the test values and predicted values of sampling points are shown in Figure 6. To intuitively analyze the data sampled from the starting point, midpoint and end point of the curve, the test data of the simulator is compared with the theoretical calculation value of the speed prediction model in the design specification, and it can be found that the overall change trend of speed is consistent, while there exists large speed deviation at the entrance of individual curves. The results of the simulator experiment are generally higher than the theoretical values. The points where the simulated value is lower than the theoretical speed appear in curves 9, 10, 14, and 16. The curve radius query results show that the radius of the curves 9 and 16 is 55 m, and the radius of the curve 14 is 150 m, demonstrating that the driver in large-radius curves is more aggressive in the driving simulator; however, the driver tends to be more cautious in the small-radius sharp curves, which is related to the lack of effectiveness provided by the simulator to drivers in sharp turns, thus complying with the conclusion from the reliability evaluation that the driver test performs with low reliability in small-radius curves.

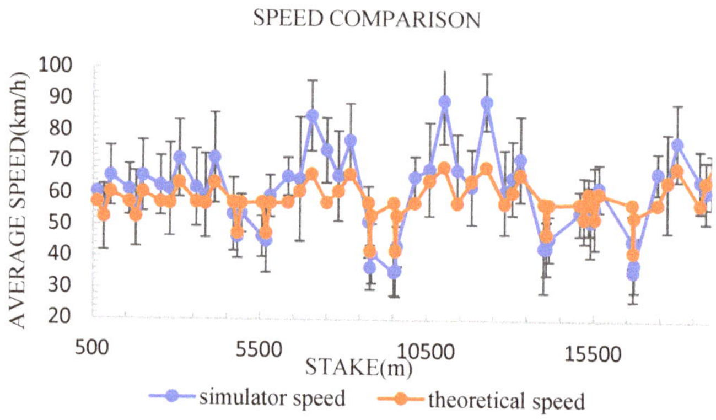

Figure 6. Speed comparisons of curve sampling point.

From a quantitative point of view, according to the criterion-related validity theory, the Pearson correlation method is employed to obtain the validity coefficient of 0.802 between the simulator data and the theoretical data. Based on the obtained correlation coefficient, the Student's t testis performed. The Student's t test is to use the t distribution theory to infer the probability of the difference, so as to compare whether the difference between two averages is significant. According to the result of the paired sample test, when the preset significance level α is 0.05, t is 0.781, and the significance level is 0.438. Hence, it can be considered that the difference between the two groups of data is not significant. The comprehensive validity and statistical test prove that the driving simulator is effective in simulating turning road conditions. Based on the validity analysis, according to the meta-analysis method, the component difference t is adopted to calculate Cohen's d index as:

$$d = \frac{2t}{\sqrt{df}} \tag{7}$$

By substituting the test value into (7), we can get $d = 0.2$. According to the Cohen's effect level classification, the effect level is low at this time, this is mainly due to the large difference in the test value of the small-radius curve based on the subjective analysis of the test value. By excluding the small-radius curve data for verification, we can get $t = 3.659$, $df = 47$ and $d = 1.07$, highlighting that the simulator test owns large effect level. Therefore, it can be explained that the simulator generates a large effect in turning conditions with a radius of greater than 150 m, but with low effect for small-radius curves. The test simulator has a high validity coefficient under turning road conditions when the radius of the curve is greater than 150 m. However, the validity on small radius curves with radius less than 55 m is low and needs to be improved.

3.3. *Verification of Vehicle Trajectory in Curve Test*

According to the designed experimental scheme, the trajectory data of the vehicle at the selected curve 6 (R = 150 m) and curve 9 (R = 55 m) is obtained using the typical vehicle trajectory planning RRT algorithm, and under the same virtual landmark, the lateral offset of the path planned by RRT algorithm at the curve is gained after transformation. The RRT algorithm is a growing tree search algorithm based on random sampling, which is widely used in robot path planning [45]. The experimental vehicle trajectory and RRT predicted trajectory for curves 6 and 9 are illustrated in Figures 7 and 8, and the trajectory similarity values are shown in Tables 2–4.

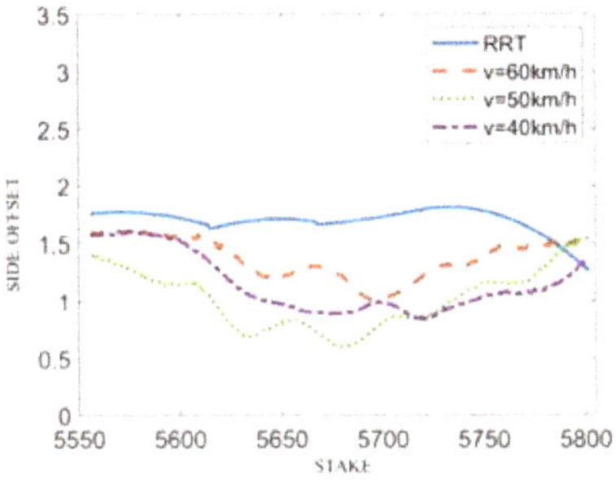

Figure 7. Curves of RRT lateral offset at different speeds in Turn 6.

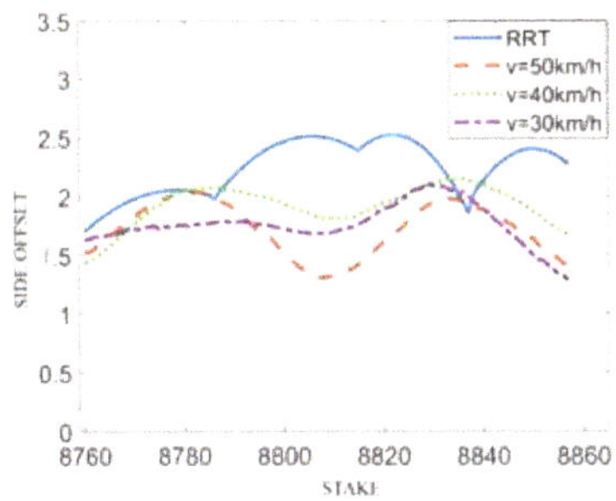

Figure 8. Curves of RRT lateral offset at different speeds in Turn 9.

Table 2. Trajectory similarity of different types of drivers.

	Cure 6 (60 km/h)	Cure 6 (40 km/h)	Cure 9 (40 km/h)
Skilled driver	0.9715	-	0.9956
Novice driver	-	0.8583	0.9785

-: No comparison possible.

Table 3. Cosine similarity of different vehicle speeds and RRT trajectory adjustment in turn 6.

	RRT	60	50	40
RRT	-	−0.2156	−0.3718	−0.0791
60	-	-	0.7909	0.8252
50	-	-	-	0.6732

Table 4. Cosine similarity of different vehicle speeds and RRT trajectory adjustment in turn 9.

	RRT	50	40	30
RRT	-	−0.539	0.2187	0.0041
50	-	-	0.6143	0.5072
40	-	-	–	0.6475

For curve 6, the vehicle with higher speed leads to larger fluctuation of the lateral offset and tends to be driven closely to the center line of the lane at the midpoint of the curve. Thus, the trend of skilled drivers driving in the middle of the road is more obvious. According to the classification of drivers with different speeds in curve 6, all those who turn at the speed of 60 km/h are skilled drivers, and those who turn at the speed of 40 km/h are basically novice drivers and a few cautious skilled drivers. The skilled drivers feature more advanced driving skills and can judge turns more accurately and turn at the speed of 60 km/h. The trajectory similarity between the skilled drivers is as high as 0.9715. The novice drivers perform unskilled and cautious driving behavior and basically control the speed at a low level of 40 km/h to pass through the turn. The trajectory similarity between novice drivers is lower than that of skilled drivers, while the overall similarity also reaches 0.8583. In the driving simulator, the trajectory similarity of vehicles with the speed of 40 km/h and 60 km/h is the largest, reaching 0.8252, proving that novice drivers with low speed can reach a trajectory similarity to that of skilled drivers with the speed of 60 km/h. In curve 9, the trajectory similarity between the novice and the skilled drivers who turn at the same speed of 40 km/h reach 0.9785 and 0.9956, respectively, and the trajectory similarity of the skilled drivers is higher than that of the novice drivers. The vehicles with the speed of 30 km/h and 40 km/h own the highest trajectory similarity of 0.6475, but from the analysis of the trajectory cosine similarity, the overall trajectory of the driving simulator in curve 9 fluctuates prominently; meanwhile, the trajectory similarity of vehicles with different speeds is markedly low. Through comprehensive analysis of the RRT trajectory calculation process under the designed curve radii, the comparison between the experimental vehicle trajectory and RRT trajectory, and between the similarity values, it can be concluded that the path point planned by RRT is quite close to the obstacle, which is the edge of the road. When planning the curve path, there emerges a phenomenon of excessive turning. The trajectory predicted by RRT in curve 6 and the trajectory predicted by the driving simulator are both negative, manifesting that the trajectories of the two are opposite. The trajectory predicted by RRT in curve 9 and the trajectory predicted by the driving simulator with the speed of 50 km/h are negative, which proves that the trajectories of the two vehicles are opposite at this speed, while the vehicle trajectories with other speeds are consistent, but the similarity is not high.

4. Improvement of Simulator Curve Track Planning Model

Through the analysis of the experimental data, it is concluded that RRT does not consider the actual driving speed, steering and other vehicle behavior factors in the curve trajectory planning process. It is necessary to establish a curve trajectory planning model driven by human-simulating operation data to provide a trajectory calibration reference for the simulator's validaton under turning conditions.

4.1. Data-Driven Modeling Method

4.1.1. Long Short-Term Memory Recurrent Neural Network

Since the lateral offset during vehicle driving is time series data, a supervised learning model can be constructed to predict this variable. The model needs to use time series data including the longitudinal displacement of the vehicle, the lateral displacement of the vehicle, the longitudinal speed of the vehicle and the steering wheel angle. The time series data itself is not only affected by the previous input characteristics, but also by the input characteristics at previous moment. In view of the above characteristics, this paper adopts a neural network with time series properties for data fitting and model construction. As shown in Figure 9, when the recurrent neural network (RNN) processes sequence data, it searches the data before the current state, saves it in the model and exploits it in the current network output, making it suitable for predicting the lateral offset of vehicles with time series. In actual calculations, in order to reduce the complexity of the algorithm and increase the inference speed of the model, it can be assumed that the current state is related to the previous states. The RNN will always receive the input x_t and loop inside the model. Each calculation will use the previous calculation information.

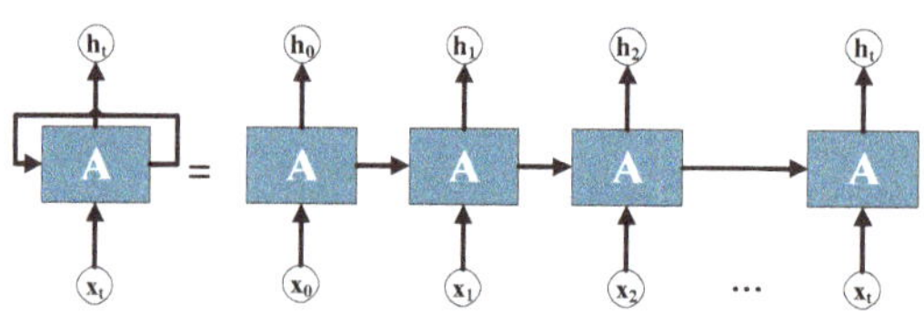

Figure 9. Schematic diagram of RNN.

From the perspective of the data set, the sampling frequency of vehicle operation is relatively high, and the current input of the model has a relationship with the input within a long period of time. When the RNN model processes long-period data, as the input pre-time sequence state is too much, the gradient will explode and disappear due to long-term dependence. Therefore, long short-term memory (LSTM) RNN is adopted to enhance the model's ability to learn long-term dependence [46]. Figure 10 presents the network structure of LSTM, where the cell state C_t represents the memory information of the model at time t. The forgetting gate f_t takes the previous sequence h_{t-1} and the current sequence x_t as input, and obtains the forgotten and retained part of the data information through the activation function. The sigmoid function is employed as activation function, which is close to 0 or 1 within the range of data values. The forgetting gate is described as:

$$f_t = \sigma(W_f \cdot [h_{t-1}, x_t] + b_f) \tag{8}$$

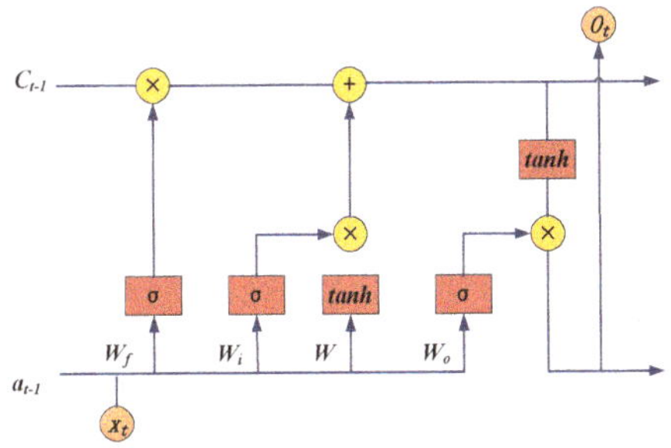

Figure 10. LSTM network structure diagram.

The input gate needs to process the input information of the current sequence, and determine the data to update the model, so as to update the model state. The sigmoid function is used to determine the information characteristics of the data which are added to the model, and the new characteristic information is converted into the added data information using the tanh function, as:

$$i_t = \sigma(W_i \cdot [h_{t-1}, x_t] + b_i) \tag{9}$$

$$\overline{C}_t = \tanh(W_c \cdot [h_{t-1}, x_t] + b_c) \tag{10}$$

The update gate updates the cell status with the forget gate and output gate information received. The update gate $f_t * C_{t-1}$ represents the information removed from the model cell, and $i_t * C_{t-1}$ represents the new data information of the model cell. The update gate is formulated, as:

$$C_t = f_t * C_{t-1} + i_t * \overline{C}_t \tag{11}$$

The output gate applies the sigmoid function to confirm the output content, and the tanh function to process the model cell content, then the output information is obtained by multiplying the two parts, as:

$$O_t = \sigma(W_o \cdot [h_{t-1}, x_t] + b_o) \tag{12}$$

$$h_t = O_t * \tanh(C_t) \tag{13}$$

where the sigmoid function is the logistic function, which is an S-shaped growth curve in biology. In practice, because of the monotonically increasing feature of the function, it is often adopted as the activation function of the neural network, which can map the input variables of the neural network to [0, 1], as:

$$\sigma = \frac{1}{1 + e^{-x}} \tag{14}$$

The tanh function is one of the hyperbolic functions. The tanh function overcomes the central asymmetry of the sigmod function, and its value range is [−1, 1], as:

$$\tanh(x) = \frac{e^x - e^{-x}}{e^x + e^{-x}} \tag{15}$$

The forget gate, update gate and output gate all adopt non-linear activation functions to enhance the learning ability of the network and enable the model to master high-dimensional complex data.

4.1.2. Bi–LSTM

The Bi–LSTM is an extension of traditional LSTM, which can improve the performance of the model to solve the time series regression problem. Based on the characteristics of the lateral offset, the prediction may need to be jointly determined by several previous inputs and several subsequent inputs, which will be more accurate. Therefore, the Bi–LSTM is employed in this study, which uses time series processing for the past and future bidirectional data, trains the front and back time series data added by the network model, and leverages forward and backward bidirectional LSTM to model the data. The forward LSTM model performs forward calculations on the data from 1 to t, saves the forward output of each sequence data time; and the backward LSTM model performs reverse calculations on the data from t to 1, and saves the backward output at each sequence data time. By synthesizing the forward and backward LSTM models to obtain the final output [47], we can yield:

$$h_t = f(w_1 x_i + w_2 h_{t-1}) \tag{16}$$

$$h'_t = f(w_3 x_t + w_5 h'_{t+1}) \tag{17}$$

$$O_t = g(w_4 h_t + w_6 h'_t) \tag{18}$$

4.1.3. Multilayer Perceptron

MLP, also a member of artificial neural networks, is based on bionics, which proposes the perception of information from the perspective of human organ perception, and the acquisition is built as a prototype. It was proposed by Frank Rosenblatt in 1958 [48].

4.1.4. Prediction Results and Model Validation

In order to evaluate the predictive performance of the model, seven error validation indicators are considered to establish a multi-indicator fusion evaluation plan [49], including:

(a) Goodness of fit (R^2)

$$R^2 = \frac{SSR}{SST} = \frac{\sum_{i=1}^{n} (\hat{y}_i - \overline{y})^2}{\sum_{i=1}^{n} (y_i - \overline{y})^2} = 1 - \frac{SSE}{SST} \tag{19}$$

(b) Mean of absolute error (Error Mean)

$$\mu = \frac{1}{n}\sum_{i=1}^{n} (\hat{y}_i - y_i) \tag{20}$$

(c) Standard deviation of absolute error (Error Std)

$$\sigma = \sqrt{\frac{1}{n}\sum_{i=1}^{n} (\hat{y}_i - y_i - \mu)^2} \tag{21}$$

(d) Mean square error (MSE)

$$MSE = \frac{1}{n}\sum_{i=1}^{n} (\hat{y}_i - y_i)^2 \tag{22}$$

(e) Root mean square error (RMSE)

$$RMSE = \sqrt{\frac{1}{n}\sum_{i=1}^{n}(\hat{y}_i - y_i)^2} \tag{23}$$

(f) Normalized root mean square error (NRMSE)

$$NRMSE = \frac{\sqrt{\frac{1}{n}\sum_{i=1}^{n}(\hat{y}_i - y_i)^2}}{\overline{y}} \tag{24}$$

(g) Rank Correlation

$$r_s = 1 - \frac{6\sum d_i^2}{n(n^2-1)} \tag{25}$$

where R^2 is the goodness of fit, y_i is the original trajectory data ($i = 1, 2, \cdots, n$), $\hat{y}_i$ is the predicted value, $\overline{y}$ is the average value, SSR is the regression sum of squares, SST is the sum of square deviations, and SSE is the residual sum of square. r_s is the rank correlation coefficient, d_i is the level difference of each pair of samples of the two variables, x_i is the difference between the variable and y_i, and n is the sample size.

4.1.5. Model Training

In order to reflect the performance advantages of Bi–LSTM model and avoid parameter overfitting, the structural performance of Bi–LSTM model is set manually to make it optimal. Twenty-four sets of crossover experiments in the form of number of hidden layers × number of traversal rounds × batch size were designed for testing as shown in Table 5. The model contains four classes of hidden layer structures, with only one hidden layer in class 1 and a number of hidden layer cells of 100, denoted as h_{100}. Class 2 has two hidden layers, both with a number of neurons of 100, denoted as $h_{100} \times h_{100}$. The remaining two classes are denoted as $h_{100} \times h_{100} \times h_{75}$ and $h_{100} \times h_{100} \times h_{75} \times h_{75}$.

Table 5. The structure of the cross-experiment.

Hidden Layers	Iteration Rounds	Lot Size	Model Number
h_{100}	100–200, Step length is 100	32–128, Step length is 32	1–6
$h_{100} \times h_{100}$			7–12
$h_{100} \times h_{100} \times h_{75}$			13–18
$h_{100} \times h_{100} \times h_{75} \times h_{75}$			19–24

The training process sets the base learning rate β to 0.005 and the dropout value is 0.5. Using the Adam iterative optimization algorithm [50], the learning rate of each parameter is dynamically adjusted using first-order moment estimation and second-order moment estimation of the gradient with the following equations.

$$m_t = \lambda \times m_{t-1} + (1-\lambda) \times g_t \tag{26}$$

$$n_t = \gamma \times n_{t-1} + (1-\gamma) \times g_t^2 \tag{27}$$

$$\hat{m}_t = \frac{m_t}{1-\lambda^t} \tag{28}$$

$$\hat{n}_t = \frac{n_t}{1-\gamma^t} \tag{29}$$

$$\theta_t = \theta_{t-1} - \frac{\hat{m}_t}{\sqrt{\hat{n}_t} + \varepsilon} \times \beta \tag{30}$$

where g_t is the gradient of the time step, m_t is the first-order moment estimate of g_t, which is the exponential moving average of g_t, n_t is the second-order moment estimate of g_t, the exponential moving average of g_t^2, λ and γ is the exponential decay rate, $\hat{m}_t$ is the deviation correction for m_t, $\hat{n}_t$ is the deviation correction for n_t, θ_t is the parameter vector for time step t, β is the learning rate, ε is the residual term, the default is taken as 10^{-8}.

After completing 24 sets of training tests, the distribution of $\overline{L}_{0.05}$ and M_{RMSE} of the model is obtained. As the number of hidden layers increases, the $\overline{L}_{0.05}$ and M_{RMSE} curves rise, and $\overline{L}_{0.05}$ and M_{RMSE} of the group 4 experiments are at a lower level, i.e., the structural performance of the group 4 model is optimal. With the same number of hidden layers, $\overline{L}_{0.05}$ and M_{RMSE} show a decreasing trend as the number of traversal rounds increases. Taking groups 1–6 as an example: the number of traversal rounds is 100 for groups 1–3 ($h_{100} \times 100 \times 32$, $h_{100} \times 100 \times 64$, $h_{100} \times 100 \times 128$) and 200 for groups 4–6 ($h_{100} \times 200 \times 32$, $h_{100} \times 200 \times 64$, $h_{100} \times 200 \times 128$). The change in the number of traversal rounds shows a significant decrease in the values of $\overline{L}_{0.05}$ and M_{RMSE}. In the case of the same hidden layer and the same number of traversal rounds, for example, with the increase of batch size, $\overline{L}_{0.05}$, M_{RMSE} also tend to increase. Based on the above analysis, group 4 (with hidden layer of, lot size of 32 and number of traversal rounds of 200) is finally adopted as the optimal Bi–LSTM model.

4.2. Model Verification and Testing

In order to compare the prediction performance of LSTM, MLP, and Mul–Bi–LSTM, 28 sets of the same vehicle curve data are used to establish trajectory planning models, respectively. The input of Mul–Bi–LSTM model is the longitudinal displacement of the vehicle, the lateral displacement of the vehicle, the longitudinal speed of the vehicle, and the steering wheel angle and the output is the lateral offset. The input and output of LSTM and MLP models are all lateral offsets. After data processing, the simulation prediction result of single variable and Mul–Bi–LSTM input is obtained. Three sets of typical data are randomly selected for comparative analysis, as shown in Figures 11–13.

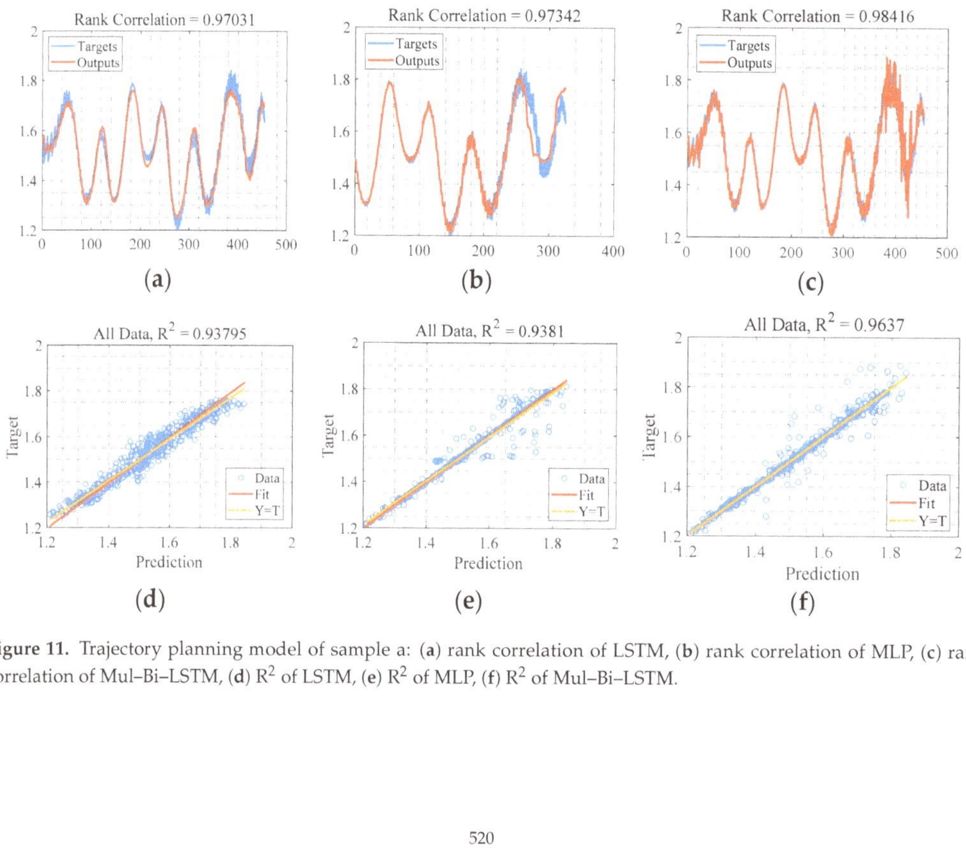

Figure 11. Trajectory planning model of sample a: (**a**) rank correlation of LSTM, (**b**) rank correlation of MLP, (**c**) rank correlation of Mul–Bi–LSTM, (**d**) R^2 of LSTM, (**e**) R^2 of MLP, (**f**) R^2 of Mul–Bi–LSTM.

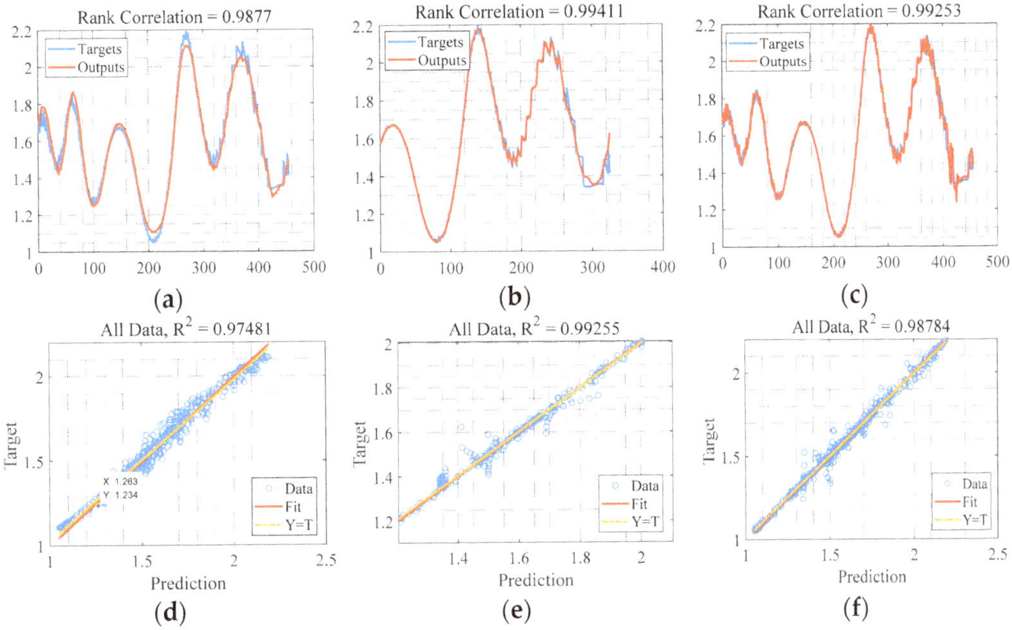

Figure 12. Trajectory planning model of sample b: (**a**) rank correlation of LSTM, (**b**) rank correlation of MLP, (**c**) rank correlation of Mul–Bi–LSTM, (**d**) R^2 of LSTM, (**e**) R^2 of MLP, (**f**) R^2 of Mul–Bi–LSTM.

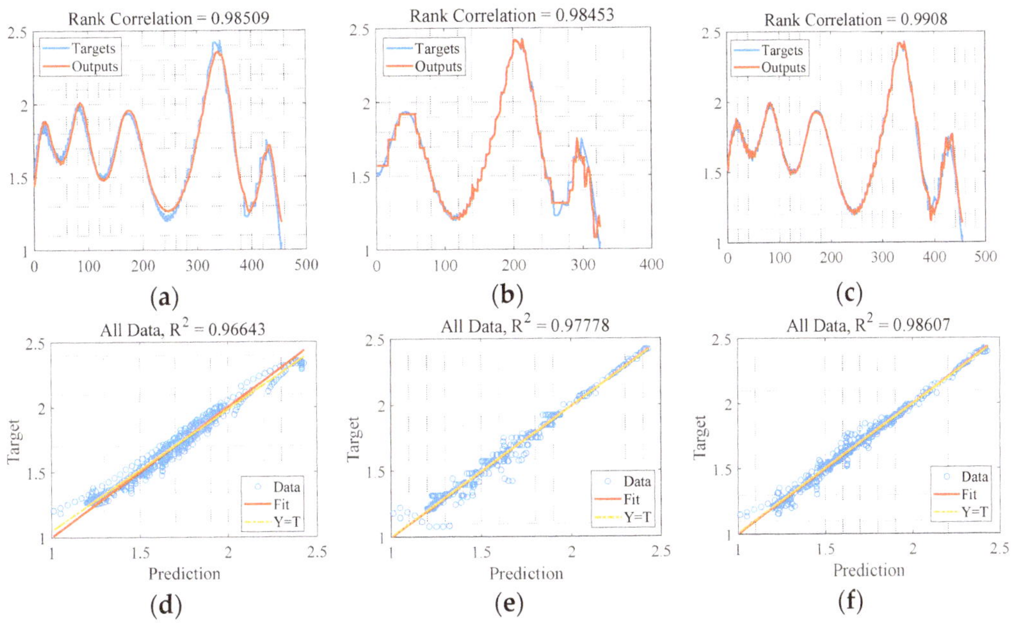

Figure 13. Trajectory planning model of sample c: (**a**) rank correlation of LSTM, (**b**) rank correlation of MLP, (**c**) rank correlation of Mul–Bi–LSTM, (**d**) R^2 of LSTM, (**e**) R^2 of MLP, (**f**) R^2 of Mul–Bi–LSTM.

As shown in Table 6, three test samples are randomly selected for model verification. In terms of Rank Correlation or R^2, the prediction results of the Mul–Bi–LSTM trajectory planning model are generally better than those of LSTM and MLP. To further verify the applicability of the LSTM model framework between different curves, curves 6, 9 and 13 are randomly selected for performance validation, and all the sample data are summarized and analyzed to obtain the planned trajectory of each curve, as shown in Table 7. It can be seen from Table 7 that in the curve trajectory planning model, the trajectory prediction value of Mul–Bi–LSTM is larger than that of LSTM. The rank correlation of Mul–Bi–LSTM trajectory prediction model in curve 6 increases by 1.3%, and R^2 increases by 11.65%, compared with the LSTM model. The Rank correlation of Mul–Bi–LSTM trajectory prediction model in curve 9 increases by 1.49, and R^2 increases by 14.35%, compared with that of LSTM model. The rank correlation of Mul–Bi–LSTM trajectory prediction model in curve 13 increases by 1.64%, and R^2 increases by 4.85%, compared with that of the LSTM model. Finally, in order to verify the portability of the LSTM model framework under the full sample condition, 80% of the samples are drawn as the training set and 20% of the samples are the test set. Mean square error (MSE), root mean squared error (RMSE) and normalized root mean squared error (NRMSE) are calculated to evaluate the prediction performance of the model. The results of each curve trajectory planning are shown in Table 8. It can be seen from Table 8 that compared with LSTM and MLP, the mean error of Mul–Bi–LSTM is reduced by 0.00275 and 0.00742, and the error Std is dropped by 0.00282 and 0.02955, and the MSE is reduced by 0.00356 and 0.00261. The results show that the value predicted by Mul–Bi–LSTM has a smaller deviation from the true value and is closer to the true value. Hence, the proposed Mul–Bi–LSTM possesses optimal prediction effect. With the introduction of human-like driving characteristic variables, the prediction performance of the model is significantly improved, and the prediction results are more accurate and closer to the true value, which is because there exists a strong recursive relationship between the trajectory data. The characteristics of driving variables can better describe and reflect the characteristics of the lateral offset, while the LSTM model separates the connections between the data, making its overall prediction performance lower, compared with that of the Mul–Bi–LSTM model.

Table 6. Comparison of samples.

	Rank Correlation			R^2		
Sample	**LSTM**	**MLP**	**Mul–Bi–LSTM**	**LSTM**	**MLP**	**Mul–Bi–LSTM**
Sample a	0.9703	0.9734	0.9842	0.9380	0.9381	0.9637
Sample b	0.9877	0.9941	0.9925	0.9748	0.9926	0.9878
Sample c	0.9851	0.9845	0.9908	0.9664	0.9778	0.9861

Table 7. Sample simulation results.

	Rank Correlation			R^2		
	LSTM	**Mul–Bi–LSTM**	**Contradistinction**	**LSTM**	**Mul–Bi–LSTM**	**Contradistinction**
Curve 6	0.9018	0.9148	1.3%	0.7294	0.8459	11.65%
Curve 9	0.8897	0.9046	1.49%	0.7216	0.8651	14.35%
Curve 13	0.9669	0.9833	1.64%	0.9192	0.9677	4.85%

Table 8. Comparison of several models.

	Error Mean	Error Std	MSE	RMSE	NRMSE
LSTM	0.00393	0.05678	0.00416	0.06235	0.03876
MLP	0.00860	0.05396	0.00321	0.05612	0.03513
Mul–Bi–LSTM	0.00118	0.02441	0.00060	0.02441	0.01599

From the previous results, it is clear that the Mul–Bi–LSTM model performed better than the MLP and LSTM models in terms of simulator data prediction. To further validate the adaptability of the Mul–Bi–LSTM model to the simulator data, a *K* fold cross-validation method was used. This method evaluates the effect of the model, with the features that each sub-sample is involved in training and testing, and can reduce the generalisation error. *K*-fold cross-validation method is based on the idea that the data set is divided into *K* regions. It selects different regions in turn as the test set. The remaining *K*-1 regions are used as the training set, and *K* times of model validation are performed. The final average of the accuracy of the *K* times validation results is taken as an estimate of the accuracy of the algorithm.The driving simulation result dataset was divided into 2 to 10 folds and the R^2 of each fold was estimated to prevent overfitting, as shown in the Figure 14. It can be seen that the test results of six-fold cross-validation have the highest R^2. Combined with the results evaluated by the *K*-fold crossover method, the Bi–LSTM model proposed in this paper has good stability and generalisability.

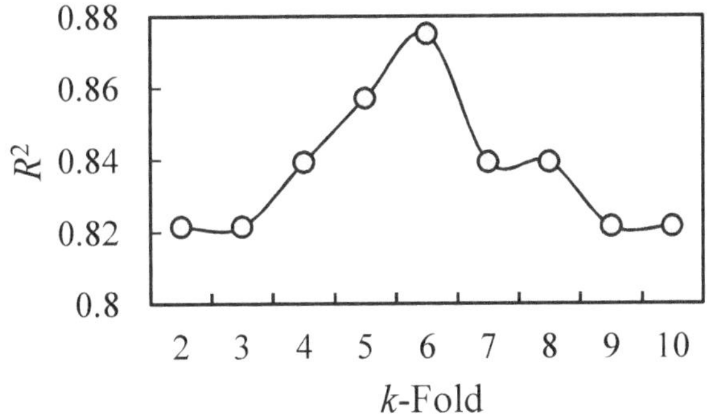

Figure 14. 2 to 10-fold cross validation R^2.

5. Conclusions

This paper mainly evaluates the effectiveness of the driving behavior characteristics of the driver under the curved road conditions in the simulator. The analysis with the speed under curves and the curve vehicle trajectory are considered as the validation indicators. The driving simulator and 27 different types of testers are selected to design the curve experiment with the radii of 55 m, 150 m, 250 m, 350 m, 450 m and 550 m for validation of the experiment simulator. In the evaluation of vehicle speed, a method for evaluating vehicle speed consistency based on reliability coefficients is proposed, and the Kronbach's α reliability, split-half reliability and r reliability are selected to conduct the reliability verification of the experimental speed results. Through sampling analysis, it is found that the overall reliability of the speed simulation for the tested driving simulator is relatively high, but the reliability of the speed simulation is low when turning in a small-radius curve. In the evaluation of speed accuracy of simulator curve, the speed values measured by the speed model at the starting point, midpoint and end point of the curve are adopted as calibration criteria, and the criterion-related validity, Cohen's d exponent method is used for validity analysis. The results reveal that the test simulator perform preferable predictive efficiency in general.

In the evaluation of the curve trajectory, the concept of lateral offset is proposed as the virtual coordinate based on the longitudinal and lateral displacement of the vehicle with the designed road form pile number. The predicted trajectory obtained by the RRT model is adopted as the evaluation standard of the curve trajectory. In the comparison of indicators, the method of revised cosine trajectory similarity is selected for analysis. The experimental results show that the overall reliability of the vehicle turning trajectory is high when the curve radius is greater than or equal to 150 m in the test simulator, while the reliability of the vehicle turning trajectory is relatively poor when the curve radius is 55 m. Based on the experimental results, in view of the low similarity of the curve trajectory between the RRT trajectory planning model and the simulator, the idea of using human-simulating operation data is proposed to establish a more complete human-simulating curve trajectory planning model by Mul–Bi–LSTM. Firstly, based on the trajectory similarity, it is concluded that the driver's curve trajectory presents strong regularity. Secondly, the trajectory planning performance of LSTM, MLP and Mul–Bi–LSTM models on the verification set and test set are compared. The results show that the Mul–Bi–LSTM model considering driving behavior factors can generate the most similar trajectory to the driver, which owns superior generalization performance. In conclusion, the speed and trajectory as the validation indicators, based on the velocity model and the Mul–Bi–LSTM trajectory model, can be used to evaluate the validity of the driving simulator under complex road conditions.

The future works to be investigated will include: (1) refinement of the validation system of the vehicle driving simulator, and (2) vehicle driving simulator development for intelligent transportation [51]. The validation of a driving simulator is a multi-level comprehensive system, which needs to be improved. The indicator system needs to be further refined, and the validation indicators need to be continuously modified through actual vehicle experiments. In addition, although the curve trajectory planning model based on human-simulating operation data driven by the thesis can be applied to emerging fields such as man-machine and autonomous driving, it is limited by experimental conditions. The effect in this process needs to be further developed for experimental scenarios and realization of related tests.

Author Contributions: Conceptualization, L.C. and F.G.; methodology, L.C.; investigation, L.C. and J.X.; data curation, W.T.; writing—original draft preparation, L.C. and S.W.; writing—review and editing, F.G. and Z.C.; supervision, F.G.; funding acquisition, F.G. All authors have read and agreed to the published version of the manuscript.

Funding: This research was funded by NATIONAL NATURAL SCIENCE FOUNDATION OF CHINA, Grant Number 71961012.

Informed Consent Statement: Informed consent was obtained from all subjects involved in the study.

Data Availability Statement: Not applicable.

Conflicts of Interest: The authors declare no conflict of interest.

Appendix A

Table A1. Detailed information of drivers.

No.	Sex	Driving Years/Kilo Miles	Age	Accidents in Recent Three Years (Y/N)	Violation in Recent Three Years (Y/N)
1	Female	9 years/50–60	35	N	N
2	Male	24 years/250	56	N	Y
3	Male	9 years/100	38	N	Y
4	Male	9 years/100	38	N	N
5	Male	10 years/80	43	N	Y
6	Male	4 years/30	28	N	Y
7	Male	8 years/140	41	Y	Y
8	Female	7 years/10	41	N	N
9	Female	10 years/50	38	N	N
10	Male	24 years/200	46	N	Y
11	Female	8 years/70	39	N	Y
12	Female	6 years/50	48	N	Y
13	Female	12 years/100	39	N	Y

Table A1. *Cont.*

No.	Sex	Driving Years/Kilo Miles	Age	Accidents in Recent Three Years (Y/N)	Violation in Recent Three Years (Y/N)
14	Female	5 years/50	32	N	N
15	Female	14 years/60	33	N	Y
16	Female	14 years/60	38	N	Y
17	Male	15 years/150	37	N	Y
18	Male	5 years/60	24	N	N
19	Male	4 years/100	23	N	N
20	Male	3 years/10	23	N	N
21	Male	20 years/300	51	N	N
22	Male	3 years/50	33	N	N
23	Male	4 years/50	23	N	Y
24	Male	3 years/20	24	N	N
25	Male	3 years/20	24	N	N
26	Male	4 years/80	22	N	N
27	Male	3 years/20	21	N	N

References

1. Wynne, R.A.; Beanland, V.; Salmon, P.M. Systematic review of driving simulator validation studies. *Saf. Sci.* **2019**, *117*, 138–151. [CrossRef]
2. Xiong, J.; Zeng, J.G.; Ding, L.; Chen, Z.L.; Shui, R.F. Application and research of vehicle driving simulator for road traffic problems. *China J. Highw. Transp.* **2002**, *15*, 117–119.
3. Broughton, K.L.M.; Switzer, F.; Scott, D. Car following decisions under three visibility conditions and two speeds tested with a driving simulator. *Accid. Anal. Prev.* **2007**, *39*, 106–116. [CrossRef] [PubMed]
4. Cox, S.M.; Cox, D.J.; Kofler, M.J. Driving Simulator Performance in Novice Drivers with Autism Spectrum Disorder: The Role of Executive Functions and Basic Motor Skills. *J. Autism Dev. Disord.* **2016**, *46*, 1379–1391. [CrossRef] [PubMed]
5. Xiong, J.; Wu, A.; Qin, Y. Experimental study on the effectiveness of driving simulator training. *Chin. J. Ergon.* **2016**, *22*, 63–67.
6. Caffò, A.O.; Tinella, L.; Lopez, A.; Spano, G.; Massaro, Y.; Lisi, A.; Stasolla, F.; Catanesi, R.; Nardulli, F.; Grattagliano, I.; et al. The Drives for Driving Simulation: A Scientometric Analysis and a Selective Review of Reviews on Simulated Driving Research. *Front. Psychol.* **2020**, *11*, 917. [CrossRef] [PubMed]
7. Bham, G.H.; Leu, M.C.; Vallati, M.; Mathur, D.R. Driving simulator validation of driver behavior with limited safe vantage points for data collection in work zones. *J. Saf. Res.* **2014**, *49*, 53–60. [CrossRef]
8. Meuleners, L.; Fraser, M. A validation study of driving errors using a driving simulator. *Transp. Res. Part F Traffic Psychol. Behav.* **2015**, *29*, 14–21. [CrossRef]
9. Dong, L.; Sun, D.; Han, G.; Li, X.; Hu, Q.; Shu, L. Velocity-free Localization of Autonomous Driverless Vehicles in Underground Intelligent Mines. *IEEE Trans. Veh. Technol.* **2020**, *69*, 9292–9303. [CrossRef]
10. Chen, X.-M.; Miao, Y.S. Driving decision-making analysis of car-following for autonomous vehicle under complex urban environment. *J. Cent. South Univ.* **2017**, *24*, 1476–1482. [CrossRef]
11. Cao, C.; Luo, Y.; Wang, J. Driving simulator validation for research on driving behavior at entrance of urban underground road. In Proceedings of the 2015 International Conference on Transportation Information and Safety, Wuhan, China, 25–28 June 2015; pp. 147–150.
12. Bella, F. Driver perception hypothesis: Driving simulator study. *Transp. Res. Part F Traffic Psychol. Behav.* **2014**, *24*, 183–196. [CrossRef]
13. Heiman, S.; Reed, N. Validation of the driver behaviour questionnaire using behavioural data from an instrumented vehicle and high-fidelity driving simulator. *Accid. Anal. Prev.* **2015**, *75*, 245–251.
14. Tozman, T.; Magdas, E.S.; MacDougall, H.G.; Vollmeyer, R. Understanding the psychophysiology of flow: A driving simulator experiment to investigate the relationship between flow and heart rate variability. *Comput. Hum. Behav.* **2015**, *52*, 408–418. [CrossRef]
15. Jesse, M.; Romain, C.; David, N.T. Driving simulator scenarios and measures to faithfully evaluate risky driving behavior: A comparative study of different driver age groups. *PLoS ONE* **2017**, *12*, e0185909.
16. Li, X.; Wang, W.; Roetting, M. Estimating Driver's Lane-Change Intent Considering Driving Style and Contextual Traffic. *IEEE Trans. Intell. Transp. Syst.* **2018**, *20*, 3258–3271. [CrossRef]
17. Zheng, R.; Nakano, K.; Yamabe, S.; Aki, M.; Nalkamura, H.; Suda, Y. Study on Emergency-Avoidance Braking for the Automatic Platooning of Trucks. *IEEE Trans. Intell. Transp. Syst.* **2014**, *15*, 1748–1757. [CrossRef]
18. Gemou, M. Transferability of driver speed and lateral deviation measurable performance from semi-dynamic driving simulator to real traffic conditions. *Eur. Transp. Res. Rev.* **2013**, *5*, 217–233. [CrossRef]

19. Groeger, J.A.; Murphy, G. Driver performance under simulated and actual driving conditions: Validity and orthogonality. *Accid. Anal. Prev.* **2020**, *143*, 105593. [CrossRef] [PubMed]
20. Zhang, Y.; Guo, Z.; Sun, Z. Driving Simulator Validity of Driving Behavior in Work Zones. *J. Adv. Transp.* **2020**, *2020*, 4629132. [CrossRef]
21. Zhou, X.; Jiang, H.; Li, A.; Ma, S. A New Single Point Preview-Based Human-Like Driver Model on Urban Curved Roads. *IEEE Access* **2020**, *8*, 107452–107464. [CrossRef]
22. Dixit, S.; Fallah, S.; Montanaro, U.; Dianati, M.; Stetvens, A.; McCullough, F.; Mouzakitis, A. Trajectory planning and tracking for autonomous overtaking: State-of-the-art and future prospects. *Annu. Rev. Control* **2018**, *45*, 76–86. [CrossRef]
23. Haigen, G. *Psychology and Educational Measurement*; Peking University Press: Beijing, China, 2008; pp. 34–40.
24. Jin, X.; Jia, Z.; Xu, W.; Yi, S. Longitudinal Acceleration Performance of Passenger Cars on Complex Mountain Highways. *China J. Highw. Transp.* **2017**, *30*, 115–126.
25. Sawada, T.; Tomori, K.; Hamana, H.; Ohno, K.; Seike, Y.; Igari, Y.; Fujita, Y. Reliability and validity of on-road driving tests in vulnerable adults: A systematic review. *Int. J. Rehabil. Res.* **2019**, *42*, 289–299. [CrossRef] [PubMed]
26. Jiang, H.; Zhou, J.; Li, A.; Zhou, X.; Ma, S. Human-like trapezoidal steering angle model on two-lane urban curves. *Int. J. Adv. Robot. Syst.* **2019**, *16*, 1729881419867614. [CrossRef]
27. Noreen, I.; Khan, A.; Habib, Z. Optimal Path Planning using RRT based Approaches: A Survey and Future Directions. *Int. J. Adv. Comput. Sci. Appl. (IJACSA)* **2016**, *7*, 142–146. [CrossRef]
28. Taylor, J.; Zhou, X.; Rouphail, N.M.; Portetr, R.J. Method for investigating intradriver heterogeneity using vehicle trajectory data: A Dynamic Time Warping approach. *Transp. Res. Part B* **2015**, *73*, 59–80. [CrossRef]
29. Knauss, A.; Berger, C.; Eriksson, H. Proving Ground Support for Automation of Testing of Active Safety Systems and Automated Vehicles. In Proceedings of the Fourth Internetional Symposium on Future Active Safety Technology toward Zero Traffic Accidents (FASTzero), Nara, Japan, 18–21 September 2017.
30. Zhang, Y.; Xiong, R.; He, H. Long Short-Term Memory Recurrent Neural Network for Remaining Useful Life Prediction of Lithium-Ion Batteries. *IEEE Trans. Veh. Technol.* **2018**, *67*, 5695–5705. [CrossRef]
31. Xiong, R.; Cao, J.; Quanqing, Y. Reinforcement learning-based real-time power management for hybrid energy storage system in the plug-in hybrid electric vehicle. *Appl. Energy* **2018**, *211*, 538–548. [CrossRef]
32. Zhang, Y.; Huang, Y.; Chen, Z.; Li, G.; Liu, Y. A Novel Learning Based Model Predictive Control Strategy for Plug-in Hybrid Electric Vehicle. *IEEE Trans. Transp. Electrif.* **2021**. [CrossRef]
33. Liu, Y.; Zhao, P.; Qin, D.; Yang, Y.; Chen, Z. Driving Intention Identification Based on Long Short-Term Memory Neural Network. In Proceedings of the 2019 IEEE Vehicle Power and Propulsion Conference (VPPC), Hanoi, Vietnam, 14–17 October 2019.
34. Chen, Z.; Zhao, H.; Shu, X. Synthetic state of charge estimation for lithium-ion batteries based on long short-term memory network modeling and adaptive H-Infinity filter. *Energy* **2021**, *228*, 120630. [CrossRef]
35. Li, A.; Jiang, H.; Li, Z.; Zhou, J.; Zhou, X. Human-Like Trajectory Planning on Curved Road: Learning From Human Drivers. *IEEE Trans. Intell. Transp. Syst.* **2019**, *21*, 3388–3397. [CrossRef]
36. Li, X.; Yan, X.; Wong, S.C. Effects of fog, driver experience and gender on driving behavior on S-curved road segments. *Accid. Anal. Prev.* **2015**, *77*, 91–104. [CrossRef] [PubMed]
37. Scialfa, C.T.; Borkenhagen, D.; Lyon, J.; Deschênes, M.; Horswill, M.; Wetton, M. The effects of driving experience on responses to a static hazard perception test. *Accid. Anal. Prev.* **2012**, *45*, 547–553. [CrossRef]
38. *Design Specification for Highway Alignment*; JTG D20-2017; Ministry of Communications of the People's Republic of China: Beijing, China, 2017.
39. Zeng, J.; Xiong, J.; Wan, H. Development and Application for 3D Scene Generated System of Driving Simulator. *J. Syst. Simul.* **2002**, *14*, 752–755.
40. Qin, Y.; Xiong, J.; Jiang, Y.; Guo, F.; Wan, H.; Jiang, L.; Jia, X. Simulator Evaluation of Drivers' Performance on Rural Highways in relation to Drivers' Visual Attention Demands. *Adv. Mech. Eng.* **2015**, *7*, 249257. [CrossRef]
41. Xie, X. γ coefficient as an estimation of reliability. *J. Chin. Psychol. Acta Psychol. Sin.* **1998**, *30*, 193–196.
42. Xie, X. Comparison of 15 equating methods. *J. Chin. Psychol. Acta Psychol. Sin.* **2000**, *32*, 217–222.
43. Li, H. The Confirmation of Measurability: Epistemological Foundation of Psychological and Educational Measurement. *Psychol. Explor.* **2015**, 304–306.
44. Huajie Engineering Consulting Co., Ltd. *Specifications for Highway Safety Audit*; JTG B05-2015; People's Communications Press Co., Ltd.: Beijing, China, 2016.
45. Kuffner, J., Jr.; Lavalle, S.M. RRT-Connect: An Efficient Approach to Single-Query Path Planning. In Proceedings of the 2000 IEEE International Conference on Robotics and Automation, San Francisco, CA, USA, 24–28 April 2000.
46. Chen, J.; Chen, Y.; Tian, F. The Method of Sample Generation for Power Grid Simulation Based on LSTM. *Proc. CSEE* **2019**, *39*, 4129–4134.
47. Alkhwiter, W.; Al-Twairesh, N. Part-of-speech tagging for Arabic tweets using CRF and Bi-LSTM. *Comput. Speech Lang.* **2020**, *65*, 101138. [CrossRef]
48. Rosenblatt, F. The perceptron: A probabilistic model for information storage and organization in the brain. *Psychol. Rev.* **1958**, *65*, 386–408. [CrossRef] [PubMed]

49. Mathematical Statistics Group, Institute of Mathematics, Chinese Academy of Sciences. *Regression Analysis Method*; Science Press: Beijing, China, 1974.
50. Li, T.; Ni, A.; Zhang, C. Short-term traffic congestion prediction with Conv–BiLSTM considering spatio-temporal features. *IET Intell. Transp. Syst.* **2021**, *14*, 1978–1986. [CrossRef]
51. Seif, H.G.; Hu, X. Autonomous Driving in the iCity—HD Maps as a Key Challenge of the Automotive Industry. *Engineering* **2016**, *2*, 159–162. [CrossRef]

energies

Article

Obsolete or Viable? Revision of Lane-Change Manoeuvre Duration Empirical Calculation

Roman Mikulec *, Marek Semela, Albert Bradáč, Stanislav Tokař, Martin Bilík, Michal Křižák, Michal Belák, Robert Kledus, Andrej Haring and Vlastimil Rábek

Institute of Forensic Engineering, Brno University of Technology, Purkynova 464/118, 612 00 Brno, Czech Republic; marek.semela@vut.cz (M.S.); albert.bradac@vut.cz (A.B.); stanislav.tokar@usi.vutbr.cz (S.T.); martin.bilik@vut.cz (M.B.); michal.krizak@vut.cz (M.K.); michal.belak@vut.cz (M.B.); robert.kledus@vut.cz (R.K.); andrej.haring@usi.vutbr.cz (A.H.); v.rabek@volny.cz (V.R.)

* Correspondence: roman.mikulec@vut.cz

Abstract: This study presents a calculation of the time required to execute a lane-change manoeuvre. Compared with other (and older) calculation methods, an analysis was conducted to determine which approach could yield the most reliable results. This study aimed to present a universal calculation method for different road surfaces, surface conditions (dry and wet road surface), and vehicle types (i.e., from small vehicles to SUVs). A total of 108 comparable manoeuvres with modern vehicles were used as a basis for statistical analysis. A new mathematical constant was found based on a regression analysis, adjusting one of the older calculation methods (so-called Kovařík equation), providing the best match between real and calculated manoeuvre duration.

Keywords: lane change; manoeuvre duration; empirical calculation; vehicle stability; lateral acceleration

Citation: Mikulec, R.; Semela, M.; Bradáč, A.; Tokař, S.; Bilík, M.; Křižák, M.; Belák, M.; Kledus, R.; Haring, A.; Rábek, V. Obsolete or Viable? Revision of Lane-Change Manoeuvre Duration Empirical Calculation. *Energies* **2021**, *14*, 8439. https://doi.org/10.3390/en14248439

Academic Editors: Guzek Marek, Rafał Jurecki and Wojciech Wach

Received: 7 November 2021
Accepted: 10 December 2021
Published: 14 December 2021

1. Introduction

A lane-change manoeuvre is part of a day-to-day commute. However, if the manoeuvre is part of a crash scenario (either in the form of crash avoidance or if the crash happened during or before the manoeuvre execution), it is necessary to determine when and where the manoeuvre began to place it in the crash scenario timeline.

Although the pre-crash analysis could be done using EDR data currently, older vehicle models usually do not possess these data, which can be problematic considering the average age of personal vehicles in the Czech Republic exceeds 15 years. Furthermore, vehicle manufacturers in Europe are not yet required to provide access to EDR data. Therefore, a calculation of the time needed to execute the avoidance manoeuvre by other methods is necessary.

Analysis of vehicle lateral movement during lane change, overtaking, or crash avoidance presents a problem in traffic accident reconstruction. The lane-change manoeuvre is not used as often as braking for crash avoidance [1,2] (the mechanism of the manoeuvre is more complex compared to braking), and there is usually less evidence left by the vehicle used to identify the vehicle's trajectory (such as yaw or skid marks). If a vehicle crash occurs on a wet road surface, a crash investigator must consider lowered adhesion while analyzing vehicle dynamics. Lane-change alteration by vehicle's adaptive cruise control was explored in [3]. As for the driving conditions, weather is one of the most common factors changing road surface conditions, especially rain, lowering tyre traction (adhesion) due to wet road surface [4].

While modern accident reconstruction methods utilize various simulation programs with integrated mathematical models, there is no guarantee that the result reflects reality. There are several unknown input parameters (delay of the steering mechanism, steering speed, driver model, and others). Therefore, using an empirical calculation based on actual manoeuvre analysis data or the simulation basis seems to be a prudent approach.

1.1. Current Calculation Methods

Many current calculation methods use simplified calculation models based on Newton's laws of motion or empirical models based on experimental measurements (driving tests). For lane-change manoeuvre, simple equations analyzing the motion of the material point can be used. These methods are described, e.g., in [1,2].

The input variables for these methods are the lateral distance, the maximum or utilized lateral adhesion (or lateral acceleration of the vehicle), and the vehicle speed. Lateral distance limits the vehicle trajectory, and lateral acceleration (or adhesion) represent and simplify vehicle (vehicle tyres) interaction with the road surface. Some variables have the same meaning across the presented formulas but differ slightly (mostly in indexes used). For unification, the variables in different formulas are defined as follows:

a_y	utilizable lateral acceleration (m/s^2)
$a_{y\,max}$	maximal lateral acceleration (m/s^2)
t	manoeuvre time (duration) (s)
v	vehicle speed (m/s)
y	lateral distance (m)
y_{max}	maximal lateral distance (m)
C	mathematical constant

Runkel [5] derived a model for a lane-change time based upon an assumption of constant acceleration inputs, resulting in a curve consisting of two circular arcs. A similar analysis was published by Jennings [6]. This model, however, does not seem to apply in noncritical situations.

$$t = C \cdot \sqrt{\frac{y}{a_y}} \quad (1)$$

Sporrer et al. in their study [7] mention that attempts were made to reflect the steering motion of a driver, which gave values of K varying from 2.51 to 2.83. Sporrer's paper then investigated the steering manoeuvre by way of driving trials. The trials showed that the manoeuvre was only symmetrical for an abrupt lane change. The results indicated that the average acceleration would not give a realistic time for the manoeuvre for a routine lane change.

In [8–10], almost identical Equation (2) is shown to be used for the theoretical calculation of the duration of the vehicle avoidance manoeuvre (t_{sin}), assuming the lateral acceleration of the vehicle is of sinusoidal shape and one-period duration.

$$t_{sin} = 2.51 \cdot \sqrt{\frac{y}{a_y}} \quad (2)$$

In the case of the actual manoeuvre, the curve of the lateral acceleration does not correspond fully with one period of the sine function. Therefore, the theoretical assumption for the deduction of the Equation (2) is not met.

Practical calculation of the manoeuvre duration (t_{rpac}) was done using the so-called Kovařík formula, described in [1] and given as Equation (3). In this formula, the mathematical constant of 2.51 was recalculated using data from experimental testing, accommodating other influences (such as delay in the steering mechanism, tyre elasticity, and others) in the overall manoeuvre duration calculation.

$$t_{prac} = 3.13 \cdot \sqrt{\frac{y}{a_y}} \quad (3)$$

A more recent form of practical calculation is the Weiss formula (Equation (4)), where K coefficient is expressed as function $K\ (a_y, v, y)$ using Equation (5), with an approximate value of 2.67, as shown in [9,11].

$$t \geq K \cdot \sqrt{\frac{y}{a_y}} \quad (4)$$

$$K = 2.2 \cdot 10^{-4} \cdot a_Y^2 + 2.6 \cdot 10^{-3} \cdot a_Y - 2.1 \cdot 10^{-2} \cdot y + 2.1 \cdot 10^{-4} \cdot v + 2.72 \quad (5)$$

The manoeuvre duration expressed by the Equations (3) and (4) is always considered to last longer than the calculated theoretical value, which is (among other things) the result of delays in the steering caused by steering wheel plays and the rigidity of the steering mechanism and elasticity of tyres.

1.2. Motivation

Most of the investigated calculation methods of lane-change duration are based on experiments carried out more than ten years ago. In contrast, the newer experiments were carried out mainly on a dry road surface. Therefore, it is necessary to explore if (and how) modern vehicle construction, vehicle stability systems, and lowered adhesion conditions affect these methods and their applicability, i.e., whether it is necessary to change the approach in the manoeuvre duration calculation or whether the empirical models used currently are still viable.

1.3. Hypothesis

The duration of the lane-change manoeuvre is conditioned by its intensity, that is, the intensity of lateral acceleration and the lateral distance. Modern electronic devices allow precise measuring of parameters relevant for the analyses of lane-change manoeuvres, that is, speed, time, lateral distance, lateral acceleration, and others. In this way, with an adequate concept of experimental test track, it is possible to analyze the influence of the intensity of lateral acceleration and the lateral distance to the time necessary for obstacle avoidance by lane-change manoeuvre.

The goal of this research is also to obtain the latest and more precise experimentally gained results of lane-change manoeuvre duration convenient for theoretical clarification and description of manoeuvre, social relevance, and legal certainty impact of participants in terms of road accident analysis

2. Methodology

The main idea of this paper is to explore the applicability of current empirical Equations (1) and (2) used for the manoeuvre duration calculation, and it was necessary to obtain all the relevant variables used in the calculations (Table 1).

Table 1. Relevant variables of selected empirical formulae.

Calculation Method	Essential Input Variables	Output Variables
Kovařík Equation (1)	y (m); $a_{y\ max}$ (m/s ^{2})	t (s)
Weiss Equation (2)	y (m); $a_{y\ max}$ (m/s ^{2}); v (m/s)	t (s)

For this purpose, four series of driving tests were carried out between the years 2016 and 2020 in various conditions (both dry and wet road surfaces). To repeat the experiment under the same conditions, the test track for vehicle lane-change manoeuvre was based on the procedure described in standard ISO 3888-2, Passenger cars—Test track for a severe lane-change manoeuvre, Part 2 Obstacle avoidance (or its modifications) [12].

In all test series, the drivers were always experienced males, between 30 and 35 years old, with 10 to 15 years of driving experience. The drivers always had several practice runs before testing to become better acquainted with both the test track and the tested vehicles (thus, the influence of the driver on the driving manoeuvre was eliminated).

All relevant data of the lane-change manoeuvre were documented using GPS Data Logger V-BOX Video HD 2 and Racelogic PERFORMANCEBOX with receiver antennas placed on the roof, roughly at the center of gravity of each vehicle tested. These instruments record data at a frequency of 10 Hz, which was proven sufficient in [13] for trajectory documentation using a similar device (Performance Box Sport).

The driver's input was monitored using a VBOX Video HD2 camera located inside the vehicle. The vehicle position was documented using the same equipment and was observed by cameras, placed around the test track.

A wide variety of vehicles was used throughout the individual test series, ranging from the supermini vehicle class to crossover SUV (see Table 2).

Table 2. Tested vehicles.

Test Series	Vehicle	Year of Manufacture	Vehicle Class	Surface Conditions
No. 1	Skoda Fabia III 1.4 TDI	2015	Supermini	dry
	Skoda Superb III 2.0 TDI	2015	Large family car	dry
No. 2	Skoda Superb III 1.4 TSI	2016	Large family car	wet
	Skoda Superb III 1.4 TSI Combi	2016	Large family car	wet
No. 3	Skoda Octavia III 1.5 TSI	2018	Small family car	wet
No. 4	Skoda Karoq 4 × 4 1.5 TSI	2019	Compact crossover SUV	wet
	Skoda Kodiaq 2.0 TDI	2019	Crossover SUV	wet

The first two test series (2016 and 2017) were carried out following standard ISO 3888-2, Passenger cars—Test track for a severe lane-change manoeuvre, Part 2 Obstacle avoidance.

The 2016 test series was carried out on a landing strip with a dry asphalt surface, with temperatures ranging from 14 to 18 °C. The 2017 test series was carried out on an area with an asphalt surface used for truck testing. During the 2017 testing, permanent rain was present, with temperatures ranging from 7 to 15 °C.

The track was modified for the next two test series (2019 and 2020) but still based on standard ISO 3888-2. The track widths (a and b) remained following the standard; the length of the track segment used for lane change (S2) was variable and shortened compared to the ISO standard to lengths of 13.5, 11.0, and 8.5 m, and the offset (p) of the exit gate to entry gate was variable as well—offsets of 1.5, 2.5, and 3.5 m were used (the diagram of a modified test track is presented in Figure 1).

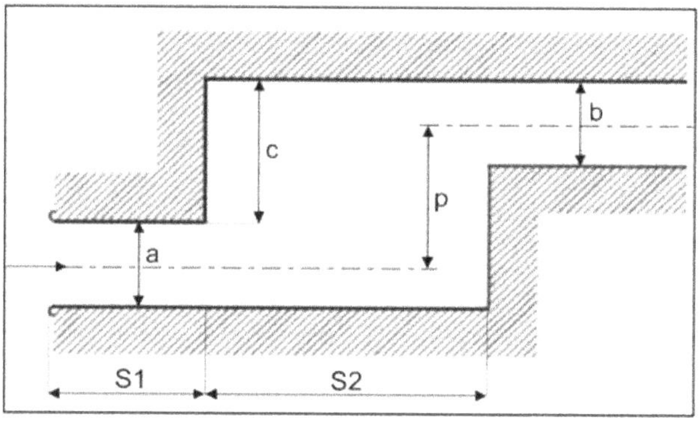

Figure 1. Adjusted test track schematics.

In all the test series, the goal was to achieve maximal vehicle speed without stability loss or exceeding the test track boundaries.

In this study, the driver's first steering intervention is regarded as the beginning of the lane-change manoeuvre and reaching maximal lateral distance (at this point, the vehicle has avoided the obstacle and changed lane at a certain width, taking a position similar to the initial one), regarded as the end of the manoeuvre, similar to [14] and as seen in Figure 2.

(a)

(b)

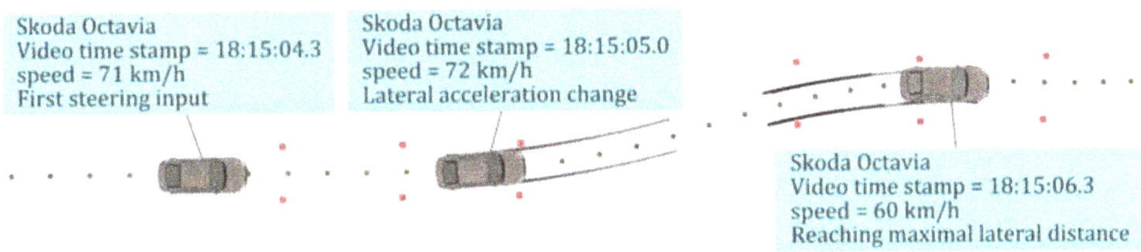

(c)

Figure 2. Example of determining the beginning and end of the manoeuvre during test series 3: (**a**) first steering input; (**b**) first discernible change of lateral acceleration and vehicle direction; (**c**) overview of the whole manoeuvre with traffic cones outlining the track and green dots representing the vehicle path.

3. Lane-Change Manoeuvre Duration Calculation

Based on the methodology chosen for the manoeuvre duration calculation, the analysis of the Kovařík and Weiss Equations (3) and (4) was carried out. Since both of these formulas were almost identical (or could be simplified into the same form), analysis of the mathematical constant C was done by using the data from 108 successfully performed manoeuvres (out of 193 manoeuvres performed in total).

A linear regression analysis was carried out. The square root value of the ratio of the maximal lateral distance y_{max} and the maximum lateral acceleration $a_{y\ max}$ achieved in the individual driving manoeuvres is plotted directly on the x-axis. The duration of the manoeuvres was again plotted on the y-axis, see Figure 3.

Based on 108 measurements from all test series and the linear regression (with correlation coefficient of 0.9899), the mathematical constant of the equation was adjusted with a value of 2.93.

$$t \geq 2.93 \cdot \sqrt{\frac{y_{max}}{a_{y\ max}}} \tag{6}$$

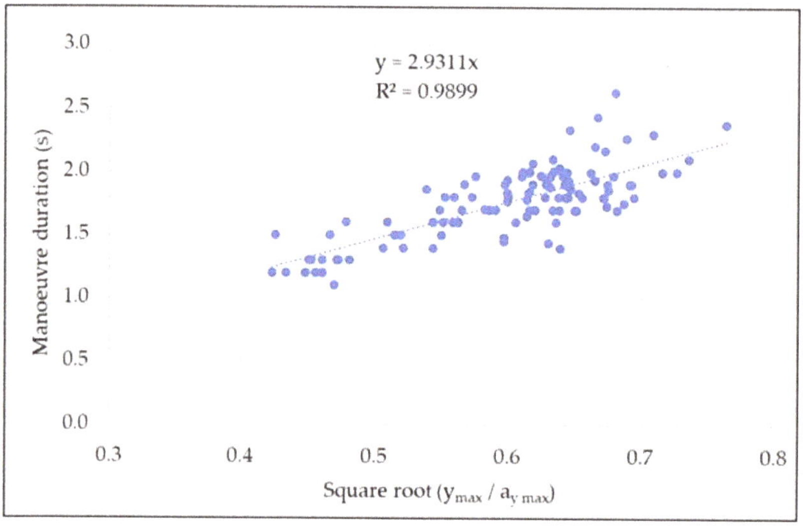

Figure 3. Data regression analysis.

4. Results

As the primary motivation of this study is to present lane-change manoeuvre duration calculation, a comparison was made between actual manoeuvre times (durations), documented during all the test series, and durations calculated based on the values of maximal lateral acceleration and lateral distance by using the original Kovařík Equation (3), the Weiss Equation (4), and the adjusted Kovařík Equation (6). An example of the calculation of manoeuvre durations using individual equations is provided in Table 3.

Table 3. A sample of actual manoeuvre times and durations, calculated using the formulas presented and data from test series 3.

Real Manoeuvre Time t (s)	Maximal Lateral Acceleration $a_{y\ max}$ (m/s^2)	Lateral Distance y_{max} (m)	Approach Speed v (km/h)	Kovařík Equation (3)	Weiss Equation (4)	Adjusted Kovařík Equations (6)
1.90	7.95	3.80	54	2.16	1.85	2.03
2.10	6.47	3.50	52	2.30	1.97	2.16
2.00	7.41	3.80	56	2.24	1.92	2.10
1.80	8.83	3.60	61	2.00	1.72	1.87
1.80	10.15	4.20	66	2.01	1.73	1.88
1.90	7.50	3.60	71	2.17	1.86	2.03
2.00	9.12	4.00	71	2.07	1.78	1.94
2.00	6.08	2.50	47	2.01	1.73	1.88
1.70	6.47	2.60	52	1.98	1.71	1.86
1.70	6.38	2.70	57	2.04	1.75	1.91

The manoeuvre durations calculated using the methods presented are compared by the box diagrams shown in Figure 4.

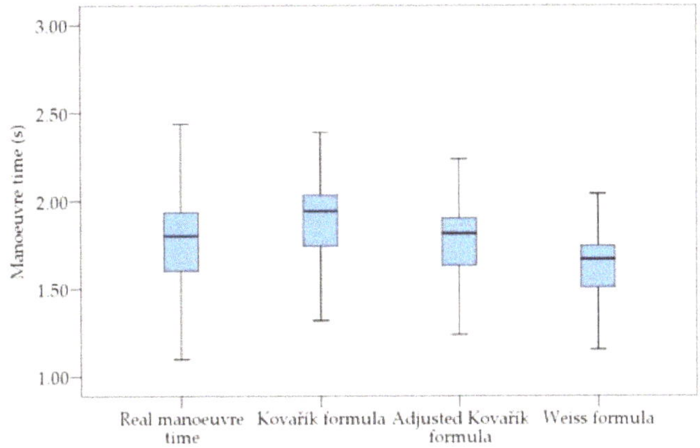

Figure 4. Equation comparison.

Using SPSS Statistics software, the Kolmogorov–Smirnov test was carried out, verifying that the data set did not have a normal distribution. The results of the test are listed in Table 4.

Table 4. Kolmogorov–Smirnov test results.

	Kolmogorov–Smirnov Test			Shapiro–Wilk Test		
	Statistics	**df**	**Sig.**	**Statistics**	**df**	**Sig.**
Real manoeuvre time	0.108	108	0.003	0.976	108	0.045
Kovařík formula	0.137	108	0.000	0.943	108	0.000
Weiss formula	0.138	108	0.000	0.942	108	0.000
Adjusted Kovařík formula	0.136	108	0.000	0.943	108	0.000

As the data did not have a normal distribution, the Mann–Whitney nonparametric test was used for comparing the equations. Tables 5 and 6 compare the results of actual manoeuvre durations, calculated by using Equation (6).

Table 5. Mann–Whitney test results.

Group	N	Mean Rank	Sum of Ranks
Real manoeuvre time	108	107.68	11,629.00
Adjusted Kovařík Equation (6)	108	109.32	11,807.00
Total	216		

Table 6. Mann–Whitney test results.

	Data
Mann–Whitney U	5743.000
Wilcoxon W	11,629.000
Z	−0.194
Asymp. Sig. (2-tailed)	0.846

The result of this analysis indicated no statistically significant difference between manoeuvre duration (p-value 0.846), calculated by the adjusted Kovařík Equation (6) and actual documented times.

By using the Mann–Whitney test to compare other calculation methods with actual manoeuvre durations, it was found there was a difference between the manoeuvre durations, calculated by the original Kovařík Equation (3) and Weiss Equation (4) compared to actual documented times, with p-values 0.000 and 0.000, respectively.

5. Discussion

Several mathematical methods of lane-change manoeuvre duration calculation were explored. These methods assume vehicle movement in a curve consisting of two circular arcs or that the vehicle's lateral acceleration is of sinusoidal shape and one-period duration. Driver's reaction time needed to perform the steering manoeuvre is not taken into consideration, nor is the response time of the steering mechanism to turning of the steering wheel.

The calculation of the manoeuvre duration presented was based on the results of experimental research carried out with a wide range of vehicles (from supermini vehicles to crossover SUVs) and in various road conditions (from dry to wet road surface). The calculation base includes only the manoeuvres that were successfully performed without vehicle stability loss or exceeding the test track borders.

Exploration of the lane-change duration calculation was introduced in [15]. Based on experimental testing with modern vehicles on a dry road surface (also described in [16]), the constant value was recalculated to 2.6.

Research determining lane-change distance during obstacle avoidance based on experimental measurements was introduced in [13]. This research has similar characteristics to the present study; however, the main goal was to present the calculation of longitudinal lane-change distance (i.e., not the overall distance travelled by the vehicle).

Table 7 lists the ranges of vehicle approach speeds and maximal documented lateral acceleration in the individual test series.

Table 7. Documented lateral acceleration in the individual test series.

Test Series	Vehicle	Approach Speed (km/h)	Maximal Lateral Acceleration $a_{y\ max}$ (m/s^2)
No. 1	Skoda Fabia III 1.4 TDI	47 to 62	6.5 to 11.5
	Skoda Superb III 2.0 TDI	42 to 65	8.5 to 12.0
No. 2	Skoda Superb III 1.4 TSI Skoda Superb III 1.4 TSI Combi	46 to 65	7.8 to 10.0
No. 3	Skoda Octavia III 1.5 TSI	47 to 79	5.5 to 10.0
No. 4	Skoda Karoq 4 × 4 1.5 TSI	45 to 68	6.0 to 10.5
	Skoda Kodiaq 2.0 TDI	48 to 73	5.5 to 9.5

The first step in the manoeuvre time calculation was to define the beginning and the end of the manoeuvre. In this paper, the driver's first steering intervention is regarded as the beginning of the lane-change manoeuvre and reaching the maximal lateral distance is considered to be the end of the manoeuvre, similar to [14]. An argument can be made that the end of the manoeuvre should include complete stabilization at the end of the manoeuvre—this, however, seems difficult to apply in actual cases, as keeping control over the vehicle depends largely on the driver's skills, vehicle speed, and severity of the manoeuvre.

The impact of tyre pressure on achievable lateral acceleration was observed during 2016 test series 1. From what was observed while testing both vehicles, when tyre pressure was set to 30% over its nominal value, the maximal achievable lateral acceleration values increased on average by 0.5 m/s^2. When the tyre pressure was set 30% below its nominal value, the maximal achievable lateral acceleration was decreased on average by 0.8 m/s^2.

A positive effect of electronic stability systems on crash occurrence was explored in numerous studies, such as [17–19]. After all, the generally accepted positive effects of these driving assistants are seen through mandatory equipment for this feature in all newly manufactured vehicles, given by the European Commission since 2014.

Compared to the older calculation methods, the method presented is based on tests of modern vehicles currently used on European roads. The proposed method also clearly defines the manoeuvre's beginning and end.

The statistical analysis found a new mathematical constant value C, adjusting the Kovařík Equation (3). This way, it is possible to apply lane-change manoeuvre time (duration) calculation with only adjustment to its mathematical constant to 2.93, giving it the form of the Equation (6).

The calculation presented is based on several simplifications; mainly, the complex interaction of the vehicle and vehicle tyres with the road surface is represented only by lateral acceleration. This simplification should not be seen as a disadvantage since, in many crash reconstruction cases, many of the variables are unknown (such as slip angle, yaw motion, and others) and cannot be precisely determined, e.g., due to the analyzed vehicle's destruction. Determination of the maximal lateral acceleration thus depends on crash the reconstructionist, who can base this value on road surface condition, vehicle technical state, and vigour of the manoeuvre. The only other unknown variable in the equation is lateral distance, based on either on-road marks or estimated vehicle trajectory.

It is important to note that all the measurements were done in a testing (safe) environment. Thus, it was possible to reach the vehicles' driving limits. To use this calculation for the driver's avoidance option calculation (as a viable option, e.g., to prevent collision with an object in the driver's lane), it is recommended to use lower values of lateral acceleration, as an "average driver" is not necessarily able to execute this manoeuvre while driving at the vehicle's stability limit (either due to fear of slipping or due to lack of driving experience). The lane-change manoeuvre is not used as often as braking for crash avoidance, as seen in [20,21] as the manoeuvring mechanism is more complex than braking.

Thus, a guideline is proposed, similar to the one used for braking on the wet road surface, using only half of the maximal achievable acceleration on the given surface.

While modern vehicles can reach a lateral acceleration of up to 10.0 m/s^2 during the lane-change manoeuvre even on the wet surface, it is the limit of the vehicle and not necessarily of the vehicle's driver.

Furthermore, it is important to note that requiring an evasive manoeuvre from the driver as possible collision prevention is complicated by manoeuvre complexity and the necessity to check not only the road ahead of the driver, but also whichever direction the driver chooses to evade, and also the immediate vicinity of the vehicle, to ensure the safety of the manoeuvre. This naturally increases the driver's reaction time (compared to "simple" braking).

6. Conclusions

Based on the analysis of the older calculation methods, it was determined that these methods were based on theory and, in some cases, outdated data.

For this reason, the applicability of these calculation methods was explored. Four series of driving tests were carried out using a wide variety of modern vehicles on the test track in dry and wet conditions. To ensure repeatability of the experiments, the research was performed according to the procedure prescribed by ISO 3888-2 standard, which defines the conditions and procedure of the obstacle avoidance manoeuvre. This procedure was later modified in the sense of variation of lateral distances.

Considering the calculation methods presented, it was concluded that the most prudent approach is to explore the value of the mathematical constant C. Based on a statistical analysis of the research results on a sample of 108 test drives, a new mathematical constant was presented, providing the most reliable results (there was no statistically discernible difference between the measured manoeuvre duration and the manoeuvre duration calculated by the modified formula).

The manoeuvre duration calculated by this formula represents the period between the driver's first steering input and reaching the vehicle's maximal lateral distance during the manoeuvre (without considering possible slight adjustments to steering in case of vigorous manoeuvring at the end of the manoeuvre).

One of the main components of the calculation is the value of maximal lateral acceleration, which encompasses both the road conditions (i.e., adhesion) and intensity of the manoeuvre, and its value should be carefully evaluated.

Future research should focus on adjusting the mathematical constant based on new data obtained by testing modern vehicles (regarding vehicle development) or providing a range of mathematical constants best suited either for different types of vehicles or for various road surfaces. Future research could also focus on developing more detailed mathematical models that would include additional parameters (such as tyre–road friction coefficient, vehicle mass, vehicle lateral stiffness) as data in this paper were insufficient for such detailed analysis.

Author Contributions: Conceptualization, R.M.; methodology, R.M., M.S., A.B. and R.K.; data analysis, R.M., M.K. and A.H.; formal analysis, M.B. (Martin Bilík) and M.B. (Michal Belák); writing—original draft preparation, R.M.; writing—review and editing, R.M., M.S., A.B., S.T. and V.R.; supervision, M.S. and R.K. All authors have read and agreed to the published version of the manuscript.

Funding: This research received no external funding.

Institutional Review Board Statement: Not applicable.

Informed Consent Statement: Not applicable.

Data Availability Statement: Data sharing not applicable.

Conflicts of Interest: The authors declare no conflict of interest.

References

1. Bradáč, A. *Soudní Inženýrství*; CERM: Brno, Czech Republic, 1999; p. 725, ISBN 80-7204-133-9.
2. Burg, H.; Moser, A. *Handbook of Accident Reconstruction: Accident Investigation, Vehicle Dynamics, Simulation, Part I, Basic*, 1st ed.; CreateSpace Independent Publishing Platform: Washington, WA, USA, 2013; ISBN 978-1492328421.
3. Freyer, J.; Deml, B.; Maurer, M.; Färber, B. ACC with enhanced situation awareness to reduce behavior adaptations in lane change situations. In Proceedings of the IEEE Intelligent Vehicles Symposium Proceedings, Istanbul, Turkey, 13–15 June 2007; pp. 999–1004. [CrossRef]
4. Pardillo, M.; José, M.; Jurado, P.R. An assessment of the skid resistance effect on traffic safety under wet-pavement conditions. *Accid. Anal. Prev.* **2009**, *41*, 881–886. [CrossRef] [PubMed]
5. Runkel, M. Die Bedeutung der Sichtbehinderung durch Lastkraftwagen für Überholvorgänge auf zweispurigen Landstraßen. Ph.D. Thesis, TH-Stuttgart, Fachbereich Wasser- u. Verkehrsw, Stuttgart, Germany, 1969; p. 85.
6. Jennings, P. Swerves and Lane Changes. *Impact J. Inst. Traffic Accid. Investig.* **1990**, *1*, 9.
7. Sporrer, A.; Prell, G.; Buck, J.; Schaible, S. Realsimulation von Spurwechselvorgängen im Straßenverkehr. *Verk. Und Fahrz.* **1998**, *36*, 69–76.
8. Bradáč, A.; Krejčíř, P.; Glier, L.; Plch, J.; Lukašík, L.; Helešic, V. *Znalecký standard č. III (Forensic Espert's Standard No. III). Technická Analýza Střetu Vozidla s Chodcem. Znalecký Standard č. IV (Forensic Espert's Standard No. IV). Technická Analýza Nárazu Vozidla na Překážku*; VUT Publishing House: Brno, Switzerland, 1991; p. 117.
9. Weiss, E. Untersuchung und Rekonstruktion von Ausweich-und Fahrspurwechselvorgängen. Ph.D. Thesis, VDI-Verlag München, Technical University, Düsseldorf, Germany, 1988; p. 135.
10. Zellner, J.W.; Weir, D.H. Development of Handling Test Procedures for Motorcycles. *SAE Trans.* **1978**, *87*, 780230–780458.
11. Weiss, E.; Woschni, G. Rekonstruktion von Überholvorgängen. *Verk. Und Fahrz.* **1986**, *24*, 303–308.
12. *ISO 3888-2 Passenger Cars-Test Track for a Severe Lane-Change Manoeuvre, Part 2, Obstacle Avoidance*, 2nd ed.; ISO CopyRight Office: Geneva, Switzerland, 2011.
13. Zovak, G.; Bogdanović, V. Empirical approach for determining lane change distance at obstacle avoidance manoeuvre. *Promet* **2016**, *28*, 267–275. [CrossRef]
14. Kledus, R. Lane change test manoeuvres: (test results and reconstructions). *Impact J. Inst. Traffic Accid. Investig.* **2005**, *14*, 63–76.
15. Panáček, V. Problematika znalecké analýzy jízdy a brzdění vozidla v obecném prostorovém oblouku při rychlostech vyšších než 50 km/h. Ph.D. Thesis, Brno University of Technology, Institute of Forensic Engineering, Brno, Czech Republic, 2015; p. 187.
16. Panáček, V.; Semela, M.; Adamec, V.; Schüllerová, B. Impact of usable coefficient of adhesion between tyre and road surface by modern vehicle on its dynamics while driving and braking in the curve. *Transport* **2016**, *31*, 142–146. [CrossRef]

17. Farmer, C.M. Effects of Electronic Stability Control: An Update. *Traffic Inj. Prev.* **2006**, *7*, 319–324. [CrossRef] [PubMed]
18. Page, Y.; Cuny, S. Is electronic stability program effective on French roads? *Accid. Anal. Prev.* **2006**, *38*, 357–364. [CrossRef] [PubMed]
19. Chouinard, A.; Lécuyer, J. A study of the effectiveness of Electronic Stability Control in Canada. *Accid. Anal. Prev.* **2011**, *43*, 451–460. [CrossRef] [PubMed]
20. Mcgehee, D.V.; Elizabeth, N.M.; Baldwin, G.H.S.; Grant, P.; Simmons, C.J.; Hankey, J.; Forkenbrock, G. *Examination of Drivers' Collision Avoidance Behavior Using Conventional and Antilock Brake Systems on the Iowa Driving Simulator*; U.S. Department of Transportation, National Highway Traffic Safety Administration: Washington, DC, USA, 1999. [CrossRef]
21. Dozza, M. What factors influence drivers' response time for evasive maneuvers in real traffic? *Accid. Anal. Prev.* **2013**, *58*, 299–308. [CrossRef] [PubMed]

energies

MDPI

Article

Analysis of the Influence of Motors Installed in Passenger Car Wheels on the Torsion Beam of the Rear Axle Suspension

Piotr Dukalski [1], Bartłomiej Będkowski [1], Krzysztof Parczewski [2], Henryk Wnęk [2], Andrzej Urbaś [3,*] and Krzysztof Augustynek [3]

[1] Łukasiewicz Research Network—Institute of Electrical Drives and Machines KOMEL, 188 Roździeńskiego Ave., 40-203 Katowice, Poland; piotr.dukalski@komel.lukasiewicz.gov.pl (P.D.); bartlomiej.bedkowski@komel.lukasiewicz.gov.pl (B.B.)

[2] Department of Combustion Engines and Vehicles, University of Bielsko-Biala, 2 Willowa Str., 43-309 Bielsko-Biala, Poland; kparczewski@ath.bielsko.pl (K.P.); hwnek@ath.bielsko.pl (H.W.)

[3] Department of Mechanical Engineering Fundamentals, University of Bielsko-Biala, 2 Willowa Str., 43-309 Bielsko-Biala, Poland; kaugustynek@ath.bielsko.pl

* Correspondence: aurbas@ath.bielsko.pl; Tel.: +48-38279360

Abstract: The influence of mounting motors in wheels' hubs and flexibility of the twist beam rear suspension on their dynamics and strength is presented in the paper. The international roughness indicator (IRI) is applied to assess the overcoming of road unevenness. This indicator is a combination of a shape of the road unevenness and of overcoming velocity. The movement of a wheel's axis during obstacles overcoming is described. For the needs of the dynamics analysis, the mathematical model of the rear suspension system with embedded motors is developed using the MSC.Adams-ANSYS interface. The discrete model of the twist beam is prepared in the ANSYS software, which is used in the next step to construct the dynamics model of the rear suspension system using the MSC.Adams program. The vertical components of displacement and acceleration of the wheel's centre, forces in the suspension's springs and dampers, as well as forces in the joints are analyzed. The analysis of the suspension beam's stress during the road unevenness overcoming is also carried out.

Keywords: road unevenness; multibody system dynamics; rear suspension system; wheel hub motor

Citation: Dukalski, P.; Będkowski, B.; Parczewski, K.; Wnęk, H.; Urbaś, A.; Augustynek, K. Analysis of the Influence of Motors Installed in Passenger Car Wheels on the Torsion Beam of the Rear Axle Suspension. *Energies* **2022**, *15*, 222. https://doi.org/10.3390/en15010222

Academic Editor: Aldo Sorniotti

Received: 24 November 2021
Accepted: 23 December 2021
Published: 29 December 2021

1. Introduction

In the era of growing market share in hybrid and electric cars, it is worth considering how mounting motors in a vehicle's wheels affects its movement and the strength of the suspension components. The application of such driving systems change many factors, which can affect vehicles' dynamics, such as safety, ease of steering, comfort and the durability of the suspension components. Due to a very wide range of issues, authors focus only on the assessment of a vehicle's behaviour during road unevenness overcoming. The analysis is carried out for motors placed in the wheels of the rear axle of a car with a semi-dependent suspension and a twist beam. Such suspension consists of two trailing arms connected by a twist beam. In a semi-dependent suspension, the vertical motion of the wheels slightly influence each other. A twist beam is located in the front of a wheel axle and is subjected to torsional forces, acting as a stabilizer. This solution does not generate high production and operating costs. It also causes, under the action of lateral forces, the wheels to tilt only slightly, which improves the grip of the car's wheels. Usually, coil springs or a torsion bar are elastic elements in this type of suspension.

Incorrect operation of a suspension can cause the premature wear of tires, suspension joints, steering system, wheel's bearings, and body mounting elements, as well as increase the braking distance.

The use of drive motors in wheels, despite the significant simplification of the drive system, is not a common solution, because the required driving force acting at the contact point of the wheel with the road results from the values of torque generated by the motor and the radius of the wheel. Car wheel diameters are usually proportional to the dimensions of the vehicle. The diameter of the wheel rim, in which the electric motor is mounted, cannot be too large either. Additionally, the required drive torque is relatively high and depends on the weight of the vehicle. Obtaining such torque requires large dimensions of the motor. Placing the motors in the wheels deteriorates the ratio of sprung to unsprung masses, which reduces driving comfort.

2. Literature Review and Paper Contributions

Installing motors in the rear wheels of a vehicle increases the weight of a suspension and its moments of inertia (this applies to the rotational motion of wheels and the vertical movement of a suspension) [1–4]. It also increases the mass of unsprung elements and generates larger loads acting on the suspension's elements.

In the literature, one can find few studies in which the analysis of the impact of mounting motors in vehicle wheels on comfort, vehicle properties, and durability of rear suspension components is comprehensively treated [5,6]. The most common are quarter rear suspension models, in which a simplified model of interaction between the suspension and the wheel is used, and unevenness on the road is modeled by kinematic inputs.

In the paper [7], X. Jiao and M. Bienvenu used the IRI (international roughness indicator) to assess the road surface condition and its impact on fuel consumption on motorway sections. It was also used to determine the rolling resistance coefficient of the vehicle. The report of the Iowa State University [8] investigated the interaction between the vehicle and the bridge during overcoming the unevenness on the bridge, which results in the dynamic loads acting between the vehicle and the bridge. The unevenness overcoming issues was described by the authors in previous works [9,10].

In this study, the IRI factor is used to evaluate the obtained results. This factor allows to refer the road unevenness to the vehicle velocity. The mechanics of the road unevenness overcoming is presented. It was proven that mounting motors in the wheels reduces the comfort of single road unevenness overcoming. A simulation model of the rear suspension with the compound twist beam axles (a typical solution used in small vehicles and the test vehicle) is built to carry out a more detailed analysis. The model is prepared in the MSC.Adams system (MSC Software, Newport Beach, CA, USA) and the obtained simulation results are used to conduct a strength analysis of the twist-beam axle, using the ANSYS system (ANSYS, Canonsburg, PA, USA). It allowed to determine the areas of stress concentration and compare them for a vehicle with motors mounted in the vehicle's wheels and without them.

3. Road Unevenness and Their Assessment

The universal testing procedure was developed at the request of the World Bank to harmonize measuring methods and qualify road surfaces. According to this procedure, the international roughness index (IRI) was introduced, which combines a description of the road surface damage condition with a safe vehicle's velocity (Figure 1). The value of this indicator depends on the standard of a road. From the computational point of view, the IRI indicator is a ratio of cumulative vertical movements of a body and a wheel to a road traveled by the vehicle during the test, determined on the basis of the analysis of the movement of a body and a wheel in a quarter car model [11,12].

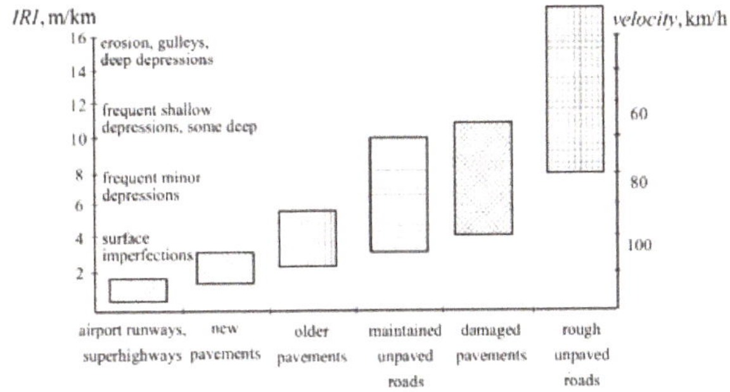

Figure 1. The *IRI* roughness scale [13].

A value of the indicator IRI_i for a single unevenness of the road is determined for a single calculation step, and it is equal to the ratio of a deflection speed of a suspension system to an assumed vehicle velocity. In practice, it determines the ratio of a deflection of a suspension system to a road traveled at that time. A surface unevenness index, as an *IRI* segmental assessment, corresponds to an average value of suspension deflections, which is calculated as the arithmetic mean $\sum_{i=1}^{n} IRI_i$ for the n measurements. In accordance with the established calculation procedure, the surface unevenness index *IRI* is determined according to the following relationship

$$IRI = \frac{1}{n}\sum_{i=1}^{n} IRI_i = \frac{1}{nV}\int_0^n \left|\dot{z}_{b,i} - \dot{z}_{w,i}\right| \mathrm{d}t, \tag{1}$$

where $\dot{z}_\alpha\big|_{\alpha\in\{b,w\}}$ are the vertical displacements and velocities of the body (b), wheel (w), and velocity (V).

In this work, the authors deal with the problem of changing wheel loads when overcoming single road unevenness.

After preliminary tests, it was found that unfavourable impact is particularly due to road unevenness in the shape of a single depression in a roadway and protuberances, similar to overcoming an obstacle. These are the most common road unevenness types that generate dynamic forces loading the suspension.

During the analysis, the simulation model described below is used. The results of previous analyses of the simpler model are described in [14–16]. In these papers, road test results were used to validate the simulation model. The paper presents an analysis of the vehicle motion for different vehicle velocities.

In the first part, an analysis of the suspension movement is performed, and the forces affecting the suspension are determined. In the second part of the paper, the strength analysis of the suspension twist beam during overcoming selected road unevenness is performed. Stresses of the twist beam suspension of the car are analyzed for the system with and without motors mounted in wheels. The modeling of the movement of the wheel axle during overcoming road unevenness is a very important issue [17–19].

4. Mechanics of Overcoming Single Road Unevenness

The problem of the wheel axis movement and forces acting on it results from the height and shape of road unevenness, wheel movement velocity, suspension characteristics, and vehicle parameters. During road unevenness overcoming, additional loading forces act on the wheel [17,20–23]. Loads arising during rectilinear wheel movement and when hitting on the road unevenness are shown in Figure 2. The rolling resistance force $\mathbf{f}_{h,0}$ is omitted in Figure 2b,c in order to improve its readability.

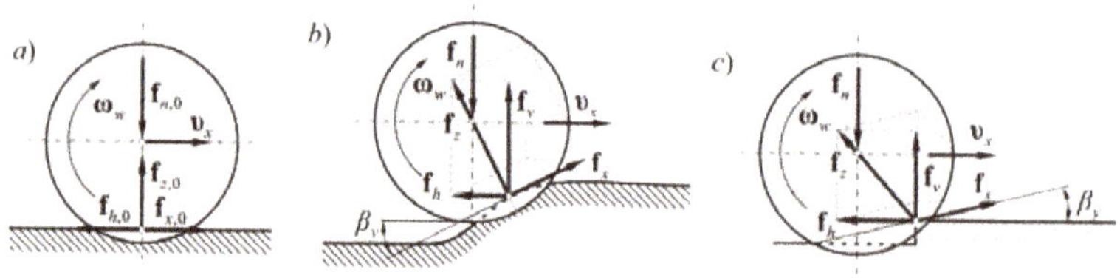

Figure 2. Diagram of a wheel load formation when overcoming road unevenness (**a**) flat road, (**b**) road unevenness, (**c**) step obstacle.

The horizontal component of the wheel load force depends on the effective angle of a road plane inclination, vehicle velocity, and shape of the unevenness. As a result of the suspension's deflection, the wheel is loaded with greater force—an additional force is generated, which loads the wheel $\mathbf{f}_n^*$, $\mathbf{f}_n = \mathbf{f}_{n,0} + \mathbf{f}_n^*$. The reaction force $\mathbf{f}_z$ is transmitted entirely by the tire. The tire deflection occurring on the edge of the unevenness largely depends on the effective angle of the road plane, the contact area of the tire with the road unevenness, and its height and shape.

Increasing the overcoming velocity through road unevenness results in a decrease in the relative amplitude, defined as the ratio of the maximum value of the vertical movement amplitude of the vehicle body to the height of the unevenness. As the vehicle velocity increases, the vertical load of the wheel increases due to the inertia of the car body when overcoming road unevenness (Figure 3).

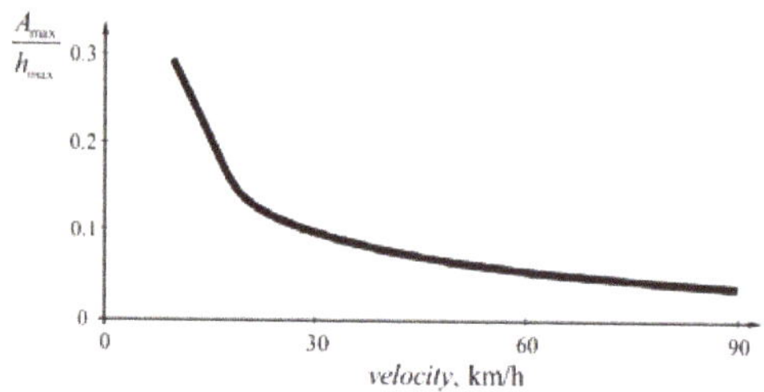

Figure 3. The relative amplitude of vertical vibrations of the body during overcoming single road unevenness with different velocities, where $A_{\max}$—the maximum value of the magnitude of the body vibrations, and $h_{\max}$—maximum height of unevenness.

Comparing the wheel movement during overcoming road unevenness to the movement on a flat horizontal road, it can be stated that, apart from forces occurring in the rectilinear motion (rolling resistance force and forces resulting from the movement conditions—driving or braking), it can be noted the following additional forces loading the wheel and suspension:

1. Additional vertical force resulting from suspension deflection (the wheel lift and body inertia);
2. Horizontal force consisting of
    (a) Horizontal resultant force component $\mathbf{f}_h$ resulting from the wheel load with force $\mathbf{f}_{n,0}$ during overcoming road unevenness;
    (b) Component of additional loading force $\mathbf{f}_n^*$ resulting from the shape of the road unevenness and driving velocity.

To model the movement of the wheel axle, the model proposed by Zegelaar using the two-point beam technique [17,18,20] is used. The possibility of using the smoothing properties of tires, described in [19], is also analyzed. Ultimately, the model of the tire's contact points of the road and unevenness is used in the analysis.

In the model proposed by Zegelaar [20], the wheel rolls on the road unevenness; however, the movement of the wheel axis is determined using a curve (sinusoid) on which a beam of a certain length moves. The beginning and the end of the beam move along a curve, and its center is determined by the trajectory of the wheel axis. In the case of overcoming road unevenness of a rectangular shape, the procedure for determining the movement of the wheel axis is supplemented with a section connecting the entry to and exit from the unevenness. The diagram for determining the wheel axis motion is shown in Figure 4.

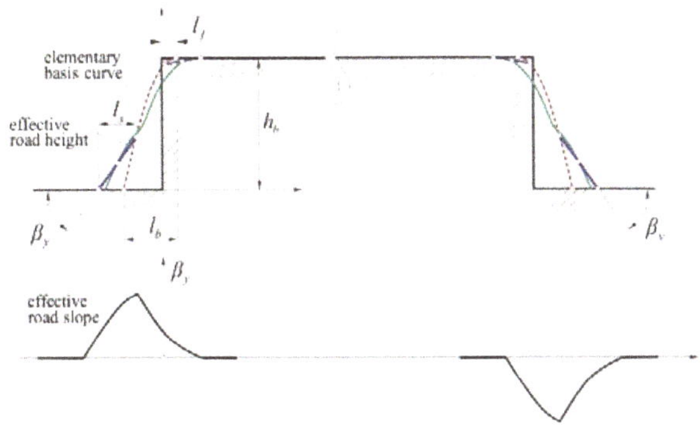

Figure 4. The use of the two-point follower technique to determine the wheel axle lift and angle of inclination of the effective road plane [20].

The methodology presented in [19] takes into account the smoothing properties of tires, and uses the fact that the tire, in the area of the contact patch with the road, averages the height of road unevenness and filters the spectrum of road unevenness in this area. This allows using the model of the point cooperation of the tire with the road

$$S_{df}(\Omega) = H^2 S_d(\Omega), \tag{2}$$

where $S_d(\Omega)$, $\mathrm{m^3rad^{-1}}$ is the power spectral density of a single longitudinal trace, $\Omega = 2\pi/L$, $\mathrm{radm^{-1}}$ is the circular road frequency, L, m is the wavelength of road unevenness, and H is the filter that allows for point cooperation of the wheel model with the road.

The analysis of the contact conditions of the wheel points with the road surface and unevenness is described below.

5. Analysis of the Dynamic's Load of the Rear Axle

Based on the tests described in [22], the results obtained for the vehicle with and without motors mounted in the rear wheels are compared. The IRI indicator calculated for a single road unevenness is applied to assess the impact of mounting motors in the wheels on the vehicle motion. The values of this indicator are determined based on the vertical movement of the body and the wheel during the rectangular unevenness overcoming. The IRI indicators obtained from road tests are compared with those obtained from simulations. Table 1 contains values of the IRI indicator for individual overcoming tests with the velocity of $\sim$12 $\mathrm{kmh^{-1}}$ for the vehicle without the load (with driver only) and with the load (driver and passengers).

Table 1. List of IRI indicators determined from the road tests ($v = 12.5$ kmh^{-1}).

Vehicle	Without Load	With Load
	Unevenness U	
without motors in wheels	17.35	17.88
with motors in wheels	19.87	20.24
	Unevenness E	
without motors in wheels	17.46	17.41
with motors in wheels	18.92	20.85

Where U—unevenness in the form of depression, E—unevenness in the form of protuberant.

Based on the results obtained and Equation (1), the maximum overcoming velocity for an individual unevenness can be determined. It means that comfortable overcoming of a short single unevenness requires a reduction in vehicle velocity. Comfortable overcoming of a rectangular obstacle requires reducing the velocity to ~30 kmh^{-1} when a vehicle is unloaded. Increasing the vehicle load allows to overcome unevenness with a slightly higher velocity. The installation of motors in the wheels of a vehicle increases values of the IRI coefficients by ~10%. Overcoming road unevenness of a larger length leads to separation of the unevenness entry and exit phases, so it is possible to overcome the obstacle at higher velocities.

Overcoming the obstacle in the shape of a step causes the wheel, in the middle phase of overcoming, to not contact the obstacle bottom but rest on its wall. Increasing the velocity shifts the lowest contact point toward the step wall in the direction of the vehicle movement. This condition causes a rapid change in the velocity direction of the wheel axis, which can cause an occurrence of greater accelerations and thus, the vertical velocity of the wheel axis. The vehicle loads generate smaller amplitudes of the wheel vibrations. Increasing the overcoming velocity can cause a decrease in the amplitude of the body vibrations, which are caused by the body's inertia, and increase the horizontal component of the reaction forces acting on the wheel [9].

6. Dynamics Model of the Rear Axle Twist Beam Suspension

The MSC.Adams package is used to elaborate the dynamics model of the rear suspension when overcoming an obstacle. This model was validated in earlier works of the authors; however, the flexibility of the suspension beam was not taken into account [14,15]. The numerical model of the rear suspension with the flexible beam is prepared using the interface MSC.Adams and ANSYS packages. The scheme of such cooperation is shown in Figure 5. The interaction between the flexible beam and the remaining suspension elements is implemented through interface points. The simulation in the MSC.Adams package allows to determine the stress distribution and the time stamps for which the stresses are the largest, but this assessment is rough. More accurate stress distribution over time can be calculated in the ANSYS package after loading the forces acting on the interface points, calculated in the MSC.Adams package.

According to the scheme shown in Figure 5, the multibody model of the rear suspension system is modeled using the MSC.Adams package. Unlike the models presented in earlier works [14–16], the method of the interconnection of the beam with the body is modified. The previous model assumed that suspension during overcoming an obstacle rotates around the axis of the spherical joint formed between the suspension beam and the body. While such assumptions are acceptable in the case when the suspension's beam is treated as rigid, in the case of the flexible beam, it can lead to excessive stresses that do not exist in the real system. In order to obtain numerical results more consistent with those measured during road tests, a sliding joint, which takes into account the vertical displacements of the beam during overcoming of the obstacle, is introduced to the system. In addition, the rigid beam is introduced to take into account the interaction between suspension springs (Figure 6).

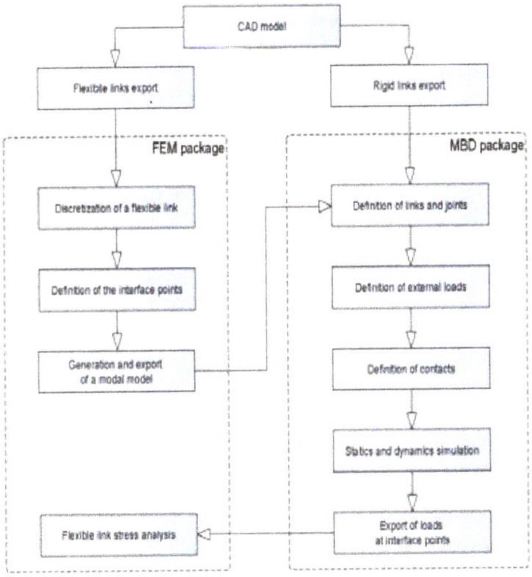

Figure 5. MSC.Adams-ANSYS interface.

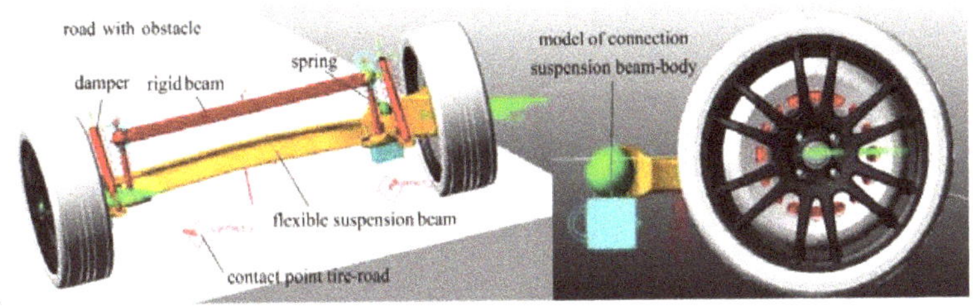

Figure 6. Numerical model of the rear suspension system of the Fiat Panda II.

In the calculations, it is assumed that the suspension's beam is made of the AISI 4130 steel. This steel, also known as chromium or chromium–molybdenum steel alloy, has admixtures of chromium (0.8 ÷ 1.1%) and molybdenum (0.15 ÷ 0.25%), which are used as reinforcing agents.

It exhibits the following mechanical properties:

- Density—7850 kgm^{-3};
- Hardness—217 (Brinell scale)–95 (Rockwell B);
- Ultimate tensile strength—540 MPa;
- Yield tensile strength—460 MPa;
- Modulus of elasticity—205 GPa;
- Machinability—72% (in annealed state).

Other advantages of this material include high plasticity, good weldability and workability.

The numerical model of the flexible beam is modeled using the ANSYS package. The suspension beam is divided into 231,039 solid and 7380 beam finite elements (Figure 7). Due to a large number of irregular and curved surfaces, making a finite element mesh is a complex task, and it requires many iterations to achieve an appropriate quality of the discrete model.

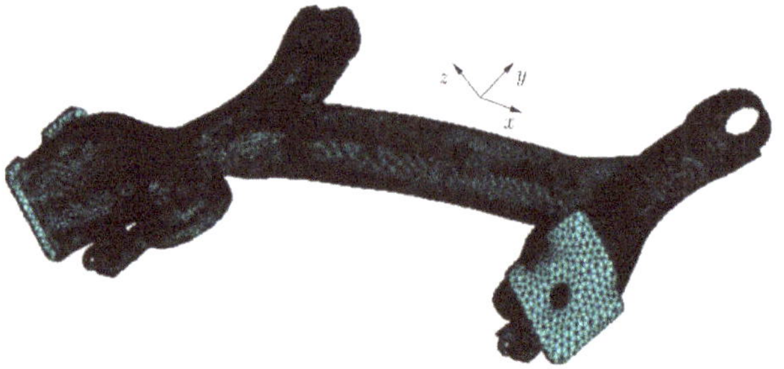

Figure 7. A discrete model of the rear suspension twist beam of the Panda II car.

Loads acting on the beam are introduced by the so-called interface points, i.e., points where forces and torques calculated in the MSC.Adams are applied (Figure 8). These points are determined in the ANSYS package and then exported with the discrete beam model to the MSC.Adams package. In the considered model, it is assumed that the interface points are located in places where the suspension beam is joined with other parts of the rear suspension system.

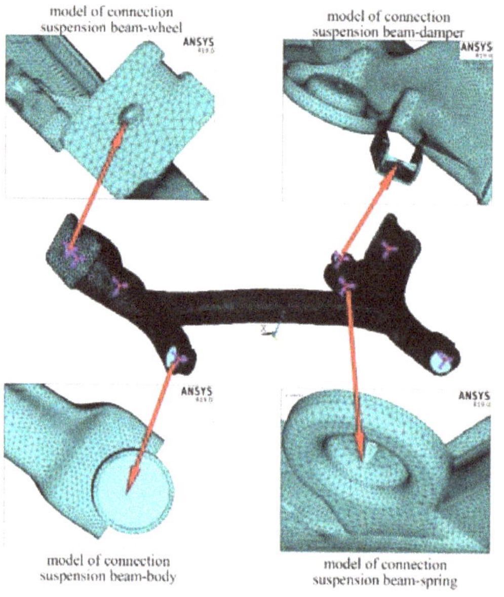

Figure 8. Location of the interface points.

It is assumed that the exported suspension beam model contains 48 natural frequencies. The number of frequencies is limited to those which are the most important when determining the dynamic response of the system. Figure 9 shows the first two frequencies of free vibrations and the corresponding forms of vibrations. Analyzing the presented forms of vibration, it can be seen that the deformations associated with bending and torsion are dominant for the analyzed frequencies.

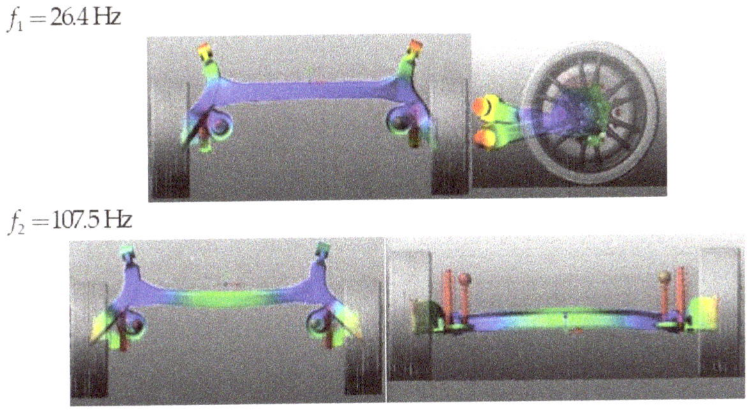

Figure 9. The suspension beam's natural frequencies and its mode shapes.

The results of the simulation of the wheel overcoming obstacles are presented in the further part of the paper. The purpose of these simulations is to determine the effect of additional unsprung masses from motors built into wheel hubs on the dynamic response of the suspension. It is assumed that the wheels overcome obstacles at speeds of 30, 60 and 90 kmh^{-1}.

The simulations analyze the dynamic response of the rear suspension system when overcoming obstacles, shown in Figure 10. These obstacles are a fairly common case of unevenness occurred on roads.

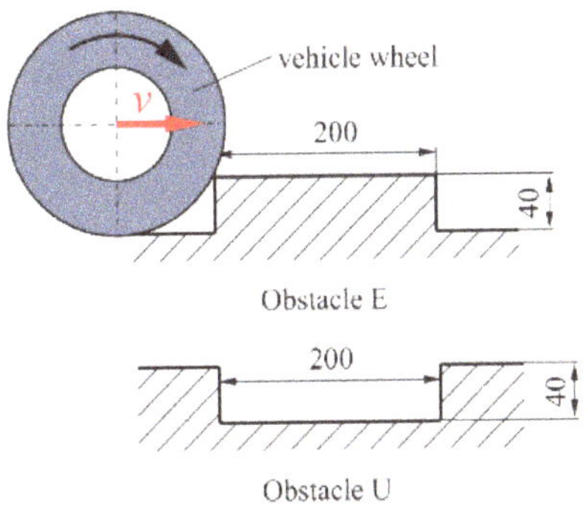

Figure 10. Obstacles types applied in simulations.

The selected time histories obtained for the suspension with the rigid beam during overcoming the U-type obstacle are presented in Figure 11. In the simulations, the effect of additional unsprung masses from motors built into the wheels on the dynamics is additionally analyzed.

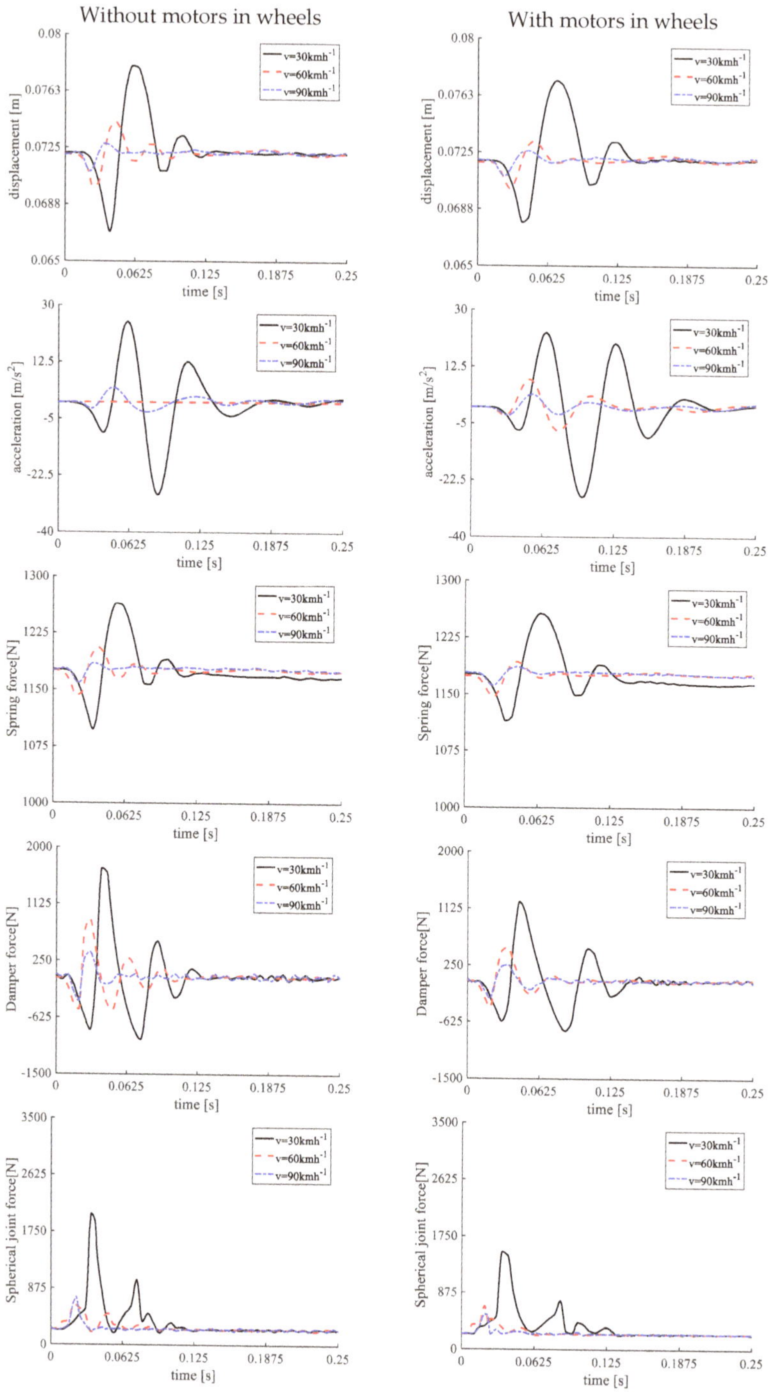

Figure 11. Selected time courses obtained for the suspension model with the rigid beam during overcoming the U-type obstacle.

Analyzing the presented results, it can be seen that the introduction of additional masses does not have a major impact on the displacement of the center of the wheel and the forces acting on the suspension spring. The influence of these masses is visible in the course of forces acting in the shock absorber and in the spherical joint, where the beam joins the car's body. Similar analyses are made for the system with the flexible beam, and their results are presented in Figure 12.

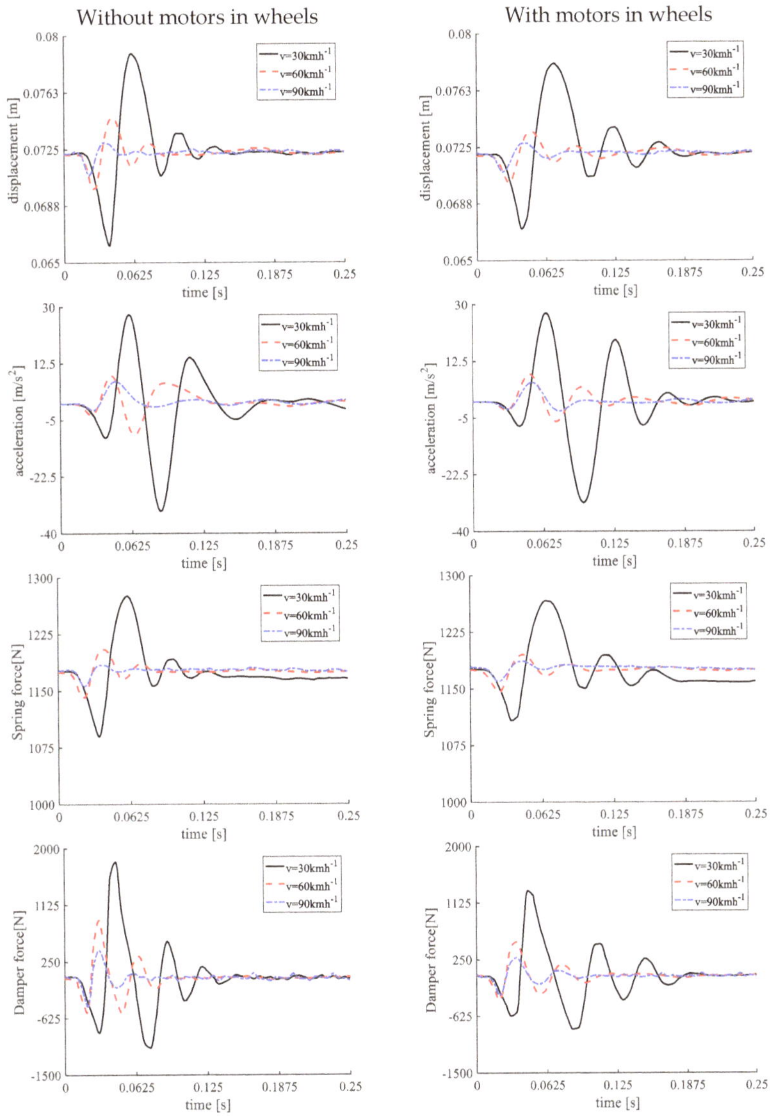

Figure 12. *Cont.*

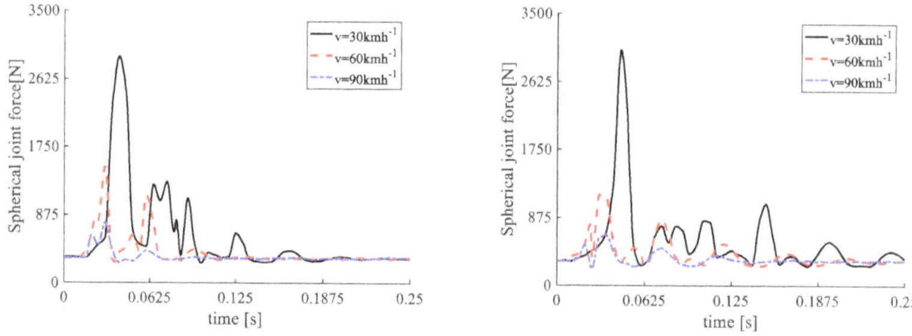

Figure 12. Selected time courses obtained for the suspension model with the twist beam during overcoming the U-type obstacle.

Analyzing the results obtained, it can be seen that the introduction of additional masses, resulting from the assembly of the motors in wheels, has no major impact on the dynamic response of the suspension. Comparing the results from Figure 11 with the results from Figure 12, it can be seen that taking into account the suspension beam's flexibility leads to a slight increase in the amplitudes of displacement and acceleration of the wheel's center as well as forces acting in the suspension spring and damper. Moreover, in the model with the flexible beam, additional oscillations appear just after obstacle overcoming. Particularly noteworthy is the force acting in the spherical joints. It can be seen that in the case of the flexible beam, these forces are greater than those occurring in the model with the rigid beam. In addition, it can be seen that as the overcoming obstacle velocity increases, the loads acting on the system are smaller.

In Figures 13 and 14, selected time courses obtained for the E type obstacle are compared. The influence of the additional masses from the motors' built-in wheels and the suspension beam's flexibility on the response of the rear suspension system are analyzed in the presented simulations results.

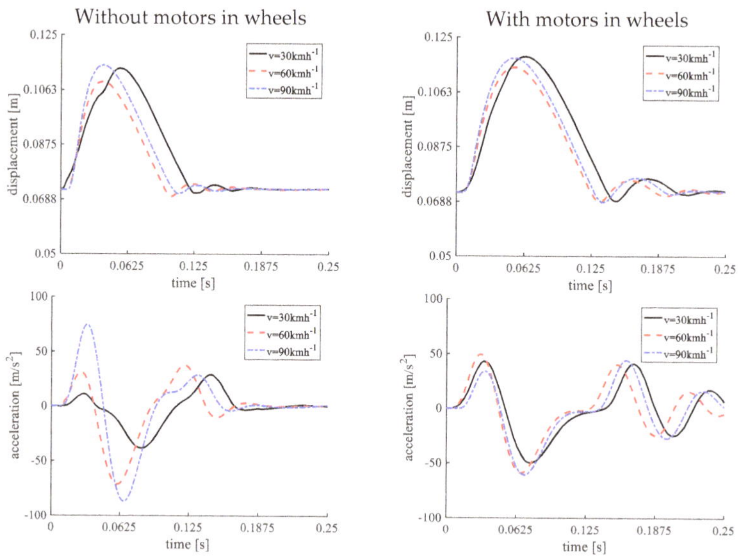

Figure 13. *Cont.*

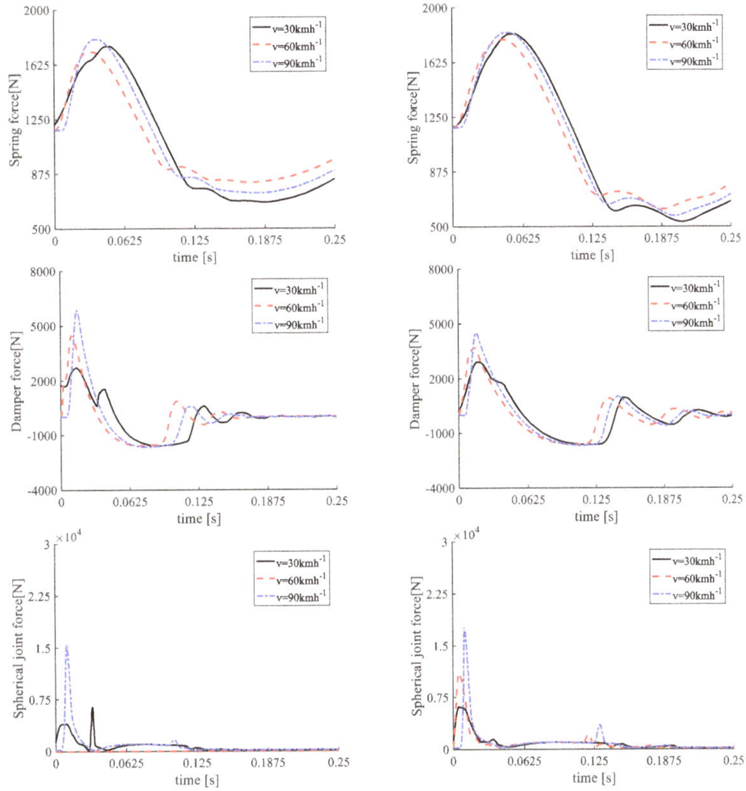

Figure 13. Selected time courses obtained for the suspension model with the rigid beam during overcoming the E-type obstacle.

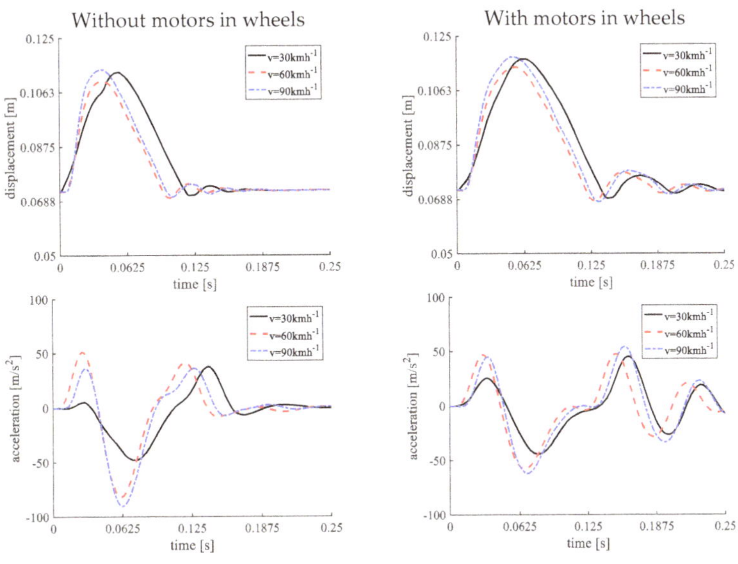

Figure 14. *Cont.*

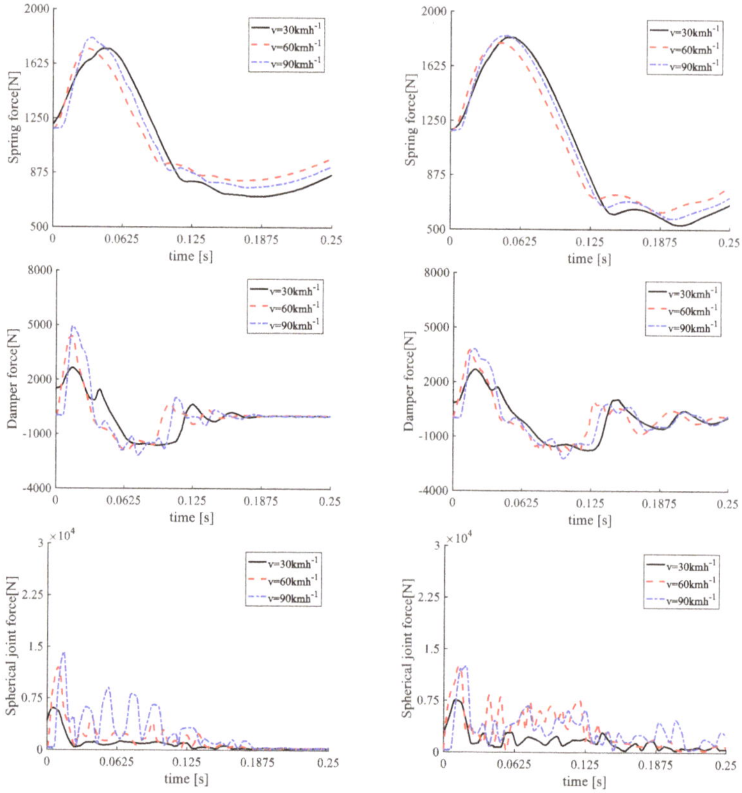

Figure 14. Selected time courses obtained for the suspension model with the flexible beam during overcoming the E-type obstacle.

Analyzing the courses presented, it can be seen that the response of the system obtained for the obstacle overcoming velocities 60 kmh^{-1} and 90 kmh^{-1} is very similar. For velocity 30 kmh^{-1}, the forces acting in the rear suspension system are significantly smaller. Examining the accelerations of the wheel's center for the system with the rigid beam and without motors in the wheels obtained for overcoming velocity equal to 90 kmh^{-1}, it can be seen that after entering the obstacle, the acceleration increases rapidly to around 80 m s^{-2}. The time courses of forces acting in the spherical joint connecting the suspension beam with the body are interesting. In the case when the beam is treated as rigid, there is observed a peak for the force in the spherical joint after overcoming the obstacle. After this peak, the force decreases to the level occurring before entering the obstacle. Forces acting in the spherical joint look different in the model with the flexible beam. In this model, the forces increase rapidly during overcoming the obstacle, after which there are damped oscillations with relatively large amplitudes.

Figure 15 shows the distribution of reduced stress obtained from the solution of the static task for the suspension system with and without motors in wheels.

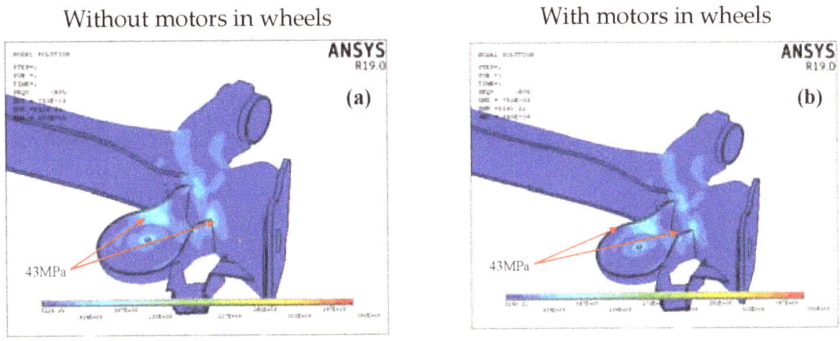

Figure 15. Contour plot of the reduced stresses obtained for the statics. (**a**) without motors in wheels, (**b**) with motors in wheels.

Analyzing the results shown in Figure 15, it can be seen that the reduced stresses in the suspension beam do not exceed 43 MPa. In the analysis, stresses above 43 MPa are rejected. They result from the simplifications used in modeling stress concentration areas (sharp edges, places of connection of mounting plates with the suspension beam) and from the applied method of modeling joints (forces from joints are transferred to the system through interface points).

Figures 16–18 show the results of the strength calculations in the form of contour plots of the reduced stresses before and during overcoming the obstacle, obtained at the time when the reduced stresses reach their highest values.

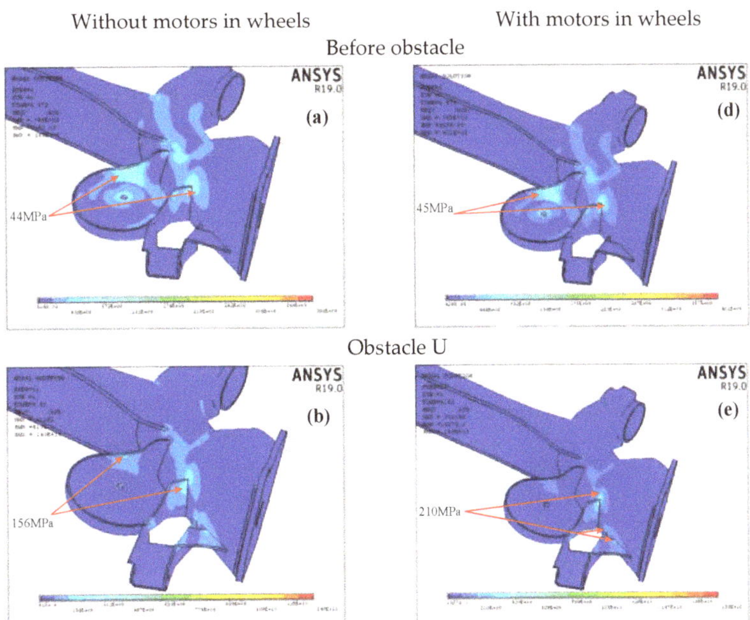

Figure 16. *Cont.*

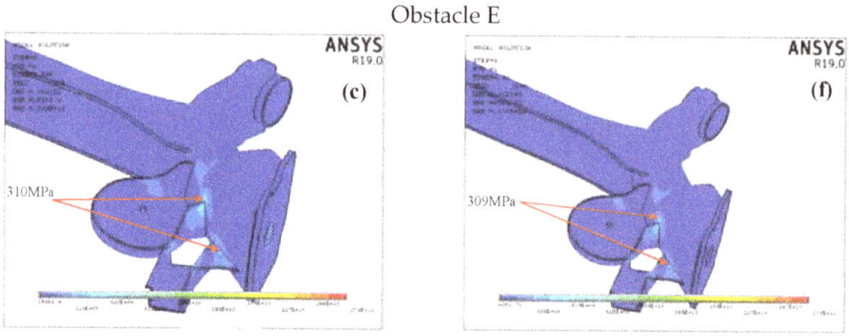

Figure 16. Contour plot of the reduced stresses obtained during overcoming an obstacle at velocity $v = 30\ \text{kmh}^{-1}$. (**a**) without motors in wheels and before obstacle, (**b**) without motors in wheels and obstacle U, (**c**) without motors in wheels and obstacle E, (**d**) with motors in wheels and before obstacle, (**e**) with motors in wheels and obstacle U, (**f**) with motors in wheels and obstacle E.

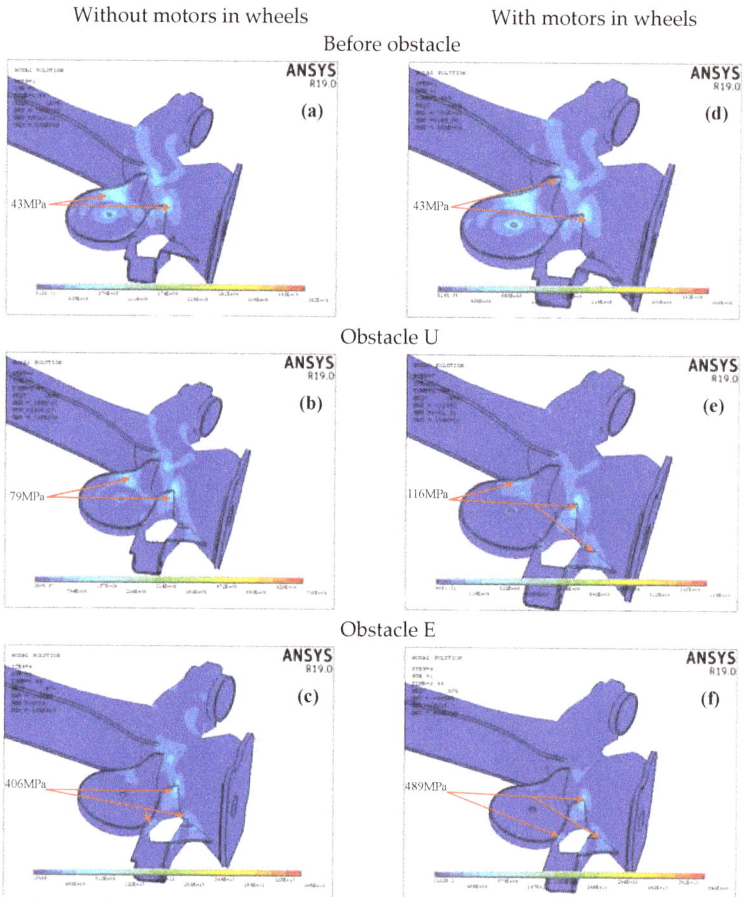

Figure 17. Contour plot of the reduced stresses obtained during overcoming an obstacle at velocity $v = 60\ \text{kmh}^{-1}$. (**a**) without motors in wheels and before obstacle, (**b**) without motors in wheels and obstacle U, (**c**) without motors in wheels and obstacle E, (**d**) with motors in wheels and before obstacle, (**e**) with motors in wheels and obstacle U, (**f**) with motors in wheels and obstacle E.

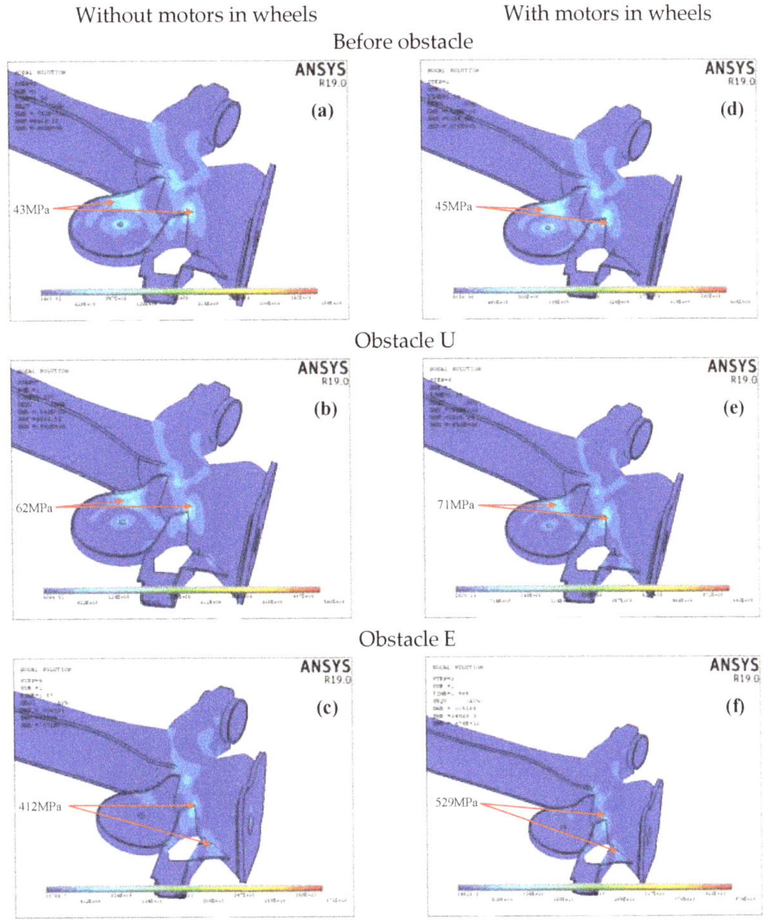

Figure 18. Contour plot of the reduced stresses obtained during overcoming an obstacle at velocity $v = 90$ kmh^{-1}. (**a**) without motors in wheels and before obstactle, (**b**) without motors in wheels and obstactle U, (**c**) without motors in wheels and obstactle E, (**d**) with motors in wheels and before obstactle, (**e**) with motors in wheels and obstactle U, (**f**) with motors in wheels and obstactle E.

Additionally, the values of the maximum reduced stresses of the system with and without motors obtained for both analyzed obstacles are calculated in order to assess the impact of the obstacle overcoming velocity on the suspension beam's strength (Figure 19).

Analyzing the results obtained, the following conclusions can be drawn:

- The highest stress concentration are located in places where the spring and damper are assembled to the beam.
- In the case of obstacle E, it can be seen that for a system without motors, the maximum reduced stresses of the suspension beam are greater than those occurring in the system with the motors built into the wheels (Figure 19). In addition, as the obstacle overcoming velocity increases, the difference between the stresses obtained for the system with and without motors in wheels increases. For velocity $v = 90$ kmh^{-1}, the difference is about 100 MPa, and the stresses occurring in the beam are close to the tensile strength (Figure 19).

- In the case of obstacle U, it can be noticed that the maximum reduced stresses decrease as the obstacle overcoming velocity increases. For $v = 90$ kmh^{-1} the maximum reduced stresses are three times smaller than those obtained for velocity $v = 30$ kmh^{-1}.

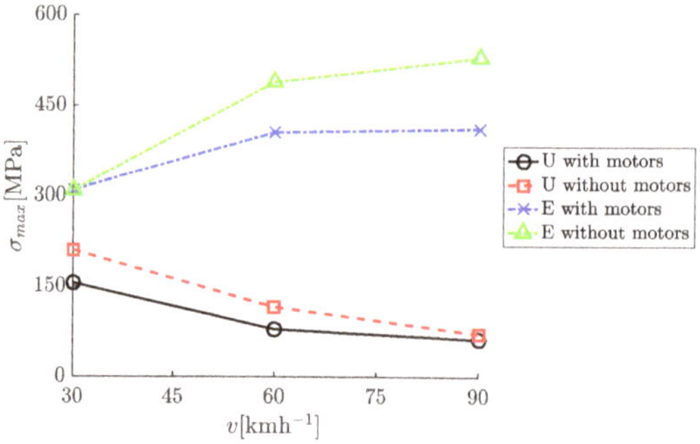

Figure 19. Relationship between maximum reduced stresses and obstacle overcoming velocity.

7. Conclusions

The first part of the paper presents the IRI indicator, combining the assessment of road surface unevenness with driving speed. This indicator was used to assess unevenness overcoming during the road tests. The movement of the wheel axis during single road unevenness overcoming was described. In the further part of the paper, the road unevenness was used as enforcement during the simulation of the dynamics of the rear suspension system. The paper presents a computational model of the rear suspension system for dynamics analysis. This model enables stress analysis of the suspension beam during obstacle overcoming for two variants: with motors built into wheel hubs and without motors. The finite element method implemented in the ANSYS commercial package was used to model the flexibility of the beam of the semi-dependent rear suspension. Thanks to the use of the MSC.Adams-ANSYS program interface, it was possible to analyze the beam strength for loads acting during unevenness overcoming and to take into account the couplings occurring between the deformable beam and other suspension elements. The results of the numerical simulations allow to formulate the following conclusions:

- Obstacles that were used during the dynamics analysis can be classified as "rough unpaved road" in Figure 1. This means that traveling at a velocity greater than $v = 80$ kmh^{-1} can lead to significant damage to the suspension. This is also confirmed by the stress values shown in Figure 19.
- In the case of the obstacle E overcoming, the reduced stresses increase with the travel velocity, and for velocity above $v = 80$ kmh^{-1}, they can exceed the allowable values. Slightly different conclusions can be formulated in the case of the U-type obstacle. For this obstacle, the reduced stresses decrease with the travel velocity. This is due to the way the shock absorber works and the wheel inertia—at higher speeds, the wheel "jumps" over a U-type obstacle.
- The simulation results show that greater dynamic forces and stresses in the beam occur during overcoming unevenness in the form of a cavity in the roadway than protuberances. In the case of a cavity, the highest loads occur at lower speeds ($v = 30$ kmh^{-1} in the analyzed case). At higher speeds, the wheel "jumps" through this cavity, as a result of which the forces and stresses are much smaller.
- It is also noticeable that smaller dynamic forces and stresses occur when the motors are built into the wheel hubs. This is due to increased wheel inertia.
- The fatigue strength of the suspension beam was not analyzed; however, it can be assumed that the final solution will require its local reinforcement.
- Experimental verification of the obtained results is planned in the future.

Author Contributions: Conceptualization, P.D., B.B., K.P., H.W., A.U., K.A.; methodology, K.P., H.W.; software, A.U., K.A.; validation, K.A., A.U.; formal analysis K.P., H.W., A.U., K.A.; investigation, P.D., B.B., K.P., H.W., A.U., K.A.; resources, P.D., B.B., K.P., H.W.; data curation, B.B.; writing—original draft preparation, K.P., H.W., A.U., K.A.; writing—review and editing, K.P., H.W., A.U., K.A.; visualization, K.A.; supervision, P.D.; project administration, P.D. All authors have read and agreed to the published version of the manuscript.

Funding: The research is a continuation of the project: "Innovative Solutions for Direct Drive of Electric Vehicles", financed by National Centre for Research and Development under the LIDER VII program, in accordance with the agreement: LIDER/24/0082/L-7/15/NCBR/2016 (POLAND). The project received the Research Award (main award) 25th Siemens Award Competition for scientists and research teams (Poland).

Institutional Review Board Statement: Not applicable.

Informed Consent Statement: Not applicable.

Data Availability Statement: The data presented in this study are available on request from the corresponding author.

Conflicts of Interest: The authors declare no conflict of interest.

References

1. Anderson, M.; Harty, D. Unsprung Mass with In-Wheel. Motors-Myths and Realities. In Proceedings of the 10th International Symposium on Advanced Vehicle Control, Loughborough, UK, 22–26 August 2010; pp. 261–266.
2. Watts, A.; Vallance, A.; Fraser, A.; Whitehead, A. Integrating In-Wheel Motors into Vehicles-Real-World Experiences. *SAE Int. J. Altern. Powertrains* **2012**, *1*, 289–307. [CrossRef]
3. Biček, M.; Gotovac, G.; Miljavec, D.; Zupan, S. Mechanical Failure Mode Causes of In-Wheel Motor. *J. Mech. Eng.* **2015**, *61*, 74–85. [CrossRef]
4. Ślaski, G.; Gudra, A.; Borowicz, A. Analysis of the influence of additional unsprung mass of in-wheel motors on the comfort and safety of a passenger car. *Arch. Automot. Eng.* **2014**, *3*, 51–64.
5. Dukalski, P.; Krok, R. Selected Aspects of Decreasing Weight of Motor Dedicated to Wheel Hub Assembly by Increasing Number of Magnetic Poles. *Energies* **2021**, *14*, 917. [CrossRef]
6. Szewczyk, P.; Łebkowski, A. Studies on Energy Consumption of Electric Light Commercial Vehicle Powered by In-Wheel Drive Modules. *Energies* **2021**, *14*, 7524. [CrossRef]
7. Phares, B.; Deng, Y. *Investigation of the Effect of Speed on the Dynamic Impact Factor for Bridges with Different Entrance Conditions*; Research Report; Institute of Transportation, Iowa State University: Ames, IA, USA, 2016.
8. Jiao, X.; Bienvenu, M. The influence of pavement–vehicle interaction on highway fuel consumption by field measurement. *Transport* **2016**, *31*, 202–210. [CrossRef]
9. Parczewski, K.; Wnęk, H. An attempt to determine the value of forces acting on the wheel while overcoming road unevenness. In Proceedings of the 2020 XII International Science-Technical Conference Automotive Safety, Kielce, Poland, 21–23 October 2020. [CrossRef]
10. Parczewski, K.; Wnęk, H. Impact of tire inflation pressure during overcoming of road unevenness. In Proceedings of the 21th International Conference Transport Means, Kaunas, Lithuania, 21–25 August 2017; Volume 1, pp. 154–157.
11. ASTM E950-98 (Reapproved 2004). *Standard Test Method for Measuring the Longitudinal Profile of Traveled Surfaces with an Accelerometer Established Inertial Profiling Reference*; ASTM International: West Conshohocken, PA, USA, 2017.
12. ASTM E1926-08. *Standard Practice for Computing International Roughness Index of Roads from Longitudinal Profile Measurements*; ASTM International: West Conshohocken, PA, USA, 2015.
13. Sayers, M.W.; Karamihas, S.M. *The Little Book of Profiling: Basic Information about Measuring and Interpreting Road Profiles*; Transportation Research Institute, University of Michigan: Ann Arbor, MI, USA, 1998.
14. Dukalski, P.; Będkowski, B.; Parczewski, K.; Wnęk, H.; Urbaś, A.; Augustynek, K. Analysis of the influence of assembly electric motors in wheels on behaviour of vehicle rear suspension system. *IOP Conf. Ser. Mater. Sci. Eng.* **2018**, *421*, 022004. [CrossRef]
15. Dukalski, P.; Będkowski, B.; Parczewski, K.; Wnęk, H.; Urbaś, A.; Augustynek, K. A simulation model of the dynamics of the rear suspension of a Fiat Panda car with motors installed in the wheels. *Electr. Mach. Probl. Noteb.* **2018**, *1*, 75–80. (In Polish)
16. Dukalski, P.; Będkowski, B.; Parczewski, K.; Wnęk, H.; Urbaś, A.; Augustynek, K. Dynamics of the vehicle rear suspension system with electric motors mounted in wheels. *Maint. Reliab.* **2019**, *21*, 125–136. [CrossRef]
17. Schmeitz, A.J.C. A Semi-Empirical Three Dimensional Model of the Pneumatic Tire Rolling over Arbitrarily Uneven Road Surfaces. Ph.D. Thesis, Delft University of Technology, Delft, The Netherlands, 2004.
18. Pacejka, H.B. *Tire and Vehicle Dynamics*; SAE: Detroit, MI, USA, 2006.
19. Lozia, Z.; Zdanowicz, P. Calculation of shock absorber asymmetry coefficient optimal value in car passive suspension using "quarter-car" model. *Proc. Wars. Univ. Technol. Transp.* **2017**, *119*, 249–265. (In Polish)

20. Zegelaar, P.W.A. The Dynamic Response of Tyres to Brake Torque Variations and Road Unevenness's. Ph.D. Thesis, Delft University of Technology, Delft, The Netherlands, 1998.
21. Wicher, J.; Więckowski, D. Influence of vibrations of the child seat on the comfort of child's ride in a car. *Maint. Reliab.* **2010**, *48*, 102–110.
22. Janczur, R. Vertical acceleration of the vehicle body when driving over the speed bump. *Arch. Automot. Eng.* **2015**, *1*, 47–60. (In Polish)
23. Haniszewski, T. Preliminary studies of vertical acceleration of a passenger car passing through the speed bump for various driving speeds. *Transp. Probl.* **2019**, *14*, 23–34. [CrossRef]

MDPI
St. Alban-Anlage 66
4052 Basel
Switzerland
Tel. +41 61 683 77 34
Fax +41 61 302 89 18
www.mdpi.com

Energies Editorial Office
E-mail: energies@mdpi.com
www.mdpi.com/journal/energies

www.ingramcontent.com/pod-product-compliance
Lightning Source LLC
Chambersburg PA
CBHW061454220926
51850CB00004B/91